Sustainable Development and Biodiversity

Volume 32

Series Editor

Kishan Gopal Ramawat, Botany Department, Mohanlal Sukhadia University, Udaipur, India

Sustainable Development Goals are best achieved by mechanisms such as research, innovation, and knowledge sharing. This book series aims to help researchers by reporting recent progress and providing complete, comprehensive, and broad subject-based reviews about all aspects of sustainable development and ecological biodiversity.

The series explores linkages of biodiversity with delivery of various ecosystem services and it offers a discourse in understanding the biotic and abiotic interactions, ecosystem dynamics, biological invasion, ecological restoration and remediation, diversity of habitats and conservation strategies. It is a broad scoped collection of volumes, addressing relationship between ecosystem processes and biodiversity. It aims to support the global efforts towards achieving sustainable development goals by enriching the scientific literature. The books in the series brings out the latest reading material for botanists, environmentalists, marine biologists, conservationists, policy makers and NGOs working for environment protection.

We welcome volumes on the themes --

Agroecosystems, Agroforestry, Biodiversity, Biodiversity conservation, Conservation of ecosystem, Ecosystem, Endangered species, Forest conservation, Genetic diversity, Global climate change, Hotspots, Impact assessment, Invasive species, Livelihood of people, Plant biotechnology, Plant resource utilization, Sustainability of the environment, Sustainable management of forests, Sustainable use of terrestrial ecosystems and plants, Traditional methods, Urban horticulture.

Sylvester Chibueze Izah • Matthew Chidozie Ogwu
Editors

Sustainable Utilization and Conservation of Africa's Biological Resources and Environment

Springer

Editors
Sylvester Chibueze Izah
Department of Microbiology
Bayelsa Medical University
Yenagoa, Bayelsa State, Nigeria

Matthew Chidozie Ogwu
Goodnight Family Department of Sustainable Development, Living Learning Center
Appalachian State University
Boone, North Carolina, USA

ISSN 2352-474X ISSN 2352-4758 (electronic)
Sustainable Development and Biodiversity
ISBN 978-981-19-6976-8 ISBN 978-981-19-6974-4 (eBook)
https://doi.org/10.1007/978-981-19-6974-4

This Springer imprint is published by the registered company Springer Nature Singapore Pte Ltd.
The registered company address is: 152 Beach Road, #21-01/04 Gateway East, Singapore 189721, Singapore

Preface

Biological resources refer to the tangible and intangible values of all micro- and macro life forms as well as their living landscape. The living landscape is the diverse environment and interactions where organisms reside and obtain sustenance. Biological and environmental resources are linked to societal detrimental and beneficial development. These resources possess intrinsic and extrinsic differences that may be exploited in their classification, exploration, utilization, and conservation. Also, some are known to have generation-defining impacts like *Phytophthora infestans*, *Batrachochytrium dendrobatidis, Agrobacterium tumefaciens, Adansonia digitata, Leo panthera,* as well as rivers Nile, Congo, and Niger, Serengeti savanna, Congo basin and the Sahara and Namib deserts. Nowhere is the connection between biological-environmental resources and the human pursuit of sustainable development so apropos in this sixth extinction era than in Africa because of the ongoing and potential exploitation of these resources. Therefore, to sustainably manage the relationship between biological-environmental resources and the human need for development, it is paramount to document the current status and undertake characterization of these resources, identify viable conservation strategies, develop sustainable utilization patterns, and minimize biological and environmental challenges from human activities. Poverty and other economic and social issues have been identified as the major threats to Africa's biological and environmental resources.

This book, entitled *Sustainable Utilization and Conservation of Africa's Biological Resources and Environment,* present a roadmap through a broad interdisciplinary collection of reviews written by researchers, intellectuals, experts, practitioners, and professionals within and outside the African continent. The book focuses on the status and value of Africa's resources, current and potential threats from human-driven environmental degradation and overexploitation, biodiversity loss and shrinking landscapes, viable conservation strategies (including the application of local knowledge and belief systems, cultural heritage, and traditional protection strategies), and the need to incorporate and address sustainability and policy issues. The book also considered the touristic value of the African environment and its connection to environmental and economic sustainability. The book encourages an active and constant conversation on the sustainable utilization and conservation of Africa's biological and environmental resources for the meaningful growth and development of the continent. Such conversation may be held through transdisciplinary workshops, forums, symposia, town hall meetings, and sensitization programs. This book is invaluable to students (undergraduates and postgraduates), academics, researchers, environmentalists, ecologists, practitioners, agricultural scientists, biodiversity experts, policymakers, conservationists, and industry professionals interested in the sustainable utilization and conservation of Africa's biological and environmental resources.

Thank you.

Yenagoa, Bayelsa State, Nigeria
Boone, North Carolina, USA
June 2022

Sylvester Chibueze Izah
Matthew Chidozie Ogwu

The original version of this book has been revised: Series Volume number 32 has been assigned. The correction to this book can be found at https://doi.org/10.1007/978-981-19-6974-4_25

Contents

1 **Overview of African Biological Resources and Environment** 1
Matthew Chidozie Ogwu, Adams Ovie Iyiola, and Sylvester Chibueze Izah

Part I Conservation of Africa's Biological Resources and Environment

2 **Biodiversity Conservation and Tourism Sustainability in Africa** . . . 35
Bukola Omotomilola Adetola

3 **Biodiversity Conservation Strategies and Sustainability** 61
Ehi Francis Okosodo and Odangowei Inetiminebi Ogidi

4 **Potentials, Threats, and Sustainable Conservation Strategies of Plankton and Macrophytes** 85
Emmanuel N. Ogamba, Adams Ovie Iyiola, Baturh Yarkwan, and Bukola Omotomilola Adetola

5 **Threats and Conservation Status of *Cercopithecus sclateri* in Akwa Ibom State, Nigeria** 119
Daniel Etim Jacob and Edem Archibong Eniang

6 **Potential Threats and Possible Conservation Strategies of Biodiversity in Niger Delta Region of Nigeria** 151
Sylvester Chibueze Izah and Ayobami Omozemoje Aigberua

7 **Exploration of Local Beliefs and Cultural Heritages as Tools for Species Conservation in Selected Sites in Africa** 177
Aroloye O. Numbere and Pasiya P. Otufu

8 **The Need to Conserve and Protect Forest Resources: African Perspective** 203
Sampson Abigha Inatimi

9 Rationale Behind Conservation of Africa's Biological Resources . . . 235
Godfrey C. Akani

Part II Utilization Patterns and Potentials of Africa's Biological Resources and Environment

10 The Value of Biodiversity to Sustainable Development in Africa . . . 269
Abiola Omolewa Saliu, Oluwatosin Olaoluwa Komolafe, Christopher Oluwasesan Bamidele, and Morufu Olalekan Raimi

11 Medicinal Potentials of Aloe Vera (*Aloe barbadensis* Miller): Technologies for the Production of Therapeutics . . . 295
Odangowei Inetiminebi Ogidi and Uchechi Emmanuella Enenebeaku

12 Sustainable Utilization of Important Medicinal Plants in Africa . . . 323
Odangowei Inetiminebi Ogidi

13 Local Food Crops in Africa: Sustainable Utilization, Threats, and Traditional Storage Strategies . . . 353
Matthew Chidozie Ogwu

Part III Challenges in the Conservation and Utilization of Africa's Biological Resources and Environment

14 Environmental Pollution: Threats, Impact on Biodiversity, and Protection Strategies . . . 377
Ayotunde Samuel Kolawole and Adams Ovie Iyiola

15 Environmental Degradation in the Niger Delta Ecosystem: The Role of Anthropogenic Pollution . . . 411
Aroloye O. Numbere, Tambeke N. Gbarakoro, and Bernard B. Babatunde

16 Climate Change and Other Environmental Factors as Drivers of Fauna and Flora Biodiversity in Africa . . . 441
Chukwudi Nwaogu, Bridget E. Diagi, Victor A. Agidi, and Simon I. Okweche

17 Contamination of African Water Resources: Impacts on Biodiversity and Strategies for Conservation and Restoration . . . 469
Baturh Yarkwan

18 Disease Outbreaks in Ex-Situ Plant Conservation and Potential Management Strategies . . . 497
Matthew Chidozie Ogwu and Moses Edwin Osawaru

Part IV Integrated Conservation and Sustainable Utilization of Africa's Biological Resources and Environment

19 Challenges of Sea Turtle Conservation in African Territorial Waters: The Way Out 519
Godfrey C. Akani, Luca Luiselli, Gijo A. Harry, Kia Tarela Jovita, and Grace N. Alawa

20 An Overview of Environmental Resources in Africa: Emerging Issues and Sustainable Exploitation 543
Adams Ovie Iyiola, Oluseun Adesola Akinsorotan, Abiola Elizabeth Ojeleye, and Abdulbasith Olabamiji Fajimolu

21 Touristic Value of African Environment: A Socio-economic Perspective 571
Opeyemi Pamela Babafemi, Adams Ovie Iyiola, and O. M. Ogundare

22 Environmental Sustainability: Relevance of Forensic Insects and Other Ecosystem Services in Africa 603
Maduamaka Cyriacus Abajue

23 Intrinsic Values of the African Environment: A Sustainable Perspective 635
Adams Ovie Iyiola and John Valbo-Jørgensen

24 Toward Sustainable Biological and Environmental Policies in Africa 665
Smith Etareri Evivie and Ejiroghene Ruona Evivie

Correction to: Sustainable Utilization and Conservation of Africa's Biological Resources and Environment C1
Sylvester Chibueze Izah and Matthew Chidozie Ogwu

Index 689

Editors and Contributors

About the Editors

Sylvester Chibueze Izah possesses a Ph.D. in Applied Microbiology and Environmental Health, an M.Sc. in Applied Microbiology, and a B.Sc. in Biological Sciences from Niger Delta University, Nigeria. Dr. Izah is a licensed Environmental Health Specialist in Nigeria. Dr. Izah has over 10 years of working and research experience in environmental studies, focusing on soil, water, and atmospheric assessment. In the course of his early career, he has carried out studies on vegetation cover and biodiversity concerning species diversity and composition. Dr. Izah is presently a lecturer at Bayelsa Medical University in Yenagoa, Nigeria, where he also serves as the Assistant Director of Academic Planning, Research, and Innovations. He is also the Sub Dean of his Faculty at the Bayelsa Medical University, Nigeria. He is also a part-time lecturer at the Department of Community Medicine, Faculty of Clinical Sciences, Niger Delta University, Wilberforce Island, Nigeria. Dr. Izah has an impressive research and publication record on water, air, and soil quality; applied microbiology; biotechnology; public and environmental health; risk assessment; bioenergy; toxicology; and biodiversity. Dr. Izah has over 250 peer-reviewed publications, including journal articles, book chapters, and an edited book. His research is now focusing on Sustainable Human-Environmental Health Interactions (covering air, soil, and water quality; toxicology; hygiene and sanitation; food science; vegetation; and wildlife and their sustainability) in the Global South. Dr. Izah has done some research with colleagues in other parts of the world. He is also an editorial and review board member of many reputable journals.

Matthew Chidozie Ogwu is an Assistant Professor of Integrated Ecology and Sustainable Development in the Goodnight Family Sustainable Development Department at Appalachian State University, USA. He obtained his Ph.D. in Biological Sciences with a research focus in the Molecular Biology of Soil Microbial and Geographical Ecology from Seoul National University. He holds an M.Sc. (Distinction) and B.Sc. (First Class) in Plant Diversity and Conservation and Plant

Biology and Biotechnology respectively from the University of Benin, Nigeria. After receiving his Ph.D., he worked as a Researcher at the Centre for Floristic Research of the Apennines, Italy where he contributed to numerous European Union-funded research including the famous NATURA 2000 project for sustainable ecosystem and biodiversity management. Before that, Dr. Ogwu was a Lecturer at the University of Benin, Nigeria, where he taught courses in plant biology and biotechnology, biometry, economic botany, cytogenetics, etc., and supervised and co-supervised honors and graduate students' theses, respectively. Dr. Ogwu is an interdisciplinary academic with transdisciplinary skills pertinent to the assessment of coupled human and natural as well as socio-ecological systems and has numerous awards, research grants, and scholarships to his name. His research and teaching revolve around sustainable biodiversity and ecosystem management by assessing the realm of coupled systems for sustainable environmental development as well as addressing sustainable development issues arising from various scales of biogeographical interactions. Some of his research interests include soil microbial and geographical ecology, molecular and genomic characterization of economic plant species for breeding, conservation, and sustainable utilization as well as biodiversity issues and policies, climate change science, gut and fecal microbiota, bioinformatics, biosystems and bioeconomy, sustainable agriculture and food production systems, ethnoscience (including traditional biotechnological practices), land use patterns, waste management, microbial diversity, environmental and social systems integrity, landscape preservation, and pollution ecology. He is spearheading some convergence works (One Health and Eco Health) in the Human Environmental and Agricultural Laboratory (HEAL lab) at Appalachian State University. Dr. Ogwu has published over 100 peer-reviewed works including several in high-impact journals and has presented his results at different international conferences. Dr. Ogwu has attended numerous professional courses and serves on the board of and as a reviewer for many peer-reviewed international journals. He continues to volunteer his time and skills to promote sustainable community development, especially in the Global South. Dr. Ogwu is fluent in several languages and loves to travel and meet new people.

Contributors

Maduamaka Cyriacus Abajue Department of Animal and Environmental Biology, Faculty of Science, University of Port Harcourt, Port Harcourt, Nigeria

Bukola Omotomilola Adetola Department of Ecotourism and Wildlife Management, School of Agriculture and Agricultural Technology, Federal University of Technology, Akure, Ondo State, Nigeria

Victor A. Agidi Department of Environmental Management, School of Environmental Sciences, Federal University of Technology, Owerri, Nigeria

Ayobami Omozemoje Aigberua Department of Environment, Research and Development, Anal Concept Limited, Elelenwo, Rivers State, Nigeria

Godfrey C. Akani Department of Applied and Environmental Biology, Rivers State University, Port-Harcourt, Nigeria

Oluseun Adesola Akinsorotan Department of Wildlife and Ecotourism Management, Faculty of Renewable Natural Resources Management, College of Agriculture and Renewable Resources, Osun State University, Osogbo, Osun State, Nigeria

Grace N. Alawa Department of Applied and Environmental Biology, Rivers State University, Port-Harcourt, Rivers State, Nigeria

Opeyemi Pamela Babafemi Global Affairs and Sustainable Development Institute (GASDI), Osun State University, Osogbo, Nigeria

Bernard B. Babatunde Department of Animal and Environmental Biology, University of Port Harcourt, Port Harcourt, Nigeria

Christopher Oluwasesan Bamidele Institute of Ecology and Environmental Studies, Obafemi Awolowo University, Ile-Ife, Nigeria

Bridget E. Diagi Department of Environmental Management, School of Environmental Sciences, Federal University of Technology, Owerri, Imo State, Nigeria

Uchechi Emmanuella Enenebeaku Department of Biological Sciences, Hezekiah University, Umudi, Imo State, Nigeria

Edem Archibong Eniang Department of Forestry and Wildlife, University of Uyo, Uyo, Nigeria

Ejiroghene Ruona Evivie Division of Plant Science, Research School of Biology, College of Science, Australian National University, Acton, Australia
Department of Plant Biotechnology, College of Life Sciences, Northeast Agricultural University, Harbin, China

Smith Etareri Evivie Department of Animal Science, Faculty of Agriculture, University of Benin, Benin City, Nigeria
Department of Food Science and Human Nutrition, Faculty of Agriculture, University of Benin, Benin City, Nigeria
Key Laboratory of Dairy Science, Ministry of Education, Northeast Agricultural University, Harbin, China

Abdulbasith Olabamiji Fajimolu Department of Wildlife and Ecotourism Management, Faculty of Renewable Natural Resources Management, College of Agriculture and Renewable Resources, Osun State University, Osogbo, Osun State, Nigeria

Tambeke N. Gbarakoro Department of Animal and Environmental Biology, University of Port Harcourt, Port Harcourt, Nigeria

Gijo A. Harry Department of Biological Sciences, Niger Delta University, Wilberforce Island, Bayelsa State, Nigeria

Sampson Abigha Inatimi Forestry Department, Bayelsa State Ministry of Environment, Yenagoa, Bayelsa State, Nigeria

Adams Ovie Iyiola Department of Fisheries and Aquatic Resources Management, Faculty of Renewable Natural Resources Management, Osun State University, Osogbo, Osun State, Nigeria

Sylvester Chibueze Izah Department of Microbiology, Bayelsa Medical University, Yenagoa, Bayelsa State, Nigeria

Daniel Etim Jacob Department of Forestry and Wildlife, University of Uyo, Uyo, Nigeria

Kia Tarela Jovita Department of Biological Sciences, Niger Delta University, Wilberforce Island, Bayelsa State, Nigeria

Ayotunde Samuel Kolawole Department of Fisheries and Aquatic Resources Management, Faculty of Renewable of Renewable Natural Resources Management, College of Agriculture and Renewable Resources, Osun State University, Osogbo, Osun State, Nigeria

Oluwatosin Olaoluwa Komolafe Institute of Ecology and Environmental Studies, Obafemi Awolowo University, Ile-Ife, Nigeria

Luca Luiselli Centre of Environmental Studies "Demetra" s.r.l. and FIZV, Rome, Italy

Aroloye O. Numbere Department of Animal and Environmental Biology, University of Port Harcourt, Port Harcourt, Nigeria

Chukwudi Nwaogu Department of Environmental Management, School of Environmental Sciences, Federal University of Technology, Owerri, Imo State, Nigeria

Emmanuel N. Ogamba Department of Biological Sciences, Faculty of Science, Niger Delta University, Wilberforce Island, Nigeria

Odangowei Inetiminebi Ogidi Department of Biochemistry, School of Applied Sciences, Federal Polytechnic Ekowe, Ekowe, Bayelsa State, Nigeria

O. M. Ogundare Department of Tourism Studies, College of Humanities and Culture, Osun State University, Osogbo, Nigeria

Matthew Chidozie Ogwu Goodnight Family Department of Sustainable Development, Living Learning Center, Appalachian State University, Boone, North Carolina, USA

Abiola Elizabeth Ojeleye Department of Agronomy, Faculty of Agricultural Production and Management, College of Agriculture and Renewable Resources, Osun State University, Osogbo, Osun State, Nigeria

Ehi Francis Okosodo Department of Leisure and Tourism, Federal Polytechnic, Ilaro, Nigeria

Simon I. Okweche Department of Forestry and Wildlife Resources Management, University of Calabar, Calabar, Nigeria

Moses Edwin Osawaru Department of Plant Biology and Biotechnology, Faculty of Life Sciences, University of Benin, Benin City, Nigeria

Pasiya P. Otufu Department of Animal and Environmental Biology, University of Port Harcourt, Port Harcourt, Nigeria

Morufu Olalekan Raimi Department of Community Medicine, Environmental Health Unit, Faculty of Clinical Sciences, Niger Delta University, Wilberforce Island, Bayelsa State, Nigeria

Abiola Omolewa Saliu Department of Plant and Environmental Biology, Kwara State University, Malete, Nigeria

John Valbo-Jørgensen Fisheries Division, Food and Agricultural Organization (FAO) of the United Nations, Rome, Italy

Baturh Yarkwan Department of Biochemistry, Joseph Sarwuan Tarkaa University, University of Agriculture, Makurdi, Nigeria
School of the Environment, University of Windsor, Windsor, ON, Canada

Chapter 1
Overview of African Biological Resources and Environment

Matthew Chidozie Ogwu, Adams Ovie Iyiola, and Sylvester Chibueze Izah

Abstract Africa has rich and diverse biological and environmental resources, but both are vulnerable to rising human population and the effects of climate change. Also, intense poverty and food insecurity are driving overexploitation of natural resources, while the absence or gross lack of technical know-how and unsustainable policies makes environmental degradation unchecked. Most of Africa's biological and environmental resources have high spatial and temporal variability. Africa's biodiversity and environmental issues require a holistic approach that incorporates sociocultural, political, and economic considerations. This chapter aims to present an overview of perspectives presented in the book toward a common agenda. This book is a collection of works on the sustainable utilization and conservation of Africa's biological resources and environment and is divided into three sections. There are six biological prime spots in Africa - Mediterranean basin forest, Guinean forest, Eastern Arc Mountain Forest, Western Indian Ocean Island, Succulent Karoo, and Cape Floristic region. African terrestrial ecosystems include deserts (e.g., Sahara Desert) and semiarid deserts (i.e., the Kalahari Desert), grasslands and savannahs, forests, and montane ecosystems, whereas aquatic ecosystems include estuaries, wetlands, mangrove swamps, lakes, rivers, and coastal and marine ecosystems. African countries are signatories to the Convention on Biological Diversity, the African Convention on Nature and Natural Resource Conservation, the Ramsar Convention, Convention on International Trade in Endangered Species of Wild

M. C. Ogwu (✉)
Goodnight Family Department of Sustainable Development, Living Learning Center, Appalachian State University, Boone, North Carolina, USA
e-mail: ogwumc@appstate.edu

A. O. Iyiola
Department of Fisheries and Aquatic Resources Management, Faculty of Renewable Natural Resources Management, Osun State University, Osogbo, Osun State, Nigeria

S. C. Izah
Department of Microbiology, Faculty of Science, Bayelsa Medical University, Yenagoa, Bayelsa State, Nigeria

S. C. Izah, M. C. Ogwu (eds.), *Sustainable Utilization and Conservation of Africa's Biological Resources and Environment*, Sustainable Development and Biodiversity 32, https://doi.org/10.1007/978-981-19-6974-4_1

Fauna and Flora, Convention to Combat Desertification, the Climate Change Framework Convention, the United Nations Convention on the Law of the Sea, and Agreement on the Trade-Relates aspects of intellectual property rights for the protection and conservation of Africa's biological resources and environment. An Integrated Conservation and Development Project approach is recommended as the most suitable conservation and utilization system for Africa. Further, to balance development with biological and environmental sustainability in Africa would also require an understanding of the capacity of the continent's resources, producing and applying this knowledge to distribute the resources for efficient utilization and management, and then modelling current use with population growth estimates and current and potential threats to produce a viable system for the future.

Keywords Biodiversity conservation · Environmental laws and policies · Resource utilization · Integrated systems approach · Community participation

1.1 Introduction

Africa has rich and diverse biological and environmental resources, but both are vulnerable to rising human population and the effects of climate change. For instance, six biodiversity hotspots are found in Tanzania alone while the horn of Africa, Cape Floristic Region, Guinean forests, coastal forests of East Africa, Congo Basin, and Afromontane regions hold significant amounts of plant biodiversity (Marshall et al. 2016; Treurnicht et al. 2017; Selemani 2020). However, intense poverty and food insecurity are driving overexploitation of natural resources, while the absence or gross lack of technical know-how and unsustainable policies makes environmental degradation unchecked (Ogwu 2009, 2010). Even though some protected areas exist for the management and conservation of natural resources for environmental sustainability and biodiversity preservation, it is affected by management inefficiencies, corruption, weak government policies, political instability, exponential population growth, and emigration (Ogwu 2019; Osawaru and Ogwu 2014; Ogwu et al. 2014, 2016). Policies, laws, and legislations geared towards sustainable approaches and strategies where they exist are mostly not well implemented. Africa remains the poorest region in the world, but the population is projected to exceed one billion in 2050. This growth is anticipated to increase the dependence on natural resources for survival, especially land, water, plants, forests, animals, renewable energy, and ecosystem services Selemani 2020).

African biodiversity supports human and ecological processes and is a good source of economic goods and services. African forests are among the richest in the world and contribute to treatment of diseases, i.e., medicinal plants (Bassey and Izah 2017; Seiyaboh et al. 2020a, 2020b; Kigigha et al. 2018a,b, 2016a,b,c, 2015a,b; Izah et al. 2019a, b, c, d, a, b, c, d; Izah and Aseibai 2018; Izah 2019; Izah and Youkparigha 2019; Youkparigha and Izah 2019; Enaregha et al. 2021), global fuel wood, non-timber, and timber forest products used for numerous industrial processes. Indigenous people rely on their environment for food and income (Izah 2018; Izah et al. 2018e; Izah et al. 2017; Izah and Seiyaboh 2018a, 2018b). And even though the adoption of science and technology innovations and their application can contribute to poverty reduction, food, and income security, and environmental sustainability objectives, much has not been achieved (Ogwu et al. 2016a, 2016b, 2016c; Ikhajiagbe et al. 2020). To meet the pressure from international and local markets, indigenous biological resources in Africa are harvested in a destructive and unsustainable manner, whereas environments are rapidly degraded. This trend is threatening the livelihood of biodiversity-dependent communities in Africa due to biodiversity loss and ecosystem degradation (Lettington 2000). The relief and slopes in the environment can be used to significantly improve livelihoods, but benefits from the tourism sector such as economic, social, cultural, employment, and others do not effectively trickle down to local communities. Monocultures are becoming more common in agricultural fields as well as the adoption of fertilizers for food production, despite attendant environmental risks and lessons from the Global North. It is high time that the interconnection and relatedness of biological and environmental resources and livelihood in the continent are recognized and prioritized in management systems. Earlier Brady (1988) mentioned that a significant amount of biodiversity and environmental resources could be lost and this would have negative effects on society and long-term development because of impaired ecological and socioecological systems like water regulation and regeneration of natural resources (including plants and animals), unchecked tourism, poaching, nutrient cycling, property values, etc. The realm of ecosystem resilience and rural livelihood is important in this era of climate change, globalization, and the mass extinction of biodiversity (Olsson and Ouattara 2013).

Little is known about Africa's biodiversity and environment, but they are been lost rapidly and their demand is constantly increasing (Sosef et al. 2017; Ogwu 2020; Osawaru and Ogwu 2020). Even areas with a relatively limited anthropogenic legacy on the continent is experiencing significant change in biological and environmental resources including marine and terrestrial systems and those within conservation areas (Chown 2010; Ikhajiagbe and Ogwu 2020). Most of Africa's biological and environmental resources have high spatial and temporal variability (McClain 2013). Some biological resources like *Bos* species (indigenous African cattle) are endangered due to uncontrolled crossbreeding and the introduction of exotic species, despite their unique adaptive traits (Mwai et al. 2015). According to Davids et al. (2018) and Wei et al. (2020), rapid rates of urbanization and population growth in Africa make it difficult for African ecosystems to provide for human well-being. Recently, Seymour et al. (2020), using the popular Delphi method, produced a list of 63 biodiversity issues with ten priority areas for parts of sub-Saharan Africa, most of which cannot be addressed by science alone. Therefore, Africa's biodiversity and environmental issues require a holistic approach that incorporates sociocultural, political, and economic considerations. Development of this all-inclusive approach would require horizon scanning, scenario planning, research insights, questions, and response, identification of priority areas, adoption of new methodologies and technologies that allow anticipation planning and ongoing responsive action together (Sutherland et al. 2010; Arts et al. 2015; Fleishman et al. 2011; Mitchell et al. 2015; Brown et al. 2016; Souza and Bernard 2019; Seymour et al. 2020). Social science has an important role to play in the renaissance and adoption of a holistic approach to biodiversity conservation and environmental sustainability in Africa. This could take the form of identifying potential conflict areas or the elucidation of best transformation methods (Reihling 2008). In the view of Chown (2010), the growing polarization highlights a disconnect between social and scientific systems to address changes in biological and environmental resources.

This book is a collection of works from diverse global scholars, academics and professionals, practitioners, experts, etc. on the sustainable utilization and conservation of Africa's biological resources and environment. It is divided into three sections—Conservation of Africa's Biological Resources and Environment (Sect. I), Utilization Patterns and Potentials of Africa's Biological Resources and Environment (Sect. II), and Challenges in the Conservation and Utilization of Africa's Biological Resources and Environment (Sect. III). This introductory chapter aims to present an overview of perspectives presented in the book towards a common agenda.

1.2 Conservation of Africa's Biological Resources and Environment

Conservation of biological resources involves the process of protecting, managing, and adequate use of resources for the sustainable achievement of various goals and activities which involves social, environmental, and cultural (Blanc et al. 2003). This benefits humans in the long run because they can, directly and indirectly, utilize these resources to achieve various individual and collective goals and objectives to improve well-being and ensure societal development. According to Allison et al. (2000) and IUCN (2001), the conservation of biological resources may be tagged around five themes:

- Clear conservation targets, goals, and objectives as well as the conservation strategies.
- Social alliance, network, and processes involved should focus on equity and effectiveness and the establishment of valuable partnerships and human participation for effective conservation of resources.
- Equitable sharing of benefits and incentivizing the conservation process were possible and appropriate in line with needs.
- Local, national, and international biodiversity policies must be supportive, complementary, and applicable in the conservation of local resources.
- Creating awareness, establishing education programmes to promote knowledge, and building capacity for effective conservation.

Biodiversity can be broadly conserved in two major ways:

- In situ conservation: is the process by which biodiversity is conserved in its native range and habitat and can be in form of protected areas or natural reserves. This

form of conservation is common in Africa. Mburugu (1995) reported that Kenya has 26 national parks and 30 national reserves.
- Ex situ conservation: is the process of conserving biodiversity outside of its natural environment such as zoos, gene banks, botanical gardens, etc. This method requires adequate and intensive knowledge about the species and huge financial investments (African Wildlife Foundation 2003).

Other conservation approaches are, *circum situ* (circa *situm*) conservation where the resources are held in farmlands, home gardens, urban gardens, botanical gardens, distant farms, and other human-managed agroecosystems, quasi situ conservation where the resources are held in natural (e.g. original or primary forests), seminatural (e.g. secondary forest) or human-created environments but their cultivation and growth is not purposely carried out by humans, and inter situ conservation where the resources are maintained in semi-wild settings to ensure species survival and restoration outside their current range. These approaches have evolved over the years, and in most cases, involve the direct or indirect participation of local communities. Presently, policies and the participatory involvement of individuals and stakeholders in Africa need to be streamlined to develop a unified strategy for the continent.

1.3 Africa's Biological Resources

Africa is rich in diverse ecosystems which range from desert to moist tropical forests. The forest is estimated to be over 650 million hectares and represents about 17% and 22% of the forest cover and African land area, respectively (FAO 2010). The continent has the world's richest savannah regions with diverse mammals which feed on the abundant grasslands in the world. Animals such as hippopotamuses, leopards, lions, elephants, chimpanzees, gorillas, hyenas, and wild dogs are found in the savannah and rainforest regions in Africa (Van Drunen et al. 2009). The climate of Africa consists of temperate and tropical conditions diversifying between 14 different types of forests (Sayer et al. 1992). The tropical forests contain about 1.5 million species and 800 plant species with numerous wildlife species for consumption by locals as a source of protein (Table 1.1).

As reported by Mittermeier et al. (2004), there are a total of 34 internationally accepted biological hotspots in the world with Africa having six of them. These locations and their identified biodiversity are presented in Table 1.2.

Table 1.1 Contribution of fauna resources in African forest

S/N	Species	Contribution (%)
1	Primates	84
2	Passerine birds	68
3	Butterfly	66

Table 1.2 The six biological locations in Africa

S/N	Name	Location	Flora species	Fauna species
1	Mediterranean Basin forest	It is situated around the Mediterranean Sea.	25,000 species and 14 endemic genera	
2	Guinean Forest	It is located along the coast of Western Africa. The forest is fragmented.	2250 species	45 mammal species, 90 bird species, 46 reptile species
3	Eastern Arc Mountain Forest	It is situated in eastern Africa and is about 30 million years old	25% of the species are endemic	
4	Western Indian Ocean Islands	It is an isolated area situated around Madagascar which has the highest endemic species in Africa		700 endemic vertebrate species and ranked the world's vertebrate endemism as the 6th
5	Succulent Karoo	It is situated between Namibia and South Africa and it is the richest desert in the world	4849 species with 40% endemic	
6	Cape floristic region	It is located in South Africa	It is the richest in the flora kingdom of the world despite its size. 8700 species are found there with 68% endemic	

Table 1.3 Relative abundance of endemic species in Eastern Africa

S/N	Species	Contribution (%)
1.	Mammals	55
2	Birds	63
3	Reptiles	49
4	Amphibians	40

The abundance of endemic species in Africa is presented in Table 1.3. It shows that Eastern Africa has more endemic species when compared with North Africa. Madagascar is the country that has the richest endemic species, the third richest in plant species diversity after South Africa and Congo DRC, and the sixth in the world (World Resources Institute 2003).

1.4 The Diverse Ecosystems in Africa

African terrestrial ecosystems include deserts (e.g., Sahara Desert) and semiarid deserts (i.e., the Kalahari Desert), grasslands and savannahs, forests, and montane ecosystems, whereas aquatic ecosystems include estuaries, wetlands, mangrove

Table 1.4 Resources in Lake Tanganyika Basin

S/N	Resources	Species abundance
1	Cichlid fish species	250
2	Non-cichlid species	145
3	Gastropods	15
4	Copepods	69
5	Leeches	20
6	Sponges	9

swamps, lakes, rivers, and coastal and marine ecosystems. Savannah ecosystems are one of the most important areas in Africa. It houses over 40 different species diversity and hoofed mammals in the 13 million km^2 land cover. Animals found there are gazelles, onyx, eland, kudu, buffalo, zebra, giraffes, elephants, warthogs, and impalas. They graze on the abundant grassland all year round around the different migratory seasons. This ecosystem is found in the West, East, and Southern subregions of Africa (Mataruka 2009). Africa holds about 15% of the global forest area and may be considered the richest biological area of the continent. The forest is about 3.5 million km^2 in the area and includes the Congo basin in East-Central Africa, the Upper Guinea forest in West Africa, the montane forest of East Africa, Coastal forests in Eastern Africa, and Madagascar and Indian Ocean Island Forest. These forests are home to chimpanzees, bonobos, elephants, monkeys, antelopes, giraffes, pygmy hippos, etc. The compositional heterogeneity of Africa's Forest is ill-understood, albeit an understanding is required for conservation and sustainability (Réjou-Méchain et al. 2021). The Atlas Mountain ecosystem in Morocco supports diverse species and possesses a unique landscape. This resource is found in every country and covers 1% of the total area in each country. Some of the countries that have the largest wetlands area in the Upper Nile are swamps in Congo DRC, Chad and Lake Victoria basins, and Niger and Zambezi floodplains. Lake Tanganyika which is the second largest lake in Africa and third by volume in the world is the most important wetland in Africa. It has the richest freshwater ecosystem in the world with over 600 endemic species in its basin (Allison et al. 2000). The Lake also has over 2000 species of plants and animals and some of the resources in the lake's basin are presented in Table 1.4. A lot of species are observed to be migratory and a large number of them have not been described.

This area is characterized by coral reefs, mangrove forests, and sea-grass beds. It contains large diversity with over 4000 species being identified and described. Some are the marlin, tuna, billfish, sea turtles, and dugongs (Ibisch et al. 2010). Seven large marine ecosystem surrounds Africa and includes Agulhas current, Benguela current, Canary current, Guinea current, Mediterranean Sea, Red Sea, and Somali coastal current (Zeller et al. 2020).

1.5 African Biological Resources and Framework for Biodiversity Management

The biological resources in Africa are diverse and contribute to livelihood, development, and global well-being.

- The African forest is a source of traditional energy sources with about 64% of the populace depending on it for their livelihood.
- It is also a source of bush meat which provides animal protein as reported by FAO (2010) in southern Ivory Coast (70%), 80–90% protein source in Liberia, and 55% in Sierra Leone.
- It is a source of income, foreign exchange, and employment at the national level.
- It is important for cultural, spiritual, or religious purposes.
- It provides timber and non-timber forest products such as latex, gum, spices, flavorings, pesticides, and dyes.

Biodiversity comprises plants, animals, and microorganisms that dwell on earth and have different functioning parts. They dwell in terrestrial, marine, and other aquatic ecosystems with diverse complexities and can meet the need of people such as basic food, health, and energy needs. Swaminathan (1996) estimated that about 2.5 million people rely on wild animals and plant species for their daily needs. Based on this, there is a need to sustain and foster management of these resources to cater to over 800 million humans who suffer from malnutrition. To foster this, various international and regional policy and legal frameworks have been developed by the local and international communities for resource utilization and conservation (Gondo 2011). Some are such as:

- The Convention on Biological Diversity
- The African Convention on Nature and Natural Resource Conservation
- The Ramsar Convention
- Convention on International Trade in Endangered Species of Wild Fauna and Flora
- Convention to Combat Desertification
- The Climate Change Framework Convention
- The United Nations Convention on the Law of the Sea
- Agreement on the Trade-Relates aspects of intellectual property rights

1.5.1 Convention on Biological Diversity

The convention in which 49 African states are a party majorly seeks to address and recognize the following:

- Promoting the conservation of biodiversity.
- Sustainable use of the components of biodiversity.

- The benefits from the use of resources.
- Promotes a global in situ cooperation for sustainable management.
- Relevance of property rights in the management of resources.
- The roles of local community dependence.
- The roles each community plays in the conservation and sustainable use of resources.

The sovereignty over natural resources is held by this convention in high esteem and it is seen as a common concern for humans. All the member states are responsible for conserving all their biological resources using measures, which can be in situ or ex situ (Gobeze et al. 2009).

1.5.2 The African Convention on Nature and Natural Resource Conservation

This convention ensures the following for its member states:

- Oblige measures for conservation and development of natural resources.
- Protects animal and plant species that are threatened with extinction.
- Establish areas of conservation to protect all species and ecosystems.
- Protection of habitats that are outside the protected areas in their development plans.

To complement the activities of the convention, it is essential to take measures that are legislative and necessary for the implementation of the provision of the convention.

1.5.3 The Ramsar Convention

The convention was ratified by 28 African countries and is concerned with the protection of natural habitats. It is also known as the Convention on Wetlands of International Importance Especially as Waterfowl Habitat. It emphasizes the following:

- Establishment of natural reserves in wetlands.
- The need to conserve wetlands.
- Sustainable utilization concerning the maintenance of natural ecosystem properties.

Currently, the conference has adopted guidelines that involve the participation of locals and indigenous people in the management of natural sites.

1.5.4 Convention on International Trade in Endangered Species of Wild Fauna and Flora

CITES provides international control over wildlife products. It emphasizes the following:

- The control of wildlife products.
- Identification of endangered species and their withdrawal from the world market.
- Creation of the CITES appendix list which is determined by the Conference of Parties and contains threatened species and species that may face extinction in the nearest future.

CITES allows states that are members of the party to notify other members of trade restrictions on species (Galaz et al. 2008).

1.5.5 Convention to Combat Desertification

This convention is applicable in countries experiencing droughts and desertification in Africa (Folke et al. 2005). Fifty-two African countries are members of the party and it emphasizes the following:

- Provision of framework against desertification.
- Migrating effects in states experiencing droughts and desertification.
- Affected parties to bear the main actions against desertification.
- Collaboration between countries, governments, locals, and groups in the community.

Members of the parties are expected to establish protocols to reduce desertification, strengthen existing legislation, and enact new laws that may not be in place.

1.5.6 The Climate Change Framework Convention

This convention addresses the global warming issues at an international level. The Kyoto convention addresses the quantification of emission, limitation, and reduction commitments by countries undergoing economic transmission. For the implementation of the convention, measures such as the provision of financial resources and the arrangement of facilities to enhance the context of the Kyoto Protocol were put in place. The convention addresses the following areas:

- Cost-effective climate change mitigation services.
- Attract new funders for countries.

- The synergy between developed and underdeveloped countries in fund investments for climate mitigation, so the objective of the convention can be achieved.
- Projects must enable sustainable developments in developing countries.

The relationship between climate change with agriculture and biodiversity is an intrinsic one because it can affect the growth and regeneration of trees which play a major role in carbon sequestration in the environment.

1.5.7 The United Nations Convention on the Law of the Sea

The convention addresses issues in the marine areas and provides measures that are related to the conservation and management of marine resources (FAO 2010). Some of the keynotes are:

- The creation of the Exclusive Economic Zone in which countries are given rights for resources within this zone.
- Provision of laws and obligations for coastal states.
- Ensure proper maintenance of living resources.
- Ensure there are no cases of overexploitation and that endangered species are preserved.

1.5.8 Agreement on Trade-Related Aspects of Intellectual Property Rights (TRIPS)

It consists of 41-member states which are parties to this agreement and are indirectly concerned with issues on biological resource conservation and management. The TRIPS Agreement highlights the following:

- Intellectual property rights and impacts on biodiversity management
- Exclusion of patentability by states in cases of human, animal, or plant protection
- Exclusion of patentability on plants, animals, and microorganisms to avoid prejudice against the environment.
- Protection of plant varieties by member states

1.6 Trends in Biological Resources Management

The issues of law and policy creation and enforcement have been an issue since time immemorial and increased compliance can be geared toward enforcing appropriate measures to safeguard the environment (Elson 2010). Some of the highlights are:

- The making of laws and policies in Africa and
- The influence of international law on member states.

During the colonial periods and immediately after in Africa, laws created at that time were basically on intense resource exploitation for exports. Conscious efforts on resource conservation, especially on wildlife, resulted in the creation of laws in the early 1900s. However, these laws were not effective in addressing issues on sustainable management of natural resources. The conclusion of the Landmark International Environmental Agreements has made the laws on natural resources conservation common in African countries (Millennium Ecosystem Assessment (MEA) 2005). It elaborates the following:

- Justification of private property holders' limitations in the interest of the general public
- The doctrine of police power in which each country based on its sovereignty must not harm the welfare of the public
- Fundamental rights and freedom must be guaranteed and stated in the constitution of member states

Countries such as Uganda, the Gambia, Guinea, Comoros, and South Africa have included in their constitutions and adopted the legislative framework on the bill of rights relating to the environment. It is explicit on:

- Protection of the environment.
- Sustainable management of the environment.
- Promotion of environmental awareness.
- Rights by everyone to clean environments.
- The use of ecologically sustainability measures on natural resources toward economic and social developments.
- The principles of environmental management.
- The setting up of environmental funds.
- Creation of an institutional framework for the management of biodiversity.

Apart from these laws and frameworks, some countries have environmental standards for the management of natural resources. A case is the Malawi Environmental Management Act which ensures the harmonization of laws and policies on the environment across the departments and environmental ministries in the country (World Bank 2010).

The environmental law, policy, and development in African countries are influenced by the international treaty and customary laws in the country. The influence of these treaties extends to the obligations and principles of each state that is a member of the party. Some of the influences on the principles are:

- Influence of international law and the
- Influence of conventions.

The first environmental law principle is aimed at ensuring that their activities do not damage the environment. This is the major principle of international

Table 1.5 Environmental Acts by countries and their principles

S/N	Act	Principles
1.	Congo Environmental Act	It states that developmental project should include an environmental impact assessment.
2	Ugandan Environmental Act	Environmental management should aim to encourage maximum participation by the populace in policy development, management plans, and processes.
3.	Angolan Environmental Act	Seek to provide a framework for cooperation with other countries and international organizations to provide solutions to common environmental problems.
4.	Mozambican Environmental Act	Activities involving environmental management must be done to avoid negative impacts on the environment.
5	Cameroon Environment Act	It is based on a practical rationale for environmental management.

environmental law and it enforces the principles of sustainable development. It has been incorporated into the general principles of national development and strikes a balance between economic growth and environmental protection. Various principles have been incorporated into domestic laws such as (Table 1.5):

- Polluter Pays Principle which states that any polluter will bear the cost of cleaning up and removing the effects of the pollutants.
- The precaution and Prevention Principle considers the fundamental principles on which rational environmental management is based.
- The need for environmental impact assessment.
- The need for public participation and cooperation.

The wildlife laws are largely influenced by the international regions and laws and the development of conservation strategies and sustainability. The African convention has provided a framework for the development of many wildlife laws in Africa. It has incorporated the principles of the 1900 and 1933 Convention which was signed by the colonial powers and is geared at flora and fauna preservation in its natural state. CITES has also influenced the wildlife laws and policies in Africa and has been ratified by 47 African states. Several countries have adopted this legislation in the trade of endangered flora and fauna species. A case is the Kenya Wildlife (Conservation and Management) Act which has banned all game animal hunting and licenses revoked in the wildlife trade. This was concerning the adoption of CITES in its activities. Recently, the levels of management have changed at international and national levels and the processes involved. In Ethiopia, the national conservation strategy has been adopted and focused on preservation, development, management, and conservative use of domesticated and wild flora and fauna diversities (Kiefer et al. 2010).

1.7 Sustainable Initiatives for Biodiversity Management

Some initiatives on the legal and institutional frameworks adopted for biodiversity conservation and management by countries in Africa are:

- Trade and management of wildlife
- Forest management using economic instruments

1.7.1 Trade and Management of Wildlife

In most African countries, measures for the management of wildlife resources are in place. They are incorporated into the national wildlife act and national wildlife policy and legislation. These policies also cover both fisheries and forest legislation act. The policies are expected to regulate the trade in the country's biological resources by emphasizing fisheries, wildlife, and forest ecosystems. Some of the management acts are highlighted in Table 1.6. The cases from Kenya and Zimbabwe as illustrated in the table show the actions based on CITES with no regard for domestic wildlife management. The wildlife management strategies in Zimbabwe are more community-based programmes that encourage management, while in Kenya, it is more of a preservationist strategy of management (Waheed et al. 2009).

Table 1.6 Conservation Acts in some African countries

S/N	Country	Act	Highlights
1.	Kenya	Wildlife Conservation and Management Act which was amended in 1989	• To control illegal access to wildlife exploitation. • Authority must be granted to individuals/institutions before wildlife extraction. • Approvals are to be granted by the Minister of Natural Resources and the Kenya Wildlife Service. • The rights do not cover all benefits from the exploited resources.
2.	Zimbabwe	Zimbabwe Parks and Wildlife Act of 1975	• Protection of indigenous fauna resources. • Permits are required before the collection, export, cultivation, or propagation of fauna species.
		Parks and Wildlife Amended Act of 1985	• The collection of indigenous plants can be prohibited by the Minister. • It does not allow foreign collectors to share benefits from the resources exploited.

1.7.2 *Forest Management Using Economic Instruments*

Most African countries have legislation on forest exploitation and management and these laws differ from place to place. In general, these laws are concerned with controlling the illegal access and exploitation of forest resources. In Cameroon, the forest legislation was established in the mid-1970s and it highlighted the following:

- Forest legislation in the trade regulation and export of *Prunus africana*
- Licensing by the Forest Administration before exploiting *P. africana*

In recent times, the issue of climate change and its convention has influenced forestry projects. This is so as a result of carbon sequestration and carbon conservation through the promoted growth of forests and protecting the existing ones. Degraded forests can be regenerated as well as expanding forest ecosystems through plantations. Legislation on the efficiency of the use of fuel wood by encouraging the use of stoves and reducing waste from wood from logging activities has been made (Van Herwijnen 2008).

The management of biological diversity is protected by the international legal system. This is used to create a framework for accessing biodiversity policies in Africa. Some of the concerns are:

- The biological resources exclusively
- Environmental issues
- Some are in agreement with laws that are concerned with biodiversity management

In most African countries, international law influences environmental laws and policies. This is seen as a positive step to an effective biodiversity management system and it can incorporate the cross-sectional approaches and policies which form part of the integrated approach system for biodiversity management. Based on this, it is essential to formulate biodiversity policies and laws and assess them with the present situation (Scherr and McNeely 2007).

1.8 Conservation and Utilization of Africa's Biological Resources and Environment

An Integrated Conservation and Development Project (ICDP) approach is recommended as the most suitable conservation and utilization system for Africa. African conservationists aim to develop approaches that can protect the natural resources of the continent. This will help to create long-term viability of flora and fauna species and preserve the natural ecosystems. This will benefit the political and economic perspectives of the communities as well as the government if such designed conservational strategies are workable. The ICDP approach has been used in recent years and has been linked to the conservation of biological resources within a protected area to social and economic development outside the protected areas. In this approach, the community is directly concerned with decisions involving employment, revenue sharing, provision of infrastructure, and exploitation of limited flora and fauna species. The approach started in Africa in the 1990s, and currently, the ICDP approach is used for conserving protected areas in Africa (Newmark and Hough 2000). This approach receives funding from major donors, but was stated that their successes are limited in conservation and developmental objectives.

The ICDP approach has increased in popularity due to the following:

- Habitat loss has declined drastically over the last 30 years. Kiss (1990) reported that about 56% of the original African wildlife habitat has been lost due to agricultural activities, overgrazing, and deforestation which has resulted from increased human population and poverty level. Based on this, conservation activities must be connected with development.
- It addresses the challenges associated with biological diversity in protected areas. These areas have become isolated in Africa as a result of human settlement, deforestation, and hunting of wildlife (Singh et al. 2009).
- The programs are effective in addressing social injustice issues. This is so because protected areas have affected indigenous African people and donors see the ICDP approach as a way of developing relationships with communities that bear a large proportion of the social costs incurred from the protected areas.
- It is attractive because it identifies and recognizes the old methods of management which were not effective and have created issues of confrontation among the locals.

However, there are two striking issues regarding the assessment of ICDP's effectiveness in Africa. One of these is the belief among workers that ICDP has not made modest progress or achieved all its aims which have limited its successes. This belief is due to assessment problems, internal constraints, and external challenges. Evaluators of projects have identified two assessment constraints that can affect the objectives and successes of ICDPs and include -

1. Project assessments are only at the implementation stage (Wells et al. 1992). It was concluded from this assessment that successes in conservation and developmental objectives were limited. It was observed that the period of 3–5 years for a project cycle was very inappropriate as compared to the project cycle time in the 1970s which was much longer and objectives were achieved.
2. Most projects lack ecological monitoring. In 1994, 36 projects were examined all over the world, out of which 23 were from Africa. It was observed that there were only two projects that had some form of ecological monitoring components. The lack of this component has prevented rigorous evaluation of non-resource exploitation, developmental activities, and biodiversity conservation.

Regarding internal constraints, project evaluators observed four constraints in many ICDPs, which include:

1. The behavior of individuals is not affected by public goods. This is important because ICDPs failed in their conservation goals because of the incentives the communities present and are insufficient to change the behavior of individuals. However, the effects of these incentives may vary based on the groups within a community.
2. The similarity between organizational structure in many ICDPs to colonial structures. This affects many local people because they are disconnected from ICDPs as a result of the state's ultimate authority over wildlife resources. They believe that the authorities should come ultimately from the local communities (Minnemeyer et al. 2011).
3. The unsustainable long-term harvesting scheme off-take. The large mammal harvesting which is rampant in most savannah areas is unsustainable because the wildlife populations are inherently variable. Therefore, most wildlife managers will want to maintain a constant flow of benefits to communities such as skins, meat, or revenues from wildlife exploitation and this becomes an issue when there is a decline in the wildlife resources. This can result in reduced community participation because of the reduced off-take.
4. The frequent conflicts between developmental activities and conservation objectives. These conflicts mostly arise from managers' ineffective control over the wildlife resource exploitation by individuals or the communities. Such was reported in ICDP in Tanzania that had fish ponds in a wetland in the East Usambara Mountains which disrupted the riparian habitat despite an additional protein source to the villagers (Nyenje et al. 2011).

Regarding challenges from external influences, three forces were identified to affect ICDPs in Africa and they are -

1. The potential sources of revenue for communities are unreliable and insufficient. This is due to fluctuation in exchange rates, political issues, and reduction in tourism activities as a result of unstable government activities. The dramatic decline in the tourism industry observed in Kenya, Uganda, Comoros Islands, and Zimbabwe was attributed to political unrest and economic issues. Some communities may not benefit from tourism revenues because management costs for the protected areas exceed what has been got from sales of gate receipts.
2. The influence of external market forces manipulates the pattern of resource use in Africa. The increased urbanization has placed a huge demand on meat, timber, and firewood in African cities and towns. These communities exploit natural resources which may limit the efficiency of ICDP activities. A situation was observed in the population of Cape buffalo in Serengeti National Park which declined between 50% and 90% as a result of commercial poaching which was encouraged by increased market forces.
3. The increased migration into the project areas was due to developmental activities. This was observed around Lake Manyara in Tanzania in the early 1980s in which there was a 40% increase in population between 1978 and 1998. This was due to the United Nations-supported irrigation project in the area. This stimulated rapid development which stimulated in-migration.

1.9 Assessing the Progress Towards African Sustainable Development

The following is an inexhaustive list of tools that may be used to assess the progress of sustainable biological resource and environmental development in Africa -

- Governance
- Demographic changes
- Poverty
- State of the economy
- Health
- Social equity and opportunities
- Agriculture, food security, and nutrition
- Education
- Climate change
- Natural resource base
- Energy
- Natural disasters

1.9.1 Governance

Africa has made progress and improvements in social, economic, and environmental sustainability (AfDB 2011). These incremental advances are improving the level of governance at various levels and institutional mandates. It has been observed to influence sustainable development efforts and participation at various governmental levels as well as the natural resource management strategies (Karembu et al. 2009; Armitage 2008). However, much is still required in terms of a stable governing structure required to enforce conservation treaties and policies. These results would be observable in the positive economic and political trends that result from peace and security (Olsson et al. 2006).

1.9.2 Poverty

It has been realized that the increase in GDP is not a reflection of economic well-being in most African countries and progress is being made to address connected issues. Successes have been reported in amenities such as drinking water and infrastructure and stagnation continued in areas like sanitation (Cundill 2010). The rural and urban populations have inadequate access to improved energy and water resources and interventions are required for the provision of improved services such as sanitation, water, energy, and decent housing to eliminate poverty and improve developmental trends (CBD Secretariat 2010).

1.9.3 Demographic Changes

The African population is expected to double by 2050 which is a significant threat to the attainment of sustainable development outcomes. This is due to the increased demand for goods and services and excessive pressures that may arise on natural resources. Africa requires an educated and right skilled population base that is meaningful to achieve the required sustainable goals. Foreseen challenges are seen in cities in terms of basic services such as water shortage, sanitation, transport, security, and adequate housing and they require specific attention.

1.9.4 State of the Economy

Africa is making progress in economic growth and macroeconomic performance. Value addition and tourism contribution are spurred by policy adoption, improved management of the macroeconomy, and governance. Many countries are sustaining economic progress and benefit from social and economic outcomes. The crisis recorded globally in 2009 has drastically affected the economy of Africa, although the rate of recovery from this has been fast. This was mainly propelled by a change in commodity pricing, remittances inflow, and an increase in foreign investments. It is necessary to increase efforts to track the positive performance and identify any necessary factors that can deplete the natural resource base and poverty reduction. Diverse opportunities in value addition and sustainable development of tourism will be required for robust and sustainable economic development (Carpenter et al. 2006).

1.9.5 Sustainable Consumption and Production (SCP)

Africa is making dramatic progress in terms of sustainable production, but sustainable consumption efforts need to be intensified so that the basic needs of the growing population are met (Brockhaus et al. 2012). This concept is relatively new to Africa and it is yet to be fully adopted. The effects of food consumption and production are:

- Impacts on transportation,
- Impacts on processing, packaging, and retailing of food and
- Food wastes at the point of consumption

The middle class is the target of most SCP campaigns and in achieving its objectives, organizational challenges must be addressed in Africa. Some of the challenges are:

- Poor institutional capacity
- Lack of monitoring
- Inadequate capacity
- Lack of consumption and production models (Romano and Reeb 2006).

1.9.6 Social Equity and Opportunities

A major challenge in Africa is inequality most especially concerning the distribution of income and the economic development outcomes. Significant progress is observed in women's empowerment and it has shown a positive trend in gender distribution of opportunities in health and education. Attention is required to women and youth participation in sustainable development and access to employment and education (Biermann et al. 2010).

1.9.7 Education

The progress in education in Africa is slow but on the path to meeting education-related targets. Adult literacy is slow and unable to meet the 2015 targets and interventions are required to link education and culture with the aim of transformation. Improvements are very clear in gender parity, basic education, education coverage, and enrolment (Olorunfemi and Raheem 2008). Policies are being designed in the educational and cultural areas for shaping the future of Africa and society. Curriculum and educational system transformation should be undertaken by Governments which can include teaching, learning materials, and contents (OECD 2011).

1.9.8 Health

The increased health issues and diseases are making developmental progress slow in Africa. Maternal and child mortality cases and major diseases like HIV/AIDS and malaria have not met the 2015 targets. It is important for the government to enhance improvement in primary health care and intensify preventive methods and control measures of diseases that are communicable or noncommunicable (Timko 2011).

1.9.9 Agriculture, Food Security, and Nutrition

Structural transformation of the agricultural sector is insufficient because raw materials are not evidenced in the agricultural sector. This has prevented agriculture from attaining its full potential. Efforts have been made to make agriculture sustainable by adopting practices that can enhance competitiveness and productivity and do not pose any risk to natural resources. The goals of food security, nutrition, and health will be achieved when there is increasing agricultural productivity on existing land and reducing food wastage.

1.9.10 Natural Resource Base

The increased human population has resulted in a continued depletion and degradation of natural resources and poverty reduction. There is also a difficulty in water resources management and people living in stressed environments. Appropriate technology is required for improved usage of water in sectors of agriculture, industry, tourism, and application domestically. Degradation of land is also a constraint in sustainable development and increased farming and agricultural land-based products. The sustainable exploitation of resources is a challenge with limited value addition, benefits to people, and environmental degradation. Efforts need to be managed and ensure economic, social, and environmental benefits presently and in the nearest future (UNEP 2007; UNEP 2006).

1.9.11 Energy

The sustainable development and diversification in the energy sector are stagnant and there is a need for modern access to services that are clean, safe, and reliable. Africa has a great potential for renewable and nonrenewable energy which remains untapped due to the limited investments in this sector and the high risks involved in its exploitation (UNESCO et al. 2011). Despite these potentials, the economy of Africa has shown no significant improvement, and energy poverty is increasing.

1.9.12 Climate Change

It is a major challenge to climate-sensitive sectors and their development. The cost of mitigation and adaptation is high and the African economy is vulnerable to its cost which has continued to escalate. The greenhouse emissions in Africa are increasing based on land use and its changes and deforestation which are major sources of these gases. Adaptation to climate change is very important as humans and natural systems need to adjust in their activities. Adaptation is more important than mitigation in Africa and is of major concern (IPCC 2007; Juma and Serageldin 2007). The support from developed countries in regards to adaptation and mitigation measures is crucial to fully integrate African countries into development planning and reduction in poverty.

1.9.13 Natural Disasters

The increased occurrence and frequency of disasters have a great impact on sustainable development in Africa. Particularly, the poor are vulnerable over time with impacts on social and environmental impacts. There should be increased efforts on monitoring and warning signs so that the responsive capacity of people can be built.

1.10 Contribution of Some Sectors to Sustainable Growth and Development in African Countries

The African natural environment is composed of various aspects whose potential may be harnessed for sustainable development. For example, the roles forest play in sustainable development are very crucial and the loss of forest cover is a major concern. Activities such as the expansion of agriculture, settlements, illegal logging, and uncontrolled bush fires can pose a threat to forests. Recently, policies and economic constraints have impeded the efforts of sustainable forest management. There is now a conscious effort to reverse the loss of forest trend and achieve sustainable forestry in Africa. Scale-up activities in afforestation, expansion of planting in riparian areas, and forest area protection should take preeminence in sustainable measures. Also, biodiversity loss indicates depletion and diminishing resources in a region and it can equate to a reduction in opportunities for addressing developmental challenges African countries are facing. The challenge majorly is the balance between development and conservation as well as the sustainable use of resources in the natural forests and environment. The incorporation of biotechnology in environmental management can contribute immensely to sustainable development in Africa. This approach has provided opportunities for the following:

- Poverty alleviation
- Enhanced food security
- Industrial competitiveness
- Promotion of sustainable use of natural resources

There is a need to adopt strategies and policies at local, community, national, and regional levels that are data and knowledge-driven. These would seek to harness economic, industrial, health, and environmental benefits. The conservation of mountains is critical in sustainable developmental strategies. If the ecosystem benefits are understood, there will be additional benefits and alternative sources of livelihood. The participation of indigenous people and the community will help to reduce all forms of resource-use conflicts and promote development in the mountain areas. There is a need to synergize the implementation of initiatives to promote sustainable development and management systems. Africa is rich in tourism activities and the developments have enabled the country to capitalize on its achievements. Although there are obstacles affecting the sector, improvement of these areas will promote sustainable development. These potentials can contribute immensely to the economy of the African regions.

1.11 Future Direction for Conservation Initiatives in Africa

Based on the present situation and challenges, a conservative initiative for the future needs to be addressed for increased economic development in the future. Some of the challenges are:

- Measures for ensuring that ICDPs address the complexity in their social and ecological environments must be developed. This will effectively monitor, analyze, and adapt to the changes as they arise in the environment.
- There is a need to assess, implement, and evaluate approaches to ICDPs that address external forces that affect them. This is possible through economic and land policy reforms, conservation and resource planning, and capacity management in protected areas.

This reform can help in reduction in environmental pressures on the protected areas. Examples of such pressures are the market forces on the resources and in-migration as a result of development. Land use activities need to be compatible with wildlife conservation strategies because most protected areas are small and becoming isolated ecologically. Activities that are incompatible with the natural environment along the wildlife corridors and wildlife dispersal zones should be avoided. In the African savannah regions, pastoralism is very compatible with the conservation of wildlife than agricultural practices, therefore efforts must be geared at maintaining the pastoral systems around the protected areas.

Dialogue is very crucial in developmental and conservation activities; it can be between managers and local communities or protected areas stakeholders. This enables the identification of areas with a common interest and necessary measures for advisory management between park managers and local people. These dialogues are inexpensive, easy to implement, and can reduce unnecessary tensions between park authorities and local community dwellers. Community-Based Natural Resources Management (CBNRM) approaches have been used in Zimbabwe, Zambia, and Namibia and have recorded great successes. This approach is used in land-use regimes around protected areas for natural resource conservation. CBNRM approach differs considerably from ICDP because it develops management responsibilities for natural resources management between wildlife and local communities rather than developmental services in exchange for conservation. It is geared toward communities seeing values in wildlife management on a long-term sustainable basis rather than exploitation on a short term or the alternative use of land.

The management capacity between protected areas and local communities is limited. The capacity of a protected area to address conflicts that may arise can be enhanced in the following ways:

(i) Planning of Park administration
(ii) Development of courses and scholarships
(iii) Development of exchange programmes and training activities
(iv) Technical assistance
(v) Focus on ecological and social monitoring of protected areas
(vi) Conflict resolution and park planning

1.12 Conclusion

Africa needs a quantifiable and scalable approach for recording and measuring the irreplaceability of its unique environments that hold significant amounts of biodiversity. The African Union should scale-up plans to ensure environmental stability and biodiversity conservation within and outside the range of protected areas. This can be done by linking conservation data and knowledge to action. Another key strategy would be to understand coupled environmental interactions and biodiversity distribution and use the knowledge of what influences them to design effective management strategies. No doubt, the unpredictability of environmental change from emergent properties and human impacts on biodiversity will make it complex to manage, but it is known that biodiversity and environmental conservation is a "crisis discipline". Although none of the issues highlighted from the horizon scan of sub-Saharan Africa's biological and environmental is chaotic, the future is unpredictable and would need strategic communication, continuous engagement, and education to understand changes (Seymour et al. 2020).

Developmental goals in Africa need to be aligned with biodiversity and environmental sustainability goals if the integrity of the system is to be maintained in the long run. At the center of the development and sustainability conflict in Africa are scientific bias, poverty, diseases, non-integration of local knowledge systems (mainly utilization and conservation), poor technology, weak policies, insufficient capacities, lack of job opportunities, food insecurities, etc. Each of these issues needs to be uncoupled and addressed for the benefit of the systems. Financial compensation and incentives offered from a distance or in isolation are no longer cutting it. A people-centric approach should be adopted for biodiversity conservation and environmental sustainability in Africa. Consultation with local communities through town hall meetings, seminars, training programmes, and workshops should be held regularly to keep the people engaged in the management of biodiversity and environmental resources. This incorporates aspects of the progressive philosophies for ethical and authentic integration of local communities in biological resources and environment conservation described in Montgomery et al. (2020). This human heritage-centered conservation framework is summarized to include using local languages to communicate conservation goals, incorporating traditional systems, supporting interdisciplinary research, collaborating with local authorities, applying heritage-based solutions, building local capacities, providing and retaining opportunities locally, degree and training incorporation for indigenes in conservation programmes, and promote local revenue systems and peer-review of conservation solutions. Further, to balance development with biological and environmental sustainability in Africa would also require an understanding of the capacity of the continent's resources, producing and applying this knowledge to distribute the resources for efficient utilization and management, and then modelling current use with population growth estimates and current and potential threats to produce a viable system for the future. This kind of approach is necessary because interactions between drivers require a robust prediction of emergent characteristics to address potential complication areas.

References

AfDB (2011) Africa Economic Brief. 2(4)

African Wildlife Foundation (2003) Conserving wildlife, protecting land, Empowering People. Annual Report 2003. http://www.awf.org

Allison EH, Paley RGT, Ntakimazi G, Cowan VJ, West K (2000) Biodiversity assessment and conservation in Lake Tanganyika: BIOSS final technical report. Pollution control and other measures to protect biodiversity in Lake Tanganyika (RAF/92/G32)

Armitage D (2008) Governance and the commons in a multi-level world. Int J Commons 2(1):7–32

Arts K, van der Wal R, Adams WM (2015) Digital technology and the conservation of nature. Ambio 44:661–673. https://doi.org/10.1007/s13280-015-0705-1

Bassey SE, Izah SC (2017) Nigerian plants with insecticidal potentials against various stages of mosquito development. ASIO J Med Health Sci Res 2(1):07–18

Biermann F, Betsill MM, Gupta J, Kanie N, Lebel L, Liverman D, Schroeder H, Siebenhuner B, Zondervan R (2010) Earth system governance: a research framework. Int Environ Agreements Politics Law Economics 10(4):277–298

Blanc JJ, Thouless CR, Hart JA, Dublin HT, Douglas-Hamilton I, Craig CG, Barnes RFW (2003) African elephant status report 2002: an update from the African elephant database. IUCN/SSC African Elephant Specialist Group. ICUN, Gland, Switzerland and Cambridge, UK. vi + 304pp

Brady NC (1988) International development and the protection of biological diversity. In: Wilson EO, Peter FM (eds) Biodiversity. National Academies Press, Washington, DC

Brockhaus M, Djoudi H, Kambire H (2012) Multilevel governance and adaptive capacity in West Africa. Int J Commons 6(2) http://www.thecommonsjournal.org/index.php/ijc/article/view/331/280

Brown MJF, Dicks LV, Paxton RJ, Baldock KCR, Barron AB, Chauzat M-P, Freitas BM, Goulson D (2016) A horizon scan of future threats and opportunities for pollinators and pollination. PeerJ 4:e2249. https://doi.org/10.7717/peerj.2249

Carpenter SR, DeFries R, Dietz T et al (2006) Millennium ecosystem assessment: research needs. Science 314:257–258

Chown SL (2010) Temporal biodiversity change in transformed landscapes: a southern African perspective. Philos Trans R Soc Lond Ser B Biol Sci 365(1558):3729–3742. https://doi.org/10.1098/rstb.2010.0274

Cundill G (2010) Monitoring social learning processes in adaptive co-management: three case studies from South Africa. Ecol Soc 15(3):28. http://www.ecologyandsociety.org/vol15/iss3/art28/

Davids R, Rouget M, Boon R, Roberts D (2018) Spatial analyses of threats to ecosystem service hotspots in Greater Durban, South Africa. PeerJ 6:e5723. https://doi.org/10.7717/peerj.5723

Elson D (2010) Investing in locally controlled forestry: reviewing the issues from a financial investment perspective. Background paper for The Forests Dialogue's initiative on investing in locally-controlled forestry conference, London, UK, 24–25 May 2010

Enaregha EB, Izah SC, Okiriya Q (2021) Antibacterial properties of Tetrapleura tetraptera pod against some pathogens. Res Rev Insights 5:1–5. https://doi.org/10.15761/RRI.1000165

FAO (2010) Global forest resources assessment 2010—Main report. FAO Forestry Paper 163. Rome, Italy. https://www.fao.org/forestry/fra/fra2010/en/

Fleishman E, Blockstein DE, Hall JA, Mascia MB, Rudd MA, Scott JM, Sutherland WJ, Bartuska AM (2011) Top 40 priorities for science to inform US conservation and management policy. Bioscience 61:290–300. https://doi.org/10.1525/bio.2011.61.4.9

Folke C, Hahn T, Olsson P, Norberg J (2005) Adaptive governance of social–ecological systems. Annu Rev Environ Resour 30:441–473

Galaz V, Olsson P, Hahn T, Svedin U (2008) The problem of fit among biophysical systems, environmental and resource regimes, and broader governance systems: insights and emerging challenges. In: Young OR, Schroeder H, King LA (eds) Institutions and environmental change: principal findings, applications and research frontiers. MIT Press, Cambridge, MA, pp 147–186

Gobeze T, Bekele M, Lemenih M, Kassa H (2009) Participatory forestry management and its impacts on livelihoods and forest status: the case of Bonga forest in Ethiopia. Int For Rev 11(3): 346–358

Gondo PC (2011) Financing of sustainable Forest Management in Africa: an overview of the current situation and experiences. Southern Alliance for indigenous resources (SAFIRE), Harare

Ibisch PL, Vega AE, Herrmann TM (eds) (2010) Interdependence of biodiversity and development under global change. Technical series no. 54. Secretariat of the convention on biological diversity, Montreal

Ikhajiagbe B, Ogwu MC (2020) Hazard quotient, microbial diversity and plant composition of spent crude oil polluted-soil. Beni-Suef University J Basic Appl Sci 9(26). https://doi.org/10.1186/s43088-020-00052-0

Ikhajiagbe, B., Ogwu, M.C. and Lawrence, A.E. (2020). Single-tree influence of *Tectonia grandis* Linn. f. on plant distribution and soil characteristics in a planted forest. Bull Natl Res Centre, 44: Article number 29. doi: https://doi.org/10.1186/s42269-020-00285-0

International Panel on Climate Change (IPCC) (2007) Climate change 2007: impacts, adaptation and vulnerability. In: Parry ML, Canziani OF, Palutikof JP, van der Linden PJ, Hanson CE (eds) Contributions of working group II to the fourth assessment report of the intergovernmental panel on climate change (IPCC). Cambridge University Press, Cambridge, UK

IUCN (2001) IUCN Red List Categories and Criteria: version 3.1. IUCN Species Survival Commission. IUCN, Gland, Switzerland and Cambridge, UK. ii + 30 pp

Izah SC (2018) Ecosystem of the Niger Delta region of Nigeria: potentials and threats. Biodiversity Int J 2(4):338–345

Izah SC (2019) Activities of crude, acetone and ethanolic extracts of Capsicum frutescens var. minima fruit against larvae of Anopheles gambiae. J Environ Treatment Techniques 7(2): 196–200

Izah SC, Aseibai ER (2018) Antibacterial and synergistic activities of methanolic leaves extract of lemon grass (Cymbopogon citratus) and rhizomes of ginger (Zingiber officinale) against Escherichia coli, Staphylococcus aureus and Bacillus subtilis. Acta Sci Microbiol 1(6):26–30

Izah SC, Seiyaboh IE (2018a) Challenges of wildlife with therapeutic properties in Nigeria; a conservation perspective. Int J Avian Wildlife Biol 3(4):252–257

Izah SC, Seiyaboh EI (2018b) Changes in the protected areas of Bayelsa state, Nigeria. Int J Mol Evol Biodiversity 8(1):1–11

Izah SC, Youkparigha FO (2019) Larvicidal activity of fresh aqueous and ethanolic extracts of Cymbopogon citratus (DC) Stapf on malaria vector, Anopheles gambiae. BAOJ Biotech 5:040

Izah SC, Angaye CN, Aigberua AO, Nduka JO (2017) Uncontrolled bush burning in the Niger Delta region of Nigeria: potential causes and impacts on biodiversity. Int J Mol Ecol Conserv 7(1):1–15

Izah SC, Uhunmwangho EJ, Dunga KE, Kigigha LT (2018a) Synergy of methanolic leave and stem-back extract of Anacardium occidentale L. (cashew) against some enteric and superficial bacteria pathogens. MOJ. Toxicology 4(3):209–211

Izah SC, Uhunmwangho EJ, Dunga KE (2018b) Studies on the synergistic effectiveness of methanolic extract of leaves and roots of Carica papaya L. (papaya) against some bacteria pathogens. Int J Complement Altern Med 11(6):375–378

Izah SC, Uhunmwangho EJ, Etim NG (2018c) Antibacterial and synergistic potency of methanolic leaf extracts of Vernonia amygdalina L. and Ocimumgratissimum L. journal of basic. Pharmacol Toxicol 2(1):8–12

Izah SC, Zige DV, Alagoa KJ, Uhunmwangho EJ, Iyamu AO (2018d) Antibacterial efficacy of aqueous extract of Myristica fragrans (common nutmeg). EC Pharmacol Toxicol 6(4):291–295

Izah SC, Aigberua AO, Nduka JO (2018e) Factors affecting the population trend of biodiversity in the Niger Delta region of Nigeria. Int J Avian Wildlife Biol 3(3):206–214

Izah SC, Chandel SS, Etim NG, Epidi JO, Venkatachalam T, Devaliya R (2019a) Potency of unripe and ripe express extracts of long pepper (Capsicum frutescens var. baccatum) against some common pathogens. Int J Pharm Phytopharmacol Res 9(2):56–70

Izah SC, Etim NG, Ilerhunmwuwa IA, Silas G (2019b) Evaluation of crude and ethanolic extracts of Capsicum frutescens var. minima fruit against some common bacterial pathogens. Int J Complement Altern Med 12(3):105–108

Izah SC, Etim NG, Ilerhunmwuwa IA, Ibibo TD, Udumo JJ (2019c) Activities of express extracts of costusafer Ker–Gawl. [Family COSTACEAE] against selected bacterial isolates. Int J Pharm Phytopharmacol Res 9(4):39–44

Izah SC, Chandel SS, Epidi JO, Venkatachalam T, Devaliya R (2019d) Biocontrol of Anopheles gambiae larvae using fresh ripe and unripe fruit extracts of Capsicum frutescens var. baccatum. Int J Green Pharmacy 13(4):338–342

Juma C, Serageldin I (2007) Freedom to innovate: biotechnology in Africa's development. A report of the high-level African panel on modern biotechnology. African union (AU) and new Partnership for Africa's development (NEPAD). Addis Ababa and Pretoria

Karembu M, Nguthi F, Ismail H (2009) Biotech crops in Africa: the final frontier. ISAAA Afri-Center, Nairobi, Kenya

Kiefer I, Lopez P, Ramiarison C, Barthlott W, Ibish P (2010) Development, biodiversity conservation and global change in Madagascar. In: Ibisch et al. (2010)

Kigigha LT, Izah SC, Ehizibue M (2015a) Activities of Aframomum melegueta seed against Escherichia coli, Staphylococcus aureus and bacillus species. Point J Bot Microbiol Res 1: 23–29

Kigigha LT, Izah SC, Prebo PE (2015b) Effect of residual antibiotics in snacks against Escherichia coli and Staphylococcus aureus. J Environ Treatment Techniques 3(4):201–203

Kigigha LT, Biye SE, Izah SC (2016a) Phytochemical and antibacterial activities of Musanga cecropioides tissues against Escherichia coli, Pseudomonas aeruginosa Staphylococcus aureus, Proteus and Bacillus species. Int J Appl Res Technol 5(1):100–107

Kigigha LT, Izah SC, Okitah LB (2016b) Antibacterial activity of palm wine against Pseudomonas, Bacillus, Staphylococcus, Escherichia, and Proteus spp. Point J Bot Microbiol Res 2(1): 046–052

Kigigha LT, Apreala A, Izah SC (2016c) Effect of cooking on the climbing pepper (Piper nigrum) on antibacterial activity. J Environ Treatment Techniques 4(1):6–9

Kigigha LT, Selekere RE, Izah SC (2018a) Antibacterial and synergistic efficacy of acetone extracts of Garcinia kola (Bitter kola) and Buchholzia coriacea (Wonderful kola). J Basic Pharmacol Toxicol 2:13–17

Kigigha LT, Izah SC, Uhunmwangho EJ (2018b) Assessment of hot water and ethanolic leaf extracts of Cymbopogon citratus Stapf (lemon grass) against selected bacteria pathogens. Ann Microbiol Infect Dis 1(3):1–5

Kiss A (ed.) (1990) Living with Wildlife:Wildlife Resource Management with Local Participation in Africa. Washington (DC): World Bank. Technical Paper no. 130

Lettington R. (2000) The protection and promotion of traditional and marginalized medicinal knowledge: romantic dream or twenty-first century necessity? Regional conference on medicinal plants, traditional medicines and local communities in Africa: challenges and opportunities of the new Millenium, Nairobi

Marshall CAM, Wieringa JJ, Hawthorne WD (2016) Bioquality hotspots in the tropical African Flora. Curr Biol 26:3214–3219. https://doi.org/10.1016/j.cub.2016.09.045

Mataruka D (2009) Role of genetically modified crops in Africa. CBI July 2009. http://www.whybiotech.com/?p=915#more-915

Mburugu, J (1995). In situ conservation and protected areas forest in Kenya. Centre for Biodiversity, National Museums of Kenya. 285–292 pp

McClain ME (2013) Balancing water resources development and environmental sustainability in Africa: a review of recent research findings and applications. Ambio 42(5):549–565. https://doi.org/10.1007/s13280-012-0359-1

Millennium Ecosystem Assessment (MEA) (2005) Ecosystems and human Well-being: a framework for assessment. Island Press, Washington, DC

Minnemeyer S, Laestadius L, Sizer N, Saint-Laurent C, Potapov P (2011) A world of opportunity. World Resources Institute, Washington, DC. http://www.wri.org/restoringforests

Mitchell M, Lockwood M, Moore SA, Clement S (2015) Scenario analysis for biodiversity conservation: a social–ecological system approach in the Australian Alps. J Environ Manag 150:69–80. https://doi.org/10.1016/j.jenvman.2014.11.013

Mittermeier RA, Robles-Gil P, Hoffmann M, Pilgrim JD, Brooks TB, Mittermeier CG, Lamoreux JL, Fonseca GAB (2004) Hotspots revisited: earth's biologically richest and most endangered ecoregions. CEMEX, Mexico City. 390 pp

Montgomery RA, Borona K, Kasozi H, Mudumba T, Ogada M (2020) Positioning human heritage at the center of conservation practice. Conserv Biol 34(5):1122–1130. https://doi.org/10.1111/cobi.13483

Mwai O, Hanotte O, Kwon YJ, Cho S (2015) African indigenous cattle: unique genetic resources in a rapidly changing world. Asian Australas J Anim Sci 28(7):911–921. https://doi.org/10.5713/ajas.15.0002R

Newmark WD, Hough JL (2000) Conserving wildlife in Africa: integrated conservation and development projects and beyond. Bioscience 50(7):585–592

Nyenje N, Kingamkono RR, Kullaya AK, Mneney EE (2011) Biotechnology for sustainable agriculture, food security and poverty reduction in Africa. In: Pasternak C (ed) Access not Excess. Pp 19 – 30

OECD (2011) Better policies for development: recommendations for policy coherence. Organization for Economic Cooperation and Development

Ogwu MC (2009) The significance and contribution of conservation education in Nigeria: an appraisal and a call for improvement. In: Babalola F (ed) Proceedings of the maiden seminar of Nigerian Tropical Biology Association. Nigerian Tropical Biology Association and Tropical Biology Association, pp 109–114

Ogwu MC (2010) Conserving biodiversity even in poverty: the African experience'. In: Babalola F (ed) Seminar proceedings of Nigerian tropical biology association. Nigerian Tropical Biology Association and Tropical Biology Association, pp 112–117

Ogwu MC (2019) Towards sustainable development in Africa: the challenge of urbanization and climate change adaptation. In: Cobbinah PB, Addaney M (eds) The geography of climate change adaptation in urban Africa. Springer Nature, Cham, Switzerland, pp 29–55. https://doi.org/10.1007/978-3-030-04873-0_2

Ogwu MC (2020) Value of *Amaranthus* [L.] species in Nigeria. In: Waisundara V (ed) Nutritional value of amaranth. IntechOpen, pp 1–21. https://doi.org/10.5772/intechopen.86990

Ogwu MC, Osawaru ME, Ahana CM (2014) Challenges in conserving and utilizing plant genetic resources (PGR). Int J Genet Mol Biol 6(2):16–22. https://doi.org/10.5897/IJGMB2013.0083

Ogwu MC, Osawaru ME, Aiwansoba RO, Iroh RN (2016) Status and prospects of vegetables in Africa. In: Borokini IT, Babalola FD (eds) Conference proceedings of the joint biodiversity conservation conference of Nigeria Tropical Biology Association and Nigeria chapter of Society for Conservation Biology on MDGs to SDGs: toward sustainable biodiversity conservation in Nigeria. University of Ilorin, Nigeria, pp 47–57

Ogwu MC, Chibuogwu AC, Osawaru ME (2016a) Trees in the service of man: University of Benin flora as a case study. UNILAG J Med Sci Technol 4(2):71–87

Ogwu MC, Osawaru ME, Iyamu M (2016b) Tree flora and their environmental services: a case study of University of Benin flora. Ilorin J Sci 3(1):40–68

Ogwu MC, Osawaru ME, Obayuwana KO (2016c) Diversity and abundance of tree species in the University of Benin, Benin City, Nigeria. Appl Trop Agric 21(3):46–54

Olorunfemi F, Raheem UA (2008) Sustainable tourism development in Africa: the imperative for tourists/host communities security. J Sustainable Development Africa 10(3)

Olsson EG, Ouattara S (2013) Opportunities and challenges to capturing the multiple potential benefits of REDD+ in a traditional transnational savanna-woodland region in West Africa. Ambio 42(3):309–319. https://doi.org/10.1007/s13280-012-0362-6

Olsson P, Gunderson LH, Carpenter SR, Ryan P, Lebel L, Folke C, Holling CS (2006) Shooting the rapids: navigating transitions to adaptive governance of social-ecological systems. Ecol Soc 11(1):18

Osawaru, M. E. and Ogwu, M.C. (2014). Conservation and utilization of plant genetic resources. In: Omokhafe, K. and Odewale, J. (eds.). Proceedings of 38th Annual Conference of The Genetics Society of Nigeria. Empress Prints Nigeria Limited. pp. 105-119

Osawaru ME, Ogwu MC (2020) Survey of plant and plant products in local markets within Benin City and environs. In: Filho LW, Ogugu N, Ayal D, Adelake L, da Silva I (eds) African handbook of climate change adaptation. Springer Nature, Chamn, Switzerland, pp 1–24. https://doi.org/10.1007/978-3-030-42091-8_159-1

Reihling HC (2008) Bioprospecting the African renaissance: the new value of muthi in South Africa. J Ethnobiol Ethnomed 4:9. https://doi.org/10.1186/1746-4269-4-9

Réjou-Méchain M, Mortier F, Bastin JF et al (2021) Unveiling African rainforest composition and vulnerability to global change. Nature 593:90–94. https://doi.org/10.1038/s41586-021-03483-6

Romano F, Reeb D (2006) Understanding forest tenure: what rights and for whom? FAO, Rome, Italy, p 3

Sayer JA, Harcourt CS, Collins NM (eds) (1992) The conservation atlas of tropical forests: Africa. IUCN. Macmillan Publishers Ltd.

Scherr SJ, McNeely JA (2007) Farming with nature: the science and practice of Ecoagriculture. Island Press, Washington, DC

Secretariat CBD (2010) Global biodiversity outlook 3. Convention of Biological Diversity, Montréal, 94 pp

Seiyaboh EI, Odubo TC, Izah SC (2020a) Larvicidal activity of Tetrapleura tetraptera (Schum and Thonn) Taubert (Mimosaceae) extracts against Anopheles gambiae. Int J Adv Res Microbiol Immunol 2(1):20–25

Seiyaboh EI, Seiyaboh Z, Izah SC (2020b) Environmental control of mosquitoes: a case study of the effect of Mangifera Indica root-bark extracts (family Anacardiaceae) on the larvae of Anopheles gambiae. Ann Ecol Environ Sci 4(1):33–38

Selemani IS (2020) Indigenous knowledge and rangelands' biodiversity conservation in Tanzania: success and failure. Biodivers Conserv 29(14):3863–3876. https://doi.org/10.1007/s10531-020-02060-z

Seymour CL, Gillson L, Child MF, Tolley KA, Curie JC, da Silva JM, Alexander GJ, Anderson P, Downs CT, Egoh BN, Ehlers Smith DA, Ehlers Smith YC, Esler KJ, O'Farrell PJ, Skowno AL, Suleman E, Veldtman R (2020) Horizon scanning for south African biodiversity: a need for social engagement as well as science. Ambio 49(6):1211–1221. https://doi.org/10.1007/s13280-019-01252-4

Singh RK, Murty HR, Gupta SK, Dikshit AK (2009) An overview of sustainability assessment methodologies. Ecol Indic 9:189–212

Sosef MS, Dauby G, Blach-Overgaard A, van der Burgt X, Catarino L, Damen T, Deblauwe V, Dessein S, Dransfield J, Droissart V, Duarte MC, Engledow H, Fadeur G, Figueira R, Gereau RE, Hardy OJ, Harris DJ, de Heij J, Janssens S, Klomberg Y, Ley AC, Mackinder BA, Meerts P, van de Poel J, Sonke B, Stevart T, Stoffelen P, Svenning J-C, Sepulchre P, Zaiss R, Wieringer JJ, Couvreur TL (2017) Exploring the floristic diversity of tropical Africa. BMC Biol 15(1):15. https://doi.org/10.1186/s12915-017-0356-8

Souza ECA, Bernard E (2019) Setting priorities in biodiversity conservation: an exercise with students, recent graduates, and environmental managers in Brazil. Ambio 48:879–889. https://doi.org/10.1007/s13280-018-1116-x

Sutherland WJ, Clout M, Côté IM, Daszak P, Depledge MH, Fellman L, Fleishman E, Garthwaite R (2010) A horizon scan of global conservation issues for 2010. Trends Ecol Evol 25:1–7. https://doi.org/10.1016/j.tree.2015.11.007

Swaminathan MS (1996) Preface, agrobiodiversity and farmers. Konark Publishers, Delhi

Timko J (2011) HIV/AIDS, forests and futures in sub-Saharan Africa, STEPS working paper 43. STEPS Centre, Brighton

Treurnicht M, Colville JF, Joppa LN, Huyser O, Manning J (2017) Counting complete? Finalising the plant inventory of a global biodiversity hotspot. PeerJ 5:e2984. https://doi.org/10.7717/peerj.2984

UNEP (2006) Africa environment outlook (AEO2): our environment our wealth. United Nations Environment Programme, Nairobi, Kenya. http://www.eoearth.org/article/Africa_Environment_Outlook_2:_Our_Environment,_Our_Wealth_(e-book)

UNEP (2007) Global Environment Outlook (GEO4) Environment for development. United Nations Environment Programme, Nairobi

UNESCO, UNECA, AUC (2011) Promoting science, technology and innovation for development in Africa—Information note: Meeting of the committee of experts of the 4th joint annual meetings of the AU conference of ministers of economy and finance and ECA conference of African ministers of finance, planning and economic development. Addis Ababa, Ethiopia 24–27 March 2011

Van Drunen M, Kuik O, Lutz C, Wiebe K (2009) Modelling future Intelinkages, INDI-LINK – indicator based evaluation of interlinkages between different sustainable development objectives. Ther Deliv D2:5

Van Herwijnen M (2008) The sustainability A-test. Chapter 4 in OECD sustainable development studies conducting sustainability assessments. Organization for Economic Cooperation and Development (OECD) Publishing, Paris

Waheed B, Khan F, Veitch V (2009) Linkage-based frameworks for sustainability assessment: making a case for driving force-pressure-state- exposure-effect-action (DPSEEA) frameworks. Sustainability 1:441–463

Wei F, Wang S, Fu B, Liu Y (2020) Representation of biodiversity and ecosystem services in East Africa's protected area network. Ambio 49(1):245–257. https://doi.org/10.1007/s13280-019-01155-4

Wells M, Brandon K, Hannah L (1992) People and parks: linking protected area management with local communities. World Bank, Washington (DC)

World Bank (2010) World development report 2010: development and climate change. The International Bank for Reconstruction and Development/The World Bank, Washington, DC. 20433. p. 192. http://go.worldbank.org/BKLQ9DSDU0
World Resources Institute (2003) EarthTrends: the environmental information portal. http://earthtrends.wri.org. World Resources Institute, Washington DC
Youkparigha FO, Izah SC (2019) Larvicidal efficacy of aqueous extracts of Zingiber officinale roscoe (ginger) against malaria vector, Anopheles gambiae (Diptera: Culicidae). Int J Environ Agric Sci 3:020
Zeller D, Hood L, Palomares MLD, Sumaila UR, Khalfallah M, Belhabib D, Woroniak J, Pauly D (2020) Comparative fishery yields of African large marine ecosystems. Environ Dev 36(20): 100543. https://doi.org/10.1016/j.envdev.2020.100543

Part I
Conservation of Africa's Biological Resources and Environment

Chapter 2
Biodiversity Conservation and Tourism Sustainability in Africa

Bukola Omotomilola Adetola

Abstract The long-term viability of the tourism industry and the preservation of its principal attraction (nature) are inextricably linked. Biodiversity conservation is a fundamental obligation of all economic sectors and a unique paradigm in the tourist sector that makes tourism more biodiversity-friendly. Assets of biodiversity are powerful forces shaping tourism's essence, including the human drive to see and experience the pristine ecological settings. Moreover, Africa boasts of rich diversity, and in this continent, biodiversity endowment offers opportunities for economic, social, and ecological development. Whether scuba diving on coral reefs or watching wildlife, most tourists enjoy the natural beauty (animals and vegetation) that surrounds them. However, the high rates of tourist expansion must be balanced against the problems that come with it, such as massive volumes of traffic, trash, a huge area of land and resource consumption associated with travel, and impacts on fragile ecosystems among others. The tourism sector appears to place a long-term emphasis on environmental conservation and protection because of the importance and uniqueness of the environmental quality and biodiversity for tourism, and they encourage and support sustainable tourism operations. Responsible tourism benefits indigenous communities and residents by improving their standard of living while also preserving culture and biodiversity.

Ecotourism, a sustainable tourism strategy, is an interface of conservation concerns and tourism interest, a synergy that jointly preserves the environment's quality while protecting nature and promoting tourism. Consequently, Africa's response to the Sustainable Development Goals (SDGs) will be ineffective unless its biodiversity is conserved and used sustainably.

Keywords Biodiversity Conservation · Sustainable Development Goals (SDGs) · Wildlife · Environmental protection · Ecotourism

---

B. O. Adetola (✉)
Department of Ecotourism and Wildlife Management, School of Agriculture and Agricultural Technology, Federal University of Technology, Akure, Ondo State, Nigeria
e-mail: boadetola@futa.edu.ng

S. C. Izah, M. C. Ogwu (eds.), *Sustainable Utilization and Conservation of Africa's Biological Resources and Environment*, Sustainable Development and Biodiversity 32, https://doi.org/10.1007/978-981-19-6974-4_2

2.1 Introduction

Biodiversity is a blend of two words, 'bio' which means life and 'diversity' means variety. These words are combined to give a literal meaning of 'variety of life'. This means that the various species and forms of plants, animals, and microorganisms make up biodiversity. Biodiversity includes various aspects such as ecological, species, and genetic biodiversity. Varieties of various ecological zones such as rainforests, savannahs, woodlands, deserts, mangrove swamps, etc. are referred to as Ecological Biodiversity. Species diversity denotes the various species of animals and plants living within an ecosystem such as elephants, squirrels, oak, baobab, teak, etc. Genetic biodiversity considers the genetic makeup and modifications of each species and its varieties across several members of the same species. Life is made up of various components and the differences which it possesses make up its varieties; this is responsible for its beauty and spontaneity.

The efficient and effective usage and protection of resources for the current and imminent use are being referred to as conservation. It is the act of guarding and keeping safe unique natural resources for mankind. Biodiversity conservation, therefore, is the act of preserving or effectively managing the flora and fauna resources for the benefit of present use and ensuring their sustenance for the sake of future generations. Biodiversity is self-sustaining by nature and regenerative if properly managed. The implication is that biodiversity can replicate itself in larger quantities and if its growth exceeds the capacity of an environment; nature helps to introduce certain factors so that equilibrium can be reached. Therefore, it can be inferred that minimal efforts are required for the sustenance of biodiversity, if their resources are properly conserved. Biodiversity conservation invariably involves creating awareness, changing perception, and changing the behavior of people towards sustainable utilization of natural resources around us.

One of the potentials of biodiversity conservation is its ability to generate revenue through tourism. This is beneficial to national development as revenue would be derived from various activities relating to nature tourism, both directly and indirectly. Nature tourism is a form of tourism in which the main attraction is focused on the variety and beauty of nature, such as plants, animals, birds, beautifying landscapes, water bodies, and so on. It would attract tourists from other countries who would visit for sightseeing, and in the process, purchase other artifacts and visit other attractions. They would also pay for services such as transportation, accommodation, travel, and hospitality during their stay, all of these sum up to give a major boost to the country's economy. This situation can be observed in Kenya's Wildlife and Tourism sector. Sustainable tourism and responsible tourism are centered on the concept of sustainability, which is defined as "meeting current expectations and needs without endangering future generations' ability to fulfill their own needs" (Steck 2004). In touristic locations, sustainable development must balance the interests of the tourism sector, tourist satisfaction, and biodiversity conservation.

2.2 The Tourism Sector

The tourism industry is frequently referred to as the largest industry in the globe, making significant changes to GDP, employment creation, and the number of services provided for clients. In 2016, more than 1.2 billion international tourists visited the United States, resulting in huge economic effects and goods and services mobility (UNWTO 2015a; 2015b). Tourism is classified as an export because customers come to the product to consume it onsite, and it is the world's largest foreign exchange earner (Eilat and Einav 2003). The United Nations World Tourism Organization defines tourism as an economic and socio-cultural phenomenon wherein tourists visit countries or regions other than their usual residence for pleasure or business purposes. Visitors could either be tourists who stay for more than 24 h in the place visited or excursionists staying for less than 24 h; visitors could also be residents (domestic) or nonresidents (foreign) are the individuals in question, and their activities, many of which require tourism expenditure (United Nations Statistics Division 2010). Tourism has been adopted and promoted by developing countries due to its potential to significantly contribute to economic growth and development, especially in Africa (Kester 2003). According to Eilat and Einav (2003), tourism is extremely important for economic growth because of the influence it has on jobs, foreign exchange earnings, infrastructural facilities, revenue collection, and world peace promotion. Africa's endowment of nature and culture is such that tourism could considerably profit from it. Christie and Crompton (2001) affirmed Africa's tourist potential as being enormous, "Africa has a lot to offer that can no longer be found elsewhere;" according to the report, Africa invokes a sense of empathy and a good romance as the world of expeditions and pleasure seekers. Africa has some of the most beautiful vistas on the planet, as well as scenic landscapes that only some places can match. The nature-based riches, as well as its traditional culture and customs, are a significant attraction. Some African countries have become the cornerstone of tourism aid, especially in Tanzania, South Africa, Zambia, Kenya, and Botswana where wildlife tourism has been projected to be a successful venture (Hottola 2009). South Africa and Morrocco with 29% and 22%, respectively, accounted for over average of all international tourists. The region welcomed 52.4 million tourists in 2012, and the UNWTO predicts that this number will rise to 134 million by 2030, from 85 million in 2020. Tourism accounts for one out of every eleven employments. Despite the encouraging gains in recent years, Africa barely accounts for 5% of worldwide tourism and 3% of receipts, totaling US $ 1 billion. The entire contribution to GDP scaled to 9% in 2012, with total employment increasing to 7.1 percent (WTTC 2013). Tourism is already a growing contributor to GDP and exports in more than half of African countries. (Christie and Crompton 2001). A huge increase in tourist arrival has been observed since the early 1990s in Africa (Chen and Devereux 1999).

The travel and tourism industry generated USD 39.8 billion in economic activity in Africa's Sub-Saharan countries in 2003, accounting for 2.4 percent of the region's GDP and 5.4 percent of all jobs in the region. The entire export value of international tourism is $1.5 trillion, according to the UNWTO (2015a), 2015b). America, the Pacific, Asia, and Africa have witnessed the biggest growth in terms of tourism dollars earned. International arrivals would in 2030 reach approximately 1.8 billion based on UNWTO's Tourism 2030 Vision study (Association of Bhutanese Tour Operators 2010). Africa's tourist potential remains untapped and unrealized, despite its excellent endowments and strong tourism growth over the last decade.

2.3 Biodiversity as Tourism Attraction

Biodiversity is a tourist attraction rather than a market niche. It has a significant economic impact on the tourism industry. While tourism is often thought of as a service business, tourist products are collections of goods and services, with the goods being at least as essential as the services. Because the environment provides tourist attractions such as scenery, health, rare plants and wildlife, and interesting cultures, tourism is described as "renting out other people's environments." Biodiversity is an important consideration when selecting a tourism product. Zoo visitation and visits to wildlife reserves, as well as other nonconsumptive tourism like game viewing, bird watching, whale watching, and dive tours, as well as consumptive tourism like hunting and fishing trips, are all good examples.

Tourists are drawn to natural environments with a diverse range of species. Swimming in clean seas surrounded by fish and coral reefs, as well as observing seabirds, are popular activities for visitors to coastal areas. Others travel to Africa to witness wild creatures. All these activities require that ecosystems remain intact and thriving. National parks rely on healthy ecosystems to provide recreation, education, culture, and entertainment to visitors (Melita and Mendlinger 2013). The findings of Adetola et al. (2021) revealed that hiking in the rainforest vegetation is the most popular tourist activity in Cross River National Park, Nigeria, because the park provides a wonderful wilderness experience with its dense canopy forests, consisting of tall trees with huge buttresses, the coolness of the environment, the freshness of the air, and the sweet songs of the forest-dwelling birds. Its diverse flora supports a wide range of wild animals, including the Cross River Gorilla (*Gorilla gorilla diehli*), the Chimpanzee (*Pan troglodytes*), Forest Elephant (*Loxodonta cyclotis*), and the bare-headed rock fowl (*Picarthates oreas*) among others. Tourists who visit scenic areas, the majority of which owe their unique character to vegetation and animals, and the underlying topography, need to know about biodiversity, the threats they face and what they can do to help for its continued existence. The rate at which humans utilize resources of nature has exceeded their natural ability to replenish themselves. Hence, the need for nonconsumptive use of these resources in the form of ecological tourism, where tourists learn in a relax mood and nonformal settings through guided and/or self-guided tour in conservation areas both in situ and ex-situ (Adetola and Akinboboye 2020).

Pristine ecosystems with diverse animal and plant communities are critical to preserving air quality, freshwater, and unpolluted environs which are important tourism pull factors in many countries (Buckley 2004). Biodiversity is a major draw for a somewhat separate and quantifiable industry segment known as nature, eco, and adventure tourism. Tourism has enormous potential to aid in the promotion and preservation of biodiversity. Tourism employs about 9% of the worldwide workforce, with 1.2 billion tourists visiting each year and producing US$1.5 trillion in revenue (third Asian Pacific Summit). It is important to note that the industry's size and rate of expansion present opportunities and challenges to biodiversity protection.

2.4 Sustainability

The first public statement on sustainability was released in 1987 by the Brundlland Commission. According to the concept of sustainable development, growth in economic terms and protection of the environment are compatible partners. You cannot have one without the other. Sustainable development, according to (Gössling et al. 2009), is "development that meets current needs without jeopardizing future generations' ability to meet their own." Then the practical way of thinking in terms of sustainability is nothing more than a guiding philosophy containing specific ideas about our interaction with the natural world, rather than as an endpoint. One of the most essential distinguishing characteristics of tourism in today's globe is sustainability. To preserve a tourism destination's biodiversity, many actions are implemented, including obtaining a good harmony among the location's environmental, social, and economic components.

Sustainable Development Goals targets that are tourism-related include:

SDG Target 8.9: devise and implement policies to promote sustainable tourism that creates jobs and promotes local culture and products;

SDG Target 12. b: develop and implement tools to monitor sustainable development impacts for sustainable tourism that creates jobs and promotes local culture and products; and

SDG Target 14.7: by 2030, increase the economic benefits to small island developing States and least developed countries from the sustainable use of marine resources, including through sustainable management of fisheries, aquaculture and tourism.

Responsible tourism is a great asset that can help control and minimize the negative economic, social, and environmental consequences of mass tourism (Fennel 2008). During Globe '90 Conference in Vancouver in 1990, organizations "discussed the challenge of applying the wider principles of sustainable development to the tourism sector". As a result, a set of five sustainable tourism goals emerged by Tourism Canada, 1990 in Miller and Twining-Ward (2005), which include the following:

1. increase in knowledge and comprehension of the substantial achievements that tourism could harness to the economy and environment.

2. encourage fairness and advancement;
3. advance the standard of living of the indigenous community;
4. provide an excellent tourist experience;
5. sustain the quality of the environment upon which preceding objectives are entirely based.

Tourism sustainability is premised on the actualization of these objectives.

Tourism that is established and maintained in a manner and large form that will continue to be economically viable indefinitely while having no negative impact on the physical and human environment that supports and nurtures it is referred to as sustainable tourism. 'Meeting existing tourism and host region demands while safeguarding and expanding future potential' (Fennel 2008).

Sustainable tourism is based on three essential ideas. The first principle is the importance of quality. Sustainable tourism should deliver a top-notch experience to visitors, while also increasing the well-being of the host community and safeguarding the environment.

Continuity is the second essential principle. This includes continuity of the resources on which tourism is founded, the continuity of the host community's culture, and the continuity of visitor support or tourist demand which are all required for sustainable tourism.

Finally, sustainable tourism is all about finding a happy medium. The goal of tourism goal is to strike a balance between the needs of the host, the needs of the guest, and the ecology of the location.

2.5 The Significant Value of Biodiversity for Sustainable Tourism

African governments have long advocated biodiversity conservation through careful resource management, and significant progress has been made in the protection and recovery of wildlife populations (Blanc et al. 2007). Eastern and southern Africa's tourism industries are built on wildlife and outdoor leisure. About 20% of South Africa's land area is dedicated to game ranching and conservation, representing a significant percentage of the tourism revenue generation in South Africa (Van der Merwe and Saayman 2003). The Zululand Birding Route, along with the established Greater Limpopo Birding Route, is expected to be worth ZAR 50 million (US$ 6.8 million) in direct economic value to the South African region each year (Birdlife International 2008).

In locations of exceptional avian richness, birding routes provide tourists with specific itineraries, skilled local guides, and birder-friendly lodging. Seventy-five percent of international tourists and 55 percent of domestic tourists participate in nature-based or outdoor activities in Australia (Tourism Australia 2009), and these visitors stay longer and spend more money than other visitors. According to market research conducted in 2000, 42 percent of European travelers surveyed included a visit to a natural park in their vacation plans (Convention on Biological Diversity Secretariat 2008). Tourism related to high-quality natural areas is projected to be valued at £5 billion annually in England (GHK Consulting Ltd. and GFA-Race Partners Ltd. 2004).

Coral reefs in the Caribbean provided tourism earnings worth an estimated US$ 4.7 billion in gross revenue and US$ 2.1 billion in net revenue in 2000 (World Resources Institute 2004). In Mexico, around 14 million domestic and foreign tourists visit federally protected areas each year, paying a total of US$ 660 million (Convention on Biological Diversity Secretariat 2008). In the United States of America, 87.5 million tourists went for wildlife watching in 2006, hunting, or fishing, spending a total of US$ 122.3 billion on these activities, including US$ 37.5 billion on food, housing, and transportation (United State Fish and Wildlife Service 2007), tourism to beaches, estuaries, and wetlands attract 85 percent of all tourist visitation in the United States, vastly outnumbering tourism to theme parks and national monuments (Houston 1996). These figures show the magnitude of biodiversity-based tourist contributions to the economy.

Most biodiversity hotspots in developing nations, such as Madagascar, Uganda, Tanzania, South Africa, Costa Rica, Ecuador, and Belize, have biodiversity as a major tourism attraction. Even though only a small percentage of visitors come to see wildlife, its vast biodiversity is a crucial influence in their decision to visit these places. Visitors may choose to prolong their stays in these areas to visit other tourist attractions such as cultural or heritage sites or leisure resorts. The image established by these places' biodiversity is thus critical for marketing all forms of tourism they provide, whether these are primarily focused on biodiversity.

The tourism industry is heavily reliant on the country's diverse and untouched stunning features, including rare flora and fauna found in parks, nature reserves, bird parks, marine parks, and neighboring coral reefs in many places. Protected areas, for example, play an essential role in biodiversity conservation, and tourism is considered as an incentive for public participation in conservation, which is central to parks' role. Protected area managers can assist people in understanding ecotourism and promoting sustainable tourism. This is accomplished through the employment of various management tactics such as providing and maintaining visitor facilities, controlling where tourists go and what they do, and providing educational programs.

2.6 Description of Tourism Trend in Africa

The arrival of international visitors in Africa has risen by 29.5 percent since 2000. With a 19.9% increase since 2000, Sub-Saharan Africa has had the fastest growth. Since 2000, North African growth has been sluggish, increasing by 9.6%. Arab Spring has had a detrimental influence on tourism outcomes in North Africa since 2010, the region's growth slowing to its lowest level in four years. Sub-Saharan Africa, on the other hand, saw a significant increase in tourists, adding to the continent's modest growth. In 2014, the region accounted for 5% of all international arrivals and 3% of all tourism receipts. North Africa rose by about 1% in 2014, with foreign visitor arrivals in the leading destination, Morocco's arrivals, increased by 2% following a strong performance in 2013, whereas Tunisia's arrivals declined by 3%. In 2014, Sub-Saharan Africa increased by 3%, according to estimates. Côte d'Ivoire (+24%) and Madagascar (+13%) both experienced a significant increase in 2014, according to available data; nevertheless, arrivals to South Africa, the sub-region's most populous destination, remained steady (WTO Barometer 2015).

2.7 Ecotourism Clustering Potentials for Conservation of Biodiversity and Sustainable Tourism

According to Tapper and Cochrane (2005), tourism can help with sustainable use and biodiversity conservation in a variety of ways:

i. promoting the economic value of biodiversity conservation and sustainable use through tourism stimulation, particularly where jobs are created and local products and services are used, may aid in the reduction of unsustainable natural resource exploitation.
ii. Biodiversity conservation promotion at the national and local levels through increased visitor awareness and enlightenment campaign.
iii. revenue generation for conservation through tourism.

The link between sustainable tourism and biodiversity is straightforward: it is centered on the idea that sustainable tourism should contribute to biodiversity protection (Convention on Biological Diversity Secretariat 2007).

Ecotourism is a sustainable tourism strategy that brings conservation concerns and tourism interests together in a synergy that protects the environment while also increasing tourism (Huybers and Bennett 2002). Ecotourism has been at the forefront in terms of growth among the tourism business, rising with more than thrice of the whole sector (TIES 2008). Consequently, there is no doubt that environmental concerns are growing in tandem with the long-standing pattern of travel as a practice of being enticed to experience nature's beauty, propelled by "the pressures of urban living that encourage people to seek solitude with nature," resulting in the heightened number of visitors to conservation areas like the parks, game reserves, wildlife sanctuaries, and other protected areas (Ceballos-Lascurain 1990).

Ecotourism is defined by the International Union for Conservation of Nature (IUCN 1997) as "environmentally responsible travel and visitation to pristine natural areas for enjoyment and admiration of scenic features (and any associated cultural traits both past and present) that promote conservation, minimize visitor impact, and delivers active socioeconomic involvement of local people."

The International Ecotourism Society (1990) describes ecotourism as "the type of tourism that requires responsible travel to natural areas that conserves the environment and improves the standard of living of local people". Minimizing impacts, increasing cultural and environmental awareness, providing positive experiences for both visitors and hosts, providing financial justification for conservation, empowerment, and financial benefits for residents, and creating awareness and enlightenment campaigns on the environmental, social, and political climates of host communities are among the principles asserted.

Ecotourism is seen as a meeting place for conservation and tourism interests, allowing for the necessary synergy to conserve environmental integrity while also encouraging tourism. It is an important strategy for the preservation of healthy ecosystems and the economic well-being of any host-area community. As a result, it has been lauded as a novel tourism strategy that strikes a balance between development and economic gains while helping both environment and tourist destinations. It should be an ideal apt for preserving biodiversity and finding nonconsumptive uses for natural assets (Kutay 1989). In many regions, the ecotourism venture is based on the country's diverse untouched beautiful nature, including rare flora and fauna in conservation areas, game reserves, wildlife parks, bird parks, coastal ecosystems, and coral reefs. Because natural assets such as landscapes, wildlife, air, water quality, flora, and other natural heritage often lure visitors to a location, tourism has a special interest in biodiversity conservation. Protecting, preserving, and restoring a place's natural ecosystems are in the best interests of the location and its people. Biodiversity, on the other hand, is not evenly spread over the globe. The abundance of these unique resources is determined by the geographical feature and climatic conditions of a region, while tropical ecological zones are major hotspots being the most diverse and abundant. These tropical forest areas, which are being destroyed at an alarming rate, are home to thousands of indigenous species and animals (Adetola and Adetoro 2014). Terrestrial biodiversity is higher near the equator due to the warm environment and high "primary productivity," as seen in Kenya, Tanzania, Borneo, Colombia, Ecuador, and the Galapagos Islands.

2.8 Africa's Biodiversity as Ecotourism Appeal

The African continent is diverse, with untapped resources, scenic wonders, native culture, historical landmarks, biodiversity, safaris, beach resorts, deserts, and even more, all of which, if appropriately developed, could enhance cultural heritage tourism, ecotourism, and adventure tourism.

South and East African countries are known for being the best destinations in the universe to see the "Big Five" (Leopard, Lion, Elephant, Buffalo, and Rhinoceros). In addition to Africa's conventional safari sites, new goods such as gorilla trekking in Central Africa are emerging as alternative or complementary safari places. These wildlife-related tourism items are only found on the African continent, making them a one-of-a-kind focal point for African travel. Aside from the aforementioned prominent wild animals, all countries in Africa provide exceptional opportunities to explore biodiversity and beautiful landscape including sightseeing, wildlife watching, bird watching, observation of marine animals, and wildlife aggregations along migration routes (UNWTO 2015a; 2015b).

2.8.1 Nigeria

Nigeria is a huge and intriguing country with a diverse range of ecosystems and natural resources. Its wide climatic differences have resulted in a comfortable North-South gradation of ecological development, which has resulted in diverse fauna and flora sustaining around 1340 species of animals, including 274 mammals, 860 birds, and about 460 plant species (FORMECU 1996). There are around 22,000 species of vertebrates and invertebrates. There are approximately 20,000 insects, 1000 birds, 1000 fish, 247 mammals, and 123 reptiles in Nigeria (First National Biodiversity Report 2001). Nigeria has around 7895 plant species, divided into families of 338 and 2215 genera, according to the most recent census on biodiversity assessment, including a considerable number of endemic species (Borokini 2014). Nigeria is one of Africa's richest countries with diverse biological resources. Nigeria's wildlife resources are conserved in the nation's protected areas among which are the prominent national parks such as Kainji Lake National Park (KLNP), Figs. 2.1 and 2.2 present Kob (*Kobus Kob)* and Baboon (*Pabio anubis)* in KLNP, Cross River National Park, Gashaka Gumti National Park, Kamuku National Park, Chad Basin National Park, Old Oyo National Park (at the national level), and Yankari game reserve (Bauchi State-owned).

2.8.2 Kenya

Kenya has a broad range of sustainable tourism offerings, such as local cultures (aboriginal communities) and biodiversity (protection of the environment) possible alternatives (hiking and adventure tourism, and wildlife safaris). According to the report of Kenya Travel Tips (2018),

Fig. 2.1 Kob (*Kobus Kob)* in Nigeria's Premier National Park- Kainji Lake National Park

Fig. 2.2 Baboon (*Papio anubis)* in Kainji Lake National Park Nigeria

1. Conservation of wildlife covers around 7.5 percent of the country.
2. Kenya has 23 National Parks, 28 Reserves, six Marine Reserves, four Marine National Parks and four National Santuaries.
3. Diverse array of wild animals, including lions (*Panthera leo)*, African elephants (*Loxodonta africana*), zebras (*Equus quagga*), and over 1070 bird species.
4. The coastal area is 536 kilometers long, with most of it consisting of sandy beaches.
5. There are six UNESCO World Heritage Sites.

Although specific statistics on ecotourism are difficult to come by, there are some statistics on tourism in general. The Kenya National Bureau of Statistics collects a lot of tourism data as presented in Fig. 2.3 on visitors' influx to Kenya and Fig. 2.4 reveals visitors' purpose of visit to Kenya between 2009 and 2016. However, for several years, the number of international tourists has remained relatively stable, albeit with a slight decrease.

Each year, these tourists are expected to contribute $1 billion to Kenya's economy, accounting for approximately 10% of the country's GDP. Kenya's main industry, tourism, employs 9.3% of the country's workers (Kenya Tourism Board 2016). With Kilimanjaro as a backdrop, the Kuku Group Ranch in Kenya provides a regular route between Tsavo and Amboseli National Parks, as well as excellent game watching for tourists at Campi ya Kanzi, including the "Big Five". The Maasai people are recognized for their colorful culture, wildlife sightings; Zebra (*Equus quagga*) / Wildebeest (*Connochaetes taurinus*) and Lion (*Panthera leo*) in Maasai

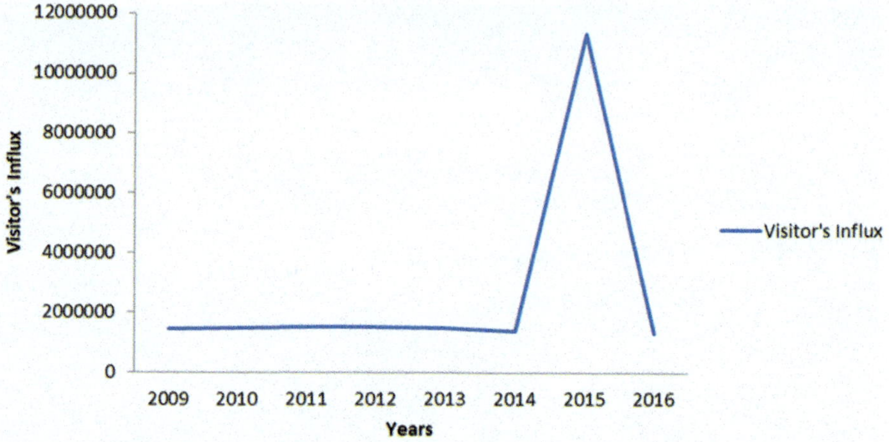

Fig. 2.3 Visitors' influx to Kenya between 2009 and 2016. Adapted from National Bureau of Statistics in Kenya 2017

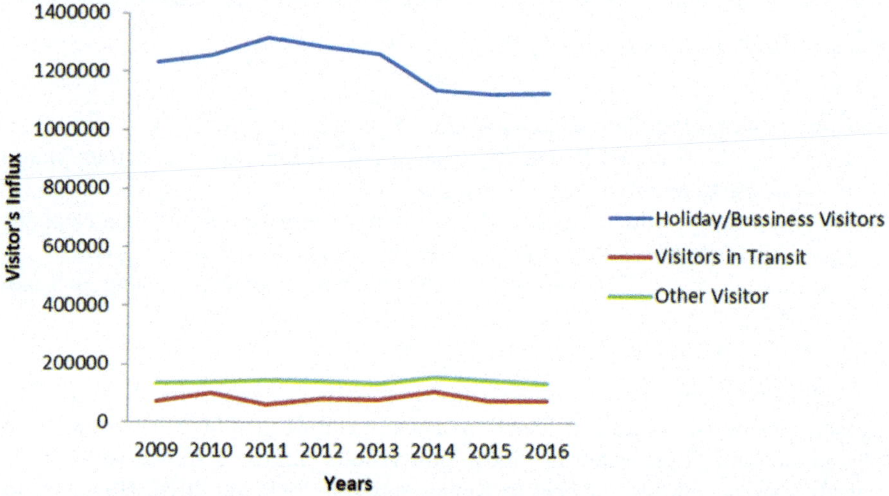

Fig. 2.4 Visitors' Purpose of visit to Kenya between 2009 and 2016. Adapted from National Bureau of Statistics in Kenya 2017

Mara National Reserve, Kenya (Figs. 2.5 and 2.6 respectively), and the dense cloud forests that supply crucial water, oxygen, and carbon sources in the Maasai's Chyulu Hills.

Fig. 2.5 Zebra (*Equus quagga*) and Wildebeest (Connochaetes taurinus) in Maasai Mara National Reserve, Kenya

Fig. 2.6 Lion (*Panthera leo*) in Maasai Mara National Reserve, Kenya

2.8.3 South Africa

South Africa is an abode of a diverse range of wildlife. Lions, leopards, cheetahs, white rhinoceroses, blue wildebeest, kudus, impalas, hyenas, hippopotamuses, and giraffes are among the massive animals. With a total of 1,659,793 visitors in the years 2014/15, the Kruger National Park (KNP), which opened in 1926, is one of South Africa's most popular tourist destinations (South African National Parks Annual Report 2015). With 14 diverse habitats to explore, KNP is South Africa's most visited wildlife conservation area. It is one of Africa's biggest game reserves, home to the "Big Five", i.e., "lion, African elephant, Cape buffalo, leopard, and rhinoceros"—as well as a varied array of other species of wildlife and birdlife.

The United Nations Educational, Scientific, and Cultural Organization designated the Grootbos Private Nature Reserve in the Cape Floral Kingdom as a UNESCO World Heritage Site in 2004. (UNESCO). The reserves are endowed with flora biodiversity, with over 9000 species and 817 plant species, as well as aquatic conservation, and as one of the few locations on earth where the "Marine Big Five" may be observed (UNESCO 2004). Since 1999, it has played an important role in the restoration and management of the ecosystem for the benefit of the natural environment and residents, cooperating with local landowners and environmentalists to protect the flora and fauna treasures.

2.9 Implication of Tourism on Biodiversity

Tourism has both positive and negative effects on biodiversity. It could be used to protect areas from more harmful sectors of development, provide economic justification for conservation and natural environment restoration investments, and create local jobs in areas where other options are limited; protected areas such as the national parks and public and private game reserves are destinations of tourisms significance. Positive consequences include economic development and employment, foreign exchange earnings, income creation for protected area management, and improving conservation awareness among tourists and local inhabitants and communities.

Poorly managed tourism has severe negative consequences such as the conversion of land for tourism, insufficient placement of tourism, pollution and waste, depletion of natural assets, and animal disruption. It can also have severe societal consequences relating to conflicts on resources, tourists' disagreements with indigenous cultural values and norms, or working conditions and limited employment possibilities for locals in tourism ventures (Drumm and Moore 2002).

Various measures, such as strategic planning controls to safeguard key biodiversity locations and sensitive areas from development of tourism, sustainable environmental management initiatives to minimize generation of waste and appropriate treatment and disposal of residual wastes, adoption of strategic purchasing schemes only to procure supplies from sustainable sources, and effective visitor management to minimize wild animal disturbance, can make valuable contributions to mitigating the negative effects of tourism.

2.10 Legislation, Governance, and Policy in Ensuring Biodiversity Sustainability

Biodiversity conservation laws are becoming more common in African countries, owing to the landmark global environmental agreements and treaties to which most of these countries have signed, among which are the underlisted according to Kameri-Mbote and Cullet (2002) and Tinker (1995).

(a) The Biodiversity Convention: The Convention of Biodiversity, with which 49 African countries subscribed and have signed, aims at promoting biodiversity sustainability, judicious use of its elements, and equitable sharing of benefits resulting from asset usages, together with acceptable access to genetic diversity and the transfer of related technological systems.
(b) The Conservation of Nature and Natural Resources African Convention: requires countries to take steps to ensure the conservation of natural resources, utilized and developed in line with guidelines and specified principles taking cognizance of the people's best interests. The convention also aims at safeguarding extinction-threatened fauna and flora species, as well as their environmental conditions. Participating countries must establish conservation areas to protect all species and ecosystems that are most reflective of their regions and are exceptional.
(c) Ramsar Convention on Internationally Significant Wetlands: It advocates for the creation of Wetland nature reserves in general while emphasizing the need for preserving wetlands for their long-term use such that they are in harmony with the ecosystem's inherent features. The Convention, on the other hand, does not constrict the judicious use of species in as much as it does not impair the natural attributes of the wetland.
(d) The Convention on International Trade in Endangered Species of wild animals and plants (CITES). This convention is the major global framework regulating wildlife trade through a listing process whereby endangered species are identified and removed from the international market. The CITES document enumerates the biodiversity that is presently threatened with extinction as well as those that have some indication of extermination shortly. The Conference of Parties decides which species should be listed.
(e) The convention to combat desertification in Africa: This convention has 52 African countries as parties, especially African countries experiencing serious drought and/or desertification. A broad framework for coordinating efforts against desertification is its goal to mitigate the consequences of drought in countries suffering from severe drought and desertification, whereas the principle of mutual but distinctive obligation is accepted; it is the directly impacted parties who carry the main responsibility for combating climate change. (Stéphane 1995).
(f) The Convention on Climate Change is an international framework agreement aimed at resolving climate change and global warming.

(g) The International Convention on the Law of the Sea: This convention oversees that the use of marine areas is governed by the United Nations Convention on the Law of the Sea. There are rules in place to govern the management of biological marine resources. Among these is the creation of Exclusive Economic Zones (EEZs), which give coastal states a larger influence in managing their biodiversity on a long-term basis. The Convention grants autonomous rights in the EEZ for utilization, preservation, and management of the natural resources, as well as imposing several obligations on coastal zone governments. They must, for example, ensure that through effective conservation and management measures, the survival of living resources in the exclusive economic zone is not compromised by overexploitation. Countries are also obligated to work collaboratively in the management and conservation of biological resources in high-seas territories.

2.11 Constitutional Laws and Policy for Sustainable Biodiversity Management in Africa

For a long time, international conventions and treaties have helped the design of national environmental laws and policies in African countries. Wildlife laws in Africa, for example, have been heavily impacted by international and regional laws on wildlife management and conservation.

The enactment of the African Convention has formed the basis for several African wildlife laws. Both the 1933 Convention and the 1900 Convention, both signed by colonial empires, incorporate the core ideas of the preservation of fauna and flora in their natural state.

These measures have served as the foundation for reserving land for wildlife management and have become a defining feature of African wildlife policy initiatives. CITES, which has also been endorsed by 47 African states, is another global instrumental document that has had a substantial impact on biodiversity policies and laws in African Continent. Most countries have entrenched in their wildlife legislative measures a federal ban on the trading activities in flora and fauna species threatened with extinction. In 1977, Kenya's Wildlife Conservation and Management Act adopted CITES regulations by prohibiting all game animal hunting and abolishing all authorizations to trade in wildlife products.

Concerns about biodiversity management have recently gained attention at the global and regional levels. Therefore, in this context, African countries have begun to develop national biodiversity strategies, plans, and programs by Article 6 of the Biodiversity Convention for effective and sustainable management of biological diversity in their custody. In most cases, this procedure has entailed incorporating the conservation and sustainable use of natural resources into appropriate cross-sectoral action plans, initiatives, and policies. Ethiopia, for example, has a national biodiversity conservation strategy in place. As the primary goal for environmental conservation, the framework outlines the conservation and restoration, development and advanced initiatives, governance and sustainable utilization of wild and domesticated flora and fauna species, as well as genetic diversity (Shibru and Martha 1998).

Most African countries have policies, laws, and administrative regulations in place to promote biological diversity conservation. These laws include measures to regulate the trade in assets, with a focus on wildlife, fisheries, and forests. They are not obtained through any single policy regime or piece of legislation. They can be found in sectoral policies and legislation such as the National Wildlife Act, National Wildlife Policy, Fisheries Act, and Forest Act. Existing wildlife and forestry laws include provisions governing resource ownership and access. For instance, the 1989 amended Wildlife Conservation and Management Act of Kenya that regulates unauthorized exploitation and access to wildlife resources. Individuals and/or organizations have no right to remove wildlife or its components that are constitutionally protected without the authorization of the national wildlife agency. According to the Act, anyone and/or institution wishing to access wildlife or sections of it must have received permission from the competent authorities; these are the Kenya Wildlife Service (KWS) and the Minister of Natural Resources. In addition, several African countries have developed and implemented forest legislation to regulate the exploitation of forest land and forest resources. The composition of these laws varies by country. Many of them include provisions to prevent unauthorized access to public forests. The Forest Act of Uganda, for example, prohibits people from entering forest reserves to obtain any forest resource or engage in activities that may harm the forest ecosystem. Other countries have also used other frameworks, such as trading activities to manage their forest resources. Cameroon, for example, has used forest legislation to regulate trade in and extract of *Prunus africana* since the mid-1970s.

Furthermore, several nations have included environmental provisions in their national constitutions. The Ugandan constitution, for example, states that Parliament is to provide protection measures, manage it sustainably, and raise environmental awareness. A lot of nations have also included environmental provisions in their bills of rights. As a result, the South African Constitution states that everybody has the right to a decent environment and the right to have it protected, as well as the right to have the environment protected through mechanisms that promote justifiable economic and social development while ensuring ecologically sustainable development and resource use.

Most countries' environmental laws have evolved on a sectoral basis. Africa is no exception to this rule, but like anywhere else, several countries have attempted to address the issue by enacting framework legislation. Such laws frequently include general principles of environmental management, established an institutional framework for biodiversity management, and defined national environmental funds. Countries such as Gambia, Guinea, Uganda, and Comoros have instituted framework legislation.

The conceptual model also specifies the mechanisms for enforcing laws. Enforcement of environmental law requirement in many African countries has relied on penal sanctions, primarily fines and imprisonment. Overall, relying on penal sanctions has proven to be an ineffective tool for ensuring compliance with the standards put in place. Specific sanctions have failed to deter likely offenders, and in some cases, flouting the standards may be more cost-effective than complying. However, some counties have attempted to use incentives to encourage action toward sustainable biological resource management. They could take the form of economic measures like taxes and subsidies. However, in the process of enacting biodiversity laws and policies, needs' evaluation and priority setting should be conducted at the early stage in specific countries to ensure that laws and policies are consistent with the situation on the ground.

2.12 Constraints of Biodiversity Conservation Policies and Laws

Although Africa has a sizable share of the global natural resources and biological diversity, the vital natural resource on which human survival depends is in severe danger (AfDB 2015). Countries in Africa are experiencing unprecedented rates of resource exploitation in recent times, due to increased raw material extraction for economic growth, land-use changes, urbanization, and weak institutional arrangements. Furthermore, the occurrence of global warming poses a major development threat to species and habitats and the future of most of the African rural populace whose livelihood strategies are directly dependent on biological resources. Many African countries have signed some international treaties, agreements, and protocols relating to the conservation and protection of biological diversity. Countries have also pledged to initiatives and declarations at the regional level to protect biodiversity. Countries are required under these pledges to establish and implement national policies, plans, or programs to promote biological variety conservation and sustainable use.

The transformation of these worldwide regimes into regional and national practices via well-defined policy initiatives, legal frameworks, and institutional structures has been a major challenge for governments (Kameri-Mbote and Cullet 2002). In African countries, biodiversity values are not adequately incorporated into national development and planning agendas under current policies and institutional frameworks. At a regional consultation discussion board, African governments revealed their failure to fulfill the Africa biodiversity targets for 2010, quoting issues such as inadequate inclusion and prioritization of biodiversity into larger sectors of the economy (UNEP 2010). Governments have expressed concern about the scientific community's failure to efficiently communicate biological diversity challenges to policymakers in ways that prioritize biodiversity in the political and development agendas (UNEP 2010). After the inability to meet the 2010 biodiversity goals, governments initiated a massive and detailed Strategic Plan for Biodiversity 2011–2020, to ensure the long-term viability of sturdy ecosystems and the provision of essential services by 2020 to curtail biodiversity loss. Achieving this gap requires policymakers to articulate policies that would slow and eventually stop the rapid loss of biodiversity. Given the crucial role of biodiversity in the advancement of Africa's economic growth and the relevance of high-quality data to inform an appropriate decision, it has also become imperative to analyze the existing government's policies, legislation, and institutional framework required for collecting, documenting, and processing biodiversity data and information in African countries to effectively meet biodiversity conservation goals.

2.13 Tourism Policy and Sustainability in Africa

A policy is, in general, a series of interconnected opinions and actions established and applied by the government and authorized institutions to address existing issues, concerns, and prospects (Airey and Chong 2011). Government policy is defined as what the apex authority, the government, chooses to do or not do in society (Dye 2008). It is a planned path of action engaged by an individual or group of individuals to address issues of worry, as well as government action in response to public concerns (Kraft and Furlong 2015). It is an action guide, a means to a goal rather than an end in and of itself, and a tool for resolving socioeconomic and political issues. What authorities in charge determine to do concerning issues and problems that require government involvement and intervention is known as public policy (Rinfret et al. 2018).

Tourism policy, on the other hand, is a system of regulations, standards, guidance, improvement, promotion, and goal-oriented techniques that form the basis for individual and group actions that have a significant influence on sustainable tourism development and daily routines activities within a destination. According to Hall (2008), tourism policy aids us in comprehending the forming of regulatory decisions and the consequences of those decisions. It gives information on real-world problems and solutions. It empowers the stakeholders and value systems involved in the decision process to be envisioned and act as a guide to understanding policy planning and implementation. Tourism policies are primarily designed to manage the growth of the tourism industry. Rules are also required to maintain constitutional control over tourism policy operations. The underlisted are some of the reasons why tourism policy is so important:

i. It establishes the ground rules or the conditions under which tourism businesses must operate.
ii. It specifies the kind of activities and behaviors that tourists and visitors are permitted to engage in.
iii. It gives all tourist stakeholders in a destination a common sense of direction and guidance.
iv. It helps people agree on precise methods and goals for a specified destination.
v. It provides a platform for public-private dialogues about the tourist sector's role and contributions to the economy and society.
vi. It makes it easier for tourism to interact with other sectors of the economy.

The tourism sector's expanding influence in Africa necessitates a detailed examination of the past, present, and future planning and policy frameworks in which tourism might be used to achieve long-term goals. According to Christie and Crompton (2001), African governments should take the leading initiative in developing the tourism sector by instituting legislative framework and institutional arrangements that guarantee the sustainable development of tourism, alleviation of poverty, and community engagement.

The United Nations World Tourism Organization (UNWTO) defines tourism governance as "the system of managing tourist destinations via complementary and orchestrated efforts by the government, at varying levels and in different capacities; civil society living in inbound tourism destinations; and the corporate sector involved in tourism operations." Multi-stakeholder participation, interaction, and integrity in determining winners and losers can all contribute to sustainable tourism governance (Bramwell 2010; Qian et al. 2016).

However, African tourism governance is based on policy dissolution, which lacks credibility, inclusion, equality, flexible power structures, and open engagement. As a result, the socio-cultural and economic impacts are weakened, particularly in rural areas with a high concentration of tourist resources. To aid the transformation of economic and financial gains in tourism into fundamental economic gains for communities, inclusive governance must be improved. This will necessitate bridging the gap between theory and practice in the industry. Justice, inclusiveness, integrity, and control interactions between indigenous communities, government authorities, and private enterprises must all be considered when promoting sustainable tourism (Kato 2018).

In addition, the consistency and stability of the African tourism industry must be guided by the principles and metrics of sustainability, with the following broad objectives in mind.

Institutional Structure

(a) Destination governance policy guidelines
(b) Framework for Guest Governance
(c) Platform for Tourist Security
(d) Planning and policy standards for various tourism activities
(e) Initiatives for employee capacity development in tourism

Economic Principles

(a) Before promoting tourism, consider the economic implications.
(b) Increase links and reduce leakages to maximize local economic advantages.
(c) Assist communities in participating in and benefiting from tourism.
(d) Facilitate the development and marketing of local goods and services.
(e) Encourage fair commercial practices and recompense with reasonable rates.

Guidelines for Social Behavior

(a) Empower host communities in the decision-making and planning process.
(b) Examine the social consequences of tourism activities.
(c) Appreciate and respect socio-cultural varieties.
(d) Take cognizance of the culture of the host community.

Environmental Practices

(a) Minimize the carbon footprint of tourism development.
(b) Utilize natural resources judiciously.
(c) Biological resources should be conserved and protected

Tourism policy and planning at all levels plays a critical role in ensuring its long-term viability. Thus, for African nations to achieve the set tourism objectives and goals, tourism policy must be frequently evaluated and reassessed to reinforce sustainable tourism in Africa.

2.14 National Conservation and Wildlife Tourism Policies: A Case Study of Botswana and South Africa

There are 2 types of policy initiatives in South Africa that regulate the conservation areas and the long-term viability of tourism. The protected areas are governed by 11 sections of national legislation and 9 sections of provincial legislation. According to Steyn and Spencer (2011), these are "the Environmental Management White Paper of 1998, the National Environmental Act of 2003, as amended, the 2003 Environmental Conservation Act, White Paper on Biological Diversity Conservation and Sustainable Use, National Environmental Management Act of 2003, the Act on Protected Areas, the National Biodiversity Act of 2004, and the National Biodiversity Strategy and Action Plan of 2006."

The Protected Areas Act and the Biodiversity Act are two essential pieces of legislation that drive the creation of management plans for South African National Parks (SANParks) protected areas to ensure that protected area management is done in collaboration with residents (Paterson 2009). As a result, SANParks has established a stakeholder engagement strategy that will be used by park management to manage natural resources while also maintaining cultural values (Paterson 2009). The political transformation in South Africa resulted in the implementation of policies aimed at promoting and developing sustainable and inclusive tourism industry.

The White Paper on Tourism Advancement from 1996, the National Tourism Strategic approach, the 2003 Tourism Act, and Black Economic Empowerment (BEE) Charter are among these measures (Steyn and Spencer 2011). The guidelines have been put in place to give business enterprises, the local communities, and the federal government more opportunities to profit from this sector of the economy. (Manwa and Modirapula 2019). In addition, the Restitution of Land Act of 1994, as amended, was critical in supporting communities in regaining land that had been seized from them throughout South Africa's colonial history. The San and Mier communities became landowners within the Kgalagadi Transfrontier Park because of this Act, as well as participants in the park's planning and management of wildlife conservation and tourism activities.

Botswana, too, has tourism and conservation policies initiatives in place. The country's tourism development is guided by the Botswana Tourism Regulation of 1994, Tourism Act of 1992, Tourism industry Master Plan of 2000, Botswana Tourism Development Framework of 2001, and the National Ecotourism Strategy (NES) of 2002 (Basupi et al. 2017).

The dearth of a regulatory framework to manage and expand the travel and tourism industry, as well as the increasing significance of tourism to the country's economy, prompted these regulations. (Basupi et al. 2017). The Community-Oriented Approach was established in 1997 to encourage community participation and the development of tourist organizations at the local level (USAID 2016). These programs, especially the NES, aim to involve the public in the country's tourism development.

The Wildlife Conservation Policy on Wildlife, Resources Conservation and Development Policy, National Conservation Policy, and the Acts on National Parks and Wildlife Conservation all developed a standard through which stakeholders such as the community could influence environmental conservation (Moswete et al. 2012). All these regulations also emphasize the relevance of conservation areas to people's quality of life, stating that societies should benefit from things like job opportunities and long-term sustainability.

Researchers emphasize the significance of visualizing conservation areas through the lens of their societal and cultural relationships with residents, emphasizing the trend of incorporating community perspectives and opinions into protected area design and management operations (Sabuhoro et al. 2017; Atanga 2019). Roy (2016) reported that wildlife tourism and conservation in protected areas are viewed as important to successful environmental protection and poverty reduction.

The significance of every tourism stakeholder, regardless of their level of influence, in long-term growth and preservation of natural resources is emphasized through collaborative management (Kossomann et al. 2016). Furthermore, this technique of protecting lands allows for the merging of traditional beliefs and traditions with modern scientific conservation concepts (Aswani et al. 2017). Community-based natural resource management (CBNRM) programs that demonstrate the efficacy of collaborative management include Communal Property Associations (CPA), Community-Based Organizations (CBOs), and Community Trusts.

2.15 Conclusion

Africa's tourist potential is almost untapped. The African continent receives only 3% of worldwide tourism receipts and 5% of global arrivals, according to the UNWTO Tourist Highlights 2015 Edition. Tourism can drive Africa's economic growth and employment creation if properly implemented because cultural endowments and natural assets may be utilized to produce opportunities for local people; the industry has the potential to make a substantial contribution to the Sustainable Development and Social Inclusion agenda. Keeping in mind that the environment is the primary tourism product, and that the tourism business relies on the environment for its continued success, environmental preservation and enhancement is crucial to the tourism industry's survival. Ecotourism is one of the most well-known terms used to represent this more responsible, "greener" style of tourism (a nonconsumptive use of natural resources and an amalgam of two words- ecology and tourism). Ecotourism promotes the preservation of human cultures for long-term continuity and sustainability, in addition to the ideals of maintaining sensitive and vulnerable ecosystems/natural environments. Ecotourism has been identified as a viable strategy and has been described as a responsible tourism industry specifically developed to accomplish the sustainable usage of resources. The participation of local people, researchers, and government officials is particularly unique and of the highest importance for ecotourism development.

As a result, conservation scientists, the tourist industry, government, and other stakeholders must make promoting a more sustainable and eco-friendly approach, as well as the need to adopt responsible tourism policies, a key priority. Integrated planning, multi-stakeholder dialogue with native peoples, zoning in land-use planning, strategic environmental assessment, standards, industry performance-recognition programs, recognized accreditation bodies, ecolabelling, codes of good practice, environmental impact assessment, and environmental imprinting are just some of the tools that can help to ensure the long-term use of biological diversity through tourism.

References

Adetola BO, Adetoro OA (2014) Threats to biodiversity conservation in Cross River National Park. Nigeria Int J Conserv Sci 5(4):547–552

Adetola BO, Akinboboye TS (2020) Visitors' connection with ex-situ conservation in zoos: catalyst for visitation intention to in-situ ecotourism destinations. Journal of Tourism Res 25:5–29

Adetola BO, Alarape AA, Ayodele IA (2021) Forest resources as ecotourism attraction: Cross River National Park, Nigeria. Ruhuna J Sci 12(1):26–39

AfDB-African Development Bank (2015) African natural resources Center (ANRC) strategy (2015–2020). Revised Version. ANRC Department

Airey D, Chong K (2011) Tourism in China: policy and development since 1949. Routledge, Oxford

Association of Bhutanese Tour Operators (2010) UNWTO Tourism Vision 2020 Forecast Released. http://www.abto.org.bt/2010/06/unwto-tourism-2020-vision-forecast-released/

Aswani S, Albert S, Love M (2017) One size does not fit all: critical insights for effective community based resource management in Melanesia. Mar Policy 81:381–391

Atanga RA (2019) Stakeholder views on sustainable community-based ecotourism: a case of the Paga crocodile ponds in Ghana. GeoJ Tourism Geosites 25(2):321–333

Basupi B, Pansiri J, Lenao M (2017) Botswana tourism policy landscape–successes and challenges. Botswana J Business 10(1):83–96

BirdLife International Press Release (2008) Avitourism 'Takes off' in South Africa (Online). http://www.birdlife.org/news/news/2008/04/SA_Birding_Routes.html. Accessed 09 April 2008

Blanc JJ, Barnes FW, Craig GC, Dublin HT, Thouless CR, Douglas-Hamilton I, Hart JA (2007) African elephant status report 2007: an update from the African elephant database. IUCN, gland, Switzerland: occasional paper series of the IUCN species survival commission, no. 33. IUCN/SSC African Elephant Specialist Group. http://www.africanelephant.org/aed/aesr2007.html

Borokini TI (2014) A systematic compilation of endemic flora in Nigeria for conservation management. J Threatened Taxa 6(11):6406–6426

Bramwell B (2010) Participative planning and governance for sustainable tourism. Tour Recreat Res 35(3):239–249. https://doi.org/10.1080/02508281.2010.11081640

Buckley R (2004) Environmental impacts of ecotourism: impacts of ecotourism on bird. Ecotourism series no 2. CAB international. 416 pp

Ceballos-Lascurain H (1990) Tourism, ecotourism, and protected areas. Paper presented at the 34th working session of the Commission of National Parks and Protected Areas, Perth, Australia

Chen LL, Devereux J (1999) Tourism welfare in sub-Saharan Africa: a theoretical analysis. J Afr Econ 8(2):209–227

Christie IT, Crompton DE (2001) "Tourism in Africa", Africa region working paper series no. 12. The World Bank, Washington DC

Drumm A, Moore A (2002) Ecotourism development: an introduction to ecotourism planning Vol.1. The Nature Conservancy, USA. 85 pp

Dye TR (2008) Understanding public policy, 12th edn. Pearson/Prentice Hall. 354pp

Eilat AT, Einav L (2003) The determinants of international tourism: a three-dimensional panel data analysis. Unpublished Working Paper

Fennel D (2008) Ecotourism. TJ International Ltd, Cornwall

First National Biodiversity Report (2001) Ministry of Home Affairs Housing and Environment, First National Report to the Conference of the Parties to the Convention on Biological Diversity. Retrieved from https://www.cbd.int/doc/world/mv/mv-nr-01-en.pdf

Forestry Management Evaluation and Coordinating Unit (FORMECU), (1996) Nigerian forestry action programme volume II: Main report Federal Ministry of agriculture and natural resources. Abuja

GHK Consulting Ltd., GFA-Race Partners Ltd. (2004) Revealing the value of the natural environment in England, a report to the Department for Environment. Food and Rural Affairs, London

Gössling S, Hall C, Weaver D (2009) Sustainable tourism futures; perspectives on systems, restructuring, and innovations. Routledge, New York; Oxon

Hall CM (2008) Tourism planning: policies, processes, and relationships. Pearson Education, London

Hottola P (2009) Tourism strategies and local responses in southern Africa. CABI, Wallingford

Houston, JR (1996). 'International Tourism and US Beaches', *Shore and Beach* (1996), quoted in 1998 *Year of the Ocean-Coastal Tourism and Recreation* (Online). http://www.yoto98.noaa.gov/yoto/meeting/tour_rec_316.html

Huybers T, Bennett J (2002) Environmental management and the competitiveness of nature-based tourism destinations. Edward Edgar Publishing. 192 pp

IUCN (1997) Ecotourism. In: Beyond fences: seeking social sustainability in conservation. IUCN, Gland. Pp 22

Kameri-Mbote P, Cullet P (2002) Biological diversity management in Africa: policy perspectives. Rev Eur Community Int Environ Law 11(1):38–52

Kato K (2018) Debating sustainability in tourism development: resilience, traditional knowledge, and community: a post-disaster perspective. Tourism Planning Dev 15(1):55–67. https://doi.org/10.1080/21568316.2017.1312508

Kenya National Bureau of Statistics (2017) Statistical abstract 2017. Accessed 27 January 2018

Kenya Tourism Board (2016) Sustainable tourism report. Available at https://magicalkenya.com/wpcontent/uploads/2020/05/Final-KTB-Sustainable-Tourism-Brochure-2016-1.pdf

Kenya Travel Tips (2018) Tourism in Kenya. Available at https://www.kenyatraveltips.com/tourism-in-kenya/

Kester IGC (2003) International tourism in Africa. Tour Econ 9(2):203–221

Kossomann CM, Behgel J, Bailey M (2016) Action and inertia in collaborative governance. Mar Policy 72:21–30

Kraft ME, Furlong SR (2015) Public policy: politics, analysis, and alternatives, 2nd edn. CQ Press, Sage. 459 pp

Kutay, K (1989) Ecotourism: travel's new wave. Vis a Vis, July, 4–80

Manwa H, Modirapula T (2019) The role of Botsalano game reserve in sustainable livelihoods and poverty alleviation through community-based tourism development in South Africa. African Renaissance 16(1):315–333

Melita AW, Mendlinger S (2013) The impact of tourism revenue on the local communities' livelihood: a case study of Ngorongoro conservation area, Tanzania. J Serv Sci Manag 6:117–126

Miller GA, Twining-Ward L (2005) Monitoring for a sustainable tourism transition. The challenge of developing and using indicators. CABI Publishing. 59 p. https://doi.org/10.1079/9780851990514.0000

Moswete N, Thapa B, Child B (2012) Attitudes and opinions of locals and national public sector stakeholders towards Kgalagadi Transfrontier Park, Botswana. Int J Sustain Dev World Ecol 19(1):67–80

Paterson AR (2009) Legal framework for protected areas: South Africa. IUCN SANParks, Gland, p 2016

Qian C, Sasaki N, Shivakoti G, Zhang Y (2016) Effective governance in tourism development – an analysis of local perception in the Huangshan mountain area. Tour Manag Perspect 20:112–123

Rinfret S, Scheberie D, Pautz M (2018) Chapter 2: the policy process and policy theories. In Public policy: a concise introduction. SAGE Publications, pp 19–44

Roy AKD (2016) Local community attitudes towards mangrove forest conservation: lessons from Bangladesh. Mar Policy 74:186–194

Sabuhoro E, Wright BA, Munanura I, Nyakabwa IN (2017) The potential of ecotourism opportunities to generate support for gorilla conservation amongst local communities Neighboring volcanoes National Park in Rwanda. J Ecotour:1–17

Secretariat of the Convention on Biological Diversity (2007) Managing tourism and biodiversity, User's manual on the CBD guidelines on biodiversity and tourism development, Secretariat of the CBD, Montreal, 12 pp

Secretariat of the Convention on Biological Diversity (2008) The value of nature: ecological, economic, cultural and social benefits of protected areas, secretariat of the CBD, Montreal

Shibru T, Martha G (1998) Biodiversity management in Ethiopia. In: Mugabe J, Clark N (eds) Managing biodiversity—National Systems of conservation and innovation in Africa. ACTS Press, Nairobi

South African National Parks Annual Report (2015). http://www.sanparks.org/assets/docs/general/annual-report-2015.pdf. Accessed 30 Jan 2022

Steck B (2004) Sustainable tourism as a development option practical guide for developers and decision makers, Eschborn. 63 pp

Stéphane DB (1995). 'La convention de Paris de 1994 sur la désertification', *in Stratégies énergétiques Biosphère & Société - Le droit international face à l'éthique et à la politique de l'environnement* 143-150

Steyn JN, Spencer JP (2011) South African tourism: an historic Evaluation of macro tourism policies. Afr J Phys Health Educ Recreation Dance 17(2):178–200

Tapper R, Cochrane J (2005) Forging links between protected areas and the tourism sector: how tourism can benefit conservation. UNEP report. https://wedocs.unep.org/handle/20.500.11822/8765

The International Ecotourism Society (TIES) (1990) What is ecotourism. https://www.ecotourism.org/what-is-ecotourism. Accessed Aug 2014

TIES (2008) The components of successful ecotourism. The International Ecotourism Society (TIES). Retrieved from https://www.scirp.org/(S(czeh2tfqyw2orz553k1w0r45))/reference/referencespapers.aspx?referenceid=2533924

Tinker C (1995) A “new breed” of treaty: the United Nations convention on biological diversity. Pace Environ Law Rev 13(1):191–218. https://digitalcommons.pace.edu/pelr/vol13/iss1/8

Tourism Australia (2009) Activities factsheet—year ending June 2009 (Online). http://www.ret.gov.au/tourism/tra/snapshots/sheets/Pages/default.aspx. Accessed 7 Sep 2010

UNEP (2010) State of biodiversity in Africa. http://www.unep.org/delc/Portals/119/State%20of%20biodiversity%20in%20Africa.pdf. Accessed 16 Dec 2015

UNESCO. (2004) Cape floral region protected areas. https://whc.unesco.org/en/list/1007/. Accessed 25 Oct 2019

United Nations Statistics Division (2010) Tourism as an internationally traded service and beyond. Newsletter of the Interagency Task Force on Statistics of International Trade in Services. No. 6, December 2010, p. 1. Retrieved from http://unstats.un.org/unsd/tradeserv/tfsits/newsletter/TFSITS_newsletter_6.pdf

US Agency for International Development (USAID) Centre for Applied Research (2016). Review of community-based natural resource management in Botswana. Report prepared for Southern African Environmental Programme (SAREP)

US Fish and Wildlife Service (2007) National Survey of fishing, hunting and wildlife-associated recreation 2006, Washington

Van der Merwe P, Saayman M (2003) Determining the economic value of game farm tourism. Koedoe 46(2):103–112

World Resources Institute (2004) Reefs at risk in the Caribbean. WRI, Washington. https://www.wri.org/research/reefs-risk-caribbean

World Tourism Organization Barometer (2015). https://tourlib.net/wto/UNWTO_Barometer_2015_01.pdf

World Tourism Organization UNWTO (2015a) Exports from International Tourism Rise 4% in 2015. http://media.unwto.org/press-release/2016-05-03/exports-international-tourism-rise-4-2015

World Tourism Organization UNWTO (2015b) Why tourism? http://www2.unwto.org/content/why-tourism

WTTC (2013) Travel and tourism. Economic impact 2013. World Travel and Tourism Council, London

Chapter 3
Biodiversity Conservation Strategies and Sustainability

Ehi Francis Okosodo and Odangowei Inetiminebi Ogidi

Abstract Biodiversity is threatened globally, but most especially in Africa due to overexploitation and despite legislation in place to address concerns about natural resources and biodiversity. The goal of this chapter is to give a critical evaluation of biodiversity conservation and consumption patterns. Some biodiversity threats identified include exponential population growth, an increase in the rate of poverty, land-use modification and fragmentation, and the introduction of harmful non-native species. Although there is an absence of a complete record and comprehendible template for evaluating the global pace of biodiversity extinction, there is compelling evidence suggesting the rapid decline of biodiversity. The chapter also reviews long-term biodiversity utilization patterns to highlight the significance of conservation areas across Africa. Accurate biodiversity data are necessary for effectively monitoring and mitigating the existing rate of biodiversity loss.

Keywords Biodiversity · Sustainable uses values · Threats · Conservation · Africa

3.1 Introduction

The term "biodiversity" is commonly used to explain the total number of biological creatures on the earth (plants, animals, fungi, and microbes). Biodiversity has been studied from three angles by biologists. They include strain, varietal, species, genetic, and environmental level diversities (Ogwu et al. 2014a, b; Osawaru and Ogwu 2014a, b; Ogwu 2020). The term "genetic" in biodiversity refers to the wide range of genes whether expressed or not within a certain locale as well as intra-and interspecies differences. The sum of species established in a given place can be

E. F. Okosodo (✉)
Department of Leisure and Tourism, Federal Polytechnic, Ilaro, Nigeria
e-mail: francis.okosodo@federalpolyilaro.edu.ng

O. I. Ogidi
Department of Biochemistry, Federal Polytechnic Ekowe, Ekowe, Bayelsa State, Nigeria

S. C. Izah, M. C. Ogwu (eds.), *Sustainable Utilization and Conservation of Africa's Biological Resources and Environment*, Sustainable Development and Biodiversity 32, https://doi.org/10.1007/978-981-19-6974-4_3

referred to as species diversity. Several millions of species exist worldwide. Some of these species play essential roles in the environmental preservation and sustenance of human socioeconomic activities (Ogwu 2019a, b; Osawaru and Ogwu 2020). Sustaining environmental interactions and slight or large-scale fluctuations over time are viewed as crucial for the diverse life forms on the planet. The establishment of nonlinear relationships among biodiversity within a given size region alters context (Ogwu et al. 2019a, b). However, many authors have expressed reservations about extrapolating figures from a particular geographic scale.

Preference for growing a diverse range of crops is gradually becoming a standing because it can serve as a backup if one crop collapses (Osawaru et al. 2013a, b, c). Modern agricultural techniques, on the other hand, are predominantly monoculture-based, with cash crops for both domestic and foreign markets receiving a lot of focus. Local food shortages, unemployment, homelessness, and drought susceptibility have resulted as a consequence (Chime et al. 2015, 2018). Irrigation, nutrients, and insecticides have all recently become very popular. However, Africa still lags in terms of industrial monoculture-based agriculture that is almost entirely reliant on agrochemical use. But predictions exist suggesting that, around 2050, the pressure from population growth might alter existing trends (Ogwu 2019b, c; Erinle et al. 2021; Evivie et al. 2020).

For ecosystems to survive, they may be reliant on the presence of a certain species known as a "keystone or indicator species" as these organisms retain information about the state of the environment. Many concrete buildings are supported by structural arches, each of which has a keystone at the top that supports the entire arch and, therefore, the building. Similarly, the survival of an ecosystem may depend on the presence of a particular species. These species are essential to the ecosystems in which they live; without them, ecosystems would become unstable and eventually change to a new equilibrium, which is often less biologically diverse and hence less capable of supporting human existence. *Orycteropus afer* is an example of a keystone species in Africa that contributes to savanna ecosystem diversity by excavating holes that serve as dens and burrows for a variety of savanna species' reproduction and protection. It should be evident that biodiversity and human well-being are inextricably linked. The fact that people need to eat species that are part of the earth's biodiversity and depend on the stability of other species and ecosystems shows that they are dependent on climate, nutrient cycles, and ecosystem stability (Ikhajiagbe and Ogwu 2020, 2021).

Based on the roles of certain species in maintaining and sustaining life forms, in addition to managing environmental degradation which is characterized by the modern society in the Global North probably due to urbanization, industrialization, and population growth, there is a need to conserve life forms with pivotal roles in Africa and other parts of the Global South. This chapter

focuses on sustainable conservation and utilization strategies of Africa's biodiversity.

3.2 Biodiversity and Sustainable Development

The Brundtland Report, published in 1987, presented the notion of sustainable development. However, the role of reconciling the framework of economic growth is frequently contradicted by the need to reverse ecological damage while simultaneously advancing human rights and poverty reduction. Meeting current demands without jeopardizing previous generations' ability to fulfill their own is a fascinating and long-term concept. It's always been challenging to transform basic sustainable strategies into practical solutions. The basic concepts of viability, such as in terms of inter and multi-equity, present a slew of tough challenges. Nonetheless, the notions of ecological sustainability may be useful in determining the objectives and consequences of ecological and sustainability initiatives. Additionally, while considering how we must live in reality, stability and progress become more than abstractions or ideals. They serve as a link between our forward-thinking and preparation and the activities and consequences rooted in natural ecosystems and contextual factors. Sustainability's materialism, or conciseness, is constantly there as a potential resistance to those who want to promote their own limited goals, and by extension limit and modify other people's choices in a variety of ways. Lockie claims that for sustainability, at least three interconnected sets of needs must be fulfilled. For starters, long-term success necessitates knowledge. As climate variability shows, the time and geographical patterns of social behavior are characterized by abrupt planned and unplanned change, interlocking impacts, and unforeseen repercussions. It will never be solely about preserving stable ecosystems, societies, or economies to provide humans with a long-term good environment. Environmental management must be established with the use of existing data and involve native people managing native environments. Our present understanding of global processes and other political and social groupings will inevitably be proven as insufficient and outpaced as species and ecosystems, and also human organizations and individuals, develop in unforeseen ways. In practice, this would imply reengineering our education system to include continuous improvement and the flexibility to adjust to changes in learning and facts. It's important to plan for the long term, but it's much more important to understand it (Ogwu 2009, 2010). Second, sustainability necessitates dialogue, i.e., information sharing and conversation about major issues that are accessible to all individuals who may be involved in the situation. More than merely people's right to vote in democratic elections is at stake. The problem of acquiring local or indigenous information isn't the only one. Contemplation is also required within sustainability since the human environment is a continually changing panorama of information, beliefs, objectives, ambitions, and alliances, all of which are equally significant (Osawaru and Ogwu 2014a, b; Osawaru et al. 2016; Ogwu et al. 2016a, 2017, Ogwu et al. 2018a, b). Deliberation, as required by sustainability and sustainable development, also means realizing that the human environment is a dynamic terrain of knowledge, beliefs, objectives, aspirations, and coalitions, all of which are critical. Each interaction or conflict with others in the context of environmental management can modify existing senses, beliefs, and goals. As new special interests emerge, areas of agreement and disagreement evolve. So, democratic debate is needed to understand and respond to the fluid reasons why social networks, interpretations, and goals are built and rebuilt by social and economic change processes.

Long-term sustainability also necessitates openness. Implementing new action plans is insufficient. It is critical to examine our strategic work and training. We must distinguish between appropriate and unsuitable, effective and failed, exceptional and dreadful efforts to make altogether possible social, now and in the past, and regulatory frameworks that draw together a diverse group of participants. During human history, several organizational frameworks have been formed to ensure such responsibility (for example, property rights and responsibilities, pollution licensing, production standards, etc.). The critical evaluation must be broadened and increased via knowledge and dialogue for these arrangements to be sustainable.

3.3 Value of Biodiversity

The value of biodiversity is often seen from the perceptive of ecotourism and its uses. Aside from the financial benefits of biodiversity preservation, there are several ways biodiversity can significantly impact our perceptions, inspiration, and originality. African biodiversity is a significant aspect of the global tourism business. National parks, sanctuaries, and resorts all over the continent attract visitors from across the world who are looking for some downtime to rest and rejuvenate. These kinds of activities promote calmer dispositions, but also help to improve our interactions with nature in terms of cultural, moral, spiritual, and ethical. This defines the sacredness and interconnectivity of all living organisms. According to the definition of ecotourism, it is "known to natural places that seek to sustain the ecosystem and local well-being." Ecotourism benefits poor countries like many of those within the African continent in several ways, including by creating alternative solutions, which have little to no environmental impacts. Residents and visitors alike are more conscious of the need of protecting natural and cultural resources. Both monetarily and in terms of foreign exchange, it is beneficial to host cities by promoting economic growth and well-being. Nature-based tourism which is often hailed as a realistic development platform for underdeveloped countries has enormous potential. For instance, the Argungu Fishing Festival in Northern Nigeria is a popular attraction as well as the Hadejia Nguru Wetlands among others that are scattered all over the African continent. In addition, these sites serve as hosts and have the capacity to sustain and welcome millions of migrating birds and other animals like wildebeests. There are a lot of promising biodiversity in Africa's natural forest (especially nature reserves), coastal habitats, and wetlands resources. Cross River gorillas *(Gorilla diehli)*, African elephants (*Loxodonta africana*), lions (*Panthera leo*), and chimps (*Pan troglodytes ellioti*) are among the most well-known big animals in the region. Several species of birds, reptiles, amphibians, and fishes abound in the region. However, due to excessive exploitation and other human activities, many are near extinction in the region.

Biodiversity has a significant influence on human health. Farming and food, healthcare, commodities, and other useful aspects are all made possible by biodiversity (Izah and Seiyaboh 2018a, b; Ogwu et al. 2016b,c,d; Izah et al. 2017). It enables humans to live in environments where they would otherwise perish. This is because many forest resources including plants and animals have been used to treat several human diseases. A lot of current medicines make use of mixtures of biologically diverse chemicals based on numerous plants (which we, therefore, label medicinal). Some of these plants have been comprehensively documented (Izah et al. 2019a,b,c, 2018a,b,c,d; Izah and Aseibai 2018; Enaregha et al. 2021; Kigigha et al. 2018a, b, 2016a, b, c, 2015a, b). Ayurvedic and Unani medicinal traditions employed several herbs to provide a range of benefits before modern medicine's arrival. Humans would be far more susceptible to illness if such plants and the myriad insects that pollinate and cross-pollinate them were not present. The ecosystem's diversity supplies forest dwellers with all they need daily, including food, building materials, fodder, medications, and a range of other items. Without biodiversity, we are without shelter, as it supplies us with materials such as timber. Biodiversity contributes significantly to the food sources of humans.

3.4 Threats and Consequences of Biodiversity Loss

This section of the chapter summarizes the threats and consequences of biodiversity loss (Fig. 3.1).

3.4.1 Extinction

The greatest evidence of the extinction of species is the loss of distinct taxonomic groupings. Most worrisome is the disappearance of African species before they are properly classified or characterized (Osawaru et al. 2014a, b, c, 2015a, b). Background extinction is ongoing, and if unchecked, might continue until no more members of a taxon survive, either in a small area of their habitat or over their whole range, and they are deemed extinct. Biodiversity decline is usually examined at the species level, but it may also be examined at the subspecies or population level. Each species has its collection of genetic, evolutionary, behavioral, and ecological characteristics that cannot be recreated after it's gone extinct. Extinction and evolution (cum speciation and diversification) occur when current species die and new ones evolve or develop respectively. Speciation has been continuous across the Paleozoic, culminating in a vast spectrum of diversity. Across the earth's natural history, biodiversity loss has been a repeating problem, but in this geologic era, humans and human activities are the driving force. According to paleontology, there

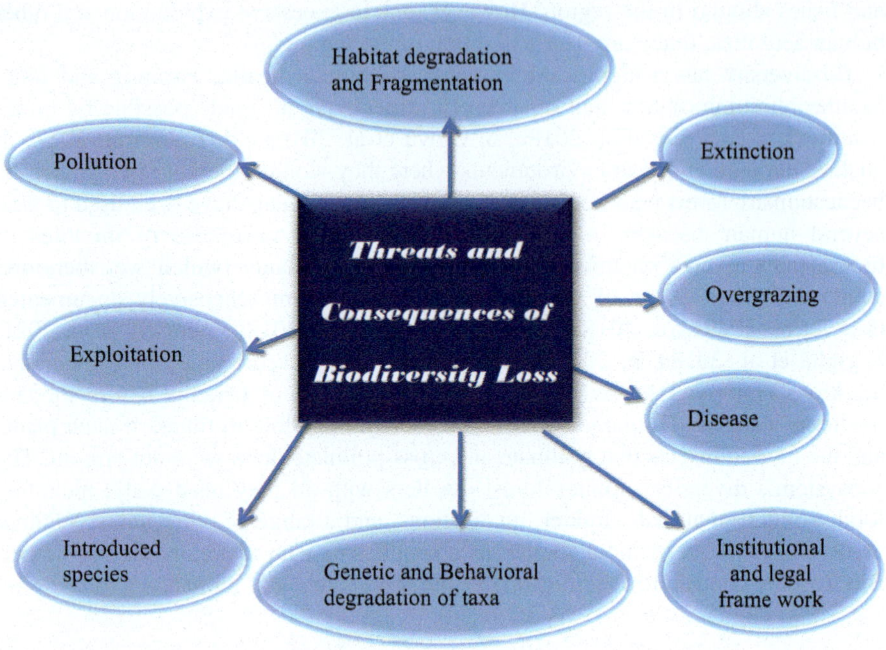

Fig. 3.1 Threats and Consequences of Biodiversity Loss

have been five cataclysmic extinctions that wiped out the majority of taxa. Habitat loss and other issues have a big influence on these areas. Extinction reduces biological diversity. The most successful strategy for expanding diversity is sustainable biological development, which requires a reassessment of extinction events in prehistoric periods. Following the massive habitat fragmentation that led to substantial biodiversity losses, habitats needed millennia to recover, resulting in a more homogeneous biosphere in terms of medicine and ecosystem.

3.4.2 Genetic and Behavioral Degradation of Taxa

The term "biodiversity" refers to the genetic and behavioral diversity observed in populations of a species. Species differences are the most evident type of variety, notwithstanding the importance of population and genetic variants in species survival. It may be difficult to discern various populations within a species when the border between them can be referred to as fuzzy—asexual reproduction or hybridization are some examples. Population loss and change as a result of human actions are the two fundamental causes of genetic and behavioral deterioration. When the majority or all of a species' surviving members are kept in captivity, as in the case of the Spix's macaw, limited populations have a significant number of both, which is necessary for diversity (*Cyanopsitta spixi*). A large number of species, which include *Procyon lotor* and *Passer domesticus*, have adapted to people and abandoned most of their native habits as a result of their modified habits in response to human sources, limiting natural variations. On the other hand, wild populations are extremely fragile. Several chimp groups including *Pan troglodytes* use twigs to remove termites from their stacks and stones to crack the nuts open. Human interaction can cause invertebrates to change or lose their habits or modify their behaviors as a form of an adaptive response strategy. The *Danaus plexippus* is not a serious global issue, but its populations migrate in extreme ways along with different countries. The disappearance of certain habits reduces total diversity, which might have lengthy consequences for other populations and species (Izah 2018; Izah et al. 2018e). Different genetic alterations occur throughout a species' populations, resulting in comprehensive biological and ecological roles in a broad range of environments.

3.4.3 Habitat Degradation and Fragmentation

The establishment of specified corridors is required to maintain population connectivity while conserving genetic diversity. When dispersal patterns in fragmented environments are altered, the genetic make-up of smaller populations is accentuated. For example, the loss of some plant species might affect the function and structure of the ecosystem. Habitat deterioration happens when a section of the natural ecosystem is destroyed or changed. While less visible, habitat damage and fragmentation are less detrimental than habitat loss. Substantial human or animal usage is another activity that has a detrimental influence on habitat quality. The *Molothrus ater* brood parasitic bird in the savanna habitat is an example of how fragmentation disrupts ecosystems by limiting the ability of species with large habitat variants or specialized ecosystems to live, while allowing species to thrive on various borders. In addition, the construction of major highways has led to the loss of biodiversity. For instance, the Nun Forest Reserve in Nigeria used to be home to much endemic wildlife but due to the establishment of Niger Delta University and the subsequent creation of a road linking the University with Yenagoa, the State capital, many communities have sprung up in the area. This has led to the loss of valuable trees and shrubs and wildlife in the area. About 10 years ago, bushmeat is common along the highway, but it is now increasingly rare to find bushmeat along the same highway.

3.4.4 Pollution

Contaminants may be released into ecosystems, putting certain creatures at risk of extinction. Pollution can be described as the destruction of the instinctive environment. Nuclear mishaps, crude oil spillage, and marine time catastrophes are all examples of severe ecological disasters. Runoff of water from agricultural activities, the application of pesticides (Inyang et al. 2020a, b, 2016a, b, c, d, e), car emissions (carbon monoxides), and erosion are all examples of human-caused persistent pollution (Vwioko et al. 2018; Ikhajiagbe et al. 2022a, b). The exposure of inorganic and organic pollutants into the air due to the industrial revolution is placing biodiversity in danger. Slight climate adjustments may have substantial consequences, but any significant local, regional, or global change might be disastrous. Acute pollution is caused by a single incident, whereas chronic pollution is caused by toxins accumulating in the environment over time. This pollutant includes solid wastes, industrial emissions, agricultural wastes, and releases.

3.4.5 Introduced Species

Blatant migration, competition killing, ecological degradation, and human-caused climate change are some of the tactics utilized to bring in new species because of human activities. In some cases, only a small percentage of species that have been relocated to new environments thrive, and even fewer add considerably to natural vegetation. Invasive plants are those that have spread beyond their natural range as a result of human interference, either directly or indirectly. Animals have successfully inhabited some islands, filling biological voids. This has been demonstrated in certain regions of the world, especially in the Pacific Islands by *Felis catus*, *Ovis aries,* and *Capra hircus* grazing on the pacific islands bereft of strong mammals. *Lythrum salicaria* has overrun several marshy regions in North America, while *Pueraria lobata* spreads along forest borders. Water hyacinth (*Eichhornia crassipes*), a worldwide invasive, exotic plant species, is wreaking havoc on African lake ecosystems (Ogamba et al. 2017, 2015a,b,c,d). Animals have effectively colonized certain islands, filling ecological gaps in the process.

3.4.6 Disease

Disorders that are characterized by genetic abnormalities, bacterial or viral infections, or parasitic illnesses are classified as such. Hosts and illnesses coevolve, allowing both hosts and viruses to survive throughout time. Viruses spread more easily when man-made variables, such as population loss, create a disparity. Living things such as chlorinated hydrocarbons can weaken the resistance to infection of living things, making them more susceptible to disease. Pollution or stress can compromise immune function, causing susceptible populations or species to become extinct. The canine morbillivirus was seen to kill numerous surviving *Mustela nigripes*, compelling the surviving animals to be caught and used in a restoration effort. Since the host-pathogen association is generally the consequence of a long history of coevolution, the pandemic is typically more lethal.

3.4.7 Exploitation

The wild game business has destroyed numerous populations and species across Africa and Asia, spurred by a cultural desire for wild animals, particularly ungulates and monkeys. Humans gather animals from the wild for nutrition, energy, and other commodities. Fishing, hunting, and logging are three examples of common exploitation methods. Even though wild-game hunting and gathering continue to be popular, cultivated flora and fauna products account for the majority of the human diet. The extinction of animals due to human activity poses a serious threat to biodiversity. Wildlife is mostly used for food in both freshwater and marine ecosystems. Wildlife has declined significantly in the majority of maritime resources, implying that exploitation puts local biodiversity at risk. Forest logging is a one-of-a-kind type of enslavement that poses the greatest threat to biodiversity in the world. Most lumbering activities are difficult even in temperate conditions, and tropical forest lumbering appears to be untenable.

3.4.8 Overgrazing

Whenever plant species are browsed quicker than they can be normally replenished, habitat loss develops, leading to permanent loss of plant cover. It can happen when there are too many animals grazing on a limited area, and whenever cattle browse greenery over longer durations with inadequate recuperation times. It devalues land and contributes to deterioration and soil erosion.

3.4.9 Institutional and Legal Framework

The National Policy on Environment is the world's foremost institutional and legal structure for environmental protection. Its duty and objective, as mentioned in the constitutions of countries all over the world, is to "ensure environmental protection and protection of the environment for sustainability." The environment ministry has specific priorities which include the coordination of environmental preservation and conservation of natural resources for long-term growth, with a focus on a person's health and well-being.

3.4.10 Legal Framework

3.4.10.1 International Legal Framework

The management, conservation, and sustainable use of biodiversity are utmost in this age of mass extinction of organisms. Therefore, national and international organizations strive to institute law and regulatory frameworks for this same purpose. International laws enable rules and regulatory mechanisms to function whereby actions on biodiversity and ecosystem conservation and sustainability can be affected and also institute a legal framework that empowers countries in an attempt to realize specific objectives of biodiversity with regards to various countries' priorities and their commitment to addressing the affairs of biodiversity locally, but in line with global efforts and goals. It gives the fundamental basis for recognized authorities' strategies and activities to sustain, manage, and conserve biodiversity and the ecosystems by setting up pertinent borders and social protections where profits come from the use of these resources which can be fairly and equitably shared for the satisfaction of all authorities. Licenses are also made available with the help of these laws to provide various incentives and acknowledge the rights and responsibilities of businesses, communities, individuals, local government areas, entrepreneurs, and other indigenous people to act decisively in preventing further damage to the ecosystem and biodiversity.

Various international legal organizations have described the implementation or execution of the National Biodiversity Strategies and Action Plans law as providing a clear blueprint for the agenda for biodiversity; incorporating general biodiversity into countries' needs or priority areas; incorporating biodiversity theories into governance structures; issuing very clear directives or orders that cuts across various biodiversity institutions; providing team members or stakeholders with assurances in conserving and sustaining biodiversity operations and incentives; acknowledging local community rights and responsibilities and pertinent customary or traditional laws; and providing fairness, justice, and equity into biodiversity-linked operations.

The Laws of the International Court of Justice instituted the foremost conventional or traditional provenance of international law, such as the international conferences or congresses in general or specific; international practice as confirmation of acceptable standards of law; fundamental ethics of law acceptable by developed nations; and legal resolutions and explanations of accomplished public relations consultants in different countries as ancillary areas for implementing the rule of law. Other forms of legal framework or law are widely recognized as "soft law". The officially irrevocable international methods approved between different nations established by the international legal framework that deals with general or individual environmental matters are referred to as multilateral environmental agreements (MEAs). MEAs, as a pact, implement international laws around the world, incorporating authorities' collaborative ability to preserve ecosystems and biodiversity. A traditional, customary, or common law action establishes responsibilities for states, regions, or other international member communities, except for those who have vehemently opposed such a legal framework. This law exists independently of treaty or concordat law. Rules and regulations may bring about the principles of concordat or treaty law. The former is to be in tune with the conservation, management, and sustainability of biodiversity, which is significant because it takes care of the implementation problems of MEAs. Customary law supplements the rule of law to strengthen the governance of biodiversity processes (ICJ 1986).

3.4.10.2 National Biodiversity Legal Framework

The acceptance of biodiversity pacts is seen as one of the major accomplishments of many nations' involvement in the United Nations Conference on Environment and Development. As a result, these countries embrace their treaty obligations according to international treaties. For instance, the Nigerian constitution contains extensive ecological provisions and recognizes serious environmental issues. The current Constitution of Nigeria's global objectives is outlined in the Forestry Ordinance and the National Parks Act, as well as the Environmental Impact Assessment Act, the National Oil Spill and Detection Agency, and the National Environmental Standards and Compliance Enforcement Agency, among other laws. Apart from the fact that several of these laws require revision, their execution has been subpar. A Biosafety Act was passed in Africa, enhancing the country's biodiversity protection initiatives. As the ecosystem improves, water, air, land, forest, and wildlife are all preserved.

3.4.10.3 Institutional Framework

The Environmental Protection Agency advises the federal government on a wide range of issues, notably forest management, management, and renewal. It is in charge of the country's overall environment protection, and also biodiversity and asset preservation, plan development with stakeholder engagement, and long-term planning. It also aids in employee training to meet various management requirements of the environment. The federal government provided the National Environmental Standards Regulation Enforcement Agency to guarantee that every policy on the environment is enforced reliably across the country. The Federal Ministry of the Environment also oversees the National Park Service and the National Oil Spill Response and Detection Agency. In Different countries in Africa, major government departments are in charge of forming effective sustainable development agreements.

3.5 Biodiversity Conservation

Biodiversity conservation necessitates the protection of all kinds of life on Earth, as well as functional and thriving natural ecosystems. This involves elements such as biological variety conservation, management, long-term usage, restoration, and development. Restoration is a subset of conservation that pertains to retaining something in its natural position without changing the actual thing. Conservation refers to the continuing application of resources such as preservation and utilization. Sustainable design is quite a difficult aspect of biodiversity preservation. This pertains to growth that fulfills current requirements without affecting the ability of the future to meet their sustainable development and entails the conservation of biodiversity by striking a balance between the environment, development, and society. It is only through the strong application of enforcement and implementation of strategies, as well as institutional environmental policies, that will allow this to happen.

As time goes on, the links between biodiversity and a long-term future become clearer, which means we must protect biodiversity because it is necessary for human survival. In light of the current situation, two basic methods of biodiversity protection have emerged: in situ and ex-situ approaches. To put it bluntly, it has to do with redistributive and intergenerational fairness.

3.5.1 in Situ Conservation

In situ conservation refers to the preservation of germplasm in wild communities of plant or animal species, such as forest genetic resources in natural populations of tree species. In situ conservation protects a species' genetic diversity as well as evolutionary adaptations that allow it to adapt to changing environmental conditions, including population shifts and climate change. As a consequence, it ensures that hosts of other related species, as well as the target species, are preserved. It is usually less costly than in-place methods (though not inexpensive). For plant species having resistant seeds, it is sometimes the only choice for conservation. In Africa, places with great biodiversity are recognized as protected areas and protected. The natural tranquility of these nature reserves, as well as the plant, animal, and bird groups that call them home, come together to make each one a one-of-a-kind destination. The pillars of in situ biodiversity conservation include national parks, forest and wildlife reserves, and sacred forests.

There are presently 17 national parks situated right across Africa, representing significant African ecosystems such as rainforests, montane forests, freshwater wetlands, and lakes, the Guinea savanna, Sudan savanna, and Sahel savanna. In Africa, the Government manages these parks, although the African National Park Service oversees them. The Gashaka-Gumti National Park (GGNP) covers 6.402 km^2 and includes savanna grassland in the north and montane forest in the south. The Mountain reedbuck (*Redunca fulvorufula*), African elephant (*Loxodonta africana*), and West African wild dog are all found in the GGNP. The red sunbird bush (*Metarungia pubinervia*), a tropical species with small villages in eastern Africa, South Africa, and Africa, is also highlighted. The Kainji National Park, located in Niger and Kwara state, is Africa's second largest after the Gashaka Gumti National Park. The Borgu game reserves, which make up the park's savanna forest component, the Kainji lake, a 136-kilometer-long reservoir that divides the Borgu game reserve in half, and the Zugurma sector, which is unreachable due to a lack of roads, are separated into three areas (*Taurotragus derbianus*). Okwango and Oban, both in Cross River state, are split into two portions of the Cross River National Park. The park lies in Guinea-Congo and is home to a wide variety of plants and animals. The Cross River National Park is Africa's most significant wildlife sanctuary and home to one of Africa's oldest rain forests. According to recent figures, deforestation, slash-and-burn farming, and hunting are now threatening both sides of the park. The ancient Oyo National Park in Africa is situated in Oyo State and extends into southern Kwara State. Flora and fauna from both the rainforest and the savanna may be found in the park. Recent events, on the other hand, demonstrate that the park has been seriously encroached upon by the rising human population. As a result, many once-common species in the park have become rare. Take the West African wild dog (*Lycaon pictus amanuensis*), which is in danger of dying out.

The Chad Basin National Park covers approximately 2258 km^2 in northern Nigeria, West Africa, in the Chad Basin. The park is split into three sections. The Chingurmi-Duguma sector of Borno State is in a Sudanian savanna ecological zone, whereas the Bade-Nguru Wetlands and Bulatura sectors of Yobe State are in a Sahel ecological zone.

Due to the park's tiny number of shared species, notwithstanding its convenient location, it has only attracted a small number of tourists. The park's success is being hampered by security issues, climate change, population increase, and environmental deterioration.

In Kaduna State, the Kamuku National Park is situated in the country's north. It has Guinea-Sudan savanna vegetation, which protects the country's most diverse plant varieties. The secretary bird (Sagittarius serpentarius), Denham's bustard (Neotis denhami), and Abyssinian ground-hornbill (Neotis denhami) are among the unique species identified in the park (Bucorvus abyssinicus). Okomu National Park in Edo State, Nigeria, is a vestige of the African lowland rainforests, which once reached from the Niger River west to Benin's Dahomey Gap. Mangrove and swampy forests divided the forest from the coastline to the south and southeast, while it became part of the Guinean Forest-Savanna Mosaic eco-region to the north. The park has downsized to less than one-third of its original size due to higher nearby intrusion.

3.5.2 *Forest and Game Reserve*

Protected areas are controlled by government tracts of land where commercial wood gathering is prohibited to gather ecological elements that may be absent in sustainably harvested regions. The plan was developed to counteract logging-related rainforest destruction by implementing a regular afforestation program. However, there is another alternative. Over 1000 forest reserves listed in the IUCN World Database on Protected Areas do not exist at the present. Owing to invasion and consequent forest transformation to plantations and farm fields, the few remaining forest reserves, such as Idanre Forest Reserve, Akure Ofosu Forest Reserve, and Oban Hills Forest Reserve, have lost much of their natural tropical forests. A wildlife sanctuary is a large area of land where wildlife may live peacefully. The vast majority of Africa's wildlife sanctuaries have been depleted and exploited for other purposes, establishing them as mere paper tiger reserves. The failure of state and local governments to adequately manage these nature reserves has resulted in significant administrative functions in the country's wildlife and forest management. The Yankari game reserves, for example, which is now under state administration, show how things may change under new management. The wildlife reserve is now just a shadow of its former glory.

3.5.3 *Sacred Groves*

Sacred groves are forest sections that have religious and cultural significance. Sacred groves are collections of trees that hold special religious importance for a specific culture, and the grove's species are protected as a function of their value. The sacred groves are in ex-situ preservation in Africa. Because traditional religion is still embraced by the bulk of Africans, sacred groves may be seen throughout Africa's many cultures. Residents of Osun State, for instance, hold the Osun-Osogbo sacred grove in high regard. Ex-situ conservation refers to the preservation of a threatened plant or animal outside of its native habitat, such as by colony relocation, which includes relocating a portion of the population to a less vulnerable site, or through daily care techniques like zoos and botanical gardens. While ex-situ preservation is beneficial in man's attempts to preserve and conserve our environment, it is rarely sufficient to prevent catastrophe. As a result, it aids in the maintenance and conservation of the biotic species it contains. As it can copy the full environment, conserve a species' entire genetic variation, its symbiosis counterparts, or those features that may help a species in adapting to changing conditions throughout time, it is often adopted as a last choice or as a supplementary to in situ conservation. Ex-situ biodiversity conservation on the African continent has mostly relied on home gardens, seed banks, and botanic and zoological parks.

3.5.4 *Home Gardens*

Home gardens are methods of production of agricultural commodities that necessitate the growing or cultivating of small parcels of land near the household or within walking distance of the residence. It provides consistent harvesting of raw vegetables every year with cost-effective and tremendous production. Home gardens might be valuable for both on-farm research and biodiversity protection. Home gardens are typically placed near houses, have a close association with the activities of the family, and have a diverse range of agricultural and animal species to suit the family's requirements (Ogwu et al. 2014a, b). Home gardens are critical to household security in rural Africa, providing food, fiber, fuel, materials, and even land ownership. However, population restrictions and increased housing demand in cities have reduced the theoretically accessible acreage necessary for home garden production. Home garden systems have made a big difference in the stability and sustainability of the environment, especially in rural parts of Africa where home gardens are common.

3.5.5 *Seed Banks*

Ex-situ seed banks are facilities for storing seeds to preserve genetic variation in rare and fragile plant species. Ex-situ biodiversity conservation is a priority for several African research organizations, including the International Institute of Tropical Agriculture (IITA), the National Centre for Genetic Resources and Biotechnology (NACGRAB), and the National Horticultural Research Institute (NHRI). Conventional breeding and researchers may utilize these resources to increase agricultural productivity and food security. Seed banks hold seed collections at consistent background temperatures and humidity levels to protect gene pools that may otherwise be lost to disease outbreaks or other natural disasters if stored in situ or field collections. Several African research institutes, notably the International Institute of Tropical Agriculture (IITA), the National Centre for Genetic Resources and Biotechnology (NACGRAB), and the National Horticultural Research Institute, have made ex-situ protecting the environment a priority (NHRI). The archives stored in these locations can help plant breeders and scientists in agriculture production and food production.

3.5.6 *Botanical and Zoological Gardens*

Botanical gardens and zoological gardens, on the other hand, house live flora and fauna collections. The Abuja Zoo in Abuja, the Audu Bako Zoo in Kano, the Enugu Zoo in Enugu, the Ibadan University Zoo in Ibadan, the Jos Wildlife Park in Jos, the

Port Harcourt Zoo in Port Harcourt, the Ogba Zoo in Benin, and the Sanda Kyarimi zoo in Maiduguri are all prominent African zoological gardens. Botanical and zoological parks serve as tourist destinations for Africans and foreign tourists, generating cash for the authorities.

3.5.7 *Ex-Situ Conservation Strategies*

For millennia, gardens, zoos, and menageries have provided sanctuaries for endangered plants and animals. While these gardens, zoos, and menageries were primarily constructed for the pleasure of the monarchs of governments and countries that financed excursions, they frequently had a more utilitarian purpose. They were converted into breeding and adaptation centers for collecting plants and animals that would be valuable to people in their new environment. In the 1900s, when agriculture and other academic institutions created independent germplasm banks, botanical and zoological parks' conservation duties were more limited to species with no obvious economic interest. Botanical gardens, zoological parks, and aquaria have taken on a new and more vital role as several flora and fauna face extinction in the natural world. Approximately 1500 gardens and components exist across the globe, including the world's most important collection of plant species outside of nature. While there has been no systematic assessment of the wide range, a botanic garden database established for the Botanic Gardens Strategy Plan offers some idea of the biological gems it possesses. Botanical gardens may accommodate approximately 80,000 species (roughly a fifth of the world's higher plants), with specific gardens containing a much larger number of species. The Royal Botanic Gardens at Kew in the United Kingdom, for example, hold over 38,000 species (around 10% of those are threatened), which is more than most countries see in their native habitat. Botanic Gardens Conservation International was established to organize floral garden environmental initiatives across the globe. Botanical gardens have the potential to aid in the recovery of threatened ecosystems. Restocking, reintegration, and restoring operations connect ex-situ and also in situ efforts to use and sensibly maintain biodiversity. All zoos and aquariums accomplish the same thing. Over 700,000 members of over 3000 species of mammals, birds, reptiles, and amphibians live in roughly 800 competently maintained zoos across the world. Zoos, like botanical gardens, are becoming increasingly conservation-focused. Many zoos now have captive farming practices for threatened animals, and some, like the National Zoo in Washington, DC, are proactively responsible for raising threatened animals and reintroducing them to their native habitats. As a result, many large zoos have invested heavily in infrastructure and the development of new technologies to assist rare and threatened species in their marine habitats. The sacred groves of India are extremely essential for the environment. Sacred groves are small patches of natural habitat that were protected by local communities for religious reasons in the past. Indigenous people maintain these groves, which are typically found in tribal areas. These are geographical "green patches" that protect biodiversity and promote conservation. The safeguarding of Khejri (Prosopis cineraria) tree groves by the Bishnoi people in Rajasthan is an example. Maharashtra, Karnataka, and Kerala have sacred groves as well. The aims of agronomic research groups for genetic resource preservation have also changed during the previous couple of decades. These institutions used to be primarily concerned with the genetic resources of a new staple crop or animal species. A range of factors, such as genetic susceptibility, the high ecological costs of using strong cultivars, and restricted or drained soil and water mineral wealth, are enabling these organizations to search for alternative uses for originally used species, as well as possible benefits for thousands of flora and fauna that are rarely known to science.

3.6 Food Security and Nutrition

By this concept, food security consists of four elements: accessible, affordability, usage, and preservation. Availability of food describes a state where everyone has continual physical, social, and economic access to sufficient, safe, and nutritional food that satisfies their nutritional needs and enables them to make appropriate food preferences. Plant species include the native plant species of the forest, as well as trees (such as agroforestry trees and other exotic trees). Agriculture provides healthy food, revenue, jobs, power, and natural ecosystems in semirural and urban areas, giving back to all the core elements of global food security. Food security and nutrition might be jeopardized by deforestation. In the short and medium term, widespread conversion of natural to other land use types, particularly agriculture, may increase food security for farming communities that rely on their own goods; however, this conversion may have long-term environmental, wellness, and food security implications for people; these implications will primarily affect forest communities, but they will also have an impact on national and global populations. Forest degradation is also anticipated to result in long-term biodiversity loss and ecological processes as a result of decreasing crop yields. Forest products such as game meat, edible insects, edible plants, mushrooms, and fish are used by around 1 billion people to some extent. Several polls indicate that they are the lowest-income households in emerging countries. Despite accounting for less than 0.6 percent of global food consumption, forest foods are critical in local communities for ensuring the provision of medically complete meals as well as necessary vitamins and minerals. Forests and trees provide cattle with pasture in the form of forage or animal feed outside of the woodlands, ensuring food security. Food security is improved by grazing in two ways. For starters, it expands the amount of food available. Livestock are important sources of animal protein and essential nutrients, and they can help third-world countries increase crop production and food security more than expected by providing grazing capacity and waste. Tiny fish, for example, are believed to be more affordable than bigger species, other animal sources of food, or crops in vast dry land areas for small-scale farmers and herders' adaptation in Africa. Goods have a critical role in maintaining food security and nutrition.

Even though forest resources provide so little to global household income, they are critical for the incomes, food production, and nutrition of the more than 80 million people who labor in the public and private forest sectors. Forest property and natural rights must be secured to reap the economic benefits of agricultural and forestry harvesting and sales, as well as to ensure forest people's choices of food and nutrition security. Regardless of the lack of female data, research implies that rural women are affected by NWFPs and fuel wood gathering and that they rely on sales revenue year-round. Despite many efforts to update NWFP data, more data are intended to explain where and for whom these foods are used to obtain a satisfactory diet and energy. Cooking has long been the most common method of ensuring that minerals from food are taken, with one-third of the world's population (2.4 billion people) utilizing traditional fuels to cook and one out of every ten heating and disinfecting water for drinking contaminated processes. Another illustration of how tree resources are employed in cuisine is that pulverized seeds of the tropical tree (Moringa oleifera) are used for household water purification. Preventive techniques like smoking and drying, which boost the accessibility of food production throughout semi-months and disperse it over a long period, also require sustainable ingredients.

3.7 Forests, Biodiversity, and Human Health

An ecosystem of diversity affects how many items or solutions it can give. Nature and variety provide a wide range of commodities and services that benefit human well-beings, such as medications, food, clean water and atmosphere, shade, and just a green place to play and relax. In addition to the nutritional and food safety net that forests and plants provide, biodiversity also supplies pharmaceuticals. Animals, plants, and microorganisms are also very vital resources for nutrition and medicinal values. Such medicinal characteristics are essential not just domestically, but also nationally and internationally, where they are traded and utilized as templates for creating new therapies. The mainstream lively mixes that were initially derivatives of forest plants are traded on local and global markets and used as prototypes for the initiation of advanced actions. Over 28,000 plant species with therapeutic characteristics have been discovered, several of which are prevalent in forest settings. Ayurvedic, ancient Chinese, and other indigenous medicinal systems employ tree remedies. Many medications utilized in conventional medicine are derived from plants that have historically been employed in forest people's ancient healthcare systems. Alternative therapies, according to the World Health Organization, are "the total of understanding, talent, and processes distinct from hypotheticals, religious views, and points of view original to diverse cultures, irrespective of whether they are easily understandable or not, used in the preferment of health as well as precautionary care, prognosis, enhanced efficiency, or patients with severe illness." As a result, forest peoples' folk medicinal systems across the world are a rich source of information. As a result, indigenous herbal systems used by forest peoples around the world are a valuable source of information. Such systems are recognized for the survivability of wilderness people around the world because they are often the most practicable, available, economical, and, in some cases, common and accepted aspects of health care. Alternative therapies seem to be "the accumulation of knowledge, talent, and methods distinct from hypotheticals, opinions, and points of view indigenous to diverse cultures, whether easily interpretable or not, used in the improvement of health and also for precautionary care, prognosis, time savings, or patients with severe illness," according to the World Health Organization. As a result, forest people's traditional medical systems across the world give vital information. Because they are typically the most realistic, available, inexpensive, and, in some cases, widespread and recognized forms of health care, such aspects have assisted in the survival of people across the globe. The knowledge base of forest herbal medicines and their correlated profits is diminishing because of the fast urbanization, significant sociocultural transformations impacting native populations, and the disappearance of the globe's biologically based variety. As the Nagoya Protocol recognizes, preserving and sustaining indigenous practices linked to species and habitat, along with safeguarding sparsely populated citizens' liberties to spread the wealth of their information and expertise, are crucial to the health and well-being of local people as well as the world community. Rural communities are losing their supply of food and medicine as a result of reserved forest, habitat destruction, and lack of information, resulting in rising food shortages, malnutrition, and sickness. Green spaces in industrialized nations and cities may increase the willingness to work out and lessen health issues associated with sedentary lifestyles, such as obesity, psychological stress, and attention fatigue (Health Council of the Netherlands). There is substantial evidence that access to natural surroundings improves overall health in people of all ages and sexes, especially in urban settings and among the poor. Some green spaces have also been developed. Reduced blood circulation and pulse rate, increased cognitive control (Berman), and even increased human immunological reactions are all physiologic impacts of forest visits. As per various studies (the propensity to develop sensitive reactions), some who live in much more naturalistic complex surroundings get a more numerous and richer microbiome and become less prone to allergy aggravation. According to mounting data, the presence of natural surroundings improves human overall health across all socioeconomic classes and sexes, especially in metropolitan settings as well as among socioeconomically underprivileged groups, according to mounting data. Forest bathing, also known as shinrin-yoku, is a Japanese technique of simply being in the environment and taking a deep breath. Extra fat, psychological stress, and cognitive exhaustion are all connected to sedentary lifestyles. Green spaces in developed countries and big cities may make people want to walk more and keep them from getting sick from sitting around all day.

A large proportion of modern human contagious diseases are cosmopolitan, indicating they originated in mammals. Modifications in forest size and growing populations in the forest regions, both of which improve human access to, and wildlife species in certain instances, bush meat-eating, might be connected to their spread. Chagas disease (sometimes referred to as American trypanosomiasis),

African trypanosomiasis (sleeping sickness), leishmaniasis, and Lyme disease are all forest-related disorders. HIV and Ebola, two zoonotic diseases that have caught global attention, have unmistakable forest roots.

3.8 Conclusion

This chapter presented the status of African biodiversity and its protection predicament. It illustrated ongoing biodiversity management approaches as well as locations in need of urgent attention. Scientists, the authorities, and other officials will use the findings to help discover and designate additional protected areas around the western nation as well as propose critical conservation initiatives for fragile species. Two major contributors should take a leadership role in sustainable biodiversity utilization and protection—government and private organizations. To start, government actions and activities must incorporate sustainability ideals. Conservationists need to make sure that the government has full access to information about biodiversity. The value of biodiversity conservation should be universally acknowledged, mostly in local communities, where the bulk of our biodiversity still exists. Native peoples' biodiversity should be preserved, and projects that offer them a sense of belonging should be promoted. Local people must be taught how to run an ecotourism business for it to be sustainable. Furthermore, tourist profits must contribute to the growth of the domestic economy, provide a reason to protect the environment, and fund local health facilities as well as grants for local children. Commercial factors are a major cause of biodiversity decline and ecological system fragility. As a consequence, biodiversity decline should be fined monetarily, while preservation should be encouraged. This will favor economic activity that does not harm the environment.

References

Chime AO, Osawaru ME, Ogwu MC, Ogunlowo EO (2015) Preservation of fresh tomato fruits (*solanum lycopersicon* L.) using leaf extracts of bitter melon (*Momordica charantia* L., Cucurbitaceae). J Phytopathol Plant Health 3:29–37

Chime AO, Aiwansoba RO, Ogwu MC (2018) Pathological status of plant germplasm and sustainable crop production and conservation. J Energy Nat Resour Manag 1:17–21

Enaregha EB, Izah SC, Okiriya Q (2021) Antibacterial properties of *Tetrapleura tetraptera* pod against some pathogens. Res Rev Insights 5:1–5. https://doi.org/10.15761/RRI.1000165

Erinle KO, Ogwu MC, Evivie SE, Zaheer MS, Ogunyemi SO, Adeniran SO (2021) Impacts of COVID-19 on agriculture and food security in developing countries: potential mitigation strategies. CAB Rev 16(16):1–16

Evivie, S.E., Ogwu, M.C., Ebabhamiegbebho, P.A., Abel, E.S., Imaren, J.O. and Igene, J.O. (2020). Packaging and the Nigerian food industry: challenges and opportunities. In: Food technology and culture in Africa. Ogunlade, C.A., Adeleke, K.M. and Oladejo, M.T. (eds). Reamsworth Publishing, Ibadan, Nigeria. 28 – 99pp

Ikhajiagbe B, Ogwu MC (2020) Hazard quotient, microbial diversity and plant composition of spent crude oil polluted-soil. Beni-Suef Univ J Basic Appl Sci 9(26). https://doi.org/10.1186/s43088-020-00052-0

Ikhajiagbe B, Ogwu MC (2021) Growth performance of ferulic acid composition of *Amaranthus* [L.] species and *Mitracarpus villosus* (Sw.) DC. In competition with weed species. Environ Sustain. doi: https://doi.org/10.1007/s42398-021-00163-z

Ikhajiagbe B, Aituae W, Ogwu MC (2022a) Morpho-physiological assessment of oil palm (*Elaeis guineensis* Jacq.) seedlings exposed to simulated drought conditions. J Oil Palm Res 34(1): 26–34. https://doi.org/10.21894/jopr.202100018

Ikhajiagbe B, Ogwu MC, Ekhator PO, Omoayena II (2022b) Implications of soil nitrogen ehnacement on the yield performance of soybean (*Glycine max*) in cadmium-polluted soil. J Pollut Effects Control 10(2):1–12. #: 1000334. https://doi.org/10.1101/2021.11.22.469589

International Court of Justice (ICJ) (1986) Military and paramilitary activities in and against Nicaragua (Nicaragua v. United States of America). Merits, Judgment. I.C.J. Reports, 14

Inyang IR, Thomas S, Izah SC (2016a) Activities of electrolytes in kidney and liver of *Clarias gariepinus* exposed to fluazifop-p-butyl. J Biotechnol Res 2(9):68–72

Inyang IR, Thomas S, Izah SC (2016b) Evaluation of activities of transferases and phosphatase in plasma and organs of *Clarias gariepinus* exposed to Fluazifop-p-butyl. J Environ Treatment Techniques 4(3):94–97

Inyang IR, Kenobi A, Izah SC (2016c) Effect of dimethoate on some selected metabolites in the brain, liver and muscle of *Clarias lazera*. Sky J Biochem Res 5(4):63–68

Inyang IR, Okon NC, Izah SC (2016d) Effect of glyphosate on some enzymes and electrolytes in *Heterobranchus bidosalis* (a common African catfish). Biotechnol Res 2(4):161–165

Inyang IR, Akio K, Izah SC (2016e) Effect of dimethoate on lactate dehydrogenase, creatinine kinase and amylase in *Clarias lazera*. Biotechnol Res 2(4):155–160

Inyang IR, Tegu DU, Izah SC (2020a) Electro-metabolic aberrations in New Zealand rabbit (Oryclotagus cuniculus) induce by Chloropyrifos. J Biochem Technol 11(1):45–48

Inyang IR, Patani DE, Izah SC (2020b) The effect of 2,4 dimethylamine salt on the blood, liver and muscle of Oryclotagus cuniculus. J Plant Animal Ecol 1(3):21–28

Izah SC (2018) Ecosystem of the Niger Delta region of Nigeria: potentials and threats. Biodiver Int J 2(4):338–345

Izah SC, Aseibai ER (2018) Antibacterial and synergistic activities of methanolic leaves extract of lemon grass (*Cymbopogon citratus*) and rhizomes of ginger (*Zingiber officinale*) against *Escherichia coli*, *Staphylococcus aureus* and *Bacillus subtilis*. Acta Sci Microbiol 1(6):26–30

Izah SC, Seiyaboh IE (2018a) Challenges of wildlife with therapeutic properties in Nigeria; a conservation perspective. Int J Avian Wildlife Biol 3(4):252–257

Izah SC, Seiyaboh EI (2018b) Changes in the protected areas of Bayelsa state, Nigeria. Int J Mol Evol Biodiver 8(1):1–11

Izah SC, Angaye CN, Aigberua AO, Nduka JO (2017) Uncontrolled bush burning in the Niger Delta region of Nigeria: potential causes and impacts on biodiversity. Int J Mol Ecol Conserv 7(1):1–15

Izah SC, Uhunmwangho EJ, Dunga KE, Kigigha LT (2018a) Synergy of methanolic leave and stem-back extract of *Anacardium occidentale L.* (cashew) against some enteric and superficial bacteria pathogens. MOJ. Toxicology 4(3):209–211

Izah SC, Uhunmwangho EJ, Dunga KE (2018b) Studies on the synergistic effectiveness of methanolic extract of leaves and roots of *Carica papaya L.* (papaya) against some bacteria pathogens. Int J Complement AlternMed 11(6):375–378

Izah SC, Uhunmwangho EJ, Etim NG (2018c) Antibacterial and synergistic potency of methanolic leaf extracts of *Vernonia amygdalina L.* and *Ocimum gratissimum L.* J Basic Pharmacol Toxicol 2(1):8–12

Izah SC, Zige DV, Alagoa KJ, Uhunmwangho EJ, Iyamu AO (2018d) Antibacterial efficacy of aqueous extract of *Myristica fragrans* (common nutmeg). EC Pharmacol Toxicol 6(4):291–295

Izah SC, Aigberua AO, Nduka JO (2018e) Factors affecting the population trend of biodiversity in the Niger Delta region of Nigeria. Int J Wildlife Biol 3(3):206–214

Izah SC, Chandel SS, Etim NG, Epidi JO, Venkatachalam T, Devaliya R (2019a) Potency of unripe and ripe express extracts of long pepper (Capsicum frutescens var. baccatum) against some common pathogens. Int J Pharm Phytopharmacol Res 9(2):56–70

Izah SC, Etim NG, Ilerhunmwuwa IA, Silas G (2019b) Evaluation of crude and ethanolic extracts of Capsicum frutescens var. minima fruit against some common bacterial pathogens. Int J Complemen Altern Med 12(3):105–108

Izah, SC, Etim, NG, Ilerhunmwuwa, IA, Ibibo, TD, Udumo JJ (2019c) Activities of express extracts of costus afer Ker–Gawl. [Family COSTACEAE] against selected bacterial isolates. Int J Pharm Phytopharmacol Res 9(4), pp.39-44

Kigigha LT, Izah SC, Prebo PE (2015a) Effect of residual antibiotics in snacks against Escherichia coli and Staphylococcus aureus. J Environ Treatment Techniques 3(4):201–203

Kigigha LT, Izah SC, Ehizibue M (2015b) Activities of *Aframomum melegueta* seed against *Escherichia coli*, *Staphylococcus aureus* and *bacillus* species. Point J Bot Microbiol Res 1(2): 1(2):23–29

Kigigha LT, Biye SE, Izah SC (2016a) Phytochemical and antibacterial activities of *Musanga cecropioides* tissues against *Escherichia coli*, *Pseudomonas aeruginosa Staphylococcus aureus, Proteus* and *Bacillus* species. Int J Appl Res Technol 5(1):100–107

Kigigha LT, Izah SC, Okitah LB (2016b) Antibacterial activity of palm wine against *pseudomonas, bacillus, staphylococcus, Escherichia,* and *proteus* spp. Point J Bot Microbiol Res 2(1):046–052

Kigigha LT, Apreala A, Izah SC (2016c) Effect of cooking on the climbing pepper (*Piper nigrum*) on antibacterial activity. J Environ Treatment Techniques 4(1):6–9

Kigigha LT, Izah SC, Uhunmwangho EJ (2018a) Assessment of hot water and ethanolic leaf extracts of Cymbopogon citratus Stapf (lemon grass) against selected bacteria pathogens. Ann Microbiol Infectious Diseases 1(3):1–5

Kigigha LT, Selekere RE, Izah SC (2018b) Antibacterial and synergistic efficacy of acetone extracts of Garcinia kola (bitter kola) and Buchholzia coriacea (wonderful kola). J Basic Pharmacol Toxicol 2(1):13–17

Ogamba EN, Seiyaboh EI, Izah SC, Ogbugo I, Demedongha FK (2015a) Water quality, phytochemistry and proximate constituents of *Eichhornia crassipes* from Kolo creek, Niger Delta, Nigeria. Int J Appl Res Technol 4(9):77–84

Ogamba EN, Izah SC, Toikumo BP (2015b) Water quality and levels of lead and mercury in *Eichhornia crassipes* from a tidal creek receiving abattoir effluent, in the Niger Delta, Nigeria. Continental J Environ Sci 9(1):13–25

Ogamba EN, Izah SC, Emaviwe D (2015c) Phytochemical assessment of *Eichhornia crassipes* from river nun, Nigeria. Res J Phytomed 1(1):24–25

Ogamba EN, Izah SC, Ogbogu MJ (2015d) Cadmium and Lead level in *Eichhornia crassipes* from river nun, Niger Delta, Nigeria. J Adv Biol Basic Res 1(1):53–56

Ogamba EN, Ebere N, Izah SC (2017) Heavy metal concentration in water, sediment and tissues of Eichhornia crassipes from Kolo Creek, Niger Delta. Greener J Environ Manag Public Safety 6(1):001–005

Ogwu MC (2009) The significance and contribution of conservation education in Nigeria: an appraisal and a call for improvement. In: Babalola F (ed) Proceedings of the maiden seminar of Nigerian Tropical Biology Association. Nigerian Tropical Biology Association and Tropical Biology Association, pp 109–114

Ogwu MC (2010) Conserving Biodiversity Even in Poverty: The African Experience'. In: Babalola F (ed) Seminar proceedings of Nigerian tropical biology association. Nigerian Tropical Biology Association and Tropical Biology Association, pp 112–117

Ogwu MC (2019a) Lifelong consumption of plant-based GM foods: is it safe? In: Papadopoulou P, Misseyanni A, Marouli C (eds) Environmental exposures and human health challenges. IGI Global, Pennsylvania, pp 158–176pp. https://doi.org/10.4018/978-1-5225-7635-8.ch008

Ogwu MC (2019b) Towards sustainable development in Africa: the challenge of urbanization and climate change adaptation. In: Cobbinah PB, Addaney M (eds) The geography of climate change adaptation in urban Africa. Springer Nature, Switzerland, pp 29–55. https://doi.org/10.1007/978-3-030-04873-0_2

Ogwu MC (2019c) Understanding the composition of food waste: an "-omics" approach to food waste management. In: Gunjal AP, Waghmode MS, Patil NN, Bhatt P (eds) Global initiatives for waste reduction and cutting food loss. IGI Global, Pennsylvania, pp 212–236. https://doi.org/10.4018/978-1-5225-7706-5.ch011

Ogwu MC (2020) Value of *Amaranthus* [L.] species in Nigeria. In: Waisundara V (ed) Nutritional value of amaranth. IntechOpen, pp 1–21. https://doi.org/10.5772/intechopen.86990

Ogwu MC, Osawaru ME, Ahana CM (2014a) Challenges in conserving and utilizing plant genetic resources (PGR). Int J Genet Mol Biol 6(2):16–22. https://doi.org/10.5897/IJGMB2013.0083

Ogwu MC, Osawaru ME, Chime AO (2014b) Comparative assessment of plant diversity and utilization patterns of tropical home gardens in Edo State, Nigeria. Scientia Africana 13(2): 146–162

Ogwu MC, Osawaru ME, Aiwansoba RO, Iroh RN (2016a) Ethnobotany and collection of west African Okra [*Abelmoschus caillei* (A. Chev.) Stevels] germplasm in some communities in Edo and Delta states, southern Nigeria. Borneo J Resour Sci Technol 6(1):25–36

Ogwu MC, Chibuogwu AC, Osawaru ME (2016b) Trees in the service of man: University of Benin flora as a case study. UNILAG J Med Sci Technol 4(2):71–87

Ogwu MC, Osawaru ME, Obayuwana KO (2016c) Diversity and abundance of tree species in the University of Benin, Benin City, Nigeria. Appl Trop Agric 21(3):46–54

Ogwu MC, Osawaru ME, Iyamu M (2016d) Tree flora and their environmental services: a case study of University of Benin flora. Ilorin J Sci 3(1):40–68

Ogwu MC, Osawaru ME, Obahiagbon GE (2017) Ethnobotanical survey of medicinal plants used for traditional reproductive care by usen people of Edo State, Nigeria. Malaya J Biosci 4(1): 17–29

Ogwu MC, Ahana CM, Osawaru ME (2018a) Sustainable food production in Nigeria: a case study for Bambara groundnut (*Vigna subterranean* (L.) Verdc. Fabaceae). J Energy Natural Resour Manag 1:68–77

Ogwu MC, Chime AO, Oseh OM (2018b) Ethnobotanical survey of tomato in some cultivated regions in southern Nigeria. Maldives Natl Res J 6(1):19–29

Ogwu MC, Kerfahi D, Song H-K, Dong K, Seo H, Lim S, Srinivasan S, Kim MK, Waldman B, Adams JM (2019a) Changes in soil taxonomic and functional diversity resulting from gamma irradiation. Sci Rep 9:7894. https://doi.org/10.1038/s41598-019-4441-7

Ogwu MC, Takahashi K, Dong K, Song H-S, Moroenyane I, Waldman B, Adams JM (2019b) Fungal elevational Rapoport pattern from a high mountain in Japan. Sci Rep 9:6570. https://doi.org/10.1038/s41598-019-43025-9

Osawaru ME, Ogwu MC (2014a) Conservation and utilization of plant genetic resources. In: Omokhafe K, Odewale J (eds) Proceedings of 38th Annual Conference of The Genetics Society of Nigeria. Empress Prints Nigeria Limited, pp 105–119

Osawaru ME, Ogwu MC (2014b) Ethnobotany and germplasm collection of two genera of cocoyam (*Colocasia* [Schott] and *Xanthosoma* [Schott], Araceae) in Edo State Nigeria. Sci Technol Arts Res J 3(3):23–28. https://doi.org/10.4314/star.v3i3.4

Osawaru ME, Ogwu MC (2020) Survey of plant and plant products in local markets within Benin City and environs. In: Filho LW, Ogugu N, Ayal D, Adelake L, da Silva I (eds) African handbook of climate change adaptation. Springer Nature, Cham, Switzerland, pp 1–24. https://doi.org/10.1007/978-3-030-42091-8_159-1

Osawaru ME, Ogwu MC, Imarhiagbe O (2013a) Biochemical characterization of some Nigerian *Corchorus* L. species. Bayero J Pure Appl Sci 6(2):69–75

Osawaru ME, Ogwu MC, Ahana CM (2013b) Current status of plant diversity and conservation in Nigeria. Nigerian J Life Sci 3(1):168–178

Osawaru ME, Ogwu MC, Imarhiagbe O (2013c) Agro-morphological characterization of some Nigerian *Corchorus* (L.) species. Biol Environ Sci J Trop 10(4):148–158

Osawaru ME, Ogwu MC, Ahana CM (2014a) Study of the distinctiveness of two genera of cocoyam (*Colocasia* [Schott] and *Xanthosoma* [Schott]) using SDS – PAGE. Nigerian Soc Exp Biol J 14(3):168–179

Osawaru ME, Ogwu MC, Emokpare AA (2014b) Preliminary assessment of the microanatomy of okra [*abelmoschus* (L.)] wood. Egypt Acad J Biol Sci 5(1):39–54

Osawaru ME, Ogwu MC, Omologbe J (2014c) Characterization of three Okra [*Abelmoschus* (L.)] accessions using morphology and SDS-PAGE for the basis of conservation. Egypt Acad J Biol Sci 5(1):55–65

Osawaru ME, Ogwu MC, Aiwansoba RO (2015a) Hierarchical approaches to the analysis of genetic diversity in plants: a systematic overview. Univ Maurit Res J 21:1–36

Osawaru ME, Ogwu MC, Imarhiagbe O (2015b) Principal component analysis (PCA) as an ideal tool for analysing on-farm research data. Biol Environ Sci J Trop 12(1):514–522

Osawaru ME, Ogwu MC, Omoigui ID, Aiwansoba RO, Kevin A (2016) Ethnobotanical survey of vegetables eaten by Akwa Ibom people residing in Benin City, Nigeria. Univ Benin J Sci Technol 4(1):70–93

Vwioko DE, Okoekhian I, Ogwu MC (2018) Stress analysis of *Amaranthus hybridus* L. and *Lycopersicon esculentum* Mill. Exposed to Sulphur and nitrogen dioxide. Pertanika J Trop Agric Sci 41(3):1169–1191

Chapter 4
Potentials, Threats, and Sustainable Conservation Strategies of Plankton and Macrophytes

Emmanuel N. Ogamba, Adams Ovie Iyiola, Baturh Yarkwan, and Bukola Omotomilola Adetola

Abstract The aquatic system is a dynamic environment which is majorly influenced by anthropogenic activities. The ecosystem houses diverse species of animals and plants and can carry out several functions and services in the ecosystem. Plankton (microplants) is composed of phytoplankton (food producers) and zooplankton (consumers) and macrophytes (macroplants) are the major components in the ecosystem. They are also found in the environment and both have been known to serve various functions such as provision of physical structure, community structuring, and affect the diversity of organisms like fish species, birds, and invertebrates in the ecosystem. The aquatic biodiversity comprises of various species composition such as fish, birds, insects, plants, and mammals and their interaction results to sustainability. Phytoplankton is key to healthy life and its significance in food production is paramount in the aquatic system. They are important in food chain, carbon cycle, and oxygen production. Conversely, the importance of plankton and macrophytes in the aquatic environment has stimulated the dire need for their sustainability, coupled with the threats to aquatic ecosystem. The global water decline owing to the ecological stressors and human-water usage constitutes an

E. N. Ogamba (✉)
Department of Biological Sciences, Faculty of Science, Niger Delta University, Wilberforce Island, Nigeria

A. O. Iyiola
Department of Fisheries and Aquatic Resources Management, Faculty of Renewable Natural Resources Management, Osun State University, Osogbo, Osun State, Nigeria

B. Yarkwan
Department of Biochemistry, University of Agriculture, Makurdi, Nigeria

School of the Environment, University of Windsor, Windsor, ON, Canada

B. O. Adetola
Department of Ecotourism and Wildlife Management, School of Agriculture and Agricultural Technology, Federal University of Technology, Akure, Ondo State, Nigeria

S. C. Izah, M. C. Ogwu (eds.), *Sustainable Utilization and Conservation of Africa's Biological Resources and Environment*, Sustainable Development and Biodiversity 32, https://doi.org/10.1007/978-981-19-6974-4_4

issue of utmost concern. This chapter has described the different types of plankton and macrophytes, their importance, potentials, and threats that affect their sustainability in aquatic systems.

Keywords Zooplankton · Phytoplankton · Macrophytes · Conservation · Management · Sustainability

4.1 Introduction

Aquatic macrophytes (also known as hydrophytes) are a diverse category of macroscopic plants with a life cycle that takes place wholly or periodically in water, freshwater bodies, watercourses, marshes, swamps, seasonally flooded areas, and brackish and marine habitats (Nakayama et al. 2017; Rejmánková 2011; Wetzel 2001; Agedah et al. 2015; Ogamba et al. 2015a,b,c,d). Green macroalgae (Chlorophyta, e.g., *Cladophora* spp.), charophytes (*Charophyceae*, e.g., *Chara* and *Nitella* spp.), and higher aquatic plants (Tracheophyta and Bryophyta) make up aquatic macrophytic vegetation (de Nie 1987; Srivastava et al. 2008). According to Chambers et al. (2008), aquatic macrophytes are Rhodophyta and Xanthophyta macroalgae.

Macrophytes include vascular flowering plants, mosses and liverworts, encrusting lichens, and a few large algae species. A multitude of factors, including light and water current, limit the presence of macrophytes in flowing rivers. The three categories of macrophytes are attached plants, rooted plants, and free-floating plants. Among the plants attached are mosses and liverworts, lichens, and various tropical blooming plants. The related macrophytes live in cool headwater streams. Mosses are special in that they require free CO_2 rather than bicarbonate as a carbon source. In dark, turbulent streams, they may outperform periphyton. In mosses, macroinvertebrates can be found in vast numbers. There are rooted plants that are submerged (e.g., Hydrocharitaceae, Ceratophyllaceae, and Halorgidaceae) and emergent (e.g., Potamogetonaceae, Ranunculaceae, and Cruciferae). Roots commonly require slow currents, moderate depth, low turbidity, and small particles. In mid-sized rivers and river borders, rooted plants lower water current velocity, promote sedimentation, and provide a substrate for epiphytic microflora. They also provide vital habitat for a range of animals. Macrophytes' adaptations to running water include tough, adaptable stems and leaves, connection by adventitious roots, rhizomes, or stolons, and vegetative reproduction. Free-floating plants (such as the Lemnaceae and Pontederiaceae) have a limited role in temperate latitude flowing streams since they rely mostly on stagnant water. They can accumulate a lot of biomass in subtropical and tropical regions. Macrophytes in lotic settings largely contribute to energy flow through decomposer food chains because few macroinvertebrates feed on living plants.

Macrophytes communities are fundamental for the functioning of many river ecosystems (Franklin et al. 2008). However, they respond to disturbances in ecosystems and are particularly sensitive to anthropogenic influences, which negatively impact their diversity and species composition (Vörösmarty et al. 2010; Allan 2004). Macrophytes association depends on various abiotic and biotic factors (Zelnik et al. 2021; Szoszkiewicz et al. 2006). Based on this, we can identify species that are reliable indicators of changes in river ecosystems and use them as a tool to assess the ecological status of rivers (Aznar et al. 2002; Mack et al. 2000; Fennessy et al. 1998). The availability and varieties of macrophytes depend on water quality, water depth, flow velocity, flow rate, hydrological conditions, water level, pH, shading, and substrate features (Yang et al. 2017; O'Hare et al. 2006; Bornette et al. 1994). In addition, they are also influenced by biotic factors, namely the properties of species, interspecific competition, grazing, and allelopathy. Shade provided by riparian habitat is an important factor in lotic ecosystems, and it also affects the distribution and abundance of aquatic plants (Mackay et al. 2003); thus, macrophyte diversity and their abundance reflect the quality of an ecosystem as a whole (Kuhar et al. 2011).

4.2 Aquatic Biodiversity

Generally, biodiversity comprises of all animals, plants species, microbes, and fungi living in different habitats on the earth. The habitat can either be aquatic (in water) or terrestrial (on land) (Padhi and Mandal 2000). Each of these organisms has their roles which are specific to them in the environment (Zeng et al. 2019; Izah 2018). As described by the Convention on Biological Diversity, biodiversity accounts for the variability of species among living organisms from the terrestrial, marine, or other ecosystems and the functions they partake in which can be within species, between species, and between species and their immediate environment (Kumar and Asija 2009) (Fig. 4.1).

Aquatic biodiversity describes the diverse life and ecosystem in fresh water, ocean, and marine regions and their interaction with biotic and nonbiotic

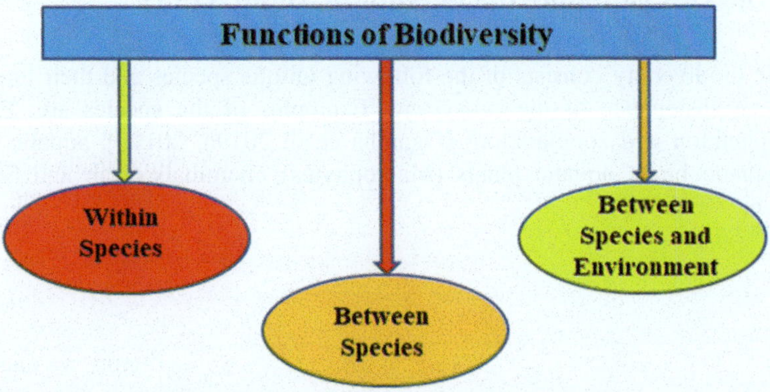

Fig. 4.1 Functions of biodiversity in an ecosystem

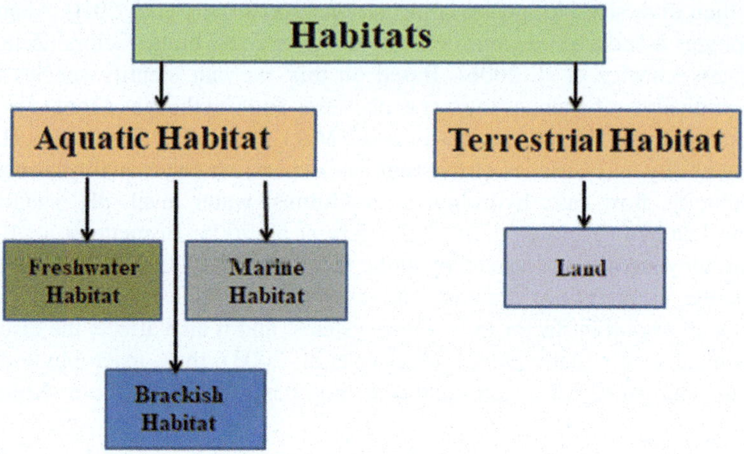

Fig. 4.2 Divisions of habitats in the environment

components (Tewari and Bisht 2021). Aquatic ecosystems consist of three main divisions as follows (Fig. 4.2):

- The freshwater ecosystem which includes the waters with salinity less than 1 part per thousand. Examples are lakes, rivers, streams, ponds, ground waters, reservoirs, and wetlands.
- The marine ecosystem includes waters having salinity above 35 parts per thousand, examples of such are estuaries, salt marshes, oceans, kelp beds, mangroves, and coral reefs (Hendrik and Martens 2005).
- The brackish ecosystem is an area of fluctuating salinity and between 1 - 34 parts per thousand. Examples are the Niger delta regions, lagoons.

4.3 Species Components of Aquatic Biodiversity

Aquatic biodiversity consists of the following unique species and their interaction results in sustenance of the ecosystem. Examples of the species are: Plankton (phytoplankton and zooplankton) (Ogamba et al. 2019a, 2019b), aquatic insects, fish, aquatic birds, aquatic plants (Macrophytes), mammals (Izah and Srivastav 2015), etc. (Fig. 4.3)

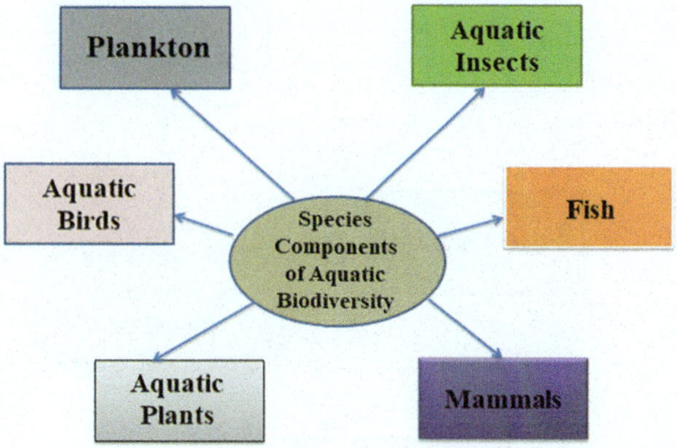

Fig. 4.3 Species components of aquatic biodiversity

4.4 Plankton as a Component of Aquatic Biodiversity

Plankton comprises of floating plants and animals in the water. They are called "wanderers" or "drifters" because their movement is dictated by the water currents. They are microscopic in nature (< 1 inch in length) and may be composed of larger organisms like jelly fish and crustaceans (Karleskint et al. 2013). They differ from nekton because the movement of the latter is not dictated by the water movement. Examples are the marine mammals, fish species, squids, etc. The presence and abundance of plankton in water is dictated by the environment. They are very sensitive to changes in parameters such as temperature, pH level, salinity, and concentration of nutrients in water. They can also be affected by the presence and abundance of other planktonic species in water. When there are excessive nutrients in water, bloom may arise which is detrimental to the water parameters (Agrawal and Gopnal 2013). Phytoplankton abundance principally dictates the abundance of zooplankton because the latter eats the former and there is a flow in the food chain.

4.5 Classification of Plankton

Plankton can be classified based on size, species, and the duration of stay in water.

4.5.1 Classification-based Trophic Levels

Based on functions, planktons are composed of five major groups including Phytoplankton (plants), Zooplankton (animals), Mycoplankton, Bacterioplankton and Virioplankton (Fig. 4.4).

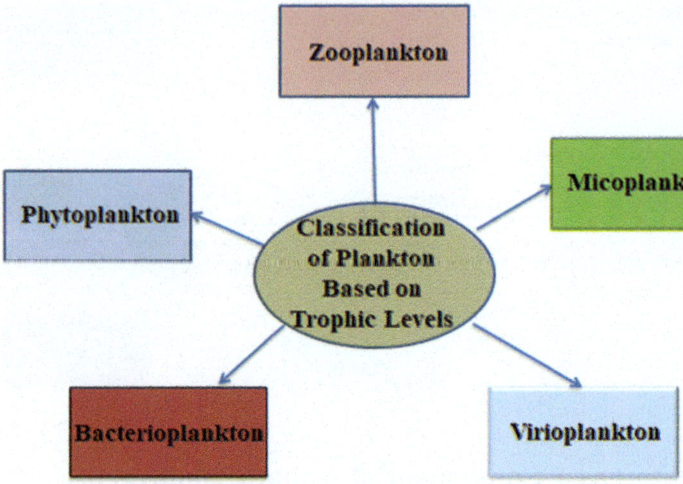

Fig 4.4 Classification of plankton based on trophic levels

4.5.1.1 Phytoplankton

They are composed of plants that are microscopic in nature and have great importance in the food web in the aquatic ecosystem. They can be autotrophic, prokaryotic, or eukaryotic algae. Their major function is to perform the process of photosynthesis through the input of the sun and carbon dioxide and oxygen is liberated. They are mostly found on the surface of water because they require large exposure to sunlight for photosynthetic activities. Phytoplankton is composed of nannoplankton and microplankton. Examples are the dinoflagellates, diatoms, blue-green and green algae, red algae, silicoflagellates, etc.

4.5.1.2 Zooplankton

They are the floating animals in the aquatic ecosystem. They are the consumers and feed on the phytoplankton in the water column. Their size can range from <60 μm in diameter (microzoans, e.g., snails, worms, krill, etc.) to large invertebrates such as fish (metazoans). They are composed of diverse taxonomic groups such as protozoans and invertebrates. They are found along with phytoplankton and can migrate to deeper areas to avoid predation. They come to the surface at night to feed on phytoplankton (Leles 2018; Steinberg and Landry 2017).

4.5.1.3 Mycoplankton

This level comprises of fungi and all fungus-like organisms which dwell in water. They are very important in the nutrient recycling and demineralization in the water (Wang et al. 2012).

4.5.1.4 Bacterioplankton

It comprises of bacteria found in the water column; they function in demineralizing organic materials in the water column and down the water. Examples are the prokaryotic phytoplankton.

4.5.1.5 Virioplankton

It comprises of virus which are the most abundant in the plankton family when compared with bacteria and archaea. They are smaller in size when compared with bacterioplankton.

4.5.2 Classification Based on Life Cycle

Planktonic species in water can be classified based on life cycle into two groups:

- **Holoplankton:** are planktonic species that spend their entire life as plankton in water column. Examples are the copepods, algae, jellyfish, etc.
- **Meroplankton;** they spend part of their life stage (the larval life stage) as plankton and metamorphose into nektons (free swimming) or benthic organisms. Examples are worms, crustaceans, most fish species, etc. (Leles 2018).

4.5.3 Classification Based on Size

Planktonic species can be grouped based on size (Fig. 4.5). These groups are described based on size as >20 cm (Megaplankton, e.g., Metazoans—jellyfish, salps, ctenophores), 2–20 cm (Macroplankton, e.g., Ostracoda, chaetognath, amphipoda), 0.2–20 mm (Mesoplankton, e.g., Large eukaryotic protists, foraminifera, tintinnids), 20–200 μm (Microplankton, e.g., Diatoms, rotifers, juvenile metazoans, copepods), 2–20 μm (Nanoplankton, e.g., Small flagellates, chlorophyta, xanthophyte), 0.2–2 μm (Picoplankton, e.g., Bacteria, chrysophyta, protists) and <0.2 μm (Femtoplankton, e.g., Marine viruses) (Karsenti et al. 2011)

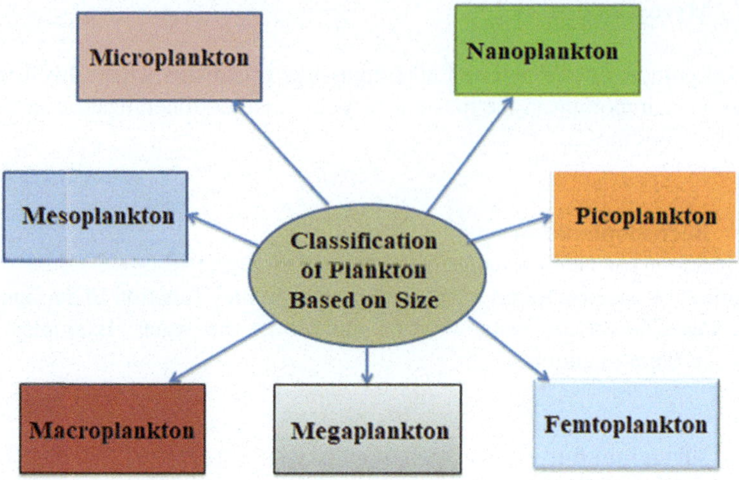

Fig. 4.5 Classification of plankton based on size

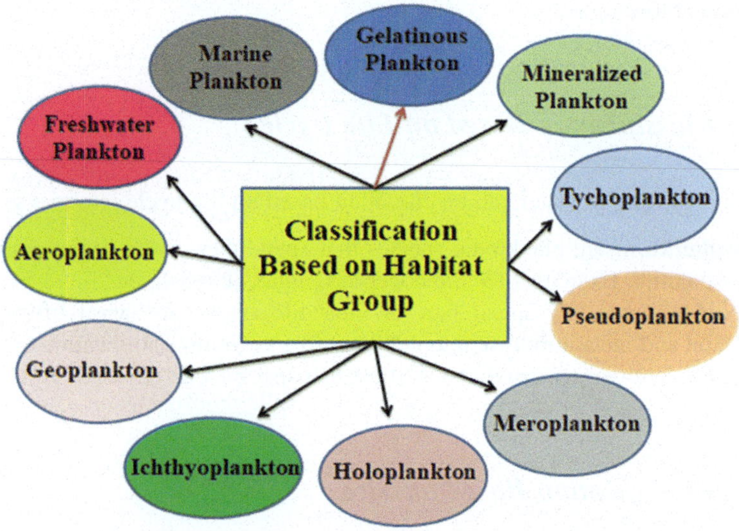

Fig. 4.6 Classification of plankton based on habitat group

4.5.4 Classification Based on Habitat Group

They are grouped into the following: Marine plankton, Freshwater plankton, Aeroplankton, Geoplankton, Gelatinous zooplankton, Ichthyoplankton, Holoplankton, Meroplankton, Pseudoplankton, Tychoplankton, and Mineralized plankton (Fig. 4.6)

4.5.4.1 Marine Plankton

They are free drifting animals or plants in the oceans, estuaries, and brackish water environments. Examples of such are the marine bacteria, archaea, protozoa, and algae. They are tolerable to high salinity.

4.5.4.2 Freshwater Plankton

These are free drifting animals or plants that inhabit freshwater environments such as rivers, streams, lakes, ponds, pools, etc.

4.5.4.3 Aeroplankton

They are found dispersed and drifting in the air as a result of the wind currents. They are mostly microscopic in size and their identification is extremely difficult because of their size. They can be collected in air using traps or sweep nets. This group of plankton is made up of several microbes, viruses, bacteria, about 40,000 species of fungi, several hundreds of algae, mosses, liverworts, and protists.

4.5.4.4 Geoplankton

These are animals that inhabit terrestrial environments by surviving on microscopic bodies of water and moisture. Examples are rotifers and gastrotrichs that lay eggs that can stay for years in dry environments. Water Bears are another example that can suspend animation during dry periods and survive for a long time. Microscopic crustaceans such as copepods, amphipods, and shrimps can go dormant when dry and can live in transient water bodies.

4.5.4.5 Gelatinous Zooplankton

These are transparent and delicate animals that live in the ocean. Examples are jellyfish, salps, medusae, and ctenophores. They are very vulnerable because they have no hard body parts which may serve as protection. Marine organisms such as arthropods, molluscs, and annelids are composed of gelatinous species and they live in the open waters of the ocean (Hartmann et al. 2012).

4.5.4.6 Ichthyoplankton

These refer to the fish eggs and larvae that are dispersed in the photic zones of fresh water. They move with the water currents within the first 200 m of the water column. They are better adapted to swimming at their latter stages of life when compared to larvae or fry stages. Both the larvae and eggs are food to large fish species.

4.5.4.7 Holoplankton

These refer to organisms that are capable of spending their entire life cycle as plankton. They dwell and move based on water currents. Examples are dinoflagellates, diatoms, amphipods, krill, copepods, gastropods, some mollusks, and radiolarians. They are found in the pelagic zones and they vary in size.

4.5.4.8 Meroplankton

These refer to organisms that live both in the pelagic zones as plankton and benthic zones at various periods in their life cycle. They start their life as holoplankton on the water surface and later grow off their larval stages to stay in the benthic zones. At this period, they become nektons. The time the organisms stay in the pelagic or benthic zones differs from species to species.

4.5.4.9 Pseudoplankton

These refer to organisms that attach themselves to substances or other plankton for survival. Examples are the Jellyella and goose barnacles. They cannot drift on their own and are referred to as false plankton. They are mostly found inside the guts of zooplankton that are filter feeders.

4.5.4.10 Tychoplankton

These are organisms that are carried into plankton by disturbance of their habitats whether by wind or erosion. The organisms can be free-living or attached to a benthic or a non-planktonic organism. They are often called accidental plankton because they spend some part of their life cycle as plankton and their reproductive portions are confined to the existence of plankton.

4.5.4.11 Mineralized Plankton

These are organisms that have shells that are mineralized. These often serve as protection for the organism (Wang et al. 2012). Examples are listed in Table 4.1.

4.6 Ecological Significance of Plankton

Plankton plays several roles in the ecosystem. The major significance in the ecosystem is in the following processes:

- Food chain
- Oxygen production
- Carbon cycle
- Absorption efficiency (Karleskint et al. 2013)

4.6.1 Food Chain

The food chain is the principal source of ecosystem sustainability in which the phytoplankton produces the food and is eaten by the zooplankton. The food chain can support various commercial fisheries and cycles.

4.6.2 Carbon Cycle

Carbon is released to the environment when they feed on phytoplankton in the food chain. This occurs during respiration which produces energy or death. The ocean is the major carbon sink in the world because organic matter tends to sink in the ocean

Table 4.1 Plankton and their mineral components

S/N	Organism	Component(s)	Functions
1	Diatoms	Glass shells called frustules	Production of oxygen for the environment
2	Microscopic marine radiolarians	Elaborate silica shells	Production of opal
3	Coccolithophores	Chalk plates called coccoliths	Production of the Cliff of Dover
4	Foraminiferans	Calcium carbonate shells	Production of limestone in the Great Pyramids.

waters away from the coastlines. This process is influenced by temperature and sea water acidification (Steinacher et al. 2010).

4.6.3 Oxygen Production

This occurs during the process of photosynthesis by phytoplankton after they absorb energy from the sun. Oxygen is a by-product and about 50% of the oxygen in the world is produced by photosynthesis from phytoplankton, while others are produced by land plants (Ward and Follows 2016).

4.6.4 Absorption Efficiency

This refers to the amount of food absorbed by plankton in relation to its demands. This is a function of the composition of the prey and feeding rate and it regulates the recycling of organic matter back into the environment. Low feeding rates translate to high absorption efficiency. Respiration rate can contribute to dissolve organic matter in water. Other factors such as light, pH, and oxygen can dictate the amount of carbon released by zooplankton (Dolan 2012).

4.7 Aquatic Macrophytes as a Component of Aquatic Biodiversity

Aquatic macrophytes are plants that live and are adapted to life in aquatic environments. They are also called hydrophytes and their life cycle is predominantly in aquatic systems. They can respond to different environmental stimulus and thrive in various habitats including wetlands, swamps, freshwaters, brackish environments, and flooded areas (Zhang et al. 2019a; Nakayama et al. 2017; Rejmánková 2011). They are basically composed of green microalgae and higher aquatic plants (Brönmark and Hansson 2017). Traditionally, it was known that aquatic vegetation evolved from terrestrial plants and research has shown the aquatic origin of several groups of angiosperms (Du and Wang 2016). Aquatic macrophytes are important components of the aquatic system and have a lot of functions, roles, ecological groups, and various adaptive features in the aquatic ecosystem (O’Hare et al. 2018)

4.7.1 Ecological Groups of Aquatic Macrophytes

Aquatic macrophytes are widely distributed around the world and their groups are closely related and cosmopolitan in nature (Zhang et al. 2019b; Santamaría 2002). They are characterized by different forms of growth, physiological and metabolic processes which are functions of the prevailing environmental conditions. They can grow in deep or shallow waters in emergent, submerged, or floating states. Some may be strictly aquatic, while some may grow in inundated environments adapting to various changes in water level and conditions (Coughlan et al. 2018; Chen et al. 2012).

The mechanism involved in the reproduction and growth of macrophytes is the vegetative and clonal propagation while subordinate methods are the sexual and genetic combination strategies (van Veen and Sasidharan 2019; Vymazal 2013). These can be achieved in three ways:

- By water—hydrochory
- By animal participation—zoochory
- By wind—anemochory

Water can disperse the pollinated pods, fruits, and seeds of angiosperms and vegetative fragments (Boedeltje et al. 2019). The dispersal by wind and animals has been reported to be vectors for aquatic angiosperm distribution (van Leeuwen 2018; Chambers et al. 2008).

4.7.2 Growth Types/Forms of Aquatic Macrophytes

Based on the large diversification of aquatic macrophytes, they can be divided into several groups based on growth and ecological traits. Arber (1920) categorized aquatic macrophytes into two groups namely:

- Rooted or attached to the bottom macrophytes
- Free-floating macrophytes (Zuo 2019; Rejmánková 2011; Arber 1920).

The rooted macrophytes are further divided into Floating Macrophytes, Submerged Macrophytes, and Emergent Macrophytes (Fig. 4.7)

4.7.2.1 Floating Macrophytes

These are plants that are found suspended on the water column. They are mostly found floating in the water and they do not have their roots buried in the bottom soil of water. They receive all their required nutrients from the water. They have less supporting tissues and their undersurfaces are waxed so as to prevent clogging of the stomata. Examples are Water Hyacinth, Giant Duckweed, and Mosquito Fern. The

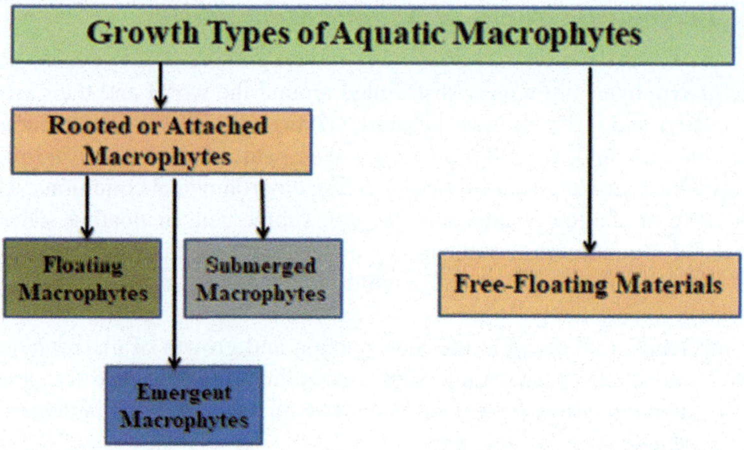

Fig. 4.7 Growth types of aquatic macrophytes

plants are highly mobile because they can drift as directed by ocean currents and are mostly invasive species. They can easily transmit diseases from one place to another and can serve as host for various pathogens (Gimenes et al. 2020).

4.7.2.2 Submerged Macrophytes

These are plants whose roots and vegetative parts are below the water. They are rooted in the soil and absorb required nutrients from the soil. Some of the plants may reach the surface of water and have leaves spread on water. They can provide shade for fish species and avoid predation. Examples of submerged macrophytes are pond weed and coontail. They can reduce fishing activities and oxygen in water at night. They absorb oxygen at night for survival and can deplete to a point where mortality of fish can occur (Wetzel 2001).

4.7.2.3 Emergent Macrophytes

These are plants that have their vegetative parts outside the water and are rooted in the soil. Most of the parts of the plants are found outside the water and tend to be more rigid and able to withstand the water currents. Examples are the Cattails, water lily, Rushes, Sedges, etc. Some may grow along shorelines or in the water. They can help in keeping the bank from being eroded and slow down wind and erosion into the water, thereby avoiding turbid conditions.

4.7.2.4 Algae

They are the oldest and confusing form of aquatic macrophytes. They do not have real roots and have a way of attaching to the substrate. They can appear as extensive mats floating on water and are highly migratory. They provide food and resources for the food web and fisheries. Examples are the Blue-green algae. The filamentous algae are the most common and easily recognizable algae type which readily grows in clusters of filaments or strands. They can cover water surface and when bloom occurs, they can deplete oxygen content in water.

4.7.3 Adaptations of Macrophytes to Aquatic Habitats

The environmental restrictions of aquatic plants are different from terrestrial plants and this had made aquatic plants develop special adaptive features to life in water. The diffusion of gas and light in aquatic environments is much lower than terrestrial environments and therefore special mechanisms are required for photosynthesis and respiratory activities by aquatic macrophytes (van Veen and Sasidharan 2019). The following mechanisms of adaptation by aquatic macrophytes are:

- The ability to switch to anaerobic mechanisms so as to generate ATP when oxygen is absent. This is to ensure adaptation to low oxygen supply (Perata and Alpi 1993).
- The development of carbon dioxide concentrating mechanisms to overcome the limitations of inorganic carbon for photosynthesis. Carbon source may be from HCO_3^- or sediment CO_2 (Mommer et al. 2005; Maberly and Madsen 2002).

The morphological and anatomical adaptations by aquatic macrophytes to submerged conditions include:

- **The formation of Aerenchyma tissues:** this consists of gas spaces for transport of oxygen and carbon dioxide between the roots and shoots (Tang et al. 2017).
- **The presence of lacunae for the transport of gases:** lacunae allow transfer of oxygen to the rhizome and roots of plants or carbon dioxide from leaves to the roots of aquatic plants (Maberly and Madsen 2002).
- **The formation of heterophylly:** plants that have more than one type of leaves on the plant are called heterophyllous plants and it can alter the morphology of aquatic plants in relation to the aquatic environment (Nakayama et al. 2017).
- The elongation of organs for contact with the atmosphere
- The poor development of mechanical and conductive tissues (Li et al. 2020, Nakayama et al. 2017; Takahashi et al. 2014; Mommer et al. 2005).

4.8 Ecological Indicators of Aquatic Biodiversity

In Inland waters, biodiversity loss of microscopic and macroscopic organisms is caused by environmental changes resulting from human activities and climate change (Yu et al. 2019). These have direct effects in the functions and services of organisms in the ecosystem. For conservation and management purposes, an insight into the effects of human activities and climate change can be gotten from the studies of succession of biotic communities on a long term. Liu et al. (2019) and Guo et al. (2019) observed the effects of these activities on the plankton biomass and water level and phytoplankton community in Lake Tai, respectively. Their results expressed how climatic factors can dictate the plankton and cyanobacteria blooms.

Anthropogenic factors have been reported to affect and modify habitats, thereby reducing all forms of homogeneity (Klamt et al. 2020; Yan et al. 2019). Some effects are presented in Table 4.2.

For aquatic biodiversity conservation and management, the study of the succession of the biotic communities for a long term is required. This will present the effects of human activities and climate change on aquatic biodiversity (Pajunen et al. 2020; Paudel Adhikari et al. 2019). These strategies have been used by some authors as presented in Table 4.3.

Table 4.2 Anthropogenic factors and their effects on aquatic biodiversity

S/N	Anthropogenic factors	Effects on biodiversity	References
1	Urbanization	It was also reported to have a negative effect on aquatic macroinvertebrates in lotic ecosystems	Gál et al. (2019), Li et al. (2019)
2	Rapid urbanization	Zooplankton were homogenized by rapid urbanization in a tropical reservoir	Liu et al. (2019)
3	Type of Land use (urban, protected, and agricultural landscapes)	It showed that the land use gradients affected the stream fish community structure	Tóth et al. (2019)
4	Human activities	It affected negatively the phytoplankton, microzooplankton, nekton, macro benthos, and zooplankton in Jiangsu offshore, China. The creation of protected areas for preservation of biodiversity was suggested.	Yu et al. (2019)
5	Hydropower operation	The spawning time of four endemic carp and Chinese sturgeon fish species in Three Gorges Dam were affected. The time was advanced by 1.3 days for the former and delayed for the latter by 2.1 days.	Zhang et al. (2021)
6	Agriculture	Eutrophication controlled the bacterial community assemblage in shallow lakes and waters	Gál et al. (2019), Zeng et al. (2019)

Table 4.3 Long-term studies of impacts of human activities on aquatic biodiversity

S/N	Lake	Duration of study	Results	References
1	Lake Poyang	3 years data set observation	Presented the influence of water level fluctuation on total biomass of phytoplankton, taxonomic composition, direct and indirect effects	He et al. (2020), Liu et al. (2019).
2	Lake Tai	1992–2017 using limnological monitoring data	It presented the variance explained by the nutrient (phosphate, nitrate, ammonium) and climatic (wind speed and air temperature) variables	Guo et al. (2019), Vilmi et al. (2019)

Table 4.4 Models for investigation of ecological status and their functions

S/N	Name of Model	Components	Functions	References
1	Macroinvertebrate-based multimetric index	Three taxonomic metrics and two functional metrics	To assess the influence of eutrophication on ecological status of shallow lakes	Zhang et al. (2019b).
2	Diatom assemblage data (predictive model)	Water parameters and climatic factors	To assess the relationship between indicator of water chemistry (pH and total phosphorus) and climatic factors (days)	Pajunen et al. (2020)

4.9 Models for Assessing Ecological Indicators

Due to the rapid urbanization and industrialization, human activities and climate change have increased and there is need to develop an assessment model/index which is reliable for investigation of the ecological status of aquatic biodiversity (Liu et al. 2019, 2020, 2021). The presence of biotic components is a widely used bio-assessment technique in freshwater habitats. Some model/index and their functions are presented in Table 4.4.

4.10 Importance of Aquatic Biodiversity

The benefits of aquatic biodiversity are enormous, having social and esthetic values.

- They are responsible for supporting the environmental health
- Humans depend on them for food, medicine, and biomaterials
- They serve for recreational (tourism purposes) and commercial purposes (fishing)
- Aquatic organisms depend on aquatic habitats for food, materials, and grounds for breeding activities (Hendrik and Martens 2005)

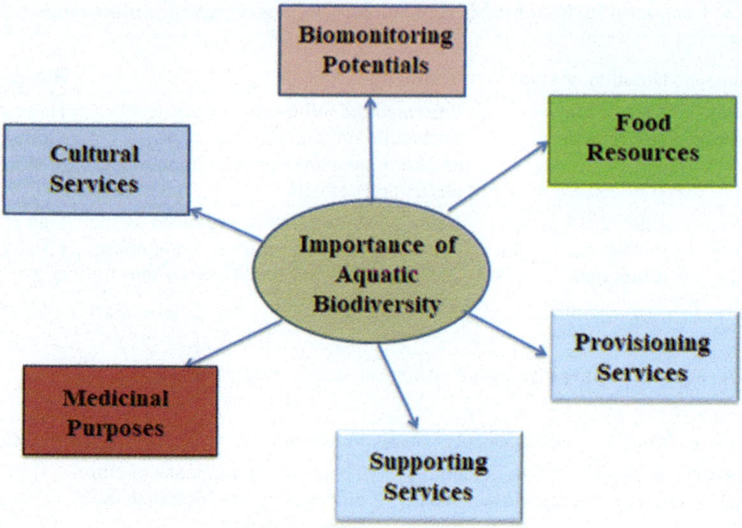

Fig. 4.8 Importance of aquatic biodiversity

Aquatic biodiversity is important in the following areas: Cultural services, Medicinal potentials, Supporting services, Provisioning services, Food resources, Research and Biomonitoring potentials (Fig. 4.8):

4.10.1 Cultural Services

This refers to the dependence on biodiversity in areas that can promote health for individuals and the society. These services include esthetics, traditional knowledge, education, inspiration, and scientific discovery associated with interaction with biodiversity. This service sustains the relationship between humans and nature as well as several health-promoting benefits biodiversity can offer (Abraham et al. 2010). The other cultural services proffered are:

- It can promote physiological health
- It can relief stress and improve cognitive functions (Gladwell et al. 2013)
- It shows improved physical health, lower mortality, and occurrence of chronic diseases
- It reduces allergies and chronic inflammation in humans (Barton and Pretty 2010)
- It reduces incidence of anxiety depression and asthma (Nieuwenhuijsen et al. 2014)
- It reduces coronary heart diseases

- The presence of aquatic fauna has helped people relax and reduce stress (Zelenski and Nisbet 2014)
- It increases mental health (Alcock et al. 2014).

4.10.2 Supporting Services

These services referred to the approach in which biodiversity supports life and its building processes. These services are very important in the existence of aquatic ecosystems. Some of these services are:

- Primary production (Chemosynthesis and Photosynthesis)
- Cycling of nutrients important for life sustenance
- Pollination

Ecosystems inadequate in support services are bound to be extinct. This is majorly sustained by primary production and plants and animals involved in this process depend on all support services to carry out this function and survival.

4.10.3 Regulating Services

Regulatory services as an importance of aquatic biodiversity go beyond the consumption of aquatic resources; it extends to the following areas:

- Disease occurrence in the population
- The life processes and influence of local climate
- Resilience against flooding and storms which are catastrophic

Principally, regulating services ensure aquatic resources are renewed and an environment that is functional and habitable. This is achieved by the following processes:

- Cleaning of air and water
- Climate services modification
- Modulation of the immune and brain functions
- Modulation of diseases that are infectious

Climate change is an important factor that can influence regulatory services in the aquatic ecosystem (Nelson et al. 2013). Natural aquifers can mitigate disasters when predicted to increase. Vector-borne diseases have also spread and expanded across the area.

4.10.4 Provisioning and Medicinal Services

Humans depend on biodiversity for food to eat and materials in form of raw materials to build their homes. Aquatic biodiversity can also be a source or material for medicinal purposes. They are important sources of:

- protein
- iron
- traditional and modern medicine from botanical products

4.10.5 Biomonitoring Potentials

The occurrence of certain species can be useful in monitoring the water quality in aquatic systems. They can be categorized into the following:

- Pollution-intolerant species: they do not thrive in areas that are polluted and the water quality is below the recommended range. They can serve as good indicators for water quality (Largo-Wight 2011).
- Pollution-tolerant species: their presence indicates the condition of the water (WHO 2014).

Despite these benefits, some factors can contribute to the decline in aquatic biodiversity. Such as:

- Introduction of exotic species
- Species overexploitation
- Pollution from human activities
- Loss of habitat as a result of dam creation
- Diversion of water (Ormond et al. 1997)

A balance in nature is required so as to promote sustainability through various strategies of conservation and protection of aquatic life forms.

4.11 Threats of Aquatic Biodiversity

Málnás et al. (2011) asserted that the extent of decline in freshwater biodiversity is progressing at a pace faster than observed in terrestrial habitats. This is because the additive effects of diverse aquatic ecosystem stressors acting within the same region create such enormous stress on the biota that results in declined individual physiological capacities and consequently species extinction. Halpern et al. (2008) investigated 17 stressors in relation to marine habitats and reported that climate change, fishing activities, and coastal habitat destruction are among the most potent stressors impacting biodiversity in marine ecosystems, just as Selig et al. (2014) later shared similar findings from their studies focusing on exclusive economic zones and areas beyond national jurisdictions. Similarly, Crowder et al. (2008) also identified environmental changes, fishing, and climate change as trio whose synergistic impacts threaten diversity in marine ecosystems. The impacts of fishing are most significant in areas with minimal natural disturbances (Jennings and Kaise 1998) tending to last longer and interfere with surface growing and benthic organisms. Thus, marine environmental stressors' impacts are aggravated due to the additive or synergistic effects of the identified factors (Halpern et al. 2008). Some of the causes of biodiversity loss are highlighted below.

4.11.1 Pollution Due to Human Activities

4.11.1.1 Oil and Gas

Oil spills are one of the most damaging sources of anthropogenic environmental pollution (Izah et al. 2017; Aigberua et al. 2017, Aigberua et al. 2016a, 2016b; Sharma et al. 2020). Chilvers et al. (2020) reported that at least 1702 acute oil spills have occurred around the world between 1970 and 2018, with 80% of the data coming from a single country. Ekperusi and Ekperusi (2021) quoted National Oil Spill and Response Agency (NOSDRA) to have recorded at least 6333 oil spills in Nigeria between 2010 and 2015. Oil spills arise from refineries, pipelines vandalization/breakage, and land transport and accidents involving marine trade ships, which accounts for the 71% reported cases of spillage (Chilvers et al. 2020). It is uncertain if refineries conceal spills to evade environmental sanctions and host community agitations for corporate environmental cleansing responsibilities, as opposed to vessel spills which are reported to balance loading and landing volumes of crude/refined oil and gas. Therefore, oil spills are a global problem and may not be localized due to flowing waters and resultant effects from pollution of both waterways and beaches. However, spillage occurs in Nigeria (Ekperusi and Ekperusi 2021) and other oil producing nations in Africa (Gdara et al. 2019) which go largely unaccounted for, with environmental impacts rarely documented.

Studies on water accommodated fractions of heavy fuel oil comprising total hydrocarbon content of 0.07 mg/L, 0.28 mg/L, and 0.55 mg/L had negative impacts on phytoplankton biomass primary production corresponding to 6, 52, and 73%, respectively; while impacts of photo toxicity worsened the reduction of primary production in the highest two concentration treatments up to 71% and 91%, respectively (Lemcke et al. 2018). This implies subacute concentrations of oil spill could drastically reduce phytoplankton survival and abundance, which could drastically affect marine food chains because of the primary position occupied by these organisms in aquatic food webs.

4.11.1.2 Dredging

Several factors could contribute to clogging waterways. When these happen, to prevent overflooding of riverbanks, management could require dredging to deepen the water bed level and increase the volume of water flow, hence saving adjacent lands (Izah et al. 2022). Moreover, profuse growth of aquatic macrophytes possesses problems globally and these are often removed via dredging. Dredging for this purpose could be to prevent clogging of hydropower plants, enhance irrigation, disease control, trade and commerce, and recreational activities such as boating, water skiing, swimming, and angling (Thiemer et al. 2021).

Dredging affects nutrient cycling in the ecosystem, sediment transport, ecosystem metabolism, and hydraulic properties, which affects macrophytes survival. Removal of submerged and free-floating macrophytes creates an increase in phytoplankton biomass and necessitates a shift towards communities where fast-growing cyanobacteria predominate (Thiemer et al. 2021; Zhu et al. 2019). The decline in population of macrophytes reduces competition for nutrients, space, and sunlight to the advantage of phytoplankton; hence the expected increase in biomass, even though a lag in growth phase may be observed before the biomass gain becomes predominantly noticeable. Tavernini et al. (2009) reported that following dredging, zooplankton communities developed towards a higher taxonomical and functional diversity. Since hydro-chemical properties are substantially altered during dredging, such changes may pave the way for more favorable environment to facilitate ecological succession of the plankton communities.

4.11.1.3 Waste Dump

Indiscriminate dumping of solid wastes and discharge of partially treated or untreated municipal/industrial wastewater effluents into aquatic bodies are often reported around several African nations (Ekperusi and Ekperusi 2021; Gdara et al. 2019; Ben-Eledo et al. 2017a, 2017b; Seiyaboh and Izah 2017a, 2017b). High concentration of plankton was recorded in the Cauvery River, Tamil Nadu, India, which is heavily polluted with household and industrial effluents (Mathivanan et al. 2007). However, a multi stressor study on a reference polluted river in Argentina showed strong decline in phytoplankton productivity due to an interplay of agro-chemicals, industrial, and household wastewater effluents (Olguín et al. 2004). The impact of pollutants on the plankton and macrophytes is a function of the composition of the pollutants, concentrations, chemical properties, additive, or synergistic effects (Hader and Gao 2015; Crowder et al. 2008). With changing seasons and associated rainfall, water quality tends to change through the year. Sewage effluents released into water bodies are usually rich in nutrients, thereby stimulating the growth of a wide range of phytoplankton and macrophytes. It is reported that nutrients ammonium and orthophosphates, high biological oxygen demand (BOD), hardness and conductivity, and high concentration of pollutants encouraged the dominance of a particular algal species, while clean water improved algal and crustacean diversity (Hader and Gao 2015). Effluents rich in atrazine from agricultural farms or production industries might inhibit the photosynthetic electron transport system, thereby reducing phytoplankton growth. Herbicides applications are thought to induce a pro-phytoplankton succession in macrophytes-dominated ecosystems (Yamamuro et al. 2012). Solid wastes such as electronics, rich in heavy metals, could exact toxic effects, affecting phytoplankton productivity which is discussed under "others" below.

4.11.1.4 Use of Chemicals for Fishing

The application of light fishing techniques limits the number of fishes that can be trapped. This often initiates the motivation to use methods aimed at immobilizing or killing fishes, which are often application of illegal chemicals such as Dichlorodiphenyltrichloroethane (DDT), Carbide, and explosives, especially dynamite, and in other instances, a mixture of garri + detergents, powdered detergent + garri + petrol have all been reported in Ghana (Afoakwah et al. 2018). Similar unhealthy practices were also reported in Cameroon where the impacts on the marine safety, ocean health, and sustainability of aquatic resources have been questioned (Beseng 2021). The highly flowing waters transport the chemicals to locations not intended, affecting other aquatic organisms. Used at high concentrations meant to achieve target results, the chemicals may be more lethal to macrophytes and zooplankton, for which there is paucity of literature to ascertain the extent of ecological harm meted on these frail aquatic lives. The accumulation of these chemicals within a target fishing location will adversely alter the ecological dynamics, facilitating a succession towards more resistant macrophytes. These practices must be checked by regulatory agencies to ensure the ill practice which does not just harm the aquatic ecosystems, but also produce fishes unfit for human consumption.

4.11.1.5 Unsustainable Use of Aquatic Resources

Coastal communities usually rely on aquatic resources for both food and economic gains. This may be mostly fishing, as a highly rated source of protein, and harvesting other aquatic resources. This indiscriminate practice entails protracted hours on the waterways, a practice that entails disrupting the quietness of water surfaces by heavy machines, fishing boats, and power engines which tend to show impacts on surface growing macrophytes (Jennings and Kaise 1998). These destructive activities constitute unavoidable consequences of fishing and harvesting aquatic resources.

4.11.1.6 Others

Metals polluting aquatic bodies of anthropogenic activities could be from mining, smelting, industrial and municipal effluents discharge, ports and harbor activities, and agricultural activities (application of phosphate fertilizers and metallo-pesticides) (Izah and Angaye 2016; Ogamba et al. 2021; Sharma et al. 2020). Dusty particles from roadsides rich in heavy metals arising from the wear of motor tires could be moved by wind and dispersed into water bodies. Being denser than water, such metals tend to deposit in the sediments and become a rich source of contamination to benthic organisms. Common heavy metals (HMs) pollutants usually found in the environment include Cr, Mn, Ni, Pb. Cu, Cd, and Zn (Khan et al. 2011). While some metals (Fe, Zn, Cu) may be of physiological value at appropriate concentrations (Sharma et al. 2020), other HMs (Hg, Pb, Cd, Cr) even at low concentrations may be toxic to organisms.

Heavy metals have been linked with causing morphological changes and reducing growth and density of phytoplankton, with high concentrations resulting in a decline in the photosynthetic capability of phytoplankton (Sharma et al. 2020). In zooplankton, HMs are known to alter physiological functioning and growth (Sharma et al. 2020, Ju et al. 2018. El-Metwally et al. (2022) reported HMs accumulation potential in zooplankton to be in the order of Cu > Fe > Mn > Zn > Cd > Ni > Pb > Co, with preferential accumulation reported for the four of Cu > Fe > Mn > Zn. However, Ju et al. (2018) reported it in the order of Zn > Cu > Pb > Ni > Cr > Cd. This could be due to the different ecosystems they studied, and the species diversity and abundance observed in the respective study locations. Water pollution by heavy metals is capable of altering the zooplankton community and subsequently the food chain in the given ecosystem (El-Metwally et al. 2022). Ju et al. (2018) reported an inverse relationship between HMs and zooplankton abundance, evidently showing the impact of HMs on zooplankton survival.

Shipping routes are associated with the release of PAHs whose toxicity on aquatic organisms including effects on plankton has been reported (Honda and Suzuki 2020). The photo-modification of PAHs, accomplished under solar irradiation, produces highly toxic chemically modified species known as oxyPAHs, which are toxic to plankton (Lampi 2005). Climate change induces increased temperature on surface water contributing to ocean warming as it is linked with increased solar and UV radiations. Sunlight is necessary for photosynthesis, but excessive sunlight is harmful. Excessive UV radiation and visible light hamper photosynthetic activity and ultimately productivity in phytoplankton, while ocean warming may alter species composition and encourage blooms of harmful eukaryotic and prokaryotic phytoplankton (Hader and Gao 2015).

4.11.2 *Pollution Due to Natural Effects*

4.11.2.1 Flooding

Terrestrial runoff may carry along sediments and dissolved organic matter into coastal waters leading to eutrophication, thus reducing UV penetration (Hader and Gao 2015). The contaminants alter physicochemical properties of water and cause death of less tolerant phytoplankton and macrophytes. The contamination of runoff waters which could be agricultural fields where agrochemicals are applied, wearing off toxic metals dumped on land surfaces, and other biological wastes would facilitate eutrophication of the water bodies, support algal blooms, and alter ecosystem dynamics in freshwaters (Whitfield et al. 2021; Zhu et al. 2019; Tavernini et al. 2009). Following flooding, floating mats of macrophytes in Mdloti Estuary of KwaZulu-Natal Province in South Africa created an adverse condition which hampered usual phytoplankton oxygen manufacture within the water column, thereby reducing oxygen diffusion through the air-water interface (Whitfield et al. 2021). The loss of productivity was felt on the phytoplankton and asphyxiation triggered the death of several fishes. The twin ills of flooding and water level recession around Africa have facilitated the loss of shoreline macrophytes and eroding of plankton. Both of these alter freshwater dynamics substantially due to the losses.

4.11.2.2 Others

Heavy metals contamination of water resources arising from natural sources could include volcanic eruptions and weathering of metal-rocks (Izah et al. 2016). Increased atmospheric concentrations of CO_2 may cause ocean acidification altering chemical properties of seawater, which reduces calcification in calcifying phytoplankton (Hader and Gao 2015), resulting in reduced growth and photosynthesis of phytoplankton and other photosynthetic marine organisms (Wu et al. 2008) and several other zoological taxa. Although phytoplankton and macrophytes constitute very important components of the aquatic ecosystem, the emergence of harmful algal blooms (HABs) poses a serious threat to other aquatic resources. Whitfield et al. (2021) reported on *Prymnesium parvum*, which first appeared in 1970 and recently in 2014 in Cape Town's Estuary and has since then been identified in other water bodies around SA, causing significant mortalities usually through gill-hemorrhaging and gonad-rupture of affected fishes. Other HAB species such as *Heterosigma akashiwo, Nannachloropsis* sp, and *Lingulodinium polyedra* have been reported due to their toxic impacts on fishes in South Africa (Whitfield et al. 2021).

Several factors capable of altering underwater transparency could reduce the radiance of sunlight reaching such beneath water surface inhabiting organisms. Zhang et al. (2019c) reported light availability as the most potent local environmental factor capable of altering freshwater macrophytes dynamics, beside water quality to which different macrophytes and plankton respond differently depending on their species' tolerance ability to stressors. They also highlighted other important natural factors to include water level, temperature fluctuations, precipitation, hours of sunlight, and water quality due to an interplay of both anthropogenic and natural factors.

4.12 Conservation Approaches to Aquatic Biodiversity

The conservation strategies must be able to support development that is sustainable by preserving ecosystems (Yan et al. 2019). To enable optimal protection of biological resources, strategies must be effective and management actions broadbased. Some approaches are highlighted:

- Areas of future research in biodiversity trend must be identified by research organizations and conferences.
- Awareness should be increased by organization of workshops, educational programs, and incentive programs because awareness is one of the public ways for biodiversity conservation (Cantonati et al. 2020).
- Wastewater discharge must be monitored and regulatory measures and actions must be prompt so as to conserve biological diversity.
- Programmes that are specialized and conservation-driven should be established. This will help to restore specific areas they are expected to address.
- The establishment of industries and plants should be away from water resources because wastes they discharge can affect the composition of natural water body, leading to biodiversity loss (Tóth et al. 2019).
- Soil erosion can be reduced by planting trees which will reduce siltation issues and biodiversity can be preserved (Vilmi et al. 2019).
- Water shed management must be carried out; waterways and exploitation should have various restrictions so as to boost aquatic resources (Gál et al. 2019).
- Degraded areas and species of animals can be restored

4.13 Conclusion

The aquatic system is composed of wide range of biodiversity and their interplay determines the sustenance of the ecosystem. Biodiversity is composed of animals, plants, fungi, and microbes having different roles in the ecosystem. They can function within species, between species, and between species and the environment. The aquatic system is composed of freshwater, brackish, and marine waters having different characteristics and species composition. Plankton is a major component in the environment and can be classified based on size, species, and length of stay on water. They are significant in the food chain, oxygen production, and carbon cycle. Aquatic macrophytes are also important components of the aquatic system and have a lot of functions, roles, ecological groups, and various adaptive features in the aquatic ecosystem. They can appear in rooted or free floating categories which can be submerged, floating, or emergent. Plankton and macrophytes have various adaptive features for their survival. Biodiversity can be assessed using various models and are important in cultural services, medicinal functions, supporting and provisioning services, food resources, and biomonitoring potentials. These functions are threatened by human activities such as oil and gas, dredging, waste dumps, and use of chemicals. Natural effects can also cause significant effects on their abundance. To this end, conservative strategies should be employed such as environmental management, awareness creation, and land usage management and environmental impact assessment of activities should be carried out to achieve sustainability of plankton and macrophytes and the aquatic environment which is the medium in which they thrive.

References

Abraham A, Sommerhalder K, Abel T (2010) Landscape and well-being: a scoping study on the health-promoting impact of outdoor environments. Int J Public Health 55:59–69

Afoakwah RO, Mensah BD, Effah E (2018) A guide on illegal fishing activities in Ghana. USAID/ Ghana Sustainable Fisheries Management Project. Narragansett, RI: Coastal Resources Center, Graduate School of Oceanography, University of Rhode Island. Prepared by the University of Cape Coast, Ghana. GH2014_SCI048_UCC 64 pp.

Agedah EC, Ineyougha ER, Izah SC, Orutugu LA (2015) Enumeration of total heterotrophic bacteria and some physico-chemical characteristics of surface water used for drinking sources in Wilberforce Island, Nigeria. J Environ Treatment Tech 3(1):28–34

Agrawal A, Gopnal K (2013) Biomonitoring of water and waste water. Springer India, p 34. ISBN 978-8-132-20864-8.

Aigberua AO, Ekubo AT, Inengite AK, Izah SC (2017) Assessment of some selected heavy metals and their pollution indices in an oil spill contaminated soil in the Niger Delta: a case of Rumuolukwu community. Biotechnol Res 3(1):11–19

Aigberua AO, Ekubo AT, Inengite AK, Izah SC (2016a) Evaluation of total hydrocarbon content and polycyclic aromatic hydrocarbons in an oil spill contaminated soil in Rumuolukwu community in Niger Delta. J Environ Treatment Tech 4(4):130–142

Aigberua AO, Ekubo AT, Inengite AK, Izah SC (2016b) Seasonal variation of nutrient composition in an oil spill contaminated soil: a case of Rumuolukwu, Eneka, Port Harcourt, Nigeria. Biotechnol Res 2(4):179–186

Alcock I, White MP, Wheeler BW, Fleming LE, Depledge MH (2014) Longitudinal effects on mental helath of moving to greener and less green urban areas. Environ Sci Tech. Pub ahead of print. https://doi.org/10.1021/es403688w

Allan JD (2004) Landscapes and riverscapes: The influence of land use on stream ecosystems. Annu Rev Ecol Evol Syst 35:257–284

Arber A (1920) Water plants. A study of aquatic angiosperms. University Press, Cambridge, 460 p

Aznar J, Dervieux A, Grillas P (2002) Association between aquatic vegetation and landscape indicators of human pressure. Wetlands 23:149–160

Barton J, Pretty J (2010) What is the best dose of nature and green exercise for improving mental health? A multistudy analysis. Environ Sci Tech 44(10):3947–3955

Ben-Eledo VN, Kigigha LT, Izah SC, Eledo BO (2017a) Water quality assessment of Epie creek in Yenagoa metropolis, Bayelsa state, Nigeria. Arch Curr Res Int 8(2):1–24

Ben-Eledo VN, Kigigha LT, Izah SC, Eledo BO (2017b) Bacteriological quality assessment of Epie Creek, Niger Delta region of Nigeria. Int J Ecotoxicol Ecobiol 2(3):102–108

Beseng M (2021) The nature and scope of illegal, unreported, and unregulated fishing and fisheries crime in Cameroon: implications for maritime security. Afr Security 14(3):262–285. https://doi.org/10.1080/19392206.2021.1982241

Boedeltje G, Sollman P, Lenssen JPM (2019) Floating ability, shoot length and abundance facilitate hydrochorous dispersal of moss and liverwort fragments. J Veg Sci 30(1):30–41. https://doi.org/10.1111/jvs.12695

Bornette G, Amoros C, Chessel D (1994) Effect of allogenic processes on successional rates in former river channels. J Veg Sci 5:237–246

Brönmark C, Hansson LA (2017) The biology of lakes and ponds, 3rd edn. Oxford University Press, Oxford. 368 p

Cantonati M, Stevens LE, Segadelli S, Springer AE, Goldscheider N, Celico F, Filippini M, Ogata K, Gargini A (2020) Ecohydrogeology: The interdisciplinary convergence needed to improve the study and stewardship of springs and other groundwater-dependent habitats, biota, and ecosystems. Ecol Indic 110:105803. https://doi.org/10.1016/j.ecolind.2019.105803

Chambers PA, Lacoul P, Murphy KJ, Thomaz SM (2008) Global diversity of aquatic macrophytes in freshwater. Hydrobiologia 198:9–26. https://doi.org/10.1007/978-1-4020-8259-7_2

Chen J, Zhang H, Han Z, Ye J, Liu Z (2012) The influence of aquatic macrophytes on microcystis aeruginosa growth. Ecol Eng 42:130–133. https://doi.org/10.1016/j.ecoleng.2012.02.021

Chilvers BL, Morgan KJ, White BJ (2020) Sources and reporting of oil spills and impacts on wildlife 1970–2018. Environ Sci Pollut Res Int 28(1):754–762. https://doi.org/10.1007/s11356-020-10538-0

Coughlan NE, Cuthbert RN, Kelly TV, Jansen MAK (2018) Parched plants: survival and viability of invasive aquatic macrophytes following exposure to various desiccation regimes. Aquat Bot 150:9–15. https://doi.org/10.1016/j.aquabot.2018.06.001

Crowder LB, Hazen EL, Avissar N, Bjorkland R, Latanich C, Ogburn MB (2008) The impacts of fisheries on marine ecosystems and the transition to ecosystems-based management. Annu Rev Ecol Evol Syst 39(1):259–278

de Nie HW (1987) The decrease in aquatic vegetation in Europe and its consequences for fish populations. FAO, Rome. http://www.fao.org/3/ac858e/AC858E00.htm

Dolan J (2012) Microzooplankton: the microsocopic (micro) animals (zoo) of the plankton. Achieved on 2016 03–04.

Du ZY, Wang QF (2016) China Phylogeny Consortium. Phylogenetic tree of vascular plants reveals the origins of aquatic angiosperms. J Syst Evol 54(4):342–348. https://doi.org/10.1111/jse.12182

Ekperusi AO, Ekperusi OH (2021) Towards transparency in oil management in Africa: Nigeria as a case study. Int Oil Spill Conf Proc 1:1141577. https://doi.org/10.7901/2169-3358-2021.1.1141577

El-Metwally MEA, Abu El-Regal MA, Abdelkader AI, Sanad EF (2022) Heavy metal accumulation in zooplankton and impact of water quality on its community structure. Arabian J Geosci 15(1)

Fennessy MS, Geho R, Elifritz B, Lopez RD (1998) Testing the floristic quality assessment index as an indicator of riparian wetland disturbance. Ohio Environmental Protection Agency, Division of Surface Water, Columbus, OH

Franklin P, Dunbar M, Whitehead P (2008) Flow controls on lowland river macrophytes: a review. Sci Total Environ 400:369–378

Gál B, Szivák I, Heino J, Schmera D (2019) The effect of urbanization on freshwater macroinvertebrates—knowledge gaps and future research directions. Ecol Indic 104:357–364

Gdara I, Zrafi I, Balducci C, Cecinato A, Ghrabi A (2019) First investigation of seasonal concentration behaviors and sources assessment of aliphatic hydrocarbon in waters and sediments from Wadi El Bey, Tunisia. Arch Environ Contam Toxicol 78(1):1–19

Gimenes LLS, Freschi GPG, Bianchini Júnior I, Cunha Santino MBD (2020) Growth of the aquatic macrophyte *Ricciocarpos natans* (L.) Corda in different temperatures and in distinct concentrations of aluminum and manganese. Aquat Toxicol 224:105484

Gladwell VF, Brown DK, Wood C et al (2013) The great outdoors: How a green exercise environment can benefit all. Extreme Physiol Med 2(3). https://doi.org/10.1186/2046-7648-2-3

Guo C, Zhu G, Qin B, Zhang Y, Zhu M, Xu H, Chen Y, Paerl HW (2019) Climate exerts a greater modulating effect on the phytoplankton community after 2007 in eutrophic Lake Taihu, China: Evidence from 25 years of recordings. Ecol Indic 105:82–91

Hader DP, Gao K (2015) Interactions of anthropogenic stress factors on marine phytoplankton. Front Environ Sci 3

Halpern BS, Walbridge S, Fujita R, Heinemann D, Lenihan HS, Madin EMP, Fox HE (2008) A global map of human impact on marine ecosystems. Science (American Association for the Advancement of Science) 319(5865):948–952

Hartmann M, Grob C, Tarran GA, Martin AO, Burkill PH, Scanlan DJ, Zubkov MV (2012) Mixotrophic basis of Atlantic oligotrophic ecosystems. Proc Natl Acad Sci U S A 109(15): 5756–5760

He F, Wu N, Dong X, Tang T, Domisch S, Cai Q, Jähnig SC (2020) Elevation, aspect, and local environment jointly determine diatom and macroinvertebrate diversity in the Cangshan Mountain, Southwest China. Ecol Indic 108:105618. https://doi.org/10.1016/j.ecolind.2019.105618

Hendrik S, Martens K (2005) Aquatic biodiversity: v. 2: the diversity of aquatic ecosystems (developments in hydrobiology). Springer

Honda M, Suzuki N (2020) Toxicities of polycyclic aromatic hydrocarbons for aquatic animals. Int J Environ Res Public Health 17(4):1363

Izah SC, Aigberua AO, Srivastav AL (2022) Factors influencing the alteration of microbial and heavy metal characteristics of river systems in the Niger Delta region of Nigeria. In: Madhav S, Kanhaiya S, Srivastav AL, Singh VB, Singh P (eds) Ecological significance of river ecosystem: challenges and management. Elsevier, pp 51–78

Izah SC, Angaye TCN, Aigberua AO, Nduka JO (2017) Uncontrolled bush burning in the Niger Delta region of Nigeria: potential causes and impacts on biodiversity. Int J Mol Ecol Conserv 7(1):1–16

Izah SC, Chakrabarty N, Srivastav AL (2016) A review on heavy metal concentration in potable water sources in Nigeria: human health effects and mitigating measures. Exposure Health 8: 285–304

Izah SC, Angaye TCN (2016) Heavy metal concentration in fishes from surface water in Nigeria: Potential sources of pollutants and mitigation measures. Sky J Biochem Res 5(4):31–47

Izah SC, Srivastav AL (2015) Level of arsenic in potable water sources in Nigeria and their potential health impacts: A review. J Environ Treatment Tech 3(1):15–24

Izah SC (2018) Ecosystem of the Niger Delta region of Nigeria: potentials and threats. Biodiversity Int J 2(4):338–345

Jennings S, Kaise MJ (1998) The effects of fishing on marine ecosystems. Adv Mar Biol 34:201–212

Ju Y-R, Lo W-T, Chen C-F, Chen C-W, Huang Z-L, Dong C-D (2018) Effect of metals on zooplankton abundance and distribution in the coast of southwestern Taiwan. Environ Sci Pollut Res Int 26(33):33,722–33,731. https://doi.org/10.1007/s11356-018-2169-x

Khan T, Muhammad S, Khan B, Khan H (2011) Investigating the levels of selected heavy metals in surface water of Shah Alam River (a tributary of River Kabul, Khyber Pakhtunkhwa). J Himalayan Earth Sci 44(2):71–79

Karleskint G, Tuner R, Small J (2013) Chapter 17: The open sea. In: Karleskint G, Small J (eds) Introduction to marine biology, (4th ed.). edn, Brooks/Cole. ISBN 978-1-133-36446-7.

Karsenti E, Acinas SG, Brok P, Bowler C, De Vargas C, Raes J, Sulivan M, Arendt D, Benzoni F, Claverie J-M, Follows M, Gorsky G, Hingamp P, Ludicone D, Jallion O, Kandels-Lewis S, Krzic U, Not F, Ogata H, Pesant S, Reynaud EG, Sardet C, Sieracki ME, Speich S,

Velayoudon D, Weissenbach J, Wincker P (2011) A holistic approach to marine eco-systems biology. PLoS Biol 9(10):e1001177

Klamt A-M, Hu K, Huang L, Chen X, Liu X, Chen G (2020) An extreme drought event homogenises the diatom composition of two shallow lakes in southwest China. Ecol Indic 108:105662. https://doi.org/10.1016/j.ecolind.2019.105662

Kuhar U, Germ M, Gaberšˇcik A, Urbaniˇc G (2011) Development of a River Macrophyte Index (RMI) for assessing river ecological status. Limnologica 41:235–243

Kumar U, Asija MJ (2009) Biodiversity: principle and conservation. Agrobios (India)

Lampi MA (2005) Environmental photoinduced toxicity of polycyclic aromatic hydrocarbons occurrence and toxicity of photomodified PAHs and predictive modeling of photoinduced toxicity. University of Waterloo, Waterloo

Largo-Wight E (2011) Cultivating healthy places and communities. Evidenced-based nature contact recommendations. Int J Environ Health Res 21(1):41–61

Leles SG (2018) Modelling mixotrophic functional diversity and implications for ecosystem function—Oxford Journals. J Plankton Res 40(6):627–642

Lemcke S, Holding J, Møller EF, Thyrring J, Gustavson K, Juul-Pedersen T, Sejr MK (2018) Acute oil exposure reduces physiological process rates in Arctic phyto- and zooplankton. Ecotoxicology (London) 28(1):26–36

Li F, Tonkin JD, Haase P (2020) Local contribution to beta diversity is negatively linked with community-wide dispersal capacity in stream invertebrate communities. Ecol Indic 108:105715

Li G, Hu S, Hou H, Kimura S (2019) Heterophylly: phenotypic plasticity of leaf shape in aquatic and amphibious plants. Plants 8:420

Liu F, Li M, Wang J, Gong Z, Liu M, Liu H, Lin P (2021) Species composition and longitudinal patterns of fish assemblages in the middle and lower Yarlung Zangbo River, Tibetan Plateau, China. Ecol Ind 125:107542

Liu J, Chen Y, Li M, Liu B, Liu X, Wu Z, Cai Y, Xu J, Wang J (2019) Water-level fluctuations are key for phytoplankton taxonomic communities and functional groups in Poyang Lake. Ecol Indic 104:470–478

Liu P, Xu S, Lin J, Li H, Lin Q, Han B-P (2020) Urbanization increases biotic homogenization of zooplankton communities in tropical reservoirs. Ecol Indic 110:105899

Maberly SC, Madsen TV (2002) Freshwater angiosperm carbon concentrating mechanisms: processes and patterns. Funct Plant Biol 29:393–405. https://doi.org/10.1071/PP01187

Mack JJ, Micacchion M, Augusta LD, Sablack GR (2000) Vegetation indices of biotic integrity (Vibi) for wetlands and calibration of the Ohio rapid assessment method for wetlands; Grant CD95276; Final report to US Environmental Protection Agency; Ohio Environmental Protection Agency: Columbus, OH, USA p. 98.

Mackay SJ, Arthington AH, Kennard MJ, Pusey BJ (2003) Spatial variation in the distribution and abundance of submerged macrophytes in an Australian subtropical river. Aquat Bot 77:169–186

Málnás K, Polyák L, Prill É, Hegedüs R, Kriska G, Dévai G, Lengyel S (2011) Bridges as optical barriers and population disruptors for the mayfly Palingenia longicauda: an overlooked threat to freshwater biodiversity? J Insect Conserv 15(6):823–832. https://doi.org/10.1007/s10841-011-9380-0

Mathivanan V, Vijayan P, Sabhanayakam S, Jeyachitra O (2007) An assessment of plankton population of Cauvery River with reference to pollution. J Environ Biol 28:523–526

Mommer L, Pons TL, Wolters-Arts M, Venema JH, Visser EJW (2005) Submergence-induced morphological, anatomical, and biochemical responses in a terrestrial species affect gas diffusion resistance and photosynthetic performance. Plant Physiol 139(1):497–508. https://doi.org/10.1104/pp.105.064725

Nakayama H, Sinha NR, Kimura S (2017) How do plants and phytohormones accomplish heterophylly, leaf phenotypic plasticity, in response to environmental cues. Front Plant Sci 8:1717. https://doi.org/10.3389/fpls.2017.01717

Nelson EJ, Kareiva P, Ruckelshaus M et al (2013) Climate change impact on key ecosystem services and the human wellbeing they support in the United States. Front Ecol Environ 11(9):483–493

Nieuwenhuijsen MJ, Hanneke K, Gidlow C et al (2014) Positive health effects of the natural outdoor environment in typical populations in different regions in Europe (PHENOTYPE): a study programme protocol. BMJ Open 4:e004951

Ogamba EN, Charles EE, Izah SC (2021) Distributions, pollution evaluation and health risk of selected heavy metal in surface water of Taylor creek, Bayelsa State, Nigeria. Toxicol Environ Health Sci 13(2):109–121. https://doi.org/10.1007/s13530-020-00076-0

Ogamba EN, Charles EE, Izah SC (2019a) Phytoplankton community of Taylor Creek in the Niger Delta using diversity indices. J Plant Animal Ecol 1(3):1–12

Ogamba EN, Charles EE, Izah SC (2019b) Application of diversity indices in the study zooplankton community of Taylor Creek in the Niger Delta, Nigeria. Sumerianz J Biotechnol 2(6):35–41

Ogamba EN, Seiyaboh EI, Izah SC, Ogbugo I, Demedongha FK (2015a) Water quality, phytochemistry and proximate constituents of *Eichhornia crassipes* from Kolo creek, Niger Delta, Nigeria. Int J Appl Res Technol 4(9):77–84

Ogamba EN, Izah SC, Toikumo BP (2015b) Water quality and levels of lead and mercury in *Eichhornia crassipes* from a tidal creek receiving abattoir effluent, in the Niger Delta, Nigeria. Continent J Environ Sci 9(1):13–25

Ogamba EN, Izah SC, Emaviwe D (2015c) Phytochemical assessment of *Eichhornia crassipes* from River Nun, Nigeria. Res J Phytomed 1(1):24–25

Ogamba EN, Izah SC, Ogbogu MJ (2015d) Cadmium and Lead level in *Eichhornia crassipes* from River Nun, Niger Delta, Nigeria. J Adv Biol Basic Res 1(1):53–56

O'Hare MT, Baattrup-Pedersen A, Baumgarte I, Freeman A, Gunn IDM, Lázár AN, Sinclair R, Wade AJ, Bowes MJ (2018) Responses of aquatic plants to eutrophication in rivers: a revised conceptual model. Front Plant Sci 9:451. https://doi.org/10.3389/fpls.2018.00451

O'Hare MT, Baattrup-Pedersen A, Nijboer R, Szoszkiewicz K, Ferreira T (2006) Macrophyte communities of European streams with altered physical habitat. Hydrobiologia 566:197–210

Olguín HF, Puig A, Loez CR, Salibián A, Topalián ML, Castañé PM et al (2004) An integration of water physicochemistry, algal bioassays, phytoplankton, and zooplankton for ecotoxicological assessment in a highly polluted lowland river. Water Air Soil Pollut 155:355–381. https://doi.org/10.1023/B:WATE.0000026538.51477.c2

Ormond RFG, Gage JD, Martin VA (eds) (1997) Marine biodiversity: patterns and processes. Cambridge University Press, New York

Padhi BK, Mandal RK (2000) Applied Fish Genetics. Fishing Chimes Publi

Pajunen V, Kahlert M, Soininen J (2020) Stream diatom assemblages as environmental indicators—a cross-regional assessment. Ecol Indic 113:106183. https://doi.org/10.1016/j.ecolind.2020.106183

Paudel Adhikari N, Liu Y, Liu K, Zhang F, Adhikari S, Chen Y, Liu X (2019) Bacterial community composition and diversity in Koshi River, the largest river of Nepal. Ecol Indic 104:501–511

Perata P, Alpi A (1993) Plant responses to anaerobiosis. Plant Sci 93:1–17. https://doi.org/10.1016/0168-9452(93)90029-Y

Rejmánková E (2011) The role of macrophytes in wetland ecosystems. J Ecol Field Biol 34(4):333–345. https://doi.org/10.5141/JEFB.2011.044

Santamaría L (2002) Why are most aquatic plants widely distributed? Dispersal, clonal growth and small-scale heterogeneity in a stressful environment. Acta Oecologica 23:137–154. https://doi.org/10.1016/S1146-609X(02)01146-3

Selig ER, Turner WR, Troeng S, Wallace BP, Halpern BS et al (2014) Global priorities for marine biodiversity conservation. PLoS One 9(1):e82898. https://doi.org/10.1371/journal.pone.0082898

Seiyaboh EI, Izah SC (2017a) Review of impact of anthropogenic activities in surface water resources in the Niger Delta region of Nigeria: a case of Bayelsa state. Int J Ecotoxicol Ecobiol 2(2):61–73

Seiyaboh EI, Izah SC (2017b) Bacteriological assessment of a tidal creek receiving slaughterhouse wastes in Bayelsa state, Nigeria. J Adv Biol Biotechnol 14(1):1–7
Sharma KV, Sarvalingam BK, Marigoudar SR (2020) A review of mesocosm experiments on heavy metals in marine environment and related issues of emerging concerns. Environ Sci Pollut Res Int 28(2):1304–1316. https://doi.org/10.1007/s11356-020-11121-3
Srivastava J, Gupta A, Chandra H (2008) Managing water quality with aquatic macrophytes. Rev Environ Sci Biotechnol 7:255–266. https://doi.org/10.1007/s11157-008-9135-x
Steinacher M et al (2010) Projected 21st century decrease in marine productivity: a multi-model analysis. Biogeosciences 7(3):979–1005. https://doi.org/10.5194/bg-7-979-2010
Steinberg DK, Landry MR (2017) Zooplankton and the ocean carbon cycle. Ann Rev Mar Sci 9: 413–414. https://doi.org/10.1146/annurev-marine-010814-015924
Szoszkiewicz K, FerreiraT KT, Baattrup-Pedersen A, Davy-Bowke J, O'Hare M (2006) European river plant communities: the importance of organic pollution and the usefulness of existing macrophyte metrics. Hydrobiologia 566:211–234
Tóth R, Czegledi I, Kern B, Eros T (2019) Land use effects in riverscapes: diversity and environmental drivers of stream fish communities in protected, agricultural and urban landscapes. Ecol Indic 101:742–748
Takahashi H, Yamauchi T, Colmer TD, Nakazono M (2014) Aerenchyma formation in plants. In: van Dongen JT, Licausi F (eds) Low-oxygen stress in plants: oxygen sensing and adaptive responses to hypoxia. Springer-Verlag, Wien, pp 247–265. https://doi.org/10.1007/978-3-7091-1254-0_13
Tang Y, Harpenslager SF, van Kempen MML, Verbaarschot EJH, Loeffen LMJM, Roelofs JGM, Smolders AJP, Lamers LPM (2017) Aquatic macrophytes can be used for wastewater polishing but not for purification in constructed wetlands. Biogeosciences 14:755–766
Tavernini S, Viaroli P, Rossetti G (2009) Zooplankton community structure and inter-annual dynamics in two sand-pit lakes with different dredging impact. Int Rev Hydrobiol 94(3): 290–307
Tewari G and Bisht A (2021) Aquatic biodiversity: threats and conservation. https://www.aquafind.com/articles/aquatic_biodiversity.php
Thiemer K, Schneider SC, Demars BO (2021) Mechanical removal of macrophytes in freshwater ecosystems: Implications for ecosystem structure and function. Sci Total Environ 782: 146,671–146,671
van Leeuwen CHA (2018) Internal and external dispersal of plants by animals: an aquatic perspective on alien interference. Front Plant Sci 9:153. https://doi.org/10.3389/fpls.2018.00153
van Veen H, Sasidharan R (2019) Shape shifting by amphibious plants in dynamic hydrological niches. New Phytol. https://doi.org/10.1111/nph.16347
Vilmi A, Tolonen KT, Karjalainen SM, Heino J (2019) Niche position drives interspecific variation in occupancy and abundance in a highly-connected lake system. Ecol Indic 99:159–166
Vörösmarty CJ, Mclntyre PB, Gessner MO, Dudgeon D, Prusevich A, Green P, Glidden S, Bunn SE, Sullivan CA, Liermann CR (2010) Global threats to humanwater security and river biodiversity. Nature 467:555
Vymazal J (2013) Emergent plants used in free water surface constructed wetlands: a review. Ecol Eng 61:582–592. https://doi.org/10.1016/j.ecoleng.2013.06.023
Wang G, Wang X, Liu X, Li Q (2012) Diversity and biogeochemical function of planktonic fungi in the ocean. In: Raghukumar C (ed) Biology of Marine Fungi. Springer, Berlin Heidelberg, pp 71–88
Ward BA, Follows MJ (2016) Marine mixotrophy increases trophic transfer efficiency, mean organism size, and vertical carbon flux. Proc Natl Acad Sci U S A 113(11):2958–2963
Wetzel RG (2001) Limnology: lake and river ecosystems, 3rd edn. Academic Press, San Diego, 1006 p. https://doi.org/10.1016/C2009-0-02112-6

Whitfield AK, Adams JB, Harrison TD, Lamberth SJ, Lemley DA, MacKay F, Weyl OLF (2021) Possible impacts of non-native plant, pathogen, invertebrate and fish taxa on the indigenous ichthyofauna in South African estuaries: a preliminary review. Biol Invasions 23(9):2729–2747
World Health Organization (WHO) (2014) Biodiversity. https://www.who.int/globalchange/ecosystems/biodiversity/en/.
Wu HY, Zou DH, Gao KS (2008) Impacts of increased atmospheric CO2 concentration on photosynthesis and growth of micro- and macro-algae. Sci China Ser C 51:1144–1150
Yamamuro M, Viaroli P, de Wit R, Mazouni N (2012) Herbicide-induced macrophyte-to-phytoplankton shifts in Japanese lagoons during the last 50 years: consequences for ecosystem services and fisheries. Hydrobiologia 699(1):5–19
Yan D, Xu H, Yang M, Lan J, Hou W, Wang F, Zhang J, Zhou K, An Z, Goldsmith Y (2019) Responses of cyanobacteria to climate and human activities at Lake Chenghai over the past 100 years. Ecol Indic 104:755–763
Yang N, Zhang Y, Duan K (2017) Effect of hydrologic alteration on the community succession of macrophytes at Xiangyang Site, Hanjiang River, China. Scientifica, 4083696.
Yu W, Zou X, Zhang D, Wang T, Wang C, Yao Y, Zhang H, Ben C, Yuan J (2019) The evaluation of marine biological value of the Jiangsu coastal zone (east of China) under the interference of human activities. Ecol Indic 102:76–83
Zelenski JM, Nisbet EK (2014) Happiness and feeling connected: The distinct role of nature relatedness. Environ Behav 46(1):3–23. https://doi.org/10.1177/0013916512451901
Zelnik I, Kuhar U, Holcar M, Germ M, Gaberščik A (2021) Distribution of vascular plant communities in Slovenian watercourses. Water 13:1071
Zeng J, Jiao C, Zhao D, Xu H, Huang R, Cao X, Yu Z, Wu QL (2019) Patterns and assembly processes of planktonic and sedimentary bacterial community differ along a trophic gradient in freshwater lakes. Ecol Indic 106:105491. https://doi.org/10.1016/j.ecolind.2019.105491
Zhang M, Molinos JG, Su G, Zhang H, Xu J (2019a) Spatially structured environmental variation plays a prominent role on the biodiversity of freshwater macrophytes across China. Front Plant Sci 10:161. https://doi.org/10.3389/fpls.2019.00161
Zhang P, Qiao Y, Grenouillet G, Lek S, Cai L, Chang J (2021) Responses of spawning thermal suitability to climate change and hydropower operation for typical fishes below the Three Gorges Dam. Ecol Indic 121:107186. https://doi.org/10.1016/j.ecolind.2020.107186
Zhang Y, Cheng L, Kong M, Li W, Gong Z, Zhang L, Wang X, Cai Y, Li K (2019b) Utility of a macroinvertebrate-based multimetric index in subtropical shallow lakes. Ecol Indic 106: 105527. https://doi.org/10.1016/j.ecolind.2019.105527
Zhang Y, Ma R, Liang Q, Guan B, Loiselle S (2019c) Secondary impacts of eutrophication control activities in shallow lakes: Lessons from aquatic macrophyte dynamics in Lake Taihu from 2000 to 2015. Freshwater Sci 38(4):802–817. https://doi.org/10.1086/706197
Zhu J, Peng Z, Liu X, Deng J, Zhang Y, Hu W (2019) Response of aquatic plants and water quality to large-scale nymphoides peltata harvest in a shallow lake. Water 11:77. https://doi.org/10.3390/w11010077
Zuo Z (2019) Why algae release volatile organic compounds—the emission and roles. Front Microbiol 10:491. https://doi.org/10.3389/fmicb.2019.00491

Chapter 5
Threats and Conservation Status of *Cercopithecus sclateri* in Akwa Ibom State, Nigeria

Daniel Etim Jacob and Edem Archibong Eniang

Abstract The primate species *Cercopithecus sclateri* is considered a sacred animal in all the villages, making up Itu Local Government Area in Akwa Ibom State, Nigeria. The communities are blessed with pockets of forests that serve as a sacred grove for this endemic species. These fragmented forests are preserved by communal laws that prohibit the extraction of specific forest resources, including the endangered primate species among other animal species in the study area. However, in recent times, there has been widespread destruction and exploitation of communal forestland due to pressure from urbanization and population growth, which has posed a severe threat to the conservation of forest resources in the area. Furthermore, studies conducted in the area have shown that anthropogenic activities in the site are high and have negatively impacted the primate species population in the study area, resulting in an aging population of the monkey species. This chapter, therefore, evaluates the threat, conservation status, and efforts undertaken to ensure species conservation in the study area. It also gives recommendations to ensure that adequate measures are urgently needed to restore and protect the remaining endemic species' habitat to ensure its conservation in the study area.

Keywords Conservation · Sclater's guenon · Community forest · Itu · Nigeria

5.1 Introduction

Majority of the world's biological diversity is found in the tropical ecosystems (Sodhi et al. 2004; Quinten 2008; Jacob et al., 2020d; Ukpong et al., 2018; Nelson et al., 2019, 2021). The region constitutes less than 7% only of the global land area and serves as a refuge for more than half of all existing flora and fauna species globally (Donohoe 2003; Jacob et al., 2015b, c). However, due to broad abuse of the

D. E. Jacob (✉) · E. A. Eniang
Department of Forestry and Wildlife, University of Uyo, Uyo, Nigeria
e-mail: danieljacob@uniuyo.edu.ng

S. C. Izah, M. C. Ogwu (eds.), *Sustainable Utilization and Conservation of Africa's Biological Resources and Environment*, Sustainable Development and Biodiversity 32, https://doi.org/10.1007/978-981-19-6974-4_5

shared assets, the locality is being converted to a place of human habitation, which has posed as a risk to neighborhood biodiversity (Lawrence 1997). Alteration of the environment by human activity has, in this manner, undermined biodiversity at a worldwide scale (Cowlishaw 1999; Cowlishaw and Dunbar 2000; Chapman and Peres 2001; Jacob, 2016). According to Ettah (2008), the tropical nations lose more than 127,000 km^2 of timberland zone every year. This does not incorporate the tremendous range being specifically logged, evaluated to cover more than 55,000 km^2 (Chapman and Lambert 2000; Bennett 2000).

In Africa, deforestation due to anthropogenic activities is a significant issue (Jacob et al., 2013a; Nelson et al., 2013, 2020). Hence, the ecosystem (e.g., swamp rainforest) is exploited at a faster rate than those of other ecosystems (Archad et al. 2002). Besides, the authors' projection has indicated that an unsustainable tropical woodland pulverization will result in the destitution of around 75% of all unique biodiversity by the next century. These scenarios of destruction are incredibly genuine concerns since the tropics are the world's most significant locale in species abundance and endemism (Mittermeier et al. 1997; Myers et al., 2000; Jacob et al., 2012), and these actions could cause the extinction of more than 42% of its biodiversity (Sodhi et al. 2004). However, biodiversity exceptionally enhances human presence because it constitutes the asset everybody depends on (Jacob et al. 2020a, b, c). Its preservation hence gets to be highly relevant (Groves, 2000; Jacob et al. 2018, 2019).

In Africa, Nigeria is the second nation in terms of being naturally diverse and positioned in primate endemism within the world (Grubb et al. 2000; Egwali et al. 2005). In show of disdain toward this vital status in primate status, the Nigerian environment is uncovered to powers of species extinction and destruction because of human actions (Eniang 2001; Eniang and Ebin 2002). Subsequently, numerous species of mammals, particularly the apes, are debilitated at different levels antagonistic to their natural selection. Consequently, the rising bushmeat trade has aggravated the problem, hence there is bound to be a forest without wildlife as in the case of *Procolobus badius waldronii,* which disappeared from some countries in West Africa (Revkin, 2000). Consequently, most primate species in Nigeria are grouped under various threat class of the IUCN Red List (IUCN 2011).

In Akwa Ibom State, Nigeria, *C. sclateri*, locally called Adiaha awah-Itam - meaning daughter of Itam deity, is a fundamentally imperiled primate of the African landmass (Egwali et al. 2005). It is evaluated as one of the foremost undermined primate species in Africa that need species preservation (Oates 1994; Tooze 1995). The species is also considered the foremost debilitated African guenons (Oates and Anadu, 1989). It was overhauled from the subspecies of *Cercopithecus erythrotis sclateri* to full species status in 1980 (Grubb et al., 2000). *Cercopithecus sclateri* is endemic to Nigeria, and has a limited run, and found only in the lower Western part of the river Niger and Cross River frameworks (Tooze 1994a, b).

Cercopithecus sclateri operates a social group that embodies the gathering of more males and females together in a group. They are effectively arborary and active in the day time. They have the ability to survive in a degraded environment due to their versatility (Egwali et al. 2005; Jacob 2017). Also, little disconnected populace of the species had been seen within the Niger Delta ecosystem (Oates et al. 1992; Tooze 1995; Rowe 1996; Oates et al. 2004). Moreover, Gadsby (1987) observed that Stubb's creek was an ecosystem with a unique and intaglio feature that favors the existence of Cercopithecus *sclateri.*

5.2 Species Description

Cercopithecus sclateri is the only monkey species that is endemic to Nigeria. The internal union for conservation of nature (IUCN) considered it as a primate species that is endangered (Cercopan.org 2011). In 1904, the first description of the species was introduced by Pocock (1904). Decades after this description, scientists still believed the primate species belonged to the subgroup of the *Cercopithecus erythrotis.* Moreover, others believed the species evolved from the species *Cercopithecus erythrotis* that was found only in the forest areas bordering Nigeria and Cameroon, and the species *Cercopithecus erythrogaster* in the forest fragments of the Niger Delta region in Nigeria. However, studies such as Hill (1953), Kingdon (1980), and Nowak (1999) had argued and voted for the full species status of *Cercopithecus sclateri* against its consideration as a subspecies of *Cercopithecus cephus* and *Cercopithecus erythrotis.* The species comparison with *Cercopithecus cephus* was attributed to its size, versatility, and being brightly colored. Consequently, the species can easily be identified by the unique color of its tail which is 30 - 50% rusty-red in coloration (Cercopan.org 2011).

Cercopithecus sclateri, a species of the guenon family, could be regarded as an exceptionally colorful monkey with a complex facial design. Its body color is dark grey with a few traces of green on its backside. It has a ruddy colored tail whose length is 75% of the total length of the species. The ruddy color on ventral proximal portion of the primate tail fades gradually till it becomes white and finally black at the distal end. The primate gag is brownish-pink with a soft white nose spot. Its confront is decorated with three distinct patches of hair, while the crown of the head and the patches on its cheek are yellow blended with darker colors. Furthermore, it has a white patch on its throat that is large and extends to its ears which have significant white tufts. Finally, there are black temporal bars which extend beyond the ears to reach the back of its head (Oates et al. 1992).

Cercopithecus sclateri, besides the other individuals of their superspecies, has a place in the littler group of guenons. The female *Cercopithecus sclateri* weighs almost 2.5 kg, though the male counterparts weigh about 4.0 kg. However, all the primate species belonging to guenons family are considered to possess sexually dimorphic canines. Furthermore, they possess longer rear appendages than their fore appendages. At the end, *Cercopithecus sclateri* possesses a unique quality that differentiates it from other species in the form of its patch on the cheek (Fleagle, 1999). In addition, *Cercopithecus sclateri* has a total body length that ranges between 80 and 120 cm (Law and Myers 2004).

There are very few literatures on the behavioral pattern of *Cercopithecus sclateri* in their natural environment. Be that as it may, within the individuals of the superspecies *Cercopitheus cephus*, social structure is of a lesser stringent characteristics compared to other individuals of *Cercopithecus* species. Particularly, those species do not have a single prevailing male as they also embrace a multi-male social structure that embodies family individuals, or all of them being only females (Kingdon 1980).

There is also paucity of information on the movement patterns of *Cercopithecus* species. Available information assume majority of the guenons species to be quadrupedal, leaping for 10 percent of their active time. However, other studies associate that their movement pattern is influenced by their eating behavior. McGraw (2002) argued that the climbing pattern of the species is contrarily related to eating habits associated with fruits, while the insectivores utilize more transitional stances than other species. Nowak (1999) also asserts that the species utilize their tails for adjusting their balance as they more often than not rest in trees.

C. sclateri is sympatric in nature. Fleagle (1999) listed a few other sympatric species to include *P. potto, A. calabarensis, C. torquatus, C. mona,* and *C. nicticans*. In Gabon, *Cercopithecus nicticans* is reported to share habitat with *Cercopithecus cephus*. They segment assets based on sort of nourishment and favored canopy nourishing level. Consequently, it can be assumed that since *Cercopithecus cephus* occupies the same biological environment with *Cercopithecus sclateri,* they are most likely to affiliate with other primate species in the area. (Jacob 2012; Tooze 1995).

The ranging pattern of *Cercopithecus sclateri* is limited by numerous components such as its body measure, diet, nourishment accessibility due to population density and dissemination, intergroup experiences, mating procedures, social structure, and shirking of predators and nourishment competition (Baker 2005; Boinski et al. 2000). Consequently, all primate species that eat fruits, for example *Cercopithecus sclateri* that depend on high-calories nourishments with inconsistent time and season accessibility, seldom have a wider habitat area and move for long distances in the day. However, species that depend on lower quality nourishments, which are readily available and accessible, have smaller feeding grounds and walk for short distances (Baker and Olubode, 2008). Scrounging models foresee that primates species ought to utilize their feeding area in a proficient way in connection to accessibility of nourishment, particularly concentrating their scrounging endeavors in zones where accessibility of nourishment is most noteworthy (Pyke et al. 1977).

In any case, *Cercopithecus sclateri* depends intensely on perpetual herbaceous vegetation that's inexhaustibly accessible where natural products can be collected. Thus, they have brief everyday distance lengths and a little yearly domestic ranges of about 2 km (Egwali et al., 2005). Be that as it may, closely related species have littler domestic ranges than other individuals of the sort. Nowak (1999) also reports that the closely related *Cercopithecus ascanius* employs a 15 ha habitat range with a 5 ha center zone, in spite of the fact that they are known to have domestic range of up to 130 ha.

Since nourishments are basically omnipresent within the rain forest eco-zones with plenitude of natural products shifting spatially on a little scale as well as among territories, *Cercopithecus sclateri* subsequently employs the rich quality nourishment of the zone over and over inside a decently brief period isolated by long interims (Watts 2000). Amid the late dry period and early onset of rain when there's continuously a significant lessening in nourishment accessibility, *Cercopithecus sclateri* relies to a great extent on natural nourishment as its major source of food at this period and would promptly attack neighboring communities for banana, plantain, and wild pear (Egwali et al. 2005; Jacob et al. 2013b; Jacob and Nelson 2015).

Cercopithecus sclateri, like other smaller species of guenons and the Cross River gorilla, are reported to be frugivorous (Egwali et al. 2005; Ettah 2008). However, other vital characteristics of the guenon's diets incorporate insects, flowers, and leaves. Hence, the guenons can be considerd as an omnivorous species. Consequently, when they occupy towns and communities with fragments or no forest land, they strike gardens and ranches for nourishment. Also, their daily calories intake reflects the regular nature of their living area, appearing dynamic in composition over the course of the year. When natural products such as fruits are limited, evidences show they switch dependence to other herbaceous vegetation for survival (Egwali et al., 2005). Specifically, the only reference to a tree species consumed by *Cercopithecus sclateri* is *Bombax buonopozense* (Butynski 2002b; Fleagle 1999; Oates et al. 1992). Other food sources (Table 5.1) recorded included *Elaeis guineensis* and *Raphia hookeri* (Egwali et al. 2005). Furthermore, C. sclateri spend a considerable length of time in a generally little range, before they continue on a longer journey to another zone. Consequently, in the fragmented forests of Itu Local Government Area, *Cercopithecus sclateri* inhabit a distinctive parts of the forest fragments, while also extending over 1.5 km radius of the fragments (Jacob 2012; Jacob et al. 2015a).

C. sclateri, just like the other individuals of the Cercopithecus species, features a unique facial design that's conceptualized to be utilized in connection associated with reproduction. Particularly, the patches on the cheek and nose are considered imperative in signaling. The design, in concurrence with exceptionally quick and complicated head waving, may suffice critical roles in keeping up connections with other individuals of the group. Sexual determination is considered to influence the advancement of facial design in this species. The profoundly colored tail is likely to be used to interact with members of the group (Kingdon 1980). Also about 22 different forms of vocalizations have been reported for the Cercopithecus species. These incorporate commotions to preserve social cohesion, caution signals, and uproarious commotions radiated by male members of the group (Nowak 1999). Hence, exteroception interaction is imperative in all primates. Prepping behaviors regularly show near connections between group members. Mothers interact with their babies by touching, as do other members. Physical animosity frequently happens, particularly between competing males (Nowak 1999).

Breeding interims for the *Cercopithecus* species vary from 1 to 5 years and no data are detailed for *Cercopithecus sclateri*. For numerous Cercopithecus species, breeding intervals occur between July and September. However, *Cercopithecus sclateri* like every other primate species that inhibit the forest is believed to exhibit

Table 5.1 Plants species consumed by *Cercopithecus sclateri*

Scientific name	Common name	Ibibio name	Status	Part (s) consumed	Utilization
Musa sapientum	Banana	*More*	C	F	***
M. paradisiacal	Plantain	*Ukom*	C	F	***
Cola argentea	Sweet cola	*Nadiya*	C	F	**
Cnestis ferruginea	Cnestis	*Utin Ewa*	W	F	**
Chrysuphyllum	African star	*Udara*	C/W	F	*
Albidum	Apple				
Aningeria robusta	Wild star apple	*Udara ebok*	W	F	**
Maesobotiga barteri	Squirrel cherry	*Nyanyatet*	W	F	**
Persea Americana	Avocado pear	*Eben mbakara*	C	F	*
Dacryodes edulis	African pear	*Eben*	C	F	***
Dacryodes kleineana	Wild pear	*Eben ikot*	W	F	***
Hippocrates Africana	–	*Mba Manang-enang*	W	F	**
Colombia lutea	Cattle stick	*Ufip-ufip*	W	L/F	**
Dennett tripetala	Pepper fruit	*Nkarika*	C/W	S/F	**
Uvaria chamae	Wild pepper fruit	*Nkarika ikot*	W	S/F	**
Pentraclethra macrophylla	African oil bean	*Ukana*	W	S	*
Carica papaya	Pawpaw	*Udia edi*	C	F	*
Zea mays	Maize	*Abakpa*	C	F/S	***
Mangifera indica	Mango	–	C	F	**
Irvingia gabonensis	Bush mango	*Uyo*	W	F	*
Elaeis guineensis	Oil palm	*Top*	W/C	F/S	***
Raphia hookeri	Raphia palm	*Ukot*	W/C	F	***

Source: Jacob et al. (2015a)

more adaptability in this aspect. Period of pregnancy for *Cercopithecus sclateri* is roughly 24 weeks, with birth happening amid the period of December - February (Jacob 2012). Baby *Cercopithecus sclateri* weighs roughly 400 g at birth and cleave to its mother's ventrum. There is paucity of information on the species period of nursing. However, like most Guenons, it is assumed to be completed at 9 months after birth. Females of the species deliver their first baby around the age of 5 to 6 years (Oates et al. 1992).

There's constrained data accessible on the reproductive nature of *Cercopithecus sclateri* since the monkey is endemic and had been rediscovered recently. The primary perceptions of the primate species within its natural habitat happened around 1988. The deferred revelation could be attributed partly to the construct that the primate species occupy a limited zone of Nigeria that had long been reported by scholars and protectionists (Jacob 2012). Within that portion of Nigeria, human population is very dense and in need of ranges that animals could be considered or examined (Oates et al. 2004). For the most part, inside the class Cercopithecus, mating season relates with the time where nourishment is readily accessible. Thus, individuals of the species are ordinarily polygynous, and it is sensible to accept that *Cercopithecus sclateri* shows this characteristic too. The mating pattern of the members of the superspecies contrasts from other Cercopithecus species within the diminished significance of group containing one male member. Consequently, female members apparently form most of the group members. They also embark on a journey regularly without a single male member in the group (Jacob et al. 2015a).

Female freedom appears to be exceptionally imperative, as they protect their domains from other bunches. Male members of the *Cercopithecus cephus* group, inclusive of *Cercopithecus sclateris,* are opportunist during coition with females instead of guarding the group (Law and Myers 2004). The male species usually flag their females counterparts before copulation. They embark on course of action with their head weaving. This form of movement by the male *Cercopithecus sclateri* has been conceptualized to be a vital romance custom utilized to console females with whom a male need to mate. Furthermore, the head weaving pattern of the species has been purported to be responsible for the complicated facial patterns of *Cercopithecus sclateri* and similar species of the *Cercopithecus cephu*s family (Fleagle 1999).

In addition, there is scarcity of information on how parents cater for their young ones. Hence, the species will likely take after other *Cercopithecus* species. The baby guenon usually stays in the mother's ventrum, cleaving on her hide and twinning its tail with hers. Like every other Cercopithecus species, the care for the young is mostly given by the mother, who cares, carries, and groom them. The newborns are by and large subordinate all their care on their mothers. The little ones also stay by their mother's side for some period after feeding. It is not unprecedented for rank of moms to influence the leader stance for their descendant. The duty of male members in caring for their young ones within a group within the group has not been fully ascertained (Nowak 1999).

5.3 Habitat and Geographical Distribution

Individuals of *Cercopithecus sclateri* can essentially be found in the rainforest region like most other guenons species. Moreover, they also exist in the auxiliary forest areas more regularly compared to other cercopithecus species. Furthermore, its other related groups appear inclined to the bottom storey of the tree cover and occasionally descend to the forest floor. The species *C. sclateri* was first thought to be lost until a few scattered populations were discovered in Nigeria on the lesser route of the Niger River and Niger delta area (Jacob 2012). Out of the five populations already known, two of them are found near communities where they are considered sacred and protected. However, each group has a population of less than 250 people. The remaining three populations are in the swamp forest in Akwa

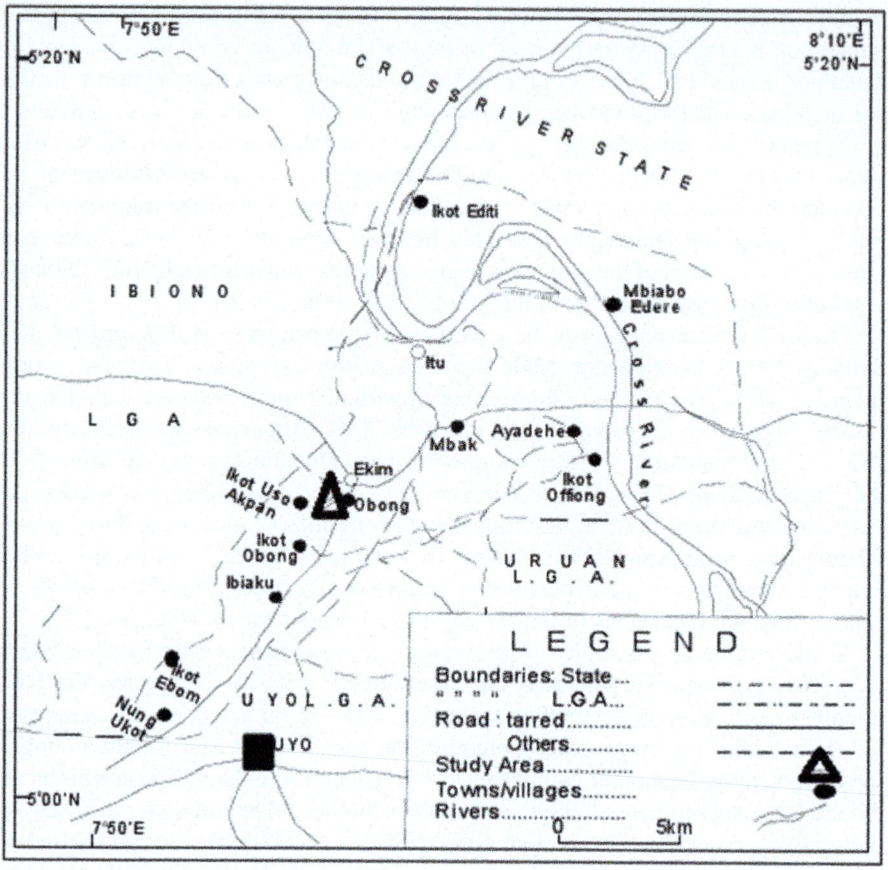

Fig. 5.1 Map of Ikot Uso Akpan, Obong Itam and neighboring villages. Source: Jacob et al. (2015a)

Ibom, and Anambra States, including the western coasts of Cross River swamp. Furthermore, a part of the species was discovered in some communities in Akwa Ibom State, with about sixty-two individuals.

Cercopithecus sclateri is confined to the forest region that lies between the rivers in the southern part of Nigeria (Fig. 5.1 and 5.2). Its degree of coverage is about 28,500 km^2. A considerable amount of the woodland all through its range covers little, regularly debased woodland areas inside a vast agricultural zone, swampy fields that are difficult to cultivate, or rows of woodland lining the waterways. *C. sclateri* groups have been reported to live within the states of Akwa Ibom, Enugu, Imo, Abia, and Cross River. The regions inside the listed states learned of *C. sclateri* presence include Utama, Stubbs Creek Forest Reserve (SCFR), Akpugoeze, Osomari, Lagwa, Blue stream, and Enyong creek/Ikpa stream (Baker 2005), including Obong Itam (Egwali et al., 2005). Furthermore, there are still three remnants' groups of Sclater's guenon reported to dwell in rural communities where

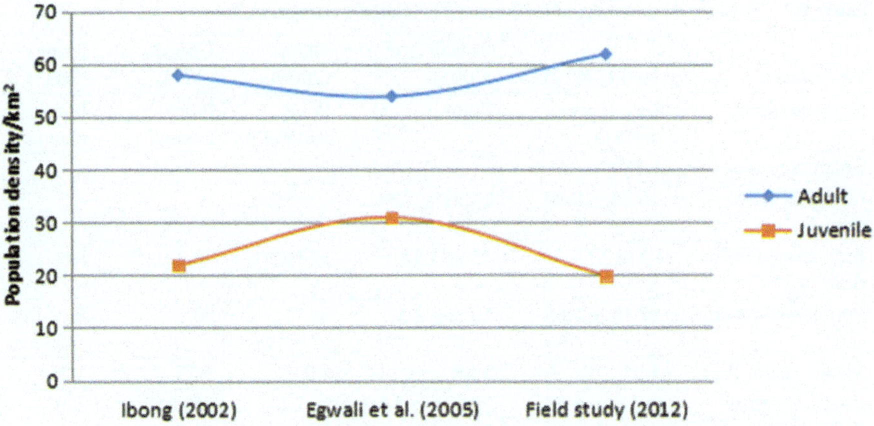

Fig. 5.2 *Cercopithecus sclateri* Population structure**.** Source: Jacob and Nelson (2015)

the people consider it sacrosanct. Even though they are not chased in these locales, the sacrosanct status of the monkey does not fundamentally ensure their survival. In 2005, about 124 individuals of *C. sclateri* moving in 15 troops were reported in Lagwa community of Imo State, with a population density averaging 15 individuals km^{-2} (Baker 2006). Similarly, Baker (2006) and Baker and Olubode (2008) reported about 193 individuals of C, sclateri moving in 20 groups in Akpugoeze, Enugu State, with a population density of about 35 individuals km^{-2}. Moreover, in Obong Itam community in Akwa Ibom Stata, Egwali et al. (2005) reported a population estimate of about 62 *C. sclateri* individuals in a group size ranging from 8 to 12.

5.4 Conservation Status and Abundance of *Cercopithecus sclateri* in Itam, Nigeria

The exploitation of woodlands and the adverse ecological consequence on the population of primates in Nigeria is usually taken with laxity. A more significant portion of the forest fragments accessible is not secured as they are found on private property and are utilized by the proprietors. Hence, these woodland layouts and contents are altered as the owners utilize them for cultivating, firewood collection, and lumber or allowed to regenerate for some time. These situations are not acknowledged because most of the research is mainly done in forestlands that have been protected (Tutin et al. 1997). Even though the hypothetical impacts of territory separation and fracture estimate on monkeys are fully documented (Hanski 1994), the impacts on *Cercopithecus sclateri* are not thoroughly examined. In a few studies that have been conducted, it is, for the most part, recognized that the ongoing

Table 5.2 Density, Abundance, and Biomass of Sclater's guenon in Itu

Parameters	Jacob (2012)	Egwali et al. (2005)	Okon (2004)	Udoedu (2004)	Ibong (2002)
Type of survey	Line transect method	Point method	Point method	Point method	Point method
Encounter rate	1.18	–	–	–	–
Group density (km^{-1})	16.86 ± 0.99	14.29 ± 2.86	–	–	–
Individual density (km^{-1})	82 ± 2.64	85 ± 3.57	82.06	70.32	80
Population density (D)	57.40 ± 1.85	59.5 ± 2.5	58	50	56
Mean weight (kg)	3.25	3.25	3.25	3.25	3.25
Biomass density (kg/km^2)	266.5 ± 8.58	276.25 ± 11.6	266.70	228.54	260
Population biomass (kg)	190.35 ± 9.81	193.38 ± 8.13	188.50	162.50	182

Source: Jacob et al. (2013b)

destruction of the species-environment poses a negative impact on the species (Egwali et al. 2005).

Cercopithecus sclateri is one of the foremost imperiled monkey species in Africa. The interplay of an excessively short range in a really crowded portion of Nigeria is leading this primate species to extinction. The region of Nigeria where *C. sclateri* is located possesses many people compared to the available land areas in Africa (Egwali et al. 2005). Moreover, a larger part of the available land has been transformed into farmlands and plantation. For Akwa Ibom State, the available populaces of *C. sclateri* are found in SCFR, Enyong Creek and Itu (Egwali et al. 2005; Jacob 2012; Jacob et al. 2013b; Jacob and Nelson 2015). For instance, a project aimed at preserving the forest and its wildlife diversity in the SCFR, Akwa Ibom State, in the 1990s was started by the Nigerian Conservation Foundation (NCF). Unfortunately, the project was abandoned due to a lack of funding from its funding agencies (Nowak 1999; Butynski 2002a, b). However, hope is not lost in the species' survival. It is related to shrines and sacrosanct forests in a few communities, as they are ensured due to penalties associated with hunting or consuming the primate species. These species are considered defenders of the sacrosanct locales in a few instances. Be that as it may, the more youthful era is considered relaxing or discontinuing a few of these rules to murdering these species. (Oates and Anadu 1989; Butynski 2002a, b).

A data set containing all survey data from 2002 to 2012 survey in Itu, Akwa Ibom State, Nigeria, was used to assess the population indices of *Cercopithecus sclateri* in the area (Table 5.2). Subsequently, the populace lists gotten in 2012 demonstrated an increment in *Cercopithecus sclateri* cluster/group density compared with the information acquired by Egwali et al. (2005). Be that as it may, the populace density of 82 individual/km^2, Total populace index of 57.40 individual/km^2, Biomass density of 266.5 kg/km^2, and Total populace biomass of 190.35 kg/km^2 of *Cercopithecus sclateri* gotten from the study were lesser than those reported by Egwali et al. (2005) and Okon (2004). However, the population indices were more significant than the indices reported by Essien (2008), Udoedu (2004), and Ibong (2002). The variation in cluster/group density, individual density, and population density of the species could be attributed to threats to their habitat due to its destruction over time as the species is considered a sacred animal and not hunted in the area.

5.5 *Cercopithecus sclateri* Population Structure

The population structure of *Cercopithecus sclateri*, as indicated in Fig. 5.2, over 10 years shows that the populace of matured *Cercopithecus sclateri* was more prominent than the adolescent populace over the study period of 2002–2012. However, within the study period, the matured populace diminished by 6.9% in 2005 compared to the 2002 survey information before increasing to 14.8% in 2012. In any case, within the same period (2002–2012), an increment of 27.27% in the populace of the adolescent *Cercopithecus sclateri* was recorded in 2005 compared to the 2002 population information. However, it later diminished in the 2012 population data by 35.48%.

The study's outcome bolsters the theory that female regenerative victory is subordinate to environmental quality and cluster estimate, suggesting that higher contest among the bigger groups is balanced by the quantity of nourishment accessible within the territory. In any case, a few other studies of fundamentally folivorous primates indicated that cluster estimate has no impact on their regenerative attainment (Steenbeek and van Schaik 2001; Robbins et al. 2007) even though such outcomes are not widespread (Borries et al., 2008; Snaith and Chapman 2008). Therefore, the grade of the *Cercopithecus sclateri* environment in Itam significantly impacts the species' childbearing, demise, movement, and resettlement rates (Fig. 5.3). Illicit lumbering exercises (Figs. 5.4 and 5.5) and the establishment of Funtumia elastica elastic plantation (Fig. 5.6) within the forest areas have negatively affected the *Cercopithecus sclateri* populace. Thus, the population of *Cercopithecus sclateri* had diminished by 3.53% in 2012 from the 2005 populace. Accordingly, an increment of 14.82% in the population of matured *Cercopithecus sclateri* was recorded, while the adolescent populace diminished by 35.48%. This result infers that logging within the species' habitat negatively impacted the regenerative capacity of *Cercopithecus sclateri*, causing a more matured populace of *Cercopithecus sclateri* than adolescence in the area.

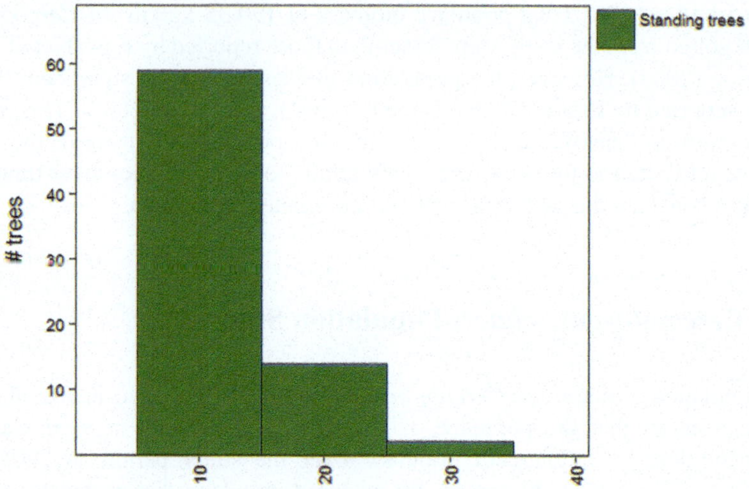

Fig. 5.3 Trees height distribution of forest fragments in Itam

Fig. 5.4 Timber logging in Ikot Uso Akpan sclater's guenon habitat

Fig. 5.5 Timber logging in Ikwat community forest sclater's guenon habitat

5.6 Threat to *Cercopithecus sclateri* Conservation in Ikot Uso Akpan, Nigeria

The worldwide dangers to biodiversity preservation are territorial destruction and overexploitation of biodiversity (Redford 1992). The risk associated with wildlife exploitation is very high within tropical woodland areas, where the efficiency of eatable wildlife is abysmal (Butynski 2002a, b). Only depending on bushmeat for protein prerequisites by the neighborhood communities without harmonizing efforts to control biodiversity abuse has caused the inflated issues of wildlife devastation with a plausibility of termination. Wild clusters of monkeys, among them *Cercopithecus sclateri*, use an exceptional portion of their time and vitality to feed. The movement to scavenge on regions with alluring nourishment makes the species very helpless to hunting (Ettah 2008; Jacob and Nelson 2015). The period spent nourishing can be equated to scattering and consistency of nourishment substances as vitality pick up / unit weight of nourishment (Fa 1988). These routines

Fig. 5.6 Rubber plantation in the community adjoining *Cercopithecus sclateri* habitat

mirror the social and natural factors they fight to outlive every day, regularly, and on a lifetime basis.

Man is the foremost imperative threat to *Cercopithecus sclateri*, and hunting is widespread over its home coverage (Cercopan.org 2011). Currently, *Cercopithecus sclateri* is not found in any government conserved area. However, there's trust for the long run because *Cercopithecus sclater*i has been able to hold on within the populated Southern region of the country, specifically due to its size, flexibility, enigmatic nature, and commonly being non-preferred as bushmeat (Baker and Olubode 2008). *Cercopithecus sclateri* are uncommon or truant over much of its assumed unique coverage area due to loss of territorial habitat. In localities where *Cercopithecus sclateri* is considered sacrosanct, the conventional convictions that bestow assurance on the species are dissolving, and the forest accessible to the monkeys is little and reducing (Jacob et al., 2015a, b).

Another danger to preserving *Cercopithecus sclateri* is environmental factors such as climate, nourishment accessibility, disease, and predation. These variables create macroevolutionary forms such as speciation, radiation, and termination; typically, given the broad awareness centered on components that impact the species cluster, e.g., healthy competition and infection (Nunn 2003). In principles, habitat-specific features vary among forestlands for some reasons, such as thermoregulatory variables related to height (Iwamoto and Dunbar 1983; Hill et al. 2000); locomotor variables related to contrasts in canopy form (Cannon and Leighton, 1994); or interspecific contest (Marshall et al. 2009). Variables such as intragroup scrambling for food and child murder are restraining cluster size (van Schaik and Janson 2000). In contrast, predation likelihood and intergroup contests are supposed to advance the usefulness of the congregation. As a result of the assumed significance of intragroup scrambling for nourishment, the health of the female species is anticipated to be lower in more prominent clusters (Borries et al. 2008). This forecast is founded on the implied suspicion that a few factors apart from nourishment availability limit cluster size, and females in bigger clusters encounter more serious nourishing competition. An elective theory is that wellness is compensated over clusters of specific measures inside a populace since females disperse themselves based on excellent free dispersion (Fretwell and Lucas Jr 1969). This point of view does not infer that healthy competition isn't vital; it implies that the impact of healthy competition fundamentally happens at the level of deciding cluster size.

Forest patches, the size of the patches, and the space between them usually pose a danger to preserving *Cercopithecus sclateri*. Indeed inside the same species, their reactions to forest patches vary (Lawes 2002). In any case, Chapman and Peres (2001) recommended that the survival of monkeys in forest patches may be decided by variables within the lattice that encompasses the forest patches. In addition to the dangers of termination related with few populaces such as *Cercopithecus sclateri*, the species populace is confronted with the peril of continued environmental destruction. The species' territory is regularly encompassed by human colonies. The bulk of the individuals are dependent on wood as fuel for cooking and construction shafts. The forest patches occupied by the species are not spared from misuse, which has caused a pernicious impact on the species' populaces.

The form of the examined forest patches denotes that the most diminutive tree height was 5.9 m, whereas the tallest tree height was 27.8 m, with a mean of 11.8 m. This demonstrates a predominance of moderately small trees in the species' habitat. Figure 5.3 underpins this exceptionally well, showing that the majority (90%) of all the examined tree heights averaged around 20 m. Compared with tropical rainforest zones like Cross River National Park, where the mean nonstop canopy forms vary from 30 to 35 m (WWF 1990), the trees of the species' habitat can be regarded as negligible. Also, dbh estimates produce an intriguing image where the least dbh is 10.2 cm and the biggest at 110.6 cm (Fig. 5.7). Despite the intermittent event of trees with colossal dbh, trees with small dbh overwhelm the species range. The relatively low mean dbh (26.9 cm) with 90% of the trees indicates the dbh of all trees is lower than 44 cm in the species range. The scenario can be credited to the significant misuse of forest assets. Also, many tall trees in Ikwat 1 can be credited to their nearness to the sacrosanct hill within the forest patches, swampy forest floor due to the sacrosanct stream that crisscrosses the forest and purges itself in the shape of a water fall. The Okuku forest patches are located on upland and are less considered sacrosanct. Hence, it is cultivated and logged than other forest patches.

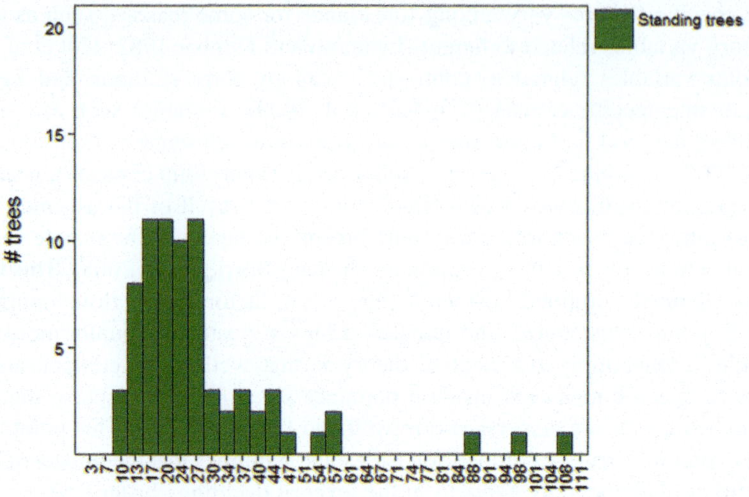

Fig. 5.7 Trees' diameter distribution of forest fragments in Itam

5.6.1 Drivers of Vegetation Changes

Cercopithecus sclateri encounters the dangers of acclimating to the elements of their environments, which are persistently transforming. They must adjust to modifications to survive, as the inability to adjust may lead to species extinction. Therefore, since most of the monkey species occur in tropical forests (Chapman et al. 2006; Lovett and Marshall 2006), the security of the forest environments ought to be a primary goal for their preservation. Be that as it may, preserving the forest patches isn't a simple undertaking for some reasons. To begin with, the forest territories are generally divided and strewed in numerous areas. Consequently, NGOs must work with the local communities to guarantee the preservation of the monkey species.

In addition, the woodland parts are found in financially underprivileged communities. Furthermore, the populace development index within the zone, especially within the Niger Delta region, is very high. Majority of the individuals rely specifically on common assets including land for farming; subsequently, they must remove the woodlands to farm (Baker 2005). This is against the concept of protecting the woodland (Chapman et al. 2006). In any case, not all the environmental changes within the region are due to anthropogenic exercises. Hence, the ecological alteration can be said to be natural and anthropogenic-induced. The biological changes incorporate changes like a wind toss of a critical nourishment plant, tree shrivel due to age group ageing, and vegetation transformation resulting from avalanches, tropical storms, and dead caused by scarcity of water. All these modifications influence *Cercopithecus sclateri* populaces adversely. Be that as it may, a few natural modifications such as forest colonization have been detailed to facilitate the populaces of a few species (Isabirye-Basuta and Lwanga 2008). In the next category, changes caused by man incorporate variables such as woodland destruction by automatic logging, forest reduction, and the introduction of non-native species.

5.6.2 *Causes of* Cercopithecus sclateri *Population Changes*

The primary cause of environmental changes in the community forest incorporates mechanized logging, woodland reduction, the introduction of non-native species, and unsustainable exploitation. Scholars have performed numerous studies to look at primates' reactions to forest destruction in tropical timberlands (Johns and Skorupa 1987; Plumptre 1996; Plumptre and Reynolds 1994; Struhsaker 1997; Mitani et al. 2000; Chapman and Lambert 2000; Chapman et al. 2003). In any case, numerous only provided negating discoveries because of contrasts in nourishing communities of the species concerned, specialized strategies utilized, unique species compositions and population, logging assertiveness, and coincidental harm to forest trees, sorts of vegetation adjoining exploited zones, age of the forest, tree composition and population prior to logging, and characteristic variety in territory sorts inside the woodland and population of large herbivores. The impacts of these variables come from studies that endeavor to screen monkeys' reactions to forest exploitation and explain why deductions from several pieces of research have been diverse indeed when the same species are evaluated.

In a perfect world, understanding how *Cercopithecus sclateri* populaces react to forest destruction needs a prelogging assessment to supply standard information against which one can estimate any alteration in their populaces after the forest destruction. Unfortunately, this has been incomprehensible for all intents and purposes since the unlawful loggers are not ready to educate the people, scholars, or conservation organizations about their activities. Moreover, gaining an excellent knowledge of the species populace in a forest to be degraded needs a more extended period of perception, which is contradictory to continuous logging routines within the range. Thus, the best viable means of evaluating the reactions of *Cercopithecus sclateri* populaces to territorial modifications induced by unlawful logging has been through comparisons of their populace over a long period earlier to the logging exercises. This may be sufficient, accepting that sometime prior to the logging activities, the degraded and undegraded forests were comparable in *Cercopithecus sclateri* populace and territorial forms. But be that as it may, this is not usually the case, thus accounting for the conflicting outcomes within the reactions of other monkey species to forest destruction. As shown in Figs. 5.4, 5.5, 5.7, unlawful forest exploitation activities have adversely affected the populace of *Cercopithecus sclateri*. As a result, there is a diminished populace of the species by 3.53% of individuals/km^2 in 2012 compared to the 2005 populace information. Subsequently, there was an increment in the adult population by 14.82%, while the adolescent populace diminished by 35.48%. This outcome suggests that forest destruction within the species ranges impacted the regenerative capability of the species.

5.6.3 *Responses of* Cercopithecus sclateri *to Forest Destruction*

While assessing reactions of *Cercopithecus sclateri* populaces to activities leading to the destruction of their habitat, it is vital to evaluate the conceivable impacts of the type of environment that encompasses their favored territories. *Cercopithecus sclateri,* like any other primates species, react to modifications in their neighboring territory, which may complicate the reactions caused by the destruction of the forest, particularly in the event that logging activities and the species' journey to or from the adjoining forest types overlap, or major floral modifications such as the colonization of the forest happen in ranges adjoining to the degraded region of the woods after the exploitation activities (Chapman and Lambert 2000).

Granted that decays within the relative abundance of a few primate species also happened within the undegraded forest, it is unsettling to trail the decrease within the populace of *Cercopithecus sclateri* to forest exploitation exercises within its range. The reduction in *Cercopithecus sclateri* in the forest patches may also be attributed to environmental discharge. As more living space is made by woodland colonization, a few species clusters might migrate out of the undisturbed forest or grow their domestic ranges to cover the area. Such transformations are cumbersome to identify from settled census courses. In this manner, future examinations of the impacts of the forest exploitation activities on the *Cercopithecus sclateri* populace at the community woodland ought to consider environmental expansion since studies have demonstrated that this type of ecological transformation can influence a few species of primate. Similarly, in a study conducted by Baker (2005) and Egwali et al. (2005) in southern parts of Nigeria and at Ikot Uso Akpan town, to be precise, the authors illustrated that *Cercopithecus sclateri* is exceptionally adaptable regarding plant species and the parts eaten for nourishment. Maybe this adaptability in dietary necessities enables them to occupy colonizing territories.

5.6.4 *Effect of Vegetation Changes on* Cercopithecus sclateri *Diets*

The decline in nourishment accessibility is one of the significant effects of forest destruction on the *Cercopithecus sclateri* populace. Hence, in circumstances where exploitation activities can invigorate or are taken after by recovery of nourishment species, the effect of logging would be negligible. However, the contrasts in nourishment tree species evacuated from the forest area may not be solely responsible for the contrasts in the *Cercopithecus sclateri* populace in the range. Since the recovery of several of the nourishment tree species often happened after logging, comparative studies in Kibale emphatically propose that browsers within the degraded ranges may smother the recovery of the plant species subsequently influencing the nourishment accessibility for *Cercopithecus sclateri* (Kasenene

1987; Nummelin 1990; Struhsaker et al. 1996; Struhsaker 1997; Paul et al. 2004; Lawes and Chapman 2006). The smothered recovery of trees may also be attributed to *Cercopithecus sclateri* populaces' moderate recuperation. Until the present rate of woodland transformation is stopped, it is inescapable that the *Cercopithecus sclateri* populace will dwell in littler confined parts until it is eventually devastated.

Contrasts in nutritional requirements among primate species may also be responsible for the need for reliability in outcomes from studies endeavoring to explore reactions of the primate populace to forest destruction. For illustration, species that are dietary specifics and generalists may not easily be influenced by forest destruction in an exact manner. Essentially, *Cercopithecus sclateri*, a frugivorous species, will react unexpectedly to forest destruction more than other folivorous species (Ettah 2008; Egwali et al. 2005). Contrasts in nourishing scores between primate groups possessing degraded and undegraded woodland result from discrepancies in nourishment accessibility regarding tree composition, quantity, and phenology (Fairgrieve and Muhumuza 2003). The study outcomes also support Johns's (1992) assertion that proposes that forest destruction adversely affects the accessibility of nourishment within the territory for *Cercopithecus sclateri*. A reduction within the growth paces and populace density of *Cercopithecus sclateri* over a period due to forest destruction activities within its range indicates that the forest destruction adversely influenced the species.

5.6.5 Effect of Forest Exploitation on Cercopithecus sclateri

The variables that permit *Cercopithecus sclateri* species to hold on in forest patches are ineffectively comprehended, thus making it cumbersome to plan measures to moderate them in a fragmented environment. Variables that scholars accept decide sustenance or termination in patches of forest, nutritional adaptation or specialized feeding, number, and area of forest patch, and how far apart is the forest patches, even though it does not appear to clarify all challenges. Indeed, among the *Cercopithecus sclateri* species, their reactions to forest destruction vary (Lawes 2002). For instance, it is recommended that the sustenance of *Cercopithecus sclateri* in forest patches could be influenced by certain variable that lies within the ecotone that encompasses the patches (Chapman and Peres 2001). Be that as it may, some of these variables are still obscured. Hence, scholars must evaluate them to ascertain if *Cercopithecus sclateri* populations that exist in forest patches are of serious concern to conservation.

Furthermore, the dangers of termination related to little populace such as *Cercopithecus sclateri* are that the species encounters the peril of persistent environmental deterioration. The environment of the species is ordinarily encompassed by human settlements. Most of the individuals depended on fuelwood for cooking and building shafts. The part of the forest where *Cercopithecus sclateri* thrives is not spared from abuse for such assets, which has a pernicious impact on the species inhabiting there. For illustration, studies on primates occupying timberland parts close to Kibale National Park uncovered that three of the 16 fragments in which primates species were inhabiting in 1995 were abused to the point that they were gone in 2000 (Chapman et al. 2003; Chapman et al. 2006).

Similarly, Gillespie and Chapman (2006), in their study, reported that the record of forest fragment debasement and the proximity of people emphatically impacted the predominance of parasitic gastrointestinal nematodes in red colobus monkey, which proposes that as forest parches ended up littler, and recurrence of connection between people and nonhuman primates gets to be higher, transmission of maladies and pathogens will be an awfully likely result (Vogel 2003; Wolfe et al. 2004; Peeters 2004; Sharp et al. 2004; Leroy et al. 2004, Rouquet et al. 2005). The hazard of illness transmission between people and nonhuman primates is, as it were, higher for the extraordinary primates since the last mentioned are phylogenetically closer to people than other primates (Ukpong et al., 2013). Nevertheless, infection transmission can happen in both ways, even though nonhuman primates have a greater danger than people. Infections in people can be identified quicker and controlled or disposed of, though, in wild nonhuman primates, it might be an inconceivable assignment (Isabirye-Basuta and Lwanga 2008). Additionally, as *Cercopithecus sclateri* thrives in little populace, a single disease episode can annihilate the whole cluster. In any case, current studies recommend that we ought to be cautious in tolerating the previously mentioned situation (Chapman et al. 2006).

Forest resources withdrawal is among the anticipated outcomes of global warming (Isabirye-Basuta and Lwanga 2008). Hence, if global warming really happens, the dangers of termination for *Cercopithecus sclateri* could be more. One conceivable clarification for particular perseverance or at slightest the deferred termination of *Cercopithecus sclateri* within the region is environmental adaptability in territory use, demeanor, or nutrient requirement (Baker, 2005; Egwali et al. 2005). *Cercopithecus sclateri* can persist in a few distinctive range types, thereby affording them the opportunity to outlive in disturbed forest areas. As tree-living species which are satisfied traversing the forest floor, the species can dodge termination through the use of branches in the forest fragments as habitat, given that much separations from patch to patch are small.

Moreover, forest degradation does not essentially yield forest areas bereft of vegetation; preferably, they are supplanted with farmlands and plantations (Fig. 5.5). *Cercopithecus sclateri* are competent in adding new plant species to their diets. Subsequently, they get a few of the nutritional needs in ranges more extensive from the fragments the species inhabit. Farm produce are usually chosen for their quality wholesome index, including their susceptibility to draw *Cercopithecus sclateri* to the area (Egwali et al. 2005). The circumstance comes about in the human-primate struggle with the monkey species due to attacks on the community farmlands to raid crops amid the planting season.

5.6.6 Effect of Climate Alteration on Cercopithecus sclateri

Although the effect of climatic change isn't clearly comprehended, especially with respect to forest arrangement and inconsistency, there's proof to propose that it is as of now influencing the species' environments and consequently their populaces (Chapman et al. 2005a, b). Scholars have anticipated global warming would be responsible for the diminished precipitation, including an extended drier period in a few tropical regions (Borchert 1998). Moreover, the interim period between El Niño occasions that are complicatedly connected with worldwide warming has been diminished (Laurence and Williamson 2001). Within the tropical region, the El Niño occasions are related to heightened cases of dieback among trees (Holmgren et al. 2001; Laurence and Williamson 2001). Hence, climatic alterations can have serious impacts on the nourishment asset base of *Cercopithecus sclateri*. On the off chance that the scenarios at last materialize, we should then anticipate diminished forest cover and expanded forest sketchiness within the forest patches that serve as living space for *Cercopithecus sclateri.*

Additionally, climate change is acknowledged to exasperate the antagonistic effects of human-initiated modifications to the forest environments, hurting the primate species further. The forest patches exploited are more defenseless to fires flare-up amid the dry seasons than the intaglio forests parts (Laurence and Williamson, 2001). This may be genuine since forest exploitation produces a quilted level of flammable fabric on the surface of the forest. Summarily, unsustainable exploitation covers a vast region, thereby suggesting that peradventure the fire starts, the heat that will be produced will slaughter numerous *Cercopithecus sclateri*. Indeed in case they elude the flame, they will likely be exterminated due to starvation since the forest is not fired safe; consequently, it will lead to a longer period of scarcity of accessible nutritional requirement. Coupled with the raised temperatures and drier periods, the number of dieback among the forest trees is anticipated to expand (Laurence and Williamson 2001).

Over 10 years, the community forest lost between 3.53 and 38.55% of its *Cercopithecus sclateri* populace as an offshoot of the specified issues. The available populace is growing towards a maturing populace regarding the region's diminished number of adolescent monkeys. These indices cover expansive environment and are cumbersome to assume the genuine cause of the decrease in population. Whereas this would as of now be more than disturbing, it is much more likely that the real circumstance is more awful. Therefore, it is of most extreme significance to require fitting activity, suggesting bypassing a progressive decrease of *Cercopithecus sclateri* species.

5.7 Conservation Measures for Sustaining Sclater's Guenon Population in Itu

5.7.1 Landuse/Cover Analysis of Cercopithecus sclateri *Habitat*

One of the fundamental factors vital to deciding the long-term possibilities of the *Cercopithecus sclateri* populace is the appropriate environment available in the community. This information, along with data on the history and existing rates of forest destruction as satisfactorily as other dangers, is essential to accurately survey the estimate and prospective patterns of the endemic *Cercopithecus sclateri* populace. Whereas the common data around the inexact degree of the different range of vegetation/forest sorts do exist, this information appears restricted, and its precision may be flawed. Nevertheless, the principal modern floral classification can be given by using Landsat 7 ETM+ satellite imagery. As the level of appraisal for this classification will be the complete area covering the entire Itu area instead of the much littler parts of the Ikot Uso Akpan community, the spatial resolution of the significant regions may confirm to be the dependable for the assurance and separation of *Cercopithecus sclateri* territory of the area. In any case, Landsat 7 satellite images possess the prospect to be utilized to produce category maps with a reasonable resolution. In this manner, relevant NGOs (e.g., Nigerian Conservation Foundation (NCF), and Biodiversity Conservation Center (BPC), etc.) ought to encourage the State Ministry of Lands and Geoinformatics to undertake and create a unique vegetation classification outline for the State or attempt to attain collective participation of all stakeholders which would permit these NGOs to utilize the Landsat information and work on such a classification in conjunction with the Ministry or individually. However, on the off-chance that all endeavors to attain this kind of cooperation stay worthless, the choice of freely procuring modern satellite imageries and endeavoring an independent classification ought to be considered. In any case, if working autonomously from each other would be quicker, less complicated, and accomplish more specific outcomes, it would moreover be an exorbitant alternative for the nongovernmental organizations since the imageries essential for these sorts of examinations are incredibly costly.

5.7.2 Coordinated Survey of Cercopithecus sclateri *Population*

Based on the earlier proposed reevaluation of the *Cercopithecus sclateri* environment, line-transect studies should be conducted in each forest patch to determine how each unique environment contributes to the entire *Cercopithecus sclateri* populace. In addition, future evaluation should be expanded to cover patches of auxiliary forests adjoining the community. When combined with available data on the degree of unique environment types, excellent gauges of the size of the *Cercopithecus sclateri* populace would be conceivable. Besides, instead of conducting such overviews in a particular area as it were once, it ought to be rehashed routinely (in interims not surpassing 5 years) since this would permit the assurance of individual populace patterns and indicate if the circumstance within the zone is moving forward or falling apart. Moreover, instead of depending on scholars who visit the area on a sporadic premise to conduct such investigations, individuals in the community ought to be instructed on relevant survey strategies that would enable them to conduct periodic assessments of any place within the forest patch and possibly in all other parts of the locality.

5.7.3 Provision of Funding for Infrastructure and Law Enforcement

As a result of its size, challenging landscape, and community taboo set on *Cercopithecus sclateri*, the community serves as a vital asylum for the species. Whereas an indigenous sanctuary was created in the community, the facilities and protection within the area are abysmal and dependent on the community. Indeed, although authoritatively forest destruction of the *Cercopithecus sclateri* habitat is banned, exploitation is still happening in the forest. Moreover, due to the increased rate of joblessness and poor livelihood standard of the people within the range, as well as their poor conservation awareness, the enforcement of guidelines within the forest patches remains frail. The Wildlife asylum that had been redundant and out of date needs financing to leave up to its mandate and to employ security for the improvement and execution of a successful program.

5.7.4 Formalizing the Protection of the Forest

The outcomes of current studies within the primate species habitat portrayed that the region is home to a sizeable populace of *Cercopithecus sclateri*, making it an awfully imperative zone for preservation. Yet, destruction and resource exploitation within the area are still unabated tragically. Thus, it is a pitiful scenario where forest exploitation persists. The forest patches will not stay untouched without any course of action or understanding between the community and the government agencies within the state. It is troublesome to predict at what degree this understanding will secure the long-term sustenance of *Cercopithecus sclateri* species. Indeed, arrangements ought to begin with the government to achieve a few legitimate assurance statuses for the community forest.

5.7.5 Connecting Forest Patches Through a Corridor Framework

Another effective approach that would back the survival of a reasonable populace of *Cercopithecus sclateri* in the area would be the creation of a natural life passage interfacing patches of the forest together in parts of Itu and the adjoining communities in the State. Such a natural corridor would offer secure connections for dispersing the monkeys that occupy the area which is threatened progressively through the isolation of forest patches because of unsustainable exploitation and territorial modification within the area. Thus, by encouraging the movement of *Cercopithecus sclateri* between the areas, the passage would serve to preserve the hereditary differing qualities inside the *Cercopithecus sclateri* populace. While this passage may not possess the sound quality of a shielded area, deforestation and territorial transformation would have to be outlawed and anthropogenic impact minimized. All endeavors pushing for such a corridor ought to be done by getting the underwriting of the State's agencies in the environmental sector. In addition, it is a vital preservation phase, and therefore, all actions aimed at achieving it should be facilitated.

5.7.6 Enhancing Education and Development Opportunities

Based on the positive effect of the outcomes, the lead actors within the development segment, such as the UNDP, UNESCO, World Bank, and numerous other organizations, advance instruction as a critical instrument for sustainable human and natural advancement (UNDP, 2004). Even though Nigeria belongs to the category of developing nations, schools were not found in the adjoining communities of the forest patches; hence, the standard of instruction is, for most people, low. The imperative construct of supportability is primarily unused for the advantage of the local people. The individuals have survived by deliberately arranging their lives or utilizing typical assets. However, their conventional local norms framework inadvertently coincidentally backed the maintainable use of assets to a few degree and assurance of *Cercopithecus sclateri*, the impact of the Christianity. Also, the developing impact of money economy has diminished the number of individuals which still follow the standard conviction of the community to protect the monkey. Thus, nearby assets are waning, and expanding numbers of exploitation have a severe effect on wildlife. Since there is no sufficient information accessible that would permit the evaluation of possible hunting levels within the range, a suitable instructive campaign is, therefore, essential to guarantee the sustenance of *Cercopithecus sclateri*, as well as the long-term victory of preservation of the forest patches. Also, there must be a little help in creating economic and financial models for the nearby individuals to secure their vocations. Orienting the community on the notion of sustenance would probably diminish the likelihood that they are attracted by the captivating short-term benefits to offer off their lands and leave the possibility to live in an intaglio environment that keeps on giving the assets they require.

5.7.7 Develop the Ecotourism Potential of the Area

With the discretion of accomplishing preservation and tourism destinations gotten simultaneously, ecotourism has become increasingly prevalent in later times and various fruitful ventures have been set up, overseeing coordinated vital parameters like financial victory, social-human advancement, and natural supportability. Ikot Uso Akpan community encompasses a vast prospect to gain from ecotourism rambles. It gives compelling motivations such as; moderately undisturbed patches of tropical rainforests, high biodiversity, inaccessible endemic zones, and indigenous people with a fascinating culture. The fruitful foundation of ecotourism ventures within the region would demand discovering suitable locations and accomplices, building the essential infrastructure, giving a holding program (itinerary/activities, etc.), and handling the publicity (notice). While innate people should continuously be the recipients, individual ventures have the potential to create more monies to support themselves similarly. The community can easily be accessible, giving a fabulous setting for ventures of this kind. Tour administrators may, on occasion, incorporate one- or two-day visits to the primate species' habitat into their programs. Sightseers would esteem the extraordinary chance to be walking and playing with *Cercopithecus sclateri*. Their payments for the tours would be shared among the various visit operators/communities and the scientific project. Such a technique would have the extra advantage that primates' research and preservation work would get progressively available to the interested person, in this way possibly driving to expanding exposure.

and more bolster.

5.8 Conlusion

The community woodland in Itu is of noticeable vital *Cercopithecus sclateri* territory for the endemic monkey species. Clusters of the species utilize this biological system, and a few of these individuals have their whole domestic coverage exterior to the boundary of the woodland patches. Whereas future evaluation ought to be carried out to affirm these outcomes, the discoveries of this current examination do not create room for any question approximating the significance of the forest patch as an environment for the preservation of *Cercopithecus sclateri*. Therefore, any exertion to effectively oversee this primate species populace should view the nearby woodland as an ecotone for the biological system in their governance techniques and schedules.

Presently, the community forest remains the only place in the state where *Cercopithecus sclateri* exists, with pockets of forest patches that function as an imperative asylum for the primate species. Also, the Cercopithecus *sclateri* species is considered sacrosanct, and the community's killing of the species is disallowed. Besides, it is overwhelmingly so since all the issues confronting Nigeria's natural life have not gone wholly neglected within the range. There have been individuals and establishments that have dedicated a part of their resources to guarantee the sustenance of the species within the region. To persist and support the initiatives, additional actions had been examined earlier in the subheads, which ought to be adopted to guarantee the long-term survival of the vegetation, animal, and particularly *Cercopithecus sclateri* species within the zone.

Unpredictable resource exploitation and agricultural transformation of the forest environment and the endemic *Cercopithecus sclateri* environment have been extreme. Although the region is still secured with little lots of better or slightly intaglio rainforest species, forest destruction has been in process within the range for a long time. Also, the cutting down of trees which were supposed to be abandoned within the neighborhood forest is still ongoing. Plates three and four make it apparent that the forest patches are being annihilated unabated into the species territory from the community, destroying even more of the environment that is fundamental for *Cercopithecus sclateri* and all other nearby natural life species. Thus, forest destruction results from logging and misuse of other forest resources and the establishment of oil palm and rubber plantations in the area. Luckily, the manors appear to have been stopped. However, apart from these activities, there is a heightened demand for affordable farmland for cultivating crops, as witnessed in the clearance of the forest patches.

Apart from the wrecking of the imperative natural living environment, trees' felling and modifying the forest patches for other intents have several other harmful effects on the forest area and *Cercopithecus sclateri* populace outstandingly by enhancing accessibility to inaccessible forest regions through felling routes. The provision of access to zones that were once cumbersome has improved the hunting pressure in the area. Moreover, the substitution of the conventional bows and arrows with advanced fire powers such as rifles has compounded the woos of wildlife exploitation. As a result, hunting is more productive nowadays. Since the customary laws utilized to control the exploitation of certain animals are no longer as effective as before, the hunt for *Cercopithecus sclateri* in the area may momentarily evolve in propensity.

References

Archad F, Eva HD, Stibig HJ, Mayaux P, Gallego J, Richards T, Mailingreau JP (2002) Determination of deforestation rates of the world's humid tropical forest. Science 297:999–1002

Baker LR (2005) Distribution and conservation status of the Sclater's guenon (Cercopithecus sclateri) in southern Nigeria. Report to Margot Marsh Biodiversity Foundation, Rufford Small

Grants, Lincoln Park Zoo Department of Conservation and Science, American Society of Primatologists, Sigma Xi, National Science Foundation, and Stanford Bay Area Charities. 23 pp

Baker LR (2006) The Nigerian guenon in southern Nigeria: outlook good or just hanging on? Unpublished Report to CENSHARE, Minneapolis, MN. 21 pp

Baker LR, Olubode SO (2008) Correlates with the distribution and abundance of endangered Sclater's monkeys (*Cercopithecus sclateri*) in southern Nigeria. Afr J Ecol 46(3):365–373

Bennett EL (2000) Is there a link between wild meat and food security? Conserv Biol 16:590–592

Boinski S, Treves A, Chapman CA (2000) A critical evaluation of the influence of predators on primates: effects on group travel. In: Boinski S, Garber PA (eds) On the move: how and why animals travel in groups. University of Chicago Press, Chicago, pp 43–72

Borchert L (1998) Responses tropical trees to rainfall seasonality and long-term changes. Clim Chang 39:381–393

Borries C, Larney E, Lu A, Ossi K, Koenig A (2008) Costs of group size: lower developmental and reproductive rates in larger groups of leaf monkeys. Behav Ecol 19:1186–1191

Butynski T (2002a) Conservation of the guenons: an overview of status, threats, and recommendations. In: Glenn C (ed) The guenons: diversity and adaptation in African monkeys. Kluwer Academic, New York, pp 411–415

Butynski T (2002b) The guenons: an overview of diversity and taxonomy. In: Glenn C (ed) The guenons: diversity and adaptation in African monkeys. Kluwer Academic, New York, pp 3–13

Cannon CH, Leighton M (1994) Comparative locomotor ecology of gibbons and macaques: selection of canopy elements for crossing gaps. Am J Phys Anthropol 93:505–524

Cercopan.org (2011) More about Sctater's guenon. http://www.cercopan.org /Primates /Guenons. Accessed 22 Sep 2011

Chapman CA, Lambert JE (2000) Habitat alteration and the conservation of African primates: a case study of Kibale National Park, Uganda. Am J Primatol 50:169–185

Chapman CA, Peres CA (2001) Primate conservation in the new millennium: the role of scientists. Evol Anthropol 10:16–33

Chapman CA, Chapman LJ, Vulinec K, Zanne A, Lawes MJ (2003) Fragmentation and alteration of seed dispersal processes: an initial evaluation of dung beetles, seed fate, and seedling diversity. Biotropica 35:382–393

Chapman CA, Chapman LJ, Struhsaker TT, Zanne AE, Clark CJ, Poulsen JR (2005a) A long-term evaluation of fruit phenology: importance of climate change. J Trop Ecol 21:35–45

Chapman CA, Chapman LJ, Zanne AE, Poulsen JR, Clark CJ (2005b) A 12-year phenological record of fruiting: implications for frugivore populations and indicators of climate change. In: Dew JL, Boubli JP (eds) Tropical fruits and frugivores. Springer, pp 75–92

Chapman CA, Lawes MJ, Eeley HAC (2006) What Hope for African primate diversity? Afr J Ecol 44:116–133

Cowlishaw G (1999) Ecological and social determinants of spacing behavior in desert baboon groups. Behav Ecol Sociobiol 45:67–77

Cowlishaw G, Dunbar RIM (2000) Primate Conservation biology. University of Chicago Press, Chicago. 134 pp

Donohoe M (2003) Causes and health consequences of environmental degradation and social injustice. Soc Sci Med 56(3):573–587

Egwali EC, King RP, Eniang EA, Obot EA (2005) Discovery of new population of Sclater's guenon (*Cercopithecus sclateri*) in the Niger Delta wetland. Nigeria Liv Sys Sus Dev 2(4):1–7

Eniang EA (2001). Effect of habitat fragmentation on the Cross River gorilla (*Gorrilla gorilla dehli*): recommendations for conservation. Unpublished report submitted to the Cross River National Park, Akamkpa, Nigeria. 30 pp

Eniang EA and Ebin CO (2002) Utilization of confiscated animals by Cross River National Park to promote *in-situ* biodiversity conservation in the rainforest of Southeastern Nigeria. In: The proceedings of pan African Association of Zoological Gardens, aquaria and botanical gardens (PAAZAB) annual conference, 28th-31st may, 2002. Johannesburg, South Africa. 17 pp

Essien IS (2008) Conservation strategies for an endangered primate species (*Cercopithecus sclateri*) in a fragmented tropical rainforest at Itu, Akwa Ibom State. Undergraduate Project submitted to the University of Uyo, Uyo. 47 pp

Ettah US (2008) Conservation strategies of Cross River gorilla (*Gorilla gorilla diehli*) in the Afi Mountain wildlife sanctuary of Cross River state, Nigeria. An Unpublished M.Sc. Thesis submitted to the University of Uyo, Uyo. 72 pp

Fa JE (1988) Time budget of Rhesus Monkeys (Macaca milatta) in a Forest Habitat in Nepal and on Cayo Santiago. In: Southwick C (ed) Ecology and behavior of food-enhanced primate group. Alan R. Liss, New York, pp 32–39

Fairgrieve C, Muhumuza G (2003) Feeding ecology and dietary differences between blue monkey (*Cercopithecus mitis stuhlmanii Matschie*) groups in logged and unlogged Forest, Budongo Forest reserve, Uganda. Afr J Ecol 41:141–149

Fleagle JG (1999) Primate adaptation and evolution. Academic Press, San Diego. 232 pp

Fretwell SD, Lucas HL Jr (1969) On territorial behavior and other factors influencing habitat distribution in birds. Acta Biotheor 19:16–36

Gadsby EL (1987) Preliminary report on the survey of drills in southern Nigeria. Unpublished Report. 13 pp

Gillespie TR, Chapman CA (2006) Prediction of parasite infection dynamics in primate metapopulations based on attributes of Forest fragmentation. Conserv Biol 20:441–448

Groves C (2000) The phylogeny of the Cercopithecoidea. In: Whitehead P, Jolly C (eds) Old World monkeys. Cambridge University Press, London, pp 92–95

Grubb P, Oates JF, White JT, Tooze Z (2000) Monkeys recently added to the Nigerian Fauna list. Nigerian Field 654:149–158

Hanski I (1994) A practical model of metapopulation dynamics. J Anim Ecol 63:151–162

Hill W (1953) Primates: comparative anatomy and taxonomy VI, Catarrhini, Cercopithecoidea. Interscience, New York. 123 pp

Hill RA, Lycett JE, Dunbar RIM (2000) Ecological and social determinants of birth intervals in baboons. Behav Ecol 11:560–564

Holmgren M, Scheffer M, Ezcurra E, Gutierrez JR, Mohrem GMJ (2001) El Nino effects on the dynamics of terrestrial ecosystems. Trends Ecol Evol 16:89–94

Ibong BU (2002) Ecology and conservation of the primate Sclater's guenon (*Cercopithecus sclateri*) in Itu local government area of Akwa Ibom state: population structure. Undergraduate Project submitted to the University of Uyo, Uyo. 48 pp

Isabirye-Basuta GM, Lwanga JS (2008) Primate populations and their interactions with changing habitats. Int J Primatol 29:35–48

IUCN (2011) IUCN Red List of Threatened Species. Version 2011.1. https://www.iucnredlist.org. Downloaded on 21 Sep 2011

Iwamoto T, Dunbar RIM (1983) Thermoregulation, habitat quality and the Behavioral ecology on gelada baboons. J Anim Ecol 52:357–366

Jacob DE (2012). Population and habitat assessment of Sclater's guenon (Cercopithecus sclateri, Pocock 1904), In Ikot Uso Akpan community Forest, Itam, Akwa Ibom state, Nigeria. An Unpublished M.Sc. Thesis submitted to the University of Uyo, Uyo, pp. 168

Jacob DE (2016) Population and Habitat Assessment of Sclater's Guenon in Itam, Nigeria. Lamber Academic Publishing (LAP), Germany. 168 pp. ISBN 9783659919602

Jacob DE (2017) Resource governance structure in selected Nigeria National Parks and impact on rural livelihood. Ph.D thesis submitted to Federal University of Agriculture, Abeokuta, 2546 pp

Jacob DE, Nelson I (2015) A survey of Sclater's guenon in Ikot Uso Akpan community Forest, Itu, Nigeria. Int J Mol Ecol Conserv 5(2):32–38

Jacob DE, Udeagha AU, Ufot IN (2012) The implication of pastoralism on the Niger-Delta ecosystem. In: Onyekwelu JC, Agbeja BO, Adekunle VAJ, Lameed GA, Adesoye PO, Omole AO (eds) De-reservation, encroachment and deforestation: implications for the future of Nigerian forest estate and carbon emission reduction. Proceedings of the 3rd Biennial National Conference of the Forest and Forest products Society held in Ibadan, pp 310–315

Jacob DE, Udoakpan, Ufot IN (2013a) The potential benefits of wildlife in food security in Nigeria. In: Etim L, Udoh JP, Etim N, Ekwu A (eds) Contemporary issues in sustainable tropical agriculture. Faculty of Agriculture, Uyo, pp 174–192

Jacob DE, Eniang EA, Udoakpan UI, Nelson IU (2013b) The conservation and status of Sclater's guenon in Ikot Uso Akpan community Forest, Itu, Nigeria. Int J Mol Ecol Conserv 3(5):27–33

Jacob DE, Eniang EA, Nelson IU, Udoakpan UI (2015a) Vegetation assessment of Sclater's guenon habitat in Ikot Uso Akpan Forest, Itu, Southeastern Nigeria. Int J Mol Ecol Conserv 2(6):32–38

Jacob DE, Nelson IU, Okoh KS, Tom AA (2015b) Trend in land use/cover changes in Stubbs Creek Forest, Akwa Ibom state, Nigeria. Int J Res Business Manag Accounting 1(1):123–130

Jacob DE, Nelson IU, Udoakapn UI, Etuk UB (2015c) Wildlife poaching in Nigeria National Parks: a case study of Cross River National Park. Int J Mol Ecol Conserv 5(4):1–7

Jacob DE, Ottong JL, Nelson IU (2018) Determinant of households' dependence on Ikot Ondo community Forest, Nigeria. J Biol Ser 1(3):062–071

Jacob DE, Eniang EA, Ukpong EE, Udoakpn UI, Nelson IU (2019) Correlates of revenue and tourist flow in old Oyo National Park. J Forestry Environ Sustain Dev 5(1):47–55

Jacob DE, Eniang EA, Ubo MO, Nelson IU (2020a) A survey of hunting practices in Ikot Ubo, Nigeria. Forestry Agric Rev 1(1):39–53

Jacob DE, Ityavyar AJ, Nelson IU (2020b) Income determinant and inequality among households around National Parks in Nigeria. Agric Stud 4(4):10–26

Jacob DE, Ityavyar J, Nelson IU (2020c) Impact National Parks on livelihood and conservation behaviours of households in Nigeria. J Forestry Environ Sustain Dev 6(1):72–85

Jacob DE, Onadeko S, Nelson IU, Shotuyo A, Ityavyar J (2020d) Determinants of income diversification among support zone communities of Nigeria National Parks. Econo Environ Stud 20(1 (53):7–23

Johns AD (1992) Vertebrate responses to selective logging: implications for the Design of Logging Systems. Philos Trans Roy Soc London Biol Sci 335:437–442

Johns AD, Skorupa JP (1987) Responses of rainforest primates to habitat disturbance: a review. Int J Primatol 8:157–191

Kasenene JM (1987). The influence of mechanized selective logging, felling intensity, and gap-size on the regeneration of moist tropical forest in Kibale Forest reserve, Uganda. Ph.D. Thesis, Michigan State University, East Lansing. 217 pp

Kingdon J (1980) The role of visual signals and face patterns in African Forest monkeys (guenons) of the genus Cercopithecus. Trans Zool Soc London 35:425–475

Laurence WF, Williamson GB (2001) Positive feedbacks among Forest fragmentation, droughts, and climate change in the Amazon. Conserv Biol 15:1529–1535

Law J, P Myers (2004) *Cercopithecus sclateri* (On-line), Animal Diversity Web. http://animaldiversity.ummz.umich.edu/site/. Accessed 16 Aug 2011

Lawes MJ (2002) Conservation of fragmented populations of *Cercopithecus mitis*. In South Africa: the role of reintroduction, corridors and metapopulation ecology. In: Glenn M, Cords M (eds) The guenons: diversity and adaptation in African monkeys. Kluwer Academic/Plenum Publishers, New York, pp 375–392

Lawes MJ, Chapman CA (2006) Does the herb Acanthus pubescens and/or elephants suppress tree regeneration in disturbed Afrotropical forest? For Ecol Manag 221:278–284

Lawrence WF (1997) Reflections on the tropical deforestation crisis. Biol Conserv 91:107–119

Leroy EM, Rouquet P, Formenty P, Souquiere S, Kilbourne A, Forment JM (2004) Multiple Ebola virus transmission events and rapid decline of central African wildlife. Science 303:387

Lovett JC, Marshall AR (2006) Why should we conserve primates? Afr J Ecol 44:113–115

Marshall AJ, Cannon CH, Leighton M (2009) Competition and niche overlap between gibbons (*Hylobates albibarbis*) and other frugivorous vertebrates in Gunung Palung National Park, West Kalimantan, Indonesia. In: Lappan S, Whittaker DJ (eds) The gibbons: new perspectives on small ape socioecology and population biology. Springer, New York, pp 161–188

McGraw W (2002) Diversity of guenon positional behavior. In: Glenn C (ed) The guenons: diversity and adaptation in African monkeys. Kluwer Academic, New York. 125 pp

Mitani JC, Struhsaker TT, Lwanga JS (2000) Primate community dynamics in old growth Forest over 23.5 years at Ngogo, Kibale National Park, Uganda: implications for conservation and census methods. Int J Primatol 21:269–286

Mittermeier RA, Myers N, Gil PR, Mittermeier CG (1997) Hotspots: earth's biologically richest and Most endangered terrestrial ecoregions. Cemex, Conservation International and Agrupacion Sierra Medre, Monterrey, Mexico. 58 pp

Myers N, Mittermeier RA, Mittermeier CG, Da Fonsa GAB, Kent J (2000) Biodiversity hotspost for conservation priorities. Nature 403:853–858

Nelson IU, Eniang EA, Jacob DE (2019) Trend and determinants of deforestation in Akwa Ibom state. J Agric Environ Resour Manag 4(1):341–347

Nelson IU, Jacob DE, Udo ES (2020) Trend and perception of Forest revenue generation in Akwa Ibom state, Nigeria. J Forest Environ Sci 36(2):122–132

Nelson IU, Jacob DE, Ver PN (2021) Temporal analysis of timber exploitation in Akwa Ibom State, Nigeria. In: Udom GN, Akpabio IA, Akpheokhai LI, Etuk UR, Ebong VO, Ekot MO (eds) A specialized compendium on agricultural and allied variables for sustainable development in Nigeria. Quickclick Printers, Uyo, pp 346–362

Nelson IU, Udoakpan UI, Daniel KS, Okon KE, Jacob DE (2013) Implications of biomass burning on the Nigerian Rainforest ecosystem. In: Popoola L, Idumah F, Ogunsanwo K, Azeez IO (eds) The green economy: balancing environmental sustainability and livelihoods in an emerging economy, Proceedings of the 36th Annual Conference of the Forestry Association of Nigeria held in Uyo, Akwa Ibom State, Nigeria. 4th–9th November, 2013. pp 300–308

Nowak R (1999) Walker's mammals of the world, 6th edn. The Johns Hopkins University Press, Baltimore and London. 234 pp

Nummelin M (1990) Relative habitat use of duikers, bush pigs, and elephants in virgin and selectively logged areas of Kibale Forest, Uganda. Trop Zool 3:111–120

Nunn CL (2003) Sociality and disease risk: a comparative study of leukocyte counts in primates. In: De Waal FBM, Tyack PL (eds) Animal social complexity. Intelligence, culture, and individualized societies. Harvard University Press, Cambridge MA, pp 26–31

Oates JP (1994) Africa's primates in 1992: conservation issues and options. Am J Primatol 34:61–71

Oates J, Anadu P (1989) A field observation of Sclater's guenon (*Cercopithecus sclateri* Pocock, 1904). Folia Primatol 52(38-42):92–96

Oates J, P Anadu, E Gadsby and J.Were (1992). Sclater's guenon: a rare Nigerian monkey threatened by deforestation. Natl Geogr Res Explor, 8(4): 476-491

Oates JF, Bergl RA, Linder JM (2004) Africa's gulf of Guinea forests: biodiversity patterns and conservation priorities. Conservation International, Center for Applied Biodiversity Sciences, Washington, DC. 26 pp

Okon AT (2004). Ecology and conservation of the Sclater's guenon (*Cercopithecus sclateri*); territorial and ranging pattern in Itu local government area of Akwa Ibom state. Undergraduate Project submitted to the University of Uyo, Uyo. 64 pp

Paul JR, Randle AM, Colin A, Chapman CA, Chapman LJ (2004) Arrested succession in logging gaps: is tree seedling growth and survival limiting? Afr J Ecol 42:245–251

Peeters M (2004) Cross-species transmission of simian retroviruses in Africa and risk for human health. Lancet 363:911–912

Plumptre AJ (1996) Changes following 60 years of selective timber harvesting in the Budongo Forest reserve, Uganda. For Ecol Manag 89:101–113

Plumptre AJ, Reynolds V (1994) The effects of selective logging on the primate populations in the Budongo Forest reserve, Uganda. J Appl Ecol 31:631–641

Pocock RI (1904) Description of a new species of spot-nosed monkey of the genus Cercopithecus. Proc Zool Soc Lond 1:433–436

Pyke GH, Pulliam HR, Charnov EL (1977) Opyimal foraging: a selective review of theory and tests. Q Rev Biol 52(2):137–154

Quinten M (2008) Survey of Primate Community of Peat Swamp Forests of Siberut, Mentawai Island, Indonesia. M.Sc./M.I.N.C. Thesis submitted to the Faculty of Biology. Georg-August Universitat Gottingen, Germany and Lincoln University, New Zealand. 84 pp

Redford KH (1992) The empty forest. Bioscience 42:412–422

Revkin AC (2000) West African monkey in extinct. The Association Press, New York, Scientist Say. 24 pp

Robbins M, Robbins A, Gerald-Steklis N, Steklis H (2007) Socioecological influences on the reproductive success of Female Mountain Gorillas (*Gorilla beringei beringei*). Behav Ecol Sociobiol 61:919–931

Rouquet P, Froment JM, Bermejo M, Kilbourn A, Karesh W, Reed P (2005) Wild animal mortality monitoring and human Ebola outbreaks, Gabon, and republic of Congo, 2001–2003. Emerg Infect Dis 11:283–290

Rowe N (1996) The pictorial guide to living primates. Poginias Press, East Hampton, New York. 263 pp

Sharp PM, Shaw GM, Hahn BH (2004) Simian immunodeficiency virus infection of chimpanzees. J Virol 79:3891–3902

Snaith TV, Chapman CA (2008) Red Colobus Monkeys display alternative behavioral responses to the costs of scramble competition. Behav Ecol 19:1289–1296

Sodhi NS, Kohl LP, Brook BW, Ng PKL (2004) Southeast Asian biodiversity: an impending disaster. Trends EcolEvol 19(12):654–660

Steenbeek R, van Schaik CP (2001) Competition and group size in Thomas's langurs (*Presbytis thomasi*): the folivore paradox revisited. Behav Ecol Sociobiol 49:100–110

Struhsaker TT (1997). Ecology of an African rain Forest. Gainesville: University Press of Florida. 136pp

Struhsaker TT, Lwanga JS, Kasenene JM (1996) Elephants, selective logging and the regeneration in the Kibale Forest, Uganda. J Trop Ecol 12:45–64

Tooze Z (1994a) Conservation status of Sclater's guenon, Akpugoeze, Nigeria. Unpublished report of the Wildlife Conservation Society, New York. 23 pp

Tooze Z (1994b) Does sacred means secure? Investigation of a sacred population of Sclater's guenon (Cercopithecus sclateri) in Southeastern Nigeria. Unpublished report. 20pp

Tooze Z (1995) Update on Sclater's guenon, *Cercopithecus sclateri*, in southern Nigeria. African Primates 1(2):38–42

Tutin CEG, White LJT, William EA, Fernandez M, McPherson G (1997) Nest building by lowland gorillas in the lope reserve, Gabon: environmental influences and implications for censusing. Int J Primatol 16(1):53–76

Udoedu UE (2004) Socio-biology of Sclater's guenon (*Cercopithecus sclateri*) in Itu local government area. Undergraduate Project submitted to the University of Uyo, Uyo. 52 pp

Ukpong EE, Jacob DE, Ibok PB, Nelson IU (2013) Nest building behaviour of chimpanzee (pan troglodytes Blumenbach 1799) at Filinga range of Gashaka Gumti National Park, Nigeria. ARPN J Sci Technol 3(7):714–717

Ukpong EE, Jacob DE, Umoh UA, Nelson IU (2018) Determinants of bushmeat traders' income in Itu, Akwa Ibom state, Nigeria. Environment 3:135–145

UNDP (2004) EFA—education for all. Global action plan: improving support to countries in achieving the EFA goals. UNDP & UNESCO, UNFPA, UNICEF, World Bank. Retrieved: 17 March 2008, http://www.unesco.org/education/GAP/GAP_v04.pdf

Van Schaik CP and Janson CH (2000). Infanticide by males and its implications. Cambridge University Press, Cambridge. 15pp

Vogel G (2003) Can great apes be saved from Ebola? Science 300:1645

Watts DP (2000) Causes and consequences of variation in the number of males in Mountain Gorilla groups. In: Kappeler P (ed) Primate males: causes and consequences of variation in group composition. Cambridge University Press, Cambridge, pp 84–93

Wolfe DN, Switzer WM, Carr JK, Bhullar VB, Shanmugan V, Tamoufe U (2004) Naturally acquired simian retrovirus infections in central African hunters. Lancet 363:932–937

WWF (1990) Cross River National Park Okwangwo Division, developing the park and its support zone. Panda House. 87 pp

Chapter 6
Potential Threats and Possible Conservation Strategies of Biodiversity in Niger Delta Region of Nigeria

Sylvester Chibueze Izah and Ayobami Omozemoje Aigberua

Abstract While much of the biodiversity of the Niger Delta is utilized as food and medicine, the region has continuously been regarded as a biodiversity hub in Africa because of its native population of diverse endemic marine and terrestrial species. This study sets forth to highlight the status of eco-diversity and their uses, some of the anthropogenic activities that are prominent across the Niger Delta region, while stating the likely ways of protecting endemic species that are native to the environment. Some of the leading factors responsible for the rapid decline in biodiversity population are the incessant episodes of flood rising from overflowing sea levels, intensive agricultural practices of seasonal bush burning, deforestation and wood lumbering, open grazing, infrastructural development, and chief among them is crude oil spillage. While the abundance of mineral resources (mainly crude oil and gas deposits) in the Niger Delta region has been responsible for its rapid industrialization, regional population expansion has continued to increase the demand for infrastructural developments such as airports, market places, industries, roads construction, among others. This in turn has led to the unabated release of several organic and inorganic pollutants into the environment, hence resulting in the exploitative utilization and destruction of forest assets. Consequently, this has led to the spatiotemporal decimation of various life forms and constant alteration in the biodiversity status of the region. The study concludes by highlighting some of the possible strategies for restoring the diversity status of the Niger Delta region which include waste minimization, reuse, or recycling, public sensitization, enforcement of existing environmental laws and legislation, and adoption of new laws in line with current global best practices.

S. C. Izah
Department of Microbiology, Faculty of Science, Bayelsa Medical University, Yenagoa, Bayelsa State, Nigeria

A. O. Aigberua (✉)
Department of Environment, Research and Development, Anal Concept Limited, Elelenwo, Rivers State, Nigeria

S. C. Izah, M. C. Ogwu (eds.), *Sustainable Utilization and Conservation of Africa's Biological Resources and Environment*, Sustainable Development and Biodiversity 32, https://doi.org/10.1007/978-981-19-6974-4_6

Keywords Biodiversity resources · Open grazing · Endemic marine and terrestrial · Seasonal bush burning · Environmental laws and legislation

6.1 Introduction

About 70,000 sq kms of Nigeria's total land mass is occupied by the Niger Delta province, being the broadest swampland territory in Africa (Ogbe 2011; Izah 2018). This region cuts across the nine member states of Cross-Rivers, Akwa-Ibom, Abia, Imo, Rivers, Bayelsa, Delta, Edo, and Ondo (Fig. 6.1). As such, these areas are characterized by the abundance of fern leaves, swamp peat moss or water, either stagnant or steadily moving surface waters of saline, brackish, or freshwater origin. The wetlands of the Niger Delta are known to host numerous aquatic plants adjusted to surviving in oxygen-deficient environments (extremophiles). Furthermore, the Niger Delta region possesses the biggest tropical mangrove in Africa as well as the third most expansive wetland estuary across the globe. In fact, this habitat occupies about half of Nigeria's shorelines (Ohimain et al. 2014a; Izah et al. 2017a; Izah 2018).

The Niger Delta of Nigeria is the oil and gas province. This oil-rich region of Nigeria plays host to a hoard of biodiversity that are globally important, most of which are endemic biotic species that inhabit the country's vast tropical and mangrove forests. The biodiversity plays key social and economic roles for the populace. As such, the Niger Delta area is divided into four primary categories namely: coastal barrier reefs, mangrove marshland, freshwater marshland, and lowland tropical forests (Izah and Seiyaboh 2018a; Ubom 2010; Asimiea and Omokhua 2013). Meanwhile, freshwater marshlands and lowland tropical forests are the most diversified environmental territories of the oil-rich region (Phil-Eze and Okoro 2009).

Estuaries of the Niger Delta area are among the principal zones in Nigeria. Also, it is classified among the prime biodiversity centers that host endemic biota across Africa. Most commonly, resources such as watersheds, climate, and biota breeding grounds are sustained by these estuaries (wetlands). Wetlands have been described as areas consisting of flowerless plants, and static or moving marshlands of freshwater, marine or brackish origin that are either native, or alien, or even short-term or long-term residents of the habitat, and they reside within vertical depths of not more than 6 meters during the low tides (Izah and Seiyaboh 2018a; Scott 1989; Ohimain and Akinnibosun 2007; Ramsar Convention Secretariat 2007; Okonkwo et al. 2015). These wetlands have been reported to host various biotic species like marine turtles, sea cows, shorebirds, and other endemic marine organisms (Izah and Seiyaboh 2018a). In addition, the Niger Delta environment hosts several vegetations that possess medicinal properties (Izah and Seiyaboh 2018a). The peculiarity of the oil-rich region in the global biodiversity sphere means that it is considered the second most fragile environment across Africa.

The Niger Delta remains a major contributor and driver of Nigeria's commonwealth. Hence, it remains the foremost producer of crude oil among the West African states. Nonetheless, it remains a hub of various biodiversity. On account of this, the

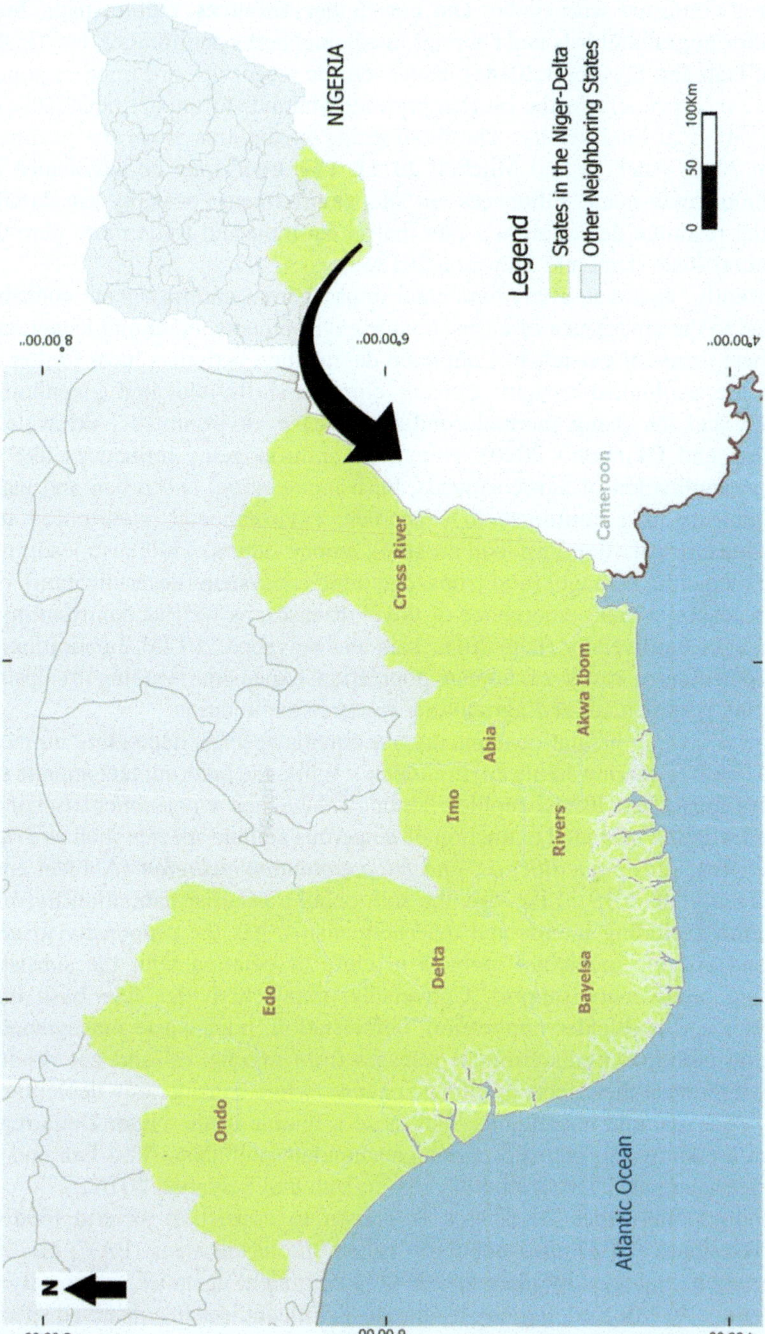

Fig. 6.1 Geographical map showing the nine (9) states of the Niger Delta region

region is conferred with timber and non-timber resources, edible snails, breeding places for migrating birds and fisheries, medicinal herbs and forages, etc. (Izah et al. 2017a; Izah and Seiyaboh 2018a). However, the recent upward trend in population expansion, especially in the oil-rich region, continues to impact negatively on the conservation of biodiversity. The biodiversity of the area is on the decline trend (Ubom 2010; Adekola and Mitchell 2011). This trend may be aggravated by the continuous man-made influences on the environment, possibly escalating into potential resource depletion at a rate that is ten-thousand folds more than that on the natural scale (Izah and Seiyaboh 2018a).

Currently, aggravated environmental impacts from anthropogenic contributions have led to the emergence of unsustainable environments, including its animate and inanimate forms of existence. Continued degradation activities have further deteriorated the ecological system, causing climatic shifts, elevated greenhouse gas releases, and the rising thermal conditions of the environment (Akinwale 2004; Aigberua and Okumoko 2020). Numerous anthropogenic influences like wood-logging, application of agrochemicals, bush incineration, and urban and industrial developments have continued to exacerbate environmental occurrences such as whirlwinds, dearth, flooding (soil erosion), among others, while also leading to the loss of medicinal herbage, food crops, declining ecosystem (desertification), climate shift, readaptation and emergence of novel diseases, as well as contributing to the depletion of biodiversity (Izah 2018; Izah and Seiyaboh 2018a). In addition, urban development is a result of human population expansion, leading to upsurge in industrial, commercial, and agricultural resource exploitation.

Nigeria's water habitat contains diverse aquatic species, depending on the water source (fresh or marine/saline environments). While the predominant aquatic species includes fin/shelled fishes, reptiles, among others, the environment is constantly plagued with the unabated extinction of numerous aquatic species such as *Trichecus senegelensis, Lutra macullicolis, and Hippopotamous pelagian* (Abowei and Hart 2008; Iwar et al. 2019). Likewise, the soil, being a peculiar habitation for different life forms, including micro- and macroorganisms, has the propensity to generate food and provide medicinal herbage in close association with the subsisting or surviving biodiversity support. Conversely, man's activities like bush burning, pesticides and herbicides application, deforestation, poor waste management, and industrial pollution (indiscriminate releases from mining, oil and gas production) tend to aggravate the effects of climate change, while considerably depleting biodiversity reserves, and marshlands, like those available in the Niger Delta region of Nigeria are known to interrupt and bioaccumulate pollutants (Phil-Eze and Okoro 2009; Ideriah et al. 2010; Izah et al. 2017a; Izah and Seiyaboh 2018a).

Although the Niger Delta area is known to consist of several biodiversity preserved zones (BPZ), also popularly called "protected areas (PA)", biodiversity conservation strategies by the responsible government agencies are grossly inadequate or poorly enforced, thereby leading to significant loss in conserved species. At times, activities such as farming, wood-logging, and even infrastructural development tend to encroach on these spots, potentially resulting in the disappearance or loss of endemic species. Even the efforts of stakeholders like Shell Petroleum

Development Company (SPDC), as well as the Niger Delta Wetland Centre (NDWC) to uphold the sustenance of such PA's, have been reasonably affected by sabotage, mainly from host communities. The poor sense of cooperation from locals mainly stems from inadequate knowledge or sensitization on the importance of biodiversity conservation. This has continued to result in the hunt-down of endemic species for food, causing scarcity of diverse species or outright extinction (FMENV 2004; NDDC 2004; Phil-Eze and Umeuduji 2004; Phil-Eze and Okoro 2009). Considering that freshwater fishes are the major protein source in the daily meals of local inhabitants of the region, any conservation strategy that restricts the local inhabitants from accessing these food resources may likely backfire and ultimately flop.

At present, the Niger Delta has 70 apportioned and publicized preserved zones (PZ) that are used as national parks, forest and wildlife, and nature parks (Izah and Seiyaboh 2018a). Of the PZ's, about one-tenth serve the purpose of protecting the various biologically diverse species, while the larger fractions are mainly used as forest reservations for producing timber. It follows that only a sparse fragment of the regions freshwaters is protected as opposed to the non-concealment of its entire marine ecosystem (Phil-Eze and Okoro 2009; Izah and Seiyaboh 2018a). This trend falls short of the global specification of $\geq$10% regional confinement for the benefit and safeguarding of biodiversity. For instance, countries like Australia and New Zealand Environment and Conservation Council (ANZECC) are governed by site selection concepts for biodiversity conservation, while the UK has adopted site selection philosophies that are mainly species-specific and adjudged on the basis of features such as rarity of characteristics, natural availability, history, diversity, among others. Other regions where similar site selection strategies have been accepted and adopted include Latin America, Carribean, Mexico, etc. (Phil-Eze and Okoro 2009). Unfortunately, these strategies are yet to gain global acceptance and adoption, especially here in the Niger Delta region.

6.2 Ecology of the Niger Delta

The maritime Niger Delta region of Nigeria serves as viable dwellings for diverse biotic macro and microorganisms among which include fin and shelled fish, sea animals, reptilian species, green algae, water hyacinths, fresh and brackish water shrimps, water-related insects, sharks, dolphins, whales, etc., which can be found across the fresh and marine habitats of the Niger Delta province. This oil-rich region of Nigeria has the largest wetlands area, accounting for up to 3% of its total landmass. Its wetlands are characterized by marsh, or swamp habitation with the potential to support different life forms within its anoxic surroundings. As such, Nigeria comprises of numerous wildlife and vegetation species, spanning over 22,000 vertebrate and invertebrate animal species, 20,000 insects, around 1000 varieties of fishes and birds, more than 100 and 200 species of reptilian and mammalian vertebrates, and about 1500 species of microorganisms (Izah 2018;

Emma-Okafor et al. 2010; FRN 2010). Reports have shown that less than 0.3 percent (<0.3%) of the identified species is either faced with existential threat or danger (Ugochukwu and Ertel 2008; Izah 2018; Izah and Seiyaboh 2018b).

The Niger Delta environment is predominated by tablelands with ground level showing the deposition of sedimentary layers washed out by the river Nun, with its estuaries of streams, rivers, and creeks which are often unloaded into the Atlantic sea. The nine (9) states of the oil-rich Niger Delta make up around 70,000 km^2 of the countries landmass, being the most extended wetland in Africa and third biggest across the globe (Izah 2018). Its marine and estuarine ecosystem is distinguished by barrier islands, tableland forests, mangroves, as well as creeks and creeklets. Reports have categorized the Niger Delta environment into six (6) diverging ecological zones (EZ), namely: tropical rainforest, mangrove or tidal forests, freshwater marshes, tableland forests, mountainous lands, and originating Savannah or grasslands (Izah 2018).

However, land use changes in the region have grossly reconstituted its ecosystem, with its three (3) prime environments consisting of mangroves (marine and estuarine), tableland rain forests, and freshwater marshlands.

6.2.1 Mangrove Swamps (Marine and Estuarine)

The mangrove habitat of the Niger Delta is among the widest in Africa. Its brackish waters are also known as estuarine because of the existence of fresh and salt water environments. The salt content of brackish waters is known to be at lower levels when compared to the marine equivalent. However, they generally possess greater saltiness with respect to freshwater environments (Izah 2018). Typical mangroves found in this region are: *Avicennia germinans* (white mangrove species), *Laguncularia racemosa*, the short and tall red mangrove species called *Rhizophora harrisonii* and *Rhizophora racemosa,* respectively. As such, the *Avicennia* and *Rhizophora* species represent the most important mangrove species within the oil-rich district (Jamabo and Chinda 2010; Ogamba et al. 2016a, Ohimain et al. 2014a; Izah 2018). Furthermore, freshwater environments constituting about 10,000 km^2, as well as mangrove areas, are partitioned by metamorphosed wetlands that have the prevailing distribution of *Machaerum lunatus*, *Dalburger escatophylum,* and *Pandanus* species (Ohimain et al. 2014a, b; Izah 2018). Other complimentary genus within the environment is *Nypa fruticans* (Izah 2018), while *Tympanotonus* and *Pachymelania* are among the most prevalent genera of shelled-fishes within the brackish water environment of the Niger Delta region (Bob-Manuel 2012). The *Tympanotonus* (*T. fuscatus* species) are often found in tidal flat zones that are rich in detritus and mud sub layers, especially where they coexist with *Neritina adansoniana* and *Pachymelania fusca var quadriseriata* (Jamabo and Chinda 2010; Ogamba et al. 2016a; Izah 2018). In addition, numerous fin fishes, benthic species, and phytoplankton and zooplankton species inhabit in the marine and estuarine environments (Izah 2018; Izah and Seiyaboh 2018b; Enerosisor et al. 2020).

6.2.2 Tableland Rainforests

Owing to the variety of species which make up the tableland rainforests of the Niger Delta environment, it is often rated among the multiplex environments of the region. The rainforest is characterized by four (4) layers/strata which are composed of murky forests of 40–50 m tall trees at the first stratum, massive trees of 20–35 m that provide cool shelters at the second stratum, 20 m towering branch plants at the third stratum, with miniature stalk bush undergrowths/herbage, and flowerless plants (ferns, moss) at the fourth stratum (Izah 2018). Most of the rainforest vegetation does have great economic value as they are lumbered for timber, firewood, pulp, and paper, electricity poles, traditional herbage, among others. The most important species of this environment include: *Lovoa trichilioides*, *Entandophragma cylindricum*, *Terminalia ivorensis*, *Gossweilerodendron balsamiferum*, etc. (Izah 2018).

6.2.3 Freshwater Forest Marshlands

The freshwater forest marshlands of the Niger Delta are found between the tableland rainforests (North of the region) and the mangrove environment (Southernmost parts of the region). Owing to its unique location, freshwater marshes contribute to the transformation shift between both ecosystems, leading to biodiversity migration. Freshwater habitats in Niger Delta are loaded with silt from the Niger River. Based on this, flooding is common during the high tide, making it an important habitation for different aquatic species. The freshwater marshlands are dotted by towering trees (*Raphia hookeri*, *Eleais guineensis*, *Musanga cecroploides*, etc.) (Izah 2018). Like in the tableland rainforests, the vegetation from freshwater forest marshlands are economically important as they are often used as wood fuel, medicinal herbage, while also serving as shoreline protection, especially during flooding (Izah 2018). The habitat also consists of freshwater resources which serve as drinking, cooking, and recreational purposes, amongst others. Apart from the freshwater environment, other common habitats of this forest zone include the riparian, coastal, and cultivable land areas. Most freshwaters in the region are connected to the estuarine habitat which further empties into the Atlantic Ocean. Some reported surface water macrophytes include *Azolla pinnata var africana*, *Nymphaea lotus*, *Salvinia nymphellula*, *Eichhornia crassipes*, among others (Agedah et al. 2015; Izah 2018; Izah and Seiyaboh 2018b).

6.3 Anthropogenic Activities Influencing the Ecology of the Niger Delta Ecosystem

6.3.1 Flooding

The Niger Delta region of Nigeria is quite frequently and severely influenced by floods. This trend can be attributed to increasing water levels. The annual rainy and dry season periods are from April to October and November to March of the subsequent year, respectively (Izah et al. 2017b). Typical relative humidity ranges between 50 and 95% with temperatures sometimes reaching up to 28 ± 8 °C all through the year (Izah et al. 2017b). Some of its creeks are below sea level. Running waters are often prone to far-reaching flow changes. Hence, their flow rates tend to be severe or mild across different seasons, or depending on river flow patterns. Often times, the most turbulent water movement occurs at sections of high tide (Davies et al. 2020). Following this, humans require protection from floods while aquatic biota (water-dwelling flora and fauna) may be washed ashore, hence causing them to lose their natural habitation. Other consequences of flooding are the water pollution problems that usually emanate from the infiltration of leachates from municipal waste dumpsites, agricultural and industrial effluents, which are readily bioaccumulated by the vulnerable aquatic organisms with further biomagnifications across the food chain (Talbot et al. 2018; Aigberua et al. 2021). When a river's drainage basin is altered, either from the direct impact of anthropogenic activities or indirect impacts from natural releases, the loss of biodiversity becomes inevitable (Omotosho 2004; Izah and Seiyaboh 2018b). For instance, when there is occurrence of aggravated loading of contaminants in surface water and sediment, it may likely result in the fluctuations of shoreline and coastal erosion, while inundating lithoral habitats and congesting river substratum, or even deterioration of floodplains (Izah and Seiyaboh 2018b).

6.3.2 Road Construction and Infrastructural Development (Airports, Markets, Industries, Etc.)

In spite of the socioeconomic benefits attached to human activities such as the building of roads or industrialization, such developments are most often associated with the loss of forest resources and its in-dwelling biodiversity (Phil-Eze and Umeuduji 2004). While different road types may include rail and motorable roads, bridges, under water tunnels, among others, infrastructural and industrial developments encompass the construction of air transportation terminals, mineral-ore mining industries, etc. These activities tend to distort or cause a fragmentation of habitats, which then aggravates invasion by non-native species, as well as leading to environmental contamination (Izah and Seiyaboh 2018b). Also, in the process of road development, a variety of mammals are hunted for food while being displaced from their natural habitation. Other times, these mammals are crushed by speeding motorists (Vaiskunaite et al. 2012; Tarvirdizadeh et al. 2014). Also, vegetative species or herbage may be affected by atmospheric deposition from vehicular emissions, suspended particulates in air (dust), soil erosion, or releases from faulty or worn-out construction equipment, among others. As a result, numerous anthropogenic activities tend to impact significantly the distribution of plant species. Constructed roads are often invaded by weeds, while waterways where construction has previously taken place are known to harbor parasitic and ravenous organisms (Izah and Seiyaboh 2018b).

6.3.3 Pollution from Industries

The increasing activities of oil and gas industries, unending acts of deforestation, continuous sand dredging (at artisanal and/or industrial exploitative scale), among others, remain the ultimate menace to the Niger Delta ecosystem. However, oil exploration remains the most negatively impacting activity plaguing host communities in the region (Ugochukwu and Ertel 2008; Okumagba 2011; Kadafa 2012; Onuoha et al. 2018). This is owing to the persistent incidence of crude-oil spillages which then lead to the contamination of mangrove forests. Also, deforestation becomes needful when creating oil pipeline right of way (ROW), installing underwater pipelines, constructing encampment for seismic workers, erecting drilling structures or platforms within forests or along a water course, etc. (Aigberua et al. 2016a). These anthropogenic activities are known to inhibit the activities of wild games and lead to the loss of biodiversity (Phil-Eze and Umeuduji 2004; Aigberua et al. 2016b; Emeudo and Emeudo 2018). Further exploitation of water resources for building construction, especially dredging operations, tends to hamper the livelihood of aquatic organisms. This is coupled with the fact that all stages of crude oil exploration are accompanied by hydrocarbon contamination and loss of wood, flora, and fauna. For example, operational releases tend to be absorbed in mangrove roots, annihilating nutrient absorption and aggravating species extinction which is identified by the discoloration and defoliation of flora (Tarawou et al. 2019). Industrial releases and their attendant pollution have been vastly reported to affect the distribution of crustaceans, mollusk, oysters, periwinkles, variant fish species, polychaetes, etc., which dwell within the mangrove environment. In the same vein, benthic species are susceptible to cessation within polluted environments (Emeudo and Emeudo 2018; Izah 2018; Izah and Seiyaboh 2018a, b).

6.3.4 Poor Waste Management

In spite of the popular practice of open waste dumps in the Niger Delta region, much like the rest of Nigeria, it remains a largely ineffective means of managing wastes on a global scale. This is because of the environmental effects linked to it (Nwosu and Pepple 2016; Aigberua and Okumoko 2020). In recent times, the incineration of solid waste materials has grown. Often times, depending on the nature of wastes, thick black clouds of smoke or soot are generated with noticeable impact on human health. The incineration of solid wastes can cause habitat disintegration, as well as lead to the loss of medicinal forages, wild species, food plant, fuel wood, plant cover, while altering the biogeochemical cycle. Similarly, leaching from open indiscriminate dumpsites tends to proliferate harmful algal blooms and kill fish foods via eutrophication (Izah et al. 2017a; Tarawou et al. 2019; Aigberua and Okumoko 2020).

6.3.5 Uncontrolled Bush Burning

Recently, bush incineration, mostly controlled and uncontrolled, has continued to be practiced in the Niger Delta area. In fact, bush incineration is a long-established, agricultural procedure among local farmers in the region, much like most third world countries. In recent times, especially during the dry season, the practice of incinerating farm weeds has continued to grow unabated. Evidently, this farming practice has continued to contribute to recent adverse alterations in the ecosystem. Air pollutants in the form of noxious gases and pyrolytic-sourced polycyclic aromatic hydrocarbons (PAH) released during the process are known to negatively affect biodiversity composition and richness. Most often, bush meat are locally hunted by setting vegetation covers ablaze. As a result, bush burning remains one of the human-induced forms of elevating atmospheric greenhouse gases (GHG) and temperature (Izah et al. 2017a; Aigberua and Okumoko 2020). In addition, the pressures from population expansion, with about 31,224,587 people across the 9 Niger Delta states, have led to persistent or exhaustive farming practices (NPC 2009). Hence, it is classified among the environmental degrading activities of the region. Furthermore, emitted gases from bush burning activities have been reported to cause cardiovascular and respiratory impairment in humans, loss of the biotic components within environmental systems, decimation of wild animals, and the disappearance of herbage, alongside the depletion of soil nutrients (Aigberua et al. 2016b; Izah et al. 2017a; Izah and Seiyaboh 2018a, b). In the same vein, biodiversity including microorganisms within vicinity of in situ burning are ill-disposed or alienated from their natural habitat.

6.3.6 Use of Chemical Fertilizers

Fertilizers can either be composed of natural or chemical substances imputed into agricultural soils for enhanced crop productivity. Some of the important microbes which control nutrient shifts include *Azobacter*, *Cyanobacteria*, *Nitrobacter*,

Nitrosomonas, *Rhizobium*, among others. Soil microbes also serve as food for wildlife. Hence, the loss of essential microorganisms often impacts the population of wild animals that require them for food (Xie et al. 2016; Izah et al. 2017a). Also, changes in flora and fauna within the ecosystem can determine the composition of its vegetation, while the absence of microorganisms limits the rate of decay of waste products. Plant nutrients such as nitrates, phosphates, etc. are found at higher concentrations in fertilized farmlands. In turn, run-offs from nutrient-rich agricultural soils tend to promote water's eutrophic condition with the possibility of causing a decline in dissolved oxygen levels, especially in the dark, and along slow-moving streams. As such, hypoxic levels of available oxygen possess the tendency to suffocate fish and other marine animals (Izah et al. 2017a; Davies et al. 2020).

6.3.7 Use of Pesticides and Herbicides

The application of pesticides for the extermination of insects that cause crop contagion has continued to gain popularity, hence being widely used. Also, a variety of herbicides are available for use in controlling the spread of unwanted vegetation or weeds. However, the continuous application of these chemicals tends to result in far-reaching environmental damages that endanger the useful plants diversity and nontarget micro and macroorganisms (Ugochukwu and Ertel 2008; Xie et al. 2016; Izah et al. 2017a). Generally, most pesticides when dispersed into the environment become unamenable by natural weathering processes. Hence, their persistent nature leads to the bioaccumulation and biomagnification of toxins across the food chain (Ugochukwu and Ertel 2008; Izah et al. 2017a; Izah and Seiyaboh 2018a; Davies et al. 2020).

6.3.8 Fuel Wood Extraction, Logging, Deforestation, and Excessive Exploitation of Forest Resource

Biodiversity, especially forest trees and plant shrubs, is known to possess social, esthetic, medicinal, tourist, ecological, economical, and cultural benefits for any nation. Sadly, human population pressure has led to an increase in activities such as wood logging, animal grazing, and the collection of woods for fuel, thereby resulting in the incessant loss of biodiversity (Phil-Eze and Umeuduji 2004). Also, urban development projects often involve the conversion of forests, or felling of trees for other purposes (farming, building, animal farmsteads, or constructing infrastructures). In addition, trees are hewn down for charcoal production, or wood timber for building construction, while the rising population continues to put pressure on food demand. Consequently, the tropical rainforests of the Niger Delta are constantly under severe threats from rapid urbanization and industrialization. Other forest resources that are critically affected are the diverse animals which are usually hunted for food. This act tends to culminate in the reduction or extinction of animals within the ecological community (Phil-Eze and Umeuduji 2004; Izah et al. 2017a; Izah and Seiyaboh 2018a).

6.4 Uses of Biodiversity in the Niger Delta

6.4.1 *Food*

The use of biodiversity for food remains one of the major challenges affecting animal and plant conservation in the Niger Delta region of Nigeria. For instance, the hunting of useful animals and the unregulated harvesting of plants are oftentimes responsible for the decimation of the community of flora and fauna species. In the Niger Delta region, much like other parts of Nigeria, bush meat is extensively sold along the highways and in the market places because it serves as an important protein source in people's diets. In addition, these animals are caught in traps so as to fend them off the destruction of cultivable food crops. Several authors have reported the sale of diverse mammalian species in markets around Edo, Delta, Bayelsa, and Rivers States (Izah 2018; Izah and Seiyaboh 2018b; Akani et al. 2015). Likewise, man's quest for food has led to unabated deforestation as forest habitats continue to be destroyed in order to provide lands for the planting of food crops. Therefore, the excessive exploitation and resulting encroachment into designated protected forests or zones continues to expose wildlife to hunting and lead to eventual migration from human-induced areas.

6.4.2 *Traditional Medicine*

The freshwater swamps of the Niger Delta are known to be migration zones for biodiversity. This is by virtue of their position between tableland forest zones and mangrove forests. As such, many of the vegetations found in this area, such as *Eleais guineensis*, *Raphia hookeri*, *Raphia vinifera,* etc., possess medicinal potential among other applications. For example, authors have reported the use of plant leaves, roots, stem, bark, and fruits for the improvement and healing of a variety of ill health conditions (Epidi et al. 2016a). Several species of plants found in the region have been reported to possess antimicrobial. Some of these plants with antimicrobial properties found in the Niger Delta region include *Alstonia boonei* (Epidi et al. 2016a), *Vitex grandifolia* (Epidi et al. 2016b), *Costus afer* (Izah et al. 2019a), Capsicum *frutescens var. minima* (Izah et al. 2019b), *Myristica fragrans* (Izah et al. 2018a), *Anacardium occidentale* (Izah et al. 2018b), *Carica papaya* (Izah et al. 2018c), Buchholzia coriacea (Izah et al. 2018d), Aframomum melegueta (Kigigha et al. 2015), *Garcinia kola* (Kigigha et al. 2018a), Cymbopogon citratus

(Kigigha et al. 2018b), *Vernonia amygdalina* and *Ocimum gratissimum* (Izah et al. 2018e), *Zingiber officinale* (Izah and Aseibai 2018), *Musanga cecropioides* (Kigigha et al. 2016), etc.

In fact, the utilization of traditional herbs is constantly gaining popularity in the rural areas while extending to the urban cities in recent times. Similarly, animals possessing zoo therapeutic or medicinal attributes are broadly dispersed across the region (Izah and Seiyaboh 2018b). Hence, zoo therapy is widely popular among Nigerian's, as it often combines psychological, spiritual, and mystic roots. Unfortunately, the increasing popularity of the practice continues to put pressure on the sustainability of species conservation (Izah 2018; Izah and Seiyaboh 2018b). Therefore, there must be a balance between the practice of zoo therapy and environmental management and sustainability.

6.5 Status of Biodiversity in the Niger Delta

Wetlands of the Niger Delta region of Nigeria remain a major repository of globally and locally endangered biodiversity species (Izah et al. 2018f). Generally, the ecosystem is a habitation of diverse plants, animals, and microorganisms that are essential for environmental sustainability. Some authors have reported different species of *Afrixalus*, *Aubria*, *Nectophryne*, *Opisthothylax*, *Bufo*, *Silurana*, *Hymenochirus*, *Hylarana*, *Ptychadena*, *Conraua*, *Hoplobatrachus*, *Phrynobatrachus*, *Arthroleptis*, *Chiromantis*, *Phrynomantis*, *Hyperolius*, *Leptopelis*, *Phlyctimantis* (Izah 2018; Akani et al. 2004). The distribution of biodiversity of the region is shown in Fig. 6.2

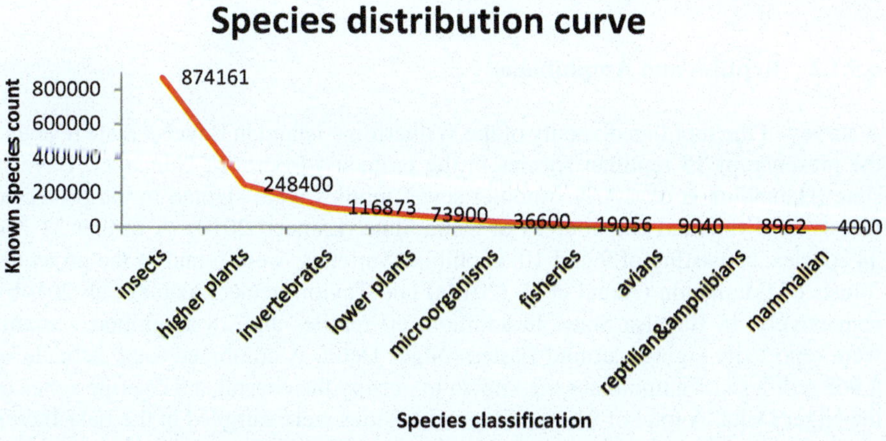

Fig. 6.2 Biodiversity species distribution within the Niger Delta ecosystem (Modified from Davies et al. 2020)

6.5.1 *Animals*

Mammals:Several studies have revealed a number of endemic mammalian species within the Niger Delta region. For instance, Sclater's guenon (*Cercopithecus sclateri*), Nigerian white-throated guenon (*Cercopithecus erythrogaster pococki*), red-capped mangabey (*Cercocebus torquatus*) and the endangered Nigeria–Cameroon chimpanzee (*Pan troglodytes ellioti*), among others (Ikemeh 2015; Izah 2018), the two-spotted civet (*Nandinia binotata*) (Amadi et al. 2016) have been found in some forest protected zones in the Niger Delta. Meanwhile, the endemic mammal Heslop's pygmy hippopotamus (*Hexaprotodon liberiensisheslopi*) can no longer be found in the Niger Delta province (Ikemeh 2015). Many other endemic species such as Sclater's guenon (*Cercopithecus sclateri*), red-capped mangabey (*Cercocebus torquatus*), Niger Delta Red Colobus monkey (*Procolobus epieni*), etc. are still spotted across forest resources in the region. Overall, there is an overall estimation of about 4,000 known species of mammalian diversity within the coastal regions (Davies et al. 2020).

6.5.1.1 Birds

About 78 avian fauna species belonging to 27 families have been identified on the Wilberforce Island area of Bayelsa in the Niger Delta region (Ohimain et al. 2014b). The predominant population of these birds inhabit forest areas, with fewer diversity been found in the villages, on the farmlands, and by river banks or beaches. Another 67 avian fauna species that were further classified into 25 families were reported within protected zones of the Nun river forest (Hamadina et al. 2007). Overall, an estimated 9,040 known species of birds have been identified across different coastal areas of Nigeria (Davies et al. 2020).

6.5.1.2 Reptiles and Amphibians

A survey of the reptilian diversity of the Wilberforce Island in Bayelsa State revealed the presence of 18 reptilian species in the protected forests of Nun river, Bayelsa state (Hamadina et al. 2007). Another seven reptiles were sighted in the protected forests in Kwale and Okpai areas of Delta State (Lameed 2010). A further 31 and 34 species, consisting of 9 and 10 amphibian varieties, were found in the protected forests of Edumanon (Akani et al. (2014a) and Taylor creeks (Akani et al. 2014b), respectively, in Bayelsa State. In addition, 60 reptiles and 28 amphibian variants were reportedly sighted around Eastern Niger Delta. A combined total estimate of 8,962 reptiles and amphibians are known to occupy the coastal and riverine zones of the Niger Delta. A total of 91 amphibian diversities were sampled in the Ijala-Ikeren wetlands of the Niger Delta, and 13 of the species belonged within 3 families

(*Bufonidae*, *Hyperoliidae* and *Ranidae*), with all species converging at the outskirts of the mangrove forest (Uwagbae et al. 2019).

6.5.1.3 Fishes

Several diversities of fish species have been identified across the Niger Delta region. Some of the species have been sighted, observed, and noted on-field. For example, 20 characteristic species of the cat and tilapia fish have been identified within the Ijala-Ikeren ecosystem in the Niger Delta (Uwagbae et al. 2019). Other species of fin fish found in the wetland area include *Tilapia zilli, Oreochromis niloticus, Chrysicthys walkeri, Chrysicthys furcatus, Arius gigas, llisha africana, Ethmalosa fimbriata, Parachana obscura* and *Clarias lazera* (Aghoghovwia et al. 2016), *Clarias garepinus* (Ogamba et al. 2016b), *Oreochromis niloticus, Clarias camerunensis* (Ogamba et al. 2016c, 2017), *Citharinus citharus* and *Synodontis clarias* (Ogamba et al. 2015), and some shell fish like *Tympanotonus fuscatus* (Ogamba et al. 2016a; Aigberua and Izah 2018). Generally, water resources around the province are known to host different species of fin and shelled fishes (Izah et al. 2017a; Izah and Angaye 2016; Uwagbae et al. 2019). Even though the current count of identified fish species in Nigeria's coastal regions is 19,056, an estimated 10% of fisheries are yet to be discovered (Davies et al. 2020).

6.5.1.4 Insects and Other Arthropods

Insects are meal for fish and avian species, amphibians, among others. However, some insects are considered inedible. Among the inedible species are *Rhinoceros oryctes, Musca domestica, Macrotermes,* and *Zonocerus* species, etc. Most often, its abundance and diversity is independent on regional or anthropogenic influences. Their presence aids in balancing the energy and nutrient flow cycle across the food web. Insects and arthropods are composed of the Niger Delta mangroves, estimated 30 million insect variants. This is based on forest canopy assessments within Nigeria's rainforests. However, only about 874,161 species of this fauna have been identified (Davies et al. 2020). In addition, insects have been sighted beneath woods, logs, and fallen leaves in the Ijala-Ikeren mangrove area (Uwagbae et al. 2019). Three (3) insect species were found in the habitat namely: *Aquarius remigis* (water strider), *Formica sp*., and *Chironomus sp*. The *Formica* species. were collected from hives draping from mangrove vegetations, being the most important insect species in the mangrove, followed by the *Chironomus* species (Uwagbae et al. 2019).

6.5.1.5 Microorganisms

There are about 36,600 known microorganism diversity in Nigeria's coastal region. Meanwhile, the scarcity of data has made determining the total number of estimated species quite difficult. The mangrove environment is rich in bacteria and fungi species. Some of the essential microbes (necessary for nutrient cycling) that have been reported include *Azobacter, Nitrobacter* (nitrogen cycle), *Nitrosomonas, Desulfovibrio* (sulphur cycle), *Beijerinckia, Begigathoa, Rhizobium, Cyanobacteria*, etc. These native microorganisms are usually destroyed during vegetation incineration processes, thereby causing nutrient and organic matter depletion. Invariably, the abundance in the flora and fauna community residing in a forest environment determines the occurrence of potential shifts in plant distribution and diversity. The microorganisms found in soil storage are needed for the sustenance and enhancement of diverse species (Uwagbae et al. 2019). Also, the zooplankton species composition of the Opobo River, Rivers State, were classified into 12 species and subclassed into 9 families (*Rotifera*, *Copepoda*, *Polychaete*, *Decapoda*, *Gastropoda*, *Echinodermata*, *Bivalve*, *Cladocera* and *Spirotricha*). Overall, *Acartia longiremis* and *Calamus finmarchicus* were particularly dominant in nearly all sections of the river (Enerosisor et al. 2020).

6.5.2 Plants

Majority of the Niger Delta ecosystem is characterized by similar mangrove species, namely: *Rhizophora harrisonii*, *Racemosa* species, *Laguncularia racemosa*, *Rhizophora mangle*, *Germinans sp.*, and *Avicennia africana*. Palm species in the region are dominated by the Nypa palm, while the most widely spread grass species is *Dalbergia ecastophylum*. Other invasive species like water hyacinth (*Eichhornia crassipes*) are found in stagnant waters. Meanwhile, *Rhizophora racemosa* is the most important of the different vegetation species scattered across the Niger Delta's mangroves. This is because it overlays about 90% of the mangrove environment (Izah 2018). More extensively, the coast of West Africa is known to be dominated by white mangroves (*Avicenniaceae sp.*), *Avicennia africana* and *Combretaceae* (*Laguncularia racemosa* and *Conocarpus erectus*). Some of the plants diversities in the region that are reported to contain effective antimicrobial ingredients include *Anthocleista vogelii*, *Musanga cecropioide*, *Vitex grandifolia*, *Anthocleista djalonensis*, among others. Again, the Raphia palm, oil palm, and rubber trees are prevalently distributed in the Southern parts of Nigeria, while kola nut plants, cocoa forest plantations, as well other commercial fruit trees can be found across the Niger Delta region. Furthermore, the vegetation resources of the Niger Delta (oil palm, banana, maize, mango, rice, coconut, beans, avocado pear, vegetables, pineapple, oranges, etc.) are important food sources of the people (Izah 2018; Uwagbae et al. 2019). The identified species of higher and lower plants in Nigeria's coastal region are about 248,400 and 73,900, respectively. Still, about 15% of total plant diversity remains undiscovered (Davies et al. 2020). For instance, the protected forest areas of Taylor creek are characterized by vegetation species such as *Raphia hookeri* (Raphia palm), *Eleais guineensis* (oil palm), *Musanga cecropioides* (umbrella tree), ferns, macrophytes, etc. Many of the trees of this protected zone are economically essential. Another protected area (Nun river forest) is a tropical rainforest characterized by layers of vegetation, while protected forest zones of the Ikibiri creek play host to important resources such as ornamental forages, timber resources, and medicinal herbages (Numbere 2018).

6.6 Impacts of Biodiversity Decline in the Niger Delta

6.6.1 Reduction in Population and Diversity Trend

Population pressure and the resultant effect of urbanization have led to the loss of forest resources. Therefore, the overexploitative acts of deforestation have continued to decimate the numerous valuable species of vegetation and animals which are required for energy balance across the food chain. Available global data reveal that Nigeria's deforestation rate has increased from 3.3% to 37.7% between 1990 and 2005. Based on this, the nation's environmental sustainability goals are been threatened (Izah and Seiyaboh 2018b). For instance, 21 of the 55 animal species used by traditional herb preparers in Ogun state, Nigeria, including the amphibian species such as toad (*Bufo regularis*) and frog (*Rana temporaria*)), reptilian species such as African python (*Python sebae*), Nile crocodile (*Crocodylus niloticus*) and cobra (*Naja* species), avian species such as African grey parrot (*Psittacus erithacus*), spotted eagle owl (*Bubo africanus*), black kite (*Milvus migrans*), red eye dove (*Streptoprelia semitorquata*) and double-spurred francolin (*Francolinus bicalcaratus*), etc.), mammalian species such as white-bellied pangolin (*Manis tricuspis*), African civet (*Civettictis civetta*), African buffalo (*Syncerus caffer*), Nile rat (*Arvicanthis niloticus*), spotted hyena (*Crocuta crocuta*), straw-colored fruit bat (*Eidolon helvum*), etc. are classified under the threatened category of Nigeria's endangered species (Decree 11, 1985 and CITES) (Soewu 2008). As such, the hunting of wildlife for food or medicine has greatly contributed to the decline in species diversity across its forestlands (Izah 2018; Izah and Seiyaboh 2018b).

6.6.2 Loss of Medically Valuable Biodiversities

Climate change stemming from activities such as deforestation, hunting and wildlife trading, bush burning, industrialization, among other environmental issues continues to portend negative impacts on the larger environment (Izah et al. 2017a). As habitats are negatively affected or destroyed, there is dwindling population of biodiversity. Hence, animals and plants of therapeutic importance are lost, or in some cases, migrate to adjoining habitats. Owing to the reliance of an estimated 80% of people around the globe to botanicals and wild games for medicine (Izah et al. 2018d), the increasing popularity of herbal concoctions of wildlife origin has continued to threaten the existence of some endangered mammalian species (Izah and Seiyaboh 2018b). As such, there is need to integrate the socioeconomic and ecological aspects of practice in a way that biodiversity remains conserved and ecological roles are left unaltered (Numbere 2018).

6.6.3 Loss of Economically Important Biodiversities

The contribution of various biodiversity to human livelihood is immense (Ogunrinola and Adepegba 2012). While insects are responsible for the pollination of plants, animals provide majority of man's food protein needs. For instance, its skin hide is utilized for leather production. Furthermore, some plants are used as medicine, while others are lumbered as timber, or utilized for paper or fabric production. Also, the presence of biodiversity aids in balancing life-sustaining energy sources. An estimated 75% of the global population is reported to use fuel wood to either generate heat energy during cold winters or for cooking. This is according to reports of the United Nation's Food and Agricultural Organization. Meanwhile, the continuous acts of deforestation and wood lumbering, bush burning, and intensive agricultural practices tend to culminate in the loss of numerous economically important flora and fauna species. Based on this, changes in forest resources have the potential to impact species composition by causing inter-habitat migration and loss or total extinction of species (Izah et al. 2017a).

6.6.4 Climate Change and Environmental Degradation

If the current environmentally unsustainable activities persist, there is a possibility that human lives will be at risk of environmental implosion from climate change effects such as increasing temperatures, rise in flood occurrences, elevated levels of greenhouse gases (GHG), among others. Consequently, most of the forest biodiversity can either be classified as threatened, endangered, critically endangered, and extinction according to the IUCN criteria. As at 2011, the decimation of forests in the Niger Delta has led to about 195 square kilometers of bare land, 132 square kilometers of surface waters, 8338 square kilometers of agricultural farms, 2898 square kilometers of urban zones, 1535 square kilometers of mangroves, and 1424 square kilometers of freshwater rainforest (IUCN 2018; Aigberua and Okumoko 2020). Of the various ecosystems that make up this region, the tableland rainforest has been most impacted, especially from intense agricultural practices. Another major contributor to climate change and environmental degradation is bush-burning which exacerbates desertification, hence contributing to the global rise in temperature. Land use and deterioration can negatively impact biodiversity abundance and cause climate shift. For example, rising carbon dioxide and atmospheric temperatures, rising water levels, and the concurrent flooding episodes tend to affect the phenology of flora and fauna species, as well as impact the composition of species population by impairing their manner of reaction, ultimately culminating to the loss of habitat (Izah et al. 2017a; Angaye et al. 2018).

6.6.5 Effects on Ecological and Nutrient Cycling

An estimated 66% and 50% of organic carbon are stored in above-ground terrestrial and forest soils, respectively. This emanates from resident organisms (flora and fauna). However, the activities of hunters and farmers (via bush burning) continue to threaten the existence of these species. For instance, bush burning significantly lowers the population of biodiversity, while leading to organic matter loss or depletion. The bush burning process tends to interrupt ecological functions, in which case animals in caves or hiding in below-ground enclaves are hunted for food or medicine. Although this act alters wildlife assemblages, it improves the overall soil quality and reconstitutes vegetation diversity, as well as creating carbon equilibrium. In addition, microorganisms act as life support and soil nutrient supplier, thus enhancing the crop productivity potential of soil. As such, the loss of microbes may impact agriculture and negatively affect humans whose survival is hinged on food plants. *Cyanobacteria, Azobacter, Nitrobacter, Nitrosomonas, and Desulfovibrio* are among the numerous nutrient-cycling microbes, but fires tend to weaken or exterminate soil fungi and animals thereby limiting the environmental degradation of waste garbage (Izah et al. 2017a).

6.6.6 Destruction of Habitats

The frequency of bush burning episodes tends to cause an imbalance in ecological functions. During the incineration process, bird's eggs, nests, rodents, and numerous food sources are destroyed, while insect and mammalian diversity are significantly affected. For instance, the loss to fire of the *Musanga cecropioides* (African corkwood tree or umbrella), a major staple of wildlife, (monkeys, etc.) could result in the migration of the mammals (Ohimain et al. 2014a, b). Similarly, the incineration of open wastes in dumpsites (preferred habitation of rats) often leads to their death, thereby leading to the absence of food for animals that prey on rodents. Some of the wildlife most affected by bush burning in the Niger Delta includes mammalian species (white-throat monkey, black house rat, house mouse, grass cutter/cane rat, antelope, wild rabbit, red-less tree squirrel, red river hog, etc.), reptilian species (black-necked cobra, viper, tortoise (*Kinixys erosa*)), West African black forest turtle (*Pelusios niger*), avian species (harrier hawk, crested guinea fowl, black kite, etc.) (Ohimain et al. 2014b), insects (crickets, maggot (*Rhynchoporus sp.*), grasshoppers, etc.), and edible mollusc species such as snail. Based on the foregoing, the incineration of the environment leads to the loss of habitat, species diversity, and abundance, while indirectly affecting humans (Izah et al. 2017a).

6.7 Possible Conservation Strategies

6.7.1 Pasture Management

The practice whereby grasses or related plants are grown for the sustenance of forage diversity or abundance and support of livestock production should be adopted. The inclusion of this method will allow for environmental balance and health. In addition, the practice will likely improve the overall quality of soil. If pastures are well-managed or structured, animal ranches will become sustainable, while ultimately enhancing environmental health (Izah et al. 2017a).

6.7.2 Minimization of Waste; Reuse or Recycle

Human and industrial activities have continued to negatively impact the environment. The Niger Delta of Nigeria is known to be a major hub of crude oil and gas production, while also prone to seasonal flood regimes due to rising water levels, soil erosion, oil spillage events, suspended particulates/smog/black soot, elevated levels of greenhouse gases and temperature, among others. Consequently, it is the most environmentally impacted region in Nigeria, with the inadequate remediation of hydrocarbon-related pollutants often leading to public health emergencies, while causing agitation among local inhabitants. Hence, there is continuous clamor for appropriate environmental remediation plan of action. A more effective waste management option that incorporates waste reuse and recycling is needed to assuage the ever-increasing and emerging waste streams that are generated from increased industrialization and urbanization (Izah et al. 2017a; Angaye et al. 2018; Aigberua and Okumoko 2020). Overall, practices that will discourage the increasing levels of waste discharged into the environment must be instituted as a way of safeguarding ecological and human health. Hazardous waste releases must be efficiently controlled or prohibited to optimize resource conservation and reuse.

6.7.3 Formulation, Implementation, and Enforcement of Strict Compliance to Environmental Laws

There is urgent necessity to conserve our nature—given ecological resources intensify the need to institute working policies and legislations. One of such elements that will make environmental laws workable is its inclusivity of all stakeholders and interest groups (Izah et al. 2017a). The government may need to utilize environmental extension workers to notify rural dwellers and farmers of the dangers of bush incineration and indiscriminate animal hunting, while delegating local vigilante associations and the police with the responsibility of enforcing government rules and regulatory stipulations as a check against defaulters.

6.7.4 Sensitization of Holiday Makers and the General Public

Foreigners or visitors should be made aware of the environmental laws guiding their host country. As such, environmental guidelines should be publicly displayed on billboards, via social media sites, and television/radio jingles, while also utilizing tour guides as alternative channels for educating holiday makers on the existing environmental laws guiding every locality. This may forest all flagrant abuse of the environment. For instance, people visiting protected areas or reserved forests should be properly oriented in order to forestall acts that may endanger the ecological biodiversity (Izah et al. 2017a). Furthermore, signs disallowing smoking in parks and games reserves can be sighted in public places in order to avoid intentional or accidental fires. Also, signs disallowing open grazing, defecation, or littering will help preserve the esthetic value of reserved parks.

6.7.5 Introduction of Biodiversity Conservation in School Curriculum

Biodiversity conservation should be introduced to children at a tender age, while being incorporated into tertiary learning for young adults and the aged (Izah et al. 2017a). That way, information regarding a person's social responsibility towards safeguarding ecological biodiversity will be adequately communicated and culturally instilled, hence inculcating a positive attitude towards future environmental sustainability.

6.8 Conclusion

The adverse effect of climate change, rising temperature and water levels, elevated concentrations of greenhouse gases owing to industrialization and intensive agricultural practices (encroachment of farmlands and cattle grazing within forest reserves, and bush burning), infrastructural development, amongst others, has continued to negatively impact the distribution of endemic species surviving in the Niger Delta area of Nigeria. Although environmental sustainability guidelines and laws exist, they are largely unenforced, partly due to the poor public dissemination of environmental information. However, a more socially responsible attitude can be inculcated among the local inhabitants of the region by imputing biodiversity conservation studies within school curriculum at all ages, thereby creating a more sustainable environment. In addition, the government may need to adopt a deliberate environmental appraisal with the aim of gauging the effectiveness of existing environmental policies, eco-diversity scheme, or initiatives. Since the Niger Delta environment by virtue of its abundant oil and gas reserves has its ecosystem constantly being threatened by the devastating effects of crude oil spills during oil production and transportation, it has become pertinent that effective eco-diversity conservation policies and strategies are considered and incorporated during the planning of reclamation or pollution containment programmes.

References

Abowei JFN, Hart AI (2008) Artisanal fisheries characteristics of the fresh water reaches of lower Nun River, Niger Delta, Nigeria. J Appl Sci Environ Manage 12(2):13–15

Adekola O, Mitchell G (2011) The Niger Delta wetlands: threats to ecosystem services, their importance to dependent communities and possible management measures. Int J Biodiver Sci Ecosyst Serv Manage 7(1):50–68

Agedah EC, Ineyougha ER, Izah SC, Orutugu LA (2015) Enumeration of total heterotrophic bacteria and some physico-chemical characteristics of surface water used for drinking sources in Wilberforce Island, Nigeria. J Environ Treat Tech 3(1):28–34

Aghoghovwia OA, Ohimain EI, Izah SC (2016) Bioaccumulation of Heavy metals in different tissues of some commercially important fish species from Warri River, Niger Delta, Nigeria. Biotechnol Res 2(1):25–32

Aigberua AO, Izah SC (2018) Evaluation of heavy metals in tissue of *Tympanotonus fuscatus* sold in some markets in Port Harcourt metropolis, Nigeria. MOJ Toxicol 4(5):334–338

Aigberua AO, Okumoko DP (2020) Distribution spread and environmental risk status of Pb, Cd, And Cr in soils of an open-air waste dumpsite along Tombia/Amassoma Road in Yenagoa Metropolis. J Plant Anim Ecol 1(3):29–43

Aigberua AO, Ekubo AT, Inengite AK, Izah SC (2016a) Evaluation of total hydrocarbon content and polycyclic aromatic hydrocarbons in an oil spill contaminated soil in Rumuolukwu Community in Niger delta. J Environ Treat Tech 4(4):130–142

Aigberua AO, Ekubo AT, Inengite AK, Izah SC (2016b) Seasonal variation of nutrient composition in an oil spill contaminated soil: a case of Rumuolukwu, Eneka, Port Harcourt, Nigeria. Biotechnol Res 2(4):179–186

Aigberua AO, Izah SC, Richard G (2021) Hazard analysis of trace metals in muscle of Sarotherodon melanotheron and Chrysichthys nigrodigitatus from Okulu River, Rivers State. Nigeria. J Environ Health Sustain Dev 6(3):1340–1356. https://doi.org/10.18502/jehsd.v6i3.7242

Akani GC, Politano E, Luiselli L (2004) Amphibians recorded in forest swamp areas of the River Niger Delta (southeastern Nigeria), and the effects of habitat alteration from oil industry development on species richness and diversity. Appl Herpetol 2(1):1–22

Akani GC, Aifesehi PEE, Petrozzi F, Amadi N, Luiselli L (2014a) Preliminary surveys of the terrestrial vertebrate fauna (mammals, reptiles, and amphibians) of the Edumanon Forest Reserve, Nigeria. Trop Zool 27(3):63–72

Akani GC, Aifesehi PEE, Petrozzi F, Amadi N, Luiselli L (2014b) Diversity of terrestrial vertebrates in Taylor creek forest reserve, an area of high environmental value in the River Niger Delta (Bayelsa State, Nigeria). Vie et milieu Life Environ 64:59–68

Akani GC, Amadi N, Eniang EA, Luiselli L, Petrozzi F (2015) Are mammal communities occurring at a regional scale reliably represented in "hub" bushmeat markets? A case study with Bayelsa State (Niger Delta, Nigeria). Folia Zool 4(1):79–86

Akinwale A (2004) The potential for climate change mitigation in the nigerian solid waste disposal sector: a case study from Lagos (MSc.). Lund University, Sweden

Amadi N, Akani GC, Micheloni P, Eniang EA, Luiselli L, Petrozzi F (2016) Distribution, habitat ecology and conservation status of the two-spotted Palm Civet Nandinia binotata (Carnivora, Nandiniidae) in south-eastern Nigeria. Small Carnivore Conserv 52–53:24–38

Angaye TCN, Daokoru-Olukole C, Abowei JFN (2018) Environmental impacts of municipal solid wastes in Yenagoa Metropolis, Bayelsa State, Nigeria. Biotechnol Res 4(1):17–23

Asimiea A, Omokhua G (2013) Environmental impact of illegal refineries on the vegetation of the Niger delta, Nigeria. J Agric and Soc Res 13(2):121–126

Bob-Manuel FG (2012) A prelimenary study on the population estimation of the periwinkles Tympanotonus fuscatus (Linnaeus, 1758) and Pachymelania aurita (Muller) at the Rumuolumeni mangrove swamp creek, Niger Delta, Nigeria. Agric Biol J North Am 3(6): 265–270

Davies OA, Esaenwi AE, Amachree D (2020) Diodiversity issues in Fresh waters of Niger Delta. Int J Agric Earth Sci 6(1):52–58

Emeudo OA, Emeudo CO (2018) Biodiversity and oil activities in the Niger Delta region of Nigeria. J Geo Environ Earth Sci Int 14(3):1–8

Emma-Okafor LC, Ibeawuchi II, Obiefuna JC (2010) Biodiversity conservation for sustainable agriculture in tropical rainforest of Nigeria. New York Sci J 3(1):81–88

Enerosisor MS, Ugbomeh AP, Miebaka M (2020) Abundance and diversity of zooplankton in the lower reach of the Opobo river, Rivers State, Nigeria. Afr J Environ Nat Sci Res 3(2):49–59

Epidi JO, Izah SC, Ohimain EI (2016a) Antibacterial and synergistic efficacy of extracts of Alstonia boonei tissues. Br J App Sci 1(1):21–26

Epidi JO, Izah SC, Ohimain EI, Epidi TT (2016b) Antibacterial and synergistic potency of tissues of Vitex grandifolia. Biotechnol Res 2(2):69–76

Federal Republic of Nigeria (FRN) (2010) Fourth National Biodivers Rep, Abuja

FMENV (2004) Site prioritization to select conservation site in selected states of the Niger Delta, Federal Ministry of Environment (FMENV), Abuja, Nigeria. Techni Rep

Hamadina MK, Otobotekere D, Anyanwu DI (2007) Impact assessment and biodiversity considerations in Nigeria: a case study of Niger Delta University campus project on wildlife in Nun River forest reserve. Manag Environ Qual Int J 18(2):179–197

Ideriah TJK, Harry FO, Stanley HO, Igbara JK (2010) Heavy metal contamination of soils and vegetation around solid waste dumps in port Harcourt, Nigeria. J Appl Sci Environ Manage 14(1):101–109

Ikemeh RA (2015) Assessing the population status of the critically endangered Niger Delta Red Colobus (Piliocolobus epieni). Primate Conser 29:87–96

IUCN Niger Delta Panel (2018) Developing a biodiversity conservation strategy for the Niger Delta: integrating biodiversity considerations into SPDC's operations. IUCN, viii+36 pp, Gland, Switzerland. https://doi.org/10.2305/IUCN.CH.2018.07.en

Iwar IM, Egwumah PO, Olufeagba SO (2019) Phylogenetic study of West African manatee Trichechus Senegalensis link (1708) along River Benue, Nigeria. J Res Wildlife Environ 11(4):70–78

Izah SC (2018) Ecosystem of the Niger Delta region of Nigeria: potentials and threats. Biodiversity Int J 2(4):338–345

Izah SC, Angaye TCN (2016) Heavy metal concentration in fishes from surface water in Nigeria: potential sources of pollutants and mitigation measures. Sky J Biochem Res 5(4):31–47

Izah SC, Aseibai ER (2018) Antibacterial and synergistic activities of methanolic leaves extract of lemon grass (Cymbopogon citratus) and rhizomes of ginger (Zingiber officinale) against Escherichia coli, Staphylococcus aureus and Bacillus subtilis. Acta Sci Microbiol 1(6):26–30

Izah SC, Seiyaboh EI (2018a) Changes in the protected areas of Bayelsa State, Nigeria. Int J Mol Evol Biodiversity 8(1):1–11

Izah SC, Seiyaboh EI (2018b) Challenges in wildlife with therapeutic properties in Nigeria; a conservation perspective. Int J Avian Wildlife Biol 3(4):256–264

Izah SC, Angaye TCN, Aigberua AO, Nduka JO (2017a) Uncontrolled bush burning in the Niger Delta region of Nigeria: potential causes and impacts on biodiversity. Int J Mol Ecol Conserv 7(1):1–16

Izah SC, Bassey SE, Ohimain EI (2017b) Assessment of heavy metal in cassava mill effluent contaminated soil in a rural community in the Niger Delta region of Nigeria. EC Pharmacol Toxicol 4(5):186–201

Izah SC, Zige DV, Alagoa KJ, Uhunmwangho EJ, Iyamu AO (2018a) Antibacterial efficacy of aqueous extract of Myristica fragrans (Common Nutmeg). EC Pharmacol Toxicol 6(4):291–295

Izah SC, Uhunmwangho EJ, Dunga KE, Kigigha LT (2018b) Synergy of methanolic leave and stem-back extract of Anacardium occidentale l. (cashew) against some enteric and superficial bacteria pathogens. MOJ Toxicol 4:209–211

Izah SC, Uhunmwangho EJ, Dunga KE (2018c) Studies on the synergistic effectiveness of methanolic extract of leaves and roots of Carica papaya L. (papaya) against some bacteria pathogens. Int J Complement Altern Med 11(6):375–378

Izah SC, Uhunmwangho EJ, Eledo BO (2018d) Medicinal potentials of *Buchholzia coriacea* (wonderful kola). Med Plant Res 8(5):27–43

Izah SC, Uhunmwangho EJ, Etim NG (2018e) Antibacterial and synergistic potency of methanolic leaf extracts of Vernonia amygdalina L. and Ocimum gratissimum L. J Basic Pharmacol Toxicol 2(1):8–12

Izah SC, Aigberua AO, Nduka JO (2018f) Factors affecting the population trend of biodiversity in the Niger Delta region of Nigeria. Int J Avian Wildlife Biol 3(3):206–214

Izah SC, Etim NG, Ilerhunmwuwa IA, Ibibo TD, Udumo JJ (2019a) Activities of express extracts of Costus afer Ker–Gawl. [Family COSTACEAE] against selected bacterial isolates. Int J Pharm Phytopharm Res 9(4):39–44

Izah SC, Etim NG, Ilerhunmwuwa IA, Silas G (2019b) Evaluation of crude and ethanolic extracts of Capsicum frutescens var. minima fruit against some common bacterial pathogens. Int J Complement Altern Med 12(3):105–108

Jamabo N, Chinda A (2010) Aspects of the ecology of *Tympanotonus fuscatus var fuscatus* (Linnaeus, 1758) in the mangrove swamps of the Upper Bonny River, Niger Delta, Nigeria. Curr Res J Biol Sci 2(1):42–47

Kadafa AA (2012) Environmental impacts of oil exploration and exploitation in the Niger Delta of Nigeria. Global J Sci Frontier Res Environ Earth Sci 12(3):18–28

Kigigha LT, Izah SC, Ehizibue M (2015) Activities of Aframomum melegueta seed against Escherichia coli, *Staphylococcus* aureus and Bacillus species. Point J Bot Microbiol Res 1: 23–29

Kigigha LT, Biye SE, Izah SC (2016) Phytochemical and antibacterial activities of *Musanga cecropioides* tissues against *Escherichia coli*, *Pseudomonas aeruginosa Staphylococcus aureus*, *Proteus* and *Bacillus* species. Int J Appl Res Technol 5(1):100–107

Kigigha LT, Selekere RE, Izah SC (2018a) Antibacterial and synergistic efficacy of acetone extracts of Garcinia kola (Bitter kola) and Buchholzia coriacea (Wonderful kola). J Basic Pharmacol Toxicol 2:13–17

Kigigha LT, Izah SC, Uhunmwangho EJ (2018b) Assessment of hot water and ethanolic leaf extracts of Cymbopogon citratus Stapf (Lemon grass) against selected bacteria pathogens. Ann Microbiol Infect Dis 1(3):1–5

Lameed GA (2010) Potential impact on biodiversity in Kwale's forest reserve by power plant establishments. Afr J Food Agric Nutr Dev 9(9):1878–1900

National Population Commission (NPC) (2009) Legal notice on publication of 2006 Census, Final Results. Federal Republic of Nigeria, Abuja. Official Gazette, Vol. 96, pp B1-42

Niger Delta Development Commission (NDDC) (2004). Biodiversity and natural resources sector study. Niger Delta regional master plan. Niger Delta Development Commission (NDDC), Abuja, Nigeria.

Numbere O (2018) Mangrove species distribution and composition, adaptive strategies and ecosystem services in the Niger River Delta, Nigeria. In Mangrove ecosystem ecology and function. Intech Open Publishers, London, UK. DOI: doi: https://doi.org/10.5772/intechopen.79028.

Nwosu EE, Pepple GT (2016). Site selection and analysis of solid waste dumpsites in Ile-Ife, Nigeria (8363). FIG Working Week 2016, recovery form disaster Christchurch, New Zealand, May 2–6, 2016.

Ogamba EN, Izah SC, Ebiowe RG (2015) Bioconcentration of Mercury, Lead and Cadmium in the bones and muscles of *Citharinus citharus* and *Synodontis clarias* from the Amassoma Axis of River Nun, Niger Delta, Nigeria. Res J Pharmacol Toxicol 1(1):21–23

Ogamba EN, Izah SC, Omonibo E (2016a) Bioaccumulation of hydrocarbon, heavy metals and minerals in *Tympanotonus fuscatus* from coastal region of Bayelsa state, Nigeria. Int J Hydrol Res 1:1–7

Ogamba EN, Izah SC, Isimayemiema F (2016b) Bioaccumulation of heavy metals in the gill and liver of a common Niger Delta wetland fish, *Clarias garepinus*. Br J Appl Res 1(1):17–20

Ogamba EN, Izah SC, Ofoni-Ofoni AS (2016c) Bioaccumulation of chromium, lead and cadmium in the bones and tissues of Oreochromis niloticus and Clarias camerunensis from Ikoli creek, Niger Delta, Nigeria. Adv Sci J Zool 1(1):13–16

Ogamba EN, Ebere N, Izah SC (2017) Levels of lead and cadmium in the bone and muscle tissues of Oreochromis niloticus and Clarias camerunensis. EC Nutr 7(3):117–123

Ogbe MG (2011) Managing the environmental challenges of the oil and gas industry in the Niger Delta, Nigeria. J Life Sci 1(1):1–17

Ogunrinola IO, Adepegba EO (2012) Health and economic implications of waste dumpsites in cities: the case of Lagos, Nigeria. Int J Econ Finance 4(4):239–251

Ohimain EI, Akinnibosun HA (2007) Assessment of wetland hydrology, hydrophytic vegetation and hydric soil as indicators for wetland determination. Trop J Environ Sci Health 10(1):1–11

Ohimain EI, Otobotekere D, Biriduba W (2014a) Unsustainable Exploitation of Freshwater Wetland Turtles and Tortoises in Central Niger Delta. Int J Environ Monitor Anal 2(2):57–64

Ohimain EI, Izah SC, Otobotekere D (2014b) Selective impacts of the 2012 water floods on the vegetation and wildlife of Wilberforce Island, Nigeria. Int J Environ Monitor Anal 2:73–85

Okonkwo CNP, Kumar L, Taylor S (2015) The Niger Delta wetland ecosystem: What threatens it and why should we protect it? Afr J Environ Sci Technol 9(5):451–463

Okumagba P (2011) Oil exploration and ethnic militia activities in the Niger Delta region of Nigeria. Int Multidisc J Ethiopia 5(5):56–67. https://doi.org/10.4314/afrrev.v5i5.6

Omotosho FO (2004) A reflection on the hazards of sea pollution: Case of Minmata disease in Japan. Paper presented at the proceeding of the 12th Annual National Conference of EBAN Abeokuta, Nigeria.

Onuoha CB, Bassey EI, Ufomba H (2018) The impact of oil exploration and environmental degradation in the Niger Delta region of Nigeria: a study of oil producing communities in Akwa Ibom State. Global J Human Soc Sci F, Pol Sci 18(3):54–69

Phil-Eze PO, Okoro IC (2009) Sustainable biodiversity conservation in the Niger Delta: a practical approach to conservation site selection. Biodivers Conserv 18:1247–1257

Phil-Eze PO, Umeuduji JE (2004) Human impact on biodiversity. In: Bell-Gam WI, Arokoyo SB, Umeuduji JE (eds) Perspectives on the human environment. Amethyst & Colleagues Publishers, Port Harcourt, Nigeria, pp 54–70

Ramsar Convention Secretariat (2007) Information Sheet on Ramsar Wetlands (RIS): Apoi Creek. https://rsis.ramsar.org/ris/1751

Scott DA (1989) A directory of Asian wetlands. IUCN, Gland and Cambridge

Soewu DA (2008) Wild animals in ethnozoological practices among the Yorubas of southwestern Nigeria and the implications for biodiversity conservation. Afr J Agric Res 3(6):421–427

Talbot CJ, Bennett EM, Cassell K et al (2018) The impact of flooding on aquatic ecosystem services. Biogeochem 141:439–461. https://doi.org/10.1007/s10533-018-0449-7

Tarawou T, Erepamowei Y, Aigberua A (2019) Assessment of heavy metals contamination of water and Cupscale Grass (*Sacciolepis africana*) along the Epie Creek in Bayelsa State, Nigeria. J Plant Anim Ecol 1(4):7–15

Tarvirdizadeh H, Nikooy M, Pourbabaei H, Naghdi R (2014) Effects of road construction on biodiversity and composition of herbaceous species cover, Asalem forest, Northern Iran. Forest Ideas 20(2):157–169

Ubom RM (2010) Ethnobotany and biodiversity conservation in the Niger Delta. Nigeria. Intern J Bot 6(3):310–322. https://doi.org/10.3923/ijb.2010.310.322

Ugochukwu CNC, Ertel J (2008) Negative impacts of oil exploration on biodiversity management in the Niger Delta area of Nigeria. Impact Assess Proj Appra 26(2):139–147. https://doi.org/10.3152/146155108X

Uwagbae MA, Dirisu, AR, Gbarakoro TN, El Surtasi EI, Agwu JE, Aisien MSO, Rotimi J (2019) Impact of disturbances on the biodiversity of Ijala-Ikeren wetland ecosystem in Niger Delta. In: Changing ecosystems and their services. IntechOpen Publishers, London, UK

Vaiskunaite R, Mierauskas P, Spakauskas V (2012) Biodiversity impact assessment in road development in Lithuania. Transport 27(2):187–195. https://doi.org/10.3846/16484142.2012.695749

Xie Y, Fan J, Zhu W, Amombo E, Lou Y, Chan L, Fu J (2016) Effect of heavy metals pollution on soil microbial diversity and bermudagrass genetic variation. Front Plant Sci 7:755

Chapter 7
Exploration of Local Beliefs and Cultural Heritages as Tools for Species Conservation in Selected Sites in Africa

Aroloye O. Numbere and Pasiya P. Otufu

Abstract The most effective means of conserving species is incorporating the human factor in the conservation plan, a "win-win" ecology situation. In Africa, cultural and traditional heritages are highly respected because they are conservation tools and play a vital role in people's daily lives. There is a strong acceptance of local cosmology, which believes in the potency of ancestral spirits and powers. Therefore, traditional laws can be used to protect sites with endangered species and would prevent the overexploitation and extinction of many species in the Niger Delta area. Sacred places can be established in localities that have endangered species across the region, such as the crocodile (*Osteolaemis tetraspis*) holy site in the Biseni and Osiama Kingdoms, which resulted over the years in the protection of thousands of this species from human hunting. If adequately documented, the protected areas can serve as conservation sites that become a tourist's attraction to generate revenue for the local community. Therefore, this chapter uniquely provides examples of how local belief has been used to preserve forests and rivers in some Niger Delta communities.

Keywords African religious · Cultural beliefs · Heritage sites · Species conservation strategies

7.1 Introduction

Conservation is the prevention of wasteful use of a resource (Sandbrook 2015). In the past, much of traditional conservation was rooted in an economical, utilitarian philosophy where the primary motivation was to maintain high yields of selected species for harvest. Contributions develop conservation biology from theoretically oriented academicians where scientific models are applied to real-world situations. Furthermore, conservation biology embraces input from nonbiologists such as locals

---

A. O. Numbere (✉) · P. P. Otufu
Department of Animal and Environmental Biology, University of Port Harcourt, Port Harcourt, Nigeria

S. C. Izah, M. C. Ogwu (eds.), *Sustainable Utilization and Conservation of Africa's Biological Resources and Environment*, Sustainable Development and Biodiversity 32, https://doi.org/10.1007/978-981-19-6974-4_7

and indigenous people who know the types of species in their community. This knowledge is mainly derived from oral history, where their forefathers told the next generation about the cultural practice of the community. However, today philosophical and spiritual beliefs hold sway in preserving some selected species in many African societies. The people's cultural opinions in seeing animals as the embodiment of spirits make them revere some animals, known as totems. It is the practice of protecting certain animals from being killed because of their cultural value and spiritual significance. Conservation biology has its historical roots in three philosophical movements, which began in America (Frank 2022). Before the nineteenth century, there was no effort to conserve wildlife, and natural resources were exploited for food, timber, fuel, and farm. Conservation of natural resources, which has spread to Africa, began its foundation in African traditional religion, which will be discussed later in this chapter. The three philosophical movements are (1) Mid 1800s-romantic-transcendental conservation ethics, (2) 1900-resource conservation ethic, and (3) early/mid-twentieth century-evolutionary ecological land ethic.

7.1.1 Mid 1800s-Romantic-Transcendental Conservation Ethics

This theory gained its root from Ralph Waldo Emerson, John Muir, and Henry Thoreau, who all stated that nature has an intrinsic value and, thus, is good for the soul (Bergstrom 2022). The soul factor shows the earth as a living entity with life that must be protected for present and future generations. In other words, they see nature in a quasi-religious sense where people worship nature as a demi-god. Muir led a national movement to preserve nature in pristine form, a situation whereby natural resources would be left alone and not exploited for the use of humans. His legacy is the setting up of the Sierra Club in the United States (Kennedy 2022). However, the above noneconomic view was countered by the 1900-resource conservation ethics that stipulate the sustainable use of natural resources for the more significant benefit of humanity. It is believed that when resources benefit humans, they will take up the responsibility of protecting them from elimination.

7.1.2 1900-Resource Conservation Ethic

This ethic holds a contrary view to the romantic-transcendental ethics by declaring that nature provides good to people, making it worthwhile. The utilitarian principle means that nature is valuable because species benefit people, and if species benefit people, they need to be protected from extinction (Sullivan et al. 2022). This ethic was made popular by Gifford Pinchot, who was the first head of the National Forest

Service (NFS) (Schweitzer 2019). He stated that nature can be exploited sustainably. Currently, NFS canvasses for multiple use policy. Gifford Pinchot stated that ": nature should be used for the greatest good for the greatest number for the largest time" (Pinchot 1947). The romantic transcendentalists believe that the anthropogenic valuing of nature is not because it is part of God's design; rather, nature is appreciated because it feeds and contributes to the material quality of life. Pinchot's approach stressed equity and fair distribution of resources for present and future generations (Banzhaf 2019). This theory led to the adoption of the "multi-use concept," which remains the mandate of the US Forest Service and the Bureau of land management. This style of land management has been adopted so far in many parts of the world including Africa to manage species. For instance, in the Niger Delta, there is a clarion call to preserve the most dominant resource, crude oil, and adopt other sources of energy such as solar, biomass, geothermal, and hydro energies. It has been predicted that in the next 50 years, the crude oil resource may vanish and will not be available for future generations.

7.1.3 Early/Mid-Twentieth Century-Evolutionary Ecological Land Ethic

This ethical model was propounded by Aldo Leopold, who stated that nature is complex because it is composed of ecosystems (Heffron 2022), and ecosystem is composed of interacting species such as predator-prey, mutualistic, and competitive interactions. It states that nature is a complicated and integrated system of interdependent processes and components. A dynamic nonequilibrium view subsequently replaced this equilibrium view. It is the philosophical foundation of conservation biology. If we want to explore nature, we will focus on one species, such as plants and cattle. This is because species interaction has a cascading effect on other problems. Leopold said that we need a better understanding of how nature works in other to exploit it sustainably, i.e., we need to do more scientific study, fieldwork to collect data, and explore more unanswered questions in science. Summing all the aspects of the historical antecedents of conservation, it is pertinent that the world has held onto different ethical models to protect and sustain the environmental resources.

7.2 Categorization of Protected Areas

Protected areas are categorized into the following groups:

(i) Ecological reserves and wilderness areas (e.g., Cassidy et al. 2022)
(ii) National parks: Namibia National Parks (Thomsen et al. 2022); Virunga National Park (Marijnen 2022)

(iii) Natural monuments and archeological sites: Greece (Mentzafou and Dimitriou 2022)
(iv) Habitat and wildlife management areas, e.g., Almasieh et al. 2022
(v) Cultural, scenic landscapes or recreation areas, e.g., Zhao et al. 2022

There has been a steady increase in protected areas worldwide after the first park (Yellowstone National Park) was established in the United States in 1872 (Smith and Bangs 2009). The growth of parks, especially in the tropical region, has equally risen since the early 1950s. The International Union for the Conservation of Nature (IUCN) has, in the past years, developed a conservation strategy to maintain ecological processes, preserve genetic diversity, and ensure sustainability (IUCN 1980). The concerted efforts of these world bodies have led to more than 530 million ha (nearly 4% of the earth's land) being designated as parks and wildlife refuges. According to the United Nations Environmental Program (UNEP), North and Central America have the most significant fraction of all protected land, i.e., 33% or nearly 10% of their land area is designated for protection of any continent (Sierra et al. 2002). Also, the largest protected area is in the tropical dry forests and savannas. Biodiversity hotspots are also found in Africa, e.g., Cape of South Africa (Forest et al. 2007), and the temperate deciduous forest in North America and Europe. Africa possesses the largest wildlife centers globally, from South Africa to Tanzania and Congo to Burundi. Most of the large, bodied animals (elephants, giraffes, rhinoceros, etc.) in the zoos of Europe and America are shipped from Africa. For instance, one of the top zoos in the United States, the Saint Louis Zoo, has 90% of its wild animals brought from different parts of Africa. This is because of Africa's favorable tropical environment that supports the survival of all kinds of beasts of the earth. According to one strange theory, it was alleged that Africans were responsible for the extinction of big-bodied animals in Europe and America. The theory stated that Africans migrated to Europe searching for animal meat through a land bridge and hunted down its big animals after exhausting the animals in Africa. Whether this theory is true is left for more scientific evidence to justify. However, the problem with the African wildlife is recurring conflict such as wars and intertribal misunderstandings, which has resulted in the decimation of most of these animals. Some countries have already developed good plans to protect 10% of their land areas in Africa. These countries include Tanzania, Rwanda, Botswana, Benin, Senegal, the Central African Republic, and Zimbabwe.

The establishment of protected areas has been a good idea, but recently, it has generated controversy between conservationists and social scientists. Most people, especially in developing countries, see parks as government policy to usurp their resources and land right. This can be traced to the original concept of parks, which was for its aesthetic, educational, and recreational values without much consideration for land ownership (Zube 1995). The landowners feel that it is a government idea to take away their socioeconomic, cultural-ancestral-and spiritual rights. The use of force by the government to evict landowners is not taken lightly in many places. For instance, in Thailand, the forceful ejection of the people from their land led to the deliberate setting of fire to the forest and the poisoning of animals that strayed into their village.

Similarly, some great apes were slaughtered in the Democratic Republic of the Congo because of anger originating from a confrontation between the local people and park rangers. Because of these skirmishes between host community bodies such as World Bank, USAID, IUCN, and WWF, other NGOs have all called for people-centered or inclusive conservation programs to reduce the anger of local people (Brown 2003). However, various opinions surfaced concerning whether more land should be protected (UNEP) or given up for agriculture (IUCN). This disagreement negatively affects the local community because it is a volatile topic. A school of thought believes keeping more land for conservation with little human interference is the best option (Kramer and van Schaik 1997).

However, others believe all areas should be open to human use, i.e., "the use advocate" (Pimbert and Pretty 1997; Janzen 1994; Wood 1995). In the face of these disagreements, more people get displaced for protection, leading to more discontent (Cernea 2002), while more people still face the risk of more displacement from their ancestral land (Geisler 2001). According to Cernea and McDowell (2000), the significant impoverishment risks of local people include landlessness, joblessness, homelessness, marginalization, increased morbidity, mortality, etc. Africans also face loss of access to common property, and social disarticulation is more common in Africa. For the sake of urbanization, many communities are being sacked of its inhabitants. Here burial ground and ancestral land are destroyed to construct roads and bridges. An example is what happened in Finima community. Finima is a coastal town in Bonny local government area, some years ago some parts of the town were evacuated of its inhabitants and its people resettled in another location because of the building of the liquified natural gas (LNG) plant. During this relocation process, burial sites, ancestral homes, and family houses were lost.

7.2.1 The Niger Delta

The Niger Delta is situated in the southern part of Nigeria and makes up a quarter of the population of Nigeria. It is an area rich in natural resources, especially crude oil, and a biodiversity hotspot (NDES 1997). The Niger Delta wetlands cover over 11,020 km^2, 12% of Nigeria's surface area. It is among 200 global ecoregions classified as critically endangered by the Worldwide Fund for Nature (WWF). It is also the second most sensitive environment in Africa. The largest mangrove forest in Africa and the Atlantic is found in this region. Conservation of its resources is essential because of the area's vulnerability to global warming, such as sea-level rise leading to flooding and erosion. The silting of rivers and channels especially in the advent of exotics with weeds (e.g., water hycenth) cause ecological problems. There is also a fear of future land tremors and earthquakes resulting from seismic activity during oil and gas exploration (Nwankwoala and Orji 2018). The conservation effort in Nigeria started in 1899 when the first forest reserve was established (Anadu 1987).

Fig. 7.1 Massive deforestation in an unprotected mangrove and Nipa palm forest at Eagle Island, Niger delta, Nigeria

By 1939, most of the existing reserves in Nigeria had been created (NEST 1991). The original policy was to conserve 25% of the total land area, but only 10% was achieved. There are thirteen (13) game reserves and six (6) national parks in Nigeria. Two of these parks are situated in the Niger Delta area: the Cross River National Park (CRNP) and the Edo National Park. There are 70 protected areas in the Niger Delta. They are distributed as follows: Strict nature reserve (1), national parks (2), game reserve (4), and forest reserve (63). Despite the already established protected area in the Niger Delta, biodiversity declines due to anthropogenic activities such as farming, deforestation, and developmental activities (Numbere 2021), and pollution from oil explorations. These activities have made the establishment of more protected areas difficult. Protected area in Nigeria is about 5.6% (Naughton-Treves and Weber 2001; Perrings and Edgar 2000), meaning that over 90% is still unprotected. Over 90 percent of the forest in Nigeria has been lost to deforestation. This loss has resulted to the loss of about 75% of species habitat. The forest is a habitat to many organisms that disappear immediately as their homes are destroyed. This situation calls for massive protection of the remaining environmental resources to prevent them from going into extinction in the next 50 years.

Many forest areas and rivers are protected in the Niger Delta because of local spiritual beliefs that forbid people from exploiting the environmental resources (e.g.,

Anwana et al. 2012) (Fig. 7.1). Other areas with no such protection are overexploited, as shown at Eagle Island coastal community, where most of the mangrove and Nypa vegetation has been cut down for firewood (Fig. 7.1).

7.3 Conservation Ethics

The conservation of resources is also founded on some ethics (Fig. 7.2), which are grouped as follows:

7.3.1 Anthropocentrism

This requires individuals, corporations, and other interest groups to reasonably consider their actions that directly affect the natural environment and indirectly affect other human beings. It may also deprive indigenous people of their homes and traditional means of subsistence, undiscovered resources, valuable ecosystem services, esthetic experience, and scientific knowledge. The anthropocentric view takes humans as the center of the universe. It is a view propounded by Plato and Aristotle (Bookchin 2022), which shows how humans relate to the biosystems, including plants and animals.

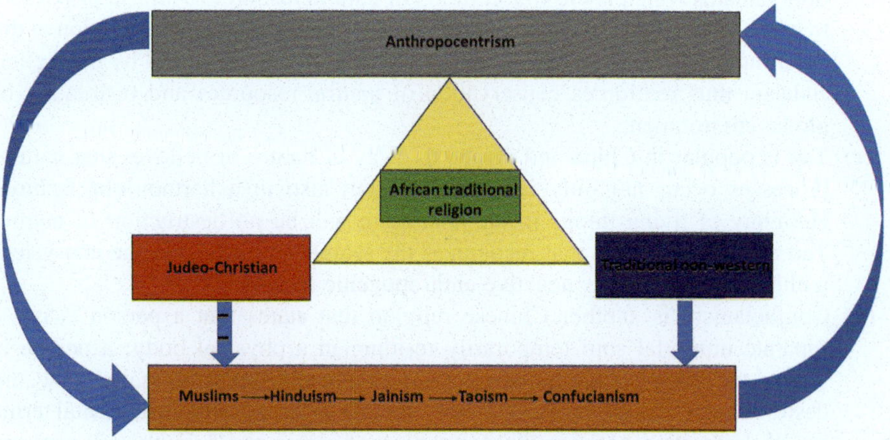

Fig. 7.2 Conservation ethics of different beliefs around the world

7.3.2 *The Judeo-Christian Stewardship Conservation Ethic*

This theory believes that God pronounced all creatures 'good'. It states that since God created the heavens, the earth is left for man to take dominion of the universe by preserving species and not driving them extinct. It thus confers objective intrinsic value on nature on the most precise and most ambiguous by divine decree. It makes human beings directly accountable to God for conserving biodiversity.

7.3.3 *Traditional Non-western Environmental Ethics*

We have different beliefs across the globe.

(i) Muslims believe that Allah founded Islam in the seventh century C.E. Muslim's belief of world view is similar to basic Judeo Christian world view. It regards man as the manager of the earth and not a proprietor, a beneficiary, not a disposer. It emphasizes the just and equitable distribution of natural resources.

(ii) Hinduism is a major religion in India and is practiced in some scattered areas of the world. It invites human beings to identify with other life forms, for all life forms share the same essence. Those that practice this religion are mostly vegetarians because they believe that animals have spirits and are next to humans. The strict vegetarians don't eat any animal or animal-based product, whereas the casual vegetarians don't eat animals but can eat animal-based products. The essence of this practice is that animals have spirit like human beings, so they are to be protected and not to be consumed as food.

(iii) Judaism has a great influence in India. It talks about *Ahimsa*, which means noninjury of all living things. This means no living thing should be hurt. It thus deplores the killing of organisms, deforestation, and wiping out of populations of organisms which is the conservation of natural resources. *Ahimsa* also refers to peace to all living things including humans. The killing and destruction of the environment through negative human activities is not supported by this belief. Judaism thus reinforces conservation of natural resources and restoration of global environment.

(iv) Tao is popular in China and means the way of nature. It believes that natural processes occur not only in an orderly but also in a harmonious fashion. Harmony of living things means that there will be no destruction of nature. Tao is a belief system that recognizes the seamless working of the ecosystem without interruption by negative anthropogenic actions.

(v) Confucianism is another Chinese religion that states that a person is not a separate immortal soul temporarily residing in a physical body; a person is instead the unique center of a network of relationships. It states that biocide, the destruction of life, is tantamount to suicide. And suicide in environmental terms is self-destruction because the destruction of plants and animals will truncate the food chain and cause food scarcity for many leading to an ultimate death, through hunger and starvation of humans.

7.3.4 African Traditional Religion

African traditional religion is the practice of declaring animate and inanimate objects as gods and the objectivation and worship of these materials. This religion believes in the supreme creator God and believes that there are lesser gods who act as intermediaries and guard spirits and is involved in human affairs. People communicate with these gods through rituals, sacrifices, and prayers. They also believe that the human condition is imperfect, so there will always be sickness, suffering, and death, which is probably caused by sins and misdeeds that offend the gods and ancestors; therefore, ritual action and sacrifices will alleviate the sufferings (Lugira 2009). Based on this knowledge, people fear and respect traditional laws because of the negative consequences that will befall any offender. This is displayed in the worship of land, river, animals, and plants. Seventy percent of Africans are traditionalists and will respect any traditional declarations intended to preserve nature. African traditional religion is therefore moving forward with the tenets of global conservation. This religion is passed from generation to generation orally. Here fathers pass on the tradition to their sons and daughters without any written record. Documentation of these oral stories that are beneficial to conservation is another aspect of research for African scientist who can use it to plan conservation strategies in the different localities.

The preservation of nature and its resources is a long-term belief by various religious and cultural groups (Kimengsi et al. 2022). For instance, one of the ideals of christianity is to take dominion over the heavens and the earth as well as the creatures inside. Christians should do their utmost best to preserve nature and prevent its destruction by having authority over living things. Furthermore, as part of the Islamic injunction, humans are expected to safeguard the natural environment to not go against Allah's (God) wish. This belief pattern in preserving nature also runs across other religious and secular organizations. In Buddhism, killing a cow is sacrilegious because it is regarded as an object of worship. Many other religions have plants and animals used as objects of worship. These religions directly or indirectly support the protection of nature. Vegetarianism is a belief in the sanctity of animals and is practiced by Buddhists, who believe that animals have souls and should not be used as food; that is why they feed on plants alone and not animals or animals' products (Minton et al. 2022).

Conservation is indirectly practiced by declaring the forest as evil forests in many communities to prevent people from plundering its resources. In African traditional religion, living things are objectified as sacred and thus worshipped. In many communities in the Niger Delta, it is taboo to cut down a giant tree because they are places of libating to the ancestral spirits and the gods of the land. Trees, land, and animals serve as totems in the communities, which regard them as deities and objects of worship. For instance, in a community in the Niger Delta, the boa constrictor snake is worshipped and therefore not killed if found. If such an animal object is killed, it is accorded full burial rights as though it is a human. The ceremony involves buying a coffin, conducting funerals, entertaining guests, and the internment of that animal in a befitting burial. A classic example of conservation practice via local beliefs is the preservation of the crocodiles of the Biseni community in Bayelsa State in the Niger Delta. There is a crocodile sanctuary in this community where crocodiles are found in large quantities because the locals or visitors do not kill them. After all, the crocodiles are regarded as "Gods" in the community. A chief priest oversees the crocodile shrine and periodically libates it to appease the land Gods for protection. Natives don't hunt down the crocodiles because of the fear of reprimand from the Gods, which may be sickness or death to the person involved or a family member. But if a crocodile is killed accidentally, a befitting burial will be conducted for the dead animal (Anwana et al. 2012). The practice of declaring crocodile pond sacred is also observed in many other communities across the Niger Delta such as the Ogonis.

In modern-day, those religious and traditional beliefs can be used as tools to protect nature from wanton destruction by humans. These traditional beliefs have existed for a long time, even when the destruction of nature has not worsened. Thus, this chapter recommends that more traditional means of protecting rare and endangered species in every community be encouraged. Traditional declarations and bans should be placed on killing endangered species to prevent their extinction. People in Africa are more afraid of traditional laws than government laws because of their potency and ability to punish offenders and administer retributive justice swiftly.

Many evil forests can be established to protect the trees by conserving and managing them for future generations (John and Enang 2022; Madongonda and Gudhlanga 2022). Presently, there is the ongoing destruction of pristine mangrove forests in the Niger Delta to produce firewood and build houses (Fig. 7.3). Thoughtfully, threatened patches of mangrove forest scattered in fragments across the Niger Delta can be declared a protected zone or a spiritual forest to deter persons from destroying the forest in the various communities. The remaining forest fragments, if not protected, may be wiped out with time because of human development, such as the building of houses in the cleared forest, which has led to the elimination of thousands of hectares of forest in the Niger Delta. As the population of people increases, the need to acquire more land spaces to construct houses increases in various communities leading to the destruction of pristine forests. The government can resolve this aspect by establishing low-cost housing in areas that have little or no impact on the environment. Building of houses will discourage people from acquiring and building houses within the forest.

Fig. 7.3 Houses built inside cleared mangrove forest at Abalama, Asari-Toru Local Government Area of Rivers State, Niger Delta, Nigeria

7.4 African Religious and Cultural Beliefs in Conservation

Some African traditional beliefs that support biodiversity conservation include the following:

(i) Establishment of virgin forest.
(ii) Establishment evil forest (Onor et al. 2022).
(iii) Worship of some aspects of nature and its components, e.g., trees, animals, rivers, land, snake, bird, etc.
(iv) Use of animals as totems and Gods.
(v) Establishment of sacred land, river, or forest.

The establishment of an evil forest is an ancestral belief that people who do evil or go against the norms of the land or nature are thrown into the woods, dead or alive (Wild and Walters 2022). A forest so designated forbids entry and the exploitation of its resources. Anyone who dares to go into the forest may get missing forever or face instant death. The declaration of the evil forest is beneficial to the tenets of conservation, which is a means of preservation (Seile et al. 2022). In the past, areas designed as "evil forests" were found in many communities of the Niger Delta region. Studies have shown that the forest has its trees and wildlife population still intact because of the prevention of human exploitation. The forest has robust growth with large tree canopies that serve as a habitat for a host of biodiversity ranging from primates, birds, reptiles, rodents, mammals, and arthropods. The designation of some areas as "sacred" had preserved them from being exploited by the local people (Reyes-García et al. 2022; Sinthumule 2022). For example, some forest in the riverine areas of the Niger Delta serves as the dumping ground for those who die by drowning or mysteriously. This forest is not usually entered or exploited for resources, but left alone and not allocated to people to build houses. This action shows that the people's traditional beliefs indirectly support the course of conservation of natural resources. However, this practice is outdated. In most communities, such traditional beliefs had been stopped to give more freedom to the people to exploit their resources, most time in an unsustainable manner, which has resulted in the overexploitation and plundering of the natural resources in such areas (Verma and Sadguru 2022). Former virgin forests had been destroyed, the trees used for timber and firewood, and the animals hunted down and killed for food. Hardly will one find an unexploited forest in the various communities. Instead, the only forest spared is those that are found in rugged terrain that is not easily accessible to intruders. However, some desperate people still reach this interior forest and exploit their resources for survival. Human exploitation for wood and firewood has led exploiters to approach the deeper parts of the interior forests of the Atlantic Ocean in the Niger Delta area. In areas around the Bonny River, large population of mangroves had been removed, thereby paving the way for invasive Nypa palm (*Nypa fruticans*) to invade and take over large hectares of the coastal territory.

Some communities still practice the culture of setting aside sacred locations to preserve their forest and animal populations (Table 7.1). These communities have been practicing the protection of such forests for centuries, with the tradition passed down from generation to generation (e.g., Chakrabarty et al. 2022). A classic example is the crocodile river of Biseni (Anwana et al. 2015). Over the years, this river had remained untouched because of the fear of incurring the wrath of the Gods of the land. It has thus become a conservation and tourist destination for people from far and wide. The river has become a cynosure of all eyes because of the large population of giant crocodiles (*Crocodylus niloticus*) that are in the river. In population dynamics, increase in population in a small area can lead to an increase in competition and a fight for survival. Thus, the crocodiles of Biseni are so many that they became a threat to fisherfolks in the nearby towns. The climax came when a community member was killed and eaten by the crocodile, and this angered the community members who hunted and killed the crocodile carrying this cowardly act.

Table 7.1 Some cultural beliefs and practices that promote the conservation of environmental resources in the Niger Delta, Nigeria

Communities/ Tribe	Local beliefs	Environmental resources/organisms	Local laws	Conservation value
Biseni	Local deity	Crocodile ponds	Forbidden	Protected area
Osiama	Local deity	Crocodile ponds	Forbidden	Protected area
Kula	Totem	Snake	Forbidden	Protected species
Ikwerre	Cultural value	Giant trees	Forbidden	Protected species
Ogoni (Bogono. Kaani)	Mysticism	River and forest	Forbidden	Protects trees and rivers (Nyoma)
Ogoni (Kwei-yang, Neko)	Sacred	Forest	Forbidden	Wildlife and trees protected
Ogoni (Kwawa)	Local deity (Nyorkebe)	Forest, river, and crocodile	Forbidden	Protected river and crocodiles
Ogoni (Neko)	Sacred	Forest	Forbidden	Protected forest
Okrika	Water spirits	River	Forbidden	River
Ogba	Local deity	Forest/crops	Forbidden	Plants
Etche	Local deity (*Ajuala*)	Python	Forbidden	Protected species
Kalabari	Local deity (Akaso)	Python	Forbidden	Protected species
Urhobo	Local deity (*Orogun*)	Iguana lizard	Forbidden	Protected species
Ogbomosho	Cultural value	Dog	Forbidden	Protected species
Nembe	Cultural value	Python	Forbidden	Protected species

Later, because of the large population of the crocodile, the community started killing some of them occasionally to reduce the number of crocodiles to a sizable limit. The planned killing of the crocodile later became a norm and is now made at a particular time in a year to help reduce the crocodile population from exploding beyond a manageable number. Some tribes glorify animal images in plays and poems. For instance, in the snake song of the Itshekiri people where there is a dramatization of snake movement (Eyituoyo 2022). The Ndoni community in the Niger Delta revere the monitor lizard (*Varanus niloticus*), which they regard as a local deity.

7.5 Identification of Communities that Practice Local Conservation

The awareness of the conservation of the environment has been gaining momentum over the years. Many communities have developed their means of protecting and preserving their natural resources. These communities have local mechanisms of protecting forests, rivers, and land from overexploitation or intrusion by unwanted persons. Two such communities, one of which was mentioned earlier, are Biseni and

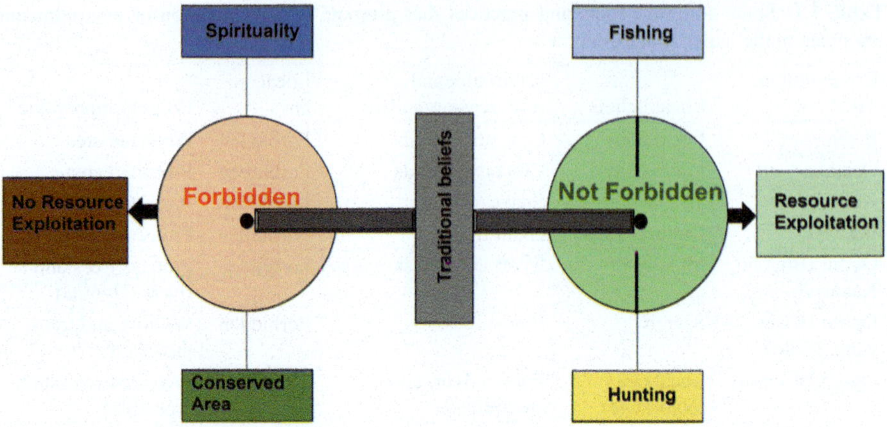

Fig. 7.4 Traditional customs and beliefs influence resource use (not forbidden) and nonuse (forbidden) in Nigeria's Niger Delta

Osiama, which are Ijaw-speaking groups in the Niger Delta. The belief system divides the natural landscape into *aweye* (forbidden) and *aweaya* (not forbidden). This belief system is demonstrated in Fig. 7.4. Forbidden freshwater lakes such as Lake Esiribi and Lake Adigbe contain sacred species representing the Gods and the ancestral spirits. One of these species is the ecologically threatened crocodile (*Osteolaemus tetraspis*), a respected species and, therefore, protected from hunting by locals. The belief is that the crocodile is the people's brother, which necessitates its protection within the *aweye* (forbidden) lake. If mistakenly killed, the crocodile is given a befitting burial like a human being. Still, when killed intentionally, the culprit is made to replace the dead crocodile with a live one (Anwana et al. 2015). Sacred groves, a protected area with cultural significance, can be used to protect small animals in Africa (Decher 1997).

Joint protection of lakes eliminates the "island effect" by creating a metapopulation of fishery resources that can be managed sustainably for the more significant benefit of the people. The lake is so sacred that fishing activities are restricted and only done during specific dates and seasons using fishing instruments approved by the group that is nonintrusive and selective. There is also cooperation among neighboring communities that are ancestrally related to practice similar protection of lakes through interconnectivity. The government can complement this effort by indigenous groups through the provision of legal backing.

Another community, Kula, a town in the Kalabari-speaking area of the Niger Delta, reveres the snake python and prevents anyone from killing it. This belief in the snake's power deifies it and makes it protected in the community. If mistakenly found near human habitation, it is carefully taken back to the forest. However, calamity befalls anybody who knowingly kills it.

7.5.1 Worshipping and Libating at the Foot of Big Trees

The practice of not cutting big trees and libating at its foot is an old practice in many communities of the Niger Delta, especially the Ikwerre-speaking tribe of the Niger Delta. Many big and giant trees are regarded as sacred sites by villages which create a barrier around them and use the site as their shrine for invoking the spirits of their ancestors. The people libate near the big trees during festivals, e.g., yam festivals, wrestling matches, and community cleaning ceremonies. The holding of these ceremonies is an annual event in many communities and helps preserve such trees. A red and white piece of cotton material is tied around the base of the giant tree. The clothes signify a sacred site.

Similarly, some shrines have life fowl tied around the tree, used to make a blood sacrifice. A built-in box is also constructed near the trees where materials used for libations are kept (Fig. 7.5). The sacred trees remain protected for years without being cut down even when other neighboring trees are cut for urban renewal projects. They do this as an act of tree worship (Rao 2022).

However, if, for urban development, the trees must be cut down to create a right of way passage, an animal would be used as a sacrifice to appease the gods. If this sacrifice is not made and an attempt is made to bring down the trees, it will lead to the havoc that may cause injury or death. There are many cases where trees fall and kill drivers of bulldozers who attempt to mow down large trees without necessary sacrifice. The preservation of big trees has become the norm in many Ikwerre-speaking areas and some other tribes in the Niger Delta such as the Ogoni. Some

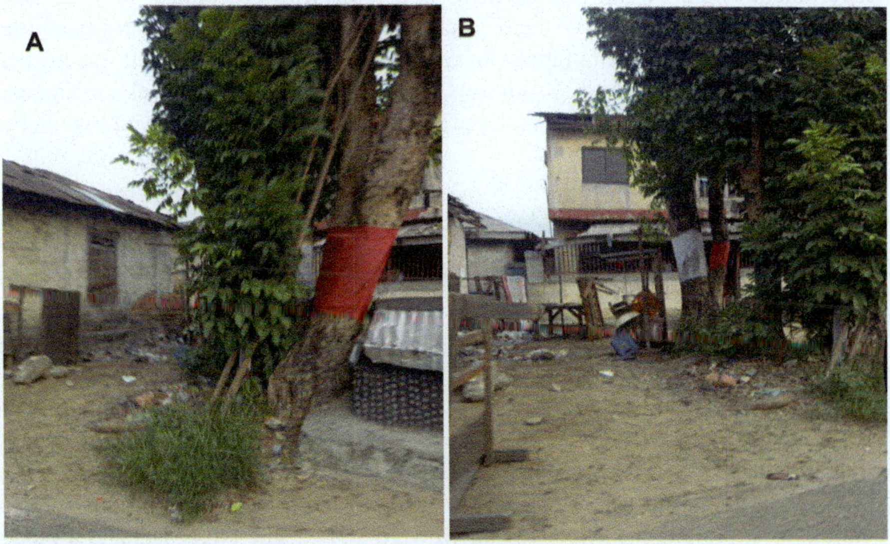

Fig. 7.5 Sacred giant trees in the Diobu area of the Niger Delta shown in Fig. 7.5 A and B. The tree is preserved and used as a site for libations to the gods of the land. At the foot of the tree is a box where native concoctions are kept while white and red clothes are tied around the tree

of these trees may be as tall as 50 feet from the ground, with large canopies and imposing branches (Eniang and Luiselli 2002). The Ogonis have big trees in many of their communities that are regarded as sacred and have not been cut for centuries. However, because of modernization many of those trees have been cut down to build houses. For instance, a sacred forest in Gwara town in Khana Local Government Area of Ogoni was cut down by some churches after much intercessory prayers and the areas allocated to people for building houses.

7.5.2 *Mystical Forest and Rivers in Ogoni*

In the Bogono community of Kaani in the Khana Local Government of the Ogoni Kingdom of the Niger Delta, two rivers flow in the same direction but do not mix. It is a sacred place where people go to take refuge. There is also a thick forest in the Kaani area called Kwe alegere, regarded as a mystical forest where people are not allowed to hunt or fish because of the negative consequences that will befall anybody that dares that. This forest has water called Nyoma, which is seen as holy water, and people are not allowed to go into that area. Anybody that goes in unauthorized may never come out of the forest alive. Some researchers believe mystical powers do exist in Ogoni (e.g., Babatunde 2022). In another community in Ogoni called Luawi, the crocodile is sacred and thus is not killed or captured, but if the crocodile migrates to the neighboring community, called Gwara, it can be killed and eaten without any problem. There is another area in the Kwawa community in the Khana Local Government Area; by the bridge, there is a river with crocodiles. Anyone who kills these crocodiles will give them a befitting burial; if they fail to bury them, the Gods will wipe out the entire family forever. Any child born will not survive in such a home. A deity called *Nyorkebe* controls this forest. Kweiyang in Nueko in Gokana Local Government of Ogoni has a sacred forest where nobody is allowed to hunt the wildlife to preserve its resources.

7.6 Factors that Cause Resource Overexploitation

Overexploitation of resources is a gradual process that leads to the depletion of environmental resources, which means the amount removed is greater than the amount added in each area. Overexploitation leads to a reduction in the population of species. A reduced number of species leads to a small population, which is vulnerable to stochastic events that can drive the population to zero or extinction. And over time, this situation leads to the loss of species due to habitat loss due to continuous exploitation (Ayanlade and Howard 2017). The Niger Delta is a biodiversity hot spot endowed with a significant wealth of natural resources that serve as a source of livelihood for the local people and foreign exchange for the country. Some plants and animals serve as food and a source of income for the people, while the crude oil resource is refined to produce petroleum and petrochemical products that are sold abroad to earn foreign exchange for the country. However, there is a problem of resource overexploitation, which pushes some aspects of the ecosystem to extinction. The following points should be noted to identify the causes of overexploitation correctly.

(i) Poverty
(ii) Lack of regulations
(iii) Lack of compliance with rules of resource conservation
(iv) Lack of implementation of laws banning overexploitation
(v) Unemployment
(vi) Lack of enlightenment on the long-term impact of overexploitation
(vii) Lack of protection for the natural resources, i.e., forest resources
(viii) Lack of provision for an alternative source of income
(ix) Lack of compensation and payment of royalties by state and federal governments on the acquisition of oil-rich land in communities
(x) Lack of monitoring of endangered sites, i.e., sea, land, forest, etc.

Lack of money, among others, is the primary cause of resource overexploitation because people need to provide food for family members. Without being meaningfully employed, the people fall back on exploiting the resources in their environment as a form of livelihood. However, if they have jobs, they won't go into the bush to hunt for rare animals or go to the sea to scoop away rare fishery. Instead, hunting and fishing would be left to professional hunters and fisherfolks who don't possess a college degree. Another thing that encourages people to plunder the resources is the lack of punishment for offenders. Those involved in destroying the environmental resources do so because no one in their memory has ever been punished for their destructive actions, which encourages them to carry out their act of degradation without fear.

7.7 Establishment of Communal Conservation Projects

To ensure the effective conservation of natural resources, such effort needs to start from the local communities that possess these resources. There is no community in the Niger Delta that doesn't have natural resources. So, the local people need to be incorporated into the action for any conservation effort to succeed. The communities will work in conjunction with the local and state authorities. The composition of the joint conservation committee will include state, local and communal coordinators who will supervise the activities of its members. When this is done, any conservation project initiated by the state government will trickle down to members of the communities. Some conservation projects that can be initiated and executed include the following:

(i) Training of conservation volunteers (students, researchers, etc.)
(ii) Formation of conservation committees in the villages
(iii) Formation of conservation guards
(iv) Tree planting campaign
(v) Clean up forests and seas of waste products (e.g., plastic waste)
(vi) Restoration and remediation of polluted and degraded sites
(vii) Establishment of conservation clubs in secondary and tertiary institutions

7.8 The Role of Community-Government Partnership in the Conservation of Rare Species

The work of conserving rare species cannot be done by the community alone where these resources are found. The government needs to partner with the community members to achieve complete conservation. The government can use its state authority to declare endangered and rare species sanctuary protected areas legally. The state assembly should pass a bill that will empower the state government to legalize the protected areas under their control. The state will thus put in its resources, workforce, and machinery to protect, maintain, and manage protected areas in all communities in the region. Under the protected area law, the local government areas will oversee the protected areas on behalf of the state government. There are some countries that have implemented the protection laws to conserve their natural resources such as Turkey (Kaya and Raynal 2001), China (Zhang et al. 2000), and Norway (Bugge 2022). The following committees can be formed to run the daily affairs of the parks as part of the partnership between the community and the government.

7.8.1 *Research and Training*

The committee will liaise with research scientists in the state universities to conduct studies on species population, abundance, and diversity to have an accurate situation report on biodiversity in the region. This report will help the state delineate and determine species in different stages of endangerment to know their location, habitat type, and factors affecting their survival. Furthermore, the study results will provide data for the future monitoring and planning of parks and protected areas in the region. Scientists can be awarded research grants to conduct full-scale studies in all regions' forests. The studies will help conservation efforts by producing a species checklist, and community members are trained on sound environmental practices that will help boost the overall conservation effort (Mushonga 2022). For instance, community members can be trained as forest or river guards and conservation monitors with the central aim of helping to protect forests and rivers in their communities (Anagnostou et al. 2022).

7.8.2 Safety and Security of Forest

Forest guards will be recruited from each community after their training and given the power to patrol and protect the perimeter of the protected area (Moreto et al. 2017; O'leary Simpson et al. 2022). They are to be placed on a monthly salary to take the job more seriously. The position will also enable them to cater for their families while safeguarding the environment. This act will be a form of job creation for the unemployed youths in the villages. Forest guards protect the park against intruders and poachers bent on exploiting the forest resources (Cronin et al. 2016). Offenders can be identified and reported to the police for possible arrests. Arrested persons will be made to face the law and given a fine or prison sentence to deter would-be offenders. Patrol vans are to be procured and given to the guards to facilitate their movement from one location to another within their operational area. Members of the conservation groups are to be provided with arms and ammunition to defend themselves when attacked by poachers (Ramutsindela et al. 2022; Duffy and Brockington 2022).

7.8.3 Forest Resource Utilization

This group is responsible for the selective conversion of the forest products into valuable commercial commodities to help in generating income for the local authority (Milley 2022; Saxena et al. 2022). For example, non-timber resources can generate wealth (Shanley 2022). The products harvested will be sold to the public at a subsidized rate to help boost the local economy. Protected areas can also be used to generate income via ecotourism (Rahman et al. 2022). Tourists will be charged some fees in the form of tickets to enable them to enter the protected forest for sightseeing. Low-level exploitation can be conducted through agroforestry, cultivating agricultural products such as cash crops, food crops, poultry, piggery, cattle ranch, etc. They can be allowed if they won't impact the forest negatively. Two forms of protection can be practiced, such as limited and strict protection, depending on the presence of rare and endangered species. Suppose there are no rare or endangered species. In that case, the limited form of protection can be practiced where an allowance is given for resource exploitation, ecotourism, etc. If the forest has rare and endangered species, a strict form of protection will be adopted. In this method, no form of resource extraction is allowed. There will also be a restriction on entry, even for ecotourism. It is necessary to balance the need for ecotourism and conservation so that there will be no conflict of interest (Birendra 2022).

7.8.4 Sensitization and Enlightenment

This group will be responsible for informing the community members about the need to protect the parks. They will do a house-to-house campaign in the community and newspaper and radio adverts and announcements to publicize the importance of protecting the forest (Soaga and Adeleye 2022). In each community, local announcers and town criers should be hired to sensitize the people on the need to conserve their natural resources. Billboards should be set up at strategic locations in the communities to sensitize the people on the importance of natural resources to the human environment. In addition, sensitization handbills and fliers can be distributed around the neighborhoods.

7.9 The Significance of Cultural Beliefs in Species Conservation

Cultural beliefs and the worship of local deities are a potent force in the Niger Delta region. Many people who are orthodox Christians fear and respect local deities more than the God of their faith (Minton et al. 2022). Anybody who goes contrary to the local laws is visited with instant retribution or justice. They fear the punishment of the deities than their all-loving heavenly Father, Jesus Christ for the Christians, Mohammed for the Muslims, and Buddha for the Hindus. The fear and respect for the deities of those who practice African traditional religion can be redirected and utilized in conservation biology to help protect rare and endangered species in the various communities in the Niger Delta. The communities protect forests, rivers, animals, and plants by declaring them cultural symbols or totems. For instance, some communities regard the snake as their totem, represented in their cultural symbol, so they don't kill it if found. The Okrika people of the Niger Delta celebrate a festival called "Odum festival" every ten- or twenty-year interval, which is represented and preserved as an embodiment of the local deity. In this community, water snake is not killed but respected and preserved. Local rites are performed in the river to appease the Gods before the festival begins. In the Kula community of Kalabari, the boa constrictor snake serves as the totem, and members of the community do not kill them when found within or outside the community. In the Biseni community, the African crocodile (Seki) is their totem. Other organisms that serve as totems include bird or feni (Okrika), lion or Odum (Ikwerre, Okrika), Shark or Ofrima (Ijaw speaking communities such as Kalabari, Ibani, Okrika), and giant trees (Ikwerre). Apart from having specific organisms as cultural totems in communities, there are also places isolated as shrines in thick forests or rivers for worshipping local deities. These shrines are often a no-go area for ordinary members of the community who are not members of the traditional society that serves that deity. For those who are not members visiting such sacred locations is suicidal because it can result in severe consequences. Culturally important forests, animals, and rivers under the protection of the community remain untouched for years without exploitation. Members of the community are not to enter to farm, hunt, or exploit any of its resources except such bans are lifted. In many communities in the Niger Delta, people tend to obey and abide by the rules given by traditional rulers and chiefs rather than the local or state government. The traditional institution is more critical, and cultural laws are more respected and given much regard than the secular laws of the government. Cultural laws are respected because there is a belief that supernatural beings make the laws of the traditional institutions and must be respected to deter curses and calamities from the Gods.

Respect for tradition can be used to help conserve the environment for the benefit of present and future generations. Cultural values can promote wildlife's survival, especially if the community knows that it is forbidden to kill any wildlife (Krause and Tilker 2022). For instance, many wild animals are endangered because of the activities of humans, such as Ilama, elephants, antelope, lions, cheetahs, etc. For example, in Andoni, a Niger Delta community, an elephant has been sighted in the past. Still, today they are rarely found because of overhunting, which has made the local authority place a total ban on all forms of exploitation. The Andoni Forest is one of the communities with the highest concentration of biodiversity in the Niger Delta region. There is a drop in the wild animal population because of their use for food. However, the declaration of these animals as cultural and traditional totems will help save them from imminent destruction. Cultural beliefs thus help to promote conservation efforts in the Niger delta and other biodiversity-rich communities in Africa. Many communities in Africa place women in the background in decision-making on ecological matters. Still, studies have shown that the involvement of women in conservation efforts can help promote conservation. Irrespective of gender, conservation strategies that provide economic benefits can come from males or females, not from a particular gender (Massey et al. 2022).

7.10 Future Potentials in Local Conservation Strategies

Local conservation is the key to conserving the environmental resources in the Niger Delta area. It offers a win-win situation, where the environment wins, and people win. The participation of the local people will ensure the success of any conservation effort by the central government and NGOs. They are making the people the custodians of the environment by putting the environment in their hands to protect. They, in turn, will be compensated by being allowed to carry out low-level exploitation of resources to feed their families. Other financial incentives include the provision of stipends for forest guards, which can be given to them as additional funds to help build cooperation between the government and the community. The principle is that if the environment is secure, the community will be safe because the bastardization of the natural resources can affect the well-being of members of the community. Destruction of environmental resources can also impact the ecosystem services of the forest. For instance, mangroves clean the environment by absorbing carbon emissions (Zhu and Yan 2022). The forest also serves as a flood and erosion barrier that may help prevent the wiping out of communities during a hurricane or flooding event. In the past, conservation effort has failed in many African communities because of the use of force to implement protection by the authorities in charge. The use of force may not work effectively in some communities when the people resist government environmental policy as an expression of dissatisfaction and anger. The use of force infuriates the locals leading to more collateral damages to the ecological resources. For instance, in Congo, the enforcement of strict poaching rules through the crackdown on poachers led to the deliberate killing of some great apes by members of the communities as a show of anger.

Therefore, future conservation efforts should be people-centered rather than environment-centered. Resources found in each community are for its members' benefit, so when taken away forcefully, it can lead to negative consequences. Therefore, people should be hired from the community as forest guards and wildlife volunteers to ensure communal participation. They are to be trained and placed on monthly salaries to motivate them to take their work seriously. For instance, in the Niger Delta area, the firewood business is very lucrative. Firewood is derived from mangrove stems, so telling the people not to cut mangrove trees will not work out because for most persons it is their only source of livelihood. They will thus fight against any policy that will stop them from utilizing the mangroves. Therefore, the best method is to teach them sustainable forestry. They will be sustainably exploiting the resource by rationing the cutting to prevent the destruction of large stands of mangrove trees. The government can also intervene by regulating and monitoring mangrove tree cutting. Those involved in the firewood business can be registered by the government and given a "tree cutting license". As part of the license, rules concerning tree cutting will be given to persons involved in the firewood business. Thus, only licensed members will be allowed to embark on the firewood business to prevent reckless cutting of mangrove trees. The whole mangrove forest in the Niger Delta will be mapped, and sections for cutting trees will be done and delineated and apportioned to registered persons. The forest mapping will enable the government to determine areas that are near extinction and prevent further cutting. Spaces with fewer tree populations will be allowed to lie fallow so that the trees will recuperate. There would also be aggressive afforestation to replace cut trees. The government will also provide an alternative source of cooking energy to reduce pressure on firewood, such as solar cookers, stoves, and gas cookers to reduce the stress on the use of the trees for firewood production.

Hunting too can be regulated to prevent overexploitation of wild animals by members of the local communities who use them as food. Presently, the hunting laws are not strictly enforced, which has provided a loophole for reckless hunting. The existing laws are not being followed because most local people are not even aware of such laws.

Future conservation efforts should include the following actions:

(i) reduction of coastal development
(ii) Stoppage of deforestation activities

(iii) Reduction or stoppage of industrial pollution
(iv) Implementation of poverty alleviation programs such as skills acquisition and increased employment
(v) Establishment of many protected areas in all communities
(vi) Initiation of conservation education and enlightenment program in primary, secondary, and tertiary institutions
(vii) Enforcement of conservation laws and regulation
(viii) Involvement of local conservation volunteers
(ix) Involvement of traditional leaders such as kings, chiefs, and title holders
(x) Application of GIS technology to map reserves, parks, and conservation areas (Numbere 2022).

7.11 Conclusion and Recommendations

The believe in the potency of spirits and the superiority of traditional values and proclamations can be harnessed to promote conservation and protection of the environmental resources, as revealed in this chapter. In Africa, customary rules and regulations are given much respect compared to secular laws. The traditional declaration of some aspects of the environment preserves them for centuries, whereas environmental laws established by the government are flouted and disregarded at will. Many communities in the Niger Delta have declared some land, river, or forests as protected areas, which forbids exploitation leading to the preservation of a specific or a group of endangered species. A classic example is the crocodile ponds in the Biseni and Osiama communities of the Niger Delta, where traditional authorities have declared the ponds forbidden for hunting. The people are free to fish and hunt in the unforbidden areas of the community without any consequences. Poverty is a significant cause of overexploitation of natural resources; therefore, the government can reduce the destruction of the resources by providing alternative livelihood opportunities such as white-collar jobs and the recruitment of local people as forests guards in different communities. These jobs will provide an opportunity for community members to earn a living which will prevent them from exploiting the environment.

References

Almasieh K, Mohammadi A, Alvandi R (2022) Identifying core habitats and corridors of a near threatened carnivore, striped hyaena (Hyaena hyaena) in southwestern Iran. Sci Rep 12(1):1–11
Anadu PA (1987) Progress in the conservation of Nigeria's wildlife. Biol Conserv 41:237–251
Anagnostou M, Gunn V, Nibbs O, Muntaner C, Doberstein B (2022) An international scoping review of rangers' precarious employment conditions. Environ Syst Decisions:1–25
Anwana ED, Cheke RA, Martin AM, Obireke L, Asei M, Otufu P, Otobotekere D (2012) The crocodile is our brother: sacred lakes of the Niger Delta, implications for conservation management. In Sacred natural sites. Routledge, pp. 155–164

Anwana ED, Nyah G, Mbong EO (2015) Valuation of social relevance of plants amongst the Annangs in South-South Nigeria. Eur J Biomed Pharm Sci 2:328–337

Ayanlade A, Howard MT (2017) Understanding changes in a Tropical Delta: a multi-method narrative of landuse/landcover change in the Niger Delta. Ecol Model 364:53–65

Babatunde AO (2022) Reimaging women ritual space: gender and power dynamics in African religion. In The Palgrave handbook of Africa and the changing global order. Palgrave Macmillan, Cham, pp. 969–986

Banzhaf HS (2019) The environmental turn in natural resource economics: john krutilla and "conservation reconsidered". J Hist Econ Thought 41(1):27–46

Bergstrom JC (2022) Reflections on the historical development of natural resource and environmental economics. In Teaching environmental and natural resource economics. Edward Elgar Publishing

Birendra KC (2022) Complexity in balancing conservation and tourism in protected areas: contemporary issues and beyond. Tour Hospitality Res 22(2):241–246

Bookchin M (2022) The philosophy of social ecology: essays on dialectical naturalism. AK Press

Brown K (2003) Three challenges for a real people-centred conservation. Glob Ecol Biogeogr 12(2):89–92

Bugge HC (2022) Environmental law in Norway. Kluwer Law International BV

Cassidy L, Perkins JS, Bradley J (2022) Too much, too late: fires and reactive wildfire Management in Northern Botswana's forests and woodland savannas. Afr J Range Forage Sci:1–15

Cernea MM (2002) For a new economics of resettlements: a sociological critique of the compensation principle. In: Cernea MM, Kanbar R (eds) An exchange on the compensation principle in resettlement. A working paper. Cornell University, Ithaca

Cernea MM, McDowell (2000) Risks and reconstruction: experiences of resettlers and refuges. World Bank, Washington, DC

Chakrabarty SP, Tanoue M, Penteado A (2022) The trouble is, you think you have time: traditional knowledge of indigenous peoples in Japan and India, the reality of biodiversity exploitation. Environ Manage, 1–13

Cronin DT, Riaco C, Linder JM, Bergl RA, Gonder MK, O'Connor MP, Hearn GW (2016) Impact of gun-hunting on monkey species and implications for primate conservation on Bioko Island, Equatorial Guinea. Biol Conserv 197:180–189

Decher J (1997) Conservation, small mammals, and the future of sacred groves in West Africa. Biodivers Conserv 6(7):1007–1026

Duffy RV, Brockington D (2022) Political ecology of security: tackling the illegal wildlife trade. J Pol Ecol 29(1)

Eniang EA, Luiselli L (2002) Ikpan wetland rainforest: an area of high biodiversity importance in South-Eastern Nigeria. Revue d'Ecologie Terre et Vie 57(1):19–28

Eyituoyo OJM (2022) Performance in Itsekiri moonlight games and songs. Chukwuemeka Odumegwu Ojukwu J Folklore Cult Stud 1(1)

Forest F, Grenyer R, Rouget M, Davies TJ, Cowling RM, Faith DP et al (2007) Preserving the evolutionary potential of floras in biodiversity hotspots. Nature 445(7129):757–760

Frank DM (2022) Science and values in the biodiversity-ecosystem function debate. Biol Philos 37(2):1–22

Geisler C (2001) Your park, my poverty: the growth of greenlining in Africa. Paper presented at a Cornell University Conference on Displacement

Heffron JM (2022) Soka education and the land ethic: educational leadership toward the 'creative co-existence of nature and humanity'. Environ Edu res 1-15

IUCN (1980) World conservation strategy. IUCN (World Conservation Union), Gland, Switzerland

Janzen W (1994) Old testament ethics: a paradigmatic approach. Westminster John Knox Press

John EO, Enang NR (2022) Revisiting the discourse on human-nature relationship in African traditional religion and the responses to the environmental change. Bangladesh J Bioethics 13(1):39–48

Kaya Z, Raynal DJ (2001) Biodiversity and conservation of Turkish forests. Biol Conserv 97(2): 131–141

Kennedy S (2022) This land is not our land, this land is their land: returning National Park Lands to their rightful protectors. Am Indian Law J 10(1):3

Kimengsi JN, Owusu R, Djenontin IN, Pretzsch J, Giessen L, Buchenrieder G et al (2022) What do we (not) know on forest management institutions in sub-Saharan Africa? A regional comparative review. Land Use Policy 114:105931

Kramer RA, van Schaik CP (1997) Preservation paradigms and tropical rain forests. In: Kramer RA, van Schaik CP, Johnson J (eds) Last stand: protected areas and the defense of tropical biodiversity. Oxford University Press, New York

Krause T, Tilker A (2022) How the loss of forest fauna undermines the achievement of the SDGs. Ambio 51(1):103–113

Lugira AM (2009) African traditional religion. Infobase Publishing

Madongonda AM, Gudhlanga ES (2022) Wreaking mother earth: environmental intrusions and land dispossession in Musaemura Zimunya's kingfisher, Jikinya and other poems (1982). African Literature, Mother Earth and Religion, 49

Marijnen E (2022) The coloniality of crisis conservation: the transnationalization and militarization of Virunga National Park from an historical perspective. In The violence of conservation in Africa. Edward Elgar Publishing

Massey LM, Camerden PM, Gaos AR, Liles MJ, Seminoff JA, Ahern ALM (2022) Challenging gender inequity in wildlife conservation: a women's group leading sea turtle conservation efforts in El Salvador. Local Environ 27(1):1–15

Mentzafou A, Dimitriou E (2022) Hydrological Modeling for flood adaptation under climate change: the case of the ancient Messene archaeological site in Greece. Hydrology 9(2):19

Milley P (2022) Opportunities for forest-based biorefining to reverse decline in Canada's forest products sector (Doctoral dissertation)

Minton EA, Tan SJ, Tambyah SK, Liu RL (2022) Drivers of sustainability and consumer well-being: an ethically-based examination of religious and cultural values. J Bus Ethics 175(1): 167–190

Moreto WD, Brunson RK, Braga AA (2017) 'Anything we do, we have to include the communities': law enforcement rangers' attitudes towards and experiences of community–ranger relations in wildlife protected areas in Uganda. Br J Criminol 57(4):924–944

Mushonga T (2022) Violent forests, local people and the role of the state in Zimbabwe. In The violence of conservation in Africa. Edward Elgar Publishing

Naughton-Treves L, Weber W (2001) Human dimensions of the African rain forest. In Webber W, White LJT, Vedder A, Naughton-Treves N (eds), African Rain Forest Ecology and Conservation

NDES (1997) Niger delta environmental survey: biodiversity, Phase 1 Report, vol. 1v

NEST (1991) Nigeria's threatened environment: a national profile NEST, Ibadan

Numbere AO (2021) Impact of urbanization and crude oil exploration in Niger Delta mangrove ecosystem and its livelihood opportunities: a footprint perspective. In Agroecological footprints management for sustainable food system. Springer, Singapore, pp. 309–344

Numbere AO (2022) Application of GIS and remote sensing towards forest resource management in mangrove forest of Niger Delta. In Natural resources conservation and advances for sustainability. Elsevier, pp. 433–459

Nwankwoala HO, Orji OM (2018) An overview of earthquakes and tremors in Nigeria: occurrences, distributions and implications for monitoring. Int J Geol Earth Sci 4(4):56–76

O'Leary Simpson F, Lwaboshi R, Ikobo Y, Mulume P (2022) The structuration of armed mobilisation in eastern DRC's Kahuzi-Biega National Park. Discussion paper/University of Antwerp. Institute of Development Policy and Management; Université d'Anvers. Institut de politique et de gestion du développement.-Antwerp, 2002, currens

Onor K, Ityonzughul TT, Atayero SA (2022) Harnessing Nigeria's Forest resources for sustainable development in the 21st century. Afr Renaissance:177–194

Perrings C, Edgar E (2000) The economics of biodiversity conservation in sub-saharan Africa: mending the ark. Edward Elgar Publishing Ltd.

Pimbert MP, Pretty JN (1997) Parks, people and professionals: putting 'participation 'into protected area management. Social Change Conserv 16:297–330
Pinchot G (1947) Breaking new ground. Harcourt, Brace, & Co., New York
Rahman MK, Masud MM, Akhtar R, Hossain MM (2022) Impact of community participation on sustainable development of marine protected areas: assessment of ecotourism development. Int J Tour Res 24(1):33–43
Ramutsindela M, Matose F, Mushonga T (2022) Conservation and violence in Africa. In The violence of conservation in Africa. Edward Elgar Publishing
Rao MM (2022) A systematic study on Jaina iconography in Andhra from Nagabharana: Recent Trends in Jainism Studies, 379
Reyes-García V, Fernández-Llamazares A, Aumeeruddy-Thomas Y, Benyei P, Bussmann RW, Diamond SK et al (2022) Recognizing indigenous peoples' and local communities' rights and agency in the post-2020 biodiversity agenda. Ambio 51(1):84–92
Sandbrook C (2015) What is conservation? Oryx 49(4):565–566
Saxena A, Buettner WC, Kestler L, Kim YS (2022) Opportunities and barriers for Wood-based infrastructure in urban Himalayas: a review of selected National Policies of Nepal. Trees, Forests and People, 100244
Schweitzer CJ (2019) History, highlights, and perspectives of southern upland hardwood silviculture research. J For 117(1):55–64
Seile BP, Bareetseng S, Koitsiwe MT, Aremu AO (2022) Indigenous knowledge on the uses, sustainability and conservation of African ginger (Siphonochilus aethiopicus) among two communities in Mpumalanga Province. South Afr Diversity 14(3):192
Shanley P (2022) As the forest falls: the changing use, ecology and value of non-timber forest resources for caboclo communities in eastern Amazonia (Doctoral dissertation, University of Kent)
Sierra R, Campos F, Chamberlin J (2002) Assessing biodiversity conservation priorities: ecosystem risk and representativeness in continental Ecuador. Landsc Urban Plan 59(2):95–110
Sinthumule NI (2022) Conservation effects of governance and Management of Sacred Natural Sites: lessons from Vhutanda in the Vhembe region, Limpopo Province of South Africa. Int J Environ Res Publ Health 19(3):1067
Smith DW, Bangs EE (2009) Reintroduction of wolves to Yellowstone National Park: history, values, and ecosystem restoration. Reintroduction of top-order predators. Oxford: Wiley-Blackwell, pp 92–125
Soaga JA, Adeleye A (2022) Ecological Services of a Peri-Urban Recreation Centre in Abeokuta. Ogun State, Nigeria
Sullivan LM, Manfredo MJ, Teel TL (2022) Technocracy in a time of changing values: wildlife conservation and the "relevancy" of governance reform. Conserv Sci Pract 4(2):e545
Thomsen JM, Lendelvo S, Coe K, Rispel M (2022) Community perspectives of empowerment from trophy hunting tourism in Namibia's Bwabwata National Park. J Sustain Tour 30(1):223–239
Verma AK, Sadguru P (2022) Anthropogenic activities and biodiversity threats. IJBI 4(1)
Wild R, Walters G (2022) The forest is clothing for the ancestors: a rapid cultural assessment tool for forest landscape restoration policy processes. For Ecol Manag 504:119825
Wood D (1995) Conserved to death: are tropical forests being overprotected from people? Land Use Policy 12(2):115–135
Zhang P, Shao G, Zhao G, Le Master DC, Parker GR, Dunning JB Jr et al (2000) China's forest policy for the 21st century. Science 288(5474):2135–2136
Zhao N, Wang H, Zhong J, Sun D (2022) Assessment of recreational and cultural ecosystem services value of islands. Land 11(2):205
Zhu JJ, Yan B (2022) Blue carbon sink function and carbon neutrality potential of mangroves. Sci Total Environ, 153438
Zube EH (1995) Greenways and the US national park system. Landsc Urban Plan 33(1–3):17–25

Chapter 8
The Need to Conserve and Protect Forest Resources: African Perspective

Sampson Abigha Inatimi

Abstract The tropical rain forests in Africa are rapidly vanishing because of constant increase in the demand for forest resources (timber and non-timber products) and forest lands for agricultural expansion and infrastructural development. These forest resources provide man with materials that aid in the satisfaction of our basic needs. It is very important to retain these benefits for future utilization in a sustainable way to meet increasing human population demands and ensure a stable eco-friendly environment, conserving, protecting, and maintaining the remaining forests biodiversity from further unsustainable exploitation. The methods of conserving these genetic resources are global concerns to nations of tropical and subtropical forests lands including Nigeria, where the rate of forests degradation is high. Nations have adopted biodiversity conventions to enhance sustainable approach in ensuring the conservation of forest resources. Some of the strategies in Nigeria include NBSAPs (National biodiversity strategies and action plans) which is aimed at encouraging the government of participating countries with an integrated template for effective biodiversity conservation approach and National REDD+ Strategy (Reducing Emissions from deforestation and forest degradation) project that will oversee the implementation and enforcement of new tactical approaches for sustainable management and conservation of the forests while augmenting carbon stock in Nigeria. Therefore, this research is aimed at identifying the need for the conservation of forest resources to nations and highlighting the effects of deforestation, the application of possible biodiversity conservation method needed for future utilization, and availability of these natural resources through effective enforcement of the biodiversity laws by the governments of participant countries from various conventions relating to biodiversity.

Keywords Plant protection · Animal conservation · Environmental sustainability · Human activities

---

S. A. Inatimi (✉)
Forestry Department, Bayelsa State Ministry of Environment, Yenagoa, Bayelsa State, Nigeria

S. C. Izah, M. C. Ogwu (eds.), *Sustainable Utilization and Conservation of Africa's Biological Resources and Environment*, Sustainable Development and Biodiversity 32, https://doi.org/10.1007/978-981-19-6974-4_8

8.1 Introduction

Africa being surrounded by the oceans is known for its vast land and forest vegetations predominated by the presence of tall canopy trees that support wildlife coexistence and is home to varied biodiversity. Human dependence for food and provision of materials for our existence is largely derived from the forest resources available to us. The economy of some tropical and subtropical countries in Africa depends immensely on forest resources which include wood fuel, timber sales, medicinal herbs, edible fruits, palms especially Raffia and Elea's species, and nuts, and animal remains are useful raw materials for production processes. These products are also exported for commercial and industrial purposes, creating instant employment for the inhabitants of the area enhancing the increasing demands for these forest products. These products play significant roles in the lifestyle of the indigenous people, meeting the basic nutritional requirement of present and future generations and improving the livelihoods of the rural inhabitants as well as defining their cultural craftiness.

Forest resources are commercially useful materials that enhance the survival of humans on this planet. They are of two categories based on the valuable items obtained: timber and non-timber forest products. Timbers are woods (both hard and soft) obtained from fallen forest trees that are used as wood fuel in cooking and heating, house construction, furniture production, domestic equipment such as boats, baskets, farm tools, animal ranch fence, ply-woods, and wood-pulp for paper and textile production, etc. Non-timber forest products, on the other hand, are a diverse range of economic resources derived from the forest other than lumber; example includes food from plants such as food additives (nuts, wild fruits, herbs, spices, mushrooms, aromatic plants); plant materials (fronds, fibers, creepers, and flowers); plant derivatives (bamboo, raffia, rattan, cork, and essential oils); and animal products (honey, fur, silk, and other animal products), as highlighted by Adekola and Mbalisi (2015). Tropical and subtropical biomes receive more annual rainfall and fair temperature most likely between 20 °C and 35 °C. Thus, forests from these regions have qualities that sustain diverse range of species, which is referred to as high biodiversity. The warm, humid environment promotes the growth of a diverse range of plant species, as well as the survival of a diverse range of animals, birds, and insects.

In general, there are three types of forests in Nigeria: tropical rain forests with dense canopy tree communities in the south, mangrove swamp forests in the coastal Niger Delta, and dispersed savanna forests in the north. This forest also provides a wide range of dynamic equilibrium to our ecosystem functioning such as; a balance to respiration, pollination, and photosynthetic activities, water and nutrient cycling within the aquatic and terrestrial habitats to the atmosphere, natural air cleaner of air pollutants from oxides of nitrogen, carbon, and sulfur, while releasing oxygen and stabilizing our ecological climate change also. The tropical forests are the main target of infrastructural developments for oil exploration and exploitation, logging concessions, or dam construction which require the expansion of the road network and the construction of roads in pristine locations (Kaimowitz and Angelsen 1998).

The management and conservation of forest resources has in recent years been threatened by the indiscriminate exploitation and mismanagement of the forest resources in Africa mainly by urbanization and developmental projects (Izah 2018; Izah et al. 2017, 2018a; Izah and Seiyaboh 2018a, b). This is partly so because forests are seriously undervalued and many of their environmental benefits are not considered by the market values.

In Nigeria, the deliberate exploitation of forests for economic or sociopolitical reasons is very common and major losses have been recorded in vegetation with forest diversity been reduced to few species. Within the last four decades, the population trends of some animals in the forests have decreased while several others are threatened, an example is the Niger Delta Red colobus monkey, *Piliocolobus epieni*, which is critically threatened by hunting and habitat loss, yet they depend, adapt, and rely on certain tree species for habitation which is cleared and cut for logs. There has been a great reduction in the population of the Red colobus monkey within the past three decades so it is declared as an endangered primate of the Niger Delta area of Nigeria. A lot of animals depend on the conditions available in the forests for their food, mates, space, light, breeding, etc.

The high rates of deforestation or vegetation loss within the tropics of Africa with dense forests resources removed daily have created serious concern on biodiversity conservation. This makes forest conservation techniques critically needed since their removal has posed threats to environmental stability and resulted in increased pollutions, soil erosion, and flooding concerns in the southern and eastern regions of the Nigeria and there are limited trees to shed off flooding and gully erosion. Therefore, regulating and adopting contemporary tree cutting measures such as clear cutting, selective cutting, and shelter wood cutting is a step to mitigate the rate of deforestation in this area.

Governments of tropical countries are concerned about the conservation of her forests and their biological resources. Therefore, they have enacted laws and policies to safeguard forests, but the implementation has not been impressive. There is a call for government agencies to enforce the Forestry Conservation and Management Acts in order to preserve the potential of our forest resources for future utilization and sustainability as well as to educate traditional institutions such as family (kinship) religions, town unions, clan and kindred, among others, to shift the trend away from traditional forest management methods with the application of technical approach and focus on the need to conserve forests resource generally in a way that will ensure greater benefit flow to everyone within the communities.

8.2 The Role of Forest in National Development

Forest resources, particularly woods, are important valuable gem for a country's economic development. The resources that come from forest cannot be thoroughly measured, but they are extremely valuable to such countries because they are regarded as their natural heritage and play an integral part of their economic earnings. Forests supply a variety of goods that are used as raw materials in different sectors. For example, forests wood is used as a source of energy for heating and cooking by rural families and the paper industries for production. These forests also provide rubber, cotton, textiles, and other raw materials for further development. Furthermore, local households earned greater forest revenue and were more reliant on forest products income than households in a distinct biotic settlement (Asfaw et al. 2013). Forest is a renewable resource whose products may be replenished or conserved if it is recognized as key contributor to the country's economic growth and to the residents whose livelihoods are shaped by what is accessible for utilization. The undervalued forests contribute to job creation, poverty reduction, industrialization, and increase in the sales of forest products. The current global foreign earnings, on the other hand, are focused on conserving, protecting, and exploiting forest resources. Fuel wood, food, medicine, forage, alcohol, industrial and culinary oil, spices, honey, gum and resin, as well as crafts like mat weaving and construction materials, all contribute significantly to forests income. Frogs, insects, snails, reptiles, mammals, birds, and fish are among the fauna species used for sustenance. Forests are home to a variety of wildlife species that are plainly beneficial to humans. As tropical forests dwindle, ecologists around the world decry the decline in wildlife populations which could occasionally lead to extinction of animal species within such forests' lands. Our forests play certain important roles to us which are described in the following.

8.2.1 Rendering Environmentally Friendly Services

Forests render environmentally friendly services such as soil stability, nutrient cycling, air and water purification, watershed protection to limit runoff, and carbon sequestration (storage) to the environment (Adekola and Mbalisi 2015). The forests help in preventing the actions of gully erosion and flooding by shedding off river water from entering into the land to result in further environmental hazards. Forests also recycle nutritional minerals made available for plants and animals, as well as the natural environment, and provide ecological stability to the ecosystem.

8.2.2 Forests are Ideal for Recreational Activities

Recreational facilities, such as game reserves, zoos, national tourism parks, sanctuaries, and biosphere reserves, not only attract tourists from all over the world, but also produce jobs for the locals and generate revenue for the maintenance of those places. Tourism encompasses a variety of activities involving people traveling to and fro, staying in locations other than their usual residence for leisure, business, entertainment, and other hospitalities required for the recreation of a new atmosphere of balance to the psychological trauma they face daily from work and other commitments while on leave. As a result, the core of recreation popularizes tourism (Ijeomah and Enaing 2018). In Nigeria, the Cross River National Park is home to the Agbokim and Kwa waterfalls, which are the state's most famous and richest forest falls. Also, the Oli camp of Kainji Lake National Park, Okomu National Park, and Yankari Wildlife Park, among others, have provided jobs for many people while also facilitating infrastructure development such as paved roads, airports, hotels, power, and railroads, as well as revenue collection (Ijeomah and Enaing 2018).

8.2.3 Increasing Nation's Gross Domestic Product

The sum of market values contributed to all producers of products and services inside a nation's borders within a given period, generally a year, is referred to as gross domestic products (GDP). The value added by industry is generally calculated at basic prices, while the overall GDP is measured at purchaser prices (Ogunbadejo and Oladipo 2017). Economic growth is necessary for sustainable development, likewise a high degree of poverty leads to poor growth and low growth leads to a high level of poverty (World Bank 2006). The United Nations Forum on Forests (UNFF) emphasized the monetary contributions of forests to global economies and it was taken into consideration. It was more than twice as much as the total output of gold and silver resources combined. Despite the lack of attention paid to agriculture, forestry, and fisheries, Nigeria's productive value added (percent of GDP) was 21.21 in 2016, with the maximum value of 48.57 in 2002 and the lowest value of 20.24 in 2014. As a result, forest and fisheries have contributed steadily to Nigeria's GDP during the last three decades (Rotowa et al. 2019). According to Sonone (2018), the forestry industry contributed 1.7 percent of India's GDP in 2002. Economic, ecological, and social functions of natural resources all have a noticeable trade income.

8.2.4 International Trade

Forest products help countries balance their foreign exchange and financial equality by bringing in a lot of money. Forest resources can be exported and used domestically as fiber, wood, medicinal, ornamental, industrial raw materials, and energy products, among other things. As forested nations, notable nations such as India, Canada, Italy, Malaysia, and others use timber as a significant source of finance (Sonone 2018).

8.2.5 *Job Creation*

For many individuals living in forest areas, the forest is their primary source of income and this population of individuals runs the home-made producing industries. According to the Forest Connect Report on Nepal by Sonone (2018), one-third of Nepal's rural population collect and trade forest products, benefiting over 70,000 people. Forest resource sales are a well-known way for the government to interact with the public through economic and socio-cultural activities. Timber contractors, tree takers, saw-millers, timber lorry drivers, machine operators, log rolling crew, timber clerks, and gatherers of non-wood forest products all benefit from these socio-cultural activities (FAO 2001). Wood products are sold for construction and industrial reasons, as well as for domestic cooking and heating. Forests dwellers made available some seasonal food crops, fruits, and nuts which were carried to non-forest locations in need of it as commerce, particularly urban sites (Sonone 2018).

8.2.6 *Medicine*

Medicinal plants for the treatment of various types of ailments can be found in forests around the world, with some of these medicinal plants growing only in specific environments, such as *Capsicum frutescens* (Izah et al. 2019a, b), *Costus afer* (Izah et al. 2019c), *Carica papaya*, *Vernonia amygdalina*, *Ocimum gratissimum*, *Myristica fragrans*, *Anacardium occidentale*, *Cymbopogon citratus* and *Zingiber officinale* (Izah and Aseibai 2018), *Vitex grandifolia* (Epidi et al. 2016a), *Alstonia boonei* (Epidi et al. 2016b), *Cymbopogon citratus* (Kigigha et al. 2018a), *Buchholzia coriacea* (Kigigha et al., 2018b), and *Musanga cecropioides* (Kigigha et al., 2016).

8.2.7 *Infrastructure Building*

Wooden timbers are used in construction, furniture, wooden boats, and miniature bridges, among other things. For building dwellings and fencing farms, most villagers employ forest trees, prickly bushes, and bamboos. Some farming implements are made of wood, and the bullock cart is an ancient mode of transportation constructed primarily of wood (Sonone 2018).

8.2.8 Food Security

Our forests provide us with wide range of food crops to aid the poor forest dwellers' survival and the commercial sale of these seasonal non-timber products brings income to them. Some of the classes of plant resources obtained from the forest could either be edible crops or materials for further productions which include plant-based food additions (nuts, wild fruits, herbs, spices, mushrooms, fragrant plants); plant materials (fibers, creepers, and flowers); plant derivatives (bamboo, raffia, rattan, cork, and essential oils); and animal products (honey, silk), as noted by the USDA (Adekola and Mbalisi 2015).

8.2.9 Others

Supplying and distribution of commercial forest materials such as food, biomass, pulp and paper, rayon fibers, wooden articles, and medicinal plants locally and internationally.

8.3 Factors Leading to the Destruction of Forests

The increasing growth in population of humans and the need to satisfy our consumption rate are the biggest causes of forest destruction due to the immeasurable amounts of resources, products, and services derived from the forests to meet up with this large population. According to Rainforest Concern (2021), half of the world's rainforests have been cleared within a century; at this rate there exists an extinction threat. This calls for an urgent need for drastic action so that these forests, its fauna and flora, and human beings who depend on them will continue to survive. Loss of vegetation is considered the second major compelling cause of climate change (even more than the entire universal transport sector), responsible for 18-25% of world annual carbon dioxide emissions by an article Rainforest Concern (2021). The factors leading to the destruction of the forest are discussed in this section.

Deforestation is the intentional removal of forest and other types of vegetative cover from a location without replacement, altering the natural arrangement of trees, the flora, and wildlife within the forest (Adekola and Mbalisi 2015). It's also known as a process in which trees are felled for various reasons, but no trees are replanted to replace the ones that are destroyed (Omofonmwan and Osa-Edoh 2008; Nzeneri 2010). Most forest areas are cleared for unsustainable human uses such as agricultural crop lands and cattle ranching, logging, urbanization and industrialization activities, and dam-building for power generation and this results in great loss of forests resources. The determining factors of deforestation are shown in Figs. 8.1 and 8.2, respectively. Since 1960, the uses have had negative impact on the stability of

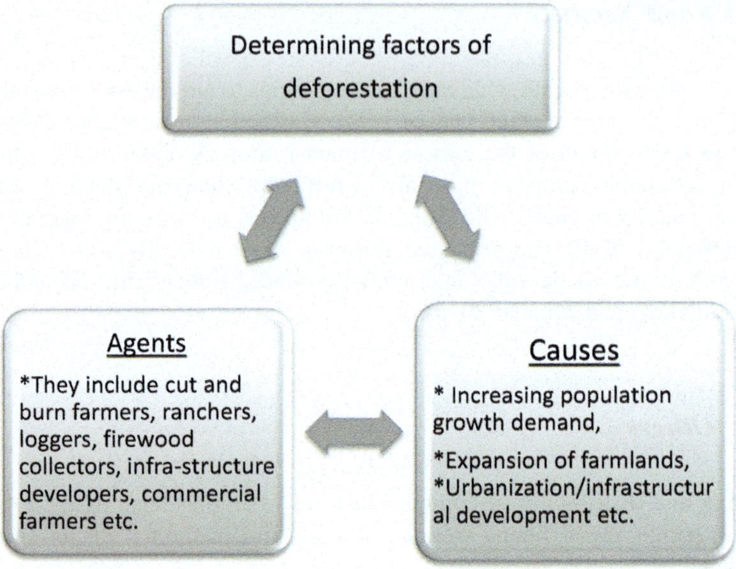

Fig. 8.1 Determining factors of deforestation

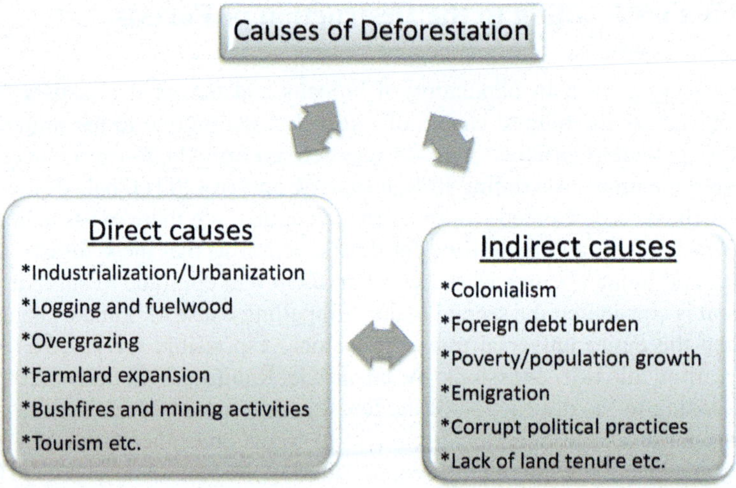

Fig. 8.2 Causes of deforestation

our natural ecosystems, biodiversity, and climate. According to an article by Youmatter, the UN's Food and Agriculture Organization estimates an annual pace of deforestation of roughly 1.3 million km^2 per decade (Youmatter 2020). According to Mongabay.com (2010), Nigeria is forested on 9.9% of its land, or over nine million hectares, with 2.38 percent (about 400 hectares) lost each year between 1990 and 2010. Between 1990 and 2010, Nigeria lost approximately 40% of its forest (about eight million hectares) in total. However, the reason why forests are destroyed will become evident if only we could distinguish between the agents of deforestation and their causes, which is crucial to comprehend the primary determining factors of forest loss (Sumit et al. 2012).

The agents of deforestation are those who are cutting down the forests, such as cut and burn farmers, ranchers, loggers, firewood collectors, infrastructure developers, commercial farmers, etc. (Fig. 8.1). The causes of deforestation are the forces or conditions that compel the agents to clear the forests which could be the increasing food demand from the population growth, leading to unsustainable agricultural expansion of farmlands, urbanization, and industrial demands, etc. These causes and agents of forests loss are interconnected and collectively attribute to the determining factors of deforestation mentioned earlier. For the sake of clarity, we can categorize the causes to be either direct or indirect cause of deforestation.

The direct and immediate causes of forest loss (Panayotou 1990; Barbier et al. 1994; Caviglia 1999) are relatively easy to identify, but the indirect causes are usually the main drivers or purpose of forest loss and the root of most discrepancies leading to destructions that are not easily calculated (Bhatnagar 1991; Mather 1991; Humphreys 2006; Sands 2005).

8.3.1 Deforestation's Direct Causes

The following are some of the direct causes of deforestation that may be seen in our environment (Fig. 8.2):

8.3.1.1 Industrialization/Urbanization/Infrastructure

Increasing population growth rates and rural-urban relocation are causing industrialization and urbanization to become inextricably connected problems. Cleared forests provide the needed land for increasing towns and cities to build infrastructure for the support of the growing population (Mather 1991; Sands 2005). Urbanization brings about an increase in human population in a particular industrialized locality and necessitates a large amount of land for infrastructural development (such as electric power stations, worship centers, health centers, houses, roads, schools, and so on) to meet the needs of the people living in the densely populated area (Adekola and Mbalisi 2015). As a result, forest vegetation in this area is destroyed to make way for these infrastructures. For smooth establishment of firms and industries, there is need for deforestation activities to build access roads, railways, bridges, airports, buildings and offices, parking lots, and other infrastructure required for these enterprises to travail. Forest resources are also explored and utilized by most enterprises as raw materials for the development of goods and services that are beneficial to them. As a result, urbanization and industrialization rely on unsustainable exploration and exploitation activities that devastate the forest and its resources. Petroleum exploration, exploitation, and oil spills, according to Onuche (2010), are destroying vast swaths of Nigerian swamp forest. One of the main reasons of infrastructure developments for oil exploration, logging concessions, or dam construction is the tropical woods, which facilitate the extension of the road network and the construction of roads in remote areas for these woods to be readily accessible (Kaimowitz and Angelsen 1998).

8.3.1.2 Logging and Fuel Wood

Logging is the removal of trees from the forest for the goal of producing, using, and exporting timber while fuel wood is the removal of trees from the forest for the purpose of heating and cooking as energy source (Adekola and Mbalisi 2015). This method diminishes the number of tree species in the forest and damages or injures unlogged trees, causing them to die (Fig. 8.3). The tall treetops and canopies of these logged trees protect and sustain the non-timber resources in the forests and numerous plants and animals' species depend on them for coexistence. However, when destroyed for logs, it results in their mortality and subsequently leads to the endangering of some plant and animal species depending on these trees directly, thus logging activities severely degrade forests (Putz et al. 2001).

Logging provides access roads to settlers to follow on and sold logs bring income to the dwellers. Gathering of fuel woods is often concentrated in the tropical dry forests and forest areas that have been degraded already (Repetto 1988, 1990; Rowe et al. 1992; Anonymous 1994). Although fuel wood is not usually the major cause of

Fig. 8.3 Logs from cleared forests

forests' loss in the humid tropics, it could be in some populated regions with reduced forest lands. Fuel wood gathering was reported to be the main cause of vegetation loss and forest degradation in El Salvador (Repetto 1990). Fuel wood gathering can be a major cause of deforestation and degradation when the demand for it is high in the drier areas of the tropics.

8.3.1.3 Overgrazing

Our forests are being cleared to aid the rising need for agricultural farmlands expansion and grazing for more animal products, thus leading us to the term overgrazing. The grazing by livestock animals to a point where the grasslands and other vegetations in our forests are completely degraded and left bare is the result of overgrazing. These activities create room for soil erosion and other ecological disasters when water and wind action invade the degraded soil. These overgrazing activities are more common to dry areas of the tropics with reduced soil fertility and few plant species mainly of shrubs and grasses available for grazing by animals. Clear cutting and overgrazing have turned huge landmass into a desert eventually, and the grazing lands which used to be so green and rich have now become sand areas. According to Sumit et al. (2012), enormous flocks of livestock like cattle, sheep, goats, etc. strip the vegetation through grazing because of its high soil fertility. Lands could be shared by the people living within, but animals are raised for themselves as wealth generation. There is need for effective control of grazing activities by animal rangers to avoid further degradation of our forests' vegetations.

8.3.1.4 Expansion of Farmland for Agriculture

Agricultural projects begin with forest clearing activities, which is the modification of the initial vegetation. According to Ogunleye et al. (2004), this occurs in places where people rely heavily on forest lands for subsistence and shifting agriculture, which is mostly found in tropical developing countries. These cleared areas provide vast lands for subsistence farming to the world's most vulnerable people (Myers 1991). In addition to logging and other factors such as urbanization, agricultural arrangements account for about 60% of cleared forests in tropical wet forests, while fuel wood accounts for the remaining 40% (Myers 1994; Anonymous 1994) The principal sources of deforestation in Nigeria, according to Ogunleye et al. (2004) and Salau (1993), include farming activities such as shift cultivation, bush clearance, and burning. However, when the condition of the land deteriorates, people are obliged to seek out fresh forest, resulting in more deforestation (Wilkie et al. 2000; Amor 2008; Amor and Pfaff 2008). As a result, a slash and burn strategy is employed where agricultural practice becomes clearing wooded area for the purpose of cultivating crops and left to fallow by weeds to increase soil fertility.

8.3.1.5 Bushfires

In both rural and urban areas, we have seen how an uncontrolled bushfire can burn into an area of flammable plants. These flames are among the most regular natural disasters in some areas, such as Siberia, California, and Australia, where they have lately destroyed enormous hectares of forest and wildlife and are attributed to climate change being a major contributor to the size of the fires that have occurred. Fires are the most essential instrument utilized in clearing the forest for agriculture, urbanization, and other infrastructure operations, aside from the harm caused by naturally occurring wildfires in our woods. According to Sumit et al. (2012), fire is a nice servant, but it has an unlucky master in the form of forest destruction contributors. According to Anonymous (2010), forest fires affected an average of 19.8 million hectares or 1% of all forests in 118 nations, accounting for 65 percent of the world's forest area.

8.3.1.6 Mining and Oil Company Activities

The need for profitable mining for gold and diamonds, coal, aluminum, copper, and other economic minerals and metals, as well as oil corporations looking for fresh oil resources, has significant impact on rainforests. As a result of these activities, massive road networks are constructed through these pristine forests to provide miners access to dig beneath the ground for minerals and metals, as well as to construct pipelines for oil extraction. Oil companies' exploration activities require the clearing and destruction of the forests in most cases within the Niger Delta swamp for oil extraction to travail conveniently. The local dwellers within these forests are left with no other choice than to encourage themselves in relocating deep into areas as a result of the negative effects of the exploration activities, especially the noise and other chemical pollution that may spill into their habitations and they are compelled to start cutting more timbers to make new homes, sell, or to produce charcoal as occupation for their sustenance. The activities from mining and oil extractions affect the biodiversity availability and cause wildlife migrations due to the disturbances they bring to the forests and the forests are contaminated eventually with ruptured oil pipelines spouting gallons of oil into the surrounding forest, which are later washed into nearby rivers affecting aquatic organisms.

8.3.1.7 Wars and Role of the Military

The impact of military operations during wars on forest destruction was seen during Guinea, Liberia, and Sierra Leone civil war in Africa (Burgess et al. 2015) and these wars further encourage armed groups to extract important forest resources to fund their activities. However, in Nigeria, the Sambisa Forest has become a hostile hide out for insurgents and their hostages causing chaos and unrest to the forest.

Sometimes, they engaged in combat confrontation with the military soldiers destroying the forests and wildlife within. Some strategic combat activities that degrade the forests are wartime bombing, chemical spraying, building of sheltering camps or hideouts, or setting straps for enemies as such that were used by the US military in the VietnamWar (Burgess et al. 2015; Mather 1991; Sands 2005) and it is expected that forests cover might be decreased in war areas. Most military camping and training of newly recruited officers are done in the forests, with the firing of various ammunitions into forests destroying and causing disturbances to the wildlife and other organisms which could die because of this unrest. On the foreign scene recently, Myanmar government sells timber to the Thailand to fund its civil war against the Karen hill tribe and deforestation in El Salvador has been attributed to war (Sumit et al. 2012). It was documented that the military plays role in huge forest loss in Southeast Asia and South America (Mather 1991; Sands 2005).

8.3.1.8 Tourism

Having national parks, forest and game reserves, sanctuaries, etc. within certain tropical forests environment brings about local employment and infrastructural development such as good roads, airports, hotels, power, railways, and evidently revenue generation to that nation (Ijeomah and Enaing 2018), but destroys the natural forests to achieve these infrastructural aims by excessive exploitations. Perhaps, you may want to say national parks and sanctuaries protect the forests, well to a little extent, but the inappropriate opening of these areas to the public for tourism is detrimental (Sumit et al. 2012). Regrettably, most governments of tropical environment adopt tourism for making profit and sacrifice the need for conservation of these forests as recreational attractions. In addition, some companies and resorts claim to be eco-tourist establishments and are exploiting the forests for profit also and some curious tourists cause repeated injuries to the plants and animals within (Shukla 2010). Using eco-tourism as a disguise, infrastructure development is what is seen today, attracting people other than tourists also, causing massive deforestation particularly deep into the forest.

8.3.2 Indirect Causes of Deforestation

The bid to ascertain economic and infrastructural development in countries of tropical forest regions globally has led to the uncontrolled forests removal majorly by the poor socioeconomic policies made by the government of these nations and this is the reason for the World Rainforest Movement calling for actions on forests protection (Anonymous 1990). Development is the main reason while irrational deforestation has taken place in the form of direct factors (agents and causes) highlighted earlier such as indiscriminate logging, need for agricultural farmlands, urbanization, and industrialization, increase in urban–rural migrations, etc.

According to Sumit et al. (2012), our wasted forest resources are found in the industrialized nations where most of our resources end up, mainly the tropical timbers. These industrialized countries influenced us by advertising their own folkways to us which seems free and easy going and later making us to give away our culture for easily explorations and exploitations. Therefore, the indirect causes of deforestation are considered below:

8.3.2.1 Colonialism

Most underdeveloped tropical rainforest countries were colonies by European first world countries who made policies that allow them to acquire full or partial political control over them and break their indigenous powers while exploiting their available resources economically. These tropical countries have local populations who had their own traditional system of land management in these areas years before the advent of explorers from rich industrialized countries. In this process, previously self-sustaining local economy of the people was turned into export sites for commercial exploitation of indigenous genetic resources for international production (Sumit et al. 2012). The colonial migrants encourage wood cutting from our natural forest reserves for the development of cocoa and palm oil plantations of which the logs were exported out of West Africa and until now, this process continues in different forms of influences and colonists encourage the incessant exploitation (Colchester and Lohmann 1993). These wealthy countries with colonial powers have deficit of their own natural resources and rely on the resources of the financially poorer countries. Now you can see why colonialism is a driver of deforestation because of the political influence it had on their colonized countries.

8.3.2.2 Heavy International Debt

Most Third World countries are poorer countries with rich natural forests resources and yet already undergoing heavy foreign debt burden all in bid to meet up with the global development agenda of sustainability, and now with the recent fluctuations in the interest rates, call for urgent way of repaying these enormous debts. Obviously, they are compelled to exploit forests resources and other minerals available to them as foreign exchange services for their rising debts (Sumit et al. 2012). Most rich nations assist poorer countries to attain visible infrastructural development (such as building or supplying materials for the construction of electric power stations, bridges, railways, airports, tarred roads, health centers, houses, schools, and so on) and in return pay with the exploitation of their rich natural resources for a period to cover for this indebtedness. However, the inconsistent decision enforcements from these poor, tropical countries are also attributed to issues with corruption in government, the military, and struggle for economic powers which encourage more debt burdens to them.

8.3.2.3 Poverty and Population Growth

The increased rate of exploration and exploitation of forest resources (timber and non-timber forest products) is necessary to meet the demands of a growing human population and the high poverty rate within forested settlements. The people's needs are satisfied via exploitation of forest resources to improve their survival and meet their demands. As a result of the constant strain placed on the forest and its resources, overexploitation occurs. Fuel wood and other forms of wood for construction reasons are in high demand, putting a lot of strain on the forest (Salau 1993). The indigenous people use fuel wood for cooking and heating rather than petroleum products as their energy source (Onuche 2010). According to the FAO and other international organizations, poverty and overpopulation are the main causes of deforestation (Sumit et al. 2012). These organizations often feel that fostering development while attempting to decrease population growth mitigates these difficulties. However, the people desire more food and space for the need of additional lands for unsustainable agriculture and housing which leads to the loss of more forests. According to Purnamasari (2010), most people do not have the financial resources to spend in improving soil quality and increasing yields on existing cleared property. Usually, majority of the farmlands are owned by powerful influential persons which displaces unfortunate farmers into rainforest areas to clear and make new farm settlements. As far as these influential persons take hold of power, it will remain difficult to achieve lasting land reform and deforestation continues (Colchester and Lohmann 1993). However, sustainability can be promoted if the government of these tropical countries would devise a sustainable management system for individual's life support and controlling population growth as a positive step towards curbing the link between poverty and overpopulation role in our deforestation.

8.3.2.4 Emigration

One of the noticeable indirect causes of deforestation is the allowance of foreign nationals who leave their resident country perhaps with the intent to settle elsewhere or explore the land and subsequently colonize the forest (Mather 1991; Colchester and Lohmann 1993; Sands 2005). The government and people thought that by encouraging emigrants and their colonization programs into rainforest areas could lessen poverty from these financially poor countries with rich natural forest resources. Well, these programs sadly failed and caused harm to the indigenous people and the forests environment. According to Levang (2002), landless migrants raise population pressure and demand for food and other valuable forest resources, causing local farmers to enhance production by cutting deep into the forests to extend cropland or habitations. Foreign migrants, on the other hand, may be unconcerned about forest conservation in their new communities, leading to more deforestation.

8.3.2.5 Traditional Land Tenure is Absent

In most tropical countries, the government has complete control over forests and is somewhat lax in efficiently regulating their use, especially when they have a loosely defined tenure system; it is usually detrimental to the locals and the forests (Chomitz et al. 2007). Majority of forest residents and landowners lack evidence of ownership or property rights to hold on and are displaced by others; the powerful influential few would acquire tenure over their properties (Mather 1991; Deacon 1999; Sands 2005). This implies that they would have to relocate their settlements and cut deep into the forest for food, space, and shelter, resulting in further depletion of forest resources.

8.3.2.6 Undervaluation of the Forests

The forests have provided us with valuable resources for economic sustainability and meeting the increasing human population needs. However, it is given less economic value, probably because the economic value most times is seen on a very small extent of benefits on the terms of raw materials and products which are traded in formal markets and available for human consumption and production (Emerton 2003). However, these direct uses only represent a small fraction of the total value of the forests overlooking the excess benefits far above the physically marketed products. Mather (1991) and Sands (2005) identify forests to gain value only when cleared for obtaining official improvement title. When non-wood products are extracted from the forests, it is also suggested to add value to the forest but not economical when measured up to clearing alternatives. Therefore, clearing of forests for logs, fuel wood, and further wood allied productions is encouraged and paying less attention to the environmental benefits the people derive from the non-wood forest products too. If only government would see these unquantifiable economic benefits and place high value to every product from the forest, then it will be protected and preserved from indiscriminate degradation.

8.3.2.7 Corrupt and Political Practice

Corruption is one of the noticeable main causes of deforestation as identified by FAO in its 2001 report, which advised that urgent concern needs to be given to these illegal forest practices and corruption in many tropical countries. These illegal forest practices were highlighted by Contreras-Hermosilla (2000, 2001) to include: forest products being smuggled across borders, permitting illegal logging and selling of harvesting permits, forestry officers give approval to illegal private contractors or enterprises, harvesting of high economic valued trees (protected trees) by commercial dealers and unlicensed processing of forest raw materials, and so on. All these corrupt political practices are common within the tropical and subtropical nations with rich forest environment. If corruption and illegal practices are allowed to continue, then indiscriminate deforestation is the result and biodiversity is endangered.

8.4 Effects of Forest Destruction

The destruction of the tropical forest causes enormous natural consequences for our ecosystem and poses significant threats to global stability while playing an important role in the national development of most countries' economy. The effects of deforestation could either be beneficial or harmful depending on how sustainably it is utilized. Forests loss has the following primary effects on us, according to Onuche (2010), Omofonmwan and Osa-Edoh (2008), and Salau (1993).

8.4.1 Biodiversity Threat

There is a serious threat to biodiversity and wildlife as deforestation continues, affecting the complex mixing of distinct species of plants, animals, and microorganisms, as well as their genetic constitutions, habitats, and ecological niches which is referred to as biodiversity. The beauty of this world is based on its biodiversity, which can be seen everywhere, a drop in wildlife and biodiversity because of deforestation will result in the loss of many forest products that support a large proportion of dependents. The destruction of vegetations can be a threat to living organisms because there will be instability in our ecosystem, especially the oxygen – carbon (IV) oxide balance within the components of the ecosystem and this could result in global crises. The complexity of this world shows how interconnected and interdependent we are to different organisms for survival; tall canopy trees provide shades to support animals and smaller trees which may not survive heat intensity of the sun directly. In addition, trees also provide animals with food (fruits, nuts, leaves, etc.) and shelters they need to adapt and survive on.

8.4.2 Depletion of Soil Fertility

Loss of soil fertility occurs when the components which sustain the fertility of the soil are removed and not replaced or maintained leading to poor crop yields and scarcity of food results in hunger and death to dependent organisms. The removal of the forests is a major cause of soil infertility because needed nutrients are not replaced and this foster other destructive ecological agent like floods, gully erosions, and even drought in some cases causing degradations. When the nutrient-rich organic topsoil is eroded or depleted of its original organic contents, this will eventually damage the soil structure and texture embedded in it.

8.4.3 Lead to Soil Erosion

The degradation of the upper layer of the soil by agents, runoff water, and wind action is possible when the trees meant to shield off these effects are removed for infrastructural purposes. The forest acts as a network between the soil and the atmosphere by increasing the permeability of the soil to rainwater which reduces runoff and binding the roots of plants to the soil firmly to control wind action. So, the removal of vegetation overtime would increase the rate of surface erosion and eventually degraded the soil leading to destruction of crops, soil, and animals and affect human farmlands and other properties (Sumit et al. 2012). Thus, forest and vegetation help to reduce this adverse destruction by water and wind erosion by forming a firmer anchorage to the soil.

8.4.4 Pollution

The extraction, collection, and distribution of forest products within the forests and on our municipal environments litter series of wastes remains which could be washed into water bodies to cause blockage of water flow within the river system. The process of transporting wood materials to processing industrial mills for furniture and construction works also litters unfriendly solid wastes to our environment. In some cases, hazardous chemicals are likely used to separate the waste from economic minerals mixed. Often these hazardous chemical separation techniques are done within the forest site where these minerals are extracted and eventually washed into rivers, disturbing the river's ecosystem. The production of hazardous chemicals and its use in the forests imposes major threat to biodiversity globally. According to Imarhiagbe and Egboduku (2019), the major perpetrator of environmental pollution has been human beings itself; our exploration, extraction, and exploitation activities such as site clearing, emissions from refineries through flared gases, industrial processes, burning of liquid and solid waste, releasing wastes of solids (plastics), liquids (oil spills, pesticides), and inorganic fertilizers into the air, soil, and water. These non-easily degradable chemicals accumulate into the toxic level of the food chain of faunas and flora within. In return, affecting the survival and adaptation of species by this disruption results in ecosystem instability and eventually causes species extinctions and loss of biodiversity (Egboduku and Olorunfemi 2016). Hence, understanding humans' negative roles toward pollution of our environment within the forests area and how to mitigate it is of crucial importance for setting priorities for the current ecological challenges we see today.

8.4.5 Loss of Local Medicinal Plants

Any plant which protects health or prevents certain illnesses or possesses curative characteristics is referred to as a medicinal plant. Majority of the poor people living in rural and urban settlements of the tropics use medicinal plants as their only available treatments for minor and severe ailments (Elisabetsky and Wannamaacher 1993). Most traditional medicines' use in the treatment of injuries and other illnesses depends largely on the herbs or parts of plants like their leaves, flowers, roots, fruits, peels, inner and outer barks of stems and exudates, etc. that are taken from the forests and useful for the treatment of illnesses like worms, urinary infections, coughs, burns, cuts, sprains, fatigue, etc. This medicinal importance is demanded by phyto-medical and pharmaceutical industries for commercial productions leading to the collection and harvesting of these plants for sale. There have been reports in Africa that the overharvesting and destruction of specific selected species of medicinal plants is because of the collection for export and increased urbanization demands leaving local inhabitants without medicinal remedies (Cunningham 1993).

8.4.6 Apparent Reduction of Foreign Earnings from Timber Export

The importance of timber products to a nation's economy is enormous and placed topmost priority in the country's exchange market. Most tropical countries' GDP depends largely on the earnings from timber exports to other nations in need of it and is used to support the economy and foster infrastructural development in that nation. However, there have been foreign income decline from exporting timber products recently; this might be attributed to the overexploitation of timber products in our forests and lack of conservation schemes (reforestation or afforestation) after tree felling, leading to the vanishing of selected timber trees of high-quality demand in the forests.

8.4.7 Contribution to Climate Change and Global Warming

A common contributor to climate change is deforestation and the others being the combustion of fossil fuels and decomposition of organic matters which emit atmospheric carbon (IV) oxide. The removal of world's forests is responsible for about 11% of global greenhouse gas emission. Forest is one of the two natural reservoirs of carbon, the other being the ocean. This forests' removal comes in the form of wildfire, logging for wood, clear cutting, and livestock ranching for agriculture, among others. The increase in concentration of greenhouse gases results in climate change (Salau 1993). Some of the visible impacts of climate change include extreme weather conditions, drought, pest infestation, rise in sea level and storms, time changes of seasonal events, collapse of glaciers, wildfires, land degradation and pollution, and global warming. This subsequent rise of temperature will strongly affect natural biological systems, and migratory species of organisms move towards Polar region. Deforestation contributes to the warming of the earth when trees are destroyed, burnt, and allowed to decay and the stored carbon is released into the air as carbon (IV) oxide to cause a greenhouse effect resulting to global warming of the earth crust.

8.4.8 Leads to Desertification

Vegetation plays a vital role in the composition of the soil and its fertility which determines the biodiversity and wildlife within the land. However, the continuous use of land for agriculture, overgrazing, and deforestation rapidly degrade the nutrients of the forests' soil leading to gradual desertification. Lands void of biodiversity sustainability eventually affect economic activities to travail such area.

8.4.9 Forests Loss Could Lead to Flooding

The importance of vegetations in reducing floods has been well-known over time in the coastal areas. When a river cannot handle the water rise and there are no forest trees to watershed it off or tree roots to absorb water from the soil, it results into flood events in lowlands and coastal areas, leading to loss of life, properties, plants, and animals.

8.5 Options for Conserving Forest Resources

It is necessary to protect and preserve the forest and its resources from further destruction and indiscriminate exploitation to attain infrastructural developments with no consideration of the effects it will cause to our ecosystem's stability. Therefore, the conservation of the natural rainforest would ensure the continuity and availability of the benefits derived from them, such as the provision of food and habitation for wildlife, provision of raw materials for construction and industrial uses, watersheds from trees to control flood and erosion, stabilization of soil's fertility, and reduction of effects of climate change (Bassey 2003). These benefits are so significant to man that they ensure that our survival and adaptation on earth is made possible. The effort to preserve the world's rainforests and other forests keeps on growing global concern about how this issue can be managed. The good part is that it is a renewable natural resource and conservation is the best tool in use with the capacity to renew the forest resources. The decline of forested area in Nigeria has resulted in a general concern for nationwide conservation.

The reasons for conserving our forest were highlighted by World Wildlife Fund (1980) and summarized as follows:

- To sustain life support and maintain the necessary ecological processes.
- To maintain biodiversity and wildlife.
- To guarantee a sustainable management of species and ecosystems.

In other words, to achieve these reasons, every individual, country, and region must contribute to stop deforestation with the following approaches being taken:

8.5.1 Community Enlightenment

Educating the people and tourists is necessary about the need to develop or protect the forests and be part of educative schemes and activities for forest conservation and stability of our ecosystem. Educative programmes and skills are necessary so that the rural dwellers can be enlightened on the environment and its related problems caused by our activities. Environment education is the ability to acquire a conscious knowledge and awareness of a vase range of environmental context. These programmes can be used to build up the rural people and educate them on how to read and write issues concerning their surrounding environment, and thus, understand the nature and components of their forests and procedures they can utilize to protect and preserve their forests. Community enlightenment schemes will take away the ignorance from the public and policy makers on forests conservation, its aims, and methods through forestry extension aids.

8.5.2 Enforcement of Forest Laws and Policies

For deforestation to decrease, it is essential for the government of tropical countries to have a strong and stable enforcement system to curb the existing practices. The government should ensure they enforce the laws and policies they create to guarantee the protection and restoration of forests. The policy and regulatory procedures for forests have already been made to protect forests but enforcement is the problem. This calls for modification and amendment of these laws such that it will persuade the public's participation in forestry conservation and management and protect the people's traditional rights and tenure (Sumit et al. 2012). Therefore, in order to stop further deforestation and overexploitation of the forest resources, these actions are necessary; negotiation between the government and stakeholders of the settlements within the forests, warnings regarding forest laws, reserves and protected areas should be made known to everyone, violation notice be issued to offenders of these laws, arrests be made to violators if they neglect warnings, fines and environmental court actions be carried on for various offenses indiscriminately violating the stakeholders of the forests. FAO (2010) said that half of the deforestation globally in the tropical countries could have been stopped if their governments were determined to do so.

8.5.3 Reforestation and Afforestation

Indeed, the deforested areas need to be reforested and the deliberate growing of trees as replacement for the felled ones should be enforced by policy makers and dealers of forests materials. This will increase the area of forest vegetations and use of deforested lands or used marginal lands (such as railway tracts, along roadside, on contours, avenues, boundaries, etc.) or degraded lands (not suitable for agricultural production) for replanting will have a profitable sustainability (Sumit et al. 2012). Trees planted outside forest areas will decrease pressure on forests demand for fuel woods and timbers. Although efforts to replant deforested lands take place annually, in most cases, replanting is done with the aim of hastily growing trees to be cut down by the logging industry soonest. While in some other countries, replanting is done by replacing the deforested areas with some economic valued trees such as raffia palm, bush mango, moringa tree, rubber tree, etc. to enhance local economy.

8.5.4 Management of Protected Areas and Reserves

It is fundamental to conserve forests and its biodiversity by managing some protected areas or reserves (Myers 1994). Portions of land restricted by government and banned from commercial activities for the purpose of afforestation and preserving biodiversity from exploitation and extinction are referred to as forest reserves (Imarhiagbe and Egboduku 2019). However, in Nigerian forest reserves, there exists a great threat as a result of lack of consistent management and monitoring of these protected areas by policy makers, and also, increasing population and economic activities have also contributed to encroachments.

8.5.5 Decrease in Population Growth

It is evident that the increasing growth in human population is a major driver of deforestation which results in other factors like unsustainable expansion of agriculture farmland into the forests, urbanization, and industrialization to meet up with the sustainability and economic development of these tropical nations. The uncontrolled increase in the growth rate of people living within a country brings poverty to that nation and loss of income meant for development used for sustaining them. These make the inhabitants living within the forests to mount pressure on them as folkways available for exploit and habitations. However, the benefits of reduced population will be an increase in our rich forest resources for sustainability because of the decrease on our total dependents on the forest products and subsequent increase of our foreign capita earnings as a country when the genetic resources are effectively managed for commercialization and infrastructural developments and foster biological diversity conservation and availability for future utilization by the people.

8.5.6 Use of Less Wood Products

We should encourage the use of substituted materials for wood works more often. If only we can stop using timber for most constructions, then the market for wood product demands will gradually reduce the alarming rate of deforestation. However, many countries are discussing on how to encourage environmentally friendly timbers' substitute for infrastructural development and construction.

8.5.7 Adequate Monitoring and Information System

Adequate monitoring and information system should improve globally for effective biodiversity conservation, distribution, and utilization of biological resources. There should be adequate information about the nature of our forests, where they are located, and the components of the forests. We can correctly manage a forest ecosystem with the understanding of remote sensing technologies, making it possible and logical to identify flash points of deforestation. Monitoring efforts can be done by international communities organizing the necessary monitoring schemes on the locations, causes, and extent of global deforestation, and making interventions which were possible to conserve the biodiversity in our forests.

8.6 Role of Biodiversity Law on the Conservation of Forest Resources

Biological diversity is the difference among living organisms of the same species and that of the species of other organisms within the ecosystem. It is the collection of different varieties among species of plants, animals, and microorganisms in the various ecosystems which they are part of. The inhabitants of forest lands benefit from the huge diversity of biological resources within their domain through basic ecosystem interactions, goods and services provided such as food, fiber, medicine, climate regulation, erosion and flood control, nutrient cycling, air, and water purification, etc. Some people depend directly on the usable land, water, plants, and animals available to support their families and others dwell on the commercial needs from the forest resources by collecting and distributing them to places of demands as means of sustenance.

Biological diversity can be categorized into genetic diversity, species diversity, and ecosystem diversity which are mutually dependent of each other. The observable difference within species of a particular population of organism is enhanced by the variation in the inherited traits of living things and the ecological factors present that make them function within that geographical area. Therefore, the variety of species population of organisms in an area changes when the genetic make-up of that organisms changes with corresponding difference on the various ecosystems (terrestrials, arboreal, marine, freshwater, estuarine, marshes, swamps, arid, etc.) in which they survive and adapt to the physical factors present for habitations.

The legal framework, instruments, and mechanisms regulating the achievement of certain biodiversity's conservation aims and objectives in our society is called "Biodiversity law". This Biodiversity law could include legislation, Acts, Decrees, and customary by-laws for national, regional, and international laws. Biodiversity law cannot stand alone, but is an integration of the international laws on issues concerning the conditions for biological diversity conservation, management, and sustainable use of its components, usually made and adopted by participating countries during the Biodiversity Conventions held to provide rules and established mechanisms for the line of actions on biodiversity and ecosystem, and also set up a structure that supports countries' effort to achieve this goals. According to International Development Law Organization (2016), this framework provides countries with the basis for governmental policies and actions in the establishment and safeguarding of the environment and its genetic resources, likewise the effective sharing of benefits rising out from the utilization of biological resources in a fair and equitable approach. This law also acts as a tool to create incentives, empower and recognize the rights and responsibilities of the people at all levels and to act for biodiversity. The biodiversity law is no different from the forestry laws and the environmental conservation laws of a nation where it hails from because diversity is within the forests of our environment.

However, in June 1992, Nigeria joined 153 countries at the United Nations Earth Summit to sign the Convention on Biological Diversity. The Convention calls for adoption international agreement on biodiversity conservation and articulate planning strategies and foster partnership between countries and among government organizations, nongovernmental organizations, and private sectors. The five key objectives of the Convention are:

- Biodiversity conservation at every level of the ecosystem;
- Biodiversity must be sustainable to maintain the continuity life support systems.
- Biodiversity benefits should be fairly and equitably shared to support conservation and sustainable development.
- Significant technological advancement for sustainable development should be shared; and

- Biodiversity conservation should have an established universal financial mechanism.

From Article 6 of the Convention on Biological Diversity National Biodiversity Strategy and Action Plans (NBSAPs) are adopted as the main tools for translating the procedures set out in the Convention on Biological Diversity line of actions and pathway in achieving these objectives within a nation. As said in the Convention on the Conservation of Migratory Species of Wild Animals (2015), this NBSAPS is an integrated instrument for countries to plan the biodiversity conservation, use of its components in a sustainable manner, and the effective sharing of the benefits gotten out of the utilization of biological resources and concentrating on the threats of biodiversity resources. It is also intended for NBSAPs to prioritize and identify national targets and the various actions required to meet up these targets to achieve this common national biodiversity and ecosystem objectives.

United Nations Environment Programme (2018) highlighted some biodiversity laws and principles used by National Biodiversity Strategies and Action Plans as tool for conservation and protection of biological diversity and the environment which are summarized in Table 8.1.

These principles are applicable to the support of biodiversity conservation laws and protection of our forests, yet can only be effective if nations enforce it. Often, forest conservation policies are frequently implemented poorly because of inadequate legal support by the government of participating countries and the noninvolvement of forest dealers' institution stakeholders and wood-based industries, who are the key players of the commerce of the forests industry. Thus, leading to the inability of the forest conservation, protection, and regeneration schemes to work effectively affects the biodiversity and wildlife conservation goals, especially when no replacement with economic tree planting exercise is carried out due to lack of legal backing by government and the wood-based firms not willing to do it on their own also (Bassey 2003). The biodiversity law is the integral legal instrument that can help policymakers and other stakeholders of biodiversity examine the effectiveness of their nation's biodiversity processes with the aim of developing and enforcing legal knowledge and practical line of action on biodiversity conservation policies in forests areas. These will help institute a lawsuit against any violator of laws on forest conservation, maintenance, and protection.

8.7 Role of International, National, and Local Agencies in the Conservation of Biodiversity

There have been various approaches used for the conservation of biodiversity by most tropical countries' agencies contributing to the stability of the ecosystem and mitigating climate change globally while fostering economic development. These agencies (regional, international, national, or local) are saddled with the responsibilities of providing an exceptional approach for effective forest governance and

Table 8.1 Biodiversity laws and applicable environmental principles

Biodiversity laws	Principle applicable
Those held accountable for activities that cause or may likely cause damage to biodiversity should bear the cost of pollution.	The polluter pays principle
Parties and stakeholders should promote suitable demographic policies and eliminate or reduce unsustainable models of production and consumption.	The sustainable use principle
We should avoid or prevent adverse effects on biodiversity at all costs.	The safeguarding of biodiversity principle
The needs of the present and future generations will be met if all activities shall take biodiversity into account.	The sustainable development principle
Biodiversity precautions should be taken by parties and stakeholders for effective protection and conservation;	The principle to take precautionary action
Uncertain scientific approach should not be used to defer cost-effective measures to prevent biodiversity degradation, especially in situations where there are risks of serious or irreversible damage;	
We should take the various needs of stakeholders having an interest in biodiversity into consideration.	
Biodiversity conservation requires the essential participation of women.	The principle of nondiscrimination
We should take the courage of the youth and the ideals, creativity, and knowledge of indigenous people into consideration;	
Consideration should be given to all relevant stakeholders and to the identity, culture, and interests of indigenous people.	
We should avoid or prevent adverse effects on biodiversity at all costs.	The non-degradation principle
Stakeholders and institutions should guarantee that any actions they fund, authorize, or perform do not degrade biodiversity.	
There should be effective participation of all concerned citizens as the best way to handle biodiversity issues.	The public participation and access to information and justice principle
Biodiversity information will be widely available to all if nations facilitate and encourage public awareness and participation.	
Biodiversity conservation, protection, and restoration need the cooperation of nations	The principle of common but differentiated responsibilities
Developed nations should accept the responsibility that they bear in the conservation and sustainable use of biodiversity, taking into consideration the demands which their societies, technologies, and financial resources place on biodiversity.	
The conservation of biodiversity shall form an integral part of the development process and should not be considered in isolation from it;	The integration principle
Biodiversity conservation laws and policies should be incorporated into other related policies.	

(continued)

Table 8.1 (continued)

Biodiversity laws	Principle applicable
The significant impact on biodiversity shall be evaluated, assessed, halted, reversed, and minimized in advance and in a timely manner.	Preventive principle
The causes of biodiversity loss or degradation shall be projected, identified, prevented, and attacked at the source.	The Source principle

conservation of biodiversity through influencing policies, planning, budgeting, and taking joint actions to promote sustainable land use in an environmentally convenient way.

The United Nations Environment Programme on Law and NBSAPs was to promote environmentally sound practices worldwide and meet up with the objectives made during the various Biodiversity Conventions held, and adopted by participating countries, to save the world's forest resources.

Recently, the Government of Nigeria has joined in the fight to curb deforestation activities and reduce forest emissions that contribute largely to the global climate change with the launching of a National REDD+ Strategy (Reducing Emissions from deforestation and forest degradation) project that will oversee the implementation and enforcement of new tactical approaches for sustainable management and conservation of the forests while augmenting carbon stock. The demand for hundreds of thousand hectares of cleared forestlands annually across countries is for the purpose of meeting the economic developments, but there is no sustainable management plan for the ecosystem's stability. Therefore, the Government of Nigeria has given key priority to the recent happenings launching National REDD+ Strategy project to combat deforestation and other forest-related emissions in the country with the support from the World Bank's Forest Carbon Partnership Facility (FCPF). This strategy involves stakeholders' participation (inclusively the related government parastatals, forest reliant community, civil society group, and private sector) to provide needed tactical framework to mitigate deforestation and emissions in Nigeria. National REDD+ Strategy programme is not just aimed at tackling forest degradation and deforestation problems only, but also promoting conservation through afforestation (tree planting) and rehabilitation of lands degraded, creating a supporting network for sustainable management of the forest resources and developmental trajectory for low forest emissions, as well as monitoring and reporting the progress in reaching the objective of having an eco-friendly society for the citizens of the country.

8.8 Conclusion

The importance of conserving, preserving, and protecting biodiversity and wildlife in our forests from indiscriminate exploitation and destruction could be sustained if the government and other policymakers can collaborate with stakeholders of the forests to identify, proffer, and enforce policies for forests' resource protection to avoid the extinction of threatened species and ensure biodiversity availability, sustainability, and effective utilization. Therefore, there should be frequent urban-rural awareness and enlightening schemes, like the REDD+ strategic activities which will enable indigenous forest dwellers have a sense of ownership to protect their forests' biodiversity and cooperate with ecotourism initiatives for the well-being of the country. With recent happenings in climate change, urgent actions are required to protect and conserve our local forests for future utilization and ecosystem stability. So there should be legal actions against any violator of biodiversity laws or forest laws on forest conservation and protection. Also, eco-management activities such as regenerating already utilized resources or planting specific economic valued trees (basically food trees) as replacement for felled trees, positioning of forest guards in reserves or protected forests areas, and identifying alternative resources for use in construction processes rather than total dependence on timber products. These will surely help in reducing deforestation and dependence on forest resources in our society.

References

Adekola G, Mbalisi OF (2015) Conserving and preserving forest and forest resources in nigerian rural communities: implications for community education. Int J Res Agric Forest 2(5):42–52. Sryahwa Publications

Amor D (2008) Road impact on deforestation and jaguar habitat loss in the Selva Maya. Ph.D. Dissertation. Ecology Department, Nicholas School of the Environment, Duke University.

Amor D, Pfaff A (2008) *Early history of the impact of road investments on deforestation in the Mayan forest*. Working Paper, Nicholas School of the Environment and Sanford School of Public Policy, Duke University, Durham, NC, USA.

Anonymous (1990) Rainforest destruction: causes, effects and false solutions. World Rainforest Movement, Penang Malaysia

Anonymous (1994) Deforestation technical support package. Third international conference on environment enforcement, Oaxaca Mexico April 25–28, 1994. World Wildlife Fund; U. S. Environmental Protection Agency and U. S. Agency for International Development.

Anonymous (2010) Global forest resources assessment, 2010-Main Report. FAO Forestry Paper 163. Rome, Italy. 340 p.

Asfaw A, Lemenih M, Habtemariam K, Ewnetu Z (2013) Importance, determinants and gender dimensions of forest income in eastern highlands of Ethiopia: The case of communities around Jelo Afromontane forest. Forest Policy Econ 28:1–7. https://doi.org/10.1016/j.forpol.2013.01.005

Barbier EB, Burgess JC, Folke C (1994) Paradise lost? The ecological economics of biodiversity. Earthscan

Bassey EE (2003) Conservation of the Nigerian rain forest: reasons, obstacles and strategies. XII World Forestry Congress. 0139-B1
Bhatnagar P (1991) The problem of afforestation in India. International Book Distributors, Dehra Dun
Burgess R, Miguel E, Stanton C (2015) War and deforestation in Sierra Leone. IOP Publishing Ltd
Caviglia J (1999) Sustainable agriculture in Brazil. Economic development and deforestation. Edward Elgar
Chomitz KM, Buys P, Luca GD, Thomas TS, Wertz-Kanounnikoff S (2007) At loggerheads? Agricultural expansion, poverty reduction and environment in the tropical forests. World Bank Policy Research Report. World Bank, Washington, DC
Colchester M, Lohmann L (1993) The Struggle for land and the fate of forest. Zed Books, London
Contreras-Hermosilla A (2000) The underlying causes of forest decline (occasional paper no. 30). Centre for international forestry research (CIFOR), Jakarta
Contreras-Hermosilla A (2001) Illegal activities and corruption in the forest sector. State of the World's forests, pp 76–89
Convention on the Conservation of Migratory Species of Wild Animals (2015) CMS and NBSAPS. Available at _WS_Philippines2015.pdf.
Cunningham AB (1993) African medicinal plants: Setting priorities at the interface between conservation and primary health care. UNESCO, Paris. People and Plants Working Paper 1
Deacon RT (1999) Deforestation and ownership: evidence from historical accounts and contemporary data. Land Econ 341–359
Egboduku WO, Olorunfemi DI (2016) Effects of ballast water and temperature variation on the growth and development of capsicum frutescens. Nigerian J Sci Environ 14(1):117–123
Elisabetsky E, Wannamaacher L (1993) The status of ethnopharmacology in Brazil. J Ethnopharmacol 38:137–143
Emerton L (2003) Tropical forest valuation: has it all been a futile exercise?
Epidi JO, Izah SC, Ohimain EI, Epidi TT (2016a) Phytochemical, antibacterial and synergistic potency of tissues of Vitex grandifolia. Biotechnol Res 2(2):69–76
Epidi JO, Izah SC, Ohimain EI (2016b) Antibacterial and synergistic efficacy of extracts of alstoniaboonei tissues. Br J Appl Res 1(1):0021–0026
FAO (2001) Forest finance: the forest revenue system and government expenditure on forestry in Nigeria
FAO (2010) Global forest resources assessment 2010 - main report. FAO forestry paper 163. Food and agriculture organization of the United Nations, Rome, Italy
Humphreys D (2006) Forest politics. Earthscan Publications Ltd., London
Ijeomah HM, Enaing EA (2018) Ecotourism and national development in Nigeria: prospects and challenges. Proceedings of 6th NSCB biodiversity conference; Uniuyo, pp 1–12.
Imarhiagbe O, Egboduku WO (2019) Conservation and utilization of biodiversity—implications to the Nigerian environment. JOJ Wildl Biodivers 1(4):555567
International Development Law Organization (2016) "Review of Post-2010 national biodiversity strategies and action plan: legal preparedness for biodiversity mainstreaming". http://www.idlo.int/sites/default/files/pdf/initiatives/NBSAP%20Review%20of%20Legal%20Preparedness%20for%20Biodiversity%20Mainstreaming%20%20-%20FINAL.pdf
Izah SC (2018) Ecosystem of the Niger Delta region of Nigeria: potentials and threats. Biodiversity Int J 2(4):338–345
Izah SC, Aseibai ER (2018) Antibacterial and synergistic activities of methanolic leaves extract of lemon grass (Cymbopogon citratus) and rhizomes of ginger (Zingiber officinale) against Escherichia coli, Staphylococcus aureus and Bacillus subtilis. Acta Sci Microbiol 1(6):26–30
Izah SC, Seiyaboh IE (2018a) Challenges of wildlife with therapeutic properties in Nigeria; a conservation perspective. Int J Avian Wildlife Biol 3(4):252–257
Izah SC, Seiyaboh EI (2018b) Changes in the protected areas of Bayelsa state, Nigeria. Int J Mol Evol Biodiversity 8(1):1–11

Izah SC, Angaye CN, Aigberua AO, Nduka JO (2017) Uncontrolled bush burning in the Niger Delta region of Nigeria: potential causes and impacts on biodiversity. Int J Mol Ecol Conserv 7(1):1–15

Izah SC, Aigberua AO, Nduka JO (2018a) Factors affecting the population trend of biodiversity in the Niger Delta region of Nigeria. IJAWB 3(3):206–214

Izah SC, Chandel SS, Etim NG, Epidi JO, Venkatachalam T, Devaliya R (2019a) Potency of unripe and ripe express extracts of long pepper (Capsicum frutescens var. baccatum) against some common pathogens. Int J Pharm Phytopharm Res 9(2):56–70

Izah SC, Etim NG, Ilerhunmwuwa IA, Silas G (2019b) Evaluation of crude and ethanolic extracts of Capsicum frutescens var. minima fruit against some common bacterial pathogens. Int J Complement Altern Med 12(3):105–108

Izah SC, Etim NG, Ilerhunmwuwa IA, Ibibo TD, Udumo JJ (2019c) Activities of express extracts of Costusafer Ker–Gawl. [Family COSTACEAE] against selected bacterial isolates. Int J Pharm Phytopharmacol Res 9(4):39–44

Kaimowitz D, Angelsen A (1998) Economic models of tropical deforestation. A review. Center for International Forestry Research, Bogor, Indonesia

Kigigha LT, Biye SE, Izah SC (2016) Phytochemical and antibacterial activities of Musanga cecropioides tissues against Escherichia coli, Pseudomonas aeruginosa Staphylococcus aureus, Proteus and Bacillus species. Int J Appl Res Technol 5(1):100–107

Kigigha LT, Izah SC, Uhunmwangho EJ (2018a) Assessment of hot water and ethanolic leaf extracts of Cymbopogon citratus Stapf (Lemon grass) against selected bacteria pathogens. Ann Microbiol Infect Dis 1(3):1–5

Kigigha LT, Selekere RE, Izah SC (2018b) Antibacterial and synergistic efficacy of acetone extracts of Garcinia kola (Bitter kola) and Buchholzia coriacea (Wonderful kola). J Basic Pharmacol Toxicol 2:13–17

Levang P (2002) People's dependencies on forests. In Technical report, phase I 1997–2001. ITTO Project PD 12/97 Rev. 1 (F)—Forest, Science and Sustainability: the Bulunganmodel forest, pp 109–130. CIFOR, Bogor Indonesia.

Mather AS (1991) Global forest resources. International Book Distributors, Dehra Dun

Mongabay.com (2010) Nigeria forest information and data. http://rainforests.mongabay.com/deforestation/2000/Nigeria.htm. Accessed 12 May 2011

Myers N (1994) Tropical deforestation: rates and patterns. In: Brown K, Pearce D (eds) The causes of tropical of tropical deforestation. The economic and statistical analysis of factors giving rise to the loss of the tropical forest. UCL Press, pp 27–40

Nzeneri PU (2010) Investigating into the existing approaches for creating environmental awareness among community members in Rivers State. M.Ed Thesis; University of Port Harcourt.

Ogunleye AJ, Adeola AO, Ojo LO, Aduradola AM (2004) Impact of farming activities on vegetation in Olokemeji forest reserve, Nigeria. Global Nest Int J 6(2):131–140

Ogunbadejo HK, Oladipo AE (2017) Impact of agricultural output volatility on economic growth in Nigeria: Egarch analysis. IOSR J Agric Veter Sci. e-ISSN: 2319-2380

Omofonmwan SI, Osa-Edoh GI (2008) The challenges of environmental problems in Nigeria. J Hum 23(1):53–57

Onuche U (2010) Impact of poverty on the sustainability of forests in Nigeria: implication for sustainable forests and reduction in global warming. J Sustain Dev Afr 12(6):208–215

Panayotou T (1990) The economics of environmental degradation: problems, causes and responses, HIID Development discussion papers 335. Harvard University

Purnamasari RS (2010) Dynamics of small-scale deforestation in Indonesia: examining the effects of poverty and socio-economic development. Unasylva 61(1–2):14–20

Putz FE, Blate GM, Redford KH, Fimbel R, Robinson J (2001) Tropical forest management and conservation of biodiversity: an overview. Conserv Biol 15:7–20

Rainforest Concern (2021) Why are rainforests being destroyed? UK Registered charity: 1028947 Website designed by Mud

Repetto R (1988) The forest for the trees? Government policies and the misuse of forest resources. World Resource Institute, Washington DC

Repetto R (1990) Deforestation in the tropics. Scientific American April, p. 37. https://www.intechopen.com

Rotowa OJ, Adekunle EA, Adeagbo AA, Nwanze OL, Fasika OO (2019) Economic analysis of agriculture, forestry and fisheries to the economic development of Nigeria. Int J Res Stud Sci Eng Technol 6(6)

Rowe R, Sharma NP, Browder J (1992) Deforestation: problems, causes and concerns. In: Sharma NP (ed) managing the world's forests: looking for balance between conservation and development. Kendall/Hunt Publishing Company, Iowa, pp 33–46

Salau AT (1993) Environmental crisis and development in Nigeria. Inaugural lecture series, 13; University of Port Harcourt

Sands R (2005) Forestry in a global context. CABI Publishing

Shukla G (2010)Biomass production and vegetation analysis of Chilapatta Reserve Forest Ecosystem of West Bengal. Ph.D. Thesis, Uttar Banga Krishi Viswavidyalaya, Pundibari. Unpublished.

Sonone OA (2018) Facts of economic contribution of forest and wildlife

Sumit C, Ghosh SK, Suresh CP, Dey AN, Shukla G (2012) Deforestation: causes, effects and control strategies, global perspectives on sustainable forest management, Clement A. Okia (Ed.), ISBN: 978-953-51-0569-5

United Nations Environment Programme (2018) Law and national biodiversity strategies and action plans, Nairobi, Kenya

Wilkie D, Shaw E, Rotberg F, Morelli G, Auzel P (2000) Roads, development, and conservation in the Congo Basin. Conserv Biol 14(6):1614–1622

World Bank (2006) Poverty reduction and growth: Virtuous and vicious circles. World Bank, Washington D.C.

World Wildlife Fund (1980) World conservation strategy: living resource conservation for sustainable development, vol 1. IUCN, Gland, Switzerland

Youmatter (2020) What is deforestation? Definition, causes, consequences, solutions.

Chapter 9
Rationale Behind Conservation of Africa's Biological Resources

Godfrey C. Akani

Abstract This chapter identifies the rationale, justification, and significance of diverse conservation approaches as well as the challenges facing biological resources in Africa. It upholds that due to poverty and population growth, biological resources are experiencing accelerated depletion including within biodiversity hotspots. It sees conservation of biological resources as a proactive measure to guarantee continuous sustenance of life, by ensuring continuous availability, preserving the original nature of the ecosystem that supports them, avoiding shortage or wastage through indiscriminate use, ensuring food security for man, as well as a continuous source of income and employment for the present and future generations. The strategy for ensuring the sustainability of these vital resources is through effective and efficient in situ and ex-situ conservation methods, which can yield fruitful results. In situ methods preserve biological resources in their natural environment at a cheaper rate and provide opportunities for evolution. Ex-situ conservation preserves endangered species and encourages their breeding and multiplication in captivity. Long-term preservation of seed banks, gene banks, and DNA molecules is currently trending, enhancing the availability of biological resources.

Keywords Conservation · Biological resources · Rationale · In situ and ex-situ · Cryopreservation · Africa

9.1 Introduction

The term conservation, which came into use in the late nineteenth century, took its origin from the Latin word, "conservatio", meaning—to keep, or keep intact or save or safe keeping. It has been defined by various authorities in diverse ways.

G. C. Akani (✉)
Department of Animal and Environmental Biology, Rivers State University, Port Harcourt, Nigeria

S. C. Izah, M. C. Ogwu (eds.), *Sustainable Utilization and Conservation of Africa's Biological Resources and Environment*, Sustainable Development and Biodiversity 32, https://doi.org/10.1007/978-981-19-6974-4_9

Sandbrook (2015) defines conservation as the protection and prevention of a resource from wasteful use. The Merriam-Webster dictionary (2022) defines it as the planned management of natural resources to prevent waste, destruction, damage, or neglect. Others define it as official care or sustainable management of natural resources for the benefit of the present and future generations (Ogwu et al. 2014; Osawaru and Ogwu 2014a, b). Succinctly, the National Geographic Society (2022) defines it as the act of protecting earth's natural resources for current and future generations. Central to these definitions is the fact that it entails wise use, abhors wastage, and professes judicious management or preservation, protection, restoration of natural resources and their environment so that future generations would also benefit from them. Although conservation is similar to preservation, both are geared towards the protection of nature, but they accomplish the task in slightly different ways. While conservation allows the controlled use of nature by humans, preservation has stricter restrictions or zero tolerance for human use.

In some localities, the term conservation has been grossly misconstrued. Many people understand it to mean the protection of forest resources from conversion and unsustainable utilization (Song et al. 2019; Ikhajiagbe et al. 2020). Consequently, they see no justification for such protection. Extremists rather see forest workers as enemies who do not allow them a free hand to use resources endowed to them by nature, whereas in fact conservation permits the wise or judicious use of resources, when they arc abundant, but kicks against the destruction of limited resources, as their extinction is forever (Ogwu 2009; Osawaru et al. 2013). It rather entails preservation, maintenance, sustainable utilization, restoration, avoidance of waste, and enhancement of the environment which support the development of the resources.

Conservation practice, which today is gaining momentum in civilized economies, and some developing countries, has emerged as a panacea to stem the increasing pressures on the natural environment and widespread loss of natural habitats and species, due to human impingement on ecosystems (Ogwu 2019).

This chapter identifies the rationale, justification, and significance of diverse conservation approaches as well as the challenges facing biological resources in Africa. These resources and the environment that sustain them are experiencing accelerated depletion due to poverty and population growth within the continent.

9.2 Biological Resources

Biological resources refer to living plants, animals, and microorganisms and the habitat and ecosystem in which they thrive. They are of great economic importance to society because of the various services they provide, as well as the problems some precipitate. Biological resources are grouped into three categories: (i) those that affect agriculture such as cultivated plants (ii) those that are sources of scientific input such as crop varieties (and their wild relatives) that provide genetic resources for variety improvement, and (iii) those that provide natural goods and services. Examples of biological resources include not only cultivated plants, but also pollinators, pests and pest predators, wildlife, fish, and scenic attractions (Ogwu et al. 2016a, b; Mougin et al. 2018).

Biological resources also include a substance or object in the environment required by an organism for normal growth, maintenance, and reproduction. For plants, key resources are sunlight, nutrients, water, substrate, pollinators, space, and place to grow; and for animals, the key resources include oxygen, water, light, food, shelter, mates, etc. The variety of biological resources endowed to a place is often and justifiably equated to its biodiversity. This means that the supporting substances, medium of existence, and environments equally deserve conservation.

9.2.1 Significance of Biological Resources Conservation

It is important to safeguard our biological resources, especially those that are essential for economic growth and poverty reduction. Biological resources support the continuity of various ecosystems (Ikhajiagbe et al. 2021). Furthermore, they have esthetic, ethical, scientific, and medicinal values. They are sources of industrial raw materials; some industries will fold if they disappear. They provide ecosystem services worth billions of dollars, as well as various inestimable social and spiritual benefits (Kremen 2015; Diaz et al. 2018; Teixeira et al. 2019). The significance of ecosystem services and examples are summarized in Fig. 9.1.

9.2.2 Environmental Conservation

The environment in which biological resources like wildlife, vegetation, and fishes thrive needs to be conserved for sustainability. By definition, the environment of an organism refers to all abiotic and biotic factors in its surrounding that influences its health and well-being (Kerfahi et al. 2019; Erinle et al. 2021). As determinants of their survival, the environment or conditions under which the biological resources grow must be optimally maintained. Activities that culminate in habitat loss, habitat fragmentation, habitat alteration, and human disturbance have to be adequately curtailed to enrich biodiversity conservation measures. The air, water, soil, and vegetation on which organisms subsist must be conserved and protected against all forms of pollution (Ogwu et al. 2018a; Vwioko et al. 2018). Should there be any devastating natural disaster, (such as wildfire, flood, drought, volcanic eruption, etc.) an appropriate restoration of the ecosystem must be planned and executed. In the view of the Best Environmental Conservation Volunteer Program, upholding

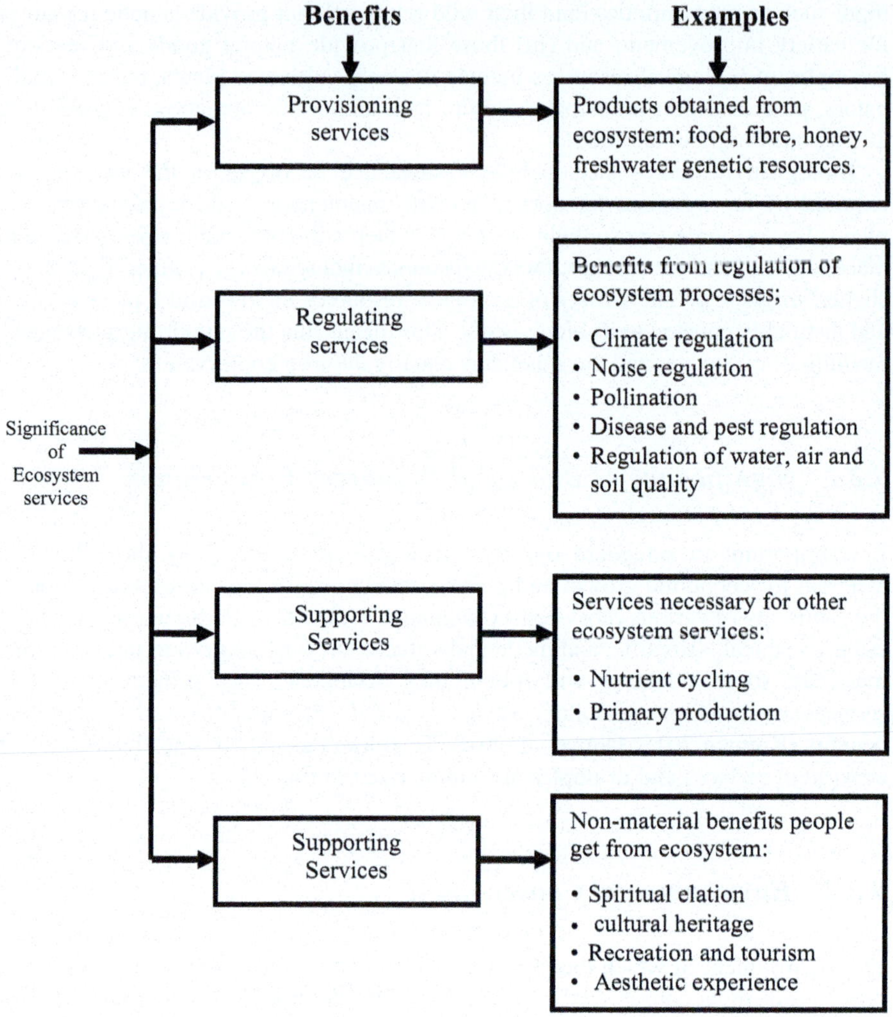

Fig. 9.1 Summary of Ecosystem services; benefits and examples (After Kremen 2005, 2015, Diaz et al. 2018)

environmental conservation is the practice of preserving the natural world to prevent it from collapsing under the pressure of human activities such as unsustainable agriculture, deforestation, and the burning of fossil fuel. The object of environmental conservation is to: minimize air, water, and land pollution; facilitate the conservation of natural resources; ensure the protection of biodiversity; ensure sustainable development; restore the ecological balance of destabilized ecosystems; and to save our planet from the deleterious consequences of global warming (Team Leverage Edu 2022).

9.2.3 *Traditional Practices that Enhance Biological Resources Conservation*

Several African traditional practices (Fig. 9.2) have been identified which promote biological resource conservation. Among them are:

- Sacred grooves—small chunks of forests where shrines are located, for pagan worship. Because it is a no-go spot except for very few worshippers, most animals that have lost their habitat find refuge in these grooves.
- Control of exploitation of iconic tree species—in some culture there is restrictions on the exploitation of iconic trees such as Iroko, Iron-wood, Ebony, Opepe, mahogany, obeche, rain tree, etc. as some expensive rituals may precede their felling.
- Cultural prohibition on killing venerated animals like Python, crocodile, Nile monitor, hippopotamus, tortoise, snails, etc., because such animals have some historical attachment to the survival as a group—family, clan, or tribe.
- Prohibition of trade on totemic species such as bushmeat.
- Imposing closure—periodic restrictions of fishing in some swamps/streams / banning the use of some fishing gears.

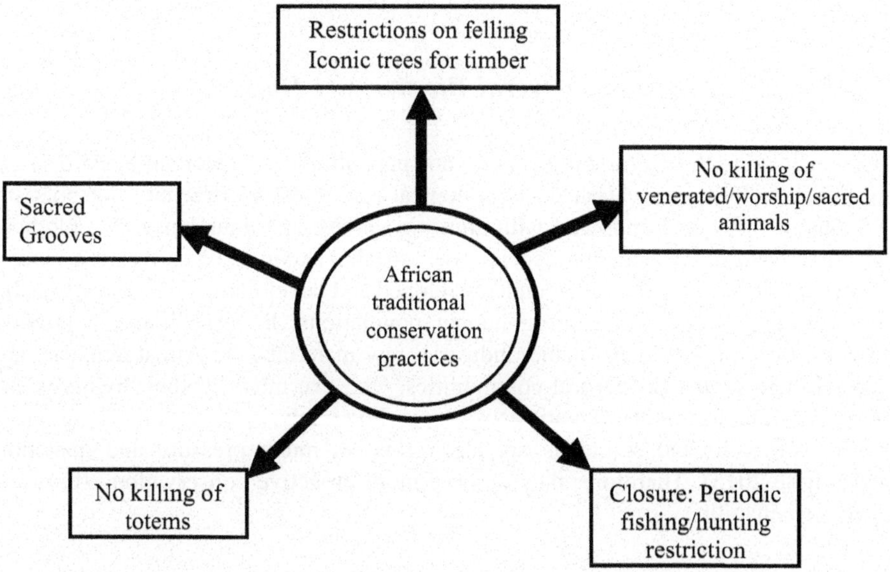

Fig. 9.2 Methods of African traditional conservation

9.3 Suitability of African Environment for Conservation

The rugged landscapes of sub-Saharan Africa and islands present a veritable environment that supports various kinds of organisms. Consequently, they deserve serious conservation efforts to ensure the sustainability of the various populations and biological resources, as the human population soars. This is due to the following -

9.3.1 Diversity of Habitats

Tropical rainforest, galloping forest, wetlands—seasonal swamp, riparian forest, rivers streams, lakes, mangrove swamps, as well as savannah, montane vegetation, Sahel, which harbor a great diversity of flora and fauna and also provide ecological niches and breeding sites for many species. The great species' richness implies that Africa holds a tremendous gene bank. Hybridization and speciation are quite promising, and the ecosystem is stable and shows great potential for resilience in the event of perturbations. Bizarre or iconic animals are often available which can be used as flagship species to garner support for conservation. A flagship species is any species that is the reason for the conservation of a forest or aquatic ecosystem.

9.3.2 Africa Includes Several Biodiversity Hotspots

The inclusion of several biodiversity hotspots in Africa (Marchese 2015) is a justification for conservation of its biological resources. Biodiversity hotspots are regions of the world endowed with amazing variety and abundance of biological resources and many endemic species (i.e., species exclusively native to a place). They happen to occur mainly in underdeveloped and developing countries within the tropical forest zone, with soring human populations. Because hotspots provide sources of food, water, medicine, industrial raw materials, etc., great demands are placed there by the poor, rural communities, who eke much of their living exclusively from the forest and aquatic ecosystems. Invariably, areas endowed with great variety of biological resources are also prone to much pressure and depletion (Marchese 2015). Therefore, they deserve more proactive conservation efforts for their sustainability.

9.3.3 Marine Biological Resources

Africa also has a great diversity of marine biological resources including every phyla of the animal kingdom. Seabirds, marine turtles, and mammals abound in virtually every region, which necessitates the conservation of the marine environment. In recognition of these sensitive sites and bizarre species, several protected areas are now delineated to salvage threatened species such as sea turtles, manatees, and Dugongs (Redaction Africanews and Campbell 2022) as seen in Gabon, Tanzania, Seychelles, and the Mediterranean coasts. This is because Africa by harboring biodiversity hotspots is prone to losing many species at alarming rates due to great demands placed on useful species.

9.4 Justification for Biological Resources Conservation

The clarion call for biodiversity conservation was orchestrated by the Earth Summit in Rio de Janeiro, Brazil, in 1992 and was informed by the alarming rate of extinction and the invaluable roles and services provided by biological resources. The significance of having a wide variety of species, genes, and ecosystems lies in the fact that it supports human livelihood and welfare in many ways especially in the form of food supplies, medicines, shelter, income employment, and cultural integrity (Cunningham and Cunningham 2015). In most cases, these biological resources provide indispensable industrial raw material bases and services, e.g., final sources, fibers, etc. (Barnes 2015). The justifications or reasons for conserving biodiversity and particularly endangered species have been identified as essentially utilitarian, ecological, esthetic, and ethical (Blicharska and Grandin 2015; Botkin 2021).

9.4.1 Utilitarian Reasons

This is based on the consideration that many wild species might be useful to mankind now or in future, and it is imprudent and unwise to destroy them before we have a chance of discovering or testing their uses. The bulk of the utilitarian justifications are unraveled from their useful genetic characteristics, chemical and tradomedical uses, potential for new crops and products, used by indigenous people for survival, use in pollution control, importance in tourism, and use in scientific and medical research (Osawaru and Ogwu 2014a, b; Ogwu et al. 2016a, b, 2018b; Ogwu et al. 2017; Osawaru et al. 2016; Collins 2020).

9.4.2 Genetic Characteristics

An amazing variety of wild strains of grains and other crops is often found in the wild state. The need to conserve the gene pools becomes necessary, since modern agricultural production of crops such as wheat, corn, millet, etc. depends on the continued introduction of fresh genetic characteristics from the wild, to create or improve new genetic hybrids. It is believed that with more varieties of genes, plant and animal breeders have a pretty wider range of genes from which they can make a combination of characters for the desired hybrid, than in situations of a limited number of genes. Again, disease organisms that attack our crops evolve continually, changing their genetic characteristics. As new disease strains develop, crops become vulnerable (Botkin 2021) and by introducing fresh genetic traits from the wild, new hybrid strains can be developed that are disease-resistant.

9.4.3 Chemical and Medical Uses

Biological resources deserve conservation because of many important chemical compounds and drugs extracted from them or were first discovered in wild organisms. For example, Digitalis, an important drug in the treatment of certain heart ailments, comes from the purple foxglove, a small flowering plant native to Europe and Morocco. Aspirin, a popular analgesic, and caffeine, a popular stimulant, are derived from willow bark and kola nut, respectively. The rosy periwinkle, from the tropical forest of Madagascar and steroids from Mexican yams, produces chemicals used for the treatment of certain cancer (Botkin 2021). Also derived from tropical plants are drugs used to treat high blood pressure, Hodgkin's disease, leukemia, multiple sclerosis, and Parkinson's disease (Botkin 2021). The potential medicinal values of plants both as unprocessed herbal material and as complex biochemical extracts are numerous, and more are still unfolding, which have helped to improve health care services (Barnes 2015). The active ingredients in at least a third of the prescription drugs used by civilization come directly from or were derived from chemical compounds found in wild plants and other organisms especially in tropical forests—digitalis, morphine, quinine, and antibiotics being the most familiar (Ehrlich 2008).

It is also evidentially clear that the availability of considerable biological resources makes it possible for mankind to adapt to changing needs and health problems in our environment. As our planet earth ages, man is more often than not plagued with disease epidemics new to science. For instance, sometime ago, it was influenza, yaws, gonorrhea, syphilis, chicken pox, tuberculosis, jaundice, and recently HIV/AIDS (Human Immunodeficiency Virus/ Acquired immunodeficiency syndrome), SARS (severe acute respiratory syndrome), and COVID-19 (Coronavirus-2019) pandemic. The control of these diseases depends on the availability of biological species for drugs. And today, the most dreaded disease is AIDS. Presently, this disease has no cure, but the scientific world is looking forth to the extraction or synthesis of drugs from biological resources for a breakthrough. If by chance, a potent drug is extracted from an endangered species, then humanity will be confronted with the problem of generating a very high population of that species for commercial and sustainable production of the drug. For this singular reason, therefore, there is every justification for mankind to embark on biodiversity conservation, since no one knows which plant or animal species would be the source of antidotes for diseases that might erupt in the future.

Many organisms produce useful medical compounds that are yet unknown and untested. For instance, some Nigerian plants such as *Vetivera zizanioides, Cassia tora, Garcinia kola, Tetrapleura tetraptera, Schumanniophyton magnificum*, etc. are commonly used as snake repellents (Etukudo 2003), but their active ingredients are subject to investigations. Recent phytochemical studies of *Tetrapleura tetraptera* indicate that the fruit (which contains saponins and molluscicides) is useful for the management and/or control of a wide variety of human ailments, including arthritis, other inflammatory conditions, asthma, diabetes mellitus, hypertension, epilepsy, schistosomiasis and good for fertility as it significantly increases sperm count, motility, daily sperm production, etc. (Larbie 2021). Extracts from the seeds of *Cassia tora* used for the treatment of Psoriasis—a chronic skin disorder caused by keratinocyte hyperproliferation—have been shown to contain Anthracene derivatives like Chryophanic acid, Chrysophenol, emodin, aloe-emodin, and rhein (Malhotra et al. 2007). Various scientists are now testing marine organisms for possible use in pharmaceutical drugs. For instance, Jack et al. (2006) extracted a ketone from the abdomen of the Hermit crab, *Clibanarius africanus,* from the Niger Delta. Among the marine animals, coral reefs have been identified to offer a particularly promising area of study for pharmaceutical drugs, because many coral reef species are known to produce self-defensive toxins; over 20,000 useful drugs have been extracted from coral reefs (NOAA 2021).

Components of snake venoms have been found useful in the treatment/management of certain diseases such as cancer, arthritis, thrombosis, multiple sclerosis, pains, neuromuscular disorders, blood, and cardiovascular disorders, infections, and inflammatory diseases, including high blood pressure, heart attack, stroke, Alzheimer's disease, Parkinson disease, etc. Snake venoms contain several neurotoxic, cardiotoxic, cytotoxic, nerve growth factors, disintegrins, hemorrhagins, and many other different enzymes (Ferraz et al. 2019). Among the pharmaceuticals made from snake, venoms are Captoprils (Enalapril), Integrillin (Eptifibatide), and Aggrastat (Tirofiban), all of which have been approved by the Food and Drug Administration (FDA). Oils from snakes like pythons and the Gaboon viper, *Bitis gabonica,* are popularly cherished for massaging, and python oil is used to suppress keloids (Akani 2019).

The Antivenin used in the treatment of bites from poisonous snakes is manufactured with the help of snake venoms. It is prepared by immunizing horses against common poisonous snakes. The horse is injected frequently with snake venom in small and increasing doses which toxicity the horse can tolerate. After 6 months such horses become immunized to the snake venom and the horses show no symptoms of snake bite, even if they receive a heavy dose of the venom (Shukla and Upadhyay 2019). The blood extracted from the view of such immunized horses is allowed to clot, and after clotting the serum, now the antivenin is packed in vials for injection purposes. These, therefore, are ample justification that snakes also deserve conservation for the sustainability of their ecosystem services.

9.4.4 Potential for New Crops and Products

Another utilitarian justification for biodiversity conservation is the great potential for developing new crops or other commercial products from wild plants and animals (World Bank 2019). Many horticultural crops, vegetables, nuts, and products such as honey, tannin, waxes, oils, resins, and latex come from tropical rain forests (Botkin 2021) and hopes are high that new products will be found in the wild (Cunningham and Cunningham 2015).

9.4.5 Use by Indigenous People for Survival

Biodiversity has direct economic values and supports human livelihood in many ways (World Bank 2019). In many economies and cultures, it provides man and other species a wide spectrum of foodstuff of both plant and animal origin including meat, fish, vegetables, nuts, fruits, flavors, and spices (World Bank 2019). In the rural communities, in particular, biodiversity means much in the continued sustenance of life, as the diversity of forests and wildlife also provide wood for shelter, fuel, tools, furniture, utensils, artwork, and materials for clothing and medicine. Consequently, a reduction in biological diversity would directly or indirectly increase poverty among the already impoverished people, as there are no alternative sources of sustenance or means of eking a living.

Various breeding and domestication literature data show that the bulk of what we cherish today as food crops, e.g., wheat, rice, and maize, were once wild species before they were developed by selective breeding, just as domestic animals like dogs, cats, goats, sheep, cow, donkeys, horses, etc. were once wild animals (Cunningham and Cunningham 2015).

9.4.6 Tourism

Wildlife-based tourism is a major source of foreign exchange in many developing countries in the tropics that have a variety of animals to showcase. Kenya, for example, has wildlife tourism as the mainstay of her economy and rakes in several millions of dollars revenue from this business, annually. Ab initio tourism was based on the larger game animals that inhabit open savanna lands; increasing attention is now being given to ecotourism, which includes an interest in the whole array of forest plants and animals (Akani 2019).

9.4.7 Medical and Scientific Research

The significance of biological resources in medical and scientific research provides another utilitarian justification for biodiversity conservation. Wild animals such as monkeys, gorillas, chimpanzees, rabbits, etc. are often used to test new drugs, as no human would offer himself as a test organism for such an experiment. The Armadillo, one of only two animal species known to contract leprosy, is an important specimen for the study of this disease to find a cure (Friedrich 2018). Studies in the social behavior of wild animals have contributed immensely to a better understanding of the ontogeny of human social behavior. Also, the ecological studies of wildlife populations have thrown better light on the dynamics and balances of the natural ecosystem, including farmlands, forests, and pastureland (Odum 2021). Furthermore, a wide variety of plants have been found useful as a source of medicine in many African countries, which need to be cultivated (Tanga et al. 2018).

9.4.8 Ecological Justifications

The variety of biological species within ecosystems is known to maintain the stability and functions of ecosystems. Thus, ecosystems characterized by high species richness are likely to be more stable and would function more effectively (Asthana and Asthana 2012). Every species performs a specific function as a member of the ecosystem, which others cannot perform. Thus, the services of every individual species are very important to the stability of the ecosystem; hence, the need to conserve every species, irrespective of whether it is dangerous, harmful, or poisonous. Loss of any species means loss of the services provided by them to the ecosystem and mankind. The individual species, entire ecosystems, and the biosphere provide public service functions essential or important to the sustenance of life, and as such, they are indirectly necessary for our survival (Asthana and Asthana 2012). Wild vertebrates, invertebrates, and microorganisms play vital roles in the pollination of wild and farm crops, germination, dispersal of seeds, and other reproductive structures of plants, soil processes, and nutrient cycling. All these functions are vital not only to the maintenance of the ecosystems of which the organisms are a part, but also to human welfare.

Natural forest ecosystems are invariably repositories of useful timber and other wood products frequently needed for furniture, building construction, carving, boat building, and pulp and paper. Forest ecosystems also have the potential of providing a suitable environment for many timber species. Ehrlich (2008) observed that trees in managed forests and tree farms have never been known to produce a better quality of a variety of timber than in natural forests. Natural forests are also valued for the diversity of non-timber products they produce. These include rubber, any kind of oil, organic chemicals, vegetables, spices, mushrooms, fruits, honey, medicinal herbs, roots, and tree backs. Others are wild animals namely—mammals, birds, reptiles, amphibians, fishes (in freshwater swamp forests), insects, snails, etc. Invariably, man benefits insensibly, from the ecosystem functions (Odum 2021).

There is no gain saying the fact that many sectors of national economies depend largely on the diversity of natural ecosystems and the ecological functions they protect. In many jurisdictions, freshwater ecosystems provide water for agriculture, hydroelectric power, inland fisheries, and inland water transport. It is also obvious that a crucial proportion of the protein in our diet comes straight from nature in the form of fishes and other animals harvested from the sea (Ehrlich 2008), namely shrimps, crabs, oysters, whelks, clams, cockles, mussels, etc.

9.4.9 *Moral and Ethical Justification*

The moral justification of biodiversity conservation is based on the belief that species have a moral right to exist, independent of our need for them, and human behaviors should be directed positively to coexist with them (Vucetich et al. 2021). It is therefore a human obligation to assist the continued existence of these species, irrespective of our religious or cultural inclinations. The right of existence by organisms was stated in the U.N General Assembly World Charter for Nature in 1982.

9.5 Drivers of Biological Resource Loss in Africa

Large volumes of literature have been accumulated that address the worrisome rate at which species are getting endangered or extinct over the millennium. In Africa, most of these problems are anthropogenically induced factors and include:

- Habitat destruction: refers to places where the original habitat of organisms has been lost. It is now known that the new threat to 85% of all species that have been listed as endangered is traceable to habitat destruction. Factors responsible for habitat loss include deforestation, bushfire, urbanization, and excessive land-take for agriculture and infrastructure developments. This is noticeable in places where several hectares of forest, grassland, and woodland have been converted to farmland, pastureland, estates, and cities and where wetlands have been reclaimed for residential accommodations (Akani 2019).
- Habitat fragmentation: this is the splitting or reduction of habitats into smaller and smaller scattered blocks or fragments, as seen in places where a network of roads is constructed, as well as areas of seismic operation, and where pipeline Right -of -Way crisscrosses a region. Such fragmentation culminates in biodiversity reduction, as many trees and shrub species are lost in the process and large-sized animals like the big cats, elephants, antelopes, and buffaloes which require large continuous territories to subsist are prompted to migrate. Other species such as

forest interior birds that reproduce successfully only in the serene deep forest far from the habitat edges and human settlement are denied optimal nesting sites and so they flee. Habitat fragmentation separates populations into isolated small groups, which may not have the normal sex ratio and age ratio of their original population. Very small populations may not have enough breeding adults which invariably will affect the rate of recruitment into the population, by birth. Besides, fragmentation creates access for nearby communities to further exploit forests, and fishing grounds which, ab initio, were inaccessible.

- Overexploitation of biological resources (wildlife, forest, and fisheries) for commercial purposes without replacement considerations. In most African countries, there is open access to fishing and hunting, without any closures or gear control. Unorthodox methods of fishing and hunting are also practiced in some localities; poisonous chemicals such as gammalin-20 and poisonous herbs are used to kill fish, *en masse*. Alcohols (native dry gin, distilled from fermented palm wine) are also used to attract monkeys and elephants to be killed. When intoxicated, they are helpless and can easily be maneuvered.
- Human population growth: In many African countries, the population is growing exponentially, thereby placing an overwhelmingly high demand on biological resources, as consumption rates also increase.
- African traditional medicine—An amazing variety of plant and animal species are usually in great demand for African ethnomedicine and spiritual healing. Most of them are sold in fetish market stalls at exorbitant rates, and they are often endangered species parts (e.g., leopard skin, lion teeth, tree barks, nuts, vulture eggs, etc.) difficult to find. Because they are highly prized, hunters go to any length to make them available to customers.
- Chronic environmental pollution—this is the frequent release of substances or energy into the environment by man to levels where they are harmful or deleterious to living things and undermine amenities and quality of the environment. Many African countries engage in the use of agrochemicals (pesticides, herbicides, fungicides, fertilizers, etc.) to improve their agricultural yield annually. The unbridled use of these substances has killed many animals and plants for a long term without detection. Such form of pollution is described as 'latent' as they are hidden or unnoticeable. Many African countries are endowed with crude oil and in the course of oil exploration and production, there is a lot of biodiversity loss through seismic operation, pipeline construction, and oil spillage (Akani 2019).
- Exotic/Invasive species—these are species that have established and spread in places where they are not native. Among the most threatening invasive species are rats, cats, green crabs, fish, algae, zebra mussels, the African tulip tree, and the brown tree snake (Cunningham and Cunningham 2015; Yede et al. 2016). The exotic species may be introduced deliberately to check a pest population or any other useful purpose. Inadvertent introduction can occur; for example, by organisms "hitch-hiking" in containers, ships, cars, trains, etc. In their new ecosystem, they flourish and eliminate the native species by outcompeting them in resource acquisition or infecting them with their parasite (Yede et al. 2016). An example is Nypa palm; *Nypa fruticans* which was introduced to check coastal erosion in

Oron, Nigeria, and eventually became invasive, displacing the more useful native mangroves—*Rhizophora* and *Avicennia* species (Akani 2019).

- Climate change—since biological processes are optimal at certain temperature ranges, global warming/climate change will upset the processes and ecosystems and lead to the following; (i) changes in the distribution of plants and animals; population sizes, growth rates, the timing of plant flowering, and timing of animal migration, (ii) changes in intensity and frequency of storms, flood, drought, and fires; in Africa, increased temperature and decreased rainfall mean that some lakes might dry out. With the rise in ocean temperature, the sea will become more acidic, resulting in the widespread degradation of tropical coral reefs. Sintayehu (2018) and many workers reported that climate change could decrease the genetic diversity of populations due to directional selection and rapid migration, which could in turn affect ecosystem functioning and resilience. Some species could disappear following the extinction of their associated species. Melting of ice in the polar region and rise in sea level will affect African coastlines causing floods in low-lying cities and wetlands. Salt intrusion into the hinterland (due to the rise in sea level) will disrupt freshwater ecosystems, killing several freshwater plant species and oligohaline fishes. Furthermore, protected habitats and their wildlife may no longer have the right climate that supported them, and the ecosystem will be jeopardized as the food webs get disrupted.

  The elevated temperature will affect sea turtles immensely. This is because the sex of sea turtle hatchlings is dependent on the temperature of the beach; at higher temperatures, the hatchlings are predominantly females, while males are hatched in cooler incubation temperatures. Thus, if the temperature rise persists for decades, beach temperatures will rise and males may become extinct, leaving a monosexual population of female sea turtles. Then sea turtles may become extinct because of the lack of males to fertilize the eggs (Akani 2019)

9.6 Methods of Conserving Biological Resources/Biodiversity

Over the years, various techniques and strategies have evolved for the effective conservation of biological resources. The techniques can be grouped into two major complementary strategies: In situ and Ex-situ conservation (Institute of Biodiversity Conservation 2005).

9.6.1 In Situ Conservation

In situ conservation refers to the conservation of species in their natural habitats. It is a vast ecosystem of several hectares established and protected by law, so that all the constituent species, small or big, dangerous or harmless, known or unknown, are encouraged to flourish, to benefit us. Thus, any trespassing/encroachment or illegal harvesting in such conserved areas as sanctuaries, Nature reserve, or Biosphere reserves is a punishable offence called poaching. Although they are given legal "muscle", all protected areas are scientifically managed like an ecosystem, where every organism has right to thrive, because of the unique ecological niche each occupies (Akani 2019). As defined in the Convention for Biological Diversity, "*in situ* conservation means the conservation of ecosystems and natural habitats and the maintenance and recovery of viable populations of species in their natural surroundings and, in the case of domesticated or cultivated species, in the surrounding where they have developed their distinctive properties." In situ conservation aims at (i) enabling biodiversity to be sustained naturally, (ii) saving the integrity of the ecosystem, species, and genetic resources, (iii) development—promoting economic development, while maintaining cultural, social, and ecological identity, and (iv) scientific research—providing support for research related to monitoring and education, local, national, and global issues (Zegeye 2017). In situ conservation, however, has its advantages and disadvantages (Fig. 9.3).

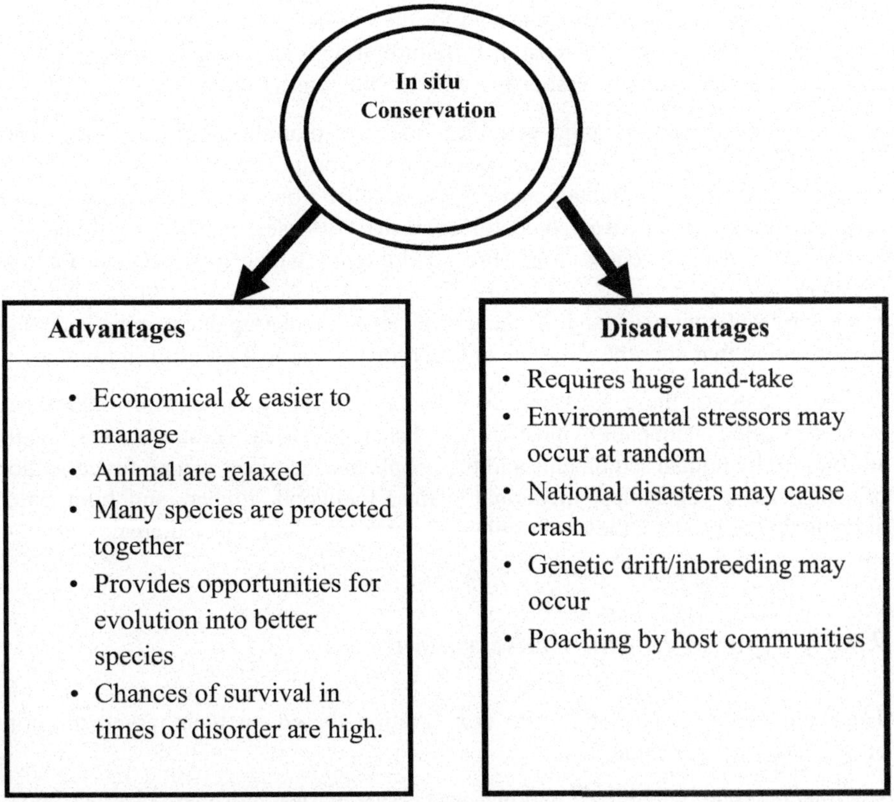

Fig. 9.3 Advantages and disadvantages of in situ conservation

9.6.2 Advantages and Disadvantages of In Situ Conservation

The advantages of conserving species in their natural habitat are that (i) it is a cheaper and easier way of conserving biodiversity, we only provide the enabling environment for species to multiply. The wildlife species are in a free state, fending for themselves (providing food, shelter, water, sex mates, etc., they need to survive) and do not suffer the boredom experienced by caged animals in zoos. (ii) Being a natural ecosystem, a large number of species (autotrophs, herbivores, carnivores, and decomposers), known or unknown to science, are protected and maintained at the same time. (iii) It allows the interplay of natural agencies (e.g., drought, fire, pathogens, flood, etc.) which predispose organisms to evolve to better species, through adjustment to adverse environmental conditions. (iv) In times of disorder, such as war, species in an in situ conservation system have higher chances of escaping and surviving than caged or tethered animals (Akani 2019).

However, a crucial disadvantage of in situ conservation is that it entails acquisition of large hectares of land, in order to preserve a full spectrum of the biodiversity of a region. The huge land take can constitute a bottleneck, in an era when there is growing demand for space, as it deprives the locals of the cultural use of their land. Although they are usually compensated for the release of land, the compensation is never commensurate with the cost of forfeiture of their land forever. Zegeye (2017) identified four natural risks associated with in situ conservation, namely:

- Alterations in demographic variables (natality, mortality, etc.) resulting from random events in the survival and reproduction of individuals.
- Unexpected environmental changes in weather conditions food supply, and populations of competitors, predators, parasites, diseases, and other biotic factors.
- Natural disasters such as flood, fire, or droughts, which may occur at random intervals.
- Genetic uncertainty or random changes in genetic make-up due to genetic drift or inbreeding that alter the survival and reproductive probabilities of individuals.

The greatest of these stressors, however, is anthropogenic encroachments and poaching which is rampant in most protected areas of Africa. These areas are usually surrounded by human settlements and the people use the forest around for collection of fruits, nuts, vegetables, medicinal plants, fuelwood, timber, and hunt game animals. In the process, they gain unlawful entry into the protected area.

9.6.3 Types of In Situ Conservation

There are various types of in situ conservation, based on their targets/mission/purpose and these include:

- Protecting natural habitats: national parks, sanctuaries, biosphere reserves, etc.
- Protecting cultivated crops including entire agroecosystems and the maintenance of domesticates and nurseries.

- Preservation of keystone species (e.g., white and black rhinoceroses, wild horse, etc.) in their natural or seminatural habitats.
- Conservation of genetic diversity.
- Programmes for the recovery of species, and
- Restoration of habitats.

9.6.3.1 Protected Areas and Management

Protected areas are natural ecosystems delineated and protected by law against human impingement and destruction, because of their important biological diversity, ecological or cultural values. The International Union for Conservation of Nature (IUCN) defines a protected area as a clearly defined geographical space, recognized, dedicated, and managed, through legal or other effective means, to achieve the long-term conservation of nature; an enormous progress has been achieved in the global number of protected areas as proposed by the Aichi targets of Convention on Biological Diversity (IUCN 2021). Examples of protected areas include nature reserves, wilderness areas, biosphere reserves, national parks, sanctuaries, forest reserves, and natural monuments, etc. These areas have varying levels of protection.

Protected areas are accorded protection because their wealth of biological resources helps to mitigate extreme weather conditions, enhance carbon sequestration (storage), and provide clean air and water, industrial raw materials, healthy soils, wild foods and medicines, etc. Besides, protected areas provide socioeconomic benefits such as wrecking in revenues to societies and economies through recreation and tourism.

Thus, for the sustainability of biological resources and ecosystem services, effective management of protected areas is key. In active management, protected area practitioners are responsive to set goals, undertake many other tasks, such as species management (taking statistics of health status and treatment of species), anti-poaching patrolling, and bringing poachers to book; providing tourist services, and engage in research and monitoring, and restoration project; as well as resisting encroachments by landlord communities (IUCN 2021).

Areas in the globe that have attracted protection include areas which are high repositories of biodiversity such as—tropical rain forests, wetlands, mangrove forests, coral reefs, world Heritage sites, etc.

9.6.3.2 Features of Protected Areas

Right from 1981 UNEP-WCMC through its Protected Areas, programme has been maintaining the world Database on Protected Areas (including terrestrial and marine habitats) and making it available to the global community IUCN (2016). Presently, the database records show that there are 202,467 protected areas covering 20 million square kilometers or 14. 7% world's land excluding Antarctica (IUCN 2021). The bulk of them are national parks, sanctuaries, and biosphere reserves, whose features are described below.

National Parks

National Parks refer to areas under state (national) control and the boundaries of which may not be altered or any portion alienated except by the competent legislative authority. They are specifically delineated to ensure multiplication, protection, preservation, and sustainable management of living organisms and their habitat for the benefit of man. However, harvest of plants and animals are only permitted for scientific and conservation purposes. National parks usually have a surrounding buffer zone or outskirt, where tourist activity as well as minimal biotic interference may be permitted but no interference is allowed in the deeper interior or the core zone of the park. In spite of the frantic efforts, National Parks in Africa have been subjected to myriads of human impingement, including—illegal settlement, hunting, fishing, grazing, logging, mining, farming, water poisoning, bush-burning, Forest product collection, and fuel food extraction (Osunsina and Osunsina 2018).

Sanctuaries

Sanctuaries are usually meant for special protection of nationally significant species or group of species, biotic communities, or physical features of the environment, where the species in question require anthropogenic intervention to survive or perpetuate. It is different from a zoo. While a zoo contains animals brought from different parts of a country and other countries for exhibition, a wildlife sanctuary contains endangered species found locally in that area. Because it is quite cumbersome to relocate the animals from their natural habitat, the best option therefore is to protect them in their natural home or sanctuary, where they are easily monitored and catered for, for the rest of their life. Sanctuaries are expensive to maintain and are owned by government, some organizations, or the affluents in the society. Consequently, no human exploitative activity or poaching is tolerated. Example is the Afi Mountain Wildlife Sanctuary established in 2000 with the concept of preserving the most endangered animal species, such as the Cross River gorilla, the Nigeria-Cameroon chimpanzee, the wild drills, and gray-necked rock fowl, all of which are native to Afi mountain (Edet and Akinyemi 2017).

Biosphere Reserves

Are large protected areas, an ecosystem with bizarre plants and animals of scientific and natural interest, delineated by law and usually more than 5000 km^2. They are designated not exclusively for protection of unique biotypes, ecosystems, or significant wetlands, but for a range of objectives including research, monitoring, capacity building, and demonstration, as well as conservation. In most cases, the human component is vital to the functioning of the biosphere reserve (Stoll-Kleemann and O'Riordan 2017). Thus, biosphere reserves have in recent times been described as "learning places for sustainable development" because they provide sites for testing interdisciplinary approaches to understanding interactions between social and ecological systems and coping with changes, while managing biodiversity. National governments nominate them, but they are designated under the intergovernmental MAB (Man & Biosphere) programme by the Director-General of UNESCO following the decision of the MAB International Coordinating Council (MAB ICC) (Stoll-Kleemann and O'Riordan 2017)

9.6.4 Protected Area Management Categories

To ensure their effective management, the IUCN categorized PAs into six groups "that define PAs according to their management objectives which are internationally recognized by various national governments and United Nations". For instance, Category Ia is reserved for Strict Nature Reserves and is managed under strict protection, while Category II is reserved for National Parks and is managed and conserved in such a way as to conserve and protect ecosystems.

Although National Parks, Sanctuaries, and Biosphere reserves are ranked very high as protected areas in most countries and regions, there are lesser and yet important protected areas established by edicts or culture at local government levels. These include (i) Forest reserves (where protection and conservation of vegetation and soil are emphasized and controlled by agencies like Forestry department), (ii) Game reserves (where the emphasis is on the protection of wildlife and their habitat and anthropogenic activity is highly controlled), and (iii) Sacred grooves—small forested area, lakes, or shrine controlled by pagan worshippers, where some wildlife species find refuge, after loosing their main habitat.

Throughout the sub-Saharan region of Africa, there have been an impressive increase in number and geographic spread of protected areas over the years, to protect its great diversity of biological resources. There are, today, more than 1100 national parks and related reserves in sub-Saharan Africa, of which 36 have been designated World Heritage Sites Records of WCMC (2004) also show that since 1970, the protected area coverage in Africa has increased almost twofold, including an area of 3.06 million km^2 of terrestrial and marine habitats.

In Kenya, where there are at least 18 National Parks, 4 Marine National parks, 4 Nature Reserves, 23 National Reserves, 5 Marine National Reserves, 5 Biosphere Reserves, and 1 Ramsar Wetland wildlife, tourism has been so developed that it now constitutes the mainstay of its economy. Other countries like South Africa, Tanzania, and Madagascar also eke considerable revenue from wildlife tourism, because of their numerous protected areas. South Africa is known to have the largest number of protected areas in Africa, with over 1500 protected areas, while Tanzania has the largest total area of 360,000 km^2 under protection. South Africa is also home to

23.420 species of vascular plants, making it the sixth most species-rich country in the world and the most species-rich country of Africa.

The Guinean Forests of west Africa—which extend through the equatorial zones of Nigeria, Cameroon, Central African Republic, etc.—is a hotspot with high levels of biodiversity, including numerous endemic species, making it a conservation priority at the global scale. The hotspot is ranked among the world's foremost regions for mammalian diversity. This explains why a total of 1936 nationally protected areas have been located in West Africa. The bulk of the protected areas of Africa belong to IUCN management categories II and IV.

9.6.5 Rationale for Delineation of Protected Area

There is no gainsaying the fact that we live in a time of unprecedented environmental change, with climate change, habitat alteration, pollution, invasive species, and overexploitation, all exacerbated by population explosion and resulting in species decline (Millennium Ecosystem Assessment 2005). In their own way, Asthana and Asthana (2009) also unraveled the rationale behind establishment of protected areas: "today, natural habitats are rapidly shrinking being replaced by human settlements, agriculture fields, or much of land is rendered barren and degraded, of no use to many of the living organisms. Wilderness has become more or less like islands in the sea of man dominated landscape which many species cannot cross to reach the locality with suitable conditions or other climatic optima. Natural barriers such as mountain ranges or an arm of the sea may also check the movement of many species. No one exactly knows, how the biosphere shall react to the climatic changes brought about by global warming. Many species could become extinct, many others could survive in reduced numbers while many may flourish with a changed geographical distribution. As different species shall respond to changes in climatic patterns in different ways, the entire biotic spectrum over the globe is expected to change and an inevitable outcome of this upheaval could be drastic reduction in biological diversity." Perhaps the only option left for mankind to circumvent these looming problems in future is to establish as many protected areas as possible to serve as biological resource "banks" to fall back to, when the "capital" is gone.

9.6.6 Conservation in the Midst of Poverty

A major problem confronting most host communities of protected areas of Africa is how to participate in conservation of biological resources, which they direly need now for sustenance of life. Having been dispossessed of their land and biological resources forever, they find it difficult to comply with the tenets of forest conservation. However, it has been suggested that this mentality could only be reversed through poverty alleviation strategies: First there should be a robust awareness programme on the benefits of conserving biological resources. Every youth should be encouraged to train in any chosen alternative livelihood support business venture sponsored by Government and Park management. Such business ventures should divert their attention from exploiting forest resources; e.g., computer operation, fashion-designing, hair dressing, tailoring, draught-manship, nursing, etc. At the end of their training, they should participate in revegetation of the buffer zone, and be paid a business start-off remuneration, and finally sign a legal agreement not to engage in poaching.

9.7 Ex-Situ Conservation

According to Wilson and Primack (2022), off-site or Ex-situ Conservation is a technique of conserving biological resources outside their natural habitat, targeting all levels of biodiversity such as genetic, species, and ecosystems. This is generally practiced as an additional measure or "saving grace" to augment, where in situ conservation has failed. Usually to prevent imminent extinction of a critically endangered species or the demise of an orphaned individual, the only option that may be left is ex-situ conservation, where the species is given intensive care to enable it survive (Wilson and Primack 2022). Broadly, ex-situ conservation includes a variety of activities ranging from managing captive populations, education, and raising awareness, supporting research initiatives to collaborating with in situ efforts. It is essentially the maintenance and breeding of endangered plant and animal species under partially or wholly controlled conditions as in zoos, gardens, nurseries, and laboratories (Asthana and Asthana 2009). A zoological garden or zoo (which is often the offsite) is a park in which wild animals are confined and maintained within enclosures or seminatural/open areas. Depending on the purpose of the zoo, the animal may be exhibited to the public for ecotourism and research, or screened off from public view and disturbance, if it is a breeding zoo, encourage mating of endangered species. Two important strategies are necessary in ex-situ conservation considerations: (a) Identification of species that need to be conserved and (b) The method of ex-situ conservation.

9.7.1 Identification of Species to Be Conserved

Because there are multitude of species threatened with extinction today, it is important to select species with more urgent need for preservation, since we cannot conserve all of them. This means that we have to have some criteria for selecting species that deserve accelerated ex-situ conservation efforts. According to Asthana and Asthana (2009), the important criteria to be considered include; (i) Vulnerability of the species to extinction and (ii) Economic, ecological, or esthetic importance of the species.

With the continuous urbanization and encroachment of human settlement into wildlife habitats, it is expected that many natural habitats will be fragmented.

Some species are more vulnerable or prone to extinction in a relatively short time than others. Such species qualify to be given priority attention for conservation. These include:

- Very rare species—critically endangered species (e.g., Black and white rhinoceros)
- Massively built species (e.g., elephants, rhinoceros, hippopotamus, Giraffe, whales, etc.)
- Species with valuable parts (e.g., elephants, crocodiles, marine turtles, leopard, cheetah, pangolins)
- Clustering species/with no migratory or dispersal drive (e.g., crocodile, python, tortoise, terrapins).
- Animals at the higher tiers of the food web (e.g., lion, leopard, cheetah, etc.)
- Animals whose adults have low survival rate. (e.g., fish)
- Altruistic animals whose grieving behavior predispose them to hunting (e.g., elephant, zebra, otters, whales, etc.)
- Animals with shorter longevity or life span. (e.g., Mayfly, fish)

After assessing the species that have a high probability of going extinct, the choice of the endangered species for ex-situ conservation will be guided also by ranking them in the order of importance to man in future. This is because there are often many species in need of conservation and the limited resources will not support ex-situ conservation of all the endangered species at the same time.

9.7.2 *Methods of Ex-Situ Conservation*

9.7.2.1 Long-Term Captive Breeding

As commonly practiced in zoos and botanical gardens. This entails capturing, maintenance, and breeding animals/plants, especially endangered species in captivity on a long-term basis. This technique is of immense use for species that have lost their habitats permanently or facing incessant perturbation disallowing resilience, such as overhunting, poaching, logging, and agricultural land-take. There are several examples of animals which owe their existence today to the benefits of long-term captive breeding technology in zoos, because their original habitats are no longer conducive for their survival. Among them are—Przewalski's Horse, *Equus przewalskii*, which once roamed the plains of Central Asia and Europe, the Addax, *Addax nosomaculatus,* of the Sahelo-Sahara region which hunters never spared because of its tasty meat and hide, the Siberian Tiger, *Panthera tigris altaica*, which are mercilessly poached because of its highly prized skin and bones. Others are the Grevy zebra, Scimitar horned oryx, Slender horned gazelle, Madagascar radiated tortoise, etc.

9.7.2.2 Short-Term Propagation and Release

This is a method of ex-situ conservation adopted when an endangered species decline is due to some temporary perturbation in the natural environment, and the animal involved would survive in the same habitat if the factor causing the decline is eliminated. While waiting for restoration of the habitat, the endangered species is maintained and bred in captivity under human control before release into its natural home. This type of ex-situ conservation efforts is known to have released some species from the brink of extinction. Among them are—Cheetah, Arabian oryx, American bison, red wolf, Bald Eagle, European eagle, etc.

9.7.2.3 Animal Translocation

Animal translocation is an intentional capture and release of an animal population into a new destination where it is not native, in order to save it from the stress that has developed in its home. For example, South African giraffes were translocated to Senegal, West Africa, when they were being nudged into extinction by hunting and habitat loss. Another example is the successful translocation of white rhinoceroses to Akagera National Park in Rwanda, which relieved the species of ruthless poaching in Uganda, Southern Sudan, and the Democratic Republic of Congo (Knight 2021). In practice, translocation of animals safely entails sedating the animal with appropriate dose of wildlife tranquilizer (Wolfe and Miller 2016) (e.g., acepromazine, Chlorpromazine diazepam) from a dart-gun and transporting it to a new destination in a truck with cage.

A number of factors may warrant or necessitate animal translocation. (i) when the species is threatened with extinction due to habitat destruction, unbridled hunting, incessant chemical or noise pollution, etc. (ii) when the population size is overwhelmingly high, exceeding the carrying capacity the habitat can support or sustain. Translocation becomes necessary at this level before an unexpected crash follows due to environmental resistance or catastrophic event. (iii) When an endemic species in its restricted locality is threatened by some natural disaster. Instead of allowing the disaster to wipe them off, translocation to a safer place, not too distant from the area, becomes the only option. (iv) If the health status or vigor of a population begins to deteriorate, the healthy ones can be saved by translocating the unhealthy individuals to a quarantine, for treatment. (v) Where competition for a biological resource becomes so keen, for two populations then the weaker competitor may be saved by translocating it to a new destination where it will be relieved of the pressure. (vi) In circumstances where animal populations like elephant, monkeys, lion, etc. constitute a nuisance to humans; threatening life, health, properties, and livelihood.

9.7.2.4 Animal Reintroduction

Animal reintroduction is the release of orphaned animals or animals born in captivity or victims of environmental pollution, etc. back to their natural habitat, after rehabilitation. Such a broken or disadvantaged animal lacks the learning behavior that should have been learnt from the parents, older individuals, or peers which enables the individuals to survive in the wild. Thus, they have to be trained or socialized in captivity before being introduced to their natural habitats. The process of training such inexperienced juveniles is called rehabilitation. It is a necessary prerequisite for reintroduction, especially for intelligent animals like primates in which learning plays an important role in their ontogeny (development from infancy to adulthood). Equally important is the post-release monitoring, which is necessary to evaluate their success (Valente et al. 2017). For animals like the crocodile, which is controlled by instinctive behavior patterns, rehabilitation is unnecessary.

9.7.2.5 Botanical Gardens

Botanical gardens are parks devoted for the conservation of threatened species of living plants. They are grown especially for research purposes and to prepare nurseries for afforestation and reforestation projects (Heywood 2017; Entwisle et al. 2017). Functional botanical gardens require infrastructures such as herbarium, laboratories, etc. The plant species to be cultivated are usually selected based on the interest of the researchers and could spread across native and non-native species as well as those with medicinal or economic benefits or ornamental appeal (Entwisle et al. 2017). These gardens cater for the needs of several agro-based industries. They are important in ecotourism and in the preservation of biodiversity. According to Vertucci et al. (2010), "botanical gardens give opportunity for arable plants to be grown under relatively modified environmental conditions (intense cultivation, relatively high fertility and high levels of disturbance)". Botanical gardens collaborate with each other, exchanging seeds, pollen, and other genetic materials to preserve threatened species.

9.7.2.6 Gene Banks

Another management technique adopted in biological resources conservation is genome resource banking. Gene banks could conserve the genetic information of seeds, plant cells, living plants, and animal tissues and organs (Muller et al. 2018). For plants, the banking is done by in vitro storage, freezing cuttings from the plant of interest, or stocking the seeds (as in seed banking). In the case of animals, it is accomplished by freezing of the gametes (sperm and eggs) until when needed. Gene banks make available desired genes for researchers and farmers to withdraw and rebuild populations or rare plant varieties and animal breeds to help increase genetic diversity (Engels and Ebert 2021). Although genebank development in Africa is upcoming, some Gene banks have developed better in East Africa, which distribute vegetable seeds to farmers in Tanzania, Kenya, and Uganda (Stoilova et al. 2019).

9.7.2.7 Seed Banking

This is a type of genebank which preserves dried and viable seeds of endangered plants for a long time in a special storage facility called seed bank; usually placed in jars seeds of different plant species protected from environmental damage. Seed banking is significant because it is capable of preserving genetic diversity for even thousands of years or more. Thus, it serves as a preventive measure to avert extreme undesirable scenario in the event of natural disaster, such as flood, nuclear fallouts, and outbreak of diseases (Walters and Pence 2020).

Three major factors influence seed longevity in the seed banks, which are checked and these include temperature, seed moisture content, and relative humidity in the storage facility. For short-term conservation, seeds are typically conserved at moisture content of 3-7% and stored at 4 °C, and at lower temperatures between −18 and −20 °C for long-term conservation (Walters and Pence 2020). Current research has revealed that the type of seed in question, whether endospermic or non-endospermic or with intraspecific variation, may affect longevity. Recalcitrant seeds (i.e., desiccation- or cold-resistant seeds) are rather conserved as live plants in ex-situ "field gene bank" which, however, present myriads of logistic challenges- large areas, high costs, vulnerability to pests and diseases, natural disaster, etc. (Walters and Pence 2020).

9.7.2.8 Cryopreservation

This is a process of cooling and storing biological materials—cells, tissues, or organs—at very low temperatures (−196 °C) in liquid nitrogen or liquid nitrogen vapor (−160 °C) (Benelli (2021). At very low temperatures as these, all enzymatic or chemical activities which might cause damage to the biological material in question are effectively stopped to maintain their viability. The purpose of plant cryopreservation is to store cells and tissues indefinitely by halting the cell's metabolism and permitting high rates of survivability of the cells upon thawing. Benelli (2022) opined that cryopreservation method can offer greater security for long term, cost-effective conservation of plant genetic resources, as well as orthodox seeds.

In the same vein, cryopreservation techniques are of immense use in extending the life span or preserving the integrity of oocytes, spermatozoa, tissues, ovarian tissues, preimplantation embryos, organs, etc. (Silva et al. 2015; Vladimirov et al. 2018). The preservation is accomplished by some minute molecules which penetrate cells and prevent dehydration and formation of intracellular ice crystals, which can cause cytolysis (or cell death) during the freezing process. Whaley et al. (2021) pointed out two popular cryoprotective agents used, namely: dimethyl sulfoxide

(DMSO) for protection of most cells and tissues, and Glycerol, primarily for protection of red blood cells. A sugar known as trehalose, present in organisms capable of tolerating extreme dehydration, is used for freeze-drying. This sugar stabilizes cell membranes; hence it is especially useful in the preservation of sperm, stem cells, and blood cells (Shen et al. 2019).

Cryopreservation technique has been commonly practiced because of its numerous benefits. Much space and labor are not needed, fertility treatment is enhanced, and it preserves biological samples for a very long time and safeguards the germplasm of endangered species. Furthermore, it preserves the genetic constitution of gametes and provides safety against genetic contamination (Jang et al. 2017).

9.7.2.9 Tissue Culture Banks

Tissue culture is a method of biological resources conservation in which a disease-free tissue fragment from an animal or plants is transferred into an artificial environment in which it can continue to thrive and function. The cultured tissue may consist of a single tissue or a whole organ or a fragment of an organ. In developing countries of Africa, many important plants have successfully been grown in tissue cultures (Abebaw et al. 2021), namely—oil palm, plantain, banana, date, eggplant, jojoba, pineapple, rubber tree, cassava, yam, sweet potato, and tomato.

9.8 Advantages and Disadvantages of Ex-Situ Conservation

Conservation of species outside their natural habitats has its advantages and disadvantages as summarized below.

9.8.1 Advantages

This method of conservation offers protection from anthropogenic factors that could lead to population depletion in the wild. The well-being of the organisms can be examined over time and medications readily administered if required. The genomes of organisms conserved using this method can be readily sequenced and advantageous traits reproduced. Critically endangered species can be bred to increase their population size (Asthana and Asthana 2012). Since the organisms are maintained in an enclosure, it is relatively easier to conduct research into their biology than if they were in the wild.

9.8.2 *Disadvantages*

A major bottleneck of keeping animals in zoos is the high maintenance cost in providing their food, medication, and shelter. Second, when an animal arrives in a zoo, it battles with a number of psychological problems, which the keeper must be conscious of. In confinement, they suffer boredom and turn rebellious, displaying repetitive self-destructive behaviors, while seeking the least chance to escape. Such psychological disorder called zoochosis includes head-bobbling, biting cage bars, pacing, playing with excrement, etc. (Mellor et al. 2015; Sawe 2017). The third oddity with zoo conservation is that by confining them in enclosures, we exclude them from the natural agencies that predispose species to evolutionary processes. Thus, in ex situ systems the germ plasms are denied opportunities for speciation (Mellor et al. 2015).

9.9 Conclusion

The foregoing review has highlighted the rationale and justification of conserving the wealth of biological resources Africa is endowed with. The impetus for conservation of biological resources was sequel to the clarion call for this action during the 1992 earth summit in Rio-de-Janiero, Brazil. Among the justification for conservation are their useful genetic characteristics, chemical and medicinal uses, and potentials for new crops and products. Also inclusive are their ecological relevance, touristic values as well as their moral and ethical justifications. It has also brought to focus the possible conservation techniques that could be practiced (for both fauna and flora) to ensure sustainability of these resources. It has presented the array of in situ and ex-situ conservation options, including examples of species to which they are applicable. Elucidated too are the advantages and disadvantages of both in situ and ex-situ conservation, and graphic examples of wildlife that have successfully been relieved from the brink of extinction were also focused.

References

Abebaw YM, Tobiaw DC, Abate BA, Eshete BK, Seymour SK, Tesfaye K (2021) Plant culture research and development in Ethiopia: a case study on current status opportunities and challenges. Adv Agric. https://www.hindawi.com:9979549

Akani GC (2019) Wildlife and biodiversity conservation: an indispensable strategy for sustainable development. An inaugural lecture series no.64, presented at the faculty of law auditorium, Rivers State University, Port Harcourt Nigeria on 31 July 2019

Asthana DK, Asthana M (2009) Environment problems and solution. S. Chand & Company Ltd., (An ISO 9001 – 2000 Company) Ram Nagar, New Delhi, 434 pp

Asthana DK, Asthana M (2012) A textbook of environmental studies Chand New Delihi, 397 pp

Barnes M (2015) Protect biodiversity, not just area. https://www.nature.com

Benelli C (2021) Plant cryopreservation: a look at the present and future. Plants 10(12):2744 pub online MDPI. https://doi.org/10.3390/plants10122744. Date of retrieval 20 May 2022
Benelli C (2022) Plant cryopreservation. MDPI books doi: https://doi.org/10.3390/books978-3-0365-2789
Blicharska M, Grandin U (2015) Why protect biodiversity ? Perspective of conservation professionals in Poland. Int J Biodiver Sci Ecosyst Serv Manag 11(4)
Botkin DB, (2021). Environmental science—earth as a living planet 9th ed Wiley 625 pp
Collins J (2020) Imperial drug economies, development, and the search for alternatives in Asia, from colonization to decolonization. International Development Policy. OpenEdition journals
Cunningham W, Cunningham M (2015) Environmental science– a global concern, 15th edn. McGraw-Hill Companies Inc.
Diaz S, Pascual U, Stenseke M, Martin-Lopez (2018) Assessing nature's contributions to people. Science 359(6373):270-272
Edet DI, Akinyemi AF (2017) Evaluation of human encroachment in Afi mountain wildlife sanctuary, Cross River State, Nigeria. Project; conservation of forest resources of the Cross River rainforest
Ehrlich PR (2008) Where does biodiversity go from here. A grim business
Engels JMM, Ebert AW (2021) A critical review of the current global ex situ conservation system for plant agrobidiversity. 1. History of the development of the global system in context of the political/legal framework and its major conservation components. Plants 10(8):1557. https://doi.org/10.3390/plants10081557
Entwisle TJ, Cole C, Symes (2017) Adapting the botanical landscape of Melbourne Gardens (Royal Botanic Gardens Victoria) in response to climate change. Plant Divers 39(6):338–347
Erinle KO, Ogwu MC, Evivie SE, Zaheer MS, Ogunyemi SO, Adeniran SO (2021) Impacts of COVID-19 on agriculture and food security in developing countries: potential mitigation strategies. CAB Rev 16(16):1–16
Etukudo IG (2003) Ethnobotany – conventional and traditional uses of plants, vol 1. The Verdict Press, Uyo, Akwa Ibom, 191 pp
Ferraz CR, Arrahman A, Xie C, Casewell NR, Lewis RJ, Kool J, Cardoso FC (2019) Multifunctional toxins in snake venoms and therapeutic implications: from pain to hemorrhage and necrosis. Front Ecol Evol 2019. https://doi.org/10.3389/fevo.2019.00218
Friedrich MJ (2018) Armadillos bring risk of leprosy. JAMA 320(11):1097. https://doi.org/10.1001/jama.2018.12741
Heywood VH (2017) Plant conservation in the Anthropocene—challenges and future prospects. Plant Divers 39(6):314–330
Ikhajiagbe B, Ogwu MC, Lawrence AE (2020) Single-tree influence of *Tectonia grandis* Linn. f. on plant distribution and soil characteristics in a planted forest. Bull Natl Res Centre 44:29. https://doi.org/10.1186/s42269-020-00285-0
Ikhajiagbe B, Ogwu MC, Ogochukwu OF, Odozi EB, Adekunle IJ, Omage ZE (2021) The place of neglected and underutilized legumes in human nutrition and protein security. Crit Rev Food Sci Nutr. https://doi.org/10.1080/10408398.2020.1871319
Institute of Biodiversity Conservation (2005) National biodiversity strategy and action plan. Institute of Biodiversity Conservation, Federal Democratic Republic of Ethiopia, Addis Ababa
IUCN (2021) World met target for protected area coverage on land, but quality must improve.
Jack IR, Braide SA, Akani GC, Okorosanye-Orubite K (2006) Isolation and characterization of ethyl piperononyl acetate-a and b-keto Ester from Hermit Crab, *Clibanarius africanus*. Niger Delta Biologia 5(2):110–114
Jang TH, Park SC, Han J (2017) Cryopreservation and its clinical applications. Integrative medicine research. Korea Institute of Oriental Medicine
Kerfahi D, Ogwu MC, Ariunzaya D, Balt A, Davaasuren D, Enhmandal O, Purevsuren T, Batbaatar D, Tibbett M, Undrakhbold S, Boldgiv B, Adams JM (2019) Metal tolerant fungal communities are delineated by high zinc, lead and copper concentrations in metalliferous Gobi desert soils. Microb Ecol. https://doi.org/10.1007/s00248-019-01405-8

Knight M (2021) Moving African rhinos: what it takes to translocate an endangered species. DownToEarth

Kremen C (2005) Managing ecosystem services: what we need to know about their ecology. Environ Sci. https://doi.org/10.1111/j.1461-0248.2005.00751.x

Kremen C (2015) Pollination services from field-scale agricultural diversification may be context dependent. Agric Ecosyst Environ 207:17–25

Larbie C (2021) *Tetrapleura tetraptera* of Ghanaian origin: Phytochemistry, antioxidant and antimicrobial activity of plant parts. SSRN papers.ssrn.com. Date of retrieval 20 May 2022

Marchese C (2015) Biodiversity hotspots: a shortcut for a more complicated concept. Global Ecol Conserv 3:297–309

Mellor DJ, Hunt S, Gusset M (2015) Caring for wildlife: the world zoo and aquarium animal welfare strategy. WAZA Executive Office, Gland Switzerland

Merriam-Webster (2022) Conservation. Merriam-Webster.com Dictionary https://wwwmerriam-webstercom/dictionary/conservation. Accessed 31 May 2022

Millennium Ecosystem Assessment MEA (2005) Ecosystem and human well-being biodiversity synthesis. World Resources Institute, Washington, DC

Mougin C, Artige E, Marchand F, Mondy S (2018) BRC4Env, a network of biological resource centres for research in environmental and agricultural sciences. Environ Sci Pollut Res 25(34): 33,849–33,857

Muller T, Schierscher-Viret B, Fossate D, Brabant C, Schori A, Keller B, Krattinger SG (2018) Unlocking the diversity of genebanks: whole genome marker analysis of Swiss bread wheat and Spelt. Theor Appl Genet 31(2):407–416. https://doi.org/10.1007/S00122-017-3010-5. Epub 2017 Nov 4: PMID 29103142

National Geographic Society (2022) Conservation. https://www.nationalgeographic.org. Date of retrieval 6th June 2022

NOAA (2021) What does coral have to do with medicine. NOAA. https://oceanservice.noaa.gov

Odum EP (2021) Fundamentals of ecology. 5th ed. Co-authored by Odum's Protégé Gary Barret. W.B Saunders, Philadelphia

Ogwu MC (2009) The significance and contribution of conservation education in Nigeria: an appraisal and a call for improvement. In: Babalola F (ed) Proceedings of the maiden seminar of Nigerian Tropical Biology Association. Nigerian Tropical Biology Association and Tropical Biology Association, pp 109–114

Ogwu MC (2019) Towards sustainable development in Africa: the challenge of urbanization and climate change adaptation. In: Cobbinah PB, Addaney M (eds) The geography of climate change adaptation in urban Africa. Springer Nature, pp 29–55. https://doi.org/10.1007/978-3-030-04873-0_2

Ogwu MC, Osawaru ME, Ahana CM (2014) Challenges in conserving and utilizing plant genetic resources (PGR). Int J Genet Mol Biol 6(2):16–22. https://doi.org/10.5897/IJGMB2013.0083

Ogwu MC, Osawaru ME, Aiwansoba RO, Iroh RN (2016a) Status and prospects of vegetables in Africa. In: Borokini IT, Babalola FD (eds) Conference proceedings of the joint biodiversity conservation conference of Nigeria Tropical Biology Association and Nigeria chapter of Society for Conservation Biology on MDGs to SDGs: toward sustainable biodiversity conservation in Nigeria. University of Ilorin, Nigeria, pp 47–57pp

Ogwu MC, Osawaru ME, Aiwansoba RO, Iroh RN (2016b) Ethnobotany and collection of West African Okra [*Abelmoschus caillei* (A. Chev.) Stevels] germplasm in some communities in Edo and Delta States, Southern Nigeria. Borneo J Resour Sci Technol 6(1):25–36

Ogwu MC, Osawaru ME, Obahiagbon GE (2017) Ethnobotanical survey of medicinal plants used for traditional reproductive care by Usen people of Edo State, Nigeria. Malaya J Biosci 4(1): 17–29

Ogwu MC, Ahana CM, Osawaru ME (2018a) Sustainable food production in Nigeria: a case study for Bambara groundnut (*Vigna subterranean* (L.) Verdc. Fabaceae). J Energy Natural Resour Manag 1:68–77

Ogwu MC, Chime AO, Oseh OM (2018b) Ethnobotanical survey of tomato in some cultivated regions in southern Nigeria. Maldives Natl Res J 6(1):19–29

Osawaru ME, Ogwu MC (2014a) Conservation and utilization of plant genetic resources. In: Omokhafe K, Odewale J (eds) Proceedings of 38th Annual Conference of The Genetics Society of Nigeria. Empress Prints Nigeria Limited, pp 105–119

Osawaru ME, Ogwu MC (2014b) Ethnobotany and germplasm collection of two genera of cocoyam (*Colocasia* [Schott] and *Xanthosoma* [Schott], Araceae) in Edo State Nigeria. Sci Technol Arts Res J 3(3):23–28. https://doi.org/10.4314/star.v3i3.4

Osawaru ME, Ogwu MC, Ahana CM (2013) Current status of plant diversity and conservation in Nigeria. Nigerian J Life Sci 3(1):168–178

Osawaru ME, Ogwu MC, Omoigui ID, Aiwansoba RO, Kevin A (2016) Ethnobotanical survey of vegetables eaten by Akwa Ibom people residing in Benin City, Nigeria. Univ Benin J Sci Technol 4(1):70–93

Osunsina IOO, Osunsina J (2018) Assessment of protected area management effectiveness in four national parks in Nigeria: In Ogunjinmi, AA, Oyeleke OO, Adeyemo AI, Ejidike BN, Orimaye JO, Ojo VA, Adetola BO, and Arowosafe FC (eds) Paper presented at proceedings of second Wildlife Society of Nigeria (WISON) Conference, "Achieving sustainable development goals: the role of wildlife", held at Federal University of Technology, Akure, Ondo State, Nigeria, 16th–19th September, 2018, pp 136– 43

Redaction Africanews, Campbell M (2022) Which African country is providing the world with a blueprint for ocean protection. Afrinews.com. Accessed 26 May 2022

Sandbrook C (2015) What is conservation? Oryx 49(4):565–566. https://doi.org/10.1017/S0030605315000952

Sawe BE (2017) What is zoochosis? Environment. https://www.worldatlas.com. Accessed 25 May 2022

Shen Y, Du K, Zou L, Zhou X, Lv R et al (2019) Rapid and continuous on-chip loading of trehalose into erythrocytes. Biomed Microdevices 21(1):5

Shukla GS, Upadhyay VB (2019) Applied and economic zoology. Rastogi Publications, Meerut. 487 pp.

Silva AR, Lima G, Peixoto G, Souza AL (2015) Cryopreservation in mammalian conservation biology: current applications and potential utility. Dove Press. https://doi.org/10.2147/RRBS.S54294

Sintayehu DW (2018) Impact of climate change on biodiversity and associated key ecosystem services in Africa: a systematic review. Ecosystem Health Sustain 4(9). https://doi.org/10.1080/20964129.2018.1530054

Song H-K, Singh D, Tomlinson K, Yang X, Ogwu MC, Slik JWF, Adams JM (2019) Tropical forest conversion to rubber plantation in Southwest China results in lower fungal beta diversity and reduced network complexity. FEMS Microbiol Ecol:fiz092. https://doi.org/10.1093/femsec/fiz092

Stoilova T, Zonneveld MV, Roothaert R, Schreinemachers R (2019) Connecting genebanks to farmers in East Africa through the distribution of vegetable seed kits. Published online by Cambridge University Press

Stoll-Kleemann S, O'Riordan T (2017) The challenges of the Anthropocene for biosphere reserves. PARKS 2017 Vol 23.1

Tanga M, Lewu FB, Oyedeji OA, Oyedeji OO (2018) Cultivation of medical plants in South Africa: a solution to quality assurance and consistent availability of medicinal plant materials for commercialization. Acad J Med Plants 6(7):168–177. https://doi.org/10.15413/ajmp.2018.0133

Team Leverage Edu (2022) Environmental conservation, Wings

Teixeira H, Lilebo A et al +15 authors (2019) Linking biodiversity to ecosystem services supply: patterns across aquatic ecosystems. Environmental science. The science of the total environment

UNEP-IUCN (2016) World database on protected areas. https://www.iucn.org. Accessed 20 May 2022

Valente A, Valente J, Fonseca C, Torres R (2017) The success of species reintroductions: a case study of red deer in Portugal two decades after reintroduction. Int J Biodiver Sci Ecosyst Serv Manag 13(1)

Vertucci VA, Walters C, Ballestros D (2010) Potentials for monitoring gene level biodiversity, using Sweden as an example. Biodivers Conserv 17(4):893–910

Vladimirov IK, Tacheva D, Dobrinov V (2018). The present and future of embryo cryopreservation

Vucetich JA, Macdonald EA, Burnham D, Bruskotter JT (2021) Finding purpose in the conservation of biodiversity by the commingling of science and ethics. Animals 11(3):837. https://doi.org/10.3390/ani11030837

Vwioko DE, Okoekhian I, Ogwu MC (2018) Stress analysis of *Amaranthus hybridus* L. and *Lycopersicon esculentum* Mill. Exposed to Sulphur and nitrogen dioxide. Pertanika J Trop Agric Sci 41(3):1169–1191

Walters C, Pence V (2020) The unique role of seed banking and cryobiotechnologies in plant conservation. Plants People Planet 3(1):83–91

Whaley D, Damyar K, Witek RP (2021) Cryopreservation: an overview of principles and cell-specific considerations. Cell Transplantation, SAGE Journals https://journals.sagepub.com. Accessed 25th May 2022

Wilson JW, Primack RB (2022) Ex situ conservation strategies Open book publishers. libretexts

Wolfe LI, Miller MW (2016) Using tailored tranquilizer combinations to reduce stress associated with large ungulate and translocation. J Wildl Dis 52(2):5118–5124

World Bank (2019) This is what It's all about: protecting biodiversity in Africa. https://www.worldbank.org

Yede SW, Kalyankar VB, Shinde VD (2016) Effect of exotic species on local flora and fauna in and around Sengaon. Int JLife Sci Res 2(5):541–543

Zegeye H (2017) In situ and ex situ conservation: complementary approaches for maintaining biodiversity. Int J Res Environ Stud IJRES 4(2017):1–12

Part II
Utilization Patterns and Potentials of Africa's Biological Resources and Environment

Chapter 10
The Value of Biodiversity to Sustainable Development in Africa

Abiola Omolewa Saliu, Oluwatosin Olaoluwa Komolafe, Christopher Oluwasesan Bamidele, and Morufu Olalekan Raimi

Abstract Biodiversity provides ecosystem services, which are the basis for healthy ecosystems that support life. These services could either be supporting, provisioning, cultural, or regulating services, and they are crucial to sustainable development as they contribute, in one way or the other, to the achievement of the Sustainable Development Goals (SDGs). Africa is a continent with rich biodiversity; it has a great number of endemic species and global natural resources; yet the continent is being threatened due to human activities including habitat loss, overexploitation of species, invasive species as well as disease, changing climate, and pollution. Africa is a rapidly urbanizing continent and projections suggest a significant population increase by the year 2050. This will further put pressure on the region's biodiversity and natural resources, with great implications for the poorest and most vulnerable who depend directly on these resources. There is, therefore, the need to use Africa's biodiversity sustainably to realize the long-term benefits that the ecosystem provides and to ensure its continuous benefits to future generations. The sustainable use of Africa's biodiversity would involve putting the right institutional mechanisms (Government coordination structures, Stakeholder engagement, inputs, etc.) in place; policies governing biodiversity conservation would not only be developed but also strictly enforced. There is also a need to put the right mechanisms in place for the continuous data collection on the status of the continent's biodiversity. This is necessary for appropriate stock-taking and to make informed decisions. There is a necessity for increased awareness about the importance of biodiversity at all levels and the bottom-up approach of community involvement in conservation should also

A. O. Saliu (✉)
Department of Plant and Environmental Biology, Kwara State University, Malete, Nigeria
e-mail: abiola.ilesanmi@kwasu.edu.ng

O. O. Komolafe · C. O. Bamidele
Institute of Ecology and Environmental Studies, Obafemi Awolowo University, Ile-Ife, Nigeria

M. O. Raimi
Department of Community Medicine, Environmental Health Unit, Faculty of Clinical Sciences, Niger Delta University, Wilberforce Island, Bayelsa State, Nigeria

S. C. Izah, M. C. Ogwu (eds.), *Sustainable Utilization and Conservation of Africa's Biological Resources and Environment*, Sustainable Development and Biodiversity 32, https://doi.org/10.1007/978-981-19-6974-4_10

be put in place. Finally, it is important to mainstream biodiversity across all sectors in each country.

Keywords Biodiversity · Africa · Sustainable development · Ecosystem services

10.1 Introduction

Biodiversity, also referred to as biological diversity, consists primarily of numerous varieties of living things existing on the earth's surface, the innumerable number of animals, plant species, and microorganisms, their genetic factor, history of their evolution, ecological characteristics as well as the habitats in which they live in. It is defined, according to the Convention on Biological Diversity (CBD), as *"the variability among living organisms from all sources including, inter alia, terrestrial, marine and other aquatic ecosystems and the ecological complexes of which they are part; this includes diversity within species, between species and of ecosystems"* (UN 1992 Article 2). Biological diversity provides the necessary life support structure and the resources that living organisms thrive on. Biodiversity comprises three distinct components, namely, ecosystem biodiversity, species biodiversity, and genetic biodiversity (Fig. 10.1). Biological diversity should be used sustainably in such a way that negative effects are minimized, and its components are preserved to meet the needs and ambitions of current and future generations (WCED, 1987), which is what sustainable development is aimed at, ensuring the economic well-being of humans, their social and human development, and achieving environmental sustainability and regeneration (Dalal-Clayton and Bass, 2002) and these can be achieved if we have healthy ecosystems.

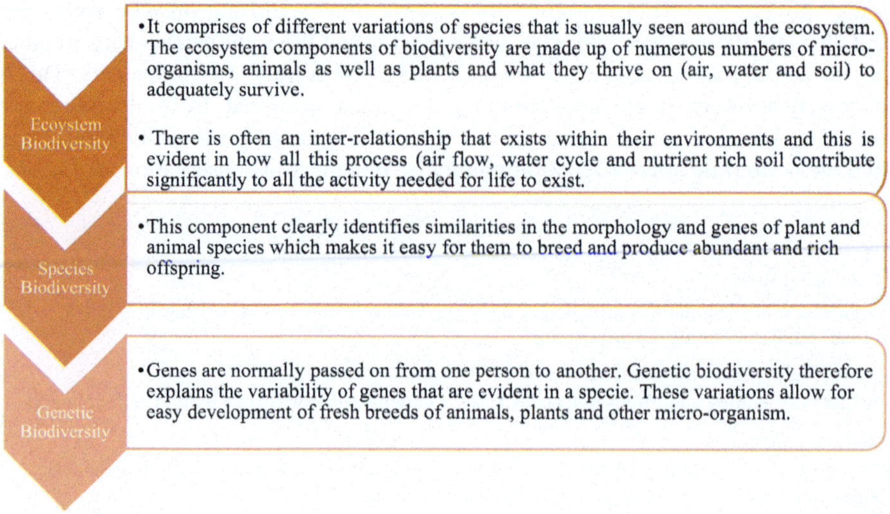

Fig. 10.1 Biodiversity components

Cadman et al. (2010) reported that biodiversity creates the foundation for socio-economic livelihood including the health and well-being of the African society. It plays a significant function in sustaining the development of any nation and it is very important for the overall health of the ecosystem which can withstand the present and incoming generations. It is, however, not a new knowledge that biological diversity globally is generally being threatened with gradual species extinction (Jenkins, 2003; Loreau et al. 2006; Olalekan et al. 2019). Biodiversity loss at present is a great concern because of the negative impacts it has on the survival of humans; it results in the shrinking of economic profits, decrease in efficiency, and ability to produce a different mass of organisms within the ecological system, decompose organic matter, and recycle required nutrients (Cardinale et al. 2012). The Millennium Ecosystem Assessment (MEA) established from its findings, that for more than five decades, humans have been responsible for changing the state of the ecological system much faster compared to any time in the history of humans (Walter et al. 2005). The report of the United Nation Millennium Development Goals (2013) also revealed that our environment is severely threatened as there is an increase in the loss of biodiversity and environmental degradation due to unsustainable exploitation of natural resources. Too often, it has been discovered that biodiversity loss is mostly fueled by climate change and unsustainable development like rapid urbanization. This, in a way, may create trade-offs that have to be properly managed and understood within the context of biodiversity development goals and conservation which in itself may be utilized as a decisive tool required to drive sustainability.

Biodiversity needs to be utilized sustainably and conserved properly; otherwise, a huge loss is imminent and can cause a knock-on effect on the goods and services available in the ecosystem. Conservation cannot be complete without having to emphasize the appropriate usage of natural resources in the ecosystem. The conservation movement is seen in how nations around the world call to attention the significance of sustaining its ecological diversities. Efforts made at conserving biodiversity include the aspect of controlled use of resources and maintaining the ecosystem processes which provide the natural resources and services required for human development. The importance of sustainable conservation of biodiversity cannot be overemphasized which is why there is a need for countries and continents to map out strategies that would help to maintain and conserve their bioresources alongside reducing any severe threat that may be encountered. Conservation practices are required to sustain, protect, restore, and rehabilitate biodiversity. The increasing knowledge of biodiversity loss and the lasting effect it creates on humans has drawn the attention of the world, which has resulted in having conventions, declarations, and codes of conduct that are geared toward supporting and encouraging nations to be firm in curbing the impending danger of decline in biodiversity. Also, this has allowed for the creation of world organizations, state institutions as well as research institutes that are working alongside stakeholders and donor groups to expound on the magnitude of biodiversity decrease and come up with sustainable strategies, policies, interventions, and plans required to tackle this imminent problem (Butchart 2010).

One of such goals is the comprehensive plan for biodiversity 2011–2020 as well as the 20 Aichi biodiversity targets, a ten-year plan which was developed to conserve biodiversity and ensure that its benefits are fully realized. However, a recent assessment of these targets reveals that none of the targets have been fully achieved (CBD, 2020); hence, the development of the post-2020 Global Biodiversity Framework, which builds on the biodiversity strategic plan 2011–2020. The framework, with 21 targets and 10 milestones, is to be achieved by the year 2030 with appropriate stock-taking. African countries also came up with a vision known as "Agenda 2063: The Africa we want' at the African Union summit in 2013 to achieve a peaceful continent by the year 2063. It is important to note, however, that there are areas of convergence; part of the 2063 agenda seeks to ensure that African citizens are healthy and well-nourished (Goal 3). It also seeks to achieve '*environmentally sustainable as well as climate-resilient economies and communities through sustainable natural resource management, biodiversity conservation, and water conservation'*. To ensure that we bend the curve of biodiversity decline in Africa and ensure sustainable development, Africa's 2063 Agenda and the post-2020 Global Biodiversity Framework must be integrated with the 2030 Agenda for sustainable development, to tackle the root cause of biodiversity loss and achieve the "*5 Ps of sustainable development; 'people, planet, partnership, peace, prosperity* to ensure by 2050, the vision of '*living in harmony with nature*".

This chapter presents an overview of the value of biodiversity to sustainable development in Africa. Africa is a rapidly urbanizing continent and projections suggest a significant population increase by the year 2050. This will further put pressure on the region's biodiversity and natural resources, with great implications for the poorest and most vulnerable who depend directly on these resources. However, indigenous biodiversity in the continent can provide ecosystem services that are essential to sustainable development.

10.2 Drivers and Threats Influencing Loss of Biodiversity

Stuart et al. (1996) explained that biodiversity is continuously been lost at an increasing rate (from 100 to 1000) compared to natural rates of extinction. Table 10.1 shows some drivers and threats to biodiversity. Many studies have

Table 10.1 Drivers and threats to biodiversity

Indicators	Threats
Socioeconomic and political components	Climate changes resulting from the increase in the fossil fuel market which has led to temperature increases and environmental pollution.
Scientific and technical components	Parasites and species that invade habitats.
Demographic components	This includes loss of habitat and deforestation arising from increasing urbanization and agricultural purposes.
Religious and cultural factors	Some identified threats include unsustainable exploitation.

indicated that biodiversity loss has harmed the environment. Loss of biodiversity could reduce the productivity level of ecosystem services, hence endangering the sustainability of such an ecological community. Many consider biodiversity as a common resource, which has led to its overexploitation globally.

Overexploitation of biodiversity has threatened and endangered wildlife (fishes, reptiles, plants, etc.). The exploitation of natural resources in the last decades has been unprecedented. The rate at which humans exploit biodiversity is more than the regenerative capacity of many ecosystems (Everard et al. 2020). Humans overexploit biodiversity through overfishing, logging, and overhunting. For instance, the ocean has seen a rapid decline in its sharks and ray population, with about one-third on the brink of extinction (Dulvy et al. 2021). This situation could lead to a cascading effect on the natural aquatic ecosystem (Ferretti et al. 2010). The same can be said about the forest ecosystem. Overexploitation of forest resources could lead to the destruction of habitats, and hence low quality of life for rural dwellers. Crowther et al. (2015) in their study estimated that about 15 billion trees are felled each year. They further adjudged that about 46% of trees have been lost since the advent of human civilization. The implication of this indiscriminate felling is increased greenhouse gases and distortion of the global ecosystem. Studies have also revealed that overexploitation of biodiversity can lead to negative economic growth in the long term (Lampert 2019). This could lead to massive unemployment and low economic productivity. A wholesome number of individuals rely on animals for their meat and income. These animals usually provide adequate micronutrients, fat, and proteins, especially for aboriginal communities in sub-Saharan regions of Africa. According to Fa et al. (2002), about 4.9 billion kilograms of wild animals have been hunted annually, particularly for their meat. As human populations expand, there is more demand for natural resources. Humans are devising better means of hunting the wildlife; however, the overexploitation of these resources is gradually depleting the reserves which are likely to hurt unborn generations. Unsustainable exploitation of biodiversity can create an increase in food scarcity and insecurity. Biodiversity may be preserved by recognizing ecological components which are constantly endangered by overexploitation. The introduction of invasive species by humans has also hurt biodiversity. Invasive species have increased the rate of biodiversity loss. It is considered to be the second major cause of biodiversity loss globally (habitat loss is the first factor). As per the National Wildlife Federation (one of the world's leading conservation groups), invasive species threaten 'approximately 42 percent of endangered wildlife.' The devastating effect of invasive species on an economy could be significant. Africa has had a fair share of the negative effect of invasive species; the introduction of invasive species into Africa has led to significant losses—both environmentally and economically (Linders et al. 2019). Many of the invasive species have suppressed the growth, led to the extinction of native plants, caused the destruction of crops, and have caused significant changes to some ecosystem processes. An example is the introduction of *Lates niloticus* (also called Nile perch) into other African lakes. The introduction of Nile Perch from Lake Albert (one of its original homes) into other lakes led to a significant reduction of native fishes. Not only that, but the dynamics of the aquatic environment and the surrounding terrestrial habitat were also changed. It is also estimated that invasive species cost Africa about $65.58 billion a year – a major economic loss to the continent (Eschen et al. 2021). There are a host of harmful parasites which pose a significant threat to the ecosystem as they feed on other species for their survival. Control measures must be put in place to identify and eradicate such predators that are likely to reduce species abundance. More so, the handling, utilization, and transfer of organisms that are modified genetically should be properly regulated and measures necessary to conserve biodiversity outside the natural environments should be strengthened as well.

Habitat loss/destruction is another factor that impedes biodiversity and its contribution to sustainable development. Habitat loss is considered to be the first major cause of biodiversity loss (Chase et al. 2020). Habitat destruction or loss is the removal of essential conditions necessary for the survival of living organisms (Singh et al. 2021). The implications of habitat loss include: a reduction in species richness and abundance and decreases in species distribution while lowering species distribution (Lituma et al. 2022). Since habitat loss is not wholly driven by humans, substantial habitat losses are caused by humans (Shrestha et al. 2021). Agriculture has been a major driver in the destruction of many habitats. Globally, agriculture has also led to the decline of biodiversity in many rural areas. According to UNEP (United Nations Environmental Programme), 'agriculture has indeed been recognized as a danger to 24,000 of the 28,000 (86 percent) species threatened with extinction'. Due to the increased population, there are demands for increased agricultural production. Many ecological systems/biomes are converted to farmlands, which has led to increased greenhouse gases. In regions such as sub-Saharan Africa, farm expansion has been a major cause of biodiversity loss while intensive agriculture in developed regions of the world has led to a reduction in beneficial organisms in the soil (El Mujtar et al. 2019). Increasing food production is a necessity; however, biodiversity loss is also detrimental to sustainable development.

Another factor that limits the contribution of biodiversity to sustainable development is pollution. Externalities from industries and agriculture have endangered many ecosystems. The release of toxic substances from pollution sources into an ecosystem has adverse effects on the biota. A case in point will be the release of acidic oxides (such as sulfur dioxide and nitrogen dioxide) by industries into an ecosystem. Such gases, if not well managed, could lead to the formation of acidic rain. This could lead to the removal of essential nutrients and the leaching of aluminum from the soil, hence limiting or preventing the growth of flora. If acidic rain runoffs into an aquatic environment (such as lakes), it could harm the food web. Biodiversity loss through pollution also has far-reaching effects on humans; causing health challenges and increasing death rates (Karlsson et al. 2021). Hence, persistent degradation of the ecosystem through pollution needs to be stopped. The challenge, therefore, lies in: how do we use Africa's resources sustainably to enjoy the benefits that healthy ecosystems provide?

10.3 African Biodiversity: Current Status

The continent of Africa boasts of having lush biodiversity with different organisms summing up to about one-quarter of the world's total (FAO 2020b). African Development Bank (2015) established in its findings that Africa has a large chunk of biological diversities and global natural resources; with a large assemblage of mammals (UNEP-WCMC, 2016). IPBES (2018) reported that a quarter of the world's mammal species are found in Africa, with a diverse species of large mammals in East and South Africa. One-sixth of the global plant species and one-fifth of global bird species are found in Africa (IPBES, 2018). Africa has 405 endemic plants; 38 of which are threatened. South Africa has the highest number of endemic plants (371), with the cape floral region as one with rich plant diversity. Madagascar has the highest number of endemic animal species, with 1028 out of a total of 2765 vertebrates, and 170 of 569 invertebrates (Fig. 10.2a–e). Currently, 8 of

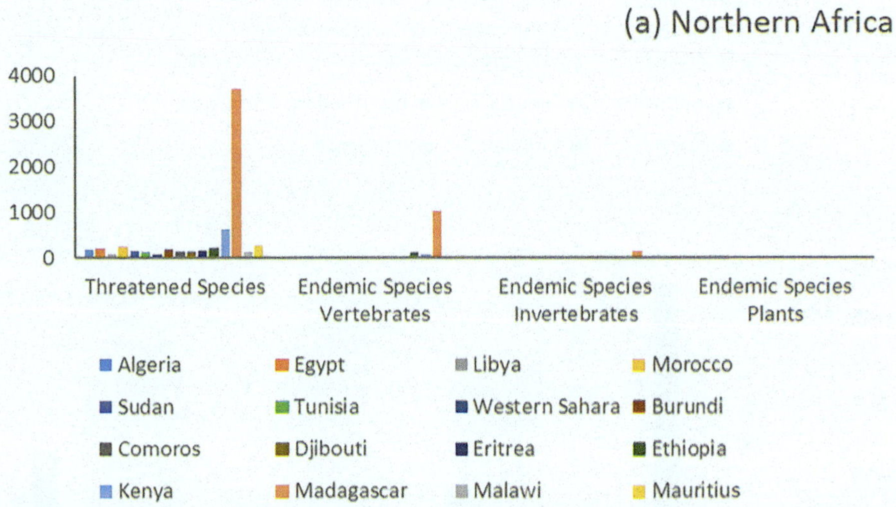

Fig. 10.2 (**a**) Data on threatened and endemic species in northern African countries (2021). (**b**) Data on threatened and endemic species in eastern African countries (2021). (**c**) Data on threatened and endemic species in southern African countries (2021). (**d**) Data on Threatened and Endemic Species in Western African Countries (2021). (**e**): Data on Threatened and Endemic Species in Middle African Countries (2021). (**a**–**e**) were calculated and modified from the IUCN Red List (2021). Note that according to IUCN, endemic species are '*species that are known to occur naturally within one country only*' and threatened species are '*species assessed in any of the three threatened Red List categories (Critically Endangered, Endangered, Vulnerable)*'

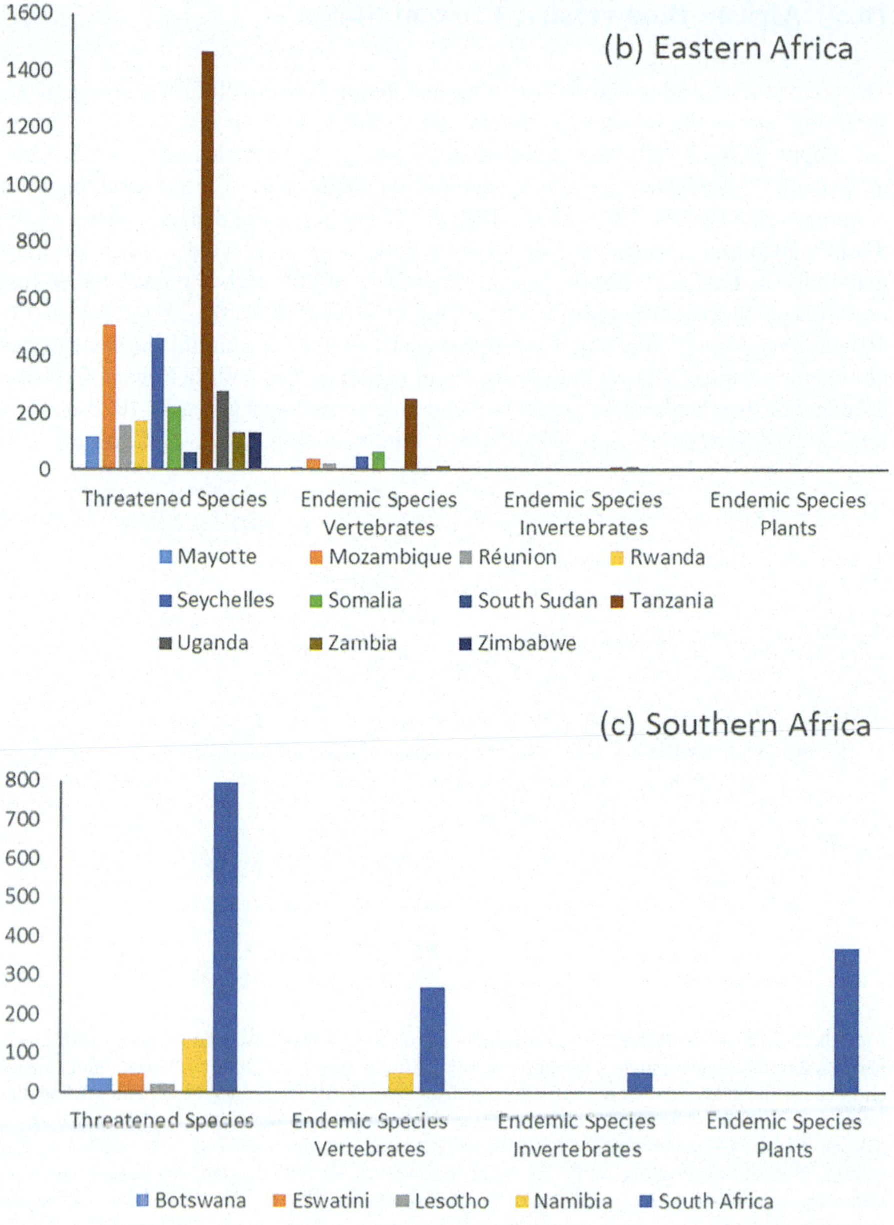

Fig. 10.2 (continued)

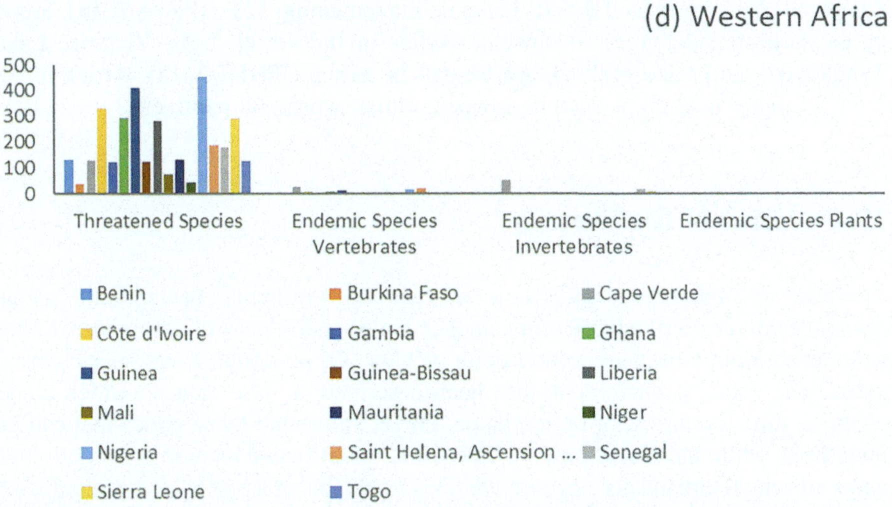

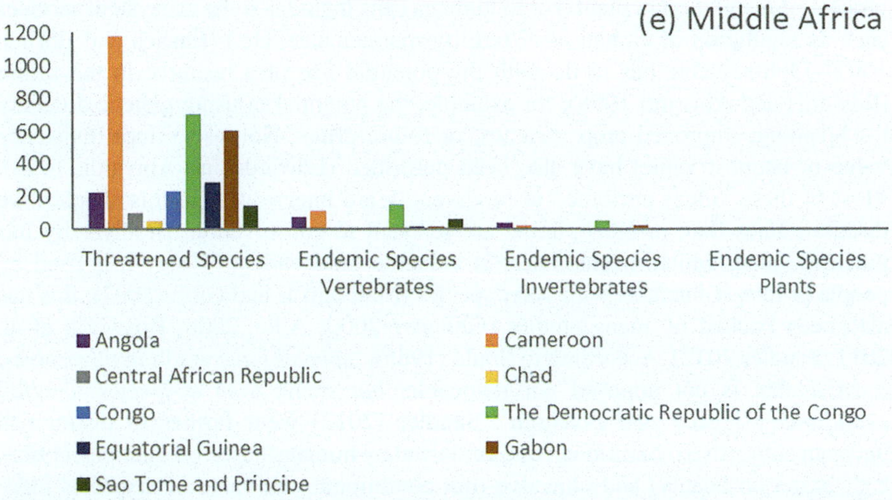

Fig. 10.2 (continued)

the 37 hotspots of the world are found in Africa; "the Coastal Forests of Eastern Africa, the Cape Floristic Region the Eastern Afromontane, the Guinean Forests of West Africa, Madagascar and the Indian Ocean Islands, the Mediterranean Basin, the Horn of Africa, the Maputaland-Pondoland-Albany, as well as the Succulent Karoo"; yet Africa's biodiversity faces serious threats from human-induced pressures. Africa has 8 major physical regions, each with its unique and diverse plants and animals. It has diverse ecosystems, which may be a result of varying climates, arising from its wide range of latitude (Lock 2001). The continent's land area is made up of 23% of forests and woodlands and 27% of arable land. Grasslands,

Savannah, Arid zones, and deserts make up the remaining 50% (IPBES 2018). Some of the largest and longest freshwater bodies in the world, Lake Victoria, Lake Tanganyika, and Lake Malawi, are located in Africa (IPBES 2018). Africa is the home of many food crops such as cowpea, millet, wheat, oil palm, etc.

10.4 Value of Biodiversity

The value of biodiversity depends on different points of human perception. Various researchers have ascribed different values to biodiversity; most of whom were either from an economic or conservation point of view. Of no doubt, every species has a value. The value of biodiversity has been described as 'Use values', which could either be direct, indirect, or option value. Direct values are those goods that can be quantified, while indirect values are difficult to quantify and include not-so-obvious ways in which organisms support life that could be of benefit to the organisms themselves (e.g., the use of the chemical energy from photosynthesis to drive the process of respiration in plants) or to humans (this includes some ecosystem services such as regulation of climate, soil fertility maintenance, etc.) (Ehrlich and Ehrlich 1997). Option value has to do with the potential use of a resource in the future (Edwards and Abivardi 1998); for example, the potential of using genetic diversity for breeding, improved crop varieties, or future drugs. Non-use values (existence value or bequest value) have also been described (Edwards and Abivardi, 1998). Most of these values attributed to biodiversity are interwoven. In this chapter, we discuss values that, in our opinion, are relevant to conservation. Biodiversity has intrinsic value, which means that it is valuable independent of how it is used by people or how it interacts with other species (Malcolm Jr and Gibbs 2007); this has also been backed by many studies (Ghilarov 2000; Alho, 2008; Bosworth et al. 2011; Sandler 2012). According to Soulé (1985), '*species have* worth *in themselves, a value* that is not imposed nor revocable, *but* stems *from a species'* lengthy *evolutionary* legacy *and potential*'. Sandler (2012) went further to distinguish between subjective (conditional; one conferred by humans based on their perception, e.g., based on beauty) and objective (not conditional; the value of a species independent of people's perception) intrinsic value. The intrinsic value of biodiversity involves conserving biodiversity because of its existence because they have a right to continue to exist (Callicott 2006; Alho 2008). This is related to the ethical/moral worth of biodiversity, which is based on people's cultural views that species have inherent value and thus should be protected from extinction by humans, albeit this varies per species. Viewing biodiversity from an ethical point helps us to protect them.

Biodiversity also has instrumental value (Malcolm Jr and Gibbs 2007; Bosworth et al. 2011) which is its relevance to other species, including humans. White (2013) refers to this as 'extrinsic value'. Economic value, esthetic value, spiritual value, scientific/educational value, ecological value, and strategic value are all examples of instrumental value. Economic value has to do with the use of biodiversity by humans for direct benefits; either through its use as food, shelter, and clothing, as a source of medicine, or as fuel. The esthetic value of biodiversity refers to its beauty. All organisms exhibit the '*beauty of design*'. The esthetic value of biodiversity helps us to appreciate biodiversity more; it is immediate and does not rely on uncertainties in the future (Kiester, 1996). The esthetic value of species supports economic activity and serves as a source of income for the government, through the establishment of national parks and zoos where tourists can go bird-watching, butterfly watching, and watching animals in a natural environment. This will also help to promote ecotourism. The esthetic value of biodiversity contributes to human well-being (Millennium Ecosystem Assessment, MEA, 2005). Hale et al. (2005) stated that there is a positive relationship between esthetic value and biodiversity when they observed that the diversity of Avian species increased significantly in locations that practiced esthetic landscape planning. According to Tribot et al. (2018), increased understanding of biodiversity builds our esthetic experience and motivates us to aspire to conserve biodiversity as humans are generally interested in preserving beautiful things. This will ultimately lead to increased diversity, which guarantees the provision of ecosystem services and further increase our esthetic experience.

Biodiversity has spiritual/religious value as many religions identify themselves with certain plants and animals. The educational value of species helps us to understand the place of species in the ecosystem. It also helps to increase our understanding of biodiversity which informs our curiosity and investigation in aspiring for more scientific knowledge about them. Species have an ecological value (Malcolm Jr and Gibbs, 2007) as each specie has a unique role it plays in the ecosystem. Every species is part of an ecosystem and therefore has an ecological role to play.

Species can also be of strategic value to conservationists by serving as flagship, umbrella, or indicator species. Malcolm Jr and Gibbs (2007) also described the uniqueness value and ascribed more importance to species with no closely related species compared to those with closely related species. It is interesting to note that researchers like White (2013) and Morton and Hill (2014) have also reported the negative value of biodiversity, a view towards biodiversity based on fear or hostility. For example, the mosquito or the influenza virus that spread disease, or snakes and crocodiles that are considered harmful.

10.5 Link Between Sustainable Development and Biodiversity

Biodiversity involves all living things: plants, animals (which include humans), and even microbes (Ranjalkar and Shelke 2022). Each living organism plays a functional role in the global ecosystem. These roles range from food provision to climate regulations; to water purification and nutrient cycling; while some provide medicinal benefits (Creamer et al. 2022). The connection between biodiversity and sustainable development is closer than many would think as it is a key factor in sustainable development. Biodiversity links the flora with the fauna, and without biodiversity, there is no life. The variability of life ensures continued existence between each category of living organisms. Humans cannot continue to exist if the resources provided by the ecosystem become depleted and out of supply. If biodiversity keeps declining, it will result in economic consequences such as increased unemployment and health consequences due to lower medicinal benefits. Apart from the utilitarian value it gives, biodiversity also provides cultural values (Toepfer 2019; Ciftcioglu 2021). Hence, any loss of biodiversity threatens the existence of life in an ecosystem. The relationship between biodiversity and sustainability is reflected through its relationship with practically all of the sustainable development goals (SDGs). It boosts the economy, hence reducing poverty rates and hunger (SDGs 1, 2, and 8); its contribution to the medicine and pharmaceutical industries is immense and crucial to human well-being (SDGs 3). Investing in biodiversity also makes human settlements inclusive and sustainable (SDG 8). It's been proven that biodiversity protection aids in the fight against climate change by absorbing as well as storing carbon (SDG 13, 14, and 15). A healthy ecosystem means social, cultural, and peaceful existence targets will be met (SDG 8, 10, and 16). However, human impacts have been a major challenge in achieving these goals.

Biodiversity is important to achieving food security and promoting sustainable agriculture (SDG Goal 2). Biological diversity provides the genetic material with which plant and animal breeders can develop improved varieties with desirable traits (Govindaraj et al. 2015; Lockie and Ransan-Cooper 2015). Genetic diversity is important to achieving sustainable agriculture as it increases the chances of a species' survival and adaptability when environmental conditions change; it also increases their resistance to pests and diseases (CBD 2020). Farmers can make use of this advantage to provide different kinds of food with improved nutritional benefits and considerable efforts have been made in Africa. Alene et al. (2015) reported the release of 367 improved varieties of cassava and 202 improved varieties of cowpea in sub-Saharan Africa (SSA) between 1970 and 2010 with 68% of cassava and over 80% of cowpea released in West and Central Africa. Improved crop varieties in SSA have also been reported for maize (Alene et al. 2015), beans (Muthoni and Andrade, 2015), rice (Diagne et al. 2015), and other crops. Pollinating services by bees and other insects are made available through biodiversity (Klein et al. 2007).

The contribution of the fishery and aquaculture sector is also important to achieving food security as most people include fish in their daily diet. However, overexploitation poses a challenge to the fishery sector as commercial fish stocks are being depleted at a fast rate (Pauly et al. 2002) and three-quarters of the world's fisheries have been exploited (FAO 2016). Fisheries and aquaculture represent an essential source of food security in Africa, especially in low-income countries with 9% of the world's total of 59.51 million people engaged in this sector from Africa; though a greater increase in employment has been reported more in the fishing sector compared to aquaculture. Globally, in 2017, 17% of animal proteins and 7% of all proteins in people's diet were a result of fish intake. An average increase of 3.1% per year was reported in food fish consumption from 1961 to 2017 and about 1.5% per year increase in per capita fish consumption from 1961 to 2018, with an increase in per capital fish consumption in Africa at the annual rate of 1.3% since 1961. Despite this increase, Africa still has the lowest per capita fish consumption compared to other regions and continents, most especially sub-Saharan Africa. This is partly due to demand being greater than supply as a result of the increasing population and an underdeveloped aquaculture sector, resulting in the importation of 35% of fish consumed in 2017 (FAO 2020a).

Biodiversity loss has had a direct impact on human health and livelihood. A well-functioning ecosystem helps to reduce the risk of illnesses associated with pollutants in soil, air, or water (Goal 3, target 3.9) and keeps us healthy through the provision of herbal medicines (Goal 3, target 3b). Africa consists of over 5000 medicinal plants (Iwu, 2014), and many plant species have been reported to have healing properties. Kingston (2011) noted that natural substances are the basis for the development of about 25% to 50% of commercial drugs. According to WHO (2003), the protection of medicinal plants lies in their cultivation and sustainable use. Carlson (2002) reported the identification of 145 species of plants for the treatment of Type II diabetes mellitus in Central Africa. The development of antibiotics and vaccines is made available as a result of genetic materials from biodiversity (Bull et al. 2000), so also herbal medicines from plants. In Southeastern Africa, the willow plant, *Salix capensis,* is used as an antipyretic and for killing pain (Iwu, 2014). The use of *Curcuma longa* in West Africa as an ointment for skin diseases has also been reported (Iwu 2014). Other African medicinal plants like *Blepharis linariifolia* have also been documented for the use of its seeds and leaves in embrocation (Iwu 2014); leaves and twigs of *Pilea microphylla* for liver and urinary inflammation (Neuwinger 2000), roots of *Phyllanthus beillei* for stomachache (Thiombiano et al. 2012), and *Cyclopia spps* endemic to southwest and southeast of South Africa for herbal teas (Mahomoodally 2013; Iwu, 2014).

Biodiversity is also crucial to ensuring that clean water and sanitation are made available for all (Goal 6) through the conservation of water-related ecosystems such as wetlands, lakes, rivers, etc. Forested watersheds account for three-quarters of the world's accessible freshwater (FAO 2018). According to IPBES (2018), Africa has some of the world's biggest wetlands and they make up 1% of the total land surface area. Wetlands help to filter out waste from the water, hence ensuring access to fresh water. Though SSA has made remarkable progress concerning the availability of drinking water, especially in urban areas, WHO/UNICEF (2021) Joint Monitoring Report stated that in 2020, 771 million people still lack basic drinking water services, 50% of whom live in SSA and over 40% of people in Sierra Leone, Chad, Madagascar, and Nigeria still use water sources that are highly contaminated. There is a need for efficient conservation of Africa's water resources to reduce the number of deaths that may occur from low water quality as a result of water-related disasters (Goal 11, target 11.5). Also, SSA has not recorded considerable progress in sanitation services as open defecation is still widespread in SSA, and over 60% of people in South Sudan, Chad, and Niger still practice open defecation (WHO/UNICEF 2021) with an increase in the number of people who lack access to basic sanitation. The regular operation of African rivers has indeed been hampered by aquatic pollution due to the discharge of pollutants from agricultural sources, industrial waste, and home sewage (Adeyemo 2003). Sustainable use of marine ecosystems and their biodiversity is also crucial to ensuring sustainable development in Africa (Goal 14, target 14.4). Oceans and their biodiversity suffer from human-induced damage (McCauley 2015). Globally, aquaculture accounted for 46% of the total fish production in 2018 with about 2.7% of this being attributed to Africa (i.e., aquaculture accounts for 17.9 percent of total fish production in Africa); Egypt and Nigeria being the top producers (FAO 2020a). However, there was an increase in the proportion of fish stocks that are within biologically unsustainable levels from 10% in 1974 to 34.2% in 2017 (FAO 2020a).

Healthy ecosystems increase our resilience and adaptation to the adverse effects of climate change (Goal 13) and climate change is posing a threat to Africa (Berrahmouni et al. 2020). Over 50% of birds and mammals in Africa are under threat from climate change and the productivity of lakes in Africa could decrease by 20–30% by 2100 (IPBES 2018). This may have indirect effects on food security as agriculture depends on weather conditions and the incidence of pests and diseases which may arise as a result of climate change. This will have severe implications for poor and vulnerable populations who depend mainly on agriculture (CBD 2018). Forest ecosystems act as carbon sinks, and they help to mitigate the adverse effects of climate as they absorb about 2 billion tons of CO_2 annually (FAO 2018). Halting deforestation will have great implications for climate change mitigation (Goal 15). Biodiversity loss has a profound effect on climate change. Climate change has harmed the global environment in recent years, while biodiversity loss has also contributed to its increase. Climate change has led to droughts, increased desertification, and melted ice glaciers and has had a disastrous impact on human health (Mills 2020). The warmer it becomes, the higher the rate of biodiversity loss. The less biodiversity, the harder it is to achieve sustainable development goals (Baumgartner 2019). To ensure adaptation/mitigation against climate change, there is a need to preserve and increase biodiversity. Increased natural biodiversity means more capacity to absorb and store carbon (Bax et al. 2021). Woodward et al. (2014) discovered that deforestation of trees in wetlands could increase annual precipitation in such areas by 15%. This can lead to a reduced ability of such environments to control flooding (one of the major functions of wetlands) (Acreman et al. 2021).

Seventy percent of terrestrial biodiversity loss is a result of expansion due to agricultural practices. By 2050, population growth and urbanization are expected to result in a major rise in demand for arable land (CBD 2018). Africa has diverse forests; yet, it experiences a high rate of deforestation (Iwu 2014) as about three million hectares of Africa's forests are lost annually (Mansourian and Berrahmouni 2021). Desertification affects 45 percent of Africa's geographical area, while 65 percent of the continent's fertile land is degraded (Mansourian and Berrahmouni 2021). Africa recorded the highest net loss of 3.9 million ha of forest between 2010 and 2020 (FAO 2020c). Forest degradation will have indirect implications on food security as forests serve as a means of livelihood for 25% of the global population (UN 2021). Human health will also be affected indirectly as forest clearing implies the loss of medicinal plants and an increase in the emergence of zoonotic diseases

(Sen 2020; UN 2021) such as COVID-19 or Ebola (Sen 2020). Africa is, however, making efforts in forest and landscape restoration with projects and initiatives such as the 'Trees for Global Benefit' project in Uganda (Kalunda et al. 2020), 'Green Belt Movement' in Kenya, the 'African Forest Landscape Restoration (AFR100) Initiative', the 'Great Green Wall of the Sahara and the Sahel Initiative', the 'Pan-African Agenda on Ecosystem Restoration', 'Regreening Africa Programme', among others.

10.6 Sustainable Use of Biodiversity in Africa

Biodiversity provides ecosystem services that we all depend on (Folke et al. 2011; Schultz et al. 2016), and which support life; hence, humans need to learn how to use biodiversity responsibly. Some of such services include the provision of food, cycling of nutrients, purification of water, and regulation of climate (Schultz et al. 2016; OECD 2018). It can therefore be said to be the basis for earth's support system. According to the Millennium Ecosystem Assessment (2005), ecosystem services are classified into four types: regulating, cultural, provisioning as well as supporting. According to CBD (2010), species extinction is increasing at an alarming rate as about one-quarter of the earth's plant species are being threatened with extinction. In Africa, over 9,000 plants are threatened with extinction according to the IUCN red list (2021) (Fig. 10.2a–e).

Healthy and well-functioning ecosystems are needed to realize the many objectives and targets of the SDGs (Fig. 10.3). There is, therefore, an overlap between the preservation of our biological diversity and the achievement of the SDGs, i.e., the sustainable and efficient use of Africa's biodiversity is important to realize the long-term benefits that the ecosystem provides; and this guarantees significant gains in many aspects of sustainable development. This is one of the many aims as well as 'targets of the 2030 Agenda for sustainable development.

Globally, about 50 percent of people depend directly on natural resources for their livelihood, many of whom are the rural poor and the most vulnerable (UNEP 2007). The negative effects of ecosystem degradation and biodiversity loss will therefore have severe consequences for them (Schultz et al. 2016). FAO et al. (2019) reported that more than 820 million people of the population of the world are undernourished, with the highest prevalence in Africa (about 20%), most especially in sub-Saharan Africa. Of particular concern is the situation in Africa; though Africa boasts of rich diversity, the large dependence on its natural resources, ecosystem degradation, and the unsustainable use of its biodiversity have resulted in an inability to deliver the services on which people rely; hence, the detrimental impacts on the well-being of its people (Cardinale et al. 2012). The WWF (2020) Living Planet Index (LPI) showed a 65% decline in the population of reptiles, mammals, amphibians, birds, as well as fish between 1970 and 2016 in Africa. This is a result of the challenges to biodiversity, for example, changes in land and sea use, habitat loss as well as degradation (45.9%), species overexploitation (35.5%), climate change (4.1%), invasive species

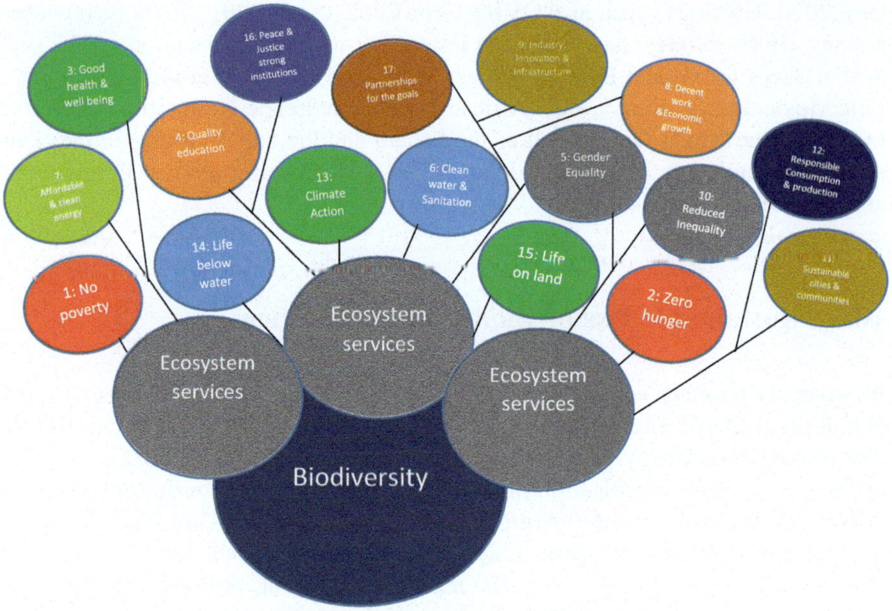

Fig. 10.3 Biodiversity provides ecosystem services that we all depend on; these services are crucial to achieving the sustainable development goals either directly or indirectly

and disease (11.6%), and pollution (2.8%). In comparison to other continents, Africa has seen a rapid pace of urbanization, with the latest estimates indicating an eight-fold increase in urban land cover between 2000 and 2030 (Schultz et al. 2016). The continent could experience a rise in population to at least 2.4 billion by 2050 (AFDB 2014). This will further put pressure on the region's biodiversity and food production systems, especially in underdeveloped areas. Conservation of biodiversity in Africa is therefore of utmost importance to avoid continuous ecosystem degradation that jeopardizes sustainable development (Stephenson et al. 2017).

Sustainable use of biodiversity means every biome will survive and flourish. Protecting a local biotic community seems to be difficult due to the various challenges attached to it; however, it is possible. Significant success has been made by various national and international organizations in ensuring biodiversity conservation. According to a mid-term review by UNEP (2016) to check the progress toward the Aichi Biodiversity Targets, Africa has made progress in fighting biodiversity loss in the last decade. However, it was also noted in the report that Africa needs to increase its efforts to ensure it meets the targets. Hence, there is a need for greater actions to ensure that suggestions on biodiversity conservation are implemented. There are guidelines to ensure natural resources are used responsibly. In 2004, there was a convention on biodiversity (CDB), where sustainable principles and strategies for managing different ecosystems in relationship with their relevant sectors were made.

The first principle involves each community identifying threats and using suggested principles/frameworks as a guide to combat biodiversity loss. This will lead to a faster renewal of natural resources. In agriculture, for example, reduced use of pesticides and chemical fertilizers will decrease the negative effect of farming on local biodiversity (Gonthier et al. 2014). Other suggestions are land sharing and land sparing strategies. The land sharing process involves the integration of conservation approaches into agricultural activities (Dudley and Alexander 2017). This will promote higher biodiversity and lower intensive farming on the same agricultural land. However, land sparing means more expanse of land is allocated for biodiversity conservation (Grass et al. 2019). Protecting the ecosystem from the impact of climate change is another way of ensuring sustainable growth and use of biodiversity. An ecosystem-based approach (ecosystem-based mitigation and adaptation strategies) has proven to be a good strategy in this regard. Examples of such approaches include diversification of livelihood for rural dwellers, reforestation of forests, and marine conservation (Chausson et al. 2020).

To enforce the sustainable use of local biodiversity, governments of African countries need to adopt these principles and suggestions for biodiversity conservation. There is also a need to have an institutional mechanism (government coordination structures, stakeholder engagement, inputs, etc.). This will ensure that they enact policies that will protect the natural ecosystem. Increasing awareness about the importance of biodiversity to human existence is another useful strategy. Many are ignorant of its importance and functions; however, enlightenment can help in ensuring the sustainable use of local natural resources (Anabaraonye et al. 2019). Akindele et al. (2021) reported the low level of awareness of Nigerians about biodiversity conservation and advocated for an increase in the level of awareness among the populace as this will ease the task of government agencies and other stakeholders concerning biodiversity conservation.

Progress has been made in many countries and across many continents with regard to sustainable use of biodiversity. Continents such as North America, South America, and Asia are very good examples. China and Brazil are good examples of countries in those continents that have increased their rate of biodiversity conservation in recent years. China, for instance, increased its protected areas from 1.48 million square kilometers (km^2) in 2008 to 1.73 million km^2 in 2018. It is also recording progress in 16 out of the 20 Aichi Biodiversity Targets (Li and Pimm 2020). Brazil has also remained the world's most diverse country in the last decade. Brazil possesses the most freshwater species, has the most mammals, and possesses the most plant species (CBD 2021). However, there is a need for more countries in Africa (such as Nigeria) to adopt biodiversity conservation strategies and ensure its sustainable use as its benefits in the long term outweigh the short-term benefits.

10.7 Policies and Registration Towards Sustainable Conservation of Biodiversity in Africa

For over 30 years, many sub-Saharan African nations have come up with national guidelines, policies, plans, legislations, and institutions with a focus on managing and conserving biodiversity. Nonetheless, there hasn't been much availability of instruments necessary to form a mechanism for producing, executing, and distributing biodiversity data required to make informed decisions. On the other hand, corruption and unending conflicts have continued to stand as hindrances to prosperity. In addition, poor governance and lack of commitment to social inclusion, equity, and diversity have led to marginalization of vulnerable groups like the poor in the society (Yap 2007).

Certain legal frameworks, government regulations, and policies can ultimately be used as a tool to encourage and control sustainable exploitation of the components of the ecosystem. This can include providing enabling rights to community members to protect the environment and manage the resources of the forest as well. Pienaar et al. (2013) suggested that communities should be allowed to play significant roles in influencing the sustainable use of land resources. For example, in Kenya, forty-seven Kaya forests of about 6000 hectares are being managed through community participation (Githitho 2003). The 'Trees for Global benefit' project in Uganda is another example of a success story of community participation (Kalunda et al. 2020). However, in some cases, even the best of laws may not necessarily be effective as a result of deficiency in the will and capacity of those required to enforce and implement these laws. Additionally, problems associated with accountability, transparency, corruption, and involvement in the policy developmental phase can greatly impact the potency of these polices and laws. It is very important to also take into consideration regulations, policies, and laws that regulate the market services, organizational plans as well as describe rights to property and tenure. Conservation policies can be sustained where there is deliberate participation among different social groups and regions in the community, particularly taking into account the natives and vulnerable groups in the area.

10.7.1 Sustainable Conservation Principles for Enacting Policies for Biological Diversity

Policies and laws governing biodiversity provide the outline required of an institution to attain its sustainable goals for conservation. Policies are enacted to improve the values fundamental to biodiversity conservation and practice which promotes strategic, focused, and planned utilization of resources. It further identifies that the progress, well-being, and living conditions of individuals are completely reliant on the healthy state of biodiversity in ecosystems. A stable development is much achievable if the ecosystem and the services it provides are securely guarded and properly maintained using pragmatic actions needed to curb the drivers and threats to biodiversity conservation.

Many African nations participate in national conventions and agree with the protocols involving the protection and sustainable conservation of biodiversity. Today, much attention is given to issues relating to climate change; however, efforts aimed at conserving biodiversity are not prioritized at all. This complacency has resulted in failure of the community in their responsibilities to highlight key issues that policy makers require in order to come up with guidelines needed to prevent biodiversity loss (UNEP 2010). There should be a strong knowledge base highlighting the essentiality of biodiversity to the achievement of the SDGs (CBD 2018). In order to halt loss of biodiversity in the coming years, it is important to identify gaps that exist within the policy and protocols needed to drive sustainable conservation. In addition, policymakers would need to put together suitable policies that will address the fast decrease in biodiversity abundance. Efforts should also be intensified to mainstream biodiversity into all sectors; agriculture, economic, financial (the inclusion of biodiversity into national budgets), educational, environmental, etc. (CBD 2018; FAO 2020b).

10.7.2 Steps Towards Formulating Sustainable Conservation Policy

1. The first step would involve setting up meetings to address the importance of civil society participation in developing the biological diversity policy. The stakeholders would involve civil society organizations, committee representative from the ministry of environment as well as the institutions responsible for biodiversity tourism and environmental affairs (DEAT 1997).
2. Management of the policy developmental phase by reflecting on the existing constitutional ethics. The committee assigned to this task would prepare the required policy documents.
3. A support group that comprises statutory boards, NGOs, and government institutions will be tasked with the responsibility of overseeing the management committee during the policy implementation phase and confirm that the policy contents reflect the identified interest and concerns of the communities.
4. The documents created from the policy implementation phase will be sent out to interested persons as well as institutions willing to share ideas that will help to sustain and conserve biodiversity.
5. It is important to allow a wider involvement and encourage participation of the general public so that important briefings by stakeholders and pertinent items raised can be properly addressed.
6. Issues identified in the forum discussion would be discussed and available policy decisions will be explored through consultative forums in order to achieve the set goals.

7. Important submissions will be taken into consideration and recommendations will be made and if there is a need for an adjustment, the support group will verify and make recommendations to the ministry in charge of tourism and environmental affairs.
8. Following the successful drafting of the policy, a detailed action plan and strategy will be established, as well as the total cost spent on the entire process plan.

10.7.3 Policy Towards Achieving Sustainable Conservation of Biodiversity

10.7.3.1 Creation of Effective Conservation Plan

Programs relating to conservation and sustainability should be strategic and properly planned to assess the potential threats to biodiversity in the ecosystem. In order to create an effective conservation strategy, programs can be developed to measure progress performance as well as generate time (USAID biodiversity policy framework, 2015).

10.7.3.2 Reduction in Emissions of Greenhouse Gases

A healthy ecological system can mitigate climate change impacts on individuals. It is therefore important to minimize gas emissions by coming up with activities to enhance good land use practices that would invariably create a stable ecosystem.

10.7.3.3 Raising Awareness on Conflict-Sensitive Programs

Effective biodiversity plans should be designed and implemented in order to raise awareness of conflict-sensitive programs in vulnerable regions, thereby reducing complaints and establishing social and organizational resistance.

10.7.3.4 Rights of Involvement in Resource Management

Sustainable conversation strategies aimed at addressing social challenges such as gender issues should be promoted to increase impact. There should be rights to accessing resources in the ecosystem, but at a controlled rate. Individuals should be given priority rights of involvement to manage resources effectively.

10.8 Conclusion

African countries need to take more proactive measures to halt biodiversity loss; it is not just sufficient to set up agendas, attend conventions, and agree with protocols, appropriate steps should be taken towards implementation so as to achieve desired results. Insights can be taken from countries who have recorded considerable success with biodiversity conservation. Policy gaps should be identified and addressed, with the involvement of the right and concerned stakeholders. It is also important to strengthen the capacity of personnel directly involved with conservation through training, so they can effectively use established systems to manage protected areas while adhering to good conservation practices as well. Degraded ecosystems should be identified and restored, if possible, to further strengthen the recovery strategies needed to develop new species and promote environmentally safe regions that are in close proximity to the protected ecosystem.

References

Acreman M, Smith A, Charters L, Tickner D, Opperman J, Acreman SF, Edwards PS, Chivava F (2021) Evidence for the effectiveness of nature-based solutions to water issues in Africa. Environ Res Lett 16(6):063007

Adeyemo OK (2003) Consequences of pollution and degradation of Nigerian aquatic environment on fisheries resources. Environmentalist 23:297–306

AfDB-African Development Bank (2014) Tracking Africa's progress in figures

AfDB-African Development Bank (2015) African Natural Resources Center (ANRC) Strategy (2015–2020). Revised Version. ANRC Department

Akindele EO, Ekwemuka MC, Apeverga P et al (2021) Assessing awareness on biodiversity conservation among Nigerians: the Aichi Biodiversity Target. Biodivers Conserv. https://doi.org/10.1007/s10531-021-02175-x

Alene AD, Abdoulaye T, Rusike J, Manyong V, Walker TS (2015) The effectiveness of crop improvement programmes from the perspectives of varietal output and adoption: cassava, cowpea, soybean and yam in sub-Saharan Africa and maize in west and central Africa. Crop improvement, adoption and impact of improved varieties in food crops in sub-Saharan Africa

Alho CJR (2008) The value of biodiversity. Braz J Biol 68(4 Suppl):1115–1118

Anabaraonye B, Chukwuma J, Eriobu C (2019) Green entrepreneurial opportunities in climate change adaptation and mitigation for sustainable development in Nigeria. J Environ Pollut Manag 2(2):102

Baumgartner RJ (2019) Sustainable development goals and the forest sector—a complex relationship. Forests 10(2):152

Bax N, Sands CJ, Gogarty B, Downey RV, Moreau CV, Moreno B, Barnes DK (2021) Perspective: increasing blue carbon around Antarctica is an ecosystem service of considerable societal and economic value worth protecting. Global Change Biol 27(1):5–12

Berrahmouni N, Hoffmann I, DelaRosa R, Kuhne I, Gbehounou G, Min K (2020) Africa's biodiversity mainstreaming across agricultural sectors. In FAO (2020b). Biodiversity: its central role in the sustainable development of Africa. Nature Faune 33(1) Accra

Bosworth A, Chaipraditkul N, Cheng MM et al (2011) Ethics and biodiversity. Ethics and climate change in Asia and the Pacific (ECCAP) project, working group 16 report.

Bull AT, Ward AC, Goodfellow M (2000) Microbiology. Mol Biol Rev 64(3):573

Butchart SHM (2010) Global biodiversity: indicators of recent declines. Science 328(5982): 1164–1168

Cadman M, Petersen C, Driver A, Sekhran N, Maze K, Munzhedzi S (2010) Biodiversity for Development: South Africa's landscape approach to conserving biodiversity and promoting ecosystem resilience. South African National Biodiversity Institute, Pretoria

Callicott JB (2006) Explicit and implicit values. In: Scott JM, Goble DD, Davis FW (eds) The Endangered Species Act at thirty: conserving biodiversity in human-dominated landscapes, vol 2. Island Press, Washington, D.C, pp 36–48

Cardinale BJ, Duffy JE, Gonzalez A, Hooper DU, Perrings C, Venail P, Narwani A, Mace GM, Tilman D, Wardle DA, Kinzig AP (2012) Biodiversity loss and its impact on humanity. Nature 486(7401):59–67

Carlson T, Foula BM, Chinnock JA, King SR, Abdourahmaue G, Sannoussy BM, Bah A, Cisse SA, Camara M, Richter RK (2002) Case study on medicinal plant research in guinea: prior informed consent, focused benefit sharing, and compliance with the convention on biological diversity. Econ Bot 55(4):478–491

CBD (2010) Global Biodiversity Outlook 3. Secretariat of the Convention on Biological Diversity, Montréal, Canada

CBD (2018) Biodiversity at the heart of sustainable development. Input to the 2018 High-level Political Forum on Sustainable Development (HLPF). Secretariat of the Convention on Biological Diversity (CBD). https://sustainabledevelopment.un.org/content/documents/18277CBD_input_to_2018_HLPF.pdf. Accessed 1 March 2022

CBD (2020) Secretariat of the Convention on Biological Diversity (2020) Global Biodiversity Outlook 5. Montreal, 211 pp

CBD (2021) Brazil-Biodiversity Facts. https://www.cbd.int/countries/profile/?country=br. Accessed 9 March 2022

Chase JM, Blowes SA, Knight TM, Gerstner K, May F (2020) Ecosystem decay exacerbates biodiversity loss with habitat loss. Nature 584(7820):238–243

Chausson A, Turner B, Seddon D, Chabaneix N, Girardin CA, Kapos V, Seddon N (2020) Mapping the effectiveness of nature-based solutions for climate change adaptation. Global Change Biol 26(11):6134–6155

Ciftcioglu GC (2021) The social valuation of agro-ecosystem services at different scales: a case study from Kyrenia (Girne) Region of Northern Cyprus. Environ Dev 39:100645

Creamer RE, Barel JM, Bongiorno G, Zwetsloot MJ (2022) The life of soils: integrating the who and how of multifunctionality. Soil Biol Biochem 166:108561

Crowther TW, Glick HB, Covey KR, Bettigole C, Maynard DS, Thomas SM, Bradford MA (2015) Mapping tree density at a global scale. Nature 525(7568):201–205. https://doi.org/10.1038/nature14967

Dalal-Clayton B, Bass S (eds) (2002) Sustainable development strategies: a resource book. Organization for Economic Co-operation and Development (OECD), Paris and United Nations Development Programme (UNDP), New York. 358 pp

Department of Environmental Affairs and Tourism, DEAT (1997) White paper on the conservation and sustainable use of South Africa's Biological Diversity. Government Gazette, 28 July 1997

Diagne A, Kinkingninhoun-Medagbe FM, Amovin-Assagba E, Nakelse T, Sanni K, Toure A (2015) Evaluating the key aspects of the performance of genetic improvement in priority food crops and countries in sub-Saharan Africa: the case of rice. In Walker TS, Alwang J (eds) Crop improvement, adoption and impact of improved varieties in food crops in sub-Saharan Africa

Dudley N, Alexander S (2017) Agriculture and biodiversity: a review. Biodiversity 18(2–3):45–49. https://doi.org/10.1080/14888386.2017.1351892

Dulvy NK, Pacoureau N, Rigby CL, Pollom RA, Jabado RW, Ebert DA, Simpfendorfer CA (2021) Overfishing drives over one-third of all sharks and rays toward a global extinction crisis. Curr Biol 31(21):4773–4787.e8. https://doi.org/10.1016/j.cub.2021.08.062

Edwards PJ, Abivardi C (1998) The value of biodiversity: where ecology and economy blend. Biol Conserv 83(3):239–246

Ehrlich PR, Ehrlich AH (1997) The value of biodiversity. In: Dasgupta P, Maler K-G, Vercelli A (eds) The economics of transnational commons, 342 pp

El Mujtar V, Muñoz N, Mc Cormick BP, Pulleman M, Tittonell P (2019) Role and management of soil biodiversity for food security and nutrition; where do we stand? Global Food Security 20: 132–144

Eschen R, Beale T, Bonnin JM, Constantine KL, Duah S, Finch EA, Taylor B (2021) Towards estimating the economic cost of invasive alien species to African crop and livestock production. CABI Agric Biosci 2(1):18. https://doi.org/10.1186/s43170-021-00038-7

Everard M, Johnston P, Santillo D, Staddon C (2020) The role of ecosystems in mitigation and management of Covid-19 and other zoonoses. Environ Sci Policy 111:7–17

Fa JE, Peres CA, Meeuwig J (2002) Bushmeat exploitation in tropical forests: an intercontinental comparison. Conserv Biol 16:232

FAO (2016) The state of world fisheries and aquaculture 2016. Contributing to food security and nutrition for all. FAO, Rome, Italy

FAO (2020a) The state of world fisheries and aquaculture 2020. Sustainability in action. FAO, Rome

FAO (2020b) Biodiversity: its central role in the sustainable development of Africa. Nature Faune 33(1) Accra

FAO (2020c) Global forest resources assessment 2020—Key findings. Rome. doi: https://doi.org/10.4060/ca8753en.

FAO (Food and Agriculture Organization of the U.N) (2018) "The state of the world's forests". http://www.fao.org/documents/card/en/c/I9535EN/

FAO, IFAD, UNICEF, WFP, WHO (2019) The state of food security and nutrition in the world 2019. Safeguarding against economic slowdowns and downturns. FAO, Rome

Ferretti F, Worm B, Britten GL, Heithaus MR, Lotze HK (2010) Patterns and ecosystem consequences of shark declines in the ocean: ecosystem consequences of shark declines. Ecol Lett. https://doi.org/10.1111/j.1461-0248.2010.01489.x

Folke C, Jansson Å, Rockström J, Olsson P, Carpenter S, Chapin F, Crépin AS, Daily G, Danell K, Ebbesson J, Elmqvist T, Galaz V, Moberg F, Nilsson M, Österblom H, Ostrom E, Persson Å, Peterson G, Polasky S, Steffen W, Walker B, Westley F (2011) Reconnecting to the biosphere. AMBIO 0044-7447. https://doi.org/10.1007/s13280-011-0184-y

Ghilarov AM (2000) Ecosystem functioning and intrinsic value of biodiversity. Oikos 90(2): 408–412. https://doi.org/10.1034/j.1600-0706.2000.900222.x

Githitho A (2003) The sacred Mijikenda Kaya forests of coastal Kenya and biodiversity conservation. In: Lee C, Schaaf T (eds) The importance of sacred natural sites for biodiversity conservation. Proceedings of the International Workshop held in Kumming and Xishuangbanna Biosphere Reserve, People's Republic of China, 2003. United Nation Educational, Scientific and Cultural Organization, Paris, pp 27–35

Gonthier DJ, Ennis KK, Farinas S, Hsieh HY, Iverson AL, Batáry P, Perfecto I (2014) Biodiversity conservation in agriculture requires a multi-scale approach. Proc Roy Soc B Biol Sci 281(1791): 20141358. https://doi.org/10.1098/rspb.2014.1358

Govindaraj M, Vetriventhan M, Srinivasan M (2015) Importance of genetic diversity assessment in crop plants and its recent advances: an overview of its analytical perspectives. Genetics Res Int:431487, 14 pages. https://doi.org/10.1155/2015/431487

Grass I, Loos J, Baensch S, Batáry P, Librán-Embid F, Ficiciyan A, Tscharntke T (2019) Land-sharing/-sparing connectivity landscapes for ecosystem services and biodiversity conservation. People Nature 1(2):262–272

Hale BW, Steen-Adams MM, Predick K, Fisher N (2005) Ecological conservation through aesthetic landscape planning: a case study of the Lower Wisconsin State Riverway. Environ Manage 35: 381–395. https://doi.org/10.1007/s00267-003-3061-z

IPBES (2018) Summary for policymakers of the regional assessment report on biodiversity and ecosystem services for Africa of the Intergovernmental Science-Policy Platform on Biodiversity and Ecosystem Services.

IUCN (2021) IUCN Red List Summary Statistics. https://www.iucnredlist.org/resources/summary-statistics. Assessed 01 March 2022
Iwu MM (2014) Handbook of African medicinal plants, 2nd edn. Taylor and Francis Group, LLC, CRC Press. 446 pages
Jenkins M (2003) Prospect of diversity. Science 302:1175–1177
Kalunda PN, Butsatsa J, Kiguli L, Kirabo A (2020) Trees for global benefits: eradicating poverty one tonne (of carbon) at a time. In FAO. Biodiversity: its central role in the sustainable development of Africa. Nature Faune 33(1) Accra
Karlsson O, Rocklöv J, Lehoux AP, Bergquist J, Rutgersson A, Blunt MJ, Birnbaum LS (2021) The human exposome and health in the Anthropocene. Int J Epidemiol 50(2):378–389
Kiester AR (1996) Aesthetics of biological diversity. Human Ecol Rev 3(2):151–157. http://www.jstor.org/stable/24707207
Kingston DGI (2011) Natural products drug discovery: Challenges and strategies in the era of the convention on biological diversity. Abstracts of Papers of the American Chemical Society; (241)
Klein AM, Vaissiere BE, Cane JH, Steffan-Dewenter I, Cunningham SA et al (2007) Importance of pollinators in changing landscapes for world crops. Proc Roy Soc B Biol Sci 274 (1608):303–313
Lampert A (2019) Over-exploitation of natural resources is followed by inevitable declines in economic growth and discount rate. Nature Commun 10(1):1419. https://doi.org/10.1038/s41467-019-09246-2
Li BV, Pimm SL (2020) How China expanded its protected areas to conserve biodiversity. Curr Biol 30(22):R1334–R1340
Linders TEW, Schaffner U, Eschen R, Abebe A, Choge SK, Nigatu L, Allan E (2019) Direct and indirect effects of invasive species: biodiversity loss is a major mechanism by which an invasive tree affects ecosystem functioning. J Ecol 107(6):2660–2672
Lituma CM, Buckley BR, Keyser PD, Holcomb E, Smith R, Morgan J, Applegate RD (2022) Effects of patch-burn grazing and rotational grazing on grassland bird abundance, species richness, and diversity in native grassland pastures of the Midsouth USA. Agric Ecosyst Environ 324:107710
Lock JM (2001) Encyclopedia of biodiversity, vol 1. Academic Press, London, UK, p 56
Lockie S, Ransan-Cooper H (2015) Biodiversity and Sustainable Development. In: Redclift M, Springett D (eds) The Routledge international handbook of sustainable development. Routledge, London
Loreau M, Oteng-Yeboah A, Arroyo MTK, Babin D (2006) Diversity without representation. Nature 442:245–246
Mahomoodally FM (2013) Traditional medicines in Africa: an appraisal of ten potent African medicinal plants. Evid Based Complement Altern Med:617459, 14 pages. https://doi.org/10.1155/2013/617459
Malcolm LH Jr, Gibbs JP (2007) Fundamentals of Conservation Biology. Third Edition by Blackwell Publishing Ltd, 515 pages
Mansourian S, Berrahmouni N (2021) Review of forest and landscape restoration in Africa. FAO and AUDA-NEPAD, Accra. https://doi.org/10.4060/cb6111en
McCauley DJ (2015) Marine defaunation: animal loss in the global ocean. Science 347(6219)
MEA (Millennium Ecosystem Assessment) (2005) Ecosystems and human wellbeing: current state and trends: findings of the condition and trends working group. Island Press, Washington, DC
Mills J (2020) The global bystander effect: moral responsibility in our age of ecological crisis. J Futures Stud 25(2):61–76
Morton SR, Hill R (2014) What is biodiversity, and why is it important? In: Morton SR, Sheppard AW, Lonsdale WM (eds) Biodiversity: science and solutions for Australia. CSIRO Publishing, Collingwood, Melbourne, pp 1–12
Muthoni RA, Andrade R (2015) The performance of bean improvement programmes in sub-Saharan Africa from the perpectives of varietal output and adoption. In Walker TS, Alwang

J (eds) Crop improvement, adoption and impact of improved varieties in food crops in sub-Saharan Africa

Neuwinger HD (2000) African traditional medicine: a dictionary of plant use and applications. Medpharm Scientific Publishers, Stuttgart, Germany, p 589

OECD (2018) Mainstreaming biodiversity for sustainable development. OECD Publishing, Paris. https://doi.org/10.1787/9789264303201-en

Olalekan RM, Omidiji AO, Williams EA, Christianah MB, Modupe O (2019) The roles of all tiers of government and development partners in environmental conservation of natural resource: a case study in Nigeria. MOJ Ecol Environ Sci 4(3):114–121. https://doi.org/10.15406/mojes.2019.04.00142

Pauly D, Christensen V, Guénette S, Pitcher TJ, Sumaila UR, Walters CJ, Watson R, Zeller D (2002) Towards sustainability in world fisheries. Nature 418(6898):689–695. https://doi.org/10.1038/nature01017

Pienaar EF, Jarvis LS, Larson DM (2013) Creating direct incentives for wildlife conservation in community-based natural resource management programmes in Botswana. J Dev Stud 49(3): 315

Ranjalkar K, Shelke K (2022) Environment conservation, challenges threats in conservation of biodiversity. Scieng Publications, p 69

Sandler R (2012) Intrinsic value, ecology, and conservation. Nature Educ Knowledge 3(10):4

Schultz M, Tyrrell TD, Ebenhard T (2016) The 2030 agenda and ecosystems—a discussion paper on the links between the aichi biodiversity targets and the sustainable development goals. SwedBio at Stockholm Resilience Centre, Stockholm, Sweden

Sen M (2020) Forests: at the heart of a green recovery from the Covid-19 pandemic. UN DESA, New York

Shrestha N, Xu X, Meng J, Wang Z (2021) Vulnerabilities of protected lands in the face of climate and human footprint changes. Nat Commun 12(1):1–9

Singh V, Shukla S, Singh A (2021) The principal factors responsible for biodiversity loss. Open J Plant Sci *6*(1):011–014

Soulé ME (1985) What is conservation biology? Bioscience 35:727–734

Stephenson PJ, Nadine B, Eugenie R, Damon S, Mallé D, Robert H, Harouna A, Tanya A, Christine A, Hilary A, Olaf B, Barthelemy B, Samuel D, Arthur D, Russell G, Cicilia W, Abdoulaye BG, David LN, Hafashimana TH, Donald H, John K, Ronald K, Martha M, Kalemba IL, Budu M, Zawadi M, Monipher M, Eric O, Aggrey R, Ahmat B, Siam AT (2017) Unblocking the flow of biodiversity data for decision-making in Africa. Biol Conserv 213:335–340

Stuart LP, Gareth JR, John LG, Thomas MB (1996) The future of biodiversity. Science 269:347–350. https://doi.org/10.1016/0006-3207(96)85981-7

Thiombiano A, Schmidt M, Dressler S (2012) Catalogue Des Plantes Vasculaires du Burkina Faso. Chambesy (Switzerland). Conservatoire et Jardin Botaniques de Laville de Geneva. Boissiera 65:371

Toepfer G (2019) On the impossibility and dispensability of defining "biodiversity". In From assessing to conserving biodiversity, pp 341–351. Springer, Cham

Tribot A-S, Deter J, Mouquet N (2018) Integrating the aesthetic value of landscapes and biological diversity. Proc Roy Soc B 285:20180971. https://doi.org/10.1098/rspb.2018.0971

UN (1992) Article 2 of the convention on biological diversity. https://www.cbd.int/doc/legal/cbd-en.pdf. Assessed 15 March 2022

UN (2021) The global forest goals report 2021. United Nations Department of Economic and Social Affairs, United Nations Forum on Forests Secretariat

UNEP (2007) Global environment outlook 4: environment for development. United Nations Environment Programme. Progress Press, Valletta

UNEP (2010) State of biodiversity in Africa. http://www.unep.org/delc/Portals/119/State%20of%20biodiversity%20in%20Africa.pdf. Accessed 1 March 2022

UNEP-WCMC (2016) The state of biodiversity in Africa: a mid-term review of progress towards the Aichi Biodiversity Targets. UNEP-WCMC, Cambridge, UK

United Nations, The Millennium Development Goals Report (2013). https://www.un.org/millenniumgoals/pdf/report-2013/mdg-report-2013-english.pdf. Accessed 29 December 2022

USAID Biodiversity Policy Framework (2015) USAID guiding principles. Biodiversity Handbook, Washington, DC

Walter VR, Harold AM, Stephen RC, Kartik C (2005) Millennium ecosystem assessment. Ecosystems and human well-being: synthesis. Island Press, Washington, DC

White PS (2013) Derivation of the extrinsic values of biological diversity from its intrinsic value and of both from the first principles of evolution. Conserv Biol 27(6):1279–1285. https://doi.org/10.1111/cobi.12125

WHO (2003) WHO guidelines on good agricultural and collection practices (GACP)for medicinal plants. Switzerland, Geneva

WHO/UNICEF (2021) Progress on household drinking water, sanitation and hygiene 2000-2020: five years into the SDGs. World, Health Organization (WHO) and the United Nations Children's Fund (UNICEF), Geneva

Woodward C, Shulmeister J, Larsen J, Jacobsen GE, Zawadzki A (2014) The hydrological legacy of deforestation on global wetlands. Science 346(6211):844–847. https://doi.org/10.1126/science.1260510

World Commission on Environment and Development (1987) Our common future. Oxford University Press

WWF (2020) Living planet report 2020—bending the curve of biodiversity loss. In Almond REA, Grooten M, Petersen T (eds). WWF, Gland, Switzerland

Yap KS (2007) Access to basic services for the poor: the importance of good governance. United Nations Economic and Social Commission for Asia and the Pacific [ESCAP], Bangkok, Thailand

Chapter 11
Medicinal Potentials of Aloe Vera (*Aloe barbadensis* Miller): Technologies for the Production of Therapeutics

Odangowei Inetiminebi Ogidi and Uchechi Emmanuella Enenebeaku

Abstract An essential category of the biological resources of Africa are plants with nutritional and therapeutic properties and they are common all over Africa. Folk remedies are relied on by almost 80% of people in underdeveloped and developing nations because of their indigenous knowledge, availability, and cost-effectiveness. *Aloe barbadensis* Miller, generally known as Aloe vera, is one of more than 400 species of *Aloe* belonging to the Liliaceae family that originated in Africa, but is considered native to arid subtropical and tropical regions of the world. More than 200 distinct biologically active compounds have been found in the plant and most of them possess one biological activity or the other. Also, more than a hundred different nutrients and beneficial chemicals make up the leaf gel's interior composition. This chapter discusses the occurrence and botanical description, processing, chemistry, historical use, options for domestication, technologies for the production of high-value therapeutics potentials, medicinal uses, and safety, toxicity, and conservation strategies of the *Aloe barbadensis*.

Keywords *Aloe barbadensis* · Therapeutic potential · Safety · Toxicity · Medicinal uses · Chemistry · Cultivation · Conservation strategies

11.1 Introduction

There has been a continuous effort to find new compounds which could be used in the fight against pathogens. About 20% of plants found in tropical and subtropical environments have been used pharmacologically and a significant number of new drugs are obtained from such natural sources (Mothana and Lindequist 2005).

O. I. Ogidi (✉)
Department of Biochemistry, School of Applied Sciences, Federal Polytechnic Ekowe, Ekowe, Bayelsa State, Nigeria

U. E. Enenebeaku
Department of Biological Sciences, Hezekiah University, Umudi, Imo State, Nigeria

S. C. Izah, M. C. Ogwu (eds.), *Sustainable Utilization and Conservation of Africa's Biological Resources and Environment*, Sustainable Development and Biodiversity 32, https://doi.org/10.1007/978-981-19-6974-4_11

As a member of the Alliaceae family, Aloe vera is one of many plant species within the family that is found around the world. It's an 80–100 cm tall succulent herb that takes four to six years to develop and may live for over five decades in the right circumstances. In support of the World Health Organization's belief that medicinal plants are the finest sources of pharmaceuticals, Aloe vera (*Aloe barbadensis* Miller) has a wide range of physiologically active chemicals like other native plants (West and Zhu 2003; Yagi et al. 2003; Joshi 1997; Santos et al. 1995; Ogwu et al. 2016a, b). Southern, Eastern Africa, and Northern (especially the upper Nile of Sudan) are considered the plant's origins; it was then brought to northern Africa, where it has now become a naturalized species; even though the plant is produced commercially in many countries throughout the world (Yeh et al. 2003). The highest-quality Aloe may be found in the California desert (Aruba, Bonaire, Haiti, India, South Africa, the United States, and Venezuela). The plant can tolerate temperatures as high as 104 degrees Fahrenheit and even freeze. On the southern coast of India, it is most often seen in the wild.

There are several therapeutic uses and benefits of Aloe vera. Adding Aloe vera to food or drink has been demonstrated to lower blood glucose levels, which helps regulate diabetes. Aloe vera was ingested by diabetics in the form of tea or a yoghurt-aloe vera combo. Antiaging and anti-wrinkle lotions and moisturizers also include it. In general, people like Aloe vera because it is nongreasy and nonsticky, and since it is quickly absorbed by the skin without leaving an unpleasant scent behind. It is used topically to soothe sunburns and other types of burns because it decreases inflammation and discomfort, allowing the burn to subside and the lesion to recover much more rapidly. Burns and wounds may be treated using the sap from the leaf or Aloe vera extract. Stomach ulcers may be treated with Aloe vera extract. The extract may be consumed straight from the bottle or mixed with a variety of foods. It aids in the healing of the ulcer lesion in the stomach by reducing inflammation. The juice of Aloe vera may be used to treat a variety of stomach illnesses, including irritable bowel syndrome, Crohn's disease, acid reflux, heartburn, and indigestion. Keeping stomach acids in check provides relief for the digestive system. Aloe vera contains both antibacterial and anti-inflammatory qualities that may help prevent gum and oral disease. Gel or toothpaste may be used to massage the gums. In addition, it treats skin conditions including eczema and ringworms. The full range of their ethnobotanical value is not yet known like other locally valued plants (Osawaru et al. 2016; Ogwu et al. 2018).

Curly and fuzzy hair may benefit from the usage of this plant as a hairstyle gel ingredient. There are a variety of other products that include it as well. Dry skin, particularly around the eyes and on the face, is well-served by Aloe vera gel. To treat a wide range of ailments, Aloe vera gel has been made as a treatment for anything from coughs and ulcers to diabetes and cancer to gastritis, headaches, and arthritis to immune system inadequacies. As a laxative, however, it is the most prevalent usage. Apply the bottom leaf's gel on the damaged region of your skin to speed up the recovery process. A drought-resistant tropical perennial plant, it is one of the crops that has been identified as a "novel plant resource with high promising possibilities in the globe," although it has yet to be fully exploited.

11.2 Occurrence, Botanical Identification, and Description of *Aloe barbadensis* Plant

Aloe was first planted in tropical Africa, but it has since spread throughout Asia, Europe, and North America due to the region's mild climates (Hernadez and Giacin 1998). In the 17th century, the *A. vera* species was introduced to China and other regions of southern Europe (Farooqi and Sreeramu 2001). Australia, Barbados, Belize, Nigeria, Paraquay, and the United States are just a few of the places where this plant has become established (Akinyele and Odiyi 2007). India, Mexico, the Pacific Rim, South America, Central America, the Caribbean, and other regions of the world are also home to the plant. Throughout the globe, it has been extensively farmed and can be found both in distant farms, and home gardens in some cases are used as an indoor ornamental plant (Osawaru et al. 2014; Ogwu et al. 2014a). Several species are currently grown for commercial reasons in India and portions of Pakistan, as well as elsewhere (Zakia et al. 2013). As a possible medicinal herb and as an esthetic plant, the species is popular with contemporary gardeners.

11.2.1 Botanical Identification and Description

***Family*: Liliaceae**

***Botanical Name*: *Aloe barbadensis* Miller**

Aloe vera is a thorny cactus-like xerophyte resembling xerophyte aloe. With a dense fibrous root system and huge basal leaves that may grow up to 16 per plant and weigh up to 1.5 kg when mature, this perennial plant forms clumps. Spreading by means of offshoots and root sprouts, the plant grows around 80–100 cm tall. Serrated margins on the green leaves give them a spongy texture. The aloe plant reaches maturity at the age of 4 years and may live for up to 12 years after that. There are saw-like teeth along the edges of the leaves, which are up to 0.5 m long and 8–10 cm wide at the base. As seen in a transverse slice, the adaxial surface is somewhat concave, whereas the lower abaxial surface is convex (Grindlay and Reynolds 1986). There is a thick cuticle covering the leaves' epidermis and mesophyll. As trosette matures, subsequent leaves have less white spots and are grey-greenish in color. Later, it is distinguished in higher chlorenchyma and lower parenchyma (Eshun and He 2004). Every 6–8 weeks, the plant may be harvested by plucking 3–4 leaves from each stem. Most of the year, a long raceme of red, yellow, purple, or pale-stripe flowers grows at the top of the flower stalk, which emerges from the center of the basal leaves. The flower stem may reach a height of 1.5 m. Numerous seeds are found inside a triangular-shaped fruit. Even though the plant is generally free of illness, fungal infections or soft rotting may sometimes generate black patches on the top surface. It is believed that a bacteria is responsible for soft rotting. A third adversary of the Aloe vera plant is frost, which it cannot tolerate (Grindlay and Reynolds 1986). Farmers protect their plantations from frost by smoking in the field on frosty nights. Over 550 types of aloe are produced across the globe. Aloe vera and aloe aborescens Miller are the only two commercially available species. Two further species, the medicinally significant *A. perry-bakeri* and *A. ferox,* are also known to exist. Some Aloe vera plants contain a hemlock-like chemical that is exceedingly dangerous (Atherton 1998). A popular home plant, Aloe variegate, is a tiny species with a diameter of just a few centimeters. Aloe vera gel is made from the gel in the aloe leaf tissue.

11.2.2 Aloe vera Leaf Structural Composition

The exterior green rind of the aloe leaf, which contains the vascular bundles, and the interior colorless parenchyma, which contains the aloe gel, may be split into two primary components. Internal aloe leaf descriptions may be a little muddled owing to the several names that are used interchangeably, such as inner pulp, mucous, mucous tissue, mucilaginous jelly, inner jelly, and parenchyma tissue of the aloe leaf's inner core. This viscous transparent liquid found within the parenchyma cells is known as "gel" or "mucilage" in scientific jargon. Pulp or parenchyma tissue is the technical word for the undamaged fleshy interior section of the leaf that includes cell walls and organelles (Ni and Tizard 2004).

The pulp of Aloe vera has three structural components: cell walls, degenerated organelles, and viscous fluids. Inner leaf pulp has been revealed to include three different morphologies and sugar compositions (Ni et al. 2004a, b). About 98.5% of *Aloe vera's* raw pulp is water, while about 99.5% of the mucilage or gel is water (Eshun and He 2004). Vitamins, minerals, enzymes, polysaccharides, polyphenols, and organic acids make up the remaining 0.5–1 percent of the solid material in the final product (Boudreau and Beland 2006). Some believe that the Aloe vera pulp's complex chemical makeup may explain the wide range of pharmacological and therapeutic effects reported in aloe gel products (Talmadge et al. 2004).

11.2.3 Cultivation Practice of Aloe vera

This plant thrives in dry tropical regions, sandy loam soils, open places, and full sun with minimal watering with adequate drainage. Seeds and head cuttings are the most common methods of propagation for *Aloe vera.* Rows and columns of plants are sown one meter apart. The plants take around 4–5 years to mature from seed to harvest. Each leaf weighs between 1.5 and 2 kg when it is harvested (Nema et al. 2013).

11.3 History of *Aloe vera*

Because of its medical capabilities and health advantages, Aloe vera has been referred to be one of the oldest plants in the world. It was seen as a blessing by ancient doctors. There are various myths and legends surrounding Aloe vera, including "wonder plant" and "nature healer." Because "*Aloe*" means "bitter sparkling material" in Arabic and "*vera*" in Latin, it's no coincidence that Arabic and Hebrew are the root words of our English name. Aloe vera is commonly referred to as a "Desert Cacti" because of its cactus-like appearance. In spite of the fact that there are over 400 *aloe* species being cultivated across the globe, it is the *aloe barbadensis miller* (also known as "*Aloe vera*" or "True *Aloe*") plant that has shown to be the most beneficial to humankind due to its therapeutic characteristics.

In many countries, Aloe vera is a popular herbal supplement. It has been described in texts dating back to the Greek, Egyptian, and Roman ages, as well as many other cultures. The Indian and Chinese early civilizations have also been discovered to contain references to this; one of the most commonly used and sought after plants in history. There have been several references to Aloe vera in ancient texts, including the Bible. Aloes and myrrh were used to dress Christ's body when he was taken down from the cross, according to the Bible (John 19 : 39). With many references to its medical properties, Aloe vera may be found in every age of history (Indu 2017).

11.4 Processing of *Aloe vera*

In the food sector, Aloe vera gel is made from the plant's leaf pulp and has become a major business. In functional foods, it is used primarily for the creation of laxative-free health beverages. Milk, ice cream, confectionary, and other food items may also benefit from the use of stevia. Some foods use Aloe vera gel as a flavor or preservation (Christaki and Florou-Paneri 2010). Thus, the development of a simple and effective processing method for the Aloe vera beverage sector is needed to preserve the bioactive compounds naturally contained in the Aloe vera leaf to increase product quality and safety (Eshun and He 2004). This will contribute to the sustainable development of Africa through reducing rural emigration and urban migration as well as promoting income security (Ogwu 2019). Aloe gel recently yielded a glycoprotein known as alprogen, which has anti-allergic effects. Aloe gel has also yielded a new class of anti-inflammatory chemicals known as C-glycosyl chromones (Hutter and Salman 1996). Saponins, the gel's soapy ingredients, make up 3% of the gel and have antibacterial characteristics, making them ideal for general cleaning (Hirat and Suga 1983). Comperterol, -sitosterol, and lupeol are among the sterols (Coats 1979). An aspirin-like chemical, salicylic acid, has analgesic effects. Aloe vera gel contains about 20 of the human body's 22 amino acids, as well as seven of the eight essential amino acids. Antioxidant properties of Aloe vera juice were investigated, and the results revealed that antioxidants were abundant in aloe extracts.

An Aloe vera juice is made by pulverizing or grinding the whole leaf of the Aloe vera plant and then filtering and stabilizing the liquid to ensure that it is safe to consume (preserving the biological integrity of active ingredient to exert the reported physiological effect upon ingestion or topical application). A medicinal, cosmetic, or culinary product may then be made by incorporating or mixing the juice with additional preparations or substances. Among other uses in the food sector, Aloe vera has been used to make functional drinks like juice and tea. The availability of topical ointments, gel preparations, pills, and capsules shows just how much *aloe vera* is used in the pharmaceutical business (Eshun and He 2004). Mucopolysaccharides—the active element in many of these Aloe vera products—are in short supply as a result of poor processing techniques. Aloe vera plant leaves are known to have a broad range of biological activities, and their widespread usage necessitates that they be processed in a way that preserves as much of the plant's bioactive components as possible. In the extraction of a liver tonic, *aloe barbadensis* has found tremendous significance. The steps involved in the processing of Aloe vera are illustrated in the flow chart below (Fig. 11.1):

11.4.1 Reception of Raw Material

When the Aloe vera leaves are harvested, they must be delivered to the processing facility in refrigerated vans. To preserve the maximum potency of the active components, the leaves must be sound, undamaged, mold-free, and ripe (3–4 years) (Lawless and Allen 2000). As soon as a leaf is cut, the gel matrix begins to decompose owing to natural enzymatic processes and the activity of microorganisms naturally present on the leaf. This is a significant role in the final product's composition. Because of this, the quality of the final product may suffer. As a result, within six hours after harvesting, the leaves are either refrigerated or immediately fed to a processing machine on the farm. The green shell of the leaf is peeled off to obtain the parenchymatous tissue known as the gel fillet during filleting (Grindlay and Reynolds 1986). Aloe gel removed from the leaf was shown to be more stable than aloe gel left in the leaf, according to one study. Within 36 hours after harvesting the leaves, the filleting procedure must be finished to prevent the loss of biological activity (Ramachandra and Srinivasa 2008).

11.4.2 Filtration and Deareation

This phase removes the fibrous material. In this procedure, Aloe vera juice is affected. The sedimentation of aloe juice in storage is caused by a lack of filtering. Vitamin C and citric acid are added to the unpasteurized aloe juice to prevent

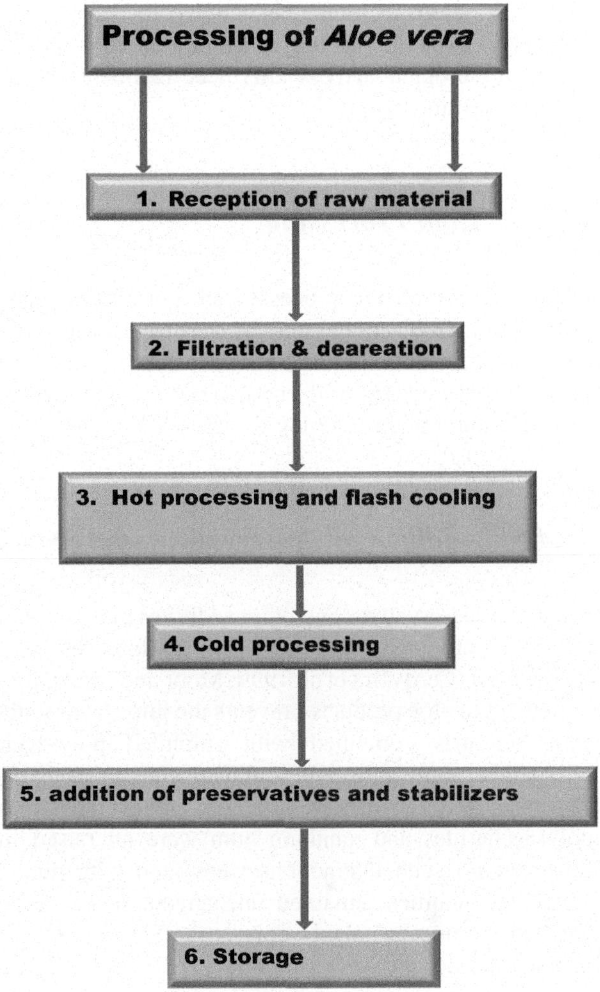

Fig. 11.1 Flow chart illustration of *Aloe vera* processing

browning reactions, enhance taste, and stabilize the juice (Kennedy et al. 1992;I Kacem et al. 1987; Tramell et al. 1986; Eison-Perchonok and Downes 1982). In order to preserve the taste of aloe vera juice, deareation is used to prevent ascorbic acid from oxidizing (Chan and Cavaletto 1986).

11.4.3 Hot Processing and Flash Cooling

By heating the aloe juice to a high temperature, activated carbon is used to sterilize it (Cerqveira et al. 1999). Aloe gel products' flavor, appearance, and biological activity may all be affected by this procedure. If Aloe vera gel is cooked to 65 °C for less than

15 min, it retains its biological activity. Increased levels of inactivity are caused by prolonged exposure to hot temperatures or higher temperatures in general. To maintain biological activity after heat treatment, the juice is rapidly cooled to 5 °C or below within 15 s.

11.4.4 Cold Processing

All of the processing processes are completed without the use of heat in cold processing. Aloe vera gel may be sterilized using enzymes such as glucose oxidase and catalase, which prevent the development of aerobic organisms inside the gel (Coats 1994). Other cold-process sterilizing procedures include ultraviolet light exposure and micron filtering (Maret 1975).

11.4.5 Addition of Preservatives and Stabilizers

Chemical preservatives and other additives may be used in any processing method to ensure the product's long-term viability. Some studies have claimed that synergism can be used (Cerqveira et al. 1999; Moor and McAnalley 1995). Adding a stabilizing ingredient to aloe products prevents the juice from sedimenting during storage. Aloe vera gel was combined with sulphited polysaccharides obtained from red microalgae, guar gum, and xantham gum in an experiment (Yaron et al. 1992). The apparent viscosities, yield points, and even hysteresis of aloe vera gel with algal polysaccharides and xantham gum were all raised in rheological investigations; however, no such interactions were seen with guar gum. During storage, these beneficial qualities remained unchanged. It was consequently hypothesized that xantham gum or algal polysaccharides might maintain fresh Aloe vera polysaccharide's network structure.

11.4.6 Storage

Like many other locally valued economic plants, there is little to no efficient packaging strategies (Evivie et al. 2020), which is even more challenging in the case of a succulemt plant like Aloe vera. The sensitive bioactive components in Aloe vera juice are protected from light exposure by being packaged in amber-colored glass bottles. Two of the most essential environmental factors that impact product quality are humidity and temperature. By altering these two variables, the packing material may absorb more of the juice's volatile ingredient, reducing both the shelf life of juice and the quality of the juice (Sadler and Braddock 1990).

11.5 Chemistry of Aloe vera Plant

Exudate from cells near to the vascular bundles and the core parenchyma tissue of *Aloe vera* leaves include a wide range of chemicals with different structures. It comprises 1,8 dihydroxyanthraquinone derivatives and their glycosides, which are largely employed for their cathartic actions in the bitter yellow exudate (Vazquez et al. 1996). Proteins, lipids, amino acids, vitamins, enzymes, inorganic chemicals, and tiny organic molecules have all been found in the aloe parenchyma tissue or pulp. A chemotaxonomic variation in the polysaccharides of aloes has been found and documented (Ni and Tizard 2004; Reynolds 2004). The chemical components of Aloe vera are explained in the following.

11.5.1 Polysaccharide Composition

The Aloe vera parenchyma is mostly composed of polysaccharides. For storage, the protoplast of parenchymal cells contains acetylated glucomannan, while the cell wall matrix contains many polysaccharides. The cell walls of the aloe leaf fillet include predominantly mannose-containing polysaccharides, cellulose, and pectic polysaccharides, whereas the epidermis of the leaf contains considerable amounts of xylose-containing polysaccharides (Femenia et al. 2003; Femenia et al. 1999).

11.5.2 Mannan

Hemicelluloses (mannans) are structural components of plants because they help to hold together the cellophane in the cellulose. As a non-starch carbohydrate reserve in seeds and vegetative tissues, they also act as a kind of storage. A signalling molecule for plant growth and development, it seems to do both of these things, as well. Linear mannans are homopolysaccharides with less than 5% galactose, consisting of linear chains of -(14D-mannopyranosyl) residua (Moreira and Filho 2008).

Acemannan was processed using the enzyme endo—D-Mannanase, and the C-4 and C-6 resonances of the fractions were examined using C13-NMR in a research examining the connections between monomers. A single-chain backbone of 14 mannose, with-(14) glucose embedded within the backbone and galactose branching from the backbone, was found in acemannan (Talmadge et al. 2004).

11.5.3 Maloyl Glucans

Three malic acid acylated carbohydrates were isolated from *A. vera* gel and characterized as 6-*O*-(1-L-maloyl)-α-,β-D-Glc*p* *i*(termed iveracylglucan A), α-d-Glcp-(1→4)-6-*O*-(1-L-maloyl)-α-,β-D-Glc*p* (termed veracylglucan B), and α-D-Glcp-(1→4)-tetra-[6-*O*-(1-L-maloyl)-α-D-Glc*p*-(1→4)]-6-*O*-(1-Lmaloyl)-*i*α-,β-D-Glc*p* (termed veracylglucan C). There were just a few hundred Da of veracylglucan A (C10H16O10) found in *Aloe vera* gel, and it was shown to be very unstable due to hydrolysis of the ester group [6-O-(1,2-Lmaloyl)-Gcp]. In terms of molecular weight, veracylglucan B (C16H26O15) is 458 Da, whereas veracylglucan C (C56H82O51) is 1570 Da (Esua and Rauwald 2006).

11.5.4 Pectins

The final major polysaccharides generally found in the plant cell walls, including in those of *Aloe*, are pectins. This ubiquitous component of all primary cell walls is also the most complex, both structurally and functionally (Caffall and Mohnen 2009; Willats et al. 2006). Pectic polysaccharides are highly hydrophilic, and with the water they attract, they can provide the cell wall with the required plasticity and flexibility. A highly plastic cell wall is especially important during cell growth where up to 65% of the wall is water. The tight connection between pectin and water allows the now pectic gel to push the cellulose microfibrils apart and thus facilitate the sideways slippage known from cell expansion.

In the cell wall, pectic polysaccharides embed the cellulose hemicellulose scaffold by cross-linking to each other via calcium bridges and hydrogen bonds. These weak bindings are easily remodeled or broken down to allow for new cell wall conformations (Albersheim 2011; Caffall and Mohnen 2009).

11.5.5 Arabinan and Arabinogalactan

Arabinogalactan contains mainly arabinose and galactose, but also other sugars including glucuronic acid and/or galacturonic acid. Certain arabinans and arabinogalactans sometimes form the neutral side chains of pectins. Arabinogalactan is present in a much lower concentration in aloe gel compared to acemannan (Ni et al. 2004a, b).

11.6 Bioactive Compounds of *Aloe vera*

11.6.1 Phytochemistry of Aloe vera

Aloe vera contains up to 200 distinct compounds (Davis 1997). About 98% of the Aloe vera gel is water content (Bozzi et al. 2007). Total solids are 0.66%, and the soluble solid percentage of Aloe vera gel fluctuates somewhat depending on the season. A dry matter basis, aloe gel is composed mostly of polysaccharides, carbohydrates, minerals, proteins, lipids, and phenolic substances (55%) (1%). Vitamins A, C, and E are found in Aloe vera gel, as well as many other nutrients. Additionally, Vitamin B1 (thiamine), niacin, Vitamin B2 (riboflavin), and choline and folic acid are found (Lawless and Allen 2000). Vitamin B12 (cyanocobalamin), which is normally found in animal sources but may be present in trace amounts in some studies, has also been proposed by some authors (Coats 1994; Coats 1979). Carbohydrates come from the plant's mucilage layer, which surrounds the parenchyma or gel inside the rind. Both monosaccharides and polysaccharides are found in them. Glucomannans [(1, 4)—connected acetylated mannan] are the most significant of the long-chain polysaccharides containing glucose and mannose. Rhamnose, galactose, and arabinose, as well as lupeol (a triterpenoid), cholesterol, campesterol, and -sitosterol, are also found in trace levels. At least four distinct partly acetylated glucomannans, linear polymers with no branching and 1,4 glycosidic connections with glucose and mannose in the ratio 1:2:8, have been identified in structural analyses of Aloe vera gel polysaccharides. Hydrolysis of these sugars diminishes the gel's viscosity. Oral carbohydrates attach to gut receptor sites and create a barrier, perhaps preventing 'leaky gut syndrome,' when consumed (Atherton 1997).

11.7 Medicinal and Therapeutic Use of Aloe vera

11.7.1 Anti-inflammatory Action

Several *in vitro* and *in vivo* researches have shown that Aloe vera gel has anti-inflammatory properties through bradykinase activity (Tyler 1994; Che et al. 1991). Bradykinin, an inflammatory molecule that causes pain, is broken down by the aloe peptidase bradykinase (Ito et al. 1993). Gel extracts yielded a new anti-inflammatory chemical called C-glucosyl chromone (Haller 1990). The cyclo-oxygenase pathway is inhibited by Aloe vera, and as a result, arachidonic acid's generation of prostaglandin E2 is reduced.

11.7.2 Anti-Ulcer Effect of Aloe vera

Borra et al. (2011) discovered an anti-ulcer activity of Aloe vera in NSAID-induced peptic ulceration in rats. Taking Aloe vera gel by mouth has been reported to be effective in the treatment of ulcerative colitis (Korkina et al. 2003). In addition, it reveals that dextran sulfate-induced ulcerative colitis in rats is protected and healed by this compound (He et al. 2013).

11.7.3 Effects on the Immune System

Chemical Alprogen from Aloe vera, an anti-allergy compound, suppresses the release of several mediators such as histamine, serotonin, SRSA, and leukotrienes from mast cells by preventing calcium influx into mast cells (Ro et al. 2000). An Aloe vera gel study in mice by Madan et al. (2008) found that it has immunomodulatory characteristics. Modified Aloe polysaccharides with the highest immunomodulatory effect were found to have the appropriate molecular size. Two new Aloe vera dihydrocoumarins, discovered by Zhang et al. (2006), showed antioxidative and immunomodulatory effects.

11.7.4 Laxative Effects

Latex contains powerful laxatives called anthraquinones. That is because they increase the volume and peristalsis of the intestine by raising water content and boosting mucus production (Ishii et al. 1994). Aloin A and B (formerly known as barbaloin) are the primary 1, 8-dihydroxyanthracene glycosides responsible for the Aloe's effects (Tyler 1994; Tyler et al. 1988). Oral aloins A and B, which are not digested in the small intestine, are hydrolyzed by intestinal bacteria in the colon, where they are converted to active metabolites (the main active metabolite is alo-emodin-9-anthrone) (Che et al. 1991; EEC 1988). This compound, like senna, stimulates and irritates the gastrointestinal tract in a similar way (Reynolds 1993). Laxative qualities of aloe latex are well-known.

11.7.5 Antimicrobial Action

Sulfur, phenols, salicyclic acid, cinnamonic acid, and urea nitrogen are all antimicrobial substances found in Aloe vera. They all have antifungal, antibacterial, and antiviral properties. Herpes and the AIDS virus cannot reproduce when Acemannan and azidothymidine (AZT) or acyclovir (Kahlon et al. 1991) are combined. In comparison to untreated animals, aloe extract therapy of Trichophyton mentagrophytes-infected guinea pig foot reduced growth by 70% (Ogidi et al. 2021a; Kigigha et al. 2018; Izah and Aseibai 2018; Kawai et al. 1998). Phyloplande microbial studies can help understand the diversity of microbes associated with plants and provide insight for their activities (Ogwu et al. 2013).

11.7.5.1 Antiviral Activity

The antiviral properties of Aloe extracts may be attributed to indirect or direct actions. The immune system is stimulated indirectly, and anthraquinones directly, to demonstrate these effects. The anthraquinone aloin inactivates different enveloped viruses such as Herpes simplex, Varicella zoster, and Influenza (Sydiskis et al. 1991).

Antiviral properties of Aloe vera gel's various components have been shown. Acemannan inhibited the growth of herpes simplex virus in two different cell lines (Kemp et al. 1990). Proliferation in cell cultures of cytomegalovirus was directly reduced by Aloe vera gel components known as lectins. All viruses, including varicella-zoster, influenza, and pseudorabies virus, could be inactivated by a pure sample of aloe emodin. It was also effective against the infectivity of herpes simplex type I and type II (Sydiskis et al. 1991).

11.7.5.2 Antibacterial Activities

Acemannan inhibited the adhesion of *Pseudomonas aeruginosa* to human lung epithelial cells in a monolayer culture after Aloe vera gel had been bactericidal (Azghani et al. 1995; Cera et al. 1980). Only 0.7% of the gel's solids are composed of carbohydrates, which make up 99% of the gel's water (El-Shemy et al. 2010). Accumulation of macrophages, stimulation of the immune system, as well as antibacterial and antiviral activities have been shown by glucomannan and acemannan (Ogidi et al. 2021b; Epidi et al. 2016; Kigigha et al. 2016; Pugh et al. 2001; Kaufman et al. 1989; Davis et al. 1987). It has been proven that Aloe vera's inner-leaf gel inhibits the development of *Streptococcus* and *Shigella species in vitro* (Ferro et al. 2003). Inhibition of Streptococcus pyogenes and Streptococcus faecalis bacteria by Aloe vera gel has been shown (Robson et al. 1982; Heggers et al. 1979). Aloe vera gel's antibacterial properties in vivo may help wound healing by removing germs that contribute to inflammation, according to a study on rats (Heggers et al. 1995). *Mycobacterium fortuitum*, *M. smegmatis*, and *M. kansasii* were all killed by aloe extract, as well as *M. tuberculosis*, *E. coli*, *S. aureus*, and *S. typhi*. The aloe extract also had antibacterial action against *P. aeruginosa*, *E. coli*, and *S. typhi*.

11.7.5.3 Antifungal Activities

Mycelia of *Rhizoctonia solani*, *Fusarium oxysporum*, and *Colletotrichum coccodes* have been tested for their ability to produce mycelium in the presence of Aloe vera, and the pulp and liquid fractions of the plant were found to inhibit F. oxysporum growth at 104 1 L-1 and to slow colony growth at 105 1 L-1 in all three species (de Rodríguez et al. 2005; Cheesbrough 1984). Candida albicans' growth was suppressed by an Aloe vera gel formulation that had been treated (Ogidi et al. 2021c; Heggers et al. 1979). Trichophyton mentagrophytes (20.0 mm) were shown to be suppressed by Aloe gel, whereas Pseudomonas aeruginosa and Candida albicans were found to be inhibited by Aloe leaf, according to Agarry et al. (2005).

11.7.6 Antitumor Activity

Researchers found that polysaccharides in Aloe reduced benzopyrene's affinity for primary rat liver cells, limiting the development of cancer-initiating benzopyrene-DNA adducts in the process. Glutathione S-transferase activation and suppression of the tumor-promoting effects of phorbol myristic acetate were observed in another investigation, indicating that Aloe gel may have a role in cancer chemoprevention (Kim et al. 1999). DMBA/croton oil-induced skin papillomagenesis in Swiss albino mice was prevented by Aloe vera, according to Saini et al. (2010). El-Shemy et al. (2010) found that Aloe vera leaf active principles obtained by supercritical carbon dioxide extraction had anticancer effects and modulated the activity of antioxidant enzymes.

11.7.7 Moisturizing and Antiaging Effect

As a result of its high concentration of muco-polysaccharides, aloe helps to retain moisture on the skin. Aloe increases the production of collagen and elastin fibers in the skin by stimulating fibroblasts. It also has a softening effect on the skin by holding together the surface peeling epidermal cells. Aloe vera gel contains amino acids that help soften tough skin. As an astringent and pore-tightening agent, zinc in this gel is present. In the treatment of dry skin caused by occupational exposure, aloe vera gel gloves enhanced skin integrity and reduced the appearance of small wrinkles and erythema, demonstrating that they had hydrating properties (West and Zhu 2003). Anti-acne properties are also provided by the gel. Aloe vera gel has a cooling and hydrating effect. It also has a function in gerontology and skin renewal as we age. It's because Aloe is made of biogenic material that has this characteristic. Cosmetic companies employ Aloe vera as a skin tonic.

11.7.8 Antiseptic Effect

Lupeol, salicylic acid, urea nitrogen, cinnamonic acid, phenols, and sulfur are the primary antiseptic agents found in Aloe vera. They all have antifungal, antibacterial, and antiviral properties (Surjushe et al. 2008).

11.7.9 Healing Properties

The healing of wounds is a dynamic process that occurs in three distinct stages. Inflammation, hyperaemia, and leukocyte infiltration characterize the initial phase. The second part involves the removal of dead tissue from the patient's body. The third phase of proliferation includes the regeneration of the epithelium and the creation of fibrous tissue (Reddy-Uma et al. 2011).

Aloe vera has been shown to be effective in the treatment of burns of the first to second degree, according to a recent study (Maenthaisong et al. 2007). Aloe vera gel's wound-healing properties have been related to Man-nose-6-phosphate, which is found in the gel (Davis et al. 1994). Actually, glucomannan and plant growth hormone gibberellins are able to boost the activity and proliferation of fibrobroblast, which in turn enhances collagen production in topical and oral aloe administration, according to Hayes' research (Hayes 1999).

Radiation burns and ulcers have been successfully treated with *aloe* gel (Syed et al. 1997), with full recovery noted in two out of three patients (Yeh et al. 2003). Lesions tended to heal more quickly with Aloe gel than with petroleum jelly gauze (18.2 days) (Visuthikosol et al. 1995) compared to burns treated with petroleum jelly gauze. *Aloe* gel has been used in a research on 27 individuals with partial thickness burns (Montaner et al. 1996).

11.7.10 Effects on Skin Exposure to UV and Gamma Radiation

Several studies have shown that aloe vera gel may protect the skin from radiation damage (Roberts and Travis 1995). Although its specific function is unknown, Aloe vera gel administration causes the skin to produce metallothionein, an antioxidant protein that scavenges free radicals and prevents the inhibition of skin antioxidant enzymes superoxide dismutase and glutathione peroxidase. Aloe reduces UV-induced suppression of delayed type hypersensitivity (Byeon et al. 1988).

11.7.11 Role of Aloe vera in Dentistry

The use of Aloe vera extracts to help with dental hygiene has been shown to be helpful (Mansourian et al. 2011; Choonhakarn et al. 2008; PDR 1998; George et al. 2009). Aloe vera gel has also been used in dentistry by Wynn (2005).

11.7.12 Anticancer Properties

There has not been enough research done on the potential carcinogenicity of Aloe vera. The use of anthranoid-containing laxatives over a long period of time may contribute to the development of colorectal cancer, although this has yet to be shown (Siegers 1993; Siegers 1992). Studies on cancer prevention have been conducted by Furukawa et al. (1991) and Fenig et al. (2004). Aloe vera juice helps the body repair itself from cancer and radiation and chemotherapy-induced damage to healthy immune cells, which are essential for the healing process. Inhibiting the formation of malignant cancer cells is one of the functions of anthraquinone, which is found in Aloe vera (Thomson 1971).

11.7.13 Stress

It helps keep the body's systems running smoothly by ingesting aloe juice (Saroj et al. 2004). Stress-related cell damage is reduced, and biochemical and physiological alterations in the body are minimized (Foster 1999). Compounds that undergo chemical processes in which their oxidative state changes are subjected to oxidative stress. Antioxidants may be found in foods, the body's own defense mechanisms, or they can be used as dietary supplements. Antioxidant-fighting Aloe vera is a wonderful example of an example of a functional food (Joseph and Raj 2010; El-Shemy et al. 2010; Barcroft and Myskja 2003).

11.7.14 Antioxidant Effects

Antioxidant properties of aloe extracts have also been discovered (Lee et al. 2000). According to Nejatzadeh-Barandozi (2013) and Baradaran et al. (2014), *Aloe vera* leaf extract has antioxidant properties. It also reveals that gentamicin-induced nephrotoxicity in male Wistar rats may be prevented by the supplement (Koul et al. 2015). It decreases the carcinogenic impact generated by cigarette smoke inhalation in the lungs of mice (Rahimifard et al. 2014). Diabetes-related oxidative stress may also be reduced by taking this supplement (Kang et al. 2014). Aloe vera

(*A. barbadensis*) gel was shown to be an effective antioxidant in both in vitro and animal studies (Kaithwas et al. 2014; López et al. 2013).

11.7.15 Cosmetic and Skin Protection Application

Pimples may be treated with aloin and its gel. Additionally, Aloe vera may be used to soothe the skin and keep it moisturized in dry weather to prevent flaky skin and a dry scalp. As well as moisturizing products, aloe sugars may also be employed (Barcroft and Myskja 2003). It is a great moisturizer, sunscreen, and a host of other beauty products when combined with essential oils. The antibacterial and antifungal properties of Aloe vera extracts have been shown to effectively cure minor skin diseases such as boils and benign skin cysts, as well as the fungi that cause tinea versicolor (Sumbul et al. 2004). The plant is being utilized in skin care, cosmetics, and as nutraceuticals (Gordon and David 2001). Several studies have shown that *aloe vera* gel may protect the skin from radiation damage (Roberts and Travis 1995).

11.8 Toxicity

The toxicity of Aloe vera extract was studied in both *in vitro* and *in vivo* experiments by Logarto et al. (2001). The median lethal concentration (LC50) of Aloe vera dried leaf extract in brine shrimp was determined to be 3.59 g/mL, while the median lethal dosage (LD50) in Swiss albino mice was 120.65 mg/kg after 24 h of acute oral exposure. When the *aloe vera* plant powder was extracted with 50% ethanol and injected intraperitoneally to adult albino mice at an initial dosage of 400–500 mg/kg, a maximum tolerated dose of 100 mg/kg body weight and LD50 of 250 mg/kg were found (Dhar et al. 1968). The half-maximal cytotoxic concentration (CC50) of Aloe vera whole-leaf material was 413.9 mg/mL in HeLa cells and 439.0 mg/mL in HepG2 cells after a 4-hour treatment (du Plessis and Hamman 2014). At doses as high as 1000 mg/mL, it induced a dose-dependent rise in apoptosis in HeLa cells.

Doses were offered to Sprague Dawley rats of 2, 4, and 8 grammes per kilogramme (2.5%, 5% and 10% *aloe* in the diet) for 90 days in a sub-chronic toxicity research (Zhou et al. 2003). In all dosed rats, there was an increase in faeces and a decrease in food efficiency and body weight among those given the highest dosages. Males exposed to 8 g/kg body weight and all females given the same dosage had considerably larger relative kidney weights than females. The pigmentation and proliferation of mesenteric lymph nodes, as well as the lamina propria of the colonic mucosa, were significantly increased in all of the exposed groups. One-fifth of the pharmacologically active dosage, 100 mg/kg Aloe vera extract per day, was shown to cause reproductive damage following a three-month period of chronic oral intake (Shah et al. 1989; Ogwu et al. 2017). Compared to control mice, severe sperm destruction, haematological abnormalities,nflammation, and death were observed.

Aloe vera non-decolorized whole leaf extract and decolorized extract were supplied to groups of four male and four female F344/N rats and the same numbers of B6C3F1 mice at 7 weeks of age as part of a 14-day drinking water research done by the NTP. In Aloe vera whole and decolorized extract solutions, the malic acid concentration ranged from 970 to 5840 g/g and aloin *aloe* content ranged from 70 to 422 g/g and 0.8 to 4.5 g/g water, respectively. On the 14th day after exposure, female rats exposed to 1.5–3.0% of decolorized extract had significantly lower blood urea nitrogen levels, while rats exposed to 3.0% of whole extract had significantly lower body weight, water consumption, and transit times in the gastrointestinal tract as well as smaller liver, heart, and kidney weights than control rats had shown. Male and female rats had considerably higher numbers of leukocytes and erythrocytes, as well as hematocrit percentages. In comparison, female mice given 2.0% non-decolorized whole extract had just a little increase in water intake.

11.9 Safety

It's difficult to know if Aloe vera products are safe and effective because of the absence of standardization in the commercial market. They are common in rural and suburban markets (Osawaru et al. 2013; Osawaru and Ogwu 2020). It is difficult to determine the correct dosage of Aloe vera formulations for various illnesses because of a lack of knowledge about the plant's active components. Topical Aloe vera gel or extract therapy is safe for mild to severe skin disorders, burns and wound healing, and inflammation (Ulbricht et al. 2008). Using it for psoriasis treatment, dermatitis treatment, and surgical wound healing has had conflicting results. Researchers determined that Aloe vera gel and Aloe latex may be safe for short-term usage, but long-term use of the latex is likely harmful because of the risk of dehydration and imbalance in electrolytes (Ulbricht et al. 2008). While some of the characteristics and applications of Aloe vera are based on legends, some are based on science. Aloe vera's usefulness must be shown in controlled trials in the future.

11.10 Conservative Measures of Aloe vera

In addition to being valued for their natural products (Grace 2011; Bosch 2006), aloes also provide nutritional benefits in food, nectar sources for honey bees (*Apis mellifera*) and nectarivorous birds, horticultural (Grace 2011), and cosmetic purposes (Sachedina and Bodeker 1999). (Sachedina and Bodeker 1999). There is a need for appropriate conservation measures and legislative approaches to ensure that real aloe may be utilized responsibly in light of their utilization value and increasing threats to plant germplasm conservation (Ogwu 2009, 2010; Grace 2011; Osawaru and Ogwu 2014b; Ogwu et al. 2014b).

Succulent plant collectors' associations and conservation organizations have as their primary goal the preservation of species variety and the long-term viability of *Aloe spp.* harvesters of succulent plant types (Osawaru and Ogwu 2014a; Chime et al. 2015; Ogwu et al. 2016a). Many of the interest groups, including the International Organization for Succulent Plant Study, British Cactus and Succulent Society, Cactus and Succulent Society of America, as well as others, promote ethical and lawful harvesting practises that are in accordance with both the Convention on Biological Diversity (CBD) and the legal requirements. However, prominent parties are well-positioned to continue encouraging the demand for plant specimens supplied in methods that respect the CBD's aims, even if this example is limited to *Aloe spp* (Crouch et al. 2008). Moreover, there is need to set up a national institute for the all round management of important plant species as recommended in Ogwu (2020)

11.11 Conclusion

Dietary supplements and herbal plants are used by more than 300 million individuals throughout the globe. Product safety, efficacy, and quality are required from dietary or herbal supplement items. It is, however, difficult to assess their effectiveness and safety because of their complicated molecular makeup. There is a lot of quality variation in herbal products due to a variety of concerns such as issues related to authenticity, adulteration, and replacement, as well as issues that arise during the cultivation, harvest, and processing phases after harvest.

Because of its many biological effects, aloe has long been utilized as a folk treatment across the globe. Various tablets, capsules, lotions, powders, and aqueous solutions include aloe extracts, which are widely available. Since certain components are susceptible to oxidation in commercial goods, stabilizers and preservatives are often included in the preparations found in Africa and other continents. Aloe vera products' varied biologic and therapeutic effects might be explained in part by their variable composition. Because Aloe vera chemical and physical makeup may vary so widely, it's almost difficult to compare one product to another. The polysaccharide substance generated from the inner gel of Aloe vera was found by the Cosmetic Ingredient Review Expert Panel to be noncytotoxic and suitable for use in foods. Due to the toxicity, mutagenicity, and carcinogenicity of anthraquinones in *aloe vera* whole leaf extract and latex, it is essential to monitor the level of these phenolic compounds. According to International Aloe Science Council standards, the maximum aloin level allowed to ben Aloe-derived material for oral intake is less than 10 ppm (parts per million); for nonmedical usage, the suggested limit is 50 ppm or less.

The leaf gel's chemical constituents can now be more easily isolated and characterized, thanks to advances in analytical chemistry technology. Aloe vera gel has lately been found to have interesting pharmacological uses, such as improving intestinal absorption and skin penetration. As an excipient in modified release matrix type tablets, the dried gel has shown promising results. More uses for this adaptable plant are being found as various perspectives on its composition and effects are studied.

References

Agarry OO, Olaleye MT, Bello-Michael CO (2005) Comparative antimicrobial activities of *Aloe vera* gel and leaf. Afr J Biotechnol 4(12):1413–1414

Akinyele BO, Odiyi AC (2007) Comparative study of the vegetative morphology and the existing taxonomic status of Aloe vera L. J Plant Sci 2:558–563

Albersheim P (2011) Plant cell walls: from chemistry to biology. Garland Science

Atherton P (1997) Aloe vera: myth or medicine. http://www.positivehealth.com. Accessed 28 Dec 2021

Atherton P (1998) First aid plant. Chem Br 34:33–36

Azghani AO, Williams I, Holiday DB et al (1995) A beta-linked Mannan inhibits adherence of *Pseudomonas aeruginosa* to human lung epithelial cells. Glycobiology 5(1):39–44. https://doi.org/10.1093/glycob/5.1.39

Baradaran A, Nasri H, Nematbakhsh M et al (2014) Antioxidant activity and preventive effect of aqueous leaf extract of *Aloe vera* on gentamicin-induced nephrotoxicity in male wistar rats. Clin Ter 165:7–11

Barcroft D, Myskja K (2003) Aloe vera: Nature's silent healer, BAAM.

Borra SK, Lagisetty RK, Mallela GR (2011) Anti-ulcer effect of *Aloe vera* in non-steroidal anti-inflammatory drug induced peptic ulcers in rats. Afr J Pharm Pharmacol 5:1867–1871

Bosch CH (2006) Aloe arborescens Mill, in: Schmelzer, G.H, Gurib-Fakim, A. (Eds.), PROTA (Plant Resources of Tropical Africa/Resources végétales de l'Afrique tropicale), Wageningen, Netherlands

Boudreau MD, Beland FA (2006) An evaluation of the biological and toxicological properties of *Aloe Barbadensis* (Miller), Aloe vera. J Environ Sci Health C 24:103–154

Bozzi A, Perrin C, Austin S et al (2007) Quality and authenticity of commercial Aloe vera gel powders. Food Chem 103(1):22–30

Byeon S, Pelley R, Ullrich SE et al (1988) Aloe Barbadensis extracts reduce the production of interleukin-10 after exposure to ultraviolet radiation. J Investig Dermatol 110:811–817. https://doi.org/10.1046/j.1523-1747.1998.00181.x

Caffall KH, Mohnen D (2009) The structure, function, and biosynthesis of plant cell wall pectic polysaccharides. Carbohydr Res 344:1879–1900. https://doi.org/10.1016/j.carres.2009.05.021

Cera LM, Heggers JP, Robson MC et al (1980) The therapeutic efficacy of *Aloe vera* Cream (Dermaide Aloe) in thermal injuries. Two case reports. J Am Anim Hosp Assoc 16:768–772

Cerqveira L, McKnight LS, Rodriguez S et al (1999) Bifurcated method to process aloe whole leaf. US Patent 5(925):357

Chan HT, Cavaletto CG (1986) Effects of deareation and storage temperature on quality of aseptically packaged guava puree. J Food Sci 51:165–168

Che QM, Akao T, Hattori M et al (1991) Isolation of human intestinal bacteria capable of transforming barbaloin to aloe-emodin anthrone. Planta Medica 57(1):15–19

Cheesbrough M (1984) Medical laboratory manual for tropical countries. The University Press, Cambridge

Chime AO, Osawaru ME, Ogwu MC, Ogunlowo EO (2015) Preservation of fresh tomato fruits (*Solanum lycopersicon* L.) using leaf extracts of bitter melon (*Momordica charantia* L., Cucurbitaceae). J Phytopathol Plant Health 3:29–37

Choonhakarn C, Busaracome P, Sripanidkulchai B et al (2008) The efficacy of *Aloe vera* gel in the treatment of oral lichen planus: a randomized controlled trial. Br J Dermatol 158:573–577

Christaki EV, Florou-Paneri PC (2010) Aloe vera: a plant for many uses. J Food Agric Environ 8(2):245–249

Coats BC (1979) Hypollergenic stabilized Aloe vera gel. US Patent number 4 178 172

Coats BC (1994) Methods of processing stabilized Aloe vera gel obtained from the whole Aloe vera leaf. US Patent 5(356):811

Crouch NR, Douwes E, Wolfson MM et al (2008) South Africa's bioprospecting, access and benefit-sharing legislation: current realities, future complications, and a proposed alternative. South Afr J Sci 104:355–366

Davis RH (1997) Aloe vera—a scientific approach. Vantage Press Inc, New York, pp 290–306

Davis RH, Kabbani JM, Maro NP (1987) *Aloe vera* and wound healing. J Am Podiatric Med Assoc 77(4):165–169

Davis RH, Di Donato JJ, Hartman GM et al (1994) Anti-Inflammatory and Wound Healing Activity of a Growth Substance in *Aloe vera*. J Am Podiatric Med Assoc 84(2):77–81

Dhar ML, Dhar MM, Dhawan BN et al (1968) Screening of Indian plants for biological activity. Indian J Exp Biol 6:232–247

EEC (1988) Council Directive No. 88/388 on the approximation of the laws of the member states relating to flavourings for use in foodstuffs and to source materials for their production. Offic J Eur Commun 184:61–66

Eison-Perchonok MH, Downes TW (1982) Kinetics of ascorbic acid oxidation as a function of dissolved oxygen concentration and temperature. J Food Sci 47:765–773

El-Shemy HA, Aboul-Soud MA, Nassr-Allah AA et al (2010) Antitumor properties and modulation of antioxidant enzymes' activity by *Aloe vera* leaf active principles isolated via supercritical carbon dioxide extraction. Curr Med Chem 17(2):129–138. https://doi.org/10.2174/092986710790112620

Epidi JO, Izah SC, Ohimain EI (2016) Antibacterial and synergistic efficacy of extracts of *Alstonia boonei* tissues. Br J Appl Res 1(1):0021–0026

Eshun K, He Q (2004) Aloe vera: a valuable ingredient for the food, pharmaceutical and cosmetic industries: a review. Crit Rev Food Sci Nutr 44:91–96

Esua MF, Rauwald JW (2006) Novel bioactive maloyl glucans from Aloe vera gel: isolation, structure elucidation and in vitro bioassays. Carbohydr Res 341:355–364

Evivie SE, Ogwu MC, Ebabhamiegbebho PA, Abel ES, Imaren JO, Igene JO (2020) Packaging and the Nigerian food industry: challenges and opportunities. In: Ogunlade CA, Adeleke KM, Oladejo MT (eds) Food technology and culture in Africa. Reamsworth Publishing, Ibadan, Nigeria, pp 28–99

Farooqi AA, Sreeramu BS (2001) Cultivation of medicinal and aromatic crops. Revised Ed. Orient Longman, p 25

Femenia A, Sanchez ES, Simal S et al (1999) Compositional features of polysaccharides from *Aloe vera* (*Aloe barbadensis* Miller) plant tissues. Carbohydr Polym 39:109–117

Femenia A, Garcia-Pascual P, Simal S et al (2003) Effects of heat treatment and dehydration on bioactive polysaccharide acemannan and cell wall polymers from *Aloe barbadensis* Miller. Carbohydr Polym 51:397–405

Fenig E, Nordenberg J, Beery E et al (2004) Combined effect of aloe-emodin and chemotherapeutic agents on the proliferation of an ad-herent variant cell line of merkel cell carcinoma. Oncol Rep 11(1):213–217

Ferro VA, Bradbury F, Cameron P et al (2003) *In vitro* susceptibilities of *Shigella flexneri* and *Streptococcus pyogenes* to inner gel of *Aloe barbadensis* Miller. Antimicrob Agents Chemother 47(3):1137–1139. https://doi.org/10.1128/AAC.47.3.1137-1139.2003

Foster S (1999) *Aloe vera:* The succulent with skin soothing cell protecting properties, Herbs for Health Magazine. http://www.healthy.net/library/articles/hfh/Aloe.htm

Furukawa F, Nishikawa A, Chihara T et al (1991) Inactivation of enveloped viruses by anthraquinones extracted from plants. Antimicrob Agents Chemother 35(12):2463–2466. https://doi.org/10.1128/AAC.35.12.2463

George D, Bhat SS, Antony B (2009) Comparative evaluation of the antimicrobial efficacy of Aloe vera tooth gel and two popular commercial toothpastes: an in vitro study. General Dentistry 57: 238–241

Gordon MC, David JN (2001) Natural product drug discovery in the next millennium. Pharm Biol 39:8–17

Grace OM (2011) Current perspectives on the economic botany of the genus Aloe L. (Xanthorrhoeaceae). South Afr J Bot 77:980–987

Grindlay D, Reynolds T (1986) The Aloe vera phenomenon: a review of the properties and modern uses of the leaf parenchyma gel. J Ethnopharmacol 16:117–151

Haller JS (1990) A drug for all seasons, medical and pharmacological history of *Aloe*. Bull New York Acad Med 66:647–659

Hayes SM (1999) Lichen planus: report of successful treatment with *Aloe vera*. General Dentistry 47(3):268–272

He ZH, Huang YQ, Weng SF et al (2013) Effect of aloe emodin on invasion and metastasis of high metastatic breast cancer MDA-MB-231 cells. Zhong Yao Cai 36:1481–1485

Heggers JP, Pineless GR, Robson MC (1979) Dermaide Aloe/*Aloe vera* gel: comparison of the antimicrobial effects. Am J Med Technol 41:293–294

Heggers JP, Kucukcelibi A, Stabenou CJ (1995) Wound healing effects of aloe gel and other topical antibacterial agents in rat skin. Phytother Res 9(6):455–457. https://doi.org/10.1002/ptr.2650090615

Hernadez RJ, Giacin JR (1998) Factors affecting permeation, sorption andmigration process in packageproducts system. In Food storage stability. CRC Press, Boca Raton, FL, pp 269–329

Hirat T, Suga T (1983) The efficiency of aloe plants, chemical constituents and biological activities. Cosmet Toiletries 98:105–108

Hutter J, Salman M (1996) Anti-inflammatory C-glucosyl chromone from Aloe barbadensis. J Nat Prod 59(5):541–543

Indu M (2017) History of Aloe vera—(a magical plant). Iosr J Human Soc Sci (iosr-jhss) 22(8): 21–24

Ishii Y, Tanizawa H, Takino Y (1994) Studies of *Aloe*. V. mechanism of cathartic effect. Biol Pharm Bull 17(5):651–653. https://doi.org/10.1248/bpb.17.651

Ito S, Teradaira R, Beppu H et al (1993) Properties and pharmacological activity of carboxypeptidase in *Aloe arborescens* Mill. var. natalensis berger. Phytother Res 7(7):S26–S29. https://doi.org/10.1002/ptr.2650070710

Izah SC, Aseibai ER (2018) Antibacterial and synergistic activities of methanolic leaves extract of lemon grass (Cymbopogon citratus) and rhizomes of ginger (Zingiber officinale) against Escherichia coli, Staphylococcus aureus and Bacillus subtilis. Acta Sci Microbiol 1(6):26–30

Joseph B, Raj SJ (2010) Pharmacognostic and phyto-chemical properties of *Aloe vera* Linn—an overview. Int J Pharm Sci Rev Res 4(2):106–110

Joshi SP (1997) Chemical constituents and biological activity of *Aloe barbadensis*—a review. J Med Aromat Plant Sci 20:768–773

Kacem B, Mathews RF, Grandall PG et al (1987) Nonenzymatic browning in aseptic packaged orange drinks: effects of amino acids, dearation and anaerobic storage. J Food Sci 52:1665–1667

Kahlon JB, Kemp MC, Carpenter RH et al (1991) Inhibition of AIDS virus replication by acemannan in vitro. Mol Biother 3:127–135

Kaithwas G, Singh P, Bhatia D (2014) Evaluation of *in vitro* and *in vivo* antioxidant potential of polysaccharides from *Aloe vera* (*Aloe barbadensis* miller) gel. Drug Chem Toxicol 37:135–143

Kang MC, Kim SY, Kim YT et al (2014) *In vitro* and *in vivo* antioxidant activities of polysaccharide purified from *Aloe vera* (*Aloe barbadensis*) gel. Carbohydr Polym 99:365–371

Kaufman T, Newman AR, Wexler MR (1989) *Aloe vera* and burn wound healing. Plast Reconstr Surg 83(6):1075–1076. https://doi.org/10.1097/00006534-198906000-00037

Kawai K, Beppu H, Shimpo K et al (1998) In vivo effects of Aloe arborescens Miller var. natalensis Berger on experimental tinea pedis in guinea pig feet. Phytother Res 12:178–182

Kemp MC, Kahlon JB, Chinnah AD et al (1990) *In vitro* evaluation of the antiviral effects of acemannan on the replication and pathogenesis of HIV-1 and other enveloped viruses: modification of the processing of glycoprotein glycoprotein precursors. Antiviral Res 3(1):83. https://doi.org/10.1016/0166-3542(90)90156-2

Kennedy FC, Rivera ZS, Loyd LL et al (1992) Lascorbic acid stability in aseptically processed orange juice in tetra brick cartons and the effect of oxygen. Food Chem 45:327–331

Kigigha LT, Biye SE, Izah SC (2016) Phytochemical and antibacterial activities of *Musanga cecropioides* tissues against *Escherichia coli*, *Pseudomonas aeruginosa Staphylococcus aureus*, *Proteus* and *Bacillus* species. Int J Appl Res Technol 5(1):100–107

Kigigha LT, Selekere RE, Izah SC (2018) Antibacterial and synergistic efficacy of acetone extracts of Garcinia kola (Bitter kola) and Buchholzia coriacea (Wonderful kola). J Basic Pharmacol Toxicol 2(1):13–17

Kim HS, Kacew S, Lee BM (1999) In vitro chemopreventive effects of plant polysaccharides (*Aloe barbadensis Miller*, *Lentinus edodes*, *Ganoderma lucidum*, and *Coriolus vesicolor*). Carcinogenesis 20:1637–1640

Korkina L, Suprun M, Petrova A et al (2003) The protective and healing effects of a natural antioxidant formulation based on ubiquinol and *Aloe vera* against dextran sulfate-induced ulcerative colitis in rats. Biofactors 18:255–264

Koul A, Bala S, Yasmeen N et al (2015) *Aloe vera* affects changes induced in pulmonary tissue of mice caused by cigarette smoke inhalation. Environ Toxicol 30:999–1013

de Rodríguez DJ, Hernández-Castillo D, Rodríguez-García R et al (2005) Antifungal activity *in Vitro* of *Aloe vera* pulp and liquid fraction against plant pathogenic fungi. Indust Crops Products 21(1):81–87. https://doi.org/10.1016/j.indcrop.2004.01.002

du Plessis LH, Hamman JH (2014) In vitro evaluation of the cytotoxic and apoptogenic properties of aloe whole leaf and gel materials. Drug Chem Toxicol 37:169–177

Lawless J, Allen J (2000) Aloe vera—natural wonder care. Harper Collins Publishers, Hammersmith 5–12

Lee KY, Weintraub ST, Yu BP (2000) Isolation and identification of a phenolic antioxidant from *Aloe barbadensis*. Free Radical Biol Med 28:261–265

Logarto PA, Silva YR, Guerra SI et al (2001) Comparative study of the assay of Artemia salina L. and the estimate of the medium lethal dose (LD50 value) in mice, to determine oral acute toxicity of plant extracts. Phytomedicine 8:395–400

López A, de Tangil MS, Vega-Orellana O et al (2013) Phenolic constituents, antioxidant and preliminary antimycoplasmic activities of leaf skin and flowers of *Aloe vera* (L.) burm. F. (syn. *A. Barbadensis* mill.) from the canary islands (Spain). Molecules 18:4942–4954

Madan J, Sharma A, Inamdar N et al (2008) Immunomodulatory properties of *Aloe vera* gel in mice. Int J Green Pharm 2(3):152

Maenthaisong P, Chaiyakunapruk N, Niruntraporn S (2007) The efficacy of *Aloe vera* for burn wound healing: a systematic review. Burns 33(6):713–718. https://doi.org/10.1016/j.burns.2006.10.384

Mansourian A, Momen-Heravi F, Saheb-Jamee M (2011) Comparison of treatment efficacy of daily use of *Aloe vera* mouthwash with triamcinolone acetonide 0.1% on oral lichen planus: a randomized double-blinded clinical trial. Am J Med Sci 342:447–451

Maret RH (1975) Process for preparing extract of Aloe vera. US Patent 3(878):197

Montaner JS, Gill J, Singer J (1996) Double-blind placebo-controlled pilot trial of acemannan in advanced humanimmune-deþciencyvirus disease. J Acquir Immune Deficiency Syndr Human Retrovirol 12(2):153–157

Moor ED, McAnalley BH (1995) A drink containing mucilaginous polysaccharides and its preparations. USPatent 5(443):830

Moreira LRS, Filho EXF (2008) An overview of mannan structure and mannan-degrading enzyme systems. Appl Microbiol Biotechnol 79:165–178

Mothana RA, Linclequist V (2005) Antimicrobial activity of some medicinal plants of the island Soqotra. J Ethnopharmacol 96(1–2):177–181. https://doi.org/10.1016/j.jep.2004.09.006

Nejatzadeh-Barandozi F (2013) Antibacterial activities and antioxidant capacity of *Aloe vera*. Org Med Chem Lett 3:5

Nema J, Shrivastava SK, Mitra NG (2013) Chemical composition of Aloe ferox under stresses of soil pH and desiccation. Int J Chem 3:44–48

Ni Y, Tizard IR (2004) Analytical methodology: the gel-analysis of aloe pulp and its derivatives. In: Reynolds T (ed) Aloes the genus aloe. CRC Press, Boca Raton, pp 111–126

Ni Y, Turner D, Yates KM et al (2004a) Isolation and characterisation of structural components of *Aloe vera* L. leaf pulp. Int Immunopharmacol 4:1745–1755

Ni Y, Yates KM, Tizard IR (2004b) Aloe polysaccharides. In: Reynolds T (ed) *Aloes the genus aloe*. Boca Raton, CRC Press, pp 75–87

Ogidi IO, Chukwudi P, Ibe AI et al (2021a) Preliminary phytochemical profile and antimicrobial potentials of white-green African garden egg (*solanum macrocarpon*) fruits obtained from Yenagoa. ASIO J Pharm Herbal Med Res 7(2):01–05. https://doi-ds.org/doilink/06.2021-76274373/10.2016-19146535/ASIO-JPHMR/2021/457

Ogidi OI, Okore CC, Akpan UM et al (2021b) Evaluation of antimicrobial activity and bioactive phytochemical properties of mango (*Mangifera Indica*) stem-bark extracts. Int J Pharmacogn 8(5):189–195. https://doi.org/10.13040/IJPSR.0975-8232.IJP.8(5).189-95

Ogidi OI, Oguoma LMO, Adigwe PC et al (2021c) Phytochemical properties and in-vitro antimicrobial Potency of Wild Edible Mushrooms (*Pleurotus ostreatus*) obtained from Yenagoa, Nigeria. J Phytopharmacol 10(3):180–184. https://doi.org/10.31254/phyto.2021.10306

Ogwu MC (2009) The significance and contribution of conservation education in Nigeria: an appraisal and a call for improvement. In: Babalola F (ed) Proceedings of the maiden seminar of Nigerian Tropical Biology Association. Nigerian Tropical Biology Association and Tropical Biology Association, pp 109–114

Ogwu MC (2010) Conserving biodiversity even in poverty: the African experience. In: Babalola F (ed) Seminar Proceedings of Nigerian Tropical Biology Association. Nigerian Tropical Biology Association and Tropical Biology Association, pp 112–117

Ogwu MC (2019) Towards sustainable development in Africa: the challenge of urbanization and climate change adaptation. In: Cobbinah PB, Addaney M (eds) *The geography of climate change adaptation in urban Africa*. Springer Nature, Switzerland, pp 29–55. https://doi.org/10.1007/978-3-030-04873-0_2

Ogwu MC (2020) Value of *Amaranthus* [L.] species in Nigeria. In: Waisundara V (ed) Nutritional Value of Amaranth. IntechOpen, pp 1–21. https://doi.org/10.5772/intechopen.86990

Ogwu MC, Chime AO, Oseh OM (2018) Ethnobotanical survey of tomato in some cultivated regions in Southern Nigeria. Maldives Natl Res J 6(1):19–29

Ogwu MC, Oforjamah IN, Ahana CM, Osawaru ME (2013) Phylloplane assessment of leaf and stalk for consumption and microbial diversity of sweet potato *Ipomoea batatas* (L.) Lam., Convolvulaceae. *Octa*. J Environ Res 1(2):107–116

Ogwu MC, Osawaru ME, Ahana CM (2014b) Challenges in conserving and utilizing plant genetic resources (PGR). Int J Genet Mol Biol 6(2):16–22. https://doi.org/10.5897/IJGMB2013.0083

Ogwu MC, Osawaru ME, Atsenokhai EI (2016a) Chemical and microbial evaluation of some uncommon indigenous fruits and nuts. Borneo Sci 37(1):54–71

Ogwu MC, Osawaru ME, Chime AO (2014a) Comparative assessment of plant diversity and utilization patterns of tropical home gardens in Edo State, Nigeria. Scientia Africana 13(2): 146–162

Ogwu MC, Osawaru ME, Obahiagbon GE (2017) Ethnobotanical survey of medicinal plants used for traditional reproductive care by Usen people of Edo State, Nigeria. Malaya J Biosci 4(1): 17–29

Ogwu MC, Osawaru ME, Obadiaru O, Itsepopo IC (2016b) Nutritional composition of some uncommon indigenous spices and vegetables. International conference on natural resource development and utilization (4th ed): positioning small and medium enterprises (SMEs) for optimal utilization of available resources and economic growth. Raw Materials Research and Development Council (Federal Ministry of Science and Technology), Maitama, Abuja Nigeria, pp 286–300

Osawaru ME, Ogwu MC (2014b) Conservation and utilization of plant genetic resources. In: Omokhafe K, Odewale J (eds) Proceedings of 38th annual conference of the Genetics Society of Nigeria. Empress Prints Nigeria Limited, pp 105–119

Osawaru ME, Ogwu MC (2014a) Ethnobotany and germplasm collection of two genera of cocoyam (*Colocasia* [Schott] and *Xanthosoma* [Schott], Araceae) in Edo State Nigeria. Sci Technol Arts Res J 3(3):23–28. https://doi.org/10.4314/star.v3i3.4

Osawaru ME, Ogwu MC (2020) Survey of plant and plant products in local markets within Benin City and environs. In: Filho LW, Ogugu N, Ayal D, Adelake L, da Silva I (eds) African handbook of climate change adaptation. Springer Nature, Switzerland, pp 1–24. https://doi.org/10.1007/978-3-030-42091-8_159-1

Osawaru ME, Ogwu MC, Ahana CM (2013) Current status of plant diversity and conservation in Nigeria. Nigerian J Life Sci 3(1):168–178

Osawaru ME, Ogwu MC, Aigbefue D (2014) Survey of ornamental gardens in five local government areas of Southern Edo State Nigeria. Bioscientist 2(1):87–102

Osawaru ME, Ogwu MC, Omoigui ID, Aiwansoba RO, Kevin A (2016) Ethnobotanical survey of vegetables eaten by Akwa Ibom people residing in Benin City, Nigeria. Univ Benin J Sci Technol 4(1):70–93

PDR (1998) PDR for herbal medicines, 1st edn. Medical Economics Company, Montvale NJ

Pugh N, Ross SA, ElSohly MA et al (2001) Characterization of aloeride, a new high-molecular-weight polysaccharide from *Aloe vera* with potent immunostimulatory activity. J Agric Food Chem 49(2):1030–1034. https://doi.org/10.1021/jf001036d

Rahimifard M, Navaei-Nigjeh M, Mahroui N et al (2014) Improvement in the function of isolated rat pancreatic islets through reduction of oxidative stress using traditional Iranian medicine. Cell J 16:147–163

Ramachandra CT, Srinivasa RP (2008) Processing of Aloe vera leaf gel: a review. Am J Agril Biol Sci 3(2):502–510

Reddy-Uma CH, Reddy SK, Reddy J (2011) *Aloe vera*—a wound healer. Asian J Oral Health Allied Sci 1:91–92

Reynolds JEF (1993) Martindale, the extra pharmacopoeia, 30th edn. Pharmaceutical Press, London

Reynolds T (2004) Aloe chemistry. In: Reynolds T (ed) Aloes the genus aloe. Boca Raton, CRC Press, pp 39–74

Ro JY, Lee B, Kim JY et al (2000) Inhibitory mechanism of aloe single component (Alprogen) on mediator release in guinea pig lung mast cells activated with specific antigen-antibody reactions. J Pharmacol Exp Ther 292:114–121

Roberts DB, Travis EL (1995) Acemannan-containing wound dressing gels reduce radiation-induced skin reactions in C3H mice. Int J Radiat Oncol Biol Physiol 32(4):1047–1052. https://doi.org/10.1016/0360-3016(94)00467-Y

Robson MC, Heggers JP, Hagstrom WJ (1982) Myth, magic, witchcraft or fact? *Aloe vera* revisited. J Burn Care Res 3(3):157–163. https://doi.org/10.1097/00004630-198205000-00005

Sachedina H, Bodeker G (1999) Wild Aloe harvesting in South Africa. J Altern Complement Med 5:121–123

Sadler GD, Braddock RJ (1990) Oxygen permealibity of low density polyethylene as a function of lomonene absorption: an approach to modeling flavour (Scalping). J Food Sci 55:587–590

Saini MR, Goyal PK, Chaudhary G (2010) Anti-tumor activity of *Aloe vera* against DMBA/croton oil-induced skin papillomagenesis in Swiss albino mice. J Environ Pathol Toxicol Oncol 29: 127–135

Santos PRV, Oliveria ACX, Tomassini TCB (1995) Controls microbiological products fitoterapices. Revista de Farmácia e Bioquímica 31:35–38

Saroj PL, Dhandar DG, Singh RS (2004) Indian aloe. Central Institute for Arid Horticulture, Bikaner

Shah AH, Qureshi S, Tariq M et al (1989) Toxicity studies on six plants used in the traditional Arab system of medicine. Phytother Res 3:25–29

Siegers CP (1992) Anthranoid laxatives and colorectal cancer. Trends Pharmacol Sci 13:229–231. https://doi.org/10.1016/0165-6147(92)90070-M

Siegers CP (1993) Anthranoid laxative abuse—a risk for colorectal cancer. Gut 34(8):1099–1101. https://doi.org/10.1136/gut.34.8.1099

Sumbul S, Ahmed SW, Azhar I (2004) Anti fungal activity of *Allium*, *Aloe*, and *Solanum* species. Pharm Biol 42(7):491–498. https://doi.org/10.3109/13880200490891845

Surjushe A, Vasani R, Saple DG (2008) *Aloe vera*: a short review. Indian J Dermatol 53:163–166

Sydiskis RJ, Owen DG, Lohr JL et al (1991) Inactivation of enveloped viruses by anthraquinones extracted from plants. Antimicrob Agents Chemother 35(12):2463–2466. https://doi.org/10.1128/AAC.35.12.2463

Syed TA, Afzal M, Ashfaq AS (1997) Management of genital herpe in men with 0.5% *Aloe vera* extracts in a hydrophilic cream: a placebo-controlled double-blind study. J Dermatol Treatment 8(2):99–102. https://doi.org/10.3109/09546639709160279

Talmadge J, Chavez J, Jacobs L et al (2004) Fractionation of *Aloe vera* L. inner gel, purification and molecular profiling of activity. Int Immunopharmacol 4:1757–1773

Thomson RH (1971) Naturally occurring quinines, 2nd edn. Academy Press, London

Tramell DJ, Dalsis DE, Malone CT (1986) Effect of oxygen on taste, ascorbic acid loss and browning for HTST pasteurized, single strength orange juice. J Food Sci 51:1021–1023

Tyler V (1994) Herbs of choice. In: Tyler V (ed) The therapeutic use of phyto medicine. Binghamton Pharmaceutical Products Press, New York, pp 131–135

Tyler VE, Bradley LR, Robbers JE (1988) Pharmacognosy, 9th edn. Lea & Febiger, Philadelphia

Ulbricht C, Armstrong J, Basch E et al (2008) An evidence-based systematic review of Aloe vera by the Natural Standard Research Collaboration. J Herb Pharmacother 7:279–323

Vazquez B, Avila G, Segura D et al (1996) Antiinflammatory activity of extracts from Aloe vera gel. J Ethnopharmacol 55:69–75

Visuthikosol V, Chowchuen B, Sukwanarat Y et al (1995) Effect of *Aloe vera* gel to healing of burn wound a clinical and histologic study. J Med Assoc Thailand 78(8):403–409

West DP, Zhu YF (2003) Evaluation of *Aloe vera* Gel gloves in the treatment of dry skin associated with occupational exposure. Am J Infect Control 31(1):40–42. https://doi.org/10.1067/mic.2003.12

Willats WG, Knox JP, Mikkelsen JD (2006) Pectin: new insights into an old polymer are starting to gel. Trends Food Sci Technol 17:97–104. https://doi.org/10.1016/j.tifs.2005.10.008

Wynn RL (2005) *Aloe vera* gel: update for dentistry. General Dentistry 53(1):6–9

Yagi A, Kabash A, Mizuno K et al (2003) Radical scavenging glycoprotein inhibiting cyclooxygenase-2 and thromboxane A2 synthase from *Aloe vera* gel. Planta Medica 69(3): 269–271. https://doi.org/10.1055/s-2003-38481

Yaron A, Cohen E, Arad SM (1992) Stabilization of Aloe vera gel by interaction with sulfated polysaccharides from micro algae and with xantham gum. J Agric Food Chem 40:1316–1320

Yeh GY, Eisenberg DM, Kaptchuk TJ et al (2003) Systematic review of herbs and dietary supplements for glycemic control in diabetes. Diabetes Care 26(4):1277–1294. https://doi.org/10.2337/diacare.26.4.1277

Zakia S, Zahid NY, Yaseen M et al (2013) Standardization of micropropagation techniques for Aloe vera: a pharmaceutically impotant plant. Pak J Pharm Sci 26:1083–1087

Zhang XF, Wang HM, Song YL et al (2006) Isolation, structure elucidation, antioxidative and immunomodulatory properties of two novel dihydrocoumarins from *Aloe vera*. Bioorg Med Chem Lett 16(4):949–953

Zhou Y, Feng Y, Wang H et al (2003) 90-day subchronic toxicity study of aloe whole-leaf powder. Wei Sheng Yan Jiu 32:590–593

Chapter 12
Sustainable Utilization of Important Medicinal Plants in Africa

Odangowei Inetiminebi Ogidi

Abstract Humans have relied on medicinal plants for food, housing, clothing, and medicine since the dawn of humanity. However, following World Health Organization's support for herbal medicine development, increased awareness of their social and economic significance, and research validation of herbal remedy efficacy claims, there appears to be a renewed increase in demand for medicinal plant resources over the last few decades. This increased exploitation has resulted in the depletion and endangerment of some of Africa's most significant medicinal plant species, making it unsustainable. This chapter focuses on traditional knowledge of medicinal plants as well as the variety of important medicinal plants in Africa, contemporary trends in their use, and issues of their unsustainable exploitation. It also proffers suggestions for remedies and a path ahead to achieve a more sustainable use of medicinal plant species in Africa.

Keywords Sustainability · Utilization patterns · Africa · Medicinal plants · WHO · Traditional medicine

12.1 Introduction

Plants were previously the world's principal source of medicines. It is likely the oldest form of plant usage known to man and dates back to periods before plant domestication began (Osawaru et al. 2012, 2016). Plants have continued to provide humans with novel treatments since then, with natural products accounting for 50% of all medications in clinical use worldwide, with higher plants accounting for 25% of the total (Van Wyk et al. 2013; Ogwu et al. 2016a; Ogwu and Osawaru 2022). In many underdeveloped nations, medicinal plants serve an important role in basic healthcare (Fullas 2007). Traditional medicines are used by about 80% of people in

O. I. Ogidi (✉)
Department of Biochemistry, School of Applied Sciences, Federal Polytechnic Ekowe, Ekowe, Bayelsa State, Nigeria

S. C. Izah, M. C. Ogwu (eds.), *Sustainable Utilization and Conservation of Africa's Biological Resources and Environment*, Sustainable Development and Biodiversity 32, https://doi.org/10.1007/978-981-19-6974-4_12

poor nations due to their cost and cultural acceptance (Maroyi 2013). Herbal medicine is becoming more popular. Herbal medicine is defined by the World Health Organization (WHO) as finished, labelled medicinal products that contain aerial or underground parts of plants, or other plant materials, or combinations thereof, as active ingredients, whether in their natural state or as plant preparations (Izah and Aseibai 2018; Gamaniel 2005).

Since the dawn of time, medicinal plants have been a vital element of Africa's healthcare system. Traditional medicine's popularity can be explained by the fact that it is an integral part of the culture of those who practise it, as well as by the economic challenge: on the one hand, pharmaceutical drugs are out of reach for the poor, while on the other hand, Africa's richness and diversity of fauna and flora provide an inexhaustible source of therapies for a wide range of ailments (Ogidi et al. 2021a, 2019a). Despite this, there is a scarcity of clinical evidence to support their efficacy and safety in people. Users of traditional medicinal herbs in Africa and elsewhere are suspicious about their efficacy without this knowledge. This limits people's ability to select plants that are potentially less expensive and more accessible (Ogidi et al. 2019b, c).

Because most medicinal and aromatic plants are edible and staple foods in most poor societies, researchers have lately analysed the notion of "food as medicine" as a factor to consider in the long-term exploitation of African medicinal plant resources (HerbFest 2015). The majority of prescription pharmaceuticals in the previous century came from natural sources, namely medicinal and aromatic plants (ISSC-MAP 2007). As a result, the issues of sustainable extraction and usage of medicinal plant resources remain a hot topic that touches on plant resource security, social and economic factors, and the accomplishment of sustainable development goals in poor countries.

The focus of this chapter is on common medicinal plants from Africa that have the potential to be developed as future phytopharmaceuticals in the short and long term, the consequences of unsustainable use of these plants, and the path forward for ensuring sustainability and conservation to treat and/or manage a wide range of infectious and chronic diseases.

12.2 Indigenous Knowledge and Medicinal Plant Harvesting

In different places of the world, different tribes and ethnic groupings have retained different versions of indigenous or traditional knowledge and practices associated with diverse plant genetic resources within their vicinity (Bruchac 2014; Ikhajiagbe et al. 2021; Ahana et al. 2022). The collection, cultivation, and harvesting of medicinal plant materials used to be a limited practise in Africa, reserved for traditional healers (Van Andel and Havinga 2008; Osawaru and Ogwu 2014). Traditional health practitioners are persons who have never had official medical training, but who are trusted by their communities to meet their healthcare requirements by employing plant, animal, and mineral ingredients (Agbor and Naidoo 2011; Ogwu et al. 2017a, b; Ogwu et al. 2018a).

Indigenous knowledge (IK), such as the knowledge gained by traditional health practitioners over many generations, which includes knowledge on medicinal plants, their uses, and methods of application, is the poor's most valuable asset in the fight for survival, food and medicine production, shelter, and self-control (Senanayake 2015). Because IK is at the heart of decision-making in their communities, it is more complex than previously thought in terms of natural resources, ecological zones, aquaculture, agriculture, game management, and forestry (Senanayake 2015). Indigenous knowledge on medicinal plant identification and usage has primarily circulated among practitioners of traditional medicine or those who benefit from such practises (Regassa 2013; Hamilton 2004).

Traditional health practitioners adhere to strict traditional values around the harvesting of medicinal plants, including taboos, superstitions, conventions, and cultural beliefs, and as a result, they have helped n the conservation of medicinal plant species (Kambizi and Afolayan 2006; Williams et al. 2000; Chime et al. 2016). They only collected plant material after it had been ordained by their ancestors and after completing particular ceremonies; (1) where the root of a plant is harvested for use as medicine, cultural tradition prevents the collecting of more than two roots of the same plant at the same time, and so on. (2) For example, *Trichilia emetica* (Natal mahogany) is saved for its fruit, despite the fact that it is also employed in traditional medicine. (3) Plants like *Siphonochilus aethiopicus* (African ginger) and Alepidea amatymbica (bigger tinsel flower) are only harvested in the winter to enable seed set and proliferation throughout the summer. (4) Plants were collected with a pointed wooden digging stick or small axe, which tended to limit the quantity of bark or roots gathered, and (5) bark for treating kidney diseases was sometimes only harvested from the eastern and western sides of the tree, traditionally resembling the kidneys and thus preventing ring-barking (Van Wyk et al. 2013; Kambizi and Afolayan 2006). However, despite their medical and healing potentials, most of these medicinal plants are underutilized (Ogwu et al. 2016b, c)

Others were purposely introduced to conserve knowledge and plant material, assuring sustainable harvesting (Kambizi and Afolayan 2006; Williams et al. 2000). Several of these procedures have become obsolete since commercial harvesters have taken over the material harvesting that was formerly done only by traditional health practitioners. Species such as Siphonochilus aethiopicus and Ocotea bullata (Black stinkwood) are now threatened with extinction in Africa as a result of these banned activities. The move from subsistence usage to commercial trading in medical plants and plant parts has resulted in an increase in the quantity and frequency of medicinal plant harvesting from natural areas (Geldenhuys and Mitchell 2006).

Poverty and a high unemployment rate as a result of Africa's economic predicament have encouraged untrained and often indifferent people from disadvantaged communities to commercial plant gathering in both urban and rural locations (Williams et al. 2000; Ogwu 2019a). According to Mander et al. (2007), South Africa has over 200,000 traditional health practitioners and 63,000 commercial harvesters. Commercial harvesting methods, in contrast to the actions of traditional health practitioners outlined above, are environmentally damaging (Delveaux et al. 2009; Grace et al. 2002).

Harvesting medicinal tree bark, for example, entails removing the maximum quantity of bark as well as any tissue exterior to the secondary xylem. Gatherers create ladders to optimize the amount of bark collected from a tree when certain tree species, for example, become rare (Cunningham 1988). Some huge trees are even toppled to collect the bark from the whole length of the trunk, and juvenile trees are being stripped for their bark when the bark of older trees has been exhausted (Chungu et al. 2007). Gatherers and sellers of medicinal plants are unconcerned about the increasing scarcity of specific plant species. Increased scarcity of medicinal plant species means higher prices for merchants, increasing their profit margin (Cunningham 1988), while higher prices for gatherers function as an incentive for increased gathering rates (Cunningham 1993). Commercial harvesters are forced to expand harvesting to produce a respectable income due of the normally low prices provided to them (Monakisi 2007). Traders become worried about medical plant scarcity only when the variety of plant material accessible in their stores is impacted, since the unavailability of particular rare species or plant parts may result in the cancellation of a client's whole purchase (Cunningham 1988).

Traditional health practitioners are now purchasing plant material from commercial harvesters, which is a source of worry for environmentalists. This is especially true for traditional health practitioners working in cities, where relevant services are typically hundreds of kilometres away (Mander 1998). Plants are seen as a common property resource by commercial harvesters in both urban and rural locations, and there are no or few incentives for resource management or conventional conservation techniques (Dold and Cocks 2002). Many of Africa's medicinal tree species are multi-purpose, meaning they are utilized for more than just medicine (Van Wyk and Gericke 2007). Because of their multiple uses, multi-purpose tree species are more vulnerable to extinction than single-use tree species, necessitating strong legal restriction as well as increased conservation efforts (Primack 2012).

12.3 Some Important Medicinal Plants in Africa

Medicinal plants that are part of an African herbal pharmacopoeia with commercial importance and those plants from which modern phytopharmaceuticals have been derived were chosen for more detailed reviews based on the following criteria: medicinal plants that are part of an African herbal pharmacopoeia with commercial importance and those plants from which modern phytopharmaceuticals have been derived.

12.3.1 *Mango (***Mangifera Indica***)*

widely known as mango in English and locally as mangoro (Yoruba), mangolo (Igbo), mangwaro (Hausa), and Ogboin (Izon), belongs to the Anacardiaceae family, which includes roughly sixty genera and 600 species of tropical trees and shrubs. It is widely utilized as a food, medicinal, and wood supply. Any mango processing industry's main by-product is mango stem bark. This waste product needs a significant financial investment in order to degrade properly and avoid contamination. As a result, converting it to generate bioactive chemicals is essential to save the food processing industries a significant amount of money. Dentifrice, antiseptic, astringent, diaphoretic, stomachic, vermifuge, tonic, laxative, and diuretic are all typical uses for the plant in Nigeria. It's also used to treat diarrhoea, dysentery, anaemia, asthma, bronchitis, cough, gastrointestinal problems, hypertension, sleeplessness, rheumatism, toothache, gastrointestinal tract infections, respiratory and urinary tract infections, sore gums, sore throats, leucorrhoea, haemorrhage, and piles (Ogidi et al. 2021b).

12.3.2 *Gum Arabic (Acacia senegal)*

This plant is often known as gum Arabic and is native to sub-Saharan Africa's semi-desert and arid areas, although it's found all across the continent, from Southern to Northern. Northern Nigeria, West Africa, North Africa, and other places of the world utilize it as a medicinal herb (Gurib-Fakim et al. 2010). Gum arabic (or gumacacia), which is formed from a bark exudate, has been used since the first Egyptian Dynasty (3400 B.C.).

It's occasionally used to treat skin and mouth infections caused by bacteria and fungi. It has been used to heal irritated skin and calm mucous membranes in the intestines (Okoro et al. 2011; Aliyu 2006). The demulcent, emollient gumis is used both internally and topically to treat inflammation of the intestinal mucosa, such as burns, painful nipples, and nodular leprosy. It's also been used as an antitussive, expectorant, astringent, and catarrh to treat colds, coughs, diarrhoea, dysentery, gonorrhoea, bleeding, sore throat, typhoid, and urinary tract problems (Okoro et al. 2011). The gum of A. senegal has been used in the pharmaceutical industry to make emulsions, tablets, and troches (as an excipient), as a demulcent for throat and stomach inflammations, as a masking agent for harsh smelling chemicals like *Capsicum*, and as a film-forming ingredient in peel-off masks. Gum arabic is extensively utilized as a component in foods such as candy and soft drinks since it possesses glue-like characteristics that are safe to consume. Gumacacia is used in commercial emulsification for the manufacturing of drinks and flavour concentrates and is frequently utilized in organic products as a natural alternative to artificial binders (Okoro et al. 2011; Brendler et al. 2010; Aliyu 2006).

A. senegal bark extracts were tested in vitro for antibacterial activity against human pathogenic isolates, according to a recent study (Escherichia coli, Staphylococcus aureus, Streptococcus pneumoniae, Klebsiella pneumoniae, Shigella dysenteriae, Salmonella typhi, Streptococcus pyogenes, Pseudomonas aeruginosa, and Proteus vulgaris). The presence of tannins and saponins in the plant was thought to be the reason for the extract's considerable antibacterial action. Following in vitro cytotoxicity testing, it was also discovered that the plant extract may not be hazardous to humans (Okoro et al. 2011).

*12.3.3 Okra (*Abelmoschus *species)*

Okra (i.e., both *A. esculentus and A. caillei*) contains significant amounts of mucilage, fibre, proteins, oil, vitamins, and phenolic compounds wh ch are abundant in leaves, fruits, and seeds, which are ingested as vegetables (Osawaru et al. 2012; Ogwu and Osawaru 2014; Ogwu et al. 2017a, b, 2018b). Mucilage is used to raise blood volume by replacing plasma, and it also binds cholesterol, lowering the risk of heart attacks. The oil in the seed is high in linoleic acid and polyunsaturated fatty acids, both of which are good for your heart.

*12.3.4 Alligator Pepper (*Piper guineensis*)*

Piperine, wisanine dihydrocubebin, guineensine, and other compounds are found in the fruits of the plant whose fruits are often used as a chilly spice in meals. Antifungal, antimicrobial, anti-tumor, hypotension, bradycardia (slow heartbeat), immunomodulatory, antiulcerogenic, contraceptive, central nervous system depression, analgesic, antipyretic, anti-inflammatory, antioxidant, and antisickling properties are all present (Kunle and Egharevba 2013).

12.3.5 Aloe ferox*: (Bitter Aloe or Cape Aloe)*

The most prevalent Aloe species in South Africa is *Aloe ferox*, which is endemic to South Africa and Lesotho. *A. ferox* has been utilized as an alternative medicine for millennia and is one of the few plants featured in San rock drawings. In Africa and Europe, bitter latex, also known as Cape aloe, is used as a laxative medication and is said to have bitter tonic, antioxidant, anti-inflammatory, antibacterial, and anticancer qualities (Chen et al. 2012; Van Wyk 2008; Jia et al. 2008; Van Wyk and Wink 2004).

The usage of *A. ferox* as a versatile traditional medicine has led to various commercial uses, and the plant is highly appreciated in the pharmaceutical, natural health, food, and cosmetic sectors. *A. ferox* is the most often collected and sold wild species in South Africa. Since 1761, when aloe bitters was first transported to Europe, the final product derived from aloe tapping has been an important South African export product. Many rural populations rely on the aloe tapping business for their livelihood, and the sector has been formalized via the formation of cooperatives and trade agreements. It has been proposed that its commerce might have a significant impact on Africa's poverty reduction (Melin 2009; VanWyk and Gericke 2000). There are several traditional and confirmed medical applications for *A. ferox*. It's well- known for its laxative properties and as a topical treatment for the skin, eyes, and mucous membranes.

*12.3.6 Bitter Cola (*Garcinia kola*)*

The seed contains antioxidants and is used to treat sickle cell anaemia. Kolaviron, biflavanones, and tannins were said to be present (Kigigha et al. 2018; Kunle and Egharevba 2013).

*12.3.7 African Garden Egg (*Solanum *species)*

One of these plants is the garden egg, often known as eggplant. It has approximately 100 species throughout Africa, most of which are utilized as vegetables, with 25 species identified in Nigeria. *Solanum aethiopicum* (green-striped round shaped garden egg) and Solanum macrocarpon (white-green striped oval garden egg) are the most commonly distributed and are eaten raw, cooked, or fried to prepare vegetable sauce (Ogwu et al. 2016c). Additionally, different portions of *S. Melongena* can be used to treat inflammatory diseases, cardiac debility, neuralgia, nasal ulcers, cholera, bronchitis, and asthma. Because of its flavour and increased proportion of vitamin B2, the fruit is a very important vegetable all over the world. The fruit can also be used to treat diabetes (Ogidi et al. 2021c).

12.3.8 Artemisia herba-alba—*White Word or Wormwood*

Asso (Mediterranean) Wormwood or desert wormwood is the common name for Artemisia herba-alba (known in Arabic as shih, and as Armoise blanche in French). It is a perennial dwarf shrub native to Northern Africa, the Arabian Peninsula, and Western Asia that is greyish and scented (Segal et al. 1987). Many civilizations have employed *A. herba-alba* in folk medicine since prehistoric times. It is used to treat arterial hypertension and diabetes in Moroccan folk medicine, and diabetes, bronchitis, diarrhoea, hypertension, and neuralgias in Tunisian traditional medicine (Ziyyat et al. 1997).

Herbal tea from A. herba-alba has been used in traditional medicine as an analgesic, antimicrobial, antispasmodic, and hemostatic agent (Zeggwagh et al. 2008; Laid et al. 2008; Tahraoui et al. 2007; Friedman et al. 1986). During an ethnopharmacological investigation among the Bedouins of the Negev desert, *A. herba-alba* was discovered to be used to treat stomach problems. This plant is also thought to be useful as a source of fodder for sheep and animals in Algeria's plateau regions, where it grows abundantly. It has also been reported that the oil of the Libyan *A. herba-alba* killed Ascaridae from pigs and ground worms in a short period of time (Mohamed et al. 2010; Benmansur et al. 1990; Fenardji et al. 1974). The antifungal activity of the constituents and biological activities of *Artemisia herba-alba* essential oils of 25 Moroccan medicinal plants, including *A. herba-alba*, were reported in another study against *Penicillium digitatum*, *Phytophthora citrophthora*, *Geotrichum citri-aurantii*, and *Botrytis cinerea* (Belhattab et al. 2012; Mohamed et al. 2010).

*12.3.9 Onion (*Allium cepa*)*

Onions have long been a staple of many civilizations' diets and have long been prized for their medical properties. Onions are monocotyledons of the Liliaceae family. The onion is one of the most widely grown vegetables in the world, and it is said to have originated in Central and Western Asia. Onions were widely consumed throughout Europe throughout the Middle Ages, and their pungent stench was subsequently supposed to protect against bad spirits and the plague. The vegetable has been linked to a variety of health advantages, including cancer prevention and cardiovascular disease prevention. Following this, research was conducted on particular chemicals discovered in onion bulbs. Onions have a unique blend of three chemical families that are thought to have health-promoting properties. They include fructans, which are tiny carbohydrate molecules that support gut health by supporting good microorganisms. Flavonoids operate as antioxidants and deactivate chemicals that are harmful to the body's cells. As well as organosulphur chemicals, which are chiefly responsible for onions' flavour and odour. These chemicals minimize diabetes mellitus symptoms, limit platelet aggregation (which causes thrombosis), and prevent asthmatic inflammatory processes (Ogidi and Julius 2021).

*12.3.10 Rooibos (*Aspalathus linearis*)*

The well-known herbal tea, also known as rooibos, is made from Aspalathus linearis, a native South African fynbos plant. Its caffeine-free and low tannin content, as well as its putative health-promoting characteristics, most notably antioxidant activity, have all led to its widespread acceptance and appeal. The use of rooibos has expanded beyond herbal tea to include intermediate value-added goods such extracts for the beverage, food, nutraceuticals, and cosmetics industries (Joubert and de Beer 2011; Marnewick et al. 2009; Van Heerden et al. 2003).

Rooibos has been utilized in a variety of ways across Africa for centuries. It's been utilized as a cooling beverage and a nutritious tea beverage (Brendler et al. 2010; Gurib-Fakim et al. 2010). Only when she discovered that giving her colicky infant a rooibos infusion relieved her child's persistent restlessness, vomiting, and stomach pains did rooibos become generally known as a "healthy" beverage, resulting in a larger user base. Since then, rooibos has been used to nourish many newborns, either in their milk or as a weak drink (Kreuz et al. 2008; Gilani et al. 2006; Khan and Gilani 2006; Van Wyk and Verdoorn 1989).

12.3.11 Vernonia amygdalina *(Bitter Leaf)*

The leaf has been utilized as an antidiabetic, antimalarial, antisickling, antioxidant, and other medicinal properties. Saponins, alkaloids, sterol-terpenes, and phenolic acids were observed to be present (Ogidi et al. 2019d; Izah et al. 2018a).

12.3.12 Centella *Species*

Centella species has been utilized as a medicinal herb from prehistoric times. It is employed in many healing cultures, including Ayurvedic medicine, Chinese traditional medicine, Kampo (Japanese traditional medicine), and African traditional medicine, and has a pan-tropical distribution (Brendler et al. 2010; Brinkhaus et al. 2000). To this day, it is being utilized in folk medicine, and it is increasingly being found at the intersection of traditional and modern scientifically oriented medicine.

C. asiatica has traditionally been used to treat wounds, ulcers, leprosy, tuberculosis, lupus, skin disorders, eye illnesses, fever, inflammation, asthma, hypertension, rheumatism, syphilis, epilepsy, diarrhoea, and mental illness, as well as being eaten as a vegetable or used as a spice. The use of *C. asiatica* in the treatment of leprosy was first described in Mauritius in 1852, and its clinical usage as a therapeutic agent suited for the treatment of leprous lesions has been documented since 1887. (Brendler et al. 2010).

Clinical effects in the treatment of chronic venous illness, wound healing, and cognitive functioning, among other things, describe the active ingredients (Brendler et al. 2010). *C. asiatica* includes many pentacyclic triterpenoids that have been investigated extensively. The two most major active chemicals employed in medication formulations are asiaticoside and madecassoside. Based on their anti-inflammatory properties, both are commercially employed primarily as wound-healing agents. The ursane-type triterpene saponin asiaticoside, which is responsible for wound healing activities (Kim et al. 2009; Shukla et al. 1999) and is known to induce type 1 collagen production in fibroblast cells, is one of the key active ingredients of *C. asiatica* (Lee et al. 2006).

*12.3.13 Black Eye Beans/Cowpea (***Vigna unguiculata***)*

In most African cultures, the plant's seed is a staple diet. It also contains saponins, which reduce carbs, lipids, and oil, as well as steroids, glycosides, and alkaloids, and is utilized in antisickling recipes (Kunle and Egharevba 2013).

*12.3.14 Clove (***Eugenia caryophyllata***)*

In sickle cell disease, the fruit, leaf, and stalk are employed. Antimicrobial, antioxidant, antifungal, and antiviral activities are present in the plant, as well as anti-inflammatory, cytotoxic, insect repellent, and anaesthetic effects. The plant had been found to produce essential oils high in eugenol, eugenyl acetate, -caryophyllene, gallotannic acid, and other compounds (Kunle and Egharevba 2013).

12.3.15 **Catharanthus roseus** *(Madagascan Periwinkle)*

Catharanthus roseus is a well-known medicinal plant with its origins in Africa. The medicinal value of this species stems from the fact that it is the sole source of the anticancer alkaloids vincristine and vinblastine, which are difficult to produce in the laboratory due to their complexity; the leaves of this species are still the only source of these alkaloids today (Pereira et al. 2010; Brendler et al. 2010). *C. roseus* is native to Madagascar, although it currently has a broad distribution across the tropics. Its traditional use may be traced back to Madagascar, where healers have used it extensively to cure a variety of diseases. It's often used as a bitter tonic, galactogogue, and emetic in traditional medicine. Rheumatism, skin ailments, and venereal infections have all been claimed to be treated with it (Gurib-Fakim et al. 2010; Abegaz et al. 2004).

The main component of *C. roseus* is vindoline, which contains a variety of phytochemicals (as many as 130 components) (up to 0.5 percent). Serpentine, catharanthine, ajmalicine (raubasine), akuammine, lochnerine, lochnericine, and tetrahydroalstonine are some of the other physiologically active chemicals. Bisindole alkaloids are abundant in the plant; vindoline and catharanthine are detected in trace amounts: vincristine (leurocristine) in up to 3 g/t of dry medicine and vinblastine (vincaleucoblastine) in slightly more (Pereira et al. 2009; Ferreres et al. 2008; Van der Heijden et al. 2004; Gurib-Fakim et al. 2010).

12.3.16 *African (White) Star Apple (***Chrysophyllum albidum***)*

In most African societies, the fruit is eaten as a snack and a vegetable. Eleagnine, found in the seed, has antioxidant, anti-inflammatory, and antibacterial properties. The seed is used as an antidiarrheal, among other things (Egharevba et al. 2015).

12.3.17 Cyclopia genistoides *(Honeybush)*

Cyclopia genistoides is a South African herbal tea that is considered a health food. To make tea, the green shoots and blossoms were traditionally fermented and dried. It has also been appreciated as a stomachic that supports poor digestion without harming the heart since ancient times for its direct favourable benefits on the urinary system. It is mostly used as a tea alternative because of the absence of hazardous chemicals such as caffeine. It's one of the few indigenous South African plants to have successfully transitioned from the wild to a commercial commodity in the last century. Over the last two decades, research has focused on propagation, production, genetic enhancement, processing, composition, and the possibility for value addition (Joubert and de Beer 2011; Brendler et al. 2010).

Honeybush decoction was employed as a restorative and expectorant in the treatment of chronic catarrh and pulmonary TB (Bowie 1830). Drinking a honeybush infusion supposedly enhances hunger as well, but no specific species is mentioned (Watt and Breyer-Brandwijk 1962). Many colonists hailed honeybush as nutritious, valuing it as a stomachic that supports poor digestion without creating any severe stimulating effects on the heart, according to Marloth 1925. It also helps with nausea and heartburn (Van Wyk et al. 2009; Van Wyk 2008; VanWyk and Smith 2004, 1996). It boosts milk production in breast-feeding mothers and cures colic in newborns, according to anecdotal data (Rood 1994).

12.3.18 Harpagophytum procumbens *(Grapple plant)*

Harpagophytum procumbens is found in red sand regions of South Africa's Transvaal, Botswana, and Namibia. It's now found all throughout the Kalahari and Savannah deserts. For years, if not millennia, the indigenous San and Khoi people of Southern Africa have utilized Devil's Claw medicinally (Andersen et al. 2004).

Harpagophytum procumbens has a long history of indigenous applications and is one among Africa's most marketed indigenous traditional medicines, with bulk exports mostly to Europe, where it is converted into a variety of health products including teas, pills, capsules, topical gels, and patches (Mncwangi et al. 2012).

Traditional uses include allergies, analgesia, anorexia, antiarrhythmic, antidiabetic, antiphlogistic, antipyretic, appetite stimulant, arteriosclerosis, bitter tonic, blood diseases, boils (topical), childbirth difficulties, choleretic, diuretic, climacteric (change of life) problems, dysmenorrhea, dyspepsia, edema, fever, and fibromyalgia., Devil's claw is used as an anti-inflammatory and analgesic in the treatment of joint ailments, back discomfort, and headaches. The widespread use of standardized Devil's claw as a moderate analgesic for joint pain in Europe is based on evidence from scientific investigations in animals and people (Mncwangi et al. 2012; Andersen et al. 2004; Chrubasik and Bradley 2004; Wegener and Lupke 2003; Ernst and Chrubasik 2000).

12.3.19 Momordica charantia *(Bitter Melon)*

Momordica charantia is a tropical vegetable produced throughout Africa. The juice derived from the various plant components (fruit pulp, seeds, leaves, and whole plant) is a fairly widespread folkloric cure for diabetes. M. charantia has a long history of usage as a folkloric hypoglycemic medication, with the plant extract dubbed "vegetable insulin" (Ahmad et al. 1999).

M. charantia has yielded a number of active chemicals, as well as some mechanistic research (Mahomoodally et al. 2004; Matsuura et al. 2002; Matsuda et al. 1998; Kimura et al. 1991). Khanna et al. (1981) isolated "polypeptidep," a 17-amino acid, 166-residue polypeptide that did not cross-react in an immunoassay for bovine insulin, from fruits, seeds, and tissue culture of seedlings. The antilipolytic and lipogenic effects of a galactose binding lectin isolated from the seeds of M. charantia were tested in isolated rat adipocytes and found to be equivalent to insulin (Dubey et al. 1987; Akhtar et al. 1981). In diverse animal models, extracts of M. charantia fruit pulp, seed, leaves, and entire plant showed hypoglycemic effects (Dubey et al. 1987; Akhtar et al. 1981). When the bitter melon fruit juice was given orally to rats, Karunanayake et al. (1984) discovered a considerable improvement in glucose tolerance and hyperglycemia. Eight Evidence-Based Complementary and Alternative Medicine In normal rats and noninsulin dependent diabetics (NIDDM), both fresh and freeze-dried M. charantia, greatly improved glucose tolerance (Jayasooriya et al. 2000; Yesilada et al. 1999). M. charantia fruit is thought to have more than one type of hypoglycemic component. An active principle known as "charantin," a homogeneous combination of -sitosterol-glucoside and 2-5-stigmatadien-3—olglucoside that may cause hypoglycemia in normal rabbits, could be one of them (Raman and Lau 1996).

12.3.20 *Laggera (*Laggera pterodonta*)*

Taba-taba (Hausa) and Taba-ebora are two more names for the same plant (Yoruba). Anti-inflammatory, antiviral, antibacterial, anti-tubercular, and hepatoprotective activities are found in the aerial component. Essential oils, sesquiterpenoids, and triterpenoids, among other things, are present (Egharevba et al. 2010b).

12.3.21 Pelargonium sidoides *(South African Geranium)*

Pelargonium sidoides is a tuberous plant endemic to South Africa's coastal areas, and existing ethnobotanical data indicate that it is an important traditional medicine with a long ethnobotanical history (Brendler et al. 2010). *P. sidoides* root extract EPs 7630 is a herbal medicine that is supposed to help with acute respiratory infections. *P. sidoides* and respiratory tract infections have been studied extensively (Agbabiaka et al. 2008; Timmer et al. 2008). These investigations suggested that *P. sidoides* might help adults with acute rhinosinusitis and the common cold, although there is still some scepticism. It may help relieve the symptoms of acute bronchitis in both adults and children, as well as sinusitis in adults (Timmer et al. 2008). By day 7, EPs 7630 has lowered bronchitis symptom ratings in patients with acute bronchitis (Agbabiaka et al. 2008). There were no major side effects noted. EPs 7630 has a beneficial effect on phagocytosis, oxidative burst, and intracellular cell death (Conrad et al. 2008; Conrad and Frank 2008; Conrad et al. 2007). The synthesis of secretory immunoglobulin A in saliva, both interleukin-15 and interleukin-6 in serum, and interleukin-15 in the nasal mucosa is all modulated by *P. sidoides* extract (Brendler and Van Wyk 2008).

12.3.22 *Pawpaw (*Carica papaya*)*

The fruit and leaves of the pawpaw (*Carica papaya* L) are edible and frequently used as vegetables. Tea produced from papaya leaves is used to treat malaria and enhance platelet counts in the blood. The alkaloid papain was extracted from the plant (Izah et al. 2018b; Kunle and Egharevba 2013).

12.3.23 Agathosma betulina *(Buchu)*

Agathosma betulina is a woody plant with the botanical name *Agathosma betulina* (Berg.) pillans (Van Wyk et al. 1997). It's only found in the Western Cape Cederburg region of South Africa, where it's suited to arid circumstances and grows on sunny rocky-sandstone slopes (Golblatt and Manning 2000). Buchu is a significant plant in the Khoi-San tradition (VanWyk and Gericke 2000), and it has a long history of use as a general health tonic, diuretic, and mild urinary antiseptic (Van Wyk et al. 1997). The antispasmodic, antibacterial, and diuretic properties of the essential oil are most likely due to this ingredient (Lis-Balchin et al. 2001; Wichtl and Bisset 2000). Antispasmodic, antipyretic, liniment, cough cure, cold and flu remedy, diuretic, treatment of kidney and urinary tract infections, haematuria, prostatitis, cholera, stomach illnesses, rheumatism, gout, bruising, calculus, and antiseptic are some of the medical applications of *A. betulina* (Simpson 1998; Watt and Breyer-Brandwijk 1962). Buchu tinctures have a long history of usage as general health tonics, stomach pains therapy, aromatic bitters, diuretics, and mild urinary antiseptics (VanWyk and Gericke 2000). To keep their skin supple and wet in the arid atmosphere, the San oiled their bodies with scented herbs combined with fat. The lubricant also worked as an antibacterial and antifungal protectant, an insect repellent, a deodorant, and a way to enhance overall bodily well-being by allowing fragrant chemicals to be absorbed via the skin (Van Wyk et al. 1997).

Buchu's composition has been thoroughly investigated, and various chemicals have been found. The compounds found in *A. betulina* were mostly concentrated in the plant's volatile fractions and included limonene, menthone, diosphenol, and l-pulegone (Fluck et al. 1961), with isomenthone and diosphenol as the major volatile compounds (Posthumus et al. 1996; Kaiser et al. 1975), which are responsible for the distinctive flavour as well as antispasmodic and antiseptic (Van Wyk 2011). The economically relevant sulphur-containing compounds (cis and trans-8-mercapto-p-methan-3-one) are a hallmark of *A. betulina*, and despite their minor levels, they are responsible for the oil's distinctive organoleptic qualities (Sandasi et al. 2010).

*12.3.24 Cocao (***Theobroma cacao***)*

The fruit and seed of the cocao tree are processed into beverages and consumed. The stem bark is utilized as an antisickling, antioxidant, and other medicinal properties. Saponins, polyphenolics, catechins, and anthocyanins were discovered in the plant (Kunle and Egharevba 2013).

*12.3.25 Neem (***Azadirachta indica***)*

Neem is a versatile medicinal plant that contains a variety of compounds with varying chemical structures and biological effects. Steroids, glycosides, alkaloids, and tannins are some of the active phytoconstituents found in neem. Neem leaves effectively treat eczema, ringworm, acne, antihyperglycemic properties, and anti-inflammation. It aids in the purification of our blood and the neutralization of free radicals. Neem leaves have anticancer properties. Neem seed extract has been shown to have a wide range of pharmacological properties, including antioxidant, antimalarial, antimutagenic, anticarcinogenic, and anti-inflammatory properties. The presence of several bioactive chemicals in various areas of the plant is credited with the biological activity. Antifungal, antiviral, anti-inflammatory, antibacterial, analgesic, and antioxidant are just a few of the therapeutic properties of neem. Every part of the neem tree is bitter, and it has a wide range of uses (Ogidi et al. 2021d).

12.3.26 *Guayava (*Psidium guajava*)*

The fruit of the guayava tree (*Psidium guajava* Linn) is edible and widely consumed. Fever, malaria, gastroenteritis, vomiting, diarrhoea, dysentery, wounds, ulcers, toothaches, coughs, and other conditions are treated with the leaf, fruit, and stem bark. Saponins, tannins, flavonoids, and glycosides were reported to be present in the plant (Egharevba et al. 2010a).

12.4 Current Trends in the Use of Medicinal Plants

With civilization, improved awareness, and the availability of more research data, the use of medicinal plants for foods, infrastructure development, and raw materials for industries (e.g. agro, chemical, textile, pharmaceutical, etc.) appears to be on the rise. Also, there is need for rebranding which can begin with improved packaging of these medicinal plant products and creating a conducive environment for their product development, market, and sale as well as understanding their long-term use implications (Ogwu 2019b, 2020; Evivie et al. 2020). More studies are being conducted in the fields of gene manipulation for increased yield and bio-production, herbal remedies, and as a source of leads/hits for the development of novel pharmaceuticals. Though the overall pattern of use stays mostly same, there appears to be a shift in the relational quantum in distinct use-schemes or regions in different parts of the world, depending on their degree of civilization. However, as shown in the Fig. 12.1, the present trend seems to serve several good purposes:

Awareness: Public awareness of medicinal plants' health and economic potential is growing, and individuals in certain nations and communities are already taking use of the prospects.

Economic Benefits: As people become more aware of herbal products and resources, more people are investing in preservation and conservation practises for some economically important crops and plants used in medicine, such as bitter leaf plantations, guayava plantations, cashew plantations, and so on.

Research: More study is currently being done on the development of medicinal plants as a commercial commodity, which will lead to standardization, safety, and evidence of efficacy, product stability and preservation, and improved storage,

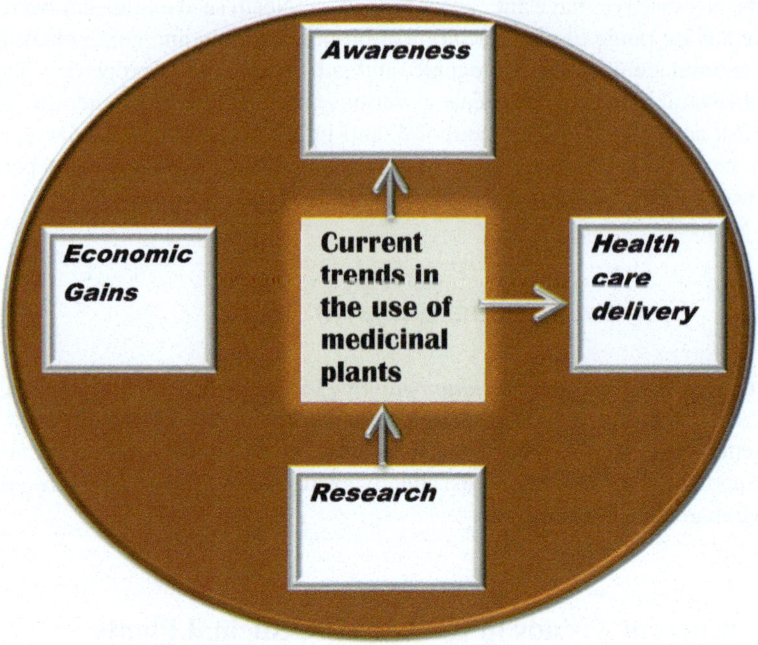

Fig. 12.1 Chart on the current trends in the use of medicinal plants

handling, and procedures (GLP, GMP, GAP). Many laboratories are also working on transforming wild species into cultivated ones for propagation.

Health-care idelivery: Scientific evidence of the usefulness of some of these plants has aided in universal acceptance, especially among the learned, and has allowed some herbal remedies to be placed on essential drug lists in several countries to support healthcare delivery. It has also backed the possibility and worldwide drive for improved integration of herbal medicine into many countries' healthcare delivery systems.

12.5 Sustainable Exploitation of Medicinal Plant Resources in Africa

The realization that humans are an intrinsic element of ecosystems is at the heart of the ecosystem approach to sustainability. As a result, one is reliant on the other. A civilization can be considered sustainable if it uses resources in such a way that both the human and ecosystem conditions remain steady or improve. Four interconnected measures must be considered when developing a sustainable harvest strategy: (1) the environment, (2) communities and ecosystems, (3) plant populations, and (4) genetic diversity (Schippmann et al. 2002b).

There is a link between resource supply, population number, and sustainable harvesting rates for every plant resource. Low stocks result in low yields, especially if the plant is slow-growing and has a low reproduction rate (Cunningham 1997). The plant's life after harvesting is determined by the plant components collected. The plant dies when bark, roots, or complete plants are taken. Leaf, flower, fruit, and seed harvesting are regarded less harmful; however, extreme pruning reduces a plant's vitality and reproductive capacity (Van Andel and Havinga 2008; Grace et al. 2002). Sustainable harvesting rates are determined by the vegetation type from which the plants or plant parts are harvested, as well as their quantity and growth rates. In general, how much can be utilized sustainably depends on the rate of regeneration following harvesting (Delveaux et al. 2009).

Plant part harvesting tolerance varies depending on environmental circumstances and the resources available to support growth and recovery (Ticktin 2004). Because research on the conservation and sustainable use of medicinal plants and their habitats has lagged far behind demand for this globally important resource, information on how much plant material constitutes sustainable harvesting is completely lacking for many countries and many plant species (Van Andel and Havinga 2008). Nonetheless, the present threats facing these resources will not go away and need the development of specific sustainable conservation and utilization programs (Ogwu et al. 2014; Osawaru and Ogwu 2014). Each species has its own ecological, socioeconomic, health, and cultural relationships, which must be understood (Schippmann et al. 2002b). According to Vermeulen (2009), controlled, experimental harvesting, and long-term monitoring are the only ways to establish yield systems and best practises for harvesting.

In the early 2000s, South Africa extracted around 20,000 tons of plant materials from wild resources each year for the local medicinal plant trade market (Mander et al. 2007). These figures do not include quantities harvested for export. In Nigeria, medinical plant species are commonly found in local rural and sub-urban markets (Osawaru and Ogwu 2020). Only 5 tons of cultivated plants are gathered, while another 40 tons of rare species are imported from neighbouring countries such as Mozambique and Swaziland (Mander et al. 2007; Mander 1998). Depending on demand and availability, annual amounts gathered for the local medicinal plant trade market may fluctuate somewhat. According to Mander et al. (2007), the following percentages of plant parts are traded in South Africa: 27% roots, 27% bark, 14 percent bulbs, 13 percent complete plants, 10% leaves and stems, 6% tubers, and 3% mixes of plant parts. Several authors have claimed that South Africa's harvesting rates are unsustainable (Van Andel and Havinga 2008; Kambizi and Afolayan 2006; Botha et al. 2004; Mander 1998), and Botha et al. (2004) have claimed that increased habitat pressure has resulted in numerous local extinctions. Due to a paucity of recent research and information, it is possible that the situation has worsened dramatically since 2007 and prior reports, emphasizing the significance of an integrated regulatory structure.

Maundu et al. (2006), Strydom and King (2013), Regassa (2013), and Cunningham (1997) identify the following as causes of unsustainable harvesting in general: (1) A high rate of global population growth, (2) competing land uses,

(3) environmental degradation, (4) loss of indigenous knowledge, (5) increasing commercialization of traditional medicine, (6) increasing demand in local and international markets, (7) lack of appropriate policies and legislation and/or failure to enforce them, (8) poverty and high unemployment (9) low medicinal plant prices, (10) invasive alien plants that threaten indigenous plant diversity, (11) unsustainable harvesting methods, (12) undue pressure on specific preferred species, and (13) slow plant growth, particularly in medicinal tree species. Unsustainable harvesting has three major consequences, according to Maundu et al. (2006): (1) local or global extinctions, (2) genetic pool shrinkage, and (3) weakened regeneration ability.

With rapidly declining medicinal plant populations, an explosive increase in the human population, and the remaining medicinal plant populations, particularly plants with roots and bulbs, and slow-growing, slow-reproducing medicinal tree populations already overexploited by injudicious harvesting, it is possible to speculate whether harvesting from the remnant medicinal plant populations, even if sustainable harvesting methods are used, could be sustainable.

12.6 Factors Related to Species Rarity of Medicinal Plants

Prior to beginning conservation efforts, species rarity is used to assess the extinction risk of medicinal plants and to identify those species that are most at risk of extinction (Figueiredo and Grelle 2009). It's crucial to figure out how uncommon each species is, as well as how rare species differ from one another. Harvesting pressures affect not all medicinal plants in the same way (Wagh and Jain 2013; Van Andel and Havinga 2008). Overexploitation, indiscriminate collection, uncontrolled deforestation, and habitat degradation all have an impact on species rarity, but they are inadequate to explain individual species sensitivity to harvest pressure. Habitat specificity, distribution range, population size, species diversity, growth rate, and reproductive system are all biological characteristics that are linked to extinction risk.

12.7 Cultivation of African Medicinal Plant Species

With the growing recognition that wild medicinal plant populations are being overharvested, several organizations, including the World Health Organization (WHO), the International Union for Conservation of Nature (IUCN), and the World Wide Fund (WWF), have recommended that wild species be introduced into cultivation systems (Schippmann et al. 2002a). Since Gerstner's suggestion in 1938, large-scale production of medical plant material for the medicinal plant trade has not been attempted due to a number of problems (Cunningham 1988). Because of the joint responsibilities between national and provincial governments, enforcing environmental regulations in Africa is difficult. As a result, implementation has become fragmented. The government's willingness and ability to take action against those who do not follow the law also limits the implementation of conservation laws (Strydom and King 2013). The lack of cultivation of medicinal plant species in Africa is due to three main factors: (1) a lack of institutional support for the production and dissemination of key species for cultivation, (2) low prices paid to plant material harvesters by herbal traders, and (3) many important medicinal species take a long time to mature (Cunningham 1997). Plants must be cultivated cheaply and in large quantities if medicinal plant material is to be successful in providing an alternative supply in herbal medicines and reducing harvesting pressure on wild stocks. Cultivation will only mask the continued exploitation of wild resources, it was stated in the 1990s, if it does not take place on a scale large enough to meet annual market demand (Cunningham 1997). Demand, plant size, and plant growth rate all influence the amount of land required for cultivation. Bulbous plants, for example, can be planted densely and rotated every six to ten years, resulting in a tiny rotational area, but trees require considerably greater land area for good root growth and development. Their rotating area would consequently be far larger, especially when the slow-growing characteristic of Africa's native trees is considered (Cunningham 1988).

Many people in rural areas think that metaphysical threats are threatening the healing ability of amayeza (isiXhosa for medicine), and this concept has ramifications for the adoption of farmed medicinal plant material. They claim that when amayeza gets in contact with "polluted humans," it loses its potency (Wiersum et al. 2006). For environmentalists, acceptance of grown medical plants is a constant source of concern, as farmed material is thought to lack the (spiritual) 'power' of healing (Cunningham 1993). Although cultivated medicinal plant material is accepted as an alternative in countries such as Swaziland, Botswana, and Ghana, conservative traditional health practitioners in most of Africa, particularly South Africa, believe that plants grown in western ways (i.e. with fertilizers and in straight rows) will not have the same healing properties as plants harvested from wild populations (Fennel et al. 2004). Factors such as season, irrigation, and fertilization of material have resulted in changed medicinal activity of grown material, which has previously been stated in this statement (Prinsloo and Nogemane 2018).

Because natural populations are older due to sluggish development, bioactive chemicals in fast developing produced stock may be lower than bioactive compounds in wild populations (Schippmann et al. 2002a). People's ideas about the cultural applications of medical plants and plant parts, on the other hand, were shown to be more essential than their concerns about the potency of cultivated plants. As a result, demand for cultivated medicinal herbs has increased (Wiersum et al. 2006).

Planting amayeza species in a separate and secluded area of a home garden, on the other hand, was seen as a reasonable option for limiting the chance of coming into touch with contaminated persons. Wiersum et al. (2006) found that the majority of those polled said they would utilize farmed material for healing or protection. Lack of water for irrigation and challenges with propagation and a lack of knowledge with suitable culture needs such as soil and light conditions, are all obstacles to home garden cultivation (Wiersum et al. 2006).

Cultivation has a variety of market benefits over wild stocks when it comes to the manufacture of plant-based medicines (Schippmann et al. 2002a). The benefits include: (1) material collected from the wild and sold in street markets is frequently adulterated with unwanted, harmful plant species to enhance potency, whereas cultivated material provides reliable botanical identification, (2) provide a steady source of raw material, (3) wholesalers and pharmaceutical companies can agree on volumes and prices with the grower, (4) controlled post-harvest handling and thus appropriate quality control, and (5) standards can be adjusted to regu (Rates 2001).

Some species may be difficult to cultivate owing to biological characteristics or ecological needs, such as sluggish growth rates, soil and water requirements, poor germination rates, and insect vulnerability (Rao et al. 2004; Schippmann et al. 2002b). Economic feasibility is the primary motivation for cultivation, but it is also a restriction. Commercial harvesters gathering material from wild populations have no input costs, therefore cultivated stock will compete with stocks obtained from wild populations at trade markets. Only a few species could be offered at high enough prices to make cultivation economical due to low pricing et al (Schippmann et al. 2002a). Cultivation for profit is thus limited to a small number of high-priced and/or fast-growing species with a specific commercial market in mind. There is ongoing competition from inexpensive plant material collected in the wild with little or no cost to the gatherers. Despite the fact that several species, such as Siphonochilus aethiopicus and Warburgia salutaris (pepper bark tree), have become extinct or vulnerable in the wild (Botha et al. 2004), low prices mean that only a few slow-growing species are grown (Cunningham 1993). Siphonochilus aethiopicus, a plant with rhizomatous roots, is easy to propagate and cultivate, and it is effectively cultivated in South Africa's warm regions (Van Wyk 2008).

Cultivation may assist certain medicinal plant material producers in rural communities by allowing them to earn money, as both men and women are equally involved (Wiersum et al. 2006). If cultivation increases to the point that foreigners with cash come in and construct large-scale monocultural plantations for export markets, rural people may only gain from plantations as a consequence of accessible employment and hence off-farm income. Large medicinal plant farms developed by wealthy and powerful individuals may manipulate market dynamics to their benefit by enforcing low salaries, limiting the social and economic growth of local people. The country's elite and the national economy will be the primary benefactors of large-scale exports (Schippmann et al. 2002a). According to Schippmann et al. (2002a), those socially disadvantaged populations that rely on traditional medicine for survival and monetary income may not have access to agricultural land and hence are unable to compete with large-scale medicinal plant production by well-established farmers.

Cultivation is a conservation alternative for vulnerable medicinal plant species because the yearly sustained production is substantially larger than the continual gathering of material from wild populations. The impact on wild populations will be lessened if demand for these species can be fulfilled from farmed sources. However, it will not boost genetic diversity. As a result, stringent protection of surviving wild populations, better ex situ security of germ cell genetic material, and investment in selection and development programmes are all critical (Schippmann et al. 2002a).

12.8 Some Challenges Associated with the Utilization of African Medicinal Plant Resources

Uncontrolled exploitation and depletion of species, particularly endangered species, is a key concern of expanding medical plant use, leading to unsustainable methods and effects. The following are the reasons behind this challenge:

- Poverty – synonymous with use of forest resources for energy, food and shelter, etc.
- Unregulated mode of use/practice (absence of protocols and SOPs)
- Level of education of practitioners
- Cultural beliefs and unsustainable land use policy
- Species displacement in cultivation or farming
- Indiscriminate collection, bad harvest, and agricultural practice
- Lack of standardization, which could have helped in determining the direction and quantum of use through monitoring and documentation.
- Environmental factors (fire, anthropologic activities, etc.)
- Inadequate or lack of plan and action for conservation and preservation.
- Absence of large-scale cultivation and lack of strategic plan to meet future demands.
- Inadequate R&D information needed to erase or modify perception, improve standardization, documentation, good agricultural practice (GAP), good manufacturing practice (GMP), and eradicate bad or unsustainable practices.

12.9 Prospects and Potentials of African Medicinal Plant Resources

Advances in tissue culture and fermentation of medicinal plants have opened new possibilities for the large-scale and highly efficient manufacture of desired bioactive chemicals, while advances in genetic engineering have made large-scale biosynthesis of natural products feasible. Tissue culture (which includes plant cell and transgenic hairy root culture) is a viable option for producing uncommon and high-value secondary metabolites with medicinal significance (Rao and Ravishankar 2002). Micropropagation through tissue encapsulation of propagules not only makes storage and transit easier, but it also boosts regeneration rates (Baker et al. 2007). Synthetic seed technology, defined as artificially encapsulated somatic embryos (or other tissues) that can be cultivated in vitro or ex vitro, is a viable option when regular seeds are insufficient for propagation (Lata et al. 2008; Zych et al. 2005). Furthermore, utilizing molecular marker-based methodologies used at the genetic level, breeding enhancements may be made and breeding time can be greatly reduced (Lata et al. 2008; Rao and Ravishankar 2002).

12.10 Conclusion

Thousands of plant species thrive in Africa, each producing distinct and important chemical compounds, with medicinal plants still the continent's most plentiful resource. More practical information on medicinal plants, on the other hand, would increase their value in agricultural landscapes by assisting farmers in improving their livelihoods while also ensuring environmental sustainability. A large percentage of Africans rely on these plants for primary health care, but rising population, increased competition for land, and unsustainable exploitation/use as an industrial raw material and for food are threatening the medicinal plants' survival for future generations. As a result, advocates and practitioners must promote and practise sustainable utilization, conservation, and exploitation of plant resources and biodiversity.

References

Abegaz BM, Ngadjui BT, Folefoc GN et al (2004) Prenylated flavonoids, monoterpenoid furanocoumarins and other constituents from the twigs of *Dorstenia elliptica* (Moraceae). Phytochemistry 65(2):221–226

Agbabiaka TB, Guo R, Ernst E (2008) *Pelargonium sidoides* for acute bronchitis: a systematic review and meta-analysis. Phytomedicine 15(5):378–385

Agbor AM, Naidoo S (2011) Knowledge and practice of traditional health practitioners in oral health in the Bui division, Cameroon. J Ethnobiol Ethnomed 7(6):8

Ahana MC, Osawaru ME, Ogwu MC (2022) Status and potentials of the genetic resources of cocoyam (*Xanthosoma* Schott., Araceae) in Nigeria. Proceedings of Cukurova 8th International Scientific Researches Conference, Adana Turkey. 15–17 April, 2022. pp. 1–11.

Ahmad N, Hassan MR, Halder H et al (1999) Effect of *Momordica charantia* (Karolla) extracts on fasting and postprandial serum glucose levels in NIDDM patients. Bangladesh Med Res Counc Bull 25(1):11–13

Akhtar MS, Athar MA, Yaqub M (1981) Effect of *Momordica charantia* on blood glucose level of normal and alloxan-diabetic rabbits. Planta Med 42(3):205–212

Aliyu BS (2006) Common ethnomedicinal plants of the semiarid regions of West Africa. Triumph Publishing, Kano, Nigeria

Andersen ML, Santos EHR, Seabra MDLV et al (2004) Evaluation of acute and chronic treatments with *Harpagophytum procumbens* on Freund's adjuvant induced arthritis in rats. J Ethnopharmacol 91(2-3):325–330

Baker DD, Chu M, Oza U (2007) The value of natural products to future pharmaceutical discovery. Nat Prod Rep 24:1225–1244

Belhattab R, Amor L, Barroso JG et al (2012) Essential oil from*Artemisia herba-albaAsso* grown wild in Algeria: variability assessment and comparison with an updated literature survey. Arab J Chem 57(4):603–619

Benmansur A, Taleb-Bendiab SA, Mashev N et al (1990) Studies on the chemical composition of *Artemisia* (*Artemisia herba-alba*). Bolgarskoi Akademii Nauk 43(8):65–67

Botha J, Witkowski ETF, Shackleton CM (2004) Harvesting impacts on commonly used medicinal tree species (*Catha edulis* and *Rapanea melanophloeos*) under different land management regimes in Mpumalanga, Lowveld, South Africa. Koedoe 47(2):18

Bowie J (1830) Sketches of the botany of South Africa. South African Q J:27–36

Brendler T, Van Wyk BE (2008) A historical, scientific and commercial perspective on the medicinal use of *Pelargonium sidoides* (Geraniaceae). J Ethnopharmacol 119(3):420–433

Brendler T, Eloff LN, Gurib-Fakim A et al (2010) African herbal pharmacopeia. AAMPS Publishing, Mauritius

Brinkhaus B, Lindner M, Schuppan D et al (2000) Chemical, pharmacological and clinical profile of the East Asianmedical plant *Centella asiatica*. Phytomedicine 7(5):427–448

Bruchac M (2014) Indigenous knowledge and traditional knowledge. In: Smith C (ed) Encyclopedia of global archaeology, pp. 3814–3844. https://repository.upenn.edu/cgi/viewcontent.cgi?article=1172&context=anthro_papers. Accessed 7 Jan 2022

Chen W, Van Wyk BE, Vermaak I et al (2012) Cape aloes—a review of the phytochemistry, pharmacology and commercialisation of *Aloe ferox*. Phytochem Lett 5(1):1–12

Chime AO, Aiwansoba RO, Danagogo SJ, Egharevba II, Osawaru ME, Ogwu MC (2016) Effects of lining with leaves of *Triclisia dictyophylla* on the fungal composition of *Cola nitida* during storage. J Indust Res Technol 5(2):128–136

Chrubasik S, Bradley PR (2004) Addendum to the ESCOP monograph on *Harpagophytum procumbens* (multiple letter). Phytomedicine 11(7-8):691–695

Chungu D, Muimba-Kankolongo A, Roux J (2007) Bark removal for medicinal use predisposes indigenous forest trees to wood degradation in Zambia. S Hem For J 69(3):157–163

Conrad A, Frank U (2008) Extract of *Pelargonium sidoides* (EPs 7630) displays anti-infective properties by enhanced phagocytosis and differential modulation of host-bacteria interactions. Planta Med 74(6):682–685

Conrad A, Hansmann C, Engels I et al (2007) Extract of *Pelargonium sidoides* (EPs) improves phagocytosis, oxidative burst, and intracellular killing of human peripheral blood phagocytes *in vitro*. Phytomedicine 14(1):46–51

Conrad A, Bauer D, Hansmann C et al (2008) Extract of *Pelargonium sidoides* (EPs 7630) improves phagocytosis, oxidative burst, and intracellular killing of human peripheral blood phagocytes *in vitro*. Zeitschrift fur Phytotherapie 29(1):15–18

Cunningham AB (1988) An investigation of the herbal medicine trade in Natal/Kwazulu. In: Investigational Report No. 29. Institute of Natural resources, Pietermaritzburg (153 pp).

Cunningham AB (1993) Setting priorities at the interface between conservation and primary healthcare. In African Medicinal Plants. People and Plants Working Paper to United Nations Educational, Scientific and Cultural Organization (UNESCO) 53 pp. https://citeseerx.ist.psu.edu/viewdoc/download?doi=10.1.1.294.2667&rep&type=pdf. Accessed 7 Jan 2022

Cunningham AB (1997) An Africa-wide overview of medicinal plant harvesting, conservation and health care. In: Bodeker G, Bhat KKS, Burley J, Vantomme P (eds) Plants for forest conservation and health care. Food and Agricultural Organization of the United Nations (FAO), pp 116–129. www.fao.org/3/a-w7261e.pdf. Accessed 7 Jan 2022

Delveaux C, Sinsin B, Darchambeau F et al (2009) Recovery from bark harvesting of 12 medicinal tree species in Benin, West Africa. J Appl Ecol 46:703–712

Dold AP, Cocks ML (2002) The trade in medicinal plants in the Eastern Cape Province, South Africa. S Afr J Sci 98:589–597

Dubey DK, Biswas AR, Bapna JS et al (1987) Hypoglycaemic and antihyperglycaemic effects of *Momordica charantia* seed extracts in albino rats. Fitoterapia 58(6):387–390

Egharevba HO, Iliya I, Ibekwe N et al (2010a) Broad spectrum antimicrobial activity of *Psidium guajava* Linn. Leaf Nature Sci 8(12):43–50

Egharevba HO, Oladosu P, Okhale SE et al (2010b) Preliminary anti-tuberculosis screening of two Nigerian Laggera species (*Laggera pterodonta* and *Laggera aurita*). J Med Plant Res 4(12): 1235–1237

Egharevba HO, Ibrahim JA, Nduaguba GA et al (2015) Phytochemical, proximate pharmacognostic analyses and thin layer chromatography of *Chrysophyllum albidum* seed. Ewemen J Herbal Chem Pharmacol Res 1(1):6–12

Ernst E, Chrubasik S (2000) Phyto-anti-inflammatories: a systemic review of randomized, placebo-controlled, double-blind trials. Rheum Dis Clin North Am 26(1):13–27

Evivie, S.E., Ogwu, M.C., Ebabhamiegbebho, P.A., Abel, E.S., Imaren, J.O. and Igene, J.O. (2020). Packaging and the Nigerian food industry: challenges and opportunities. In: Food technology and culture in Africa. Ogunlade, C.A., Adeleke, K.M. and Oladejo, M.T. (eds). Reamsworth Publishing, Ibadan, Nigeria. 28 – 99pp.

Fenardji F, Klur M, Fourlon C et al (1974) White *Artemisia* (*Artemisia herba alba* L.). Revue delevage et de medecine veterinaire des pays tropicaux 27(2):203–206

Fennel CW, Light ME, Sparg SG et al (2004) Assessing African medicinal plants for efficacy and safety: agricultural and storage practices. J Ethnopharmacol 95:113–121. https://doi.org/10.1016/j.jep.2004.05.025

Ferreres F, Pereira DM, Valentao P et al (2008) New phenolic compounds and antioxidant potential of *Catharanthus roseus*. J Agric Food Chem 56(21):9967–9974

Figueiredo MSL, Grelle CEV (2009) Predicting global abundance of a threatened species from its occurrence: implications for conservation planning. Divers Distrib 15:117–121

Fluck AAJ, Mitchell WM, Perry HM (1961) Comparison of buchu leaf oil. J Sci Food Agric 12: 290–292

Friedman J, Yaniv Z, Dafni A et al (1986) A preliminary classification of the healing potential of medicinal plants, based on a rational analysis of an ethnopharmacological field survey among Bedouins in the Negev Desert, Israel. J Ethnopharmacol 16(2-3):275–287

Fullas F (2007) The status of African medicinal plants and their future. Afr Renais 4(3/4):102–112

Gamaniel KS (2005) Ownership and sustainability issues of botanical medicines. Ethnobot Res Appl 3:017–023

Geldenhuys C, Mitchell D (2006) Sustainable harvesting technologies, a South African guide. In: Diederichs N (ed) Commercialising edible plants. Sun Press, Stellenbosch, pp 23–39

Gilani AH, Khan AU, Ghayur MN et al (2006) Antispasmodic effects of Rooibos tea (*Aspalathus linearis*) is mediated predominantly through K^+-channel activation. Basic Clin Pharmacol Toxicol 99(5):365–373

Golblatt P, Manning JC (2000) *Cape plants: a conspectus of the Cape flora of South Africa*. National Botanical Institute of South Africa, Pretoria, South Africa

Grace OM, Prendergast HDV, Van Staden J et al (2002) The status of bark in South African health care. S Afr J Bot 68:21–30

Gurib-Fakim A, Brendler T, Phillips LD et al (2010) *Green gold—success stories using southern African plant species*. AAMPS Publishing, Mauritius

Hamilton A (2004) Medicinal plants, conservation and livelihoods. Biol Conserv 13:1477–1517

HerbFest (2015) "Food as medicine: utilization and sustainable exploitation of medicinal plant. The role of NIPRD". Workshop paper presented by the Director General/CEO NIPRD at HerbFest 2015, Abuja Nigeria

Ikhajiagbe B, Ogwu MC, Ogochukwu OF, Odozi EB, Adekunle IJ, Omage ZE (2021) The place of neglected and underutilized legumes in human nutrition and protein security. Crit Rev Food Sci Nutr. https://doi.org/10.1080/10408398.2020.1871319

ISSC-MAP (2007). https://www.floraweb.de/map-pro. Britta Pätzold, Susanne Honnef (WWF Germany and TRAFFIC, Johann-Wolfgang-Goethe University Frankfurt)

Izah SC, Aseibai ER (2018) Antibacterial and synergistic activities of methanolic leaves extract of lemon grass (Cymbopogon citratus) and rhizomes of Ginger (Zingiber officinale) against *Escherichia coli, Staphylococcus aureus and Bacillus subtilis*. Acta Sci Microbiol 1(6):26–30

Izah SC, Uhunmwangho EJ, Etim NG (2018a) Antibacterial and synergistic potency of methanolic leaf extracts of Vernonia amygdalina L. and Ocimum gratissimum L. J Basic Pharmacol Toxicol 2(1):8–12

Izah SC, Uhunmwangho EJ, Dunga KE (2018b) Studies on the synergistic effectiveness of methanolic extract of leaves and roots of Carica papaya L. (papaya) against some bacteria pathogens. Int J Complement Altern Med 11(6):375–378

Jayasooriya AP, Sakono M, Yukizaki C et al (2000) Effects of *Momordica charantia* powder on serum glucose levels and various lipid parameters in rats fed with cholesterol-free and cholesterol- enriched diets. J Ethnopharmacol 72(1-2):331–336

Jia Y, Zhao G, Jia J (2008) Preliminary evaluation: the effects of *Aloe ferox* Miller and Aloe arborescens Miller on wound healing. J Ethnopharmacol 120(2):181–189

Joubert E, de Beer D (2011) Rooibos (*Aspalathus linearis*) beyond the farm gate: from herbal tea to potential phytopharmaceutical. South Afr J Bot 77(4):869–886

Kaiser R, Lamparsky D, Schudel P (1975) Analysis of buchu leaf oil. J Agric Food Chem 23(5): 943–950

Kambizi L, Afolayan AJ (2006) Indigenous knowledge and its impact on medicinal plant conservation in Guruve, Zimbabwe. Indilinga Afr J Ind Knowl Syst 5(1):26–31

Karunanayake EH, Welihinda J, Sirimanne SR et al (1984) Oral hypoglycaemic activity of some medicinal plants of Sri Lanka. J Ethnopharmacol 11(2):223–231

Khan AU, Gilani AH (2006) Selective bronchodilatory effect of Rooibos tea (*Aspalathus linearis*) and its flavonoid, chrysoeriol. Eur J Nutr 45(8):463–469

Khanna P, Jain SC, Panagariya A et al (1981) Hypoglycemic activity of polypeptide-p from a plant source. J Nat Prod 44(6):648–655

Kigigha LT, Selekere RE, Izah SC (2018) Antibacterial and synergistic efficacy of acetone extracts of Garcinia kola (Bitter kola) and Buchholzia coriacea (Wonderful kola). J Basic Pharmacol Toxicol 2(1):13–17

Kim WJ, Kim J, Veriansyah B et al (2009) Extraction of bioactive components from *Centella asiatica* using subcritical water. J Supercritical Fluids 48(3):211–216

Kimura Y, Minami Y, Tokuda T et al (1991) Primary structures of N-linked oligosaccharides of momordin-a, a ribosome-inactivating protein from *Momordica charantia* seeds. Agric Biol Chem 55(8):2031–2036

Kreuz S, Joubert E, Waldmann KH et al (2008) Aspalathin, a flavonoid in *Aspalathus linearis* (rooibos), is absorbed by pig intestine as a C-glycoside. Nutr Res 28(10):690–701

Kunle OF, Egharevba HO (2013) Chemical constituents and biological activity of medicinal plants used for the management of sickle cell disease—a review. J Med Plant Res 7(48):3452–3476

Laid M, Hegazy MEF, Ahmed AA et al (2008) Sesquiterpene lactones from Algerian *Artemisia herba-alba*. Phytochem Lett 1(2):85–88

Lata H, Chandra S, Khan IA et al (2008) Propagation of *Cannabis sativa* L using synthetic seed technology. Planta Med 74:328

Lee J, Jung E, Kim Y et al (2006) Asiaticoside induces human collagen I synthesis through TGFβ receptor I kinase (TβRI kinase)-independent Smad signaling. Planta Med 72(4):324–328

Lis-Balchin M, Hart S, Simpson E (2001) Buchu (*Agathosma betulina* and *A. crenulata*, Rutaceae) essential oils: their pharmacological action on guinea-pig ileum and antimicrobial activity on microorganisms. J Pharm Pharmacol 53(4):579–582

Mahomoodally MF, Gurib-Fakim A, Subratty AH (2004) A kinetic model for *in vitro* intestinal uptake of l-tyrosine and d(+)-glucose across rat everted gut sacs in the presence of *Momordica charantia*, a medicinal plant used in traditional medicine against diabetes mellitus. J Cell MolBiol 3:39–44

Mander M, (1998) Marketing of medicinal plants in South Africa. A case study in Kwazulu-Natal. Report Presented to the Food and Health Organization of the United Nations, Rome. 158 pp. Available from. www.fao.org/docrep/w91985e/w91985e00.htm. Accessed 7 Jan 2022.

Mander M, Nthuli L, Diederichs N et al (2007) Economics of the traditional medicine trade in South Africa. S Afr Health Rev:189–196

Marloth R (1925) The flora of South Africa with synoptical tables of the genera of the higher plants. Darter Bros & Co, Cape Town, South Africa

Marnewick JL, Van der Westhuizen FH, Joubert E et al (2009) Chemoprotective properties of rooibos (*Aspalathus linearis*), honeybush (*Cyclopia intermedia*) herbal and green and black (*Camellia sinensis*) teas against cancer promotion induced by fumonisin B1 in rat liver. Food Chem Toxicol 47(1):220–229

Maroyi A (2013) Traditional use of medicinal plants in South-central Zimbabwe. Rev Perspect J Ethnobiol Ethnomed 9(31):18

Matsuda H, Li Y, Murakami T et al (1998) Antidiabetic principles of natural medicines. III. Structure-related inhibitory activity and action mode of oleanolic acid glycosides on hypoglycemic activity. Chem Pharm Bull 46(9):1399–1403

Matsuura H, Asakawa C, Kurimoto M et al (2002) α-Glucosidase inhibitor from the seeds of balsam pear (*Momordica charantia*) and the fruit bodies of *Grifola frondosa*. Biosci Biotechnol Biochem 66(7):1576–1578

Maundu P, Kariuki P, Eyog-Matig O (2006) Threats to medicinal plant species—an African perspective. In: Miththapala, S. (Ed.), Conserving medicinal Species. Securing a healthy Future. International Union for the Conservation of Nature (IUCN), pp. 47–58. https://portals.iucn.org/library/efiles/edocs/2006-022.pdf. Accessed 8 Jan 2022.

Melin A (2009) A bitter pill to swallow: a case study of the trade & harvest of Aloe ferox in the Eastern Cape, South Africa [M.S. thesis], Imperial College

Mncwangi N, Chen W, Vermaak I et al (2012) Gericke, Devil's Claw-A review of the ethnobotany, Phytochemistry and biological activity of *Harpagophytum procumbens*. J Ethnopharmacol 143: 755–771

Mohamed AEH, El-Sayed MA, Hegazy ME et al (2010) Chemical constituents and biological activities of *Artemisia herba-alba*. Records f Nat Prod 4(1):1–25

Monakisi C (2007) Knowledge and Use of Traditional Medicinal Plants by the Setswana speaking Community of Kimberley, Northern Cape of South Africa (M.Sc. Dissertation (Ecological Assessment)). University of Stellenbosch, 162 pp

Ogidi OI, Julius JC (2021) Assessment of ethno-pharmacological compounds and antimicrobial efficacy of white onion (*Allium cepa*) bulb extracts against pathogenic microbial isolates. PhytoChem BioSub J 15(2):167–174

Ogidi OI, Esie NG, Dike OG (2019a) Phytochemical, Proximate and Mineral compositions of *Bryophyllum Pinnatum* (Never die) medicinal plant. J Pharmacogn Phytochem 8(1):629–635

Ogidi OI, George DG, Enenebeaku UE et al (2019b) Efficacy evaluation of extracts of *Brassica juncea* (Brown mustard) seeds as potential antimicrobial agent against pathogenic microbes. J Med Plants Stud 7(4):263–265

Ogidi OI, Omu O, Ezeagba PA (2019c) Ethno pharmacologically active components of *Brassica Juncea* (Brown Mustard) seeds. Int J Pharm Res Dev 1(1):09–13

Ogidi OI, George DG, Esie NG (2019d) Ethnopharmacological properties of *Vernonia amygdalina* (Bitter Leave) Medicinal plant. J Med Plants Stud 7(2):175–181

Ogidi OI, Oguoma LMO, Adigwe PC et al (2021a) Phytochemical properties and in-vitro antimicrobial potency of wild edible mushrooms (*Pleurotus ostreatus*) obtained from Yenagoa, Nigeria. J Phytopharmacol 10(3):180–184. https://doi.org/10.31254/phyto.2021.10306

Ogidi OI, Okore CC, Akpan UM et al (2021b) Evaluation of antimicrobial activity and bioactive phytochemical properties of mango (*Mangifera Indica*) stem-bark extracts. Int J Pharm 8(5): 189–195. https://doi.org/10.13040/IJPSR.0975-8232.IJP.8(5).189-95

Ogidi OI, Chukwudi P, Ibe AI et al (2021c) Preliminary phytochemical profile and antimicrobial potentials of white-green African garden egg (*solanum macrocarpon*) fruits obtained from Yenagoa. ASIO J Pharm Herbal Med Res 7(2):01–05. https://doi-ds.org/doilink/06.2021-76274373/10.2016-19146535/ASIO-JPHMR/2021/457

Ogidi OI, Ayebabogha MN, Eze PU et al (2021d) Determination of Phytoconstituents and Antimicrobial activities of aqueous and methanol extracts of neem (*Azadirachta indica*) leaves. Int J Pharmacogn Chem 2(2):60–67. https://doi.org/10.46796/ijpc.vi.155

Ogwu MC (2019a) Towards sustainable development in Africa: the challenge of urbanization and climate change adaptation. In: Cobbinah PB, Addaney M (eds) *The geography of climate change adaptation in urban Africa*. Springer Nature, Switzerland, pp 29–55. https://doi.org/10.1007/978-3-030-04873-0_2

Ogwu MC (2019b) Lifelong consumption of plant-based GM foods: is it safe? In: Papadopoulou P, Misseyanni A, Marouli C (eds) Environmental exposures and human health challenges. IGI Global, Pennsylvania, USA, pp 158–176pp. https://doi.org/10.4018/978-1-5225-7635-8.ch008

Ogwu MC (2020) Value of *Amaranthus* [L.] species in Nigeria. In: Waisundara V (ed) Nutritional value of amaranth. IntechOpen, pp 1–21. https://doi.org/10.5772/intechopen.86990

Ogwu MC, Osawaru ME (2014) Comparative study of microflora population on the phylloplane of common okra [*Abelmoschus esculentus* L. (Moench.)]. Nigerian J Biotechnol 28:17–25

Ogwu, M.C. and Osawaru, O.P. (2022). State of the genetic resources of West African Okra (*Abelmoschus caillei* [A. Chev.] Stevels.): a taxon with industrial potentials. In: Ilkim, M. and Amanzholova, A. (eds.). Proceedings of the II Başkent International Conference on Multidisciplinary Studies, Ankara Turkey, pp. 93–102

Ogwu MC, Osawaru ME, Ahana CM (2014) Challenges in conserving and utilizing plant genetic resources (PGR). Int J Genetics MolBiol 6(2):16–22. https://doi.org/10.5897/IJGMB2013.0083

Ogwu MC, Osawaru ME, Aiwansoba RO, Iroh RN (2016a) Ethnobotany and collection of West African Okra [*Abelmoschus caillei* (A. Chev.) Stevels] germplasm in some communities in Edo and Delta States, Southern Nigeria. Borneo J Resour Sci Technol 6(1):25–36

Ogwu MC, Osawaru ME, Obadiaru O, Itsepopo IC (2016b) Nutritional composition of some uncommon indigenous spices and vegetables. International Conference on Natural Resource Development and Utilization (4th Edition): Positioning small and medium enterprises (SMEs) for optimal utilization of available resources and economic growth, Raw Materials Research and Development Council (Federal Ministry of Science and Technology), Maitama, Abuja Nigeria. pp. 286–300

Ogwu MC, Osawaru ME, Atsenokhai EI (2016c) Chemical and microbial evaluation of some uncommon indigenous fruits and nuts. Borneo Science 37(1):54–71

Ogwu MC, Osawaru ME, Obahiagbon GE (2017a) Ethnobotanical survey of medicinal plants used for traditional reproductive care by Usen people of Edo State, Nigeria. Malaya J Biosci 4(1): 17–29

Ogwu MC, Osawaru ME, Onosigbere-Ohwo U (2017b) Characterization of Okra (*Abelmoschus* [Medik.]) Accessions Using Dehydrogenase Isozymes and Protein. Maldives Natl J Res 5(1): 45–62

Ogwu MC, Chime AO, Oseh OM (2018a) Ethnobotanical survey of tomato in some cultivated regions in Southern Nigeria. Maldives Natl Res J 6(1):19–29

Ogwu MC, Onosigbere-Ohwo U, Osawaru ME (2018b) Morphological characterization of okra (*Abelmoschus* [Medik.]) accessions. Makara J Sci 22(2):67–76. https://doi.org/10.7454/mss.v22i2.9126

Okoro SO, Kawo AH, Arzai AH (2011) Phytochemical screening, antibacterial and toxicological activities of *Acacia senegal* extracts. Bayero J Pure Appl Sci 5(1):163–170

Osawaru ME, Ogwu MC (2014) Ethnobotany and germplasm collection of two genera of cocoyam (*Colocasia* [Schott] and *Xanthosoma* [Schott], Araceae) in Edo State Nigeria. Sci Technol Arts Res J 3(3):23–28. https://doi.org/10.4314/star.v3i3.4

Osawaru ME, Ogwu MC (2020) Survey of plant and plant products in local markets within Benin City and environs. In: Filho LW, Ogugu N, Ayal D, Adelake L, da Silva I (eds) African Handbook of Climate Change Adaptation. Springer Nature, Switzerland, pp 1–24. https://doi.org/10.1007/978-3-030-42091-8_159-1

Osawaru ME, Ogwu MC, Chime AO, Osifo E (2012) Morphological characterization of fruits and protein profiling of nine accessions of cultivated Okra species in Nigeria. Biol Environ Sci J Tropics 6(1):156–167

Osawaru ME, Ogwu MC, Omoigui ID, Aiwansoba RO, Kevin A (2016) Ethnobotanical survey of vegetables eaten by Akwa Ibom people residing in Benin City, Nigeria. Univ Benin J Sci Technol 4(1):70–93

Pereira DM, Ferreres F, Oliveira J et al (2009) Targeted metabolite analysis of *Catharanthus roseus* and its biological potential. Food Chem Toxicol 47(6):1349–1354

Pereira DM, Faria J, Gaspar L et al (2010) Exploiting *Catharanthus roseus* roots: source of antioxidants. Food Chem 121(1):6–61

Posthumus MA, Van Beek TA, Collins NF et al (1996) Chemical composition of the essential oils of *Agathosma betulina, A. crenulata* and an *A. betulina* x crenulata hybrid (Buchu). J Essential Oil Res 8(3):223–228

Primack RB (2012) A primer of conservation biology, 5th edn, Sunderland, Massachusetts. Sinauer 363

Prinsloo G, Nogemane N (2018) The effect of water availability on chemical composition, secondary metabolites and biological activity in plants. Phytochem Rev:1–14

Raman A, Lau C (1996) Antidiabetic properties and phytochemistry of *Momordica charantia* L., (Cucurbitaceae). Phytomedicine 2:349–362

Rao SR, Ravishankar GA (2002) Plant cell cultures: chemical factories of secondary metabolites. Biotechnol Adv 20:101–153

Rao MR, Palada MC, Becker BN (2004) Medicinal plants in agroforestry systems. Agr Syst 61: 107–122

Rates SMK (2001) Plants as source of drugs. Toxicon 39:603–613

Regassa R (2013) Assessment of indigenous knowledge of medicinal plant practice and mode of service delivery in Hawassa City, Southern Ethiopia. J Med Plant Res 7(9):517–535

Rood B (1994) Uit Die Veldapteek. Tafelberg-Uitgewers Bpk, Cape Town, South Africa

Sandasi M, Kamatou GPP, Baranska M et al (2010) Application of vibrational spectroscopy in the quality assessment of Buchu oil obtained from two commercially important Agathosma species (Rutaceae). South Afr J Bot 76(4):692–700

Schippmann U, Leaman DJ, Cunningham AB (2002a) Impact of cultivation and gathering of medicinal plants on biodiversity. Global trends and issues. In: FAO. 2002. Biodiversity and ecosystem approach in agriculture, forestry and fisheries. Satellite event on the occasion of the ninth regular session of the Commission on Genetic Resources for Food and Agriculture. Rome, 12–13 October 2002. Inter-Departmental Working Group on Biological Diversity for Food and Agriculture, Rome, 22 pp

Schippmann U, Leaman DJ, Cunningham AB (2002b) A comparison of cultivation and wild collection of medicinal and aromatic plants under sustainability aspects. In: Bogers RJ, Craker LE, Lange D (eds) 2006. Medicinal and aromatic plants, Springer, Netherlands, pp 75–95

Segal R, Feuerstein I, Danin A (1987) Chemotypes of *Artemisia herba-alba* in Israel based on their sesquiterpene lactone and essential oil constitution. Biochem Syst Ecol 15(4):411–416

Senanayake SGJN (2015) Indigenous knowledge as key to sustainable development. J Agric Sci 2(1):9. https://doi.org/10.4038/jas.v2i1.8117

Shukla A, Rasik AM, Jain GK et al (1999) *In vitro* and *in vivo* wound healing activity of asiaticoside isolated from *Centella asiatica*. J Ethnopharmacol 65(1):1–11

Simpson D (1998) Buchu—South Africa's amazing herbal remedy. Scott Med J 43(6):189–191

Strydom HA, King ND (2013) Fuggle and Rabie's Environmental Management, 2nd edn. Juta, Claremont, p 1142

Tahraoui A, El-Hilaly J, Israili ZH et al (2007) Ethnopharmacological survey of plants used in the traditional treatment of hypertension and diabetes in south-eastern Morocco (Errachidia province). J Ethnopharmacol 110(1):105–117

Ticktin T (2004) The ecological implications of harvesting non-timber forest products. J Appl Ecol 41:11–21

Timmer A, Gunther J, Rucker G et al (2008) *Pelargonium sidoides* extract for acute respiratory tract infections. Cochrane Database Syst Rev 3:CD006323

Van Andel T, Havinga R (2008) Sustainability aspects of commercial medicinal plant harvesting in Suriname. For Ecol Manag 256:1540–1545

Van der Heijden R, Jacobs DI, Snoeijer W (2004) The Catharanthus alkaloids: pharmacognosy and biotechnology. Curr Med Chem 11(5):607–628

Van Heerden FR, Van Wyk BE, Viljoen AM et al (2003) Phenolic variation in wild populations of *Aspalathus linearis* (rooibos tea). Biochem Syst Ecol 31(8):885–895

Van Wyk BE (2008) A broad review of commercially important southern African medicinal plants. J Ethnopharmacol 119(3):342–355

Van Wyk BE (2011) The potential of South African plants in the development of new medicinal products. South Afr J Bot 77(4):812–829

Van Wyk BE, Gericke N (2007) People's plants. A guide to useful plants of southern Africa. Briza, Pretoria, p 351

Van Wyk BE, Verdoorn GH (1989) Alkaloids of the genera Aspalathus, Rafnia andWiborgia (Fabaceae-Crotalarieae). South Afr J Bot 55:520–522

Van Wyk BE, Wink M (2004) Medicinal plants of the world: an illustrated scientific guide to important medicinal plants and their uses. Briza Publications, Pretoria, South Africa

Van Wyk BE, Van Oudsthoorn B, Gericke N (1997) Medicinal plants of south Africa, Briza Publications

Van Wyk BE, Van Oudtshoorn B, Gericke N (2009) Medicinal plants of south Africa, 2nd edn. Briza Publications, Pretoria

Van Wyk BE, Oudshoorn B, Gericke N (2013) Medicinal plants of south Africa, 2nd edn. Briza, Pretoria, p 336

VanWyk BE, Gericke N (2000) People's plants: a guide to useful plants of southern Africa. Briza Publications, Pretoria, South Africa

VanWyk BE, Smith GF (1996) Guide to the Aloes of South Africa. Briza Publications, Pretoria, South Africa

VanWyk BE, Smith GF (2004) Guide to the Aloes of SouthAfrica. Briza Publications, Pretoria, South Africa

Vermeulen WJ (2009) The sustainable harvesting of non-timber forest products from natural forests in the southern Cape, South Africa: development of harvest systems and management prescriptions (PhD Thesis (Conservation Ecology)). Stellenbosch University 197

Wagh VV, Jain AK (2013) Status of threatened medicinal plants of Jhabua district, Madhya Pradesh, India. Ann Plant Sci 2:395–400

Watt JM, Breyer-Brandwijk MG (1962) Medicinal and poisonous plants of southern and eastern Africa, 2nd edn. E&S Livingstone, Edinburgh, UK

Wegener T, Lupke NP (2003) Treatment of patients with arthrosis of hip or knee with an aqueous extract of Devil's claw (*Harpagophytum procumbens* DC.). Phytother Res 17(10):1165–1172

Wichtl M, Bisset NG (2000) Herbal drugs and phytopharmaceuticals. CRC Press, Boca Raton, FL

Wiersum KF, Dold AP, Husselman M et al (2006) Conservation cultivation of medicinal plants as a tool for biodiversity and poverty alleviation in the Amatola Region, South Africa. In Rogers RJ, Craker LE, Lange D (eds) Medicinal and aromatic plants, Netherlands, pp. 43–57

Williams VL, Balkwill K, Witkowski ETF (2000) Unraveling the market for medicinal plants and plant parts on the Witwatersrand. South Africa Econ Bot 54(3):310–327

Yesılada E, Gurbuz I, Shibata HJ (1999) *Momordica charantia*: an overview. J Ethnopharmacol 66: 289–293

Zeggwagh NA, Farid O, Michel JB et al (2008) Cardiovascular effect of *Artemisia* herba alba aqueous extract in spontaneously hypertensive rats. Methods Find Exp Clin Pharmacol 30(5): 375–381

Ziyyat A, Legssyer A, Mekhfi H et al (1997) Phytotherapy of hypertension and diabetes in oriental Morocco. J Ethnopharmacol 58(1):45–54

Zych M, Furmanowa M, Krajewska-Patan A et al (2005) Micropropagation of *Rhodiola kirilowii* plants using encapsulated axillary buds and callus. Acta Biol Cracov Bot 47:83–87

Chapter 13
Local Food Crops in Africa: Sustainable Utilization, Threats, and Traditional Storage Strategies

Matthew Chidozie Ogwu

Abstract In Africa, local foods produced from local food crops have social, economic, cultural, religious, moral, spiritual, symbolic, ethical, environmental, psychological, physiological, ethical, and intrinsic values and continue to evolve with indigenous systems. Local food crops include many orphan or minor crops that are indigenous or naturalized and invariably grown as well as wild edible plants that likely evolved with the hunter–gatherer–forager civilization of the Stone Ages and are as old as human prehistory. This chapter aims to discuss the diversity, distribution, and classification of local food crops in Africa as well as their utilization patterns, threats, and traditional storage strategies. *Sesamum radiatum, Hibiscus sabdariffa, Vangueria infausta, Carissa macrocarpa, Tulbachia violacea, Eragrostis abyssinica, Oryza glaberrima, Eleusine coracana, Cenchrus biflorus, Echinochloa stagnina, Vigna unguiculata, V. subterranean, Detarium senegalesis, Tylosema esculentum, Parkia biglobosa, Allanblackia stuhlmannii, Dissotis rotundifolia, Coleonema pulchellum, Oxalis pes-caprae, and Agathosma apiclata* are some examples of local food crops in Africa. These local food crops can be separated into several economic food crop categories based on their dominant utilization mode. The distribution of local food crops may be determined by the extent of human influence, for instance, either *ex situ* or *circum situ* (*circa situm*) in the case of minor or orphan crops (underutilized and neglected food crops), where they are held in human-managed agroecosystems or driven by *in situ* (and *quasi in situ*) considerations as in the case of wild edible plant species, wherein they are found in natural and or semi-natural environments without humans purposely managing their cultivation and growth. The mission of local food crop sustainability is to produce safe and quality food in sufficient quantity to meet the current needs of the human population in a locale without adversely altering the capacity of the environment and food systems to meet future needs. Some factors that contribute to

M. C. Ogwu (✉)
Goodnight Family Department of Sustainable Development, Living Learning Center, Appalachian State University, Boone, North Carolina, USA
e-mail: ogwumc@appstate.edu

S. C. Izah, M. C. Ogwu (eds.), *Sustainable Utilization and Conservation of Africa's Biological Resources and Environment*, Sustainable Development and Biodiversity 32, https://doi.org/10.1007/978-981-19-6974-4_13

the sustainable utilization of local food crops are humans and other biotic components and abiotic factors such as fertile land, water, soil amendments, and stable environment and energy source. Others are market systems, economic state, and connected emergent systems. Derivable benefits from local food crops in Africa include food, nutrition, and health, (agro)biodiversity and ecological, socio-cultural, and intrinsic and economic benefits. Traditional cribs and barns, pits or underground storage units, wrapping with other plants material, exposure to sunlight, drying on a bare floor, holding in baskets or trays, adding ground plant materials, bagging or sacking, wrapping with clothes, use of wooden shelf, and warehousing are some traditional storage strategies for local food crops. There is a need to promote the adoption, expansion, regular evaluation, breeding of improved varieties, and sustainable policies, production, and marketing strategies of Africa's local food crops.

Keywords Local food crops · Crop storage · Traditional methods · Underground pits · Post-harvest treatments · Plant utilization

13.1 Introduction

Food, food systems, and resources are the core of biological existence and socio-cultural interactions (Ray et al. 2020). Local foods can be considered as any indigenously produced liquid, semi-solid, solid, dried or frozen, dehydrated or rehydrated compounds or materials that can be ingested raw or cooked for their nutritional and health benefits, taste, or physiological relief (Ogwu 2019a, b, c; Osawaru and Ogwu 2020). In Africa, local foods have social, economic, cultural, religious, moral, spiritual, symbolic, ethical, environmental, psychological, physiological, ethical, and intrinsic values and continue to evolve with indigenous systems. An example is the food production with native crops as well as packaging and processing of local foods and food products and the different storage conditions and requirements (Chime et al. 2016; Ogwu 2019b; Ogwu et al. 2019; Evivie et al. 2020; Ikhajiagbe et al. 2021a, b). Local food crops include many orphan and minor crops, i.e., crops that are indigenous and invariably grown by indigenous people mostly at subsistence scales and may not be internationally traded (Mabhaudhi et al. 2019; Tadele 2019; Talabi et al. 2022) and wild edible plant species alongside major economic crops found almost all over the world. The answer to the almost present question of whether Africa can feed itself lie in closing the gap between current yield and yield potentials of indigenous crop resources including orphan and minor crops held on existing croplands (van Ittersum et al. 2016). In addition, van der Hoeven et al. (2013) noted the importance of storage techniques and utilization patterns in minimizing loss and wastage to address local food crops and food issues in Africa.

The new green revolution program is driving agricultural intensification in Africa specifically through the conversion of smallholder agricultural processes for production, profit maximization, building socio-ecological resilience, food security, and sustainable development (Clay and Zimmerer 2020). The drive to commercialize agriculture in Africa is economic development with little consideration accorded to systems linked to agriculture (Carletto et al. 2017). This has imposed pressure on the other systems, which in turn affect local food crop availability and production because of issues linked to the farmer's crop choices (such as what to grow and family income status), environmental condition, nutrition security, food distribution, and marketing amongst other things that have a foundational influence on agricultural inputs and outputs for sustainable social and economic development. Frelat et al. (2016) provided data to support the interconnectedness of people, environment, and economic development from smallholder farmers in parts of Africa. Moreover, the transition from subsistence to commercial agriculture may exacerbate malnutrition and non-communicable diseases in non-city centres rather than their mitigation and management (van der Hoeven et al. 2013; Erinle et al. 2021).

The historical, social, economic, and cultural benefits of Africa's local food crops are well-documented from the start of domestication and early visits of European and Arabian merchants to the adoption of new value systems and modern-day wealth creation interests (Brun 1991). These crops evolved with the society and hold a historical record of social, economic, and cultural development. Therefore, the transition to modern agriculture hinged on a few exotic or naturalized crop species solely for economic benefits will further relegate native crops to the background regardless of their relevance (Chivenge et al. 2015). Moreover, these crops also contribute to Africa's (agro)biodiversity composition and richness and their abandonment will, directly and indirectly, affect linked processes. Africa's food system is already described as fragile because of the multiple inefficiencies of production and supply to meet the food and environmental needs of the continent; therefore, building resilience through local food crops is paramount (Shilomboleni 2020). Interests in local food and food crop resources by researchers and policymakers should have a broad focus to address and increase the resilience of the entire food security dimension including their stability. The small size of local food systems and local food crop resources used to address part of the needs of the system poseses a unique challenge that is often perceived as easy because of the size. However, the size and importance of the system to rural and community development might make differentiating each sustainable development dimension difficult (Stein and Santini 2021). For instance, in the review of local food climes, Caspi et al. (2012) noted the scarcity of data and measurement challenges of salient aspects of the food security dimension, which makes it difficult to assess the sustainability of the system. In addition, certain indices tied to the local food economy and politics drive demand and supply for local food crops and products with impacts on rural natural and social capitals, food security, human behaviour, and environmental outcomes and require integration for better management (Martinez et al. 2010; NRC 2015; Paul et al. 2019; Vågsholm et al. 2020). This is also the connection point between local and global food systems in the form of local data and knowledge.

The focus of this chapter is the sustainable utilization of local food crops in Africa. This chapter will begin by discussing the diversity, distribution, and classification of local food crops in Africa and proceed to enumerate utilization patterns and trends in local food crops within the continent. Next, the chapter will present some threats to the diversity and sustainable utilization of local food crops. Finally, this chapter will highlight the importance of storage and local food crop management strategies to the sustainable utilization of Africa's local food crop resources.

13.2 Local Food Crop Resources in Africa

Local food crop resources refer to indigenous plant genetic resources used to meet local food needs. Local food crop is the connector between traditional and indigenous foods. Traditional food, i.e., staples, eaten by cultural groups within the last centuries, whereas indigenous food refers to the food eaten since time immemorial (Baskarachary et al. 2015; van der Merwe et al. 2016). Local foods made from local food crop resources are cheaper, safer, and more nutritious compared to exotic ones. Furthermore, local people often adopt time-tested local cultivation methods that are driven by indigenous ecological knowledge and techniques including post-harvest processing, food preparation, preservation, and storage methods (Kuyu and Bereka 2019). Consumption of indigenous and traditional food made from local food crops has the potential to shape food, nutrition, health, and income security within the African continent. However, regardless of their connection to traditional and indigenous foods and food cultures, Africa's local food crops are not well-known and are dwindling because of the perception that introduced crops which are better in many aspects (Akinola et al. 2020). In addition, the productivity and yield from local food crops are severely affected by extreme environmental events' changes, especially by drought and floods (Ogwu 2019a, b, c; Ngcamu and Chari 2020). The work of Masipa (2017) showed that these climate-change linked events affect local food crop production, distribution, and consumption by influencing availability, accessibility, utilization, stability, and affordability. In addition, local food crops are particularly more vulnerable to these effects, since they have not been improved cultivars, unlike their exotic relatives.

African local food crop resources can be divided into two classes—(1) orphan and minor food crops and (2) wild edible plants.

Orphan and minor food crops: These are underutilized and neglected local food crop resources and they make up a significant amount of the local food crop in Africa. They are considered neglected, because they are largely ignored by researchers, policymakers, marketers, and investors within and outside the continent. Nonetheless, these crops are widely accepted by indigenous people and their genetic resources are held and maintained by local farmers (Ogwu et al. 2014; Osawaru and Ogwu 2014; Talabi et al. 2022). These crops are rich in nutritional phytochemicals and have optimal growth requirements on the relatively poor soil and extreme environmental conditions of the continent making them a significant contributor to dietary diversification as well as food and nutrient security (Mabhaudhi et al. 2019; Talabi et al. 2022). The main secondary use of these orphan and minor local food crop resources is in traditional medical practices for the treatment and management of diverse health conditions (Osawaru and Ogwu 2014; Osawaru et al. 2016; Ogwu et al. 2016a, 2017; Ogwu 2020). This is a huge sector in the continent that is driven by knowledge and practices associated with these crop resources. Mabhaudhi et al. (2019) mentioned that they are also important in addressing climate change and environmental degradation challenges from a local viewpoint as well as essential for job creation and small-scale income securities. On the other hand, van der Hoeven et al. (2013) considered orphan and minor crops as indigenous and traditionally valued plants because of the local knowledge associated with these crops as well as the pattern of usage. Other workers (Ogwu et al. 2016b; Mabhaudhi et al. 2017) referred to them as underutilized and neglected food crops because of the poorly diversified utilization pattern associated with part(s) or whole plant and little to no adoption of contemporary management and utilization approaches for their germplasms.

Wild edible plants: Wild edible plants also make up a significant amount of the local food crop resources in Africa and they likely evolved with the hunter–gatherer–forager civilization of the Stone Ages. Borelli et al. (2020) opined that their consumption is "*as old as human prehistory*" and they contributed to subsequent homestead location and crop domestication. Forests are considered the reservoir for edible wild plants on the continent. Their unique ecology and plant characteristics, i.e., growing and reproducing without human influence, make them available when other crops are out of season and they contribute significantly to maintain food security (Motti 2022). However, to avoid food poisoning, knowledge about availability, the required method of processing before consumption, and phytochemical contents are necessary. The utilization pattern for wild edible plants is linked to ethnobotanical knowledge, while availability is linked to environmental conditions (growth requirements) and phenology, especially plant habit and maturity period. The work of Ray et al. (2020) consolidates the fact that wild edible plants have long been associated with indigenous and traditional food and still provide easy access to micronutrients. However, wild edible plants are significantly undervalued, because alone they cannot fill the food demand and supply gap, yet without them, the gap would be wider (Bharucha and Pretty 2010). Some vital characteristics of wild edible plants according to Shumsky et al. (2014) and Catarino et al. (2019) include:

- High availability and use, which depends on local ecological knowledge systems and customs
- Tolerant of extreme conditions than minor and orphan crops
- A low-cost option for enhancing food and nutrient security of poor rural households with little to no inputs
- Most beneficial to households that are vulnerable to climate change effects
- Most reliable food source during famine
- Important non-timber forest products that are traded in local markets

Local food crops can be separated into several economic food crop categories based on their dominant utilization mode (Table 13.1). It should be noted that the list presented in Table 13.1 is only a very miniature representation of the thousands of indigenous minor and orphan crops and wild edible plants that make up the actual and potential local food crop resources of Africa.

Table 13.1 Some examples of local food crops and their classification into major economic food crop groups including native and naturalized minor or orphan food crops and wild edible plants

Economic food crop category	Some local food crops
Leafy vegetables and fruit crops	*Amaranthus species, Cleome gynandra, Urtica urens, Portulaca oleracea, Citrullus lanatus, Ricinodendron rautanenii, Cucurbita maxima, Lagenaria siceraria, Telfaria occidentalis, Celosia argentea, Vernonia amygdalina, Sonchus asper, S. oleraceus Abelmoschus caillei, A. esculentus Ficus species, Cucumis manni, Ensete edule, Solanum macrocarpon, S. aethiopicum, Corchorus spp., Psidium guajava, Musa paradisiaca, M. sapientum, M. balbisiana, M. acuminata, Annona senegalensis, Carica papaya, Bombax costatum, Adansonia digitata, Sesamum radiatum, Hibiscus sabdariffa, Vangueria madagascariensis, Vangueria infausta, Cocos nucifera, Uapaca kirkiana, Talinum fruticosum, Solanum scabrum, Parinari curatellifolia, Irvingia gabonensis, Gnetum africanum, Garcinia mangostana, G. livingstonei, Ensete ventricosum, Anarcadium occidentale, Diospyros mespiliformis, Dacryodes edulis, Cyphomandra batacea, Cucumis metuliferus, Crassocephalum rubens, Chrysophyllum cainito, Nasturtium offcinale, Malva sylvestris,* etc.
Stimulants and beverage crops	*Carissa macrocarpa, Tulbachia violacea, Salvia africana lutea, Sclerocarya cafra, Grewia occidentalis, Coffea Arabica, Ximenia Americana, Catha edulis, Cola nitida, C. acuminate, Saba senegalensis,* etc.
Grains and grasses	*Sorghum verticiliflorum, S. bicolor, Pennisetum americanum, Digitaria exilis, D. iburua, Eragrostis abyssinica, E. tef, Brachiaria deflexa, Oryza glaberrima, Eleusine coracana, Cenchrus biflorus, Echinochloa stagnina, Panicum laetum,* etc.
Fibre crops	*Hibiscus cananabinus Vitex doniana, Crotalaria juncea,* etc.
Pulses and leguminous crops	*Vigna unguiculata, V. subterranean, V. radiata, Vicia faba, Sphenostylis stenocarpa, Acacia Senegal, Detarium senegalesis, Tylosema esculentum, Parkia biglobosa, Phaseolus vulgaris, Macrotyloma geocarpum, Lablab purpureus, Faidherbia albidia,* etc.
Oil crops	*Elaeis guineensis, Balanites aegyptiaca, Sesamum indicum, Vitellaria paradoxa, Macadamia ternifolia, Sideroxylon spinosum,* etc.
Root and tuber crops	*Colocasia esculenta, Xanthosoma sagittifolium, Dioscorea rotundata, D. cayenensis, D. bulbifera, D. preussii, D. praehinsilis, D. sansibarensis, D. dumetorum, D. alata, Manihot esculenta, Plectranthus rotundifolius, Ipomoea batatas,* etc.
Medicinal plants	*Pelargonium culallatum, P. tomentosum, Coleonema pulchellum, Artemesia afra, Carpobrotus edulis, Mentha longifolia, Strychnos* species, *Neocarya macrophylla, Tamarindus indica, Ziziphus mauritiana, Antidesma venosum, Carpobrotus edulis, Bidens pilosa, B bipinata, Portulaca oleracea, Justicia flava, Syzygium guineense, Moringa oleifera, Lannea microcarpa, Allanblackia stuhlmannii, Dissotis rotundifolia,* etc.
Spices and condiments	*Coleonema pulchellum, Oxalis pes-caprae, Agathosma apiclata, Tulbaghia violacea, Ocimum, Ricinodendron heudelotii, Allium cepa, Zingiber officinale, Monodora myristica, Aframomum meleguetta, Piper guineense, Xylopia aethiopica, Cucuma longa,* etc.

Among these local food crops, vegetables are considered the most promising despite their complex storage and marketing requirements (Ogwu et al. 2016c). Vegetables especially wild ones are a long-standing rural food source for vitamins and proteins (Misselhorn and Hendriks 2017). This is also supported by the findings of Ojelel et al. (2019), wherein fruits and leafy vegetables account for more than 90% of collected wild edible plants in Uganda. Results of Baraka et al. (2022) show that the cultivation of local (minor and orphan and wild edible) vegetables as well as their conservation will make a significant contribution to the state of food and income security within the continent.

In terms of the distribution of local food crop resources, this may be determined by the extent of human influence or control. For instance, either *ex situ* or *circum situ* (*circa situm*) in the case of minor or orphan crops (underutilized and neglected food crops). Here, the local food crop resources are held in farmlands, home gardens, urban gardens, botanical gardens, distant farms, and other human-managed agroecosystems. In the case of wild edible plants, the distribution is driven by *in situ* considerations, wherein they are found in natural (e.g., original or primary forests), semi-natural (e.g., secondary forest) or human-created environments, but their cultivation and growth are not purposely carried out or entirely determined by humans. In some cases, this wild edible plant distribution is influenced by *quasi in situ* setups to prevent genetic erosion and conservation threats and to mitigate the effects of environmental change.

13.3 Sustainable Utilization of Local Food Crops

The focus of local food crop sustainability is to produce quality food in sufficient quantity and to meet the current needs of the human population within a locale without adversely altering the capacity of the environment and food systems to meet future needs. Factors that contribute to the sustainability of the local food system besides human and other biotic components are fertile land, water, availability of soil amendments, and a stable environment and energy source (Shennan 2008; NRC 2015; Morawicki and Díaz González 2018). Others are market systems, economic state, and connected emergent systems. A sustainable local food system should be able:

- To meet human food, fibre, and shelter needs
- To enhance environmental integrity and resource base including those needed for local food production and utilization
- To use non-renewable resources and food resources appropriately not to damage natural nutrient sinks and reservoirs
- To support informed transformation and socio-political commitment and accountability
- To maintain the equilibrium necessary for the functioning of renewable resources, such as soil

- To maximize yield from available inputs for income, ecological, and societal security

If sustainable production and utilization are not placed at the core of local food crop practices, it may result in negative environmental, economic, and social impacts (Lindgren et al. 2018). This would be in the form of reduced crop yield, less nutritious crops, poor environmental quality, loss of local food crop diversity, changing consumer behaviour, less spending power and loss of income, negative economic growth and development, etc. Refocusing the local food crop supply chain through product diversification and building infrastructure for the processing of perishable local food crops and products can foster sustainability (Nchanji and Lutomia 2021). This will help adapt the system to or encourage the adoption of reforms, valuation, reduction of waste from local food crops, standardized sustainable assessment, and practices, ensure healthy diets from robust crop diversity and agroecosystems, and strengthen governance (Delabre et al. 2021). This exemplifies the sustainability transition described in Gaitán-Cremaschi et al. (2019), wherein a productivist paradigm in local food crop management resulted in substantial amounts of inexpensive and safe food for the populace. In Gaitán-Cremaschi, they highlighted the benefits of diagnosing each aspect of the local food crop resources and food value chain to design an operational approach to characterize how and where the sustainable transition is most needed to ensure usefulness and address shortcomings. Their conceptual framework included characterization of the diversity in local food crops at a geographical scale and then the evolution of a classification system to support mainstream practices that would include and ensure the sustainable utilization of local food crop resources. In the views of Pretty and Bharucha (2014) and Rockström et al. (2017), sustainable utilization of local food crops would require a research-driven paradigm shift toward effective agricultural intensification involving the crop resources and integration of connected goals through the adoption of sustainable multidimensional practices to meet human needs while addressing ecosphere resilience. A balanced agricultural policy that gives equal attention to maximize yield from local food crops, sustainable utilization patterns, fair and equitable sharing of benefits and losses, and human nutrition, economic growth, and environmental justice will have a positive impact on the sustainable utilization of local food crops. This approach signals a win–win scenario and also meets diverse pertinent goals such as land consolidation, food production technological innovations, and management to ensure timely nutritious food availability, stability, and accessibility as well as social, ecological, and economic sustainability (Davis et al. 2019; Vargas et al. 2021; Lu et al. 2022).

Local food crops are put to diverse uses within Africa. The diversity and value of indigenous food are an expression of the contribution of local food crops. Besides the production of indigenous food and food products, Akinola et al. (2020) opined that local food crops also contribute to the economic, environmental, socio-cultural, and nutritional well-being of the continent, but most of these value systems for local food crop resources have not been widely examined. Broadly, the utilization of local

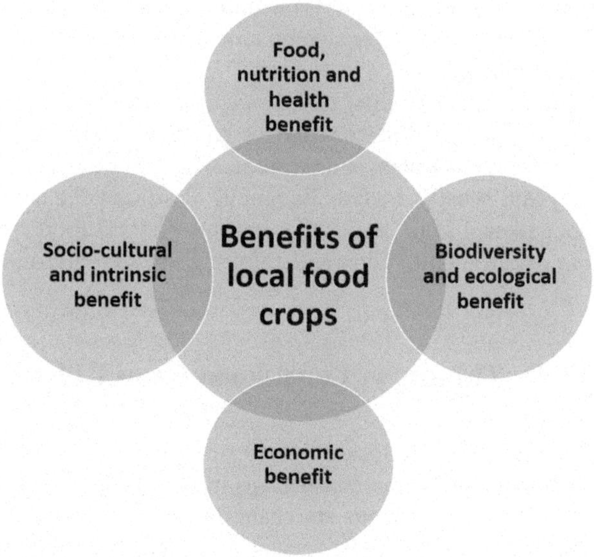

Fig. 13.1 Benefits of local food crops

food crops can be placed into one of the four categories, which are linked to their perceived, potential, actual, and derivable benefits (Fig. 13.1):

i. Food, nutrition, and health utilization benefits
ii. Biodiversity and ecological benefits
iii. Socio-cultural and intrinsic benefits
iv. Economic benefits

Food, nutrition, and health benefits: Local food crops are rich in nutrients and remain central to traditional food cultures and medicine practices or herbalism. Food is medicine and local food crops are often used in the treatment and management of diverse disease conditions. Lately, due to the poor health caused by poor diet, scientists and medical practitioners are in favour of integrating local food crops into the conventional health care system as a form of food and nutrition intervention (Verma et al. 2018; Downer et al. 2020). Local food crops have been used to address micronutrient deficiency through the "*Hidden Hunger*" and "*Nutrient Sustainability*" and other related initiatives (Burchi et al. 2011; Finley et al. 2017).

Biodiversity and ecological benefits: Besides promoting food diversity and system resilience, local food crops contribute to a biodiverse environment within the continent. Their management can be used to model environmental outcomes, such as land use impacts, patterns, and trends from their stand structure (Durazzo and Lucarini 2021). This loss of this agro-biodiversity can result in environmental susceptibility to stressors and functional losses (Shelef et al. 2017).

Socio-cultural and intrinsic benefits: Local food crops are a symbol of cultural diversity, spiritual heritage, and identity of indigenous people and ethnic groups (Akinola et al. 2020). They are important in the development and practice of socio-cultural norms and lifestyles within local communities. They are also celebrated as is the case in Nigeria for several crops such as the New Yam festival and are used in spiritual rites.

Economic benefits: Local food crops are integral to rural economic development and provide income security for growers, sellers, processors, food preparers, food vendors, consumers, and waste managers. Local markets in rural centres are where they are mostly traded. Economic development and industrialization have been documented to have negative impacts on local food crop utilization through changing policies, cultivation and consumption patterns, and acceptability.

13.4 Threats to Local Food Crop Utilization

The main threat to local food crops is the transition to commercial agriculture which is dependent on major cash crops from outside Africa. In addition, people's perception and value systems are changing and threatening local food crops with neglect (Akinola et al. 2020). For instance, the industrialization of food and food production through commercial agriculture is driving the decline in domestication, as well as in the demand and supply of local food crops. There is a rising pressure on local food crops to meet the food demands of a growing population in the face of climate change. In addition, human behaviour and unhealthy practices of overexploitation (of edible wild plant species), overconsumption, and excess food loss and waste can affect the social and intrinsic benefits of local food crops (Morawicki and Díaz González 2018). The effects of unsustainable exploitation and consumer behaviour were also discussed extensively in Nchanji and Lutomia (2021), wherein they presented results to show that it contributes to shorten availability or ensure unavailability, accessibility, affordability, reducing social connections, health and safety while increasing local food crop losses and waste and harmful environmental emission.

Urbanization and the emigration of farmers to city centres were reported in Satterthwaite et al. (2010); Ogwu et al. (2019) as a threat to the local food crop utilization and diversity of their genetic resources. According to Noort et al. (2022), expansion of urban centres and availability of disposable income are causing a shift in diet and lifestyle that were dependent on local food crops. Youths and women make a substantial amount of growers and holders of local food crop germplasms and when the effects of climate and environmental change reduce their yield and income from local food crops when they are likely to move to the city centres in search of other means of livelihood. Poverty and inequality, infrastructural development, socio-ecological resilience, agricultural practices, and environmental performance are common threats to local food crop utilization to meet local food needs. For instance, women and youth often lack access to land, credit facilities, social protection, and redress for violations which may discourage them from utilizing local food crops (Jacobi et al. 2020). To address environmental degradation-driven emigration issues, there is a need to adopt a cost-effective ecosystem-based adaptation strategy to regulate rural–urban development and promote resilience within agricultural systems as presented in the work of Zhong et al. (2022). Nonetheless, the absence of investment or institutional framework to coordinate activities related to local food crop resources also discourages utilization. Investment in transgenic technologies to create and support experimental breeding of improved cultivars alongside a legal framework to coordinate intellectual property rights would go a long way to support and promote activities related to the utilization of local food crop resources (Batur and Dedeurwaerdere 2014). The lack of gene banks and seed libraries for local food crops is also a topical issue. The establishment of these facilities would encourage the documentation and housing of genetic variations in local food crops, and then, they can be exploited in breeding improved varieties. This was highlighted in the work of Lusty et al. (2021), wherein they stated that gene banks support breeding objectives to stem the loss of agrobiodiversity. This would also be complementary to conservation efforts for local food crops.

Overexploitation of certain groups of local food crops (e.g., wild edible plants) is threatening current and future availability and use and there are few to no policies in place to ensure their conservation and sustainable use (Borelli et al. 2020). They also listed the following adapted from Heywood (1999) as topical threats to utilize local food crop resources:

- Few to no information on the use and socio-cultural and economic relevances of local food crops to rural communities and livelihoods
- Few to no data on the value of local food crop resources
- Policy neglect or weak policies that only focus on agrobiodiversity for commercial agriculture
- Presence of substitutes that are generally perceived as superior to local food crops
- Lack of infrastructure and extension support for local food crop resources and germplasm holders
- Absence of standards for assessing the quality of local food crop resources
- Lack of data on genetic diversity and distribution
- Value chain inefficiencies such as the unequal supply of local food crop resources
- Absence of inconsistencies in information about the quantity of harvest and contribution of household food and income security
- Absence of conservation efforts

According to MacFall et al. (2015), threats to the utilization of local food crop resources may be divided into natural and anthropogenic sources. Anthropogenic threats occur more frequently and often have greater effects on local food crop resources. Natural threats could be climate change and natural disasters and may be biological and ecological. The biological threats may result in the loss of germplasm, increased incidence of disease pathogens, and decline in soil organic matter as well as in the natural functionalities necessary to sustain utilization of the genetic resources, while ecological threats may be disruption of nutrients, and loss of soil fertility leading to a decline in productivity. Anthropogenic threats can be from population increase and industrialization, which may promote a cultural shift away from the use of local food crop resources and utilization patterns as well as land-related issues. Land governance and traditional leadership are central to the use, perception, access, quality, and diversity of local food crops in rural communities (Ekesa et al. 2020; Ngcamu and Chari 2020). Unfavourable land policies such as those that encourage the displacement of local farmers and the germplasm that they hold from their native land threaten the continuity and sustainable use of local food crops. Since the 2000s, significant amount of land has been removed from the lands of traditional holders and such deals affect local food production (Müller et al. 2021). Defects in the network that sustain the utilization of local food crop resources will harm the resources. In addition, the incidence of diseases, pests, and pathogens threatens the diversity of local food crops and highlights the need for improved cultivars. Water availability, environmental factors, and biogeochemical regulatory processes may affect the use of local food crops. An intergenerational shift in culture can place pressure on the utilization pattern of local food crop resources leading to their loss in some cases (Montúfar and Ayala 2019).

13.5 Traditional Food Crop Storage Strategies

Local food crop storage is an essential aspect of post-harvest processing and often incorporates local knowledge and practices. The aim is largely to make local food crops available between seasons and in times of famine or unfavourable environmental and economic conditions. Stored materials also provide planting materials for the next growing season. Storage and post-harvest processing of local food crop resources are important because of the following reasons:

- To extend the shelf life and preserve the nutritional quality of the local food crop.
- To extend the availability of the food crops.
- To encourage large-scale processing and sale.
- To extend the viability of the genetic resources and encourage multiplication and material exchanges.
- To stem extinction or disappearance of the plant genetic resources and serve as a pseudogene or seed banks.
- Preservation of nutritional quality and value of local food crops.
- To promote rural sustainability through food security and socio-economic stability.
- To control food prices and the price of cultivation materials.
- To document yield and other output indices for socioeconomic, cultural, and environmental sustainability.
- To encourage education and research founded on local food and food crop cultures.
- To preserve the socio-cultural and spiritual heritage of rural and indigenous communities and the local populace in general.
- To provide the opportunity for out-of-season sales and export markets.

Fig. 13.2 Classification of local food crop storage systems and facilities based on duration

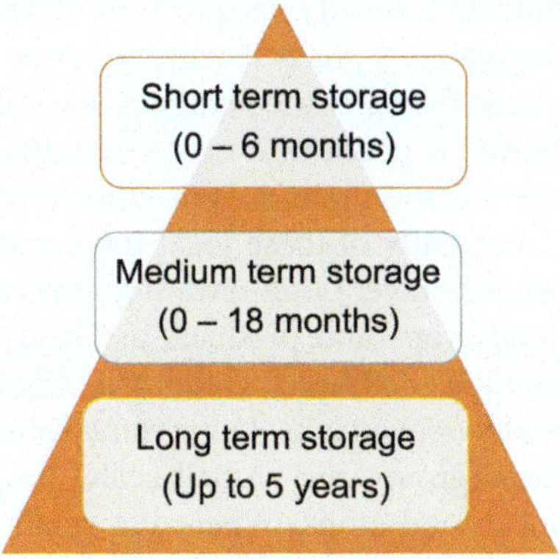

During harvest seasons of local food crops, availability and supply are significantly higher than demand and consumption, which makes the price of local food crops to plunge. In addition, storage and post-harvest processing of local food crops contribute to reduce unnecessary field losses. Traditional storage facilities and techniques only provide medium-to-short-term storage (Chime et al. 2016; Ikhajiagbe et al. 2021b). The largely subsistent farming pattern requires less intense storage requirements coupled with the small yield quantity. However, the main reason behind the short-to-medium storage period is the lack of elaborate institutional support to provide much-needed infrastructures. Environmental conditions also exacerbate storage challenges. Specifically, high rainfall and humidity during the rain and dry season in parts of West Africa. These climate conditions are not adequate for the traditional method of storage practised by most farmers. The basic storage principles to consider for each or group of local food crop resources would be the crop-specific storage requirements or principles, duration of storage, required size of the storage facility, type of plant material, and available infrastructure. The main factors considered in the storage and postharvest processing of local food crop resources are:

- Availability of pests such as rodents, ants, insects, and weevils.
- Harvesting season, i.e., prevalent environmental condition at the time immediately during and immediately after the crops are harvested.
- Moisture content as crops with high moisture contents such as fruits and vegetables are stored for a considerably shorter period compared to drier ones, such as cereals and pulses.
- Maturity or ripening stage and processes.
- Biodegradability of essential parts of the local food crops.
- Hygroscopicity of the local food crop, i.e., extent of shrinkage and swelling that may affect storage length, quality, and sale.

In terms of traditional storage systems and facilities required, local food crops may be classified based on the storage duration into three categories viz—short-term, medium-term, and long-term storage system or facility (Fig. 13.2). Short-term storage is mostly used to hold materials from 0 to 6 months and is used for fruits and vegetable. Care is usually given to control water loss during storage, and in some cases, they are dried entirely and sold dried. Medium-term local storage systems last between 0 and 12 or 18 months and care is paid to monitor the quality and deterioration rate. It is used for root and tuber crops such as cocoyam, and cassava. The quality of these local food crops may not be guaranteed after 12–18 months in local storage. Long-term storage is used for grains—cereals and pulses, and the local food crops and products are held in storage for up to 5 years depending on the storage conditions. Under the long-term storage, attention is given to prevent insect and rodent attacks. In the same vein, the scale of storage systems may be divided into three groups viz—small-scale, medium-scale, and large-scale storage systems for local food crops. Where the small-scale storage facility can hold up to 1 tonne of local plant material but not more and medium-scale storage facility can hold up to 50 tonnes, while the large-scale storage facility for local food crops can hold anything more than 50 tonnes depending on the output. The medium- and large-scale storage facilities are often community-owned and managed.

Some traditional storage structures and systems used for local food crops include traditional cribs and barns, pits or underground storage units, wrapping with other plant materials, exposure to direct sunlight, drying on bare floor, baskets or trays, pouring or adding ground plant materials, bagging or sacking, wrapping with clothes, wooden shelves and warehouse or stalls (Fig. 13.3)

- **Traditional cribs:** Used for storing grains (cereals and pulses) and may be cylindrical or rectangular shaped rhombus made from palm fronds or bamboo on raised wooden, mud, or concrete structure or simply placed on the floor. This is an example of a medium to long-term storage structure.
- **Traditional barns:** Used for yams (*Dioscorea* species), cassava (*Manihot* species). These shelf-like support structures are simply connected to wooden or bamboo poles with spaces in between them for holding the yams. The yam tubers are held on the barn with raffia or other local cordage materials. This is an example of a medium storage structure.
- **Pits or underground storage units**: Plants such as Musa species, *Colocasia, Xanthosoma* species may be stored by burying them in pits that are less than 6 feet deep.
- **Wrapping with other plants material**: Plants are held in other plants materials as a form of short-term storage. Leaf of plants such as *Thaumatococcus dannielli, Musa* species, *Xanthosoma* species, and *Colocasia* species have been used in this way.
- **Exposure to (direct) sunlight**: This short-term storage system is widely used for vegetables, such as *Telfaria occidentalis* and *Vernonia amygdalina*.
- **Drying on a bare floor**: Root crops, fruits, vegetables, and grains may be stored by spreading them on bare mud or concrete floor for short-term storage.
- **Baskets or trays**: Woven baskets and trays made from diverse materials are used in storing local fruits crops for the short-to-medium term.

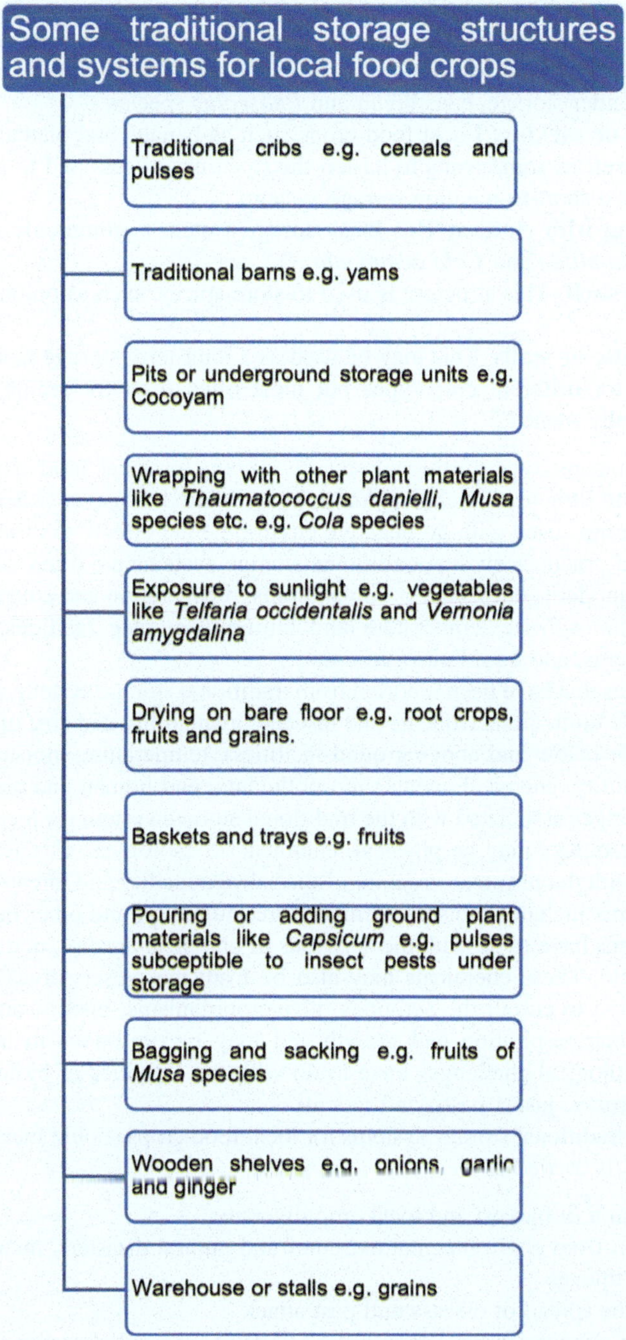

Fig. 13.3 Traditional storage systems and structures for local food crops in Africa

- **Pouring or adding ground plant materials**: Some plants are ground-like pepper (*Capsicum* species) and sprayed on other plants to extend their shelf life and quality. Used for pulses that are susceptible to attack from insect pests such as weevils under storage, e.g., *Vigna* and *Phaseolus* species
- **Bagging or sacking**: Local food crops such as banana and plantain are held in bags as well as some fruits to hasten the ripening process and to preserve their quality as a short-to-medium storage system.
- **Wrapping with clothes:** This local storage system is commonly used for cola nut—*Cola nitida* and *Cola acuminata.*
- **Wooden shelf**: This structure is used to store spices, such as onions, garlic, and ginger.
- **Warehouse or stalls**: This may be used as a long-term storage facility that may be used for different crop types but most often used for grains and may be communally owned.

The advantages of a traditional storage system for local food crops lie in the simplicity and little to no technical know-how required to use it. Others are low-to-no-cost materials used with no negative environmental effects. As traditional communities and practices evolve, traditional storage systems are often used in combination with moderns ones such as silo and bins, refrigeration and cold storage units, controlled atmospheric temperature and humidity storage facilities, evaporative coolant systems, and liquid nitrogen units.

Challenges that have been reported from traditional storage systems for local food crops include biotic pests (rodents and insects), availability and cost of construction materials (for below and above ground facilities), temperature, moisture or precipitation, humidity, chemical agents, and pollutants, and human and animal thieves. These challenges associated with the traditional storage system for local food crops, food, and resources may be physical, chemical, or biological. Physical challenges have to do with the physical structure of the storage facility and threats to the stored materials. This includes moisture, temperature, humidity, and other factors. Chemical challenges have to do with the activities of chemical agents, such as dust, wax, pollutants, etc. These chemicals may also be from the preservatives added to the storage system to curtail the activities of microorganisms, insects, and rodents by restricting their respiration and growth and may be poisonous to humans when ingested. Biological challenges have to do with the activities of biological agents, such as microbes, insect pests, and rodents.

Efficient traditional storage systems for local food crops should meet all the basic storage criteria as well as

1. Preservation of harvest and local crop diversity
2. Protection from environmental extremes and natural disasters, man-made disasters, and thieves
3. Prevent the spread of disease and pest attack
4. Serves as a means to have easy access to plant materials for consumption, sale, education, and research
5. Preservation from man-made disasters and thieves

13.6 Recommendations and Conclusion

Achieving local food sustainability requires focus to be placed on orphan and minor crops (and their crop wild relatives as well as edible wild plants) within Africa with targeted efforts to promote adoption, expansion, regular evaluation, breeding of improved varieties, and the development of sustainable policies for production, utilization and marketing. These minor crops provide avenues to address some socio-economic, cultural, and environmental woes. An agricultural research agenda that addresses the agronomy, diversity, and distribution, breeding, conservation, phytochemical composition, post-harvest handling and storage, and value addition of local food crops will boost sustainability and realign it with other pertinent developmental facets on the continent. The new climate smart-focused Africa's new green revolution can impact the local food system if integrated into local rural-focused developmental policies, which will enable smallholder farmers to contribute to the decision-making process and governance as well as to adapt and sustainably incorporate innovative strategies. Moreover, the mission of the African Orphan Crops Consortium (AOCC) should be integrated into local, national, and regional farming and food systems. The AOCC needs to build infrastructure to collect, maintain, conduct trials, and distribute minor or orphan crops as well as wild edible crops in the form of gene banks, seed libraries, and food crop sanctuaries for the continent. This would increase diversity in access, which has been shown to increase resilience and sustainability in local food crops and systems (Nchanji and Lutomia 2021).

To close the gap between the current yield and yield potentials of local food crops and meet Africa's current and future food needs, van Ittersum et al. (2016) suggested a sustainable increase in crop intensity and production area to close the gap. While bridging the yield gap is important, there is a need to also improve market access and upscale policy and development, as more land may not automatically translate into more food (Frelat et al. 2016). Specifically, wild edible plant research should focus on inventorying diversity, traditional collection, processing, and utilization of knowledge and practices as well as chemical profiles and protection strategies (Motti 2022). Africa's local food crops are in dire need of stylized facts to reinvigorate research and investment interest. According to Christiaensen (2017), research agendas and policy debates are driven by stylized facts. In the case of Africa's local food crop, this could be in the form of simply enough to adopt and follow concepts, sound methodologies to obtain reliable and interpretable data, and synthesis of findings for decision making. The decision about local food crops in Africa should incorporate information about species diversity, distribution, seed maturity period, yield quality and quantity, and conservation threats, especially genetic erosion.

References

Akinola R, Pereira LM, Mabhaudhi T, de Bruin FM, Rusch L (2020) A review of indigenous food crops in Africa and the implications for more sustainable and healthy food systems. Sustainability 12(8):3493. https://doi.org/10.3390/su12083493

Baraka E, Willis MS, Ishimwe BA (2022) What Kigali's open-air markets reveal about achieving food and nutrition security: the role of African indigenous crops. Agric Food Secur 11(1):17. https://doi.org/10.1186/s40066-022-00359-4

Baskarachary K, Yemula S, Gavaravarapu SR, Joshi AK (2015) Traditional foods, functional foods and nutraceuticals. Proc Natl Acad Sci 82(5):1565–1577

Batur F, Dedeurwaerdere T (2014) The use of agrobiodiversity for plant improvement and the intellectual property paradigm: institutional fit and legal tools for mass selection, conventional and molecular plant breeding. Life Sci Society Policy 10:14. https://doi.org/10.1186/s40504-014-0014-7

Bharucha Z, Pretty J (2010) The roles and values of wild foods in agricultural systems. Philos Trans R Soc Lond B Biol Sci 365(1554):2913–2926. https://doi.org/10.1098/rstb.2010.0123

Borelli T, Hunter D, Powell B, Ulian T, Mattana E, Termote C, Pawera L, Beltrame D, Penafiel D, Tan A, Taylor M, Engels J (2020) Born to eat wild: an integrated conservation approach to secure wild food plants for food security and nutrition. *Plants (Basel, Switzerland) 9*(10):1299. https://doi.org/10.3390/plants9101299

Brun TA (1991) The nutrition and health impact of cash cropping in west Africa: a historical perspective. World Rev Nutr Diet 65:124–162. https://doi.org/10.1159/000419470

Burchi F, Fanzo J, Frison E (2011) The role of food and nutrition system approaches in tackling hidden hunger. *Int J Environ Res Public Health 8*(2):358–373. https://doi.org/10.3390/ijerph8020358

Carletto C, Corral P, Guelfi A (2017) Agricultural commercialization and nutrition revisited: empirical evidence from three African countries. Food Policy 67:106–118. https://doi.org/10.1016/j.foodpol.2016.09.020

Caspi CE, Sorensen G, Subramanian SV, Kawachi I (2012) The local food environment and diet: a systematic review. Health Place 18(5):1172–1187. https://doi.org/10.1016/j.healthplace.2012.05.006

Catarino L, Romeiras MM, Bancessi Q, Duarte D, Faria D, Monteiro F, Moldão M (2019) Edible leafy vegetables from West Africa (Guinea-Bissau): consumption, trade and food potential. Foods (Basel, Switzerland) 8:10–493. https://doi.org/10.3390/foods8100493

Chime AO, Aiwansoba RO, Danagogo SJ, Egharevba II, Osawaru ME, Ogwu MC (2016) Effects of lining with leaves of Triclisia dictyophylla on the fungal composition of Cola nitida during storage. J Indus Res Technol 5(2):128–136

Chivenge P, Mabhaudhi T, Modi AT, Mafongoya P (2015) The potential role of neglected and underutilised crop species as future crops under water scarce conditions in sub-Saharan Africa. Int J Environ Res Public Health 12(6):5685–5711. https://doi.org/10.3390/ijerph120605685

Christiaensen L (2017) Agriculture in Africa—telling myths from facts: a synthesis. Food Policy 67:1–11. https://doi.org/10.1016/j.foodpol.2017.02.002

Clay N, Zimmerer KS (2020) Who is resilient in Africa's green revolution? Sustainable intensification and climate smart agriculture in Rwanda. Land Use Policy 97:104558. https://doi.org/10.1016/j.landusepol.2020.104558

Davis KF, Chhatre A, Rao ND, Singh D, Ghosh-Jerath S, Mridul A, Poblete-Cazenave M, Pradhan N, DeFries R (2019) Assessing the sustainability of post-Green Revolution cereals in India. *Proc Natl Acad Sci U S A 116*(50):25034–25041. https://doi.org/10.1073/pnas.1910935116

Delabre I, Rodriguez LO, Smallwood JM, Scharlemann JPW, Alcamo J, Antonarakis AS, Rowhani P, Hazell RJ, Aksnes DL, Balvanera P, Lundquist CJ (2021) Actions on sustainable food production and consumption for the post-2020 global biodiversity framework. *Sci Adv 7*(12):eabc8259. https://doi.org/10.1126/sciadv.abc8259

Downer S, Berkowitz SA, Harlan TS, Olstad DL (2020) Food is medicine: actions to integrate food and nutrition into healthcare. *BMJ Clin Res ed 369*:m2482. https://doi.org/10.1136/bmj.m2482

Durazzo A, Lucarini M (2021) Environmental, ecological and food resources in the biodiversity overview: health benefits. *Life (Basel, Switzerland) 11*(11):1228. https://doi.org/10.3390/life11111228

Ekesa B, Ariong RM, Kennedy G, Baganizi M, Dolan I (2020) Relationships between land tenure insecurity, agrobiodiversity, and dietary diversity of women of reproductive age: evidence from Acholi and Teso subregions of Uganda. Matern Child Nutr 16(Suppl 3):e12965. https://doi.org/10.1111/mcn.12965

Erinle KO, Ogwu MC, Evivie SE, Zaheer MS, Ogunyemi SO, Adeniran SO (2021) Impacts of COVID-19 on agriculture and food security in developing countries: potential mitigation strategies. CAB Reviews 16(16):1–16

Evivie, S.E., Ogwu, M.C., Ebabhamiegbebho, P.A., Abel, E.S., Imaren, J.O. and Igene, J.O. (2020). Packaging and the Nigerian food industry: challenges and opportunities. In: Food technology and culture in Africa. Ogunlade, C.A., Adeleke, K.M. and Oladejo, M.T. (eds). Reamsworth Publishing, Ibadan, Nigeria. 28 – 99pp.

Finley JW, Dimick D, Marshall E, Nelson GC, Mein JR, Gustafson DI (2017) Nutritional sustainability: aligning priorities in nutrition and public health with agricultural production. Adv Nutr (Bethesda, MD) 8:780–788. https://doi.org/10.3945/an.116.013995

Frelat R, Lopez-Ridaura S, Giller KE, Herrero M, Douxchamps S, Andersson Djurfeldt A, Erenstein O, Henderson B, Kassie M, Paul BK, Rigolot C, Ritzema RS, Rodriguez D, van Asten PJ, van Wijk MT (2016) Drivers of household food availability in sub-Saharan Africa based on big data from small farms. Drivers of household food availability in sub-Saharan Africa based on big data from small farms. Proc Natl Acad Sci U S A 113(2):458–463. https://doi.org/10.1073/pnas.1518384112

Gaitán-Cremaschi D, Klerkx L, Duncan J, Trienekens JH, Huenchuleo C, Dogliotti S, Contesse ME, Rossing WAH (2019) Characterizing diversity of food systems in view of sustainability transitions. A review. Agron Sustain Dev 39(1):1. https://doi.org/10.1007/s13593-018-0550-2

Heywood VH (1999) Use and potential of wild plants in farm households. Volume 15 FAO; Rome, Italy: 1999. (FAO Farm System Management Series)

van der Hoeven M, Osei J, Greeff M, Kruger A, Faber M, Smuts CM (2013) Indigenous and traditional plants: South African parents' knowledge, perceptions and uses and their children's sensory acceptance. J Ethnobiol Ethnomed 9(1):78. https://doi.org/10.1186/1746-4269-9-78

Ikhajiagbe B, Atoe R, Ogwu MC, Loveniers P-J (2021a) Changes in Telfaria occidentalis leaf morphology, quality and phytochemical composition under different local preservation regimes. VEGETOS – Int J Plant Res Biotechnol 31(1):29–36. https://doi.org/10.1007/s42535-021-00188-z

Ikhajiagbe B, Ogwu MC, Ogochukwu OF, Odozi EB, Adekunle IJ, Omage ZE (2021b) The place of neglected and underutilized legumes in human nutrition and protein security. Crit Rev Food Sci Nutr. https://doi.org/10.1080/10408398.2020.1871319

van Ittersum MK, van Bussel LG, Wolf J, Grassini P, van Wart J, Guilpart N, Claessens L, de Groot H, Wiebe K, Mason-D'Croz D, Yang H, Boogaard H, van Oort PA, van Loon MP, Saito K, Adimo O, Adjei-Nsiah S, Agali A, Bala A, Chikowo R, Kaizzi K, Kouressy M, Makoi JH, Ouattara K, Tesfaye K, Cassman KG (2016) Can sub-Saharan Africa feed itself? Can sub-Saharan Africa feed itself? Proc Natl Acad Sci U S A 113(52):14964–14969. https://doi.org/10.1073/pnas.1610359113

Jacobi J, Mukhovi S, Llanque A, Giger M, Bessa A, Golay C, Ifejika Speranza C, Mwangi V, Augstburger H, Buergi-Bonanomi E, Haller T, Kiteme BP, Delgado Burgoa JMF, Tribaldos T, Rist SA (2020) A new understanding and evaluation of food sustainability in six different food systems in Kenya and Bolivia. *Sci Rep 10*(1):19145. https://doi.org/10.1038/s41598-020-76284-y

Kuyu CG, Bereka TY (2019) Review on contribution of indigenous food preparation and preservation techniques to attainment of food security in Ethiopian. Food Sci Nutr 8(1):3–15. https://doi.org/10.1002/fsn3.1274

Lindgren E, Harris F, Dangour AD, Gasparatos A, Hiramatsu M, Javadi F, Loken B, Murakami T, Scheelbeek P, Haines A (2018) Sustainable food systems-a health perspective. *Sustain Sci 13*(6):1505–1517. https://doi.org/10.1007/s11625-018-0586-x

Lu Y, Zhang Y, Hong Y, He L, Chen Y (2022) Experiences and lessons from agri- food system transformation for sustainable food security: a review of China's practices. *Foods (Basel, Switzerland) 11*(2):137. https://doi.org/10.3390/foods11020137

Lusty C, Sackville Hamilton R, Guarino L, Richards C, Jamora N, Hawtin G (2021) Envisaging an effective global long-term agrobiodiversity conservation system that promotes and facilitates use. *Plants (Basel, Switzerland) 10*(12):2764. https://doi.org/10.3390/plants10122764

Mabhaudhi T, Chimonyo VGP, Chibarabada TP, Modi AT (2017) Developing a roadmap for improving neglected and underutilized crops: a case study of south Africa. Front Plant Sci 8, https://www.frontiersin.org/article/10.3389/fpls.2017.02143

Mabhaudhi T, Chimonyo VGP, Hlahla S, Massawe F, Mayes S, Nhamo L, Modi AT (2019) Prospects of orphan crops in climate change. Planta 250(3):695–708. https://doi.org/10.1007/s00425-019-03129-y

MacFall J, Lelekacs JM, LeVasseur T, Moore S, Walker J (2015) Toward resilient food systems through increased agricultural diversity and local sourcing in the Carolinas. *J Environ Stud Sci 5*(4):608–622. https://doi.org/10.1007/s13412-015-0321-1

Martinez S, Hand M, Da Pra M, Pollack S, Ralston K, Smith T, Vogel S, Clark S, Lohr L, Low S, Newman C (2010) Local food systems: concepts, impacts, and issues. Economic research report #97. Economic Research Service, United States Department of Agriculture. 87 p

Masipa TS (2017) The impact of climate change on food security in South Africa: Current realities and challenges ahead. Jamba (Potchefstroom, South Africa) 9(1):411. https://doi.org/10.4102/jamba.v9i1.411

van der Merwe J, Cloete PC, van der Hoeven M (2016) Promoting food security through indigenous and traditional food crops. Agroecol Sustain Food Syst 40(8):830–847

Misselhorn A, Hendriks SL (2017) A systematic review of sub-national food insecurity research in South Africa: missed opportunities for policy insights. PLoS One 12(8):e0182399. https://doi.org/10.1371/journal.pone.0182399

Montúfar R, Ayala M (2019) Perceptions of agrodiversity and seed-saving practices in the northern Andes of Ecuador. *J Ethnobiol Ethnomed 15*(1):35. https://doi.org/10.1186/s13002-019-0312-5

Morawicki RO, Díaz González DJ (2018) Food sustainability in the context of human behavior. *Yale J Biol Med 91*(2):191–196

Motti R (2022) Wild edible plants: a challenge for future diet and health. Plants (Basel, Switzerland) 11(3):344. https://doi.org/10.3390/plants11030344

Müller MF, Penny G, Niles MT, Ricciardi V, Chiarelli DD, Davis KF, Dell'Angelo J, D'Odorico P, Rosa L, Rulli MC, Mueller ND (2021) Impact of transnational land acquisitions on local food security and dietary diversity. Proc Natl Acad Sci U S A 118(4):e2020535118. https://doi.org/10.1073/pnas.2020535118

Nchanji EB, Lutomia CK (2021) Sustainability of the agri-food supply chain amidst the pandemic: diversification, local input production, and consumer behavior. *Adv Food Security Sustain 6*:211–229. https://doi.org/10.1016/bs.af2s.2021.07.003

Ngcamu BS, Chari F (2020) Drought influences on food insecurity in Africa: a systematic literature review. Int J Environ Res Public Health 17(16):5897. https://doi.org/10.3390/ijerph17165897

Noort MWJ, Renzetti S, Linderhof V, du Rand GE, Marx-Pienaar NJMM, de Kock HL, Magano N, Taylor JRN (2022) Towards sustainable shifts to healthy diets and food security in sub-Saharan Africa with climate-resilient crops in bread-type products: a food system analysis. Foods (Basel, Switzerland) 11(2):135. https://doi.org/10.3390/foods11020135

NRC (2015). Social and economic effects of the U.S. Food System. In Nesheim MC, Oria M, Yih PT (eds) A framework for assessing effects of the food SYSTEM. National Research Council's

Committee on a Framework for Assessing the Health, Environmental, and Social Effects of the Food System; Food and Nutrition Board; Board on Agriculture and Natural Resources; Institute of Medicine; National Research Council. Washington (DC): National Academies Press (US). Social and Economic Effects of the U.S. Food System. https://www.ncbi.nlm.nih.gov/books/NBK305168/

Ogwu MC (2019a) Lifelong consumption of plant-based GM foods: is it safe? In: Papadopoulou P, Misseyanni A, Marouli C (eds) Environmental exposures and human health challenges. IGI Global, Pennsylvania, USA, pp 158–176. https://doi.org/10.4018/978-1-5225-7635-8.ch008

Ogwu MC (2019b) Towards sustainable development in Africa: the challenge of urbanization and climate change adaptation. In: Cobbinah PB, Addaney M (eds) The geography of climate change adaptation in urban Africa. Springer Nature, Switzerland, pp 29–55. https://doi.org/10.1007/978-3-030-04873-0_2

Ogwu MC (2019c) Understanding the composition of food waste: an "-Omics" approach to food waste management. In: Gunjal AP, Waghmode MS, Patil NN, Bhatt P (eds) Global initiatives for waste reduction and cutting food loss. IGI Global, Pennsylvania, USA, pp 212–236pp. https://doi.org/10.4018/978-1-5225-7706-5.ch011

Ogwu MC (2020) Value of amaranthus [L.] species in Nigeria. In: Waisundara V (ed) Nutritional value of Amaranth. IntechOpen, pp 1–21. https://doi.org/10.5772/intechopen.86990

Ogwu MC, Osawaru ME, Ahana CM (2014) Challenges in conserving and utilizing plant genetic resources (PGR). Int J Genet Mol Biol 6(2):16–22. https://doi.org/10.5897/IJGMB2013.0083

Ogwu MC, Osawaru ME, Aiwansoba RO, Iroh RN (2016a) Ethnobotany and collection of West African Okra [Abelmoschus caillei (A. Chev.) Stevels] germplasm in some communities in Edo and Delta States, Southern Nigeria. Borneo J Resour Sci Technol 6(1):25–36

Ogwu MC, Osawaru ME, Obadiaru O, Itsepopo IC (2016b) Nutritional composition of some uncommon indigenous spices and vegetables. International Conference on Natural Resource Development and Utilization (4th Edition): Positioning small and medium enterprises (SMEs) for optimal utilization of available resources and economic growth. Raw Materials Research and Development Council (Federal Ministry of Science and Technology), Maitama, Abuja Nigeria, pp 286–300

Ogwu MC, Osawaru ME, Aiwansoba RO, Iroh RN (2016c) Status and prospects of vegetables in Africa. In Borokini IT, Babalola FD (eds) Conference proceedings of the joint biodiversity conservation conference of Nigeria tropical biology association and Nigeria chapter of society for conservation biology on MDGs to SDGs: toward sustainable biodiversity conservation in Nigeria. University of Ilorin, Nigeria, pp 47–57

Ogwu MC, Osawaru ME, Obahiagbon GE (2017) Ethnobotanical survey of medicinal plants used for traditional reproductive care by Usen people of Edo State, Nigeria. Malaya J Biosci 4(1):17–29

Ogwu MC, Chime AO, Aiwansoba RO, Emere AO (2019) Effects of storage methods and duration on the microbial composition and load of tomato (Solanum lycopersicum [L.], Solanaceae) fruits. Bitlis Eren University. J Sci Technol 9(1):1–7. https://dergipark.org.tr/download/article-file/746056

Ojelel S, Mucunguzi P, Katuura E, Kakudidi EK, Namaganda M, Kalema J (2019) Wild edible plants used by communities in and around selected forest reserves of Teso-Karamoja region, Uganda. J Ethnobiol Ethnomed 15:3. https://doi.org/10.1186/s13002-018-0278-8

Osawaru ME, Ogwu MC (2014) Ethnobotany and germplasm collection of two genera of cocoyam (Colocasia [Schott] and Xanthosoma [Schott], Araceae) in Edo State Nigeria. Sci Technol Arts Res J 3(3):23–28. https://doi.org/10.4314/star.v3i3.4

Osawaru ME, Ogwu MC (2020) Survey of plant and plant products in local markets within Benin City and environs. In: Filho LW, Ogugu N, Ayal D, Adelake L, da Silva I (eds) African handbook of climate change adaptation. Springer Nature, pp 1–24. https://doi.org/10.1007/978-3-030-42091-8_159-1

Osawaru ME, Ogwu MC, Omoigui ID, Aiwansoba RO, Kevin A (2016) Ethnobotanical survey of vegetables eaten by Akwa Ibom people residing in Benin City, Nigeria. Univ Benin J Sci Technol 4(1):70–93

Paul CJ, Paul JE, Anderson RS (2019) The local food environment and food security: the health behavior role of social capital. Int J Environ Res Public Health 16(24):5045. https://doi.org/10.3390/ijerph16245045

Pretty J, Bharucha ZP (2014) Sustainable intensification in agricultural systems. Ann Bot 114(8): 1571–1596. https://doi.org/10.1093/aob/mcu205

Ray A, Ray R, Sreevidya EA (2020) How many wild edible plants do we eat—their diversity, use, and implications for sustainable food system: an exploratory analysis in India. Front Sustain Food Syst 4, https://www.frontiersin.org/article/10.3389/fsufs.2020.00056

Rockström J, Williams J, Daily G, Noble A, Matthews N, Gordon L, Wetterstrand H, DeClerck F, Shah M, Steduto P, de Fraiture C, Hatibu N, Unver O, Bird J, Sibanda L, Smith J (2017) Sustainable intensification of agriculture for human prosperity and global sustainability. Ambio 46(1):4–17. https://doi.org/10.1007/s13280-016-0793-6

Satterthwaite D, McGranahan G, Tacoli C (2010) Urbanization and its implications for food and farming. Philos Trans R Soc Lond B Biol Sci 365(1554):2809–2820. https://doi.org/10.1098/rstb.2010.0136

Shelef O, Weisberg PJ, Provenza FD (2017) The value of native plants and local production in an era of global agriculture. Front Plant Sci 8:2069. https://doi.org/10.3389/fpls.2017.02069

Shennan C (2008) Biotic interactions, ecological knowledge and agriculture. *Philos Trans R Soc Lond B Biol Sci 363*(1492):717–739. https://doi.org/10.1098/rstb.2007.2180

Shilomboleni H (2020) COVID-19 and food security in Africa: Building more resilient food systems. AAS Open Res 3:27. https://doi.org/10.12688/aasopenres.13078.1

Shumsky SA, Hickey GM, Pelletier B, Johns T (2014) Understanding the contribution of wild edible plants to rural social-ecological resilience in semi-arid Kenya. Ecol Soc 19(4):34. https://doi.org/10.5751/ES-06924-190434

Stein AJ, Santini F (2021) The sustainability of "local" food: a review for policy- makers. Rev AgricFood EnvironStud:1–13. Advance online publication. https://doi.org/10.1007/s41130-021-00148-w

Tadele Z (2019) Orphan crops: their importance and the urgency of improvement. Planta 250:677–694

Talabi AO, Vikram P, Thushar S, Rahman H, Ahmadzai H, Nhamo N, Shahid M, Singh RK (2022) Orphan crops: a best fit for dietary enrichment and diversification in highly deteriorated marginal environments. Front Plant Sci 13, https://www.frontiersin.org/article/10.3389/fpls.2022.839704

Vågsholm I, Arzoomand NS, Boqvist S (2020) Food security, safety, and sustainability—getting the trade-offs right. Front Sustain Food Syst 4. https://www.frontiersin.org/article/10.3389/fsufs.2020.00016

Vargas AM, de Moura AP, Deliza R, Cunha LM (2021) The role of local seasonal foods in enhancing sustainable food consumption: a systematic literature review. Foods (Basel, Switzerland) 10(9):2206. https://doi.org/10.3390/foods10092206

Verma M, Hontecillas R, Tubau-Juni N, Abedi V, Bassaganya-Riera J (2018) Challenges in personalized nutrition and health. Front Nutr 5:117. https://doi.org/10.3389/fnut.2018.00117

Zhong B, Wu S, Sun G, Wu N (2022) Farmers' strategies to climate change and urbanization: potential of ecosystem-based adaptation in rural Chengdu, southwest China. Int J Environ Res Public Health 19(2):952. https://doi.org/10.3390/ijerph19020952

Part III
Challenges in the Conservation and Utilization of Africa's Biological Resources and Environment

Chapter 14
Environmental Pollution: Threats, Impact on Biodiversity, and Protection Strategies

Ayotunde Samuel Kolawole and Adams Ovie Iyiola

Abstract Africa boasts immense natural and mineral resources, as they form the skeletal structure of the economy. These resources are distributed across the continent, on land, along the shoreline, and even in the deeper regions of the oceans. Over the last decades, the increased rate of industrialization and urbanization coupled with the increasing human population has contributed in a great way to the menace being caused by pollution. Over-exploitation, in terms of excessive or indiscriminate usage of resources, causes pollution, thereby leading to unsustainability in resource management. Conversely, issues of conflicts over resources and insecurity have also been on the increase most especially along the coastline of the African countries. The Niger-Delta region of Nigeria has been plagued with pollution resulting from oil spillage which has had a grave impact on the life of aquatic and terrestrial animals likewise the plant life. Many other countries within Africa have several industries producing a variety of products such as plastics, iron, pesticides, herbicides, paper, and several others of which their wastes degrade the environment. Countries located along the West African coast and South Africa are known for emitting high amounts of mercury due to their gold mining and coal combustion activities. Discharges or wastes from these industries go into the environment causing air and water pollution. Air pollutants, present within the atmosphere, pollute the water and soil in the form of atmospheric deposition. The major sources of water pollution are the urban and industrial effluent discharges into the aquatic environment. Most countries in Africa do not have enforced standards for effluent discharge into surface waters. All these pollutants affect aquatic biodiversity, as it leads to the migration of fish species from the polluted area in search of survival. The majority of the coastal waters in Africa

A. S. Kolawole (✉)
Department of Fisheries and Aquatic Resources Management, Faculty of Renewable Natural Resources Management, College of Agriculture, Osun State University, Osogbo, Osun State, Nigeria
e-mail: ayotunde.kolawole@uniosun.edu.ng

A. O. Iyiola
Department of Fisheries and Aquatic Resources Management, Faculty of Renewable Natural Resources Management, Osun State University, Osogbo, Osun State, Nigeria

S. C. Izah, M. C. Ogwu (eds.), *Sustainable Utilization and Conservation of Africa's Biological Resources and Environment*, Sustainable Development and Biodiversity 32, https://doi.org/10.1007/978-981-19-6974-4_14

are faced with this challenge, hence the need to find an immediate solution to help conserve aquatic life. To this end, this chapter will elaborate on the various sources of pollution and their impacts on the environment and biodiversity.

Keywords Aquatic environment · Biodiversity management · Human activities · Pollution · Terrestrial environment

14.1 Introduction

The word pollution can be simply defined as the introduction of a contaminant into a naturally healthy environment that causes an undesirable change that threatens the survival of existing plants and animals. Pollution has been a major concern in our society for several decades now as it brings about the deterioration in health and also gradually wears off the beauty and aesthetic value of our ecosystem. Developing countries are experiencing unprecedented and indiscriminate growth in human population, and as such, there is a further development in urban areas and constant establishment of more industries to meet up with the demand of the masses. The majority of these industries fail to carry out an environmental impact assessment (EIA), thereby having little or no regard for the eventual effect of their waste product on the community.

Environmental pollution is one of the vital issues facing humanity today (Izah and Angaye 2016). It can be defined as "the alteration in the normalcy of the environment such that environmental processes are affected adversely." Construction, transportation, and manufacturing, as a means of development in the society, produce a large number of waste products, thereby polluting water (Ogamba et al. 2021; Izah et al. 2016), air (Richard et al. 2021a, b; Uzoekwe et al. 2021; Izah et al. 2021; Seiyaboh and Izah 2019), and soil (Izah et al. 2018a, b, 2017a, b, c), and also cause depletion in natural resources. Pollution to the atmosphere causes global warming and brings about acid rains. Human activities whether directly or indirectly cause a ripple effect of damage to the environment (Seiyaboh and Izah 2017; Ikhajiagbe and Ogwu 2020; Ikhajiagbe et al. 2022a) and on biodiversity (Osawaru et al. 2013a, b; Ogwu et al. 2014a, b; Osawaru and Ogwu 2014; Ogwu et al. 2016). An automobile mechanic, in the cause of repairing a vehicle, pollutes the soil with black engine oil from used engines and pollutes the atmosphere by releasing toxic gases, such as nitrogen oxide, carbon monoxide, carbon dioxide, black soot, and a mixture of hydrocarbons through vehicle exhaust pipes. A market trader dumps refuse either on the road or waterways, thereby leading to water pollution. In addition, municipal wastes and surface runoff from crop farms with agrochemicals pollute the aquatic ecosystems (Ojesanmi et al. 2017; Inyang et al. 2018a, b, 2019a, b, c, d, 2017a, b, c). Industrial effluent discharges contain many toxic heavy metals and emit a gaseous offensive odor. A pollutant causes alteration to the environment in an undesirable manner and is of different types. They include air, soil, water, noise, and thermal pollution.

14.2 Air Pollution

Any disturbance caused to the normal composition of the atmosphere such as the presence of gaseous chemicals and dust particles which is harmful to human, animals, and plants are known as air pollution. With the increased rate of urban development, chemicals released into the atmosphere mixes with the oxygen we breathe in and become carcinogenic to humans. Gaseous emissions from manufacturing factories and automobiles are major driving factors contributing to air pollution. Sweileh et al. (2018) categorized air particulate as a form of air pollution that is easily controlled in confined spaces. Toxic gases such as sulfur oxides, carbon monoxide, and nitrogen oxide are other forms of air pollutants (Ohimain et al. 2013; Ohimain and Izah 2013; Vwioko et al. 2018). These pollutants affect physiological processes within biological organisms and also reduce environmental quality (Ikhajiagbe et al. 2022b, c). It is extremely difficult to control air pollution, as it can occupy and spread homogeneously over a large expanse of land within a short period. Gaddi and Capello (2018) reported that air pollution causes several diseases in humans, such as respiratory difficulties (an example is asthma), skin allergies (such as rashes), cardiovascular diseases (hypertension), and neuropsychiatric diseases.

14.2.1 Categories of Air Pollutants

i. **Primary pollutants:** They are referred to as pollutants that are released directly either by daily human activities or natural events. Examples of such pollutants include but are not limited to, particulate matter, carbon monoxide, sulfur and nitrogen oxides, and unsaturated hydrocarbons.
ii. **Secondary pollutant**: The reaction of primary pollutants with atmospheric moisture brings about the formation of secondary pollutants, such as nitric acid, carbonic acid, and sulfuric acid.

14.2.2 Types of Air Pollutants

14.2.2.1 Suspended Particulate Pollutants

Dust, which is mainly emitted from flour or animal feed production companies, black soot from the exhaust pipes of vehicles with bad engines, and gaseous chemicals released from chemical industries through the chimneys all constitute the suspended particulate pollutant. They vary in size ranging from 0.001 to 500 μm in diameter. Particulate matters that are less than 10 μm float within the atmosphere and easily drift with the wind. Those that are more than 10 μm in diameter have more weight and cannot be easily carried by winds; hence, they clear off the atmosphere faster and simmer down, while tiny particles less than 0.02 μm persistently form aerosols. Railway yards, power plants, marketplaces, oil refineries, construction activities, and chemical industries are all major sources of suspended particulate matter.

14.2.2.2 Gaseous Pollutants

Industries, power plants, and vehicles all make use of either petrol or diesel as fuel, and in the process of working, they all emit gaseous wastes. These wastes, because of their ability to make the environment unsuitable to survive, are referred to as gaseous pollutants. They include carbon monoxide, nitrogen, and sulfur oxides. All of these gaseous pollutants have detrimental impacts on plants, animals, and majorly humans.

14.2.3 Causes of Air Pollution

Mentioned are a few of the numerous activities that cause air pollution (Fig. 14.1).

- **Agricultural activities:** Toxic gaseous wastes are predominantly emitted during most agricultural activities. In crop farming, fertilizers, pesticides, and herbicides used usually emit chemicals into the atmosphere during usage and thereby making the environment unsuitable to humans and animals.
- **Domestic sources:** Disinfectants used in cleaning and paints applied in finishing a building usually contain toxic chemicals that ooze out into the atmosphere after usage. Pollution caused by such chemicals leads to difficulty in breathing.
- **Industries and factories:** Chemical-producing industries are the major sources of organic compounds, several hydrocarbons, and carbon monoxide. The release of these gaseous chemicals into the atmosphere causes a detrimental effect on the quality of air humans and animals inhale.
- **Fossil fuel combustion:** The release of a large amount of sulfur dioxide into the atmosphere emanates from the combustion of fossil fuels. Incidentally, the incomplete combustion of fossil fuels brings about the release of carbon monoxide which also causes air pollution.
- **Automobiles:** The emission of gases through vehicle exhaust causes pollution to the environment. Cars, trucks, trailers, and motorcycles with faulty engines are major sources of these toxic gases which form the greenhouse gases and can lead to respiratory diseases in humans.
- **Mining:** Mineral resources underneath the earth's surface are usually mined using heavy-duty tools and machinery. Mining dust particles and gaseous chemicals emitted during the mineral excavation process contaminate the atmosphere, thereby causing severe health damage to the workers and inhabitants around the mining area.

Fig. 14.1 Causes of air pollution

14.2.4 Air Pollutants and Their Effects on Humans and Animals

14.2.4.1 Carbon Monoxide

This type of pollutant is derived from vehicle exhaust, coal burning, and biomass combustion. This type of gas is released during the incomplete burning of fossil fuel or other products. It causes severe headache, dizziness, irritation of mucous membrane, unconsciousness, heart disease (carbon monoxide combines with oxygen in the blood and reduces the affinity of hemoglobin for oxygen), and death in elongated exposure.

14.2.4.2 Carbon Dioxide

Carbon dioxide is a human respiratory by-product and its concentration increases in the atmosphere as a result of fossil fuel burning, automobile exhaust release, and volcanic eruptions, industrial and agricultural activities. CO_2 causes global warming, vision problem, severe headache, and heart strain and leads to climate change.

14.2.4.3 Nitrogen Oxides

This is mostly released from volcanic eruptions, power plants, vehicle exhausts, and industries. It reacts with moisture content present in the atmosphere giving problems to the lungs, and respiratory systems, causing asthma, bronchitis, and eye irritation.

14.2.4.4 Sulfur Dioxide

Sulfur dioxide is released from refineries involved in the mining of crude oil, volcanic blast, and chemical industries. The reaction of sulphur dioxide with moisture leads to the formation of secondary pollutants that cause severe irritation to the eye and throat.

14.2.4.5 Hydrocarbons

These are mostly released by the combustion of fossil fuels. Inhalation, by humans and animals, of hydrocarbons from the atmosphere causes kidney problems, irritation to the eyes, nose, and throat, asthma, hypertension, and carcinogenic effects on the lungs.

14.2.4.6 Lead

Tetraethyl lead is a common component of anti-knocking agents in the fuel used for the movement of vehicles. Lead particles are usually released from the vehicle exhaust and into the atmosphere. Lead compounds are also part of paints and could be washed by rain, thereby polluting the environment (Ogwu et al. 2014a, b, 2015). As a result, the presence of lead particles in the air has grave effects as it alters the development of red blood cells, and causes cancer, kidney, and liver diseases in humans on inhalation of such air.

14.2.4.7 Suspended Particulate Matter (SPM)

These include particles (solid and liquid form) suspended in the air. They are mostly released from thermal power plants, construction sites, metal industries, and automobiles. The presence of suspended particulate matter in the atmosphere causes poor visibility and breathing problem in humans. Smog (smoke and fog) formation causes poor visibility and aggravates asthma and every other respiratory disease in humans.

14.2.5 Mitigating Strategies for Air Pollution

It is always a serious issue dealing with the menace of air pollution and that is why it is important to prevent it before it happens. The preventive measure can either be government-oriented or individual actions (Ogwu 2019).

14.2.5.1 Government Oriented Prevention

- Several governments, especially in developed countries, have invested heavily in solar and wind energy. This action taken is simply to reduce to the barest minimum the burning of fossil fuels which causes air pollution.
- In addition, most governments are making it mandatory for industries to find a means to reduce the level of air pollution caused by their manufacturing machines.

14.2.5.2 Individual Actions

- Bearing in mind the level of air pollution caused by manufacturing industries, the recycling and re-using of some products will reduce the dependency on these industries to make new products.
- Families should be encouraged to constantly service their vehicles to prevent fumes from bad vehicles on the road. In addition, if possible, individuals should take more of bus, trains, or bicycles to work.
- The use of firewood should also be discouraged, because gaseous chemicals are emitted during the process, hence leading to air pollution.

14.3 Noise Pollution

This is amongst the most prevalent environmental pollutant. Any sound that is unwanted by the recipient or without value can be referred to as noise. Noise from vehicles, heavy-duty trucks, manufacturing industries, markets, and loudspeakers from religious gatherings all lead to irritation and can cause loss of temper, and an increase in blood pressure if the noise persists. It is, therefore, very important to control the amount of noise in the environment. The level of noise is usually measured in terms of decibels (dB). The World Health Organization (WHO) recommended the optimum noise level for humans to be 45 dB by day and 35 dB by night. Whatever noise that is above 80 dB is dangerous to human health.

14.3.1 Sources of Noise Pollution

The issue of noise in the environment cannot be overemphasized with several human activities constantly contributing to the creation of variable loud sounds. The sources of sound pollution are numerous and are classified to be located either indoors or outdoors.

- **Indoor sources:** Generators, radio, television, electric fans, air conditioners, other home appliances, and family engagements are all sources of noise pollution. Noise is a by-product of modern civilization, industrialization, and urbanization.
- **Outdoor sources:** As a result of the growing concentration of humans and industries, noise pollution tends to be prevalent in the cities. Activities such as the indiscriminate use of loudspeakers, product manufacturing, transportation, market hustles, political rallies, sports events religious activities, and social and cultural functions all contribute to noise pollution. In rural areas, plowing, weeding, and pumping machines are the main causes of noise pollution.

14.3.2 Effects of Noise Pollution

The health implications caused by noise pollution cannot be overstressed. It becomes annoying as it disturbs sleep, and causes high blood pressure (which leads to hypertension), mental depression, and hostility. Noise pollution also affects the performance and overall efficiency of individuals.

14.3.3 Prevention and Control of Noise Pollution

Steps to ensure the control of noise pollution are but are not limited to the following;

- The proper maintenance of vehicles will reduce road traffic noise.
- For the take-off and landing of aircraft, noise can be controlled by the enforcement of noise regulation laws at the airport.
- The use of soundproof equipment and generators will curb the amount of noise emanating from manufacturing industries.
- Loud music, the use of loudspeakers in public gatherings, high-duty vehicles, and land movers should not be permitted at night.
- Noises that are generated by trains mostly during movement can be controlled by the use of electric locomotives and the deployment of a quieter rolling stock.
- Trees are very efficient noise absorbers, hence the encouragement for tree planting initiative.

14.4 Soil Pollution

Contamination to the soil can be natural or a result of human activity. It is part of land degradation mostly caused by the presence of xenon biotic chemicals and other modifications in the natural environment. By definition, soil pollution is the contamination of the soil with toxic substances. It can also be referred to as the addition of any substance, either intentionally or by accident that lowers the soil quality, thereby making the soil inhabitable for micro- and macro-organisms living in the soil (Kerfahi et al. 2019a, b; Ogwu et al. 2019a, b). In agricultural activities through the daily application of herbicides, pesticides, fertilizers, and industrial activity, the indiscriminate disposal of waste poses greater challenges that require immediate and effective solutions (Chibuike and Obiora 2014). Contamination of the soil with heavy metals, mostly in varying concentrations, causes instability in the environment for humans and terrestrial animals. The accumulation of xenobiotic substances (which are carcinogenic) in soil is extremely harmful to humans, especially through the consumption of grown food and drinking water (Giller et al. 1998).

According to Adham et al. (2011), the rate of extraction using highly technological industrial processes increases the prevalence of xenobiotic compounds in soil. It is noteworthy to know that not all soil pollutants are xenobiotic. Due to the rate of evaporation exceeding the rate of precipitation in arid regions, salt accumulation in the soil affects crop production. Ions released during rock ablation or introduced by saline groundwater tend to accumulate within the soil and form chlorides, carbonate, sulfate, and clay minerals (Li et al. 2014; Lim et al. 2008). In general, polluted water will also pollute the soil. The use of fertilizers and the irrigation process in crop production sometimes causes salts to concentrate in the soil. Improper management causes irrigated soils to eventually become toxic (Zhao et al. 2012). Solid waste, a conglomerate of plastics, metals, fabric, glass, organic matter, municipal wastes, sewage sludge, and construction debris, mostly originating from households, commercial and industrial activities all contribute to soil contamination. Acid rain can also be classified as a principal contributor to soil pollution.

14.4.1 Cause of Degradation in Soil Quality

The causes of degradation in soil quality are discussed in this section of the chapter (Fig. 14.2)

14.4.1.1 Agricultural Sources

Chemical fertilizers contain essential micronutrients such as nitrogen, phosphorus, and potassium which help in improving the crop yield or productivity. Organic and inorganic fertilizers that are washed as part of surface run-off from crop fields usually introduce excessive nutrients into water bodies, thereby causing algal bloom in water bodies. Pesticides and herbicides used in crop fields are highly toxic chemicals that are extremely detrimental to living organisms causing

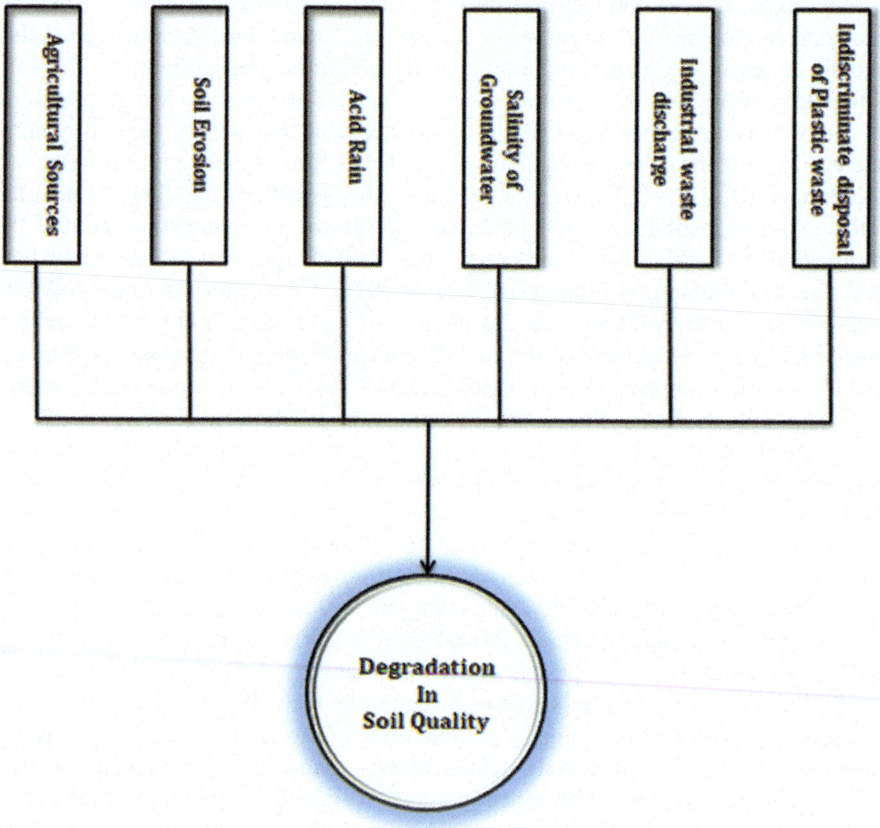

Fig. 14.2 Causes of degradation in soil quality

respiratory problems, cancer, and eventually death. Conversion of forests to plantations also changes soil microbial and physicochemical dynamics (Ikhajiagbe et al. 2020; Song et al. 2019).

14.4.1.2 Soil Erosion

The removal of the topsoil as a result of rain or flood is termed soil erosion. Soil erosion is mostly enhanced by human activities, such as deforestation, construction, mining, land purchase for agricultural purposes, livestock grazing, etc. The soil does not only become less fertile as a result of the erosion, but the soil water holding capacity is also reduced.

14.4.1.3 Acid Rain

The emission of gaseous sulfur and nitrogen oxides from several sources dissolves into the atmospheric water and produces acids in rainwater. Excessive acid rainfall will increase the level of acid penetrating the soil, thereby causing a reduction in the expected crop yield.

14.4.1.4 Water Salinity

Excessive irrigation causes a decrease in the soil quality and subsequently crop productivity as a result of a high concentration of soluble salt in the water used for irrigation. These salts accumulate on the surface of the soil, thereby disturbing the oxygen diffusion and water seepage into the soil and eventually slowing down plant growth.

14.4.1.5 Industrial Waste

Wastes from pharmaceutical companies, chemical industries, paint factories, and cosmetic industries are highly polluting the environment as most of these industries do not have proper waste disposal measures.

14.4.1.6 Plastic Bags

The disposal of waste products in terms of plastic bags (which are usually made from low-density polyethylene) has constantly been a menace to the environment. The indiscriminate plastic disposal practiced by virtually all human lead to blocked drain and sewage systems. Animals feeding on waste food and other kitchen waste may die as a result of choking on waste materials. The burning of refuse dumps causes the release of toxic and poisonous substances into the atmosphere.

14.4.2 Impacts of Soil Pollution

Anthropogenic activities such as agriculture which involves the use of chemicals, industrial activities which cause the release of pollutants in liquid, solid and gaseous forms, and the improper disposal of waste from households and market areas contribute immensely to soil pollution. Soil contamination disrupts the terrestrial ecological balance and leads to destructive health implications for humans (Richard et al. 2020). The fertility of the soil is affected by pollution; hence, crop yield is influenced. Soil pollution also affects soil microbial composition and functional gene interactions (Kyoto Protocol A 2005). However, when the crop grows, there is the possibility that the crop would have taken up; through its roots, noxious chemicals from the soil and eventually cause severe health issues when such crops are consumed by humans (Ogwu et al. 2018). The increase in the salt content of the soil inhibits crop production. In addition, contamination of the soil modifies the structure of the soil and causes the death of various valuable organisms in the soil. Furthermore, the soil becomes unbearable to predators such as birds, and they are forced to migrate in search of food and a conducive environment. In humans, researchers have reported a high incidence of stress and fatigue, extreme headaches, nausea, different types of skin diseases, and miscarriages in pregnant women living close to the contaminated lands (Pierzynski et al. 2000).

In addition, with the different concentration levels and types of contaminants found in the soil, varying accumulated health issues have been reported in humans. The long-term impacts caused by soil pollution on humans include cancer, leukemia, reproductive disorders, and damage to the kidney and liver. Such health problems are associated with purposive poisoning by the contaminated land (for instance, holding a picnic on a beach with contaminated soil) or by unintentional poisoning (such as consumption of cultivated crops on contaminated farmland, or drinking chemical polluted water) (Mishra et al. 2016).

14.4.3 Measures to Mitigate Soil Pollution

The following measures can be carried out to mitigate soil pollution:

- Tree planting should be encouraged as it prevents soil erosion to a large extent
- Industrial and domestic waste should be treated properly before being disposed of in the environment.
- Farmers should be encouraged to adopt more to the application of organic fertilizers as against synthetic fertilizers.

- Government policies regarding the ban on toxic and non-degradable products must be enforced.
- Waste materials must be recycled and reused.
- Proper awareness should be given to the public through regular extension services to enlighten the populace as to ways to prevent soil pollution.

14.5 Water Pollution

The accidental or purposeful addition of unwanted elements to water is referred to as water pollution. This is among the prevalent severe environmental issue and is triggered by a variety of anthropogenic actions which includes industrial effluents discharge which contains toxic substances, agricultural run-off carrying excess fertilizers, herbicides, and pesticides, and domestic sewage with human and animal wastes deposited into aquatic systems, thereby polluting the water. Natural phenomenon leading to water pollution includes soil erosion, leaching of mineral nutrients from rocks, and a massive amount of detritus. Water bodies (Rivers, Streams, Lakes, Seas, Oceans, and Estuaries) and groundwater can be contaminated by point or non-point sources of water pollution. The direct discharge of pollutants, such as waste pipes carrying industrial effluents, into an aquatic system symbolizes point source pollution. In contrast, non-point sources of pollution include pollutants discharged from a large expanse of the land area, such as surface run-off from crop fields, lands used for livestock grazing, roads, construction sites, and deserted mining sites.

14.5.1 Sources of Water Pollution

A majority of the human health problems and water-borne diseases are linked to water pollution. Residues or solid wastes washed by surface runoff from crop fields and discharges of partially treated or untreated effluent from municipal areas and industrial effluents into a water body are common causes of water pollution. Persistent deposition of organic matters and solid particles into the aquatic system causes an increase in the turbidity of water, hence reducing the penetration of light into the water. This, in turn, reduces the rate of photosynthesis by aquatic plants and phytoplankton.

14.5.1.1 Groundwater Pollution

The dependency, by humans, on underground water for drinking, domestic, agricultural, and industrial uses by the ever-growing human population is enormous. It is general knowledge that groundwater is usually a clean source of water. However, anthropogenic activities such as the improper disposal of treated and untreated sewage, dumping of farmyard poultry and livestock manures, agrochemicals such as pesticides and herbicides, and industrial effluents are disrupting the quality of the groundwater.

14.5.1.2 Agrochemical Pollution

Agrochemicals applied to crops contaminate water bodies. Aquatic flora and fauna are usually affected by the introduction of this pollutant into their environment. In the case of fish species and other aquatic organisms, they ingest the pesticides from the water column, and lots of damage is caused to their organs which eventually leads to death (Inyang et al. 2016a, b, c, d; Ogamba et al. 2015). Heavy metals such as lead, zinc, copper, arsenic, mercury, and cadmium emanating from industrial waste discharged into water bodies have a deleterious effect on humans and other aquatic animals, for instance, arsenic polluted water, when consumed results in the accumulation of arsenic in the human body parts like in the blood, nails, and hairs, thereby causing lesions to the skin, rough skin, dry and thickening of the skin and eventually leading to skin cancer (Izah and Srivastav 2015). The excessive concentration of mercury in freshwaters causes dropsy in fishes and Minamata disease in humans. Exposure to lead causes disorder that leads to children having difficulty in interpreting words as they grow. Crude oil pollution occurs in the sea as a result of spillage from ships, oil tankers, rigs, jetties, and pipelines used in transporting the crude oil. A large amount of crude oil gets spilled in the seas when accidents to the oil tankers occur in the sea and it becomes uncontrollable, thereby leading to the death of several marine fauna and flora (Fig. 14.3).

14.5.2 Effects of Water Pollution

The quality of water deteriorates at every contact made with contaminants hence making the water unsuitable for consumption, usage, or survival. The damage done to the environment by polluted water has a ripple effect in that its impact is felt from the species up to humans that make use of the waste or polluted water (Ogwu and Oladeji 2014). Some of the damage caused by water pollution include but are not limited to the following (Fig. 14.4).

- **Stalls economic growth:** When the biological oxygen demand which serves as an indicator used in measuring organic pollution in the aquatic system exceeds the allowable limit, it will cause a decline in the gross domestic product of areas within the same water basins by one-third.
- **Food chain poisoning:** Harvesting of fish in the contaminated water bodies and the use of this water for crop and livestock farming can introduce toxic substances into foods and can cause serious health issues when consumed.

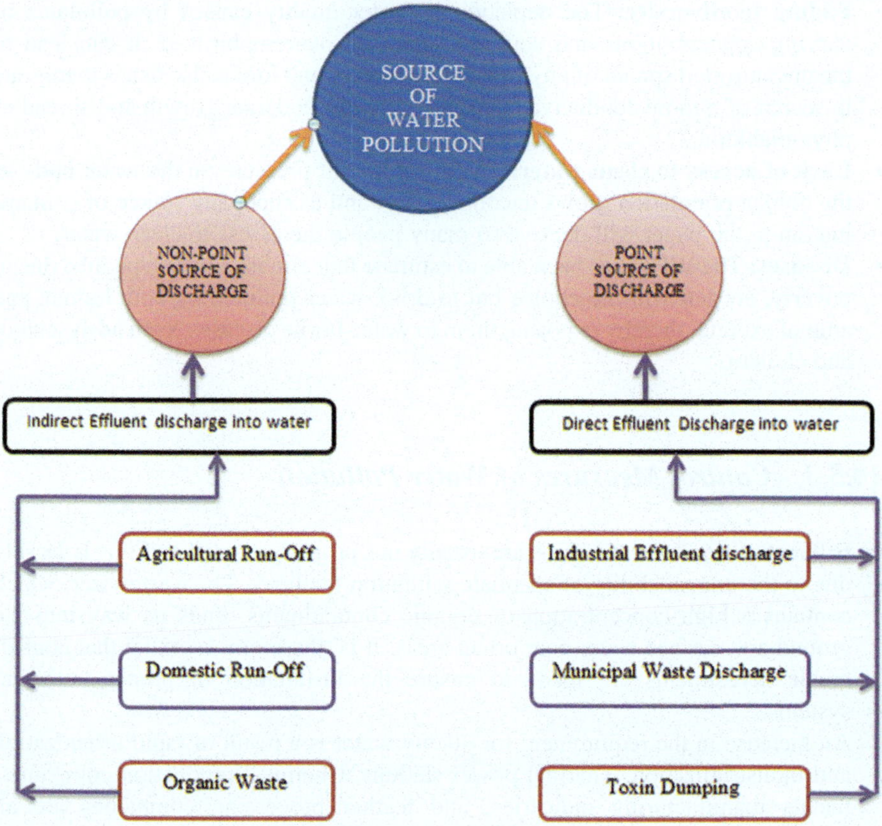

Fig. 14.3 Various sources of water pollution

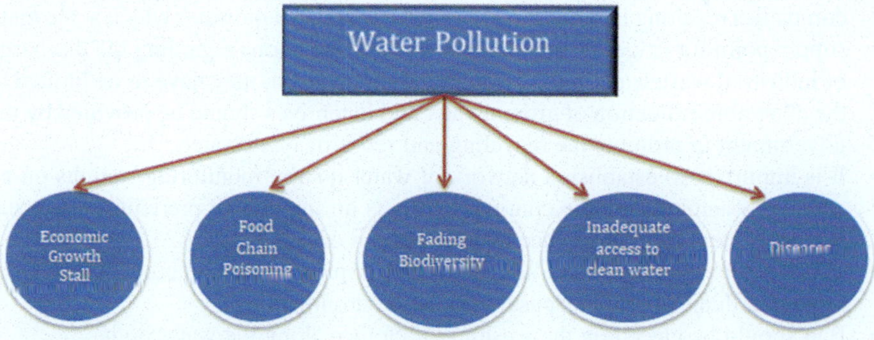

Fig. 14.4 Effects of water pollution

- **Fading biodiversity:** The depletion in water quality caused by pollutants of varying concentrations and sources cannot be overemphasized. It can lead to the uncontrolled spread of phytoplankton in lakes and turn cause fishes to migrate in search of a more conducive environment with moderate growth and spread of phytoplankton.
- **Lack of access to clean water:** There is constant pressure on the water body as the human population grows uncontrollably, and as such, any source of contamination to the water will deprive so many people the assess to clean water.
- **Diseases:** The WHO has been able to estimate that almost 2 billion people, due to poverty, are left with no choice but to drink water polluted by both human and animal excreta, thereby exposing them to water-borne diseases, such as dysentery and cholera.

14.5.3 Control Measures of Water Pollution

- Effluents from domestic areas are usually not properly treated and this is largely due to the unavailability of adequate sanitation facilities. The wastewater, which contains a high concentration of organic contaminants, finds its way into the ground and surface water near urban areas. It is, therefore, essential that considerable investments are made to ensure the installation of proper treatment systems.
- An increase in the requirement for quality water is a result of rapid urbanization and industrialization. Thermal power stations, Chemical production companies, metals manufacturing industries, and leather processing companies are all expected at the onset of their production to carry out an EIA but that does not happen most times, hence the poor environmental management system of the majority of these companies. Toxic effluents are mostly discharged from the companies resulting in pollution to the surface and groundwater which is the main supply point for irrigation and domestic use. Regulations regarding the discharge of industrial wastewater must be enforced. In addition, they have to be limited to the allowable extraction of groundwater and incentives should be provided by the government to promote the recycling and reuse of wastewater.
- It is important to establish a network of water quality monitoring stations on all aquatic systems across the continent. This is an imperative prerequisite to measure the extent of maintenance and restoration of water bodies.
- Urban settlements must have sewage treatment plants to ensure proper cleaning of sewage effluents before disposal into the environment.
- Ban should be placed on the washing of clothes along the water body.
- In addition, rainwater harvesting should be encouraged to prevent the depletion of the groundwater table by acid rain.
- Consistent awareness of the importance of water and measures of preventing pollution should be done.

14.6 Thermal Pollution

Thermal pollution, which is sometimes referred to as "thermal enrichment," is the depletion or disruption in the quality of water by any action that alters the normal temperature of the water. The common origin of thermal pollution is the use of water as a coolant by power plants and industries. Soil erosion can also cause thermal pollution. All nuclear and thermal power plants, chemical industries, and manufacturing companies make use of a large quantity of water to cool their machines and the resultant hot water is deposited into water bodies. The boiler's heating process and heat emanating from the waste produced cause an increase in the temperature of the water meant to cool the machines. The release of warm or hot water into a water body increases the temperature of the receiving water by 10–20 °C above the ambient water temperature.

In aquatic systems, dissolved oxygen becomes depleted, as the water temperature rises as a result of thermal pollution. Changes in the water temperature are usually insignificant over a short period and they occur seldomly. This instability in temperature within the aquatic system affects the metabolism of fish species and inhibits the growth of aquatic flora. This simply infers that the discharge of hot water from thermal power plants has an undesirable impact on the aquatic ecosystem. During the warm summer period, aqua flora and fauna in coastal waters within the tropics live very close to their upper-temperature threshold. The sudden increase in the water temperature as a result of hot water discharge into water bodies causes a decline in their swimming efficiency, affects fish feeding, increases their rate of metabolism, and subsequently hinders their growth. It becomes difficult for the fishes to run away from predators and they also will not be able to chase their prey. The subsequent resistance of fish to pathogenic attack decreases drastically as a result of a negative change in their environmental condition. The biological diversity of the aquatic flora and fauna is also greatly reduced. It is imperative, therefore, that in reducing the effects caused by thermal pollution, hot water from power plants should be stored in a pond, set aside to cool the water before it is discharged into any receiving water body.

14.6.1 Cause of Thermal Pollution

Natural and human activities contribute immensely to the menace of thermal pollution. Power plants and manufacturing machinery are the main cause of thermal pollution. Water is a superb and free, cooling agent. As such, many industrial operations require water from close-by water bodies to cool down the temperature of their manufacturing machines and afterward release the relatively warm water to return to the aquatic system. The natural causes of thermal pollution include the introduction of massive heat into water bodies by geothermal vents or hot springs. Deforestation is a process of clearing trees that are supposed to serve as shades for both terrestrial and aquatic animals, thereby allowing direct sunlight to hit the ground and water surface. When water gets on hot cemented surfaces, it becomes hot, and subsequently finds its way into a nearby water body, thereby causing an increase in the water temperature most times to unbearable limits to the aquatic life. Small and shallow water bodies, such as retention ponds, tend to absorb quite a bit of heat energy from the sun and this makes them a source of thermal shock. When such water is pumped directly into a river or lake, it causes an increase in temperature.

14.6.2 Impacts of Thermal Pollution

The impacts created by thermal pollution are enormous and diverse. Thermal pollution damages aquatic ecosystems and causes a decline in animal populations. Ecosystem flora and fauna respond differently to drastic significant changes in temperature. It causes survival of the fittest and organisms that cannot adapt to such a change will die or be forced to migrate out of the area. In the polluted area, biological diversity is negatively affected as potential broodstock species would have migrated off the area.

14.6.2.1 Decreased Dissolved Oxygen

An increase in the temperature of the aquatic system causes a decrease in dissolved oxygen. Warm water usually holds less oxygen than cool or normal temperature water. If there is a decline in the oxygen level, animals within the area will migrate to find a more conducive environment but those that do not migrate stand the risk of death. In deep water bodies, the accidental or intentional introduction of hot or warm water will hinder the diffusion of oxygen into the deeper regions of a water body, though beneficial to some microorganisms but detrimental to aquatic animals. The decrease in dissolved oxygen can enhance the negative impact caused by algae bloom on aquatic animals. The algae bloom menace is perhaps the most common side effect of thermal pollution.

14.6.2.2 Ecological Impacts

The sudden introduction of large amounts of warm or hot water causes thermal shock. An aquatic ecosystem is damaged by thermal pollution. Thermal shock can kill off fish, aquatic insects, and amphibians (Inyang et al. 2019d).

14.6.2.3 Biodiversity Halt

The swift temperature changes within an ecosystem cause death to organisms that are susceptible to such environmental change or force them to move away from the environment. This is one of several severe issues facing threatened and endangered plant and animal species. Such loss can be a result of organisms dying from the hot water, their inability to reproduce effectively, or simply migrating from the area. Animals are usually thought of as primary victims of water pollution, but multi-celled aquatic plants are also susceptible to the changes caused by thermal pollution.

14.6.2.4 Migration

Prolonged changes in the environmental temperature cause fish and amphibians to migrate far away from the warm water to a more suitable and conducive location. Birds that feed on aquatic organisms may also need to leave such polluted areas to survive and search for more food.

14.6.2.5 Increased Toxins

An increase in temperature elevates the potency of chemical pollution. Toxic chemicals in the water are more of a side effect of dumping wastewater than the direct impact of thermal pollution. Chemical pollution is an almost unavoidable side effect of using water for cooling in manufacturing industries. Diffused heavy metals, chemical solvents, and crude oil mostly end up in the lagoons, lakes, or sometimes rivers, where the cooling water gets deposited.

14.6.3 Ways to Prevent Thermal Pollution

- The heated water from manufacturing and production companies must be treated properly before releasing it directly into water bodies.
- In addition, cooling towers and cooling ponds can be installed to receive the heated water from several industries.
- Adequately treated water from industries can also be recycled to be used for domestic purposes.

14.7 Anthropogenic Activities and Biodiversity Conservation

The harm caused to biodiversity by humans is enormous. In light of this, the populace must have a good understanding of the effect anthropogenic activities that are having on biological diversity, bearing in mind that life itself revolves around all-around biological differences amongst all plant and animal species. If the right attitude is not in place on the way people use natural resources, there will be destruction in biodiversity until human life seize to exist. Humans influence biodiversity through their constantly rising population, which transcends into more land utilization and their ways of life, all of which can destroy natural habitats. Humans should accept that their activities have both positive and negative impacts on biodiversity; hence, humans must find the means to cope with whatever biological diversity the earth has left.

14.7.1 Definition and Importance of Biodiversity

Biodiversity simply refers to a variety of life on the planet and the patterns it may form naturally. Such variety and patterns are a result of constant development, regular cycles, and several human impacts (Secretariat 2000). Biodiversity contains a variety of potentials in an animal type, species, and the environment (Frequently 2005). Biodiversity is usually not dictated by just one element, but instead by lots of variables that are distinct spatially and transiently (Climate 2005). Albeit, the importance of biodiversity cannot be overemphasized, as it holds the key to our existence. As each day passes by, people utilize 40,000 animal and plant species, the greater part of which are mostly undetected (Eldredge 2000). Though people do not like to admit it, biodiversity provides the populace with lots of food, water, oxygen, energy, detoxification of waste, adjustment of earth's environment, medication, openings for entertainment, and the travel industry (Secretariat 2000). In short, humans, animals, and plants would not be existing without biodiversity.

14.7.2 Extinction of Species

The quantity of species in the world is the most obvious indicator of biodiversity. There are about 1.75 million identified species; although, some authors state that there are no less than 10 million species living on the planet (Eldredge 2000). To check out the deficiency of biodiversity, the quantity of species extinction should be inspected. The extinction of species has been reported to be up to 40,000 species each year (which rounds up to approximately 100 species every day or 4 species every hour) (Wood 2000). At that rate, it is 50 multiple times the normal speed at which species go into extinction and it is predicted to increase (Sherbinin 2002). The rate of extinction is of great concern, because the moment an animal variety is terminated; there is simply no way of reviving such extinct species (Izah et al. 2021).

According to Noss et al. (2005), three primary issues that cause species extinction are:

- Habitat loss occurs when a habitat is destroyed.
- Habitat degradation occurs when a habitat reduces to an extent when it is unable to sustain and support biological communities (Effects 2005).
- Habitat fragmentation occurs when a territory is degraded into smaller segments of land for expansion (Mapping 2005).

When these three problems occur in an environment, species extinction is evident and it can be related to the impacts of anthropogenic activities on habitat.

14.7.3 Human Actions Toward Species Biodiversity

The determination of human influence on biodiversity cannot be determined one way, it is a multi-facet activity from different dimensions, and nonetheless, many activities by people are causing a drastic diminishing effect on biodiversity. To determine the effects of human activities on an environment, the land and water required to cultivate a particular product to be consumed and the necessity to take a record of the wastes generated by humans should be noted according to the production and the management practices which is used throughout the production process (Wackernagel et al. 2002).

The direct and ancillary activities carried out by people have brought about the decline of biodiversity (Izah and Seiyaboh 2018; Izah 2018; Izah et al. 2018b, 2017d). The Convention of Biological Diversity expresses that there are both direct and indirect human drivers. The indirect human drivers include economic, demographic, socio-political, technological, cultural, scientific, and religious factors. Some direct human drivers comprise changes in local land use and land cover, species removals or introductions, air and water pollution, and environmental change (Climate 2005). Human movement has significantly transformed approximately 33% to one-half of the world's surface (Frequently 2005). In the next 50 years, it is expected that people will effectively affect between 50% and 90% of land in agricultural-based nations. This is a result of population growth and the utilization of regular assets (Mapping 2005). The human population is mostly considered the problem of biodiversity (Eldredge 2000). The amount of people on the planet and the consistent expansion of humans creates an issue that eventually leads to the conversion of natural habitats for animal species to land for human cultivation and use.

One way that humans have sustained their development is by changing the natural habitats to lands to be utilized for producing food. No less than 23% of the land on earth is being utilized for agriculture (31% of all land is not fit for farm practices). In the United States of America, there is an immediate connection between forest degradation to increase land for crop production (Dobson 1996). Universally, there exists a large ratio of forest to farmland. A possible risk of diminishing the remaining natural habitats is that the species are no longer available on the planet. This directly has a great influence on horticulture on the basis that large numbers of the strains that are being obliterated for crop fields would be utilized for yields with improving hereditary potentials. As such, the expansion of farming areas will continually hurt our agrarian future. Human activities have constantly assumed a part in changes evident in the environment, which is in the long run causing extraordinary risk for biodiversity. The modification in the environment is a result of expanded climatic groupings of carbon dioxide, which increases land and sea temperatures and changes in the rate of precipitation, and a rise in the ocean level. The change experienced in the environment additionally brings about an alteration in the strains or varieties of animals. It is likewise expected that the modification in the environment during the 21st century will have a lot higher rate than the beyond 10,000 years and establishes a greater effect on biodiversity (Climate 2005). Normally, 80% of naturally rich areas will experience extraordinary misfortunes of plant and creature species due to an unnatural weather change. The pace of progress of environments is relied upon to increment up to multiple times because of a worldwide temperature alteration (Sherbinin 2002).

14.7.4 Relationship Between Poverty and Biodiversity

Biodiversity has varying impacts on our lives. Individuals that lack basic needs rely extremely on nature to provide them with assets for their survival. In underdeveloped nations, logging has turned into a common action of underprivileged people. It is an immense issue in many non-industrial nations as it can destroy the natural environment, yet it is by all accounts one of the main ways that individuals can make sufficient income to assist and improve their families. The New York Times published an article about people in Malawi unlawfully chopping down trees to earn some money to purchase food. During an interview, an individual made a statement, "We have no cash to raise our families; we have no place to run, nothing else to do; so we need to cut the trees to take care of our families." However, in any event, when the people cut down trees and sell the wood, they eventually still do not get enough money to help their families. In Malawi, the pace of deforestation was at 2.8% and 23 species that were found in their forests are considered endangered species (Wines 2005). The environmental ruin will influence both poor and industrialized countries. Nonetheless, the agricultural countries will be the ones that are impacted the most by the environmental degradation by increasing poverty, destitution, decreasing work efficiency, and fueling the current economic and social crisis (Mapping 2005). Developing countries do not have the assets to help their citizens; hence, they find a choice to depend on nature for survival.

Educating the populace, especially in developing countries about the need to guard their biodiversity, is an absolute necessity for assuring human survival. Teaching people in the local community about the effect individuals are having on the environment and showing individuals how they can live in harmony with nature will assist with safeguarding biodiversity without bringing on additional persecution. People usually do not have the understanding that there are alternative methods of making money that does not place the environment in danger. For example, in Malawi's case, the people in the community could get more money from harvesting honey from bee colonies in their forests rather than cutting down the trees for sale (Wines 2005). Public awareness in developing nations is important to biodiversity regardless of whether it centers on sustainable living or not. Several types of research have shown that empowering ladies has led to a reduction in the rate of childbirth which eventually affects the population growth, particularly since developing countries have a higher birth rate when compared with industrialized countries. Educating people, all people, and not just the devastated ones, about their impacts on biodiversity is a positive development (Eldredge 2000).

14.7.5 Actions by the Government on Environmental Sustainability

There are several activities that humans are involved in and can help with sustaining biodiversity; notwithstanding, the government must take some steps that will bring about a larger impact in saving biodiversity. Former President George Bush of The United States of America has dispensed with the roadless rule, which was a standard that held logging and streets back from being available in 60 million sections of land of national forests. Bush also cut 42 million acres of critical habitat from the 83 million sections of land that are required for species that are threatened and endangered. There has been an extraordinary diminish in how much is assigned to the wild environment in the United States. President Bush assigned 530,000 sections of land as wild regions contrasted with 9.5 million under the Clinton administration, 10.6 million acres of land under the Reagan organization, and 66.3 million acres of land under the Carter organization. There was also a decrease in the number of species that were added to the endangered species list; Bush has added 31 species, Reagan 253, Bush Senior 228, and Clinton 521 (Wetsone et al. 2005).

Bush settled on one more ruling against the environment by not endorsing the Kyoto convention. The Kyoto Convention is a deal that agrees to oversee climate change on earth by decreasing how much carbon dioxide and other greenhouse gasses from gaining access to the atmosphere or to utilize trading of emissions (in which case the governments provide incentives for individuals to lessen the rate of discharges) (Kyoto Protocol A 2005). Al Gore signed the Kyoto convention in 1998, but this arrangement was held off on restriction until it was ratified (Kyoto Protocol A 2005; Kyoto Protocol B 2005). The United States did not uphold the consenting to of the arrangement, since they figured it "would bring about genuine harm to the economy of the United States" (Kyoto Protocol A 2005) The United States did not approve of the understanding and had no aim of doing so at this time (Kyoto Protocol B 2005). Bush expressed that he disagreed with the treaty, since it did exclude each country, particularly ones that were releasing huge greenhouse gases, for instance, China (World Wide Fund 2002).

14.7.6 Actions for Environmental Management and Prosperity in Africa

14.7.6.1 The African Ministerial Conference on the Environment

The African Ministerial Conference on the Environment met on the 23rd of October 2019 which was the sixteenth convention. It involved a look back on the progression, growth, and evolution of the conference in its 34 years of establishment in areas of formulation of policy and decision making. The conference on the environment has a mandate which is focused on the environment and sustainable developmental policies. The various proclamations from the conference are to inspire all stakeholders, partners, and member states. With the rate of urbanization, decisions targeted at transforming socio-economic activities must be given utmost priority. It is expected that the approach will ensure that decisions are made to address specific gaps and opportunities which are geared from real-time effects and feedback from implementation. This will prepare the member states for the Sustainable Development Goals and the Agenda 2063 by the African Union.

14.7.6.2 Review of Decision Delivery of THE Conference from Its Inception

The conference has diligently ensured that its mandate is delivered since its establishment 34 years ago. It involves ensuring the collective protection of the environment in meeting the socio-economic development needs of Africa sustainably. The conference has four key dispensations:

- The first dispensation was in 1985–1993 which was the formative period and it focused on the creation of conference organs and their program of work. The Cairo Program for African Cooperation was developed and implemented during this period and its first session was adopted in December 1985. It focused on environmental degradation and the satisfaction of needs of food and energy for the people of Africa.
- The second dispensation was between 1993 and 2000 and this period refined the gaps observed in the first dispensation. A clear policy orientation was the principal focus in this era which was generated from the Cairo Program

implementation of the Earth summit in 1992. The operations and focus of the conference were reviewed and a new policy was adopted at the fifth session held in 1993. The focus of the Conference was shifted from operational programs to programs that are focused on realizing sustainable development in Africa. The mandate during this period was the provision of political and technical leadership across the continent on sustainable development and to come up with positions agreed upon by member states on the environment and development.

- The third dispensation was between 2000 and 2010 which was a period of the revitalization of the conference which a clearer focus on the organizational structure and positioning of a high policy forum for the African Environment instead of the continent's agenda for the environment. This dispensation ensures that the issues of the environment are embedded fully in the processes of the mandate which led to socioeconomic benefits and prosperity.
- The fourth dispensation which is the modern era started in 2010 at the 13th session in Bamako to the 17th session which was held in Nairobi in 2018. The sessions in between were the 14th session in Arusha, the 15th in Cairo, and the 16th in Libreville. During this era, there was an African transformation of the environmental policy agenda from a passive one to a major influencer of global policy. Major policy initiatives which influenced the global environment and development were observed such as the Agenda 2030; African Union summits, Africa's strong voice in the Rio +20 Summit, and several Conferences of Parties, such as biodiversity and desertification, climate change, and regional economic communities. Strengthening and mobilizing member of states were also witnessed as well as sustainable financing mechanisms for the conference. This era also focused on program development and flagship programs, policies, and directions for a sustainable environment.

14.7.6.3 Environmental Sustainability and Progress Established in Africa Through the Conference

In the 14th session in 2012 which was held in Arusha, Tanzania, it was adopted at the conference that the development and implementation of programs for effective implementation of outcomes for the United Nations Conference on Sustainable Development (Rio +20) were mandated. This was in line with the debate on environmental sustainability and the new mandate was to cover the following areas:

- **Ecosystem-based adaptation program for Africa:** When undertaken in the Zambezi basin ecosystem, it showed that the country can benefit both from the environment and socioeconomically from this form of adaptation. In Malawi, the techniques used were conservation agricultural practices such as agroforestry which was able to restore over 15ha of lands that were previously degraded, and restoration of river flow by drying rivers to ensure constant water supply all year round for agriculture and domestic uses. The Baixo-Limpopo irrigation climate resilience project was aimed at climate-resilient infrastructure to increase agricultural production with a reported increase of 150% in income and poverty reduction by 42%.

- **Green economy initiatives:** It was embarked on in Burkina Faso under the Green Economy and sustainable consumption and production programs in which plastics wastes were recycled for a cleaner and healthier city and about a 400% increase in revenue was observed along the value chain for actors. The integrated resource-efficient and cleaner production principle was carried out in Kenya and 11% per annum was saved on energy and water use
- **Sustainable consumption and production program:** The notable projects are the ones involving sustainable energy, such as the development of geothermal energy capacity in the continent. This program has developed over 400 home-grown geothermal energy experts and is ready to invest and unlock the geothermal capacity of Africa of over 20 GW which can be used for industrial applications. Some countries leading in the geothermal project are Eritrea, Kenya, Ethiopia, Djibouti, Uganda, Tanzania, and Rwanda. Injecting entrepreneurship into energy actions relating to the environment was explored by blending the relationship between women and sustainable energy.
- **African program on biodiversity and ecosystems:** Several works were done on the protection of endangered wild flora and fauna by implementing the convention on the international trade on the endangered species. This was directed to support and strengthen the country in terms of governance structures to tackle the illegal activities and trade in wildlife. The motion was supported by 11 African countries that agreed to combat the issue of illegal wildlife trade.
- **Africa integrated assessment for sustainable development:** It provides Integrated Environmental Assessment tools, guidelines, cases, studies, and reports to appropriate authorities and stakeholders. This program is anchored under the African Integrated Environmental Assessment for Sustainable Development Planning Program.

A highlight of this 14th session was the reiteration of the need for the position of Africa on climate change to be supported by scientific research, findings, and analysis. An Africa Adaptation Gap Report series controlled by the United Nations Environmental Program provided a basis for the global change regime for Africa's position. The 15th session was done some months after the Paris Climate Change Agreement's adoption in Cairo in 2015. This session endorsed the African Adaptation Report findings and the environmental ministers affirmed that the natural capital provides an opening for inclusive development in Africa and this was generated from the adoption of Agenda 2030 and the SDGs. The highlights of this session were the adoption of the Cairo Declaration on Managing Africa's Natural Capital for Sustainable Development and Eradication of Poverty. This led to a basis for the environment assets to be leveraged toward the actualization of Agenda 2030 and the SDGs as well as the aspirations of the African Union's Agenda 2063.

14.7.7 *Environment and Socioeconomic Development*

The environment is seen as an indicator of socio-economic development in the world; therefore, tangible decisions to implement must be carried out. In keeping this thrust on so solutions to the environment, progress must be made at both policy and operational levels. Several actions such as the EBA-driven agriculture and clean energy, pollution control, and waste control have been undertaken to improve the management of natural capital in the African continent. This proved that innovative actions can be potential solutions to unlock the environment as illustrated in the 16th session. Some case studies are highlighted:

14.7.7.1 Uganda

In Uganda, the resource efficiency and control of pollution aspects recorded financial benefits in a bio-waste compost facility. Savings of about $8800 annually were achieved which was initially spent on fertilizer purchase. In addition, improvement in the thermo-boiler system via insulation and its automation drastically reduced the use of wood by 70%. A recycling program of plastics was set up and a savings of $23,209 was generated this act helped in the preservation of the forest ecosystem. Green investment through the industrial symbiosis of solid waste management resulted in savings of $1,080,800 annually and a repayment period of 0.58 years. This investment created green jobs for 100 males and 150 females as corroborated by SDG 5 (gender equality) and added value to solid wastes preventing the loss of over 21,600 tons annually that were supposed to be dumped on sites.

14.7.7.2 Cameroon

A women's cooperative in the country was guided to leverage solar driers to process their cassava tubers and link micro-hydro plants to provide power for milling of cassava to have a better and higher valued flour. Women now dry their products and shelf life is increased as post-harvest loss is reduced. With this, the goal of SDG 2 in terms of food security has been achieved and the processing method reduces household poverty as elaborated by SDG 1. Spoilage of cassava products has been reduced by 30%, thereby enhancing the efficiency of resources (SDG 12). Zero-emission form of electricity is achieved from the micro hydro generating plant; this protects the environment against the previous use of generators that are diesel-operated (SDG 13). Clean energy investment as elaborated by SDG 7 was achieved for the enterprise which demands clean energy for the sustenance of the environment.

14.7.7.3 Kenya

A female individual was empowered with skills in microgrid applications to start her own business. She was a member of the African Women Energy Entrepreneurs Framework, and at the start of her business, she implemented the use of a 1KW prototype that can connect up to 16 households and shop outlets (SDG 5).

14.7.7.4 Tanzania

In this country, increased productivity and generation of income by the use of efficient technology by small-scale farmers were achieved. This was done through a drip irrigation system and farmers were able to produce over 8000 kg of French beans on 2 acres. This translated to revenue of $2000/acre in 3 months; therefore, crop productivity was increased and revenue up to $8000/annum/acre of French beans was derived justifying SDG 2.

14.7.7.5 Cote d'Ivoire

The actors in the industry were assisted in the establishment of a waste-to-biofertilizer system by installing a digester in Abidjan. This process enhanced proper cleaning, sanitation, and management of wastes as illustrated by SDG 6 and the provision of organic fertilizer for use by farmers. The organic fertilizers were used on the soil and they improved the agro-ecosystems (SDG 15).

14.7.7.6 Nigeria

Several policies on environmental sustainability have been in play in Nigeria. A task force called the inter-agency policy convened the implementation of groundbreaking policies which are key. Some are the Nigerian feed-in-tariff regulations and the iconic cassava bread policy. All these policies are strategic for the realization of Nigeria's vision of 2020. The actions toward an innovative environmental solution in Africa are ongoing and their pace is limited considering the challenges. There is a need to upscale the decisions and implementation of the conference so that the beneficial impacts can be enjoyed implementation.

14.8 Conclusion

Human activities are the principal sources of pollutants in the environment. They can impact biodiversity processes in diverse ways and protective measures are required for the sustenance of the environment. The issue of environmental pollution is a major issue and countries, most especially the developing and underdeveloped ones are vulnerable to its effects. Rapid industrialization and urbanization have increased waste generation over the last decade. Increased agricultural practices, land use, waste disposal, and management have been generated by the increased human population and their threats to biodiversity are not pleasurable. Various types of pollution can arise from improper environmental management, such as air, noise, water, soil, and thermal pollution. Air pollution which can be in primary or secondary forms of pollutants hurts the environment. They can also affect humans who inhale these gases, such as carbon dioxide, carbon monoxide, Nitrogen oxide, sulfur dioxides, hydrocarbons, and suspended particulate matter. To reduce the effects of these gases, the government and individuals need to make a conscious effort to protect the environment.

Noise pollution is another form of pollutant that affects the environment. This can be sourced from indoor and outdoor sources and they have serious health implications, such as high blood pressure, insomnia, mental depression, etc. This form of pollution can be prevented and controlled by proper maintenance of vehicular objects, and the use of soundproof devices should be advocated. Trees have a natural way of sequestering noise and tree planting should be encouraged to reduce these impacts. Soil pollution may also arise from improper management of polluted waters, and high technological processes which may release xenobiotic compounds into the environment. It may also arise from sources, such as agriculture, acid rain, soil erosion, water salinity, industrial wastes, and plastic bags. Their impacts on the environment include soil modification and affects soil fertility and soil content. Issues such as stress, fatigue, nausea, skin diseases, loss of pregnancy, etc. can result in humans when exposed to contaminated soils. To reduce these effects, trees should be planted, wastes from industries should be treated before discharge, farmers should be smart at fertilizer application, and a government ban on non-biodegradable items should be enforced.

Water pollution may also arise from improper environmental usage. Pollutants in water may be from non-point or point sources and pollutants are leeched into the water. Sources of pollutants may be from the ground or agrochemical substances and have negative activities such as food chain poisoning, reducing biodiversity and assess to clean water, reduces economic growth, and diseases. This can be controlled by treating water before discharge. Thermal pollution is another form of pollutant and it reduces the dissolved oxygen concentration in water, halts biodiversity, affects the ecological system, causes migration, and increased toxins in the environment. It may be reduced by treating heated water from manufacturing and production companies be treated properly before releasing it directly into water bodies. For biodiversity conservation, humans who are the principal factor must develop a conscious and deliberate effort to sustain the environment. Government and stakeholders have a role to play in sensitization as well as investment in the management of waste for sustainability.

References

Adham KG, Al-Eisa NA, Farhood MH (2011) Risk assessment of heavy metal contamination in soil and wild Libyan jird Merioneslibycus in Riyadh, Saudi Arabia. J Environ Biol 32:813–819

Chibuike GU, Obiora SC (2014) Heavy metal polluted soils: effect on plants and bioremediation methods. Appl Environ Soil Sci 2014:1–12

Dobson AP (1996) Conservation and biodiversity. New York: Scientific American library effects: habitat degradation, loss, and fragmentation. http://chesapeake.towson.edu/landscape/impervious/all_habitat.asp.

Eldredge N (2000) Life in the balance. Princeton University Press, Princeton, NJ

Gaddi AV, Capello F (2018) Pollution and air pollution-related diseases: an overview. In: Capello F, Gaddi A (eds) Clinical handbook of air pollution-related diseases. Springer, Cham

Giller KE, Witter E, Mcgrath SP (1998) Toxicity of heavy metals to microorganisms and microbial processes in agricultural soils. Soil Biol Biochem 30(10–11):1389–1414

Ikhajiagbe B, Ogwu MC (2020) Hazard quotient, microbial diversity, and plant composition of spent crude oil polluted-soil. Beni-Suef Univ J Basic Appl Sci 9(26). https://doi.org/10.1186/s43088-020-00052-0

Ikhajiagbe B, Ogwu MC, Lawrence AE (2020) Single-tree influence of *Tectonia grandis* Linn. f. on plant distribution and soil characteristics in a planted forest. Bull Natl Res Centre 44:29. https://doi.org/10.1186/s42269-020-00285-0

Ikhajiagbe B, Ogwu MC, Ekhator PO, Omoayena II (2022a) Implications of soil nitrogen ehnacement on the yield performance of soybean (*Glycine max*) in cadmium- polluted soil. J Pollut Effects Control 10(2):1–12. 1000334. https://doi.org/10.1101/2021.11.22.469589

Ikhajiagbe B, Chukwujama CA, Ikhajiagbe B (2022b) Intrinsic restoration of a ferruginous utisol using goosegrass (*Eleusine indica* L. [Gaertn.]) obtained from different land use areas. J Nat Resour Dev 12:1–18. https://doi.org/10.18716/ojs/jnrd2022.12.01

Ikhajiagbe B, Aituae W, Ogwu MC (2022c) Morpho-physiological assessment of oil palm (*Elaeis guineensis* Jacq.) seedlings exposed to simulated drought conditions. J Oil Palm Res 34(1): 26–34. https://doi.org/10.21894/jopr.202100018

Inyang IR, Obidiozo OZ, Izah SC (2016a) Effects of Lambda cyhalothrin in protein and Albumin content in the kidney and liver of *Parpohiocephalus obscurus*. EC Pharmacol Toxicol 2(3): 148–153

Inyang IR, Thomas S, Izah SC (2016b) Activities of electrolytes in kidney and liver of *Clarias gariepinus* exposed to fluazifop-p-butyl. J Biotechnol Res 2(9):68–72

Inyang IR, Thomas S, Izah SC (2016c) Evaluation of activities of transferases and phosphatase in plasma and organs of *Clarias gariepinus* exposed to Fluazifop-p-Butyl. J Environ Treatment Techn 4(3):94–97

Inyang IR, Kenobi A, Izah SC (2016d) Effect of dimethoate on some selected metabolites in the brain, liver and muscle of *Clarias lazera*. Sky J Biochem Res 5(4):63–68

Inyang IR, Ajimmy AR, Izah SC (2017a) Organosomatic index and behavioral response of Heterobranchus bidorsalis exposed to rhonasate 360sl containing glyphosate (isopropylamine salt glycine). ASIO J Microbiol Food Sci Biotechnol Innov 3(1):4–8

Inyang IR, Ollor AO, Izah SC (2017b) Effect of Diazinon on Organosomatic Indices and Behavioural Responses of *Clarias gariepinus* (a Common Niger Delta Wetland Fish). Greener J Biol Sci 7(2):15–19

Inyang IR, Izah SC, Johnson DT, Ejomarie OA (2017c) Effects of *Lambda cyhalothrin* on some electrolytes and metabolites in organs of *Parpohiocephalus obscurus*. Biotechnol Res 3(1):6–10

Inyang IR, Izah SC, Ntaka CM (2018a) Effect of Imidacloprid on Total Protein, Albumin and Electrolytes in *Heterobranchus bidorsalis*. Environ Anal Eco Stud 4(5):EAES.000597. 2018

Inyang IR, Puanoni AR, Izah SC (2018b) Evaluation of the effect of toluene (produced water component) on some blood cells and enzymes of *Clarias gariepinus*. MOJ Toxicol 4(6): 440–444

Inyang IR, Izah SC, Okpogholor KD (2019a) Impact of aluminum phosphide on the transferases in liver and muscle of *Parophiocephalus obscurus*. J Plant Animal Ecol Special Issue 1(4):1–6

Inyang IR, Izah SC, Okoroba VR (2019b) Effect of toluene on the organosomatic indices and electrolytes in the liver and muscle of adult *Clarias gariepinus*. Sumerianz J Biotechnol 2(5): 31–34

Inyang IR, Izah SC, Suobo K (2019c) Effect of phenol on the kidney and liver biochemical and metabolites of *Clarias gariepinus*. Noble Int J Sci Res 3(3):33–40

Inyang IR, Akparata OJ, Izah SC (2019d) Evaluation of some blood cells and metabolites of *Heterobranchus bidorsalis* (Hybrid) exposed to sublethal concentrations of sarosate. Int J Res Environ Sci 5(1):17–23

Izah SC (2018) Ecosystem of the Niger Delta region of Nigeria: potentials and threats. Biodiversity Int J 2(4):338–345

Izah SC, Angaye TCN (2016) Heavy metal concentration in fishes from surface water in Nigeria: potential sources of pollutants and mitigation measures. Sky J Biochem Res 5(4):31–47

Izah SC, Seiyaboh EI (2018) Challenges of wildlife with therapeutic properties in Nigeria; a conservation perspective. Int J Avian Wildlife Biol 3(4):259–264

Izah SC, Srivastav AL (2015) Level of arsenic in potable water sources in Nigeria and their potential health impacts: a review. J Environ Treatment Techn 3(1):15–24

Izah SC, Chakrabarty N, Srivastav AL (2016) A review on heavy metal concentration in potable water sources in Nigeria: human health effects and mitigating measures. Exposure Health 8: 285–304

Izah SC, Bassey SE, Ohimain EI (2017a) Assessment of heavy metal in cassava mill effluent contaminated soil in a rural community in the Niger Delta region of Nigeria. EC Pharmacol Toxicol 4(5):186–201

Izah SC, Bassey SE, Ohimain EI (2017b) Geo-accumulation index, enrichment factor and quantification of contamination of heavy metals in soil receiving cassava mill effluents in a rural community in the Niger Delta region of Nigeria. Mol Soil Biol 8(2):7–20

Izah SC, Bassey SE, Ohimain EI (2017c) Assessment of pollution load indices of heavy metals in cassava mill effluents contaminated soil: a case study of small-scale cassava processing mills in a rural community of the Niger Delta region of Nigeria. Biosci Methods 8(1):1–17

Izah SC, Bassey SE, Ohimain EI (2018a) Impacts of Cassava mill effleunts in Nigeria. J Plant Animal Ecol 1(1):14–42

Izah SC, Aigberua AO, Nduka JO (2018b) Factors affecting the population trend of biodiversity in the Niger Delta region of Nigeria. Int J Avian Wildlife Biol 3(3):206–214

Izah SC, Uzoekwe SA, Aigberua AO (2021) Source, geochemical spreading and risks of trace metals in particulate matter 2.5 within a gas flaring area in Bayelsa State, Nigeria. Adv Environ Technol 7(2):101–118

Kerfahi D, Tripathi BM, Slick JWF, Sukri RS, Jaafar S, Dong K, Ogwu MC, Kim. H- K. and Adams, J.M. (2019a) Soil metagenome of tropical white sand heath forests in Borneo: What functional traits are associated with an extreme environment within the tropical rainforest? *Pedosphere* 29(1):12–23. https://doi.org/10.1016/S1002-0160(18)60054-2

Kerfahi D, Ogwu MC, Ariunzaya D, Balt A, Davaasuren D, Enhmandal O, Purevsuren T, Batbaatar D, Tibbett M, Undrakhbold S, Boldgiv B, Adams JM (2019b) Metal tolerant fungal communities are delineated by high zinc, lead and copper concentrations in metalliferous Gobi desert soils. Microb Ecol. https://doi.org/10.1007/s00248-019-01405-8

Kyoto Protocol A (2005) Kyoto Protocol. http://en.wikipedia.org/wiki/Kyoto_Protocol

Kyoto Protocol B (2005) Kyoto Protocol http://unfccc.int/essential_background/kyoto_protocol/status_of_ratification/items/2613.php.

Li Z, Ma Z, Van der Kuijp TJ, Yuan Z, Huang L (2014) A review of soil heavy metal pollution from mines in China: pollution and health risk assessment. Sci Total Environ 468–469:843–853. https://doi.org/10.1016/j.scitotenv.2013.08.090

Lim H, Lee J, Chon H, Sager M (2008) Heavy metal contamination and health risk assessment in the vicinity of the abandoned Songcheon Au-Ag mine in Korea. J Geochem Explor 96:223–230

Mapping (2005) Mapping human impacts on the biosphere. http://www.globio.info/impacts/.

Mishra RK, Mohammad N, Roychoudhury N (2016) Soil pollution: causes, effects and control, Vol 3. Tropical Forest Research Institute, Jabalpur

Noss R, LaRoe ET, Scott JM (2005) Endangered ecosystems of the united states: a preliminary assessment of loss and degradation. http://biology.usgs.gov/pubs/ecosys.htm.

Ogamba EN, Izah SC, Nabebe G (2015) Effects of 2, 4-Dichlorophenoxyacetic acid in the electrolytes of blood, liver and muscles of *Clarias gariepinus*. Nigerian J Agric Food Environment 11(4):23–27

Ogamba EN, Charles EE, Izah SC (2021) Distributions, pollution evaluation and health risk of selected heavy metal in surface water of Taylor creek, Bayelsa State, Nigeria. Toxicol Environ Health Sci 13(2):109–121. https://doi.org/10.1007/s13530-020-00076-0

Ogwu MC (2019) Towards sustainable development in Africa: the challenge of urbanization and climate change adaptation. In: Cobbinah PB, Addaney M (eds) The geography of climate change adaptation in urban Africa. Springer Nature, Switzerland, pp 29–55. https://doi.org/10.1007/978-3-030-04873-0_2

Ogwu MC, Oladeji TS (2014) Effect of municipal waste effluent on albino rats [*Rattus norvegicus* (Amori)]. Ital J Occup Environ Hygiene 5(1):4–10

Ogwu MC, Chime AO, Osawaru ME, Emoekpere R (2014a) Germination of watermelon [*Citrullus lanatus* (Thunberg) Matsumura and Nakai] in lead polluted petri dish. Ital J Occup Environ Hygiene 5(3–4):126–130

Ogwu MC, Osawaru ME, Ahana CM (2014b) Challenges in conserving and utilizing plant genetic resources (PGR). In J Genetics Mol Biol 6(2):16–22. https://doi.org/10.5897/IJGMB2013.0083

Ogwu MC, Aiwansoba RO, Osawaru ME (2015) Effects of indole-3-acetic acid on germination in lead polluted petri dish of *Citrullus lanatus* (Thunberg) Matsumura and Nakai, Cucurbitaceae. Aceh Int J Sci Technol 4(3):107–113. https://doi.org/10.13170/aijst.4.3.3021

Ogwu MC, Osawaru ME, Aiwansoba RO, Iroh RN (2016) Status and prospects of vegetables in Africa. In: Borokini IT, Babalola FD (eds) Conference proceedings of the joint biodiversity conservation conference of nigeria tropical biology association and Nigeria chapter of society for conservation biology on MDGs to SDGs: toward sustainable biodiversity conservation in Nigeria. University of Ilorin, Nigeria, pp 47–57pp

Ogwu MC, Ahana CM, Osawaru ME (2018) Sustainable food production in Nigeria: a case study for Bambara groundnut (*Vigna subterranean* (L.) Verdc. Fabaceae). J Energy Natural Resour Manag 1:68–77

Ogwu, M.C., Kerfahi, D., Song, H-K., Dong, K., Seo, H, Lim, S., Srinivasan, S., Kim, M.K., Waldman, B. and Adams, J.M. (2019a). Changes in soil taxonomic and functional diversity resulting from gamma irradiation. Sci Rep 9: 7894. doi: https://doi.org/10.1038/s41598-019-4441-7

Ogwu MC, Srinivasan S, Dong K, Ramasamy D, Waldman B, Adams JM (2019b) Community ecology of *Deinococcus* in an irradiated soil. Microb Ecol. https://doi.org/10.1007/s00248-019-01343-5

Ohimain EI, Izah SC (2013) Gaseous emissions from a semi-mechanized oil palm processing mill in Bayelsa state, Nigeria. Continental J Water Air Soil Pollut 4(1):15–25

Ohimain EI, Izah SC, Abah SO (2013) Air quality impacts of smallholder oil palm processing in Nigeria. J Environ Protect 4:83–98

Ojesanmi AS, Richard G, Izah SC (2017) Mortality rate of Clarias gariepinus fingerlings exposed to 2, 3-dichlorovinyl dimethyl Phosphate. J Appl Life Sci Int 13(1):1–6

Osawaru ME, Ogwu MC (2014) Conservation and utilization of plant genetic resources. In: Omokhafe K, Odewale J (eds) Proceedings of 38th annual conference of the genetics society of Nigeria. Empress Prints Nigeria Limited, pp 105–119
Osawaru ME, Ogwu MC, Braimah L (2013a) Growth responses of two cultivated Okra species (*Abelmoschus caillei* (A. Chev.) Stevels and *Abelmoschus esculentus* (Linn.) Moench) in crude oil contaminated soil. Nigerian J Basic Appl Sci 21(3):215–226
Osawaru ME, Ogwu MC, Chime AO (2013b) Assessment of growth performance of two Okra species (*Abelmoschus esculentus* [L.] Moench and *Abelmoschus caillei* [A. Chev.] Stevels) exposed to crude oil contaminated soil. Nigerian J Biotechnol 26:11–20
Pierzynski GM, Sims JT, Vance GF (2000) Soils and environmental quality, 2nd edn. CRC, Boca Raton, FL
Richard G, Izah SC, Sawyer WE (2020) Soil-borne fungi community of a major dumpsite in Bayelsa State, Nigeria. Int J Adv Res Microbiol Immunol 2(2):14–19
Richard G, Sawyer WE, Izah SC (2021a) Outdoor air quality index of biomass combustion in the Niger Delta, Nigeria: a health impact perspective. J Adv Res Med Sci Technol 8(1):19–28
Richard G, Sawyer WE, Izah SC (2021b) Health implications of air quality index of fine and coarse particulates during outdoor combustion of biomass in the Niger Delta, Nigeria. Int J World Policy Dev Stud 7(2):21–26
Secretariat of Convention of Biological Diversity (2000) Sustaining life on earth: how the convention on biological diversity promotes nature and human well-being [Brochure] Montreal, Quebec.
Seiyaboh EI, Izah SC (2017) Review of impact of anthropogenic activities in surface water resources in the Niger Delta region of Nigeria: a case of Bayelsa state. Int J Ecotoxicol Ecobiol 2(2):61–73
Seiyaboh EI, Izah SC (2019) Impacts of soil pollution on air quality under Nigerian setting. J Soil Water Sci 3(1):45–53
Sherbinin A (2002) A CIESIN thematic guide to land-use and land-cover change. The Trustees of Columbia University
Song H-K, Singh D, Tomlinson K, Yang X, Ogwu MC, Slik JWF, Adams JM (2019) Tropical forest conversion to rubber plantation in southwest China results in lower fungal beta diversity and reduced network complexity. FEMS Microbiol Ecol fiz092. https://doi.org/10.1093/femsec/fiz092
Sweileh WM, Al-Jabi SW, Zyoud SH (2018) Outdoor air pollution and respiratory health: a bibliometric analysis of publications in peer-reviewed journals (1900 – 2017). Multidiscip Respir Med 13:15. https://doi.org/10.1186/s40248-018-0128-5
Uzoekwe SA, Izah SC, Aigberua AO (2021) Environmental and human health risk of heavy metals in atmospheric particulate matter (PM_{10}) around gas flaring vicinity in Bayelsa State, Nigeria. Toxicol Environ Health Sci 13(4):323–335. https://doi.org/10.1007/s13530-021-00085-7
Vwioko DE, Okoekhian I, Ogwu MC (2018) Stress analysis of *Amaranthus hybridus* L. and *Lycopersicon esculentum* Mill. Exposed to Sulphur and nitrogen dioxide. Pertanika J Trop Agric Sci 41(3):1169–1191
Wackernagel M, Schulz NB, Deumling D, Linares AC, Jenkins M, Kapos V, Monfreda C, Loh J, Myers N, Norgaard R, Randers J (2002) Tracking the ecological overshoot of the human economy. PNAS 99(14):9266–9271
Wetsone GE, Cousin R, Warren W (2005) Rewriting the rules: the bush administration's first term environmental record. Natural Resources Defense Council, New York
Wines M (2005) Malawi is burning, and deforestation erodes economy. New York Times, A, 3.
Wood PM (2000) Biodiversity and Democracy: Rethinking Society and Nature. UBC Press, Vancouver
World Wildlife Fund (2002) Debt for Nature Swaps. http://www.worldwildlife.org/conservationfinance/swaps/sld001.htm.
Zhao H, Xia B, Fan C, Zhao P, Shen S (2012) Human health risk from soil heavy metal contamination under different land uses near Dabaoshan mine. Southern China Sci Total Environ 417-418:45–54. https://doi.org/10.1016/j.scitotenv.2011.12.047

Chapter 15
Environmental Degradation in the Niger Delta Ecosystem: The Role of Anthropogenic Pollution

Aroloye O. Numbere, Tambeke N. Gbarakoro, and Bernard B. Babatunde

Abstract Pollution is the addition of unpleasant substances into the environment that damage health. The Niger Delta is the hotbed of oil and gas exploration and exploitation. Exploratory activity serves as the primary source of crude oil, which provides foreign exchange earnings for the country. However, the reckless actions of oil industries have added a substantial quantity of pollutants to the soil, air, and water of the Niger Delta. Several studies have revealed that the environment is highly polluted, which has resulted in the contamination of food and water consumed by humans. This chapter thus provides a holistic report of the impact of environmental pollution in different areas of the Niger Delta. The work dissects the sources, fate, and effects of environmental pollution. It also proffers some solutions for tackling this hydra-headed problem in the communities of the Niger Delta.

Keywords Air pollution · Anthropogenic effects · Soil pollution · Water pollution · Environmental sustainability · Oil spill

15.1 Introduction

The Niger Delta region is a hotbed for pollution because of oil and gas exploration activities (Numbere 2018). The sedimentary nature of the area has made it crude oil-rich, which is a blessing. Still, it has become a curse to many communities whose lifestyles and livelihood opportunities have been converted for the worst (Numbere 2021). Over the years, some oil-producing community members have lost more than they have gained because of the siting of oil industry in their community. Instead, they have, in return for hosting oiling facilities: land despoliation, land fragmentation, deforestation, and marine and atmospheric pollution. Pollution has taken away vital livelihood sources such as farming and fishing. The presence of crude oil in their land has prevented their crops from growing; even when they plant and

A. O. Numbere (✉) · T. N. Gbarakoro · B. B. Babatunde
Department of Animal and Environmental Biology, University of Port Harcourt, Port Harcourt, Nigeria

S. C. Izah, M. C. Ogwu (eds.), *Sustainable Utilization and Conservation of Africa's Biological Resources and Environment*, Sustainable Development and Biodiversity 32, https://doi.org/10.1007/978-981-19-6974-4_15

eventually harvest the crops, they are exposed to pollutants when they consume the food. Different kinds of pollution exist, including water, land, and atmospheric pollution.

Water pollution destroys aquatic life and contaminates the food chain. Thus, fishery products harvested from polluted water are not fit for human consumption (Suresh et al. 2022). However, because of lack of regulation and the need to survive, people consume harvested organisms from the water body. Pollution of the marine system has a ripple effect on aquatic resources. Surface oil spills contaminate fishes that are found in the pelagic region of the river from leaking pipes caused by corrosion or sabotage. Marine organisms in the middle and benthic areas of the river are affected by sinking pollutants from the oil spill. An increase in the activities of the artisanal refinery has also accelerated the pollution of the Niger Delta environment (Obieze et al. 2022).

On the other hand, land pollution is the spillage of crude oil or other harmful substances on land. The dangerous materials could be toxic waste buried within the soil or dumped on the soil surface. Atmospheric pollutants are air-borne substances that are mainly gaseous, contaminate the air, and can harm human health (Saunders et al. 2022). Pollutants degrade all environments, including those recovering from natural and anthropogenic damage. The environment has a way of renewing itself after each disturbance. However, the persistent deterioration and degradation of the environment have impeded the natural recovery process, and humans are the primary culprit.

The word "pollution" comes from the Latin word "pollutus," which means made foul, unclean, or dirty. Human-caused air pollution can be grouped as primary or secondary pollutants (e.g., Sikhakhane 2002). Primary pollutants are those released directly from the source into the air in a deadly form. While secondary, pollutants are modified to hazardous conditions after they enter the air or are formed by chemical reactions as components of the air mix and interact (Nazaroff and Weschler 2004). Solar radiation often provides the energy for these reactions. Photochemical oxidants and atmospheric acids formed by these mechanisms are the most important secondary pollutants in human health and ecosystem damage. Fugitive emissions do not go through a smokestack, e.g., dust from soil erosion, strip mining, rock crushing, building construction, and destruction.

Pollution can include chemical, biological, organic, thermal, physical, radiation, hydrocarbons, etc. Pollution is the release into the environment of matter or energy that causes undesirable impacts on the health or well-being of humans or other organisms. Pollution is also the entry into the environment of substances detrimental to ecosystem health. According to a joint group of experts on the scientific aspects of marine pollution defined it as the introduction by man directly or indirectly of substance or energy into an environment resulting in such deleterious effects harm to living resources, hazard to human health, a hindrance to man's activities, including activities, such as fishing, use of water and use of amenity (Zielinski et al. 2009; Scheren and Ibe 2002).

Sources of pollution can be divided into two parts: (i) natural source: the one caused by nature. It includes volcanoes, forest fires, plant and animal remains, sea spray, pollen, spores, viruses, bacteria, and storms (Luo et al. 2022) and (ii) artificial or anthropogenic sources: they are the ones caused by man's impingement on the environment, such as car pollution (exhaust), toilet sewage, industrial waste, agricultural waste, synthetic organic radionuclide, trace metal, heavy metal, thermal discharges, marine shipping activities, waste dumping by ships and, river discharges (e.g., Proshad et al. 2022).

15.2 Air Pollution

15.2.1 Conventional or Criteria Pollutants

The US Clean Air Act of 1970 designated seven major pollutants [sulfur dioxide, carbon monoxide, particulates, hydrocarbons, nitrogen oxides, photochemical oxidants (tropospheric ozone), and lead] for which maximum ambient air levels are mandated (Smith 2012; Fenger 2009). These seven conventional or criteria pollutants contribute the most significant volume of air-quality degradation. They are considered the most severe threat of all air pollutants to human health and welfare.

Characteristics of each pollutant include:

15.2.1.1 Sulfur Compounds

Natural sources of sulfur in the atmosphere include evaporation of sea spray, erosion of sulfate-containing dust from arid soils, fumes from volcanoes, etc. Ninety percent of sulfur is found in cities from an anthropogenic source, such as car exhaust, which expel sulfur dioxide (SO_2) (Saleem et al. 2022). It comes from the combustion of sulfur-containing fuel (coal and oil), purification of sour sulfur-containing natural gas or oil, and industrial processes, e.g., smelting of sulfide ores. China and the United States are the largest sources of anthropogenic sulfur, which comes from coal-burning (Smith et al. 2011). However, India is overtaking China in sulfur pollution (Li et al. 2017). Sulfur dioxide is a colorless corrosive gas that directly damages both plants and animals. An atmosphere filled with sulfur trioxide (SO_3) reacts with water vapor to form sulfuric acid (H_2SO_4), which is known as "acid rain" (Abbasi et al. 2013). Small solid particles or liquid droplets can transport the acidic sulfate ion (SO_4^-) into the lungs, where it causes severe damage (Chen et al. 2022). Thus, sulfur dioxide and sulfate ions are the major causes of air pollution-related health damage. Sulfate particles and droplets cause poor visibility. Acid rain is prominent, especially in the Niger Delta atmosphere, because of soot from illegal crude oil bunkering and artisanal refinery and gas flaring. This soot increases respiratory problems, such as asthma in the surrounding cities. The soot also coats mangrove leaves and impairs photosynthetic function, resulting in withering and defoliation.

15.2.1.2 Nitrogen Compounds

Nitrogen oxides are highly reactive gases formed when nitrogen in fuel or combustion air is heated to a temperature above 650 °C (1200 °F) in the presence of oxygen or bacteria in soil or water to oxidize nitrogen-containing compounds. Nitrogen oxide is a pollutant from car exhaust, especially in cities with high traffic volumes (Wang et al. 2022). The first product, nitric oxide (NO), oxidizes in the atmosphere to form nitrogen dioxide (NO_2), a reddish-brown gas that gives a red color to photochemical smog. Nitrogen oxides combine with water to give nitric acid (HNO_3), forming acid rain (He et al. 2022). Sixty percent (60%) of reactive nitrogen compounds are caused by humans basically by fuel combustion in transportation and electric power generation. Nitrous oxide (N_2O) is an intermediate in soil denitrification that absorbs ultraviolet light and plays a vital role in climate change (Ussiri and Lal 2012). Excess nitrogen causing fertilization and eutrophication of land waters and coastal areas, where we have mangroves. It also affects terrestrial plants and semi-aquatic plants such as Nypa palm due to excess fertilization and the rapid growth of macrophytes and weeds, which crowd native species.

15.2.1.3 Carbon Oxides

The major form in the air is carbon dioxide (CO_2). It is nontoxic but increasing atmospheric levels (~0.5% per year) due to anthropogenic activities contribute to global warming and devastating impacts on both humans and natural commuters. More than 90% of the CO_2 emitted annually comes from respiration (oxidation of organic compounds by plant and animal cells). Oceans take up about 2 billion through photosynthesis by phytoplankton (algae) and shell formation by mollusks. Mangrove vegetation is also a major carbon sequester, especially in the tropics, such as Nigeria. A study by Numbere (2020) showed that higher carbon stock (66.1 ± 15.1 $Mgha^{-1}$) was found in areas with large red mangrove trees than in areas with fewer red mangrove trees (36.0 ± 12.8 $Mgha^{-1}$). The range of mangrove carbon stock in the Niger Delta region is 23.2–81.2 $Mgha^{-1}$. The carbon stock indicates that Niger Delta mangroves, the largest in Africa and the third largest globally, play a critical role in global climate stabilization by absorbing CO_2 and other atmospheric carbon pollutants.

Carbon monoxide (CO) is a colorless, odorless, non-irritating but highly toxic gas produced by incomplete combustion of fuel (coal, oil, charcoal, and gas), incineration of biomass, or solid waste, or partial anaerobic decomposition of organic material. CO inhibits respiration in animals by attaching irreversibly to hemoglobin and depriving the body of the much-needed O_2 to survive. About 90% of the CO in the air is consumed in photochemical reactions that produce ozone (Khdary et al. 2022). Land clearing fires and cooking fires are major sources.

15.2.1.4 Metals and Halogens:

Many toxic metals are mined and used in manufacturing processes or occur as trace elements in fuels, especially coal. They get in the air in the form of metal fumes or suspended particulates by fuel combustion ore smelting and disposal waste.

Lead is a metabolic poison and a neurotoxin that binds to essential enzymes and cellular components and inactivates them. Lead: lead makeup up two-thirds of all metallic air pollution. Most of this lead is from leaded gasoline called tetraethyl (Serih et al. 2022). It is added to gasoline to improve its performance. However, when the exhaust from leaded gasoline enters the air, they are inhaled or deposited on land and water. They can enter the food chain and accumulate within the body tissue and cause central nervous system malfunction and other sicknesses. An estimated 20% of all inner-city children suffer some degree of mental retardation from high environmental lead levels.

Mercury: is another dangerous neurotoxin that is widespread in the environment (Nordberg et al. 2022). The primary sources of atmospheric mercury are coal-burning power plants and waste incinerators. Mercuric fungicide in house paints was a problem formerly but has been stopped. Lead and mercury travel long distances in the atmosphere to bioaccumulate in aquatic ecosystems and mangrove and mangrove-associated organisms. In a study by Numbere (2020) in a section of the mangrove forest at Eagle Island in the Niger Delta, some organisms were found to have high metallic contents. Mercury concentrations ranging from 0.02 to 5.90 mg/kg were found in anadara (*Senilia senilis*), crab (*Callinectes Amnicola*), insect (e.g., dragonfly and wasp), tilapia (*Sarotherodon melanotheron*), and droppings of cattle egret (*Bulbus ibis*). The presence of trace amounts of mercury in these organisms makes them unfit for human consumption, because they are harmful and bioaccumulate to lethal levels in humans.

Other toxic metals of importance are nickel, beryllium, cadmium, thallium, uranium, cesium, and plutonium. A large amount of arsenic is released from metal smelters, refinery operations, coal combustion, and pesticide.

Halogens (fluorine, chlorine, bromine, and iodine): are highly reactive and toxic in elemental form. They enter the stratosphere to release chlorine and fluorine atoms that destroy the ozone shield that protects the earth from ultraviolet radiation (Marshall and Burkholder 2022). In this group is the highly persistent chlorofluorocarbon (CFCs) widely used in spray propellants, refrigeration compressors, and foam blowing.

Tropospheric ozone: although ozone in the stratosphere shields the earth against ultraviolet rays, ozone from human activities forms and accumulates low in the troposphere and acts as a pollutant. Tropospheric ozone (also called ground-level ozone) is, thus, grouped as a secondary pollutant. They are colorless gas that results from the reaction with sunlight: heat, nitrogen oxides, and volatile carbon-containing chemicals. They are a significant component of smog and are harmful to health, mainly when the free oxygen molecule engages in reactions that injure living tissues and cause respiratory problems.

Particulate materials: an aerosol is any system of the solid particle or liquid droplets suspended in a gaseous medium. They are generally referred to as particulate material. Examples of primary pollutants are dust, ash, soot, lint, smoke, pollen, spores, algal cells, and many other suspended materials. Wind-blown dust, volcanic ash, and other natural materials may contribute to more suspended particulate material. Secondary pollutants are sulfates and nitrates. Particulates are the most common form of air pollution, since they reduce visibility and leave dirt deposits on surfaces, such as windows, painted surfaces, cars, clothes, and roofs. Particulate matter is also common in urban areas, where the wind blows particles into the air (Zerboni et al. 2022). EPA classifies particulate pollution by the size of the particles. PM_{10} pollution consists of particles less than 10 microns in diameter (one-seventh the width of human hair), whereas $PM_{2.5}$ pollution consists of finer particles less than 2.5 microns in diameter. Most PM_{10} pollution is dust, while $PM_{2.5}$. Pollution results from the combustion process. Respirable particles less than 2.5 microns are among the most dangerous of this group, because they can be drawn into the lungs, where they damage respiratory tissues. Asbestos fibers and cigarette smoke are among the most dangerous respirable particles in urban and indoor air, because they are carcinogenic. For instance, the high concentration of soot in the air emanating from the burning of crude oil in Port Harcourt in the Niger Delta is putting people's health at risk.

Volatile organic compounds (VOCs) are organic chemicals that exist as gases in the air. Plants are the largest source of VOCs, releasing 350 million tons of isoprene (C_5H_8) and 450 million tons of terpenes ($C_{10}H_{15}$) each year. About 400 million tons of methane (CH_4) are produced by natural wetlands and rice paddies and by bacteria in the guts of termites and ruminant animals. These volatile hydrocarbons are generally oxidized to CO and CO_2 in the atmosphere. In addition, human activities release the following synthetic organic compounds into the air, benzene, toluene, formaldehyde, vinyl chloride, phenols, chloroform, and trichloroethylene. These products are emitted into the atmosphere or partially burned hydrocarbons from transportation, power plants, chemical plants, and petroleum refineries. These chemicals play an important role in the formation of photochemical oxidants.

The EPA has identified 33 chemical compounds considered the greatest threat to public health in urban areas (Table 15.1). In the US, air toxic standards now affect the following industries: chemical plants, oil refineries, and steel mills, as well as smaller sources, such as dry cleaners, automobiles, and trucks.

However, in the Niger Delta, it is not known if similar air standards are followed by the major oil industries, especially the Port Harcourt Refinery and Liquified Natural Gas Limited, Bonny, which spew soot into the atmosphere via gas flaring. The major problem of the Nigerian environmental system is that laid down laws, rules, and regulations are already in existence, one of such laws is the Federal Environmental Protection Agency, FEPA bill of 1988, which is not strictly followed by the operating industries and not enforced by the regulatory agencies.

Photochemical oxidants are products of secondary atmospheric reactions driven by solar energy (Bolton et al. 2001). One of the most critical reactions involves forming singlet (atomic) oxygen by splitting nitrogen dioxide (NO_2). This atomic

Table 15.1 Urban air toxics of most significant concern

Name of chemicals	Name of chemicals	Name of chemicals
Acetaldehyde	Coke oven emissions	Manganese compounds
Acrolein	Dioxins	Mercury compounds
Acrylonitrile	1, 2-dibromoethane	Methylene chloride
Arsenic compounds	1, 3-dichloropropane	Nickel compounds
Benzene	Propylene dichloride	Polychlorinated biphenyls
Beryllium compounds	Ethylene dichloride	Polycyclic organic matter
1, 3-Butadiene	Ethylene oxide	Quinoline
Cadmium compounds	Formaldehyde	
Carbon tetrachloride	Hexachlorobenzene	Tetrachloroethylene (perchloroethylene)
Chloroform	Hydrazine	Trichloroethylene
Chromium compounds	Lead compounds	Vinyl chloride

Source: USEPA 1999; Yu and Stuart 2016; Guo et al. 2014

oxygen then reacts with another molecule of O_2 to make ozone (O_3). Ozone formed in the atmosphere provides a valuable shield for the biosphere by absorbing incoming ultraviolet radiation. Nevertheless, O_3 is a strong oxidizing reagent in ambient air and damages vegetation, building materials (e.g., paint, rubber, and plastics), and sensitive tissues (e.g., eyes and lungs). Ozone has an acrid biting odor, a distinctive characteristic of photochemical smog. Hydrocarbons in the air contribute to ozone accumulation by removing NO in the formation of compounds, such as peroxyacetyl nitrate, which is another damaging photochemical oxidant.

15.2.2 *Unconventional Pollutants*

The EPA sets emission standards for certain unconventional or criteria pollutants that are known as toxic or hazardous materials regulated by emission standards are: asbestos, benzene, beryllium, mercury, polychlorinated biphenyl, and vinyl chloride (e.g., Sola et al. 2022). Most of these materials have no natural source in the environment and are, therefore, only anthropogenic in origin.

In addition, to toxic air pollutants, some other unconventional forms of air pollution include aesthetic degradation, which consists of any undesirable changes in the physical characteristics or chemistry of the atmosphere. Noise, odors, and light pollution are examples of atmospheric degradation that may not be life-threatening but reduce life quality. Deposition of soot from the artisanal refinery is rampant within the city (Fig. 15.1).

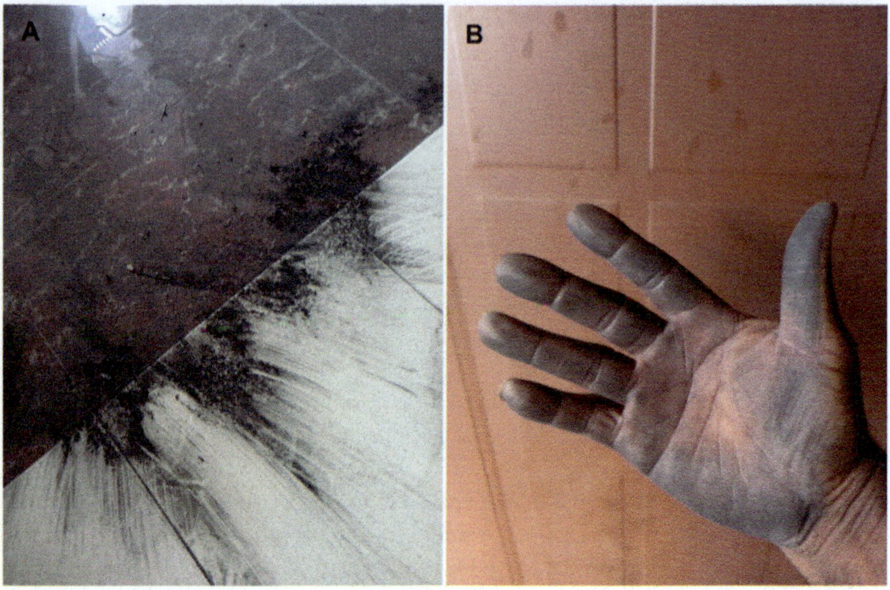

Fig. 15.1 Soot deposition in homes around Port Harcourt city in the Niger Delta, Nigeria

15.2.3 Causes of Air Pollution

It is the presence of unwanted harmful substances in the atmosphere which has a deleterious effect on human health. Atmospheric pollution is caused by natural emissions from volcanically active rocks, geologic mines, and oceanic regions. Evaporation from mine pits goes into the atmosphere to increase the pollution levels. However, the artificial addition of pollution into the atmosphere gradually becomes more than the natural emissions. Human mediated actions add pollutants to the atmosphere, including car exhaust, bush burning, incinerator emissions, smoking, and cooking with a stove or firewood.

15.3 Water Pollution

Water pollution comes in many forms and can cause diverse impacts on aquatic ecosystems and human health (Hossain et al. 2022). Some water pollution is emitted from point sources, discrete locations, such as factories or sewer pipes. On the other hand, non-point source pollution is cumulative, coming from several sources over large areas, such as farms, city streets, and residential areas. The application of fertilizers and pesticides can percolate through the soil profile into an aquatic environment, i.e., groundwater or surface water (Huang et al. 2022).

15.3.1 Causes of Water Pollution

Aquatic pollution releases poisonous substances into the marine, estuarine and freshwater environments. The cause of aquatic pollution is numerous but can be grouped into two natural and artificial causes:

Natural cause: The marine environment is sitting and surrounded by a terrestrial formation. The benthic and shoreline landforms are composed of depositional and sedimentary rocks. These rocks contain chemicals that seep into the water body.

The chemical composition of the marine, estuarine and freshwater environments is mostly based on the chemical composition of the parent rocks. Coastal chemicals, therefore, increase and influence its metallic composition. On this basis, the aquatic environment always contains metals when tested. Magmatic substances exude from oceanic ridges during sea bottom tectonic movements, e.g., during volcanoes release chemicals into the water body. Decomposed and washed-out chemicals from neighboring and surrounding rocks migrate into the water body to increase the heavy metal content; Furthermore, the metallic composition of the sea is based on the substances that flow from the base rocks and surrounding landforms, which has a great influence on the chemistry of the aquatic environment, such as the pH, salinity, sodium, chloride, phosphates, potassium, etc. A low concentration of these chemicals is not too harmful, but in high amounts may become detrimental to aquatic organisms and pose a danger to humans who consume seafood.

Artificial cause: This is anthropogenically caused by adding pollutants to the marine environment. Some activities that cause marine pollution include offshore crude oil drilling, transportation of crude oil by tankers, refueling of sea-vessel and marine craft at wharfs and jetties, offshore laying of crude oil pipelines, etc. These actions can lead to the spilling of crude oil into the water body due to accidents caused intentionally or unintentionally, resulting in the contamination of aquatic organisms. Other actions that cause spillages include sabotage, equipment failure, corrosion, lack of maintenance, and poor engineering.

15.3.2 Thermal Pollution

Moore (1958) first coined the word thermal pollution, which refers to water (rivers) pollution by heated discharges. Sources of heated water released into marine and estuarine environments are power industries situated close to the river for accessible water collection to cool hot industrial machines (Quivet et al. 2022). Water has both positive and negative effects on organisms. The positive role includes increased aquatic productivity, e.g., prolonged growth periods in bivalves and increased eutrophication due to the discharge of heated effluents. The primary sources of heated effluents in the marine environment and their percentage as recorded by the authors based on their field experiences are shown in Table 15.2.

Table 15.2 Industrial discharges of heated effluents into the aquatic environment in the Niger Delta, Nigeria

S/No	Industry	% Thermal discharge
1	Chemical/petrochemical	4.2
2	Petroleum refinery	10.1
3	Electrical generating plants	60.0
4	Iron and steel complexes	20.0
5	Paper and board	0.8
6	Food processing	1.0
7	Textiles	1.1
8	Gas and coke production	2.2
9	Soap and detergent	0.2
10	Brewing	0.5

A thermal plume is an area in the water body with artificially heated water or water above ambient temperature (García et al. 2022). Thermal plumes are influenced by the discharge (flow rate, velocity at outlet, size), receiving water (flow dynamics, tidal currents, surface waves, free turbulence, stratification, geometric characteristics, etc.), and atmosphere (wind, air, and solar).

Effluents from refineries and industries are a classic example of a cause of thermal pollution in the Niger Delta. The releases of pollutants cause physical and chemical effects, such as:

i. Decrease in viscosity which affects the buoyancy of pelagic organisms and dispersal of mangrove propagule
ii. Increase sedimentation
iii. Increase in evaporation, which increases salinity
iv. Higher salinity affects the physiology of organisms, such as fishes and mangroves
v. Increase pH with less CO_2 and hence loss in photosynthesis which affects production
vi. Loss of solubility of oxygen: reduction in the solubility of oxygen is the most crucial chemical effect. High temperature stimulates aquatic organisms' metabolic activity, which utilizes available oxygen and drastically reduces already reduced oxygen concentration. Combined with the impact of other pollutants (e.g., organic pollutants) has increased biological oxygen demand (BOD), which affects the survival of some aquatic organisms.
vii. Creation of artificial thermocline, thereby reducing nutrient circulation, affecting the vertical migration of aquatic organisms.
viii. The addition of chemicals in the cooling system to prevent corrosion in pipes or the use of biocides to prevent biological fouling (algae and Molex) add pollutants into the water

15.3.3 Pollution by Solid Organic Wastes

Organic pollution is a general name given to the harmful impact of pollutants rich in organic matter on the environment. Organic pollution is the largest and most rampant waste pollutant created by man. It makes up one of the most significant volumes of waste discharged into water and mangrove forest (Numbere 2019a). Organic waste is produced in the municipality and is exceedingly difficult to control due to the increasing number of people migrating into the cities and hinterland. Its effective management depends on human behavior, such as observance of strict environmental standards and regulations. Sources and types of organic waste that impact organisms (plants, animals, and humans) include the following:

i. Domestic activities (households): this includes food remains (garbage), metallic substances, such as empty cans, wastepaper, and peals of foodstuff, such as yam, cassava, etc. These transform into refuse dumps (solid waste), liquid waste from wash-ups, grease, and detergents, all disposed of as sewage.
ii. Market centers: this includes waste foods, shells, empty cartons, paper, cans, polythene materials, wood, and fiber materials, all converted to refuse dumps (solid waste). Sewage from gutters, waste paints, detergents, oils, and pesticides disposed of as liquid (sewage) causes a public health problem.
iii. Small industrial activities include bakeries, auto mechanic shops, and local starch/garri production from cassava. Inorganic substances, such as calcium oxides from welders (Carbide and water), paints, grease, metal, etc., are converted into refuse dumps (solid waste) and are a mixture of organic and inorganic waste (pollutants).
iv. Agricultural and allied food processing industries: this includes chemicals, pesticides, waste from industrial clean-ups, liquid foods, grease, and oil (discarded as sewage in runoff drainage). Other industrial chemicals include ammonia, fertilizer, phosphates, urea, and runoff (sewage). Solid waste food (raw materials) and refuse dumps are disposed of as solid waste.
v. Paper and allied wood industries: this includes firewood making business, wastepaper, sawdust, pulp, chemicals mixed as refuse dumps (solid waste).
vi. The pollution potential of the waste generated is also a function of the volume of effluents released, their dilution in the receiving water, and their toxicity to aquatic organisms. Most factories discharge their effluents/waste without treatment into creeks, lagoons, mangrove forests, and inshore waters, which flow into coastal waters.
vii. In the Niger Delta, there is no central municipal sewage treatment facility, which makes households, commercial establishments, and industries manage their waste without privately following environmental guidelines or disposing of it in open drains or water. Accumulation of public drains with solid waste led to blockage, and with time the contents spewed out into the adjoining river or flood roads and streets during a heavy downpour.
viii. Eighty percent of organic materials come from the economic need of humans and tend to obey the law of conservation, which states that intake (I) is

proportional to output (O). Other activities include shipping, dredging, and sand mining. These activities produce solid and liquid waste. It also result in the sedimentation of the river by washed-up soil.

$$I \propto O$$

$$I = kO$$

where k becomes constant, which is a valuable product

15.3.4 Input into Marine Estuaries and Mangrove Forest

The marine and estuarine environment are the most preferred site for liquid and solid waste disposal. In the Niger Delta, the mangrove forest is used as a site for solid waste disposal (Numbere 2019a, b). The quality of water as a universal solvent and the estuaries as a filter for wastewater have made these aquatic environments sinks for waste products. The waste disposed into the marine environment is done directly or indirectly. Private and government firms now dump waste into the marine environment. Strategic areas around Port Harcourt have been noted as sites for liquid and solid waste disposal.

Some notable areas of waste disposal around Port Harcourt city are Ntawogba Creek: this creek runs diametrically across Port Harcourt municipality, leading to neighboring streams and rivers. It has been a significant outlet point for stormwater originating from floods, homes, and internal drainages. However, because of city expansion due to the increase in population, the creek's width has been reduced due to blockages and construction work leading to an overflow during heavy rainfall. Most businesses found along this route include mechanic stores and garages, provision stores, residential quarters, small-scale industries, etc. They all discharge their waste into this creek. These discharges have clogged up the streams, creating severe flooding in the streets when their rainfall. Although the creek gets desilted occasionally but still contains assorted water products, which has facilitated the rapid encroachment of invasive Nypa palms.

Some years ago, the government reconstructed the concretized bottoms and the sides of the creek passing through the cities with evacuation points at the linking rivers to help prevent siltation from hindering the free flow of water. Still, human activities around the creek have led to the entry of sand and waste into the canal over the years. The route of the streams includes Olu-Obasanjo Road, Emekuku Street, Ogbunabali Road, Marine Base, Trans-Amadi, Marine Base, and Okrika River. These channels lack regular maintenance, and as such, the solid bottom of the creeks is filled with sand debris brought in from the land surface by flooded water, which leads to the proliferation of land weed and aquatic plants (macrophytes}. Organic waste (pollutants) has the following effect on marine estuarine and mangrove environments:

Growth of phytoplankton/algae leads to the coloration of the water body and emission of a pungent smell. Accumulation of silts brought in by floodwater. Eutrophication is the nutrient enrichment by phosphates and nitrates, which causes the growth of phytoplankton and zooplankton. Excess growth of these organisms leads to bloom (i.e., discoloration of marine/estuarine waters following overgrowth of phytoplankton or algae). These growths deprive aquatic organisms of oxygen and nutrient and eventually lead to death from asphyxiation of fishes and other marine organisms. It also causes extensive sedimentation of organic matter, depriving the water of oxygen and causing anoxic conditions. Organic pollution introduces parasitic eggs into the receiving water. The most common parasite includes nematodes, Ascaris, ancylostoma (hookworm), and tapeworms, such as *Taenia solium*. Resistant Ascaris cyst can be transferred to drinking water sources leading to a public health disaster. Mollusks are found in the polluted marine environment and, when consumed, can affect humans.

The presence of organic waste in the aquatic environment can increase dissolved organic carbon (DOC), which modifies the pH. The free radicals (carboxylic acid) and other functional groups released from the breakdown of organic waste interact in the marine environment. The interaction affects the PKa of the environment and makes the organic acid weak. They buffer the systems that span the range of pH values in natural waters (Ka is the dissociation constant of the organic acid in solution). The degree of such dissociation is temperature-dependent, which can increase the acidity of the marine environment as the organic acid addition increases. Low pH can cause reduction and loss of fish; for instance, at low pH, toxic metals such as aluminum (Al) dissolve and are absorbed by the gills leading to the disruption of the barrier of the gills. Thus, thwarting the oxygen absorption capability of fishes (i.e., respiration dysfunction). Research on the West African Red Mangrove Crab (*Goniopsis pelii*) indicates high heavy metal concentrations in the Niger Delta (Numbere 2019b).

15.4 Heavy Metal Pollution

Heavy metals are inorganic elements essential for plant growth in trace or minute quantities and are toxic and poisonous in relatively higher concentrations. Metals are biologically undegradable but easily assimilable and bioaccumulated in the protoplasm of aquatic organisms. In natural marine ecosystems, metals occur in low concentrations, usually nanogram to microgram per liter. The occurrence of these heavy metals contaminates especially more than natural levels. This situation has resulted in the rapid increases in population, urbanization, expansion of industrial activities, exploration, exploitation of natural resources, an extension of irrigation and other modern agricultural practices, and a lack of strictness in the enforcement of environmental regulations. Unlike other pollutants, heavy metals can accumulate unnoticed to a toxic level in the environment. Mercury and cadmium have been implicated in deaths in Japan and Europe and have been blacklisted in the European

Economic Community (EEC) Directive on the discharge of dangerous substances into the aquatic environment.

The EEC listed the following metals that are regarded as detrimental to health: arsenic (As), chromium (Cr), copper (Cu), lead (Pb), nickel (Ni), zinc (Zn), cobalt (Co), and vanadium (V). For example, chromium and copper can cause chromosome abnormalities and brain damage in humans. Lead accumulates in the liver, kidney, bone, and teeth of human beings who consume fish, crabs, and crayfish. It also interferes with the synthesis of haem, and so causes anemia.

15.4.1 Case Studies of Heavy Metal Pollution in the Niger Delta, Nigeria

Heavy metal contamination of mangrove and mangrove-associated organisms is rampant, because the area is highly industrialized, having chemical factories: cement factories, petroleum industry and refinery activities, food industry, and agricultural activities.

Heavy metals in soils and roots of mangrove: studies show that heavy metals have a negative linear relationship with microbial population growth. As heavy metals increase, microbial activities decrease. Furthermore, polluted soils have higher lead, cadmium, and chromium content than other heavy metals. Similarly, contaminated soil is eight times more polluted than unpolluted sites.

Heavy metals from electronic waste (mobile phone): improper electronic waste disposal is common in cities, where high life is prominent, and people have the financial ability to change their phones constantly. Accumulation of disused phones in waste dump sites is gradually becoming a problem because of the lack of proper management strategy. In a study by Numbere and George (2019), heavy metals from decomposed electronic waste migrate downward across different soil layers. This situation shows the ability of heavy metals to move from surface dumpsites into groundwater aquifer. Metal migration was higher during the rainy season. The trace amount of heavy metals is harmful to public health when people drink water from a well or borehole. Similarly, mangrove soils have higher e-waste-generated heavy metals than farm soils. Dumping old mobile phones and other electronic products directly into the mangrove forest is dangerous to the ecosystem, because it is the home to numerous organisms that depend on it directly or indirectly for survival (Numbere and George 2019). The order of heavy metal concentration is nickel > copper > zinc > lead > arsenic > cadmium.

15.4.2 Sources of Heavy Metals in Marine and Estuarine Environment

Trace metals enter the aquatic environment from natural and anthropogenic sources because of direct discharges into freshwater and marine ecosystems or indirect routes, such as dry and wet discharges and land runoff.

Natural sources: These include: (i) coastal inputs: this comes from rivers resulting from erosion produced by wave actions and glaciers. (ii) Deep-sea inputs result from metals released from deep-sea volcanic actions, geothermal activity, and magma degassing. (iii) Atmospheric inputs include metals transported into the atmosphere as dust particles (or as vapors in the case of mercury). The dust particles mainly come from ocean degassing. (iv) Nearshore: these include materials carried by glacial erosion and transported by floating ice or because of continental weathering.

Anthropogenic sources: include: (i) mining effluents, (ii) industrial effluents, (iii) domestic effluents, and urban stormwater runoff, (iv) leaching of metals from garbage and solid waste dumps (Numbere and George 2019), (v) metal inputs from rural areas, (vi) metal inputs from the farm (crop, poultry, and animals), (vii) atmospheric sources, e.g., burning of fossil fuels, crude oil, tires, incineration of municipal waste, and gas flaring, and (viii) petroleum industry activities, e.g., gas flaring, fires, and disposal of pollutants.

15.4.3 The Fate of Heavy Metals in the Aquatic Environment

When heavy metals enter the aquatic environment, they may contaminate the water, suspended solids, sediments, and biota in dissolved, particulate, or complex forms. The metals get dispersed by dilution, advection, sedimentation, and absorption/desorption). The pH, dissolved ions, and water temperature affect the distribution of heavy metals. The metal is removed from the water through absorption. In addition, microbial activity and redox reactions change the properties of the sediments and may affect their composition in intertidal water. As a result, iron and manganese oxides may be converted to carbonate or sulfides, leading to a decrease in the absorption capacity of the sediments. Heavy metals are also taken up by flora and fauna, which steadily increase in concentration if slowly excreted, leading to the bioaccumulation phenomenon.

15.5 Pesticide Pollution

Pesticides are chemicals used to control insect vectors and agricultural pests. These chemicals, especially organochlorine insecticides, have immensely benefited man. Pesticides have regulated several arthropod vectors of severe human diseases and have also increased crop yields.

15.5.1 Sources, Fate, and Effects of Pesticide Pollution on Biological Systems

Studies have shown the presence of pesticide residues in the air, rainwater, dust, rivers, and the sea. Pesticides are in the bodies of aquatic and terrestrial invertebrates, fish, birds, and mammals, including man. The most significant amounts are in organisms at or near the top of the food chain, particularly predators. Most pesticides are produced in large quantities and are persistent in the environment. They remain bound in the soil, mud, atmosphere, and biota and disappear very slowly. In the past, insecticides contained various inorganic chemicals, including Pb, Ar, S, etc., that are not too toxic to insects but are very persistent and stay long in crops from where they enter consumers. A persistent pesticide or residue is chemically different from the original substance. Residues are remnants of pesticides that are very persistent and, in some cases, more toxic than the original pesticides. They are usually changed by weathering, degradation, or metabolism within an organism. Other persistent pesticides include chlordane, endosulfan, heptachlor, and endrin. These chemicals do not occur naturally but are synthesized by chlorine and petrochemicals.

There are over 1000 organochlorine pesticides out of this number (Adjusted from Blus 2003). They are distributed, as shown in Table 15.3.

These pesticides are toxic to insects and other arthropods in small doses but are persistent in the environment (Rezende-Teixeira et al. 2022). They are not harmful to vertebrates, but the problem is that they can remain in the environment for a long time and may become toxic. Trace amounts are in plants and animal tissues, cow milk. Inadequate monitoring programs have contributed to the increase of residues in the environment. The Niger Delta happens to be the epicenter of hydrocarbon pollution due to the proliferation of oil companies, refineries, chemical industries,

Table 15.3 Number of chemicals used for different purposes

Uses	Number
Agriculture	250
Insecticides and acaricides	100
Herbicides	50
Fungicides	50
Nematocides	20
Others	530

Source: Adapted from Blus (2003)

etc. There are many pollutants in the aquatic environment, including mangrove forests and mangrove-associated organisms, e.g., crabs, periwinkle, fish, bivalve, etc. There is thus a need to establish monitoring stations in some rural areas across the zone. Most biological data on heavy metal contamination have been on crabs, mangrove parts, bivalves, and fishes (e.g., Numbere 2020; Numbere 2019a, b). The most at-risk organisms are the aquatic organisms, because they are most susceptible and concentrate these residues more readily than the terrestrial organisms. The rivers and oceans are the sinks for most toxic wastes. Instead of outrightly banning pesticides, the less harmful chemicals can be used moderately. There should be minimized usage and the incorporation of other pest control measures to mitigate environmental pollution. The use of pesticides will continue to increase because of the increase in population, making the need for more food production via rapid agricultural growth.

Sources of aquatic residues include the following:

i. Runoff from agricultural land
ii. Spraying
iii. Industrial effluents
iv. Sewage
v. Sheep dips
vi. Dust and rain
vii. River inputs

15.5.2 *Factors Influencing the Persistence of Pesticides in Water*

Even the most persistent pesticide does not remain in water unless they are carried attached to particulate matter. Usually, within a few days of contamination, concentration in waterfalls to low background levels. For this reason, most data on aquatic contamination by pesticides have been based on residue analysis made on fish, plankton, and marine invertebrates. The factors that influence their persistence in water include:

15.5.2.1 Solubility

water temperature influences the solubility of organochlorine pesticides, although they are relatively insoluble. For example, over a temperature range of 16–38 °C, the solubility of lindane increases to as much as five times and dieldrin and endrin three to five times, while DDT solubility doubles over the same temperature range. The more soluble insecticides disappear rapidly in water. Evidence indicates that they are taken up by the sediment and mud of the waterbed.

15.5.2.2 Substrate

When a persistent insecticide enters any aquatic system, very few remain longer than 1 week. Studies have shown that about 50% of the toxic insecticides get into the soil after 26 h and 22% after 24 h of contamination. The concentration of some pesticides in an aquatic environment is as follows: water > sediment > plant > fish. Some soils can absorb more pesticides than others; similarly, some types of mud can bind more pesticides. The largest concentration is typical in the sediments with the smallest particle size (e.g., clayey muds).

15.5.2.3 Organic Matter

Pesticides are highly attractive to both living and dead organic matter, especially the lipid portion. If the organic material is floating, the pesticides will be in suspension, but if the organic matter is at the bottom, there will be no pesticides in the water.

15.5.2.4 Temperature

Temperature affects the solubility and volatility of pesticides. At some hydrogen ion concentrations, some level of stability is achieved more than others. Therefore, both temperature and pH are essential in soils and thus play a key role in the pollution of the aquatic systems by pesticides.

15.6 Oil Pollution

The impact of oil pollution on aquatic and terrestrial environments in the Niger Delta has been documented for over 50 years by some research scientists (e.g., Gundlach et al. 1981; Snowden and Ekweozor 1987). Crude oil is the mainstay of Nigeria's economy. Its export to foreign countries injects billions of dollars into the nation's economy. The influx of oil wealth began after the first oil well was struck in Oloibiri, an Ijaw settlement in the Niger Delta region, in 1956. Since then, millions of crude oil barrels have been produced from that region, which has placed Nigeria on the global map as one of the world's top oil-producing nations, i.e., the third largest in the world and the largest producer in Africa. During oil exploration, a sizable amount of crude oil is spilled into the environment, often not cleaned but allowed to contaminate the terrestrial and aquatic habitat. The good things about the climate of the tropics are that any pollutant discharged into the environment soon becomes degraded by the hot weather and the numerous microbes that live in the soil and water. Nevertheless, the destroyed ecosystem cannot be fully recovered, and when attempts to recover are made, it takes a long time. Nigeria's crude oil type is unique,

Table 15.4 Predictive categorization of oil spillages in terms of quantity and type of environment in Niger Delta, Nigeria

Category	Quantity	Types of environments
Minor	Less than 50 barrels Less than 500 barrels	Inland waters on land, offshore or coastal waters and mangrove swamp and forest, mudflats, intertidal zones, wetlands
Medium	Between 50 and 500 barrels Between 500 and 5000 barrels	Inland waters on land, offshore or coastal waters and mangrove swamp and forest, mudflats, intertidal zones, wetlands
Major	Over 500 barrels Over 5000 barrels	Inland waters on land, offshore or coastal waters and mangrove swamp and forest, mudflats, intertidal zones, wetlands

because it has less sulfur content and is light. It is also preferable because of its water-soluble components, which are less dense than heavier crude oil from some South American nations such as Venezuela. Oil industry operations in Nigeria at offshore and onshore locations. Some oil terminal is found in Bonny, Qua Iboe, Brass, Forcados, Penington, Warri, etc. there are also refineries at Port Harcourt, Warri, and Kaduna. An increase in global demand for crude oil has led to increased production and the need to establish more privately-owned refineries. Most of the oiling activities in the country do occur in the Niger Delta region because of its substantial natural resources, which include oil and gas. Other resources within the mangrove forest and coastal areas are under threat by oiling activities; these include fishery, which is impacted by marine pollution, coral reefs, intertidal sand, sabotage, accidents, and other unknown causes.

Transportation and marketing cause spillage at the port facilities and terminals, which can be: (i) frequent spillages: this involves a few liters of crude oil. It occurs within and around a port or terminal and (ii) infrequent spillages: involves up to 5 tons or more of oil. This results to mechanical failure of pipelines. According to Ifeadi and Nwankwo (1989), oil spills can be categorized with respect to the quantity of oil spilled and the type of environment in which the spill occurs as minor, medium, and major. Spill categorization has increased by 50% in the last decade due to an increase in oiling activities by legitimate oil industries and illegal artisanal refineries, which has led to the encroachment of oil spillage to formerly safe zones shown in Table 15.4.

15.7 Land Pollution

Then soil serves as a source and sink of pollution. The soil is a source when metallic compounds in soil are blown or washed to contaminate the aquatic environment. The soil is a source when metal-laden soil particles are blown into the atmosphere to cause haze, fog, and smog. The airborne particles foul the air and, when breathed in, cause respiratory diseases (e.g., asthma and bronchitis). An example is the northeast trade wind called harmattan that blows sand particles from the Sahara Desert in the north to the southern part of Nigeria. The harmattan carries dust particles that impact visibility and settles on exposed objects. The dust particles are made up of harmful chemicals, so when the chemicals dissolves in water, they turn acidic and wear away the surface of any object they settle on. They reduce the aesthetic value of buildings. The soil also serves as a source of pollution to groundwater because of the percolation of poisonous chemicals into the groundwater aquifer.

Soil also serves as the sink of pollution by becoming the medium for burying waste and other pollutants. The numerous pore spaces in the ground absorb contaminants from the surface and transport them to the groundwater. When surface and sub-surface crude oil pipes puncture, they release their content into the soil. Similarly, buried pipes discharge their content when they become old and corrode or are broken through an act of sabotage. The spilled crude sinks into the soil profile to affect the groundwater aquifer. Municipal waste is also buried in landfills, decomposing, and migrating into groundwater or the surface. The soils also serve as a burial site for harmful waste from hospitals, factories, laboratories, and industries. When carried by erosion, fertilizers from farms also contaminate the soil and the river. Land pollution can also be caused naturally or artificially. For instance, soils get contaminated through the weathering of the rocks via chemical processes. The weathering process occurs chemically by solution, hydration, decomposition, and hydrolysis. Natural weathering breaks down the soil and reacts with the air to produce toxic pollutants that can flow into surface water or percolate into the groundwater aquifer. The artificial means is when humans deliberately bury poisonous substances in the soil. Examples of such actions include oil and gas exploration, construction, dredging, sand filling, agriculture, and aquaculture.

15.8 Radiation

Radiation consists of electromagnetic waste or parts of atoms moving at high speeds, which can damage living cells. Radioactive substances in the environment may give out gamma rays which tend to affect external tissue or beta rays which penetrate a few cell layers, or alpha rays which deposit a large amount of energy over a short path in the body. X-rays are biologically the most damaging.

15.8.1 Sources, Fate, and Effects of Radiation in Marine and Estuarine Environment

The emission of radiation from a substance is not continuous. Instead, it can occur within a few seconds or thousands of years, which has to do with the half-life. Half-life is the number of years it would take for a substance to decompose or degrade by half. For instance, Strontium 90 will give out half of its radiation in 28 years. Therefore, its half-life is 28 years, which means that after 28 years, the level of radiation it gives out has fallen to half; by 56 years, it has dropped to 1/4, and after 112 years, it has fallen by 1/8th of the original level. Plutonium 129 has a half-life of 24,000 years, so when released into the environment, it becomes a long-term hazard.

The effect of radioactivity, i.e., radiation, which is always related to those on humans, depends on its intensity. Radiation of about 100 gy is extremely high and fatal to humans. A substantial dose of 2–6 cy (200–600) rads may cause immediate burns and other acute effects. Six weeks of exposure to this dose leads to death. However, low exposure can lead to critical skin, hair, or bone marrow changes, but individuals exposed can recover. Those exposed to a small dose may have delayed damage, but the most significant damage in increasing concentration causes damage to reproductive organs and genetic effects. There is some background radiation from natural sources such as cosmic rays, rocks, and naturally occurring substances from outer space. The tissues of organisms always have some harmless quantity that comes from radioactive substances. Studies show that over 70% of the radiation reaching the human body comes from background levels, which is assumed to be harmless. Addition to background level can cause future harm. Therefore, all steps should be taken to stop or reduce external additions.

15.9 Industrial Activities that Promote Pollution in the Niger Delta

The Niger Delta is a fast-growing industrial city rich in natural resources, most especially crude oil, which attracts a lot of foreign and local industries to establish their branches in the town. The Source of these companies is the unregulated practice of emitting harmful substances into the environment, such as liquid and gaseous products. Industrial activities that promote pollution include the following:

15.9.1 Oil and Gas Industry

The cause of pollution is not generated by an individual event but a chain of events, which involve multiple adverse anthropogenic activities as listed above. They cause air, water, and land pollution. The oil industries, including refineries and their subsidiaries, pollute the environment before, during, and after exploration activities. The transportation and sale of crude oil products in offshore and onshore areas also spill and pollute the environment. Offshore pollution occurs from leaking pipes, which transport fuel from flow stations on the sea to ocean-going tankers and land-based depots. The refining of crude oil into petrochemical products leads to the emission of harmful methane gases into the atmosphere.

Fig. 15.2 Effect of pollution on aquatic environment (**a**) Soot on the water surface at Eagle Island and (**b**) Crude oil film on water surface Buguma in the Niger Delta Nigeria

Privately owned oil companies with oil wells also pollute the environment, because the authorities and the regulating agencies do not thoroughly monitor their activities, since they are often bribed not to do a thorough inspection. Recently, there has been a proliferation of artisanal refineries, illegally operated refineries that use crude means to refine stolen crude oil. The crude means of refining the oil lacked the requisite expertise and technology, leading to a lot of wastage of the product via spillages and gas flaring. The emission of gases has caused the proliferation of black soot in some cities, especially Port Harcourt and neighboring communities, which are the most polluted places in the country. The black soot generated from burning crude settles on almost every surface within the city. For example, a column of black soot was found during the morning hours during a field trip on a pool of water at Eagle Island (Fig. 15.2). The stagnant pool contains fingerlings that were being reared for consumption. This is a classic example of the transfer of pollutants from lower to higher organisms across trophic levels through bioaccumulation and biomagnification.

15.9.2 The Sawmill Industry

It is also a growing business in the Niger Delta because of the region's rich supply of rainforest and mangrove forests. The large quantities of trees in the area have made it a very lucrative business in many coastal communities in the area. The cutting of trees to feed the sawmills and wood industries is a major cause of deforestation. Similarly, cutting the trees into plywood generates sawdust waste dumped near the river in many sawmills.

Sawdust pollution in the Niger Delta is hardly mentioned among the types of pollution, but it is a significant form of pollution for aquatic life. Sawdust dumped in an aquatic environment causes collateral damage to marine life by changing the environmental condition of the pelagic and benthic regions of the river. Field observation has revealed that most sawmill industries are situated close to the river to easily transport the logs of wood used for the manufacturing process. The coastal environment is also preferred, because the logs of wood are often left on the river's surface to get "cured" to improve their quality before being sawed into plywood. The sawdust can be recycled as a form of biomass energy, poultry bedding, or organic manure for farms. Still, the problem is that this form of re-use is not greatly explored, resulting in heaps of sawdust being found abandoned as piles of waste at sawmills around the city. The utilization of sawdust in the manufacture of other products can be a lucrative business that can provide jobs for more persons.

15.9.3 Cement Industry

Crushed rocks are used to produce cement that is used for structural work, and the dust from the processing facility gets airborne to increase the harmful particulate matter around the city. The dust also becomes airborne to increase the particulate matter in the town, which is detrimental to human health.

15.9.4 Aquaculture

Aquaculture, the practice of rearing fish and other aquatic organisms to earn some income, is often done in the coastal environment, where the natural water environment serves as the pond for breeding a variety of fishery products for economic gain. The problem with this practice is the pollution of the aquatic environment with waste from the pond such as wastewater and solid waste from the materials used to build the pond that is harmful to aquatic life. Wastewater from the pond increases the BOD of the river. It also deprives marine organisms of oxygen, leading to their death. Those that can migrate move out to other areas far from the reach of the fisher folks, which impacts their source of livelihood for members of the community. Wastewater from fishpond is directly flushed into the open river and causes fish diseases and death due to the polluted state of the water.

15.9.5 *Beverage Industry*

This industry is set up mainly at the city's fringes or near coastal areas. They manufacture drinks and use different chemicals and dye products during the manufacturing process. This industry also utilizes a large quantity of water and produces liquid waste channeled via drainages to the aquatic environment. This liquid waste from this industry leads to eutrophication, which means nutrient enrichment. Eutrophication results in the growth of phytoplankton and zooplankton, which deprives the water of oxygen, leading to the death of fish and other aquatic organisms. The development of these water plants or macrophytes prevents the survival of beneficial marine organisms. The polluted water breeds harmful invasive species with allelopathic properties.

15.9.6 *Sand Mining*

Sand mining is rampant in the coastal region of the Niger Delta, a process of moving sand from the bottom of the sea onto land manually or mechanically. Sand mining has other harmful associated activities detrimental to the coastal environment. For instance, before sand mining, there was massive deforestation of the mangrove trees to pave the way for the mining activity. The coastal land is also fragmented under the weight of heavy-duty bulldozers and machinery that moves in to evacuate the mined sand. All these vehicles spew pollutants into the environment via smoke and the emission of engine oil.

15.10 Anthropogenic Activities that Promote Pollution

Humans are at the center of the environment and control what happens to it, whether negative or positive. Most environmental degradations are manmade and include some of the following: (i) oil and gas explorations, (ii) artisanal refineries, (iii) burning of firewood and waste materials in the open, and (iv) dumping of spent fuel in rivers or land

Exploratory activities involve searching for crude oil underneath the earth's crust using explosives. Discovered oil pocket, once found, is drilled out via pipes, and during this process, some amount spills out to contaminate the environment. Oil and gas exploration involves many parts that negatively impact the environment. For instance, before, during, and after exploration, the environment suffers a lot of degradation through deforestation, soil degradation, air, land, and water pollution resulting in contamination and eventual death of organisms. The proliferation of artisanal refineries has gathered steam for the past 10 years with devastating effects on the environment. The archaic drilling and refining crude oil method has resulted in collateral damage to the environment due to increased wastage of crude oil expelled into water and land and harmful gases and soot emissions. Firewood cooking has also contributed to the proliferation of toxic gases. Roadside cooking involving firewood creates thick smoke in outdoor and indoor environments. Cooking with a kerosene stove has also contributed to indoor and outdoor pollution, thus increasing respiratory diseases among residents in the city.

15.11 Impact of Pollution on the Environment

Pollution harms flora and fauna by contaminating or killing them outrightly (Fig. 15.1). Pollution impacts flora by contaminating the soil and water that serve as a growth medium. For example, an oil spill covers young seedling and prevent them from germinating. Even when they grow, they get deformed or mutate. Furthermore, the roots of plants get asphyxiated and die from oxygen shortage. The death of roots ultimately leads to the death of the tree. Pollution causes litterfall via defoliation and death. Plant absorption of oil pollutants impacts other organisms in the food chain through biomagnification and bioaccumulation. Pollution affects fauna directly or indirectly, e.g., when they consume contaminated materials. Marine pollution caused by oil spills impacts fish by causing deformities, mutations, and deaths (Fig. 15.3). Thermal pollution affects physiology, reproductive ability, spawning behavior, swimming ability, and other chemical reactions in their body. Contaminated fishes are harmful to humans when consumed.

15.12 Solutions to Pollution in the Niger Delta

The resolution of the pollution problem in the Niger Delta is multifaceted. First constant, oil spillage by oil companies needs to be reduced or stopped drastically. Clean-up and remediation cannot resolve the pollution problem, because if the source of pollution is not removed, the clean-up operation will be a wasted effort, and the detrimental effect of the food chain being contaminated will remain. Humans may be at the receiving end when consuming food items that have accumulated pollutants from the environment. The good thing is that there are already regulations governing good environmental practices, one of which is avoidance of pollution. However, the problem is that these regulations are not strictly enforced by the Department of Petroleum Resources (DPR) under the environment ministry. The oil companies, too, do not comply with the already existing regulations. Therefore, to resolve the pollution problem in the Niger Delta, there needs to be compliance, enforcement, and punishment, which will mark the beginning of the end of reckless and detrimental environmental practices.

In other consolidated existing laws, new regulatory rules need to be established to make polluters of the environment pay heavily and deter infringement of the law.

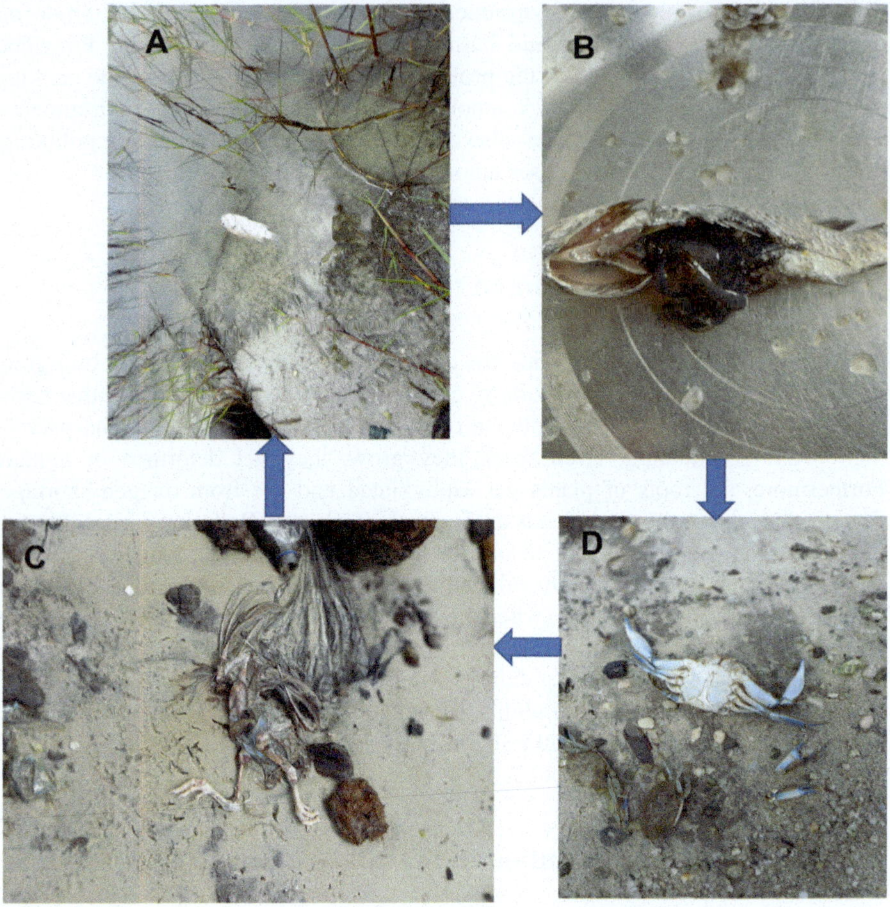

Fig. 15.3 Impact of hydrocarbon pollution on mangrove fauna affects (**a**, **b**) fish (Tilapia), (**b**) bird (heron), and (**d**) crab at Eagle Island, Niger Delta, Nigeria. Picture **b** depicts a decomposed internal organ because of pollution

The companies will determine whether to remain in business by not polluting or go out of business because of loss of profit from paying fines for polluting. Monitoring agencies such as DPR need to be empowered to do their job better and be incorruptible, especially when the companies look out for ways to monitor agents after polluting the environment. The effect of pollution is shown in Fig. 15.2.

15.13 Conclusion and Recommendation

Anthropogenic pollution is a major cause of environmental degradation globally. Nigeria is the largest producer of crude oil in Africa because of the oil-rich Niger Delta, which owns 90% of the oil wells at offshore and onshore locations. The discovery of more oil wells across the zone is directly responsible for increasing oil spillages in terrestrial and marine environments. Other causes of pollution include industrial activities, which emit pollutants into the atmosphere, water, and land. These pollutants include radioactive materials from nuclear plants, pesticides from farms, and industrial waste, which increase the heavy metal concentration along the food chain. Humans eventually become recipients of these pollutants when they consume contaminated food. Indeed "what goes around comes around," and humans become the victims of their actions. To stop the pollution problem, there needs to be a demonstration of good environmental behaviors and the respect for industrial rules and regulations that stipulate good industrial practice that outlines the stoppage of gas flaring, oil spillage, and other health and safety dangers to the environment. Implementing these rules will reduce or end the wanton pollution of the environment.

References

Abbasi T, Poornima P, Kannadasan T, Abbasi SA (2013) Acid rain: past, present, and future. Int J Environ Eng 5(3):229–272

Blus LJ (2003) Organochlorine pesticides. Handbook of. Ecotoxicology 2:313–340

Bolton JR, Bircher KG, Tumas W, Tolman CA (2001) Figures-of-merit for the technical development and application of advanced oxidation technologies for both electric-and solar-driven systems (IUPAC Technical Report). Pure Appl Chem 73(4):627–637

Chen LC, Maciejczyk P, Thurston GD (2022) Metals and air pollution. In Handbook on the toxicology of metals. Academic Press, pp. 137–182

Fenger J (2009) Air pollution in the last 50 years–From local to global. Atmos Environ 43(1):13–22

García Gil A, Garrido Schneider EA, Mejías Moreno M, Santamarta Cerezal JC (2022) Management and governance of shallow geothermal energy resources. In: Shallow geothermal energy. Springer, Cham, pp 237–272

Gundlach ER, Finkelstein KJ, Sadd JL (1981) Impact and persistence of Ixtoc I oil on the south Texas coast. In International oil spill conference (Vol. 1981, No. 1, pp. 477–485). American Petroleum Institute.

Guo SJ, Chen M, He XL, Yang WW, Tan JH (2014) Seasonal and diurnal characteristics of carbonyls in urban air in Qinzhou, China. Aero and Air Qua Res 14(6):1653–1664

He C, Wang J, Fu L, Zhao C, Huo J (2022) Associative vs. dissociative mechanism: electrocatalysis of nitric oxide to ammonia. Chin Chem Lett 33(2):1051–1057

Hossain S, Anik AH, Khan R, Ahmed FT, Siddique M, Bakar A, et al (2022) Public health vulnerability due to the exposure of dissolved metal (oid) s in tap water from a Mega-City (Dhaka, Bangladesh): source and quality appraisals. Exposure and Health, 1-20.

Huang C, Zhao D, Fan X, Liu C, Zhao G (2022) Landscape dynamics facilitated non-point source pollution control and regional water security of the Three Gorges Reservoir area, China. Environ Impact Assess Rev 92:106696

Ifeadi CN, Nwankwo JN (1989) Oil spill accidents in Nigeria petroleum industry. A critical analysis. NAPECTOR 8(4):5–11

Khdary NH, Alayyar AS, Alsarhan LM, Alshihri S, Mokhtar M (2022) Metal oxides as catalyst/supporter for CO2 capture and conversion, review. Catalysts 12(3):300

Li C, McLinden C, Fioletov V, Krotkov N, Carn S, Joiner J et al (2017) India is overtaking China as the world's largest emitter of anthropogenic sulfur dioxide. Sci Rep 7(1):1–7

Luo H, Wang Q, Guan Q, Ma Y, Ni F, Yang E et al (2022) Heavy metal pollution levels, source apportionment and risk assessment in dust storms in key cities in Northwest China. J Hazard Mat 422:126878

Marshall P, Burkholder JB (2022) Comment on 'Extremely rapid self-reactions of hydrochlorofluoromethanes and hydrochlorofluoroethanes and implications in destruction of ozone'. Chem Phy Lett:139411

Moore EW (1958) Thermal "pollution" of streams. Ind Eng Chem 50(4):87A–88A

Nazaroff WW, Weschler CJ (2004) Cleaning products and air fresheners: exposure to primary and secondary air pollutants. Atmos Environ 38(18):2841–2865

Nordberg GF, Nordberg M, Costa M (2022) Toxicology of metals: overview, definitions, concepts, and trends. Handb Toxicol Metals:1–14

Numbere AO (2018) The impact of oil and gas exploration: invasive nypa palm species and urbanization on mangroves in the Niger River Delta, Nigeria. In Threats to mangrove forests. Springer, Cham, pp. 247–266

Numbere AO (2019a) Municipal solid waste disposal in mangrove forest: environmental implication and management strategies in the Niger Delta, Nigeria. Municipal solid waste management

Numbere AO (2019b) Bioaccumulation of total hydrocarbon and heavy metals in body parts of the west African red mangrove crab (Goniopsis pelii) in the Niger Delta, Nigeria. Int Lett Nat Sci 75

Numbere AO (2020) Total hydrocarbon and heavy metal concentrations in body parts of fiddler crab (Uca tangeri) (Ocipodidae) in the Niger Delta, Nigeria. Int J Mar Sci 10

Numbere AO (2021) Impact of urbanization and crude oil exploration in Niger delta mangrove ecosystem and its livelihood opportunities: a footprint perspective. In Agroecological footprints management for sustainable food system. Springer, Singapore, pp. 309–344

Numbere AO, George G (2019) Assessment of heavy metal percolation from electronic waste (mobile phones) through different soil layers in the Niger Delta, Nigeria. Int J Sci Eng Invent 5(8):132–139

Obieze CC, Chikere CB, Adeleke R, Selvarajan R, Ntushelo K, Akaranta O (2022) Field-scale biostimulation shifts microbial community composition and improves soil pollution recovery at an artisanal crude oil refining site. Int J Environ Stud:1–20

Proshad R, Kormoker T, Al MA, Islam MS, Khadka S, Idris AM (2022) Receptor model-based source apportionment and ecological risk of metals in sediments of an urban river in Bangladesh. J Hazard Mat 423:127030

Quivet E, Höhener P, Temime-Roussel B, Dron J, Revenko G, Verlande M, Lebaron K, Demelas C, Vassalo L, Boudenne JL (2022) Underestimation of anthropogenic Bromoform released into the environment? Environ Sci Technol 56(3):1522–1533

Rezende-Teixeira P, Dusi RG, Jimenez PC, Espindola LS, Costa-Lotufo LV (2022) What can we learn from commercial insecticides? Efficacy, toxicity, environmental impacts, and future developments. Environ Pollut:118983

Saleem H, Zaidi SJ, Ismail AF, Goh PS (2022) Advances of nanomaterials for air pollution remediation and their impacts on the environment. Chemosphere 287:132083

Saunders D, Carrillo JC, Gundlach ER, Iroakasi O, Visigah K, Zabbey N, et al (2022) Analysis of polycyclic aromatic hydrocarbons (PAHs) in surface sediments and edible aquatic species in an oil-contaminated mangrove ecosystem in Bodo, Niger Delta, Nigeria: bioaccumulation and human health risk assessment. Sci Total Environ 154802.

Scheren PAGM, Ibe, AC (2002) 22 Environmental pollution in the Gulf of Guinea: a regional approach. In Large marine ecosystems, vol. 11, pp. 299–320. Elsevier

Serih KH, Abd MAA, Sarhan WM (2022) Determination of tetraethyl lead in gasoline fuel. Al-Kufa Univ J Biol 14(1)

Sikhakhane SS (2002) Cost-benefit analysis of the environmental impacts of Darvill Wastewater Works, Pietermaritzburg, KwaZulu-Natal (Doctoral dissertation)

Smith WH (2012) Air pollution and forests: interactions between air contaminants and forest ecosystems. Springer Science & Business Media

Smith SJ, van Aardenne J, Klimont Z, Andres RJ, Volke A, Arias SD (2011) Anthropogenic sulfur dioxide emissions: 1850–2005. Atmos Chem and Phy 11(3):1101–1116

Snowden RJ, Ekweozor IKE (1987) The impact of a minor oil spillage in the estuarine Niger Delta. Mar Pollut Bull 18(11):595–599

Sola MC, Santos R, Nascimento AG, da Rocha GO, de Andrade JB (2022) Occurrence, sources, and risk assessment of unconventional polycyclic aromatic compounds in marine sediments from sandy beach intertidal zones. Sci Total Environ 810:152019

Suresh JI, Janani MS, Sowndharya R (2022) Bacterial diseases in fish with relation to pollution and their consequences—a global scenario. In Bacterial fish diseases. Academic Press, pp. 113–131

United States. Environmental Protection Agency Office of Water Programs Operations (1999) Alternative disinfectants and oxidants guidance manual (Vol. 99, No. 14). US Environmental Protection Agency, Office of Water

Ussiri D, Lal R (2012) Soil emission of nitrous oxide and its mitigation. Springer Science & Business Media

Wang J, Alli AS, Clark S, Hughes A, Ezzati M, Beddows A et al (2022) Nitrogen oxides (NO and NO2) pollution in the Accra metropolis: Spatiotemporal patterns and the role of meteorology. Sci Total Environ 803:149931

Yu H, Stuart AL (2016) Exposure and inequality for select urban air pollutants in the Tampa Bay area. Science of the Total Environ 551:474–483

Zerboni A, Rossi T, Bengalli R, Catelani T, Rizzi C, Priola M et al (2022) Diesel exhaust particulate emissions and in vitro toxicity from Euro 3 and Euro 6 vehicles. Environ Pollut 297:118767

Zielinski O, Busch JA, Cembella AD, Daly KL, Engelbrektsson J, Hannides AK et al (2009) Detecting marine hazardous substances and organisms: sensors for pollutants, toxins, and pathogens. Ocean Sci 5(3):329–349

Chapter 16
Climate Change and Other Environmental Factors as Drivers of Fauna and Flora Biodiversity in Africa

Chukwudi Nwaogu, Bridget E. Diagi, Victor A. Agidi, and Simon I. Okweche

Abstract Changes in climate, land use, and other environmental factors have been and will continue to be potential threats to global biodiversity and the distribution of species. The severe danger that climate change poses to biodiversity at the local, regional, and global scale as well as species' responses has been widely studied especially in developed countries. Research on the climate change—biodiversity nexus, as well as applied models, and the implications in Africa have been low. This work is aimed at reviewing the impacts of climate change and other environmental factors (such as topography, soil, pests, and invasive species) as key drivers of biodiversity and distribution. To achieve this aim and justify the work, the following specific objectives were developed, including (i) examining the impacts of climate change on plants and animals' diversity and distribution and (ii) assessing the responses of different species to some environmental factors, such as soil, topography, invasions, pests, and diseases. This paper used case studies and relevant literature in reviewing the impacts of changes in climate, soil, topography, and other environmental factors on biological diversity and distribution. The findings from this work show that flora and fauna diversity and distribution have been significantly influenced by recent changes in climate and other environmental parameters, especially in the tropical African regions. This paper will create more awareness among the experts and the stakeholders in nature conservation on the rising threats to biodiversity in Africa. Measures to reduce the adverse effects of the environmental changes on the species are also discussed and recommended.

C. Nwaogu (✉) · B. E. Diagi · V. A. Agidi
Department of Environmental Management, School of Environmental Sciences, Federal University of Technology, Owerri, Nigeria
e-mail: chukwudi.nwaogu@futo.edu.ng

S. I. Okweche
Department of Forestry and Wildlife Resources Management, University of Calabar, Calabar, Nigeria

S. C. Izah, M. C. Ogwu (eds.), *Sustainable Utilization and Conservation of Africa's Biological Resources and Environment*, Sustainable Development and Biodiversity 32, https://doi.org/10.1007/978-981-19-6974-4_16

Keywords Climate change · Soil · Topography · Pests · Invasive species · Biodiversity · Nigeria · Africa

16.1 Introduction

The present warming of the earth–atmosphere system resulting from the continuous discharge of greenhouse gases is causing alteration in other factors that make up the earth system (Diagi and Nwagbara 2017). However, this continuous discharging of greenhouse gases into the earth is expected to cause additional warming of the earth's atmosphere and sustained alteration in other factors of the earth system, thereby creating an opportunity for increased severity, widespread, and permanent effects on humans and the ecosystem environment at large (IPCC 2013). Presently, the warming of the earth's atmosphere is largely attributable to anthropogenic activities of man emanating from the continuous emission of greenhouse gas into the atmosphere. It is the dominant greenhouse gas from most industrial activities. Carbon dioxide as a gas occurs in nature on the earth's atmosphere, it is an essential part of the carbon cycle of the earth, i.e., the natural movement of carbon amidst the atmosphere, oceans, soil, flora, and fauna. However, the increased demand of human, especially since the industrial revolution is changing the state of carbon circulation with the addition of a higher CO_2 amount to the atmosphere and also influencing negatively the capacity of natural vegetation and oceans that serves as a sink (Diagi and Nwagbara 2017). The mean temperature of the earth's surface and the rate of severe weather events like the heat wave is on the increase due to man-made change in climate (Hansen et al. 2012; McKechnie and Wolf 2010; Meehl 2007). These anthropogenic activities that lead to an increase in climate change include unsustainable farming practices, land use and transformation of land into other uses, the release of poisonous particles from industries, discharge of greenhouse gas, and use of fertilizers, such as phosphorus and nitrogen, emission from light, and noise or adverse practices during harvesting of crops (SCBD 2010). These anthropogenic stressors tend to impact negatively on other components that ensure the proper functioning of the environment, an example is the effect caused by the emission of acidifying substances on the pH of soil and water (Bouwman et al. 2002); part of the cultivated land portion is impacted by agricultural activities and the increased average global temperature resulting to global warming is influenced further by anthropogenic GHG emissions (Stocker et al. 2013) also a decline in biodiversity and extinction. Consequently, biodiversity is over-exploited by man at a rate faster than it has ever been, with severe adverse effects on human sustainability and livelihood (Turner et al. 1990).

In Ghana, for example, recent evidence from research suggests that there is an increasing rate of environmental degradation (Gyasi et al. 1995); as forests are known for abundant richness, it has been converted to savanna woodland, and existing savanna woodlands converted into near-deserts (Hawthorne and Abu-Juan 1995). Ghana's high forest is estimated to have reduced from 8.2 million hectares to 1.7 million hectares by the mid-1980s (Hall and Swaine 1981) and in the mid-1990s to about one million hectares (Forest Services Division 1996). This calls for the serious attention of all stakeholders in the environmental sector.

16.2 Biodiversity in Tropical Africa

Biodiversity refers to "the totality (numbers) and variability (types) of living organisms in the ecosystem, region and the environment" (Butler 2006). The relationship, as well as interactions that exist between species diversity and ecosystem, is essential for the welfare and health of humans (GRI 2007; UNEP 2002). Biodiversity supplies several benefits to man in the areas of economic, environment, and even the way we live, i.e., our culture (GRI 2007; UNEP 2002; European Commission 2008). In terms of economical provision, biodiversity is a source of natural resources for many industries enhancing economic growth, especially in countries that depend on oil and gas explorations as the mainstay of the economy. Biodiversity is also a means for the provision of food, fresh water, shelters, wood and fiber, fuel, and even unpolluted or fresh air to man.

Anthropogenic activities are socioeconomic activities carried out by man in the environment-like farming, deforestation, bush burning, industrial activities, etc. Biodiversity is also referred to as the diverse and changeable nature of all life on earth. As a result of the progress made in evolution, thousands of species of fauna and flora are now available on the earth's planet. Biodiversity consists of a large range of genes, species, and the ecosystem that can be seen on the planet, which interconnect with one another and impact each other. Biodiversity contributes to numerous environmental functions which include regulation of climate, controlling pollution and soil erosion, balancing the ecology, conservation of the environment, and supporting differences in species and pollination of various plants. Culturally, biodiversity contributes significantly to the support of cultural identity, gives spiritual encouragement and solace as well as plays a significant function in the mental and physical welfare of man (European Commission 2008). Besides, biodiversity also provides educational values and recreational distinction that promotes the development of ecotourism and conservation. Therefore, it is important based on the various roles played by biodiversity in the sustenance of mankind, to understand the relationship that exists between biodiversity, services provided by the ecosystem, and climate changes. Continuous changes in the earth's surface have resulted in the decline of earth's biodiversity either directly or indirectly, as climate change affects man, vegetation and crops, fishes, incidences of pests and disease, insects, animals, soil, and the overall ecosystem of man's environment. It is apparent that several species of plants and animals are dying, while some others are adapting to these changes; however, there seems to be a severe threat to the food we eat and our environment as a lot of biodiversity is gradually going into extinction arising from these changes. Although to a limited extent, evidence of the current extinction of biodiversity resulting from climate change is scarce for Africa. Research, however, attests to the fact that climate change could create a greater threat to the destruction of the dwelling place of biodiversity in the next few decades (Leadley et al. 2010). Numerous explanations have been given as to why plants and animals are less likely to adapt to the current pace of climate change. Such explanation is the rapid pace at which the change is taking place; hence, it is anticipated that over the next century, the rise in average global temperature will be faster than anything that the planet has experienced for at least the last 10,000 years. Due to the rapid increase in temperature globally, many types of plants and animals would find it difficult to adjust quickly to their new condition or move fast enough to a more suitable region for survival.

The degree of extinction of plants and animals has been considerably higher in the last 100 years than it was 500 years before the industrial era (Ceballos 2015). This, however, is driven by numerous man-made activities that continuously put pressure on biodiversity, including the loss of habitat, change in climate, and rivals from non-indigenous species (Butchart 2010). Nevertheless, a lot of research has shown that the most significant contributor to the current loss in biodiversity is majorly those resulting from climate change (MEA 2005; Brook et al. 2008; Guo et al. 2017). Globally, an increase in temperature and CO_2 levels has been attributed to a change in the climate. Effects from these changes lead to significant variation in hydrological cycles (precipitation and evaporation) leading to an increase in the extent and frequency of severe weather events such as floods, cyclones, and droughts that could produce unfavorable effects on biodiversity (Adler et al. 2009; Rinawati et al. 2013). In addition, as this century progresses, climate change is expected to put more pressure on several species (Thomas et al. 2004; Bellard et al. 2012). This is because climate change is an essential driving force on how species of plants and animal functions and is distributed in the natural environment (Parmesan and Yohe 2003).

In addition to the stress associated with an increase in temperature, terrestrial animals would most probably experience a greater level of desiccation and starvation, which is a direct impact of drought periods on the supply of water and as well as quality and availability of food plant also emanating from climate change (Clusella-Trullas et al. 2011; Hoffmann et al. 2013). These alterations outlined are regarded as a considerable risk to biodiversity (Pimm et al. 2014) in the sense that they are capable of creating a significant impact on biodiversity. Nevertheless, while some species of plant and animal may suffer, others may benefit from the present changes in climate, especially the warm adapted species, as the manifestation of these changes is different. Therefore, biodiversity response to changes in climate is probably related to individual species depending on how such species can tolerate severe temperature, desiccation, and stress from lack of food (Anderson et al. 2003; Coumou and Rahmstorf 2012). Several terrestrial and aquatic animals already are manifesting changes due to elevational depth as a reaction to the adverse effect of climate change, i.e., those associated with a water body and shifts in latitude and distribution extent, tracking of shifting isotherms (Brito-morales et al. 2018; Lenoir et al. 2020; Pecl et al. 2017; Pinsky et al. 2019; Steinbauer et al. 2018). In Equatorial latitude, there seems to be a reduction in marine species richness, as there is a redistribution of species due to climate change (Chaudhary et al. 2021; Yasuhara et al. 2020. Nevertheless, the changes in climate are not uniform the world over so also are their manifestation and impact (Diagi and Nwagbara 2017). The effects created by the change in climate and other man-made drivers of biodiversity loss differ from one geographical location to another and between niche and taxa (Blowes et al. 2019; Bowler et al. 2020).

Africa's biodiversity has been experiencing a decline during the last decades, resulting in a severe decline of species and habitats. According to (UNEP-WCMC 2016), the loss in biodiversity in Africa is attributed to man-made activities that result from cutting down forests and degradation and changes in the use of natural land cover. However, the adverse impact created by the change in climate on species and ecosystems in Africa is aggravating the already bad situation. Separate figures published for Africa indicate that a 22% loss in the forest over tropical Africa has occurred since 1900 and another estimate shows up to 35–55% to which the larger scale degradation must be included (Aleman et al. 2018); hence, close to one-third of tropical African flora is at risk and facing annihilation (Stévart et al. 2019). Therefore, the most important step to take is for effective conservation and sustainable distribution of biodiversity is documentation of its natural resources to ascertain the level of changes that have taken place. Tropical Africa provides habitat to certain of the most important species-rich biodiversity zone in the world (Linder 2001) extending from the second largest area of continuous rain forest in the world, the Congo basin, to the Namib Desert. Tropical Africa is a land that shows strong biodiversity differences (Linder 2014), besides a great extent of its wilderness has already been lost (Watson et al. 2016). In addition, climate change is anticipated to hurt sub-Sahara Africa ecosystems, with about 90% of species partly or wholly losing their areas of favorable climate by 2085 (McClean et al. 2005). These results are frightening and hence calls for immediate international policies to be put in place (Watson et al. 2016; Wen et al. 2015; Perrings and Lovett 1999) to safe guide the biodiversity of the world in general and tropical Africa in particular.

Africa is immensely rich in biodiversity. Africa boasts of two out of the five global higher biodiversity areas and nine biodiversity hotspots. In addition, it comprises about a quarter of the world's biodiversity and it helps in keeping a substantial number of large animals together in the same assembly which could have wandered about in many nations. The rich biomass of Africa spread across the mangroves in the Mediterranean zone, deserts, tropical forests, temperate, mountainous grasslands, and savannahs, including mountains in the ice cap regions.

Africa's population exceeded one billion in 2009 and is expected to grow at the rate of 2.3% yearly during 2010–2015 (World Bank 2011). This rapid growth in the population of Africa has increased the growing demand for more natural resources, leading to considerable changes in land use and the disprovable use of many species of plants and animals. Therefore, considerable pressure is placed on the natural habitat, biodiversity, and services provided. The human race relies on plants for food, medicines, crop protection, and materials for clothing and building. They also provide ecological services, cleaning our air, water, and soil. Recently, in the quest to achieve a better life for the increasing population, biodiversity has become easy prey for human overexploitation resulting from advancements in science and technology (Wuver and Attuquayefio 2006).

16.3 Climate Change Impacts on Plant and Animal Diversity and Distribution in Africa

Predictions on Africa have shown that it is extremely sensitive to the impact created by changes in climate. Projection of Africa shows that a large expanse of its region has warming above 2 °C at the turn of the twenty-first century with a possible fall in precipitation (Collier et al. 2008; Niang et al. 2014).

Overall, Africa is a continent that experienced warming of 0.7 °C during the twentieth century, with this warming increasing from 0.2 °C per decade (low level) to over 0.5 °C per decade (high level) (Hulme et al. 2001; IPCC 2001).

Consequently, these changes in two key climate elements (temperature and precipitation are having adverse negative impacts on the ecosystem and biodiversity (Biggs et al. 2008) showing signs of a shift in various species and changes to ecosystems (Garcia et al. 2012, 2014) (Fig. 16.1).

In addition to African weather, climate, and climate change, another problem is the frequency and intensity of serious events such as the El Nino Southern Vibration, the outbreak of El Nino and La Nino (Korcha and Sorteberg 2013; Midgley and Bond 2015). In addition, the anthropogenic activity resulting from changing land-use patterns, ecosystem degradation, alteration, fragmentation, species exploitation,

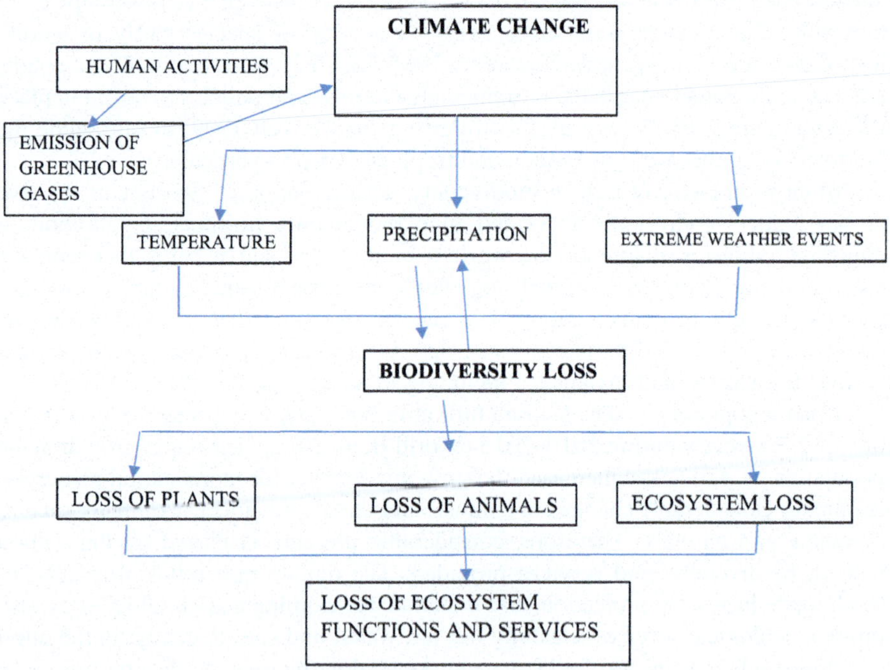

Fig. 16.1 Link between human activities and their impact on climate change, biodiversity loss, and the impacts on services provided by the ecosystem

and the introduction of alien species exacerbates the effects of climate change (Mooney et al. 2009). Climate change is projected to intensify in the next century and becomes one of the leading contributors to Africa's biodiversity loss (Sala et al. 2000; Bellard et al. 2012; Midgley and Bond 2015).

The most significant observable drivers responsible for the present decline in biodiversity are changes in habitat, excessive use of natural resources, change in climate, nonindigenous species, and overexploitation, jointly referred to as the main threat to biodiversity (Brook et al. 2008; Ogwu et al. 2014a, b; Osawaru and Ogwu 2014; Guo et al. 2017; Sonwa et al. 2017). According to the IPCC (2013), global warming and precipitation are projected to increase, and the resulting climate change is expected to contribute to the extinction of all types of biodiversity over the next 50–100 years. Despite increasing awareness that biodiversity is extremely vulnerable to climate change, the areas of biodiversity change in Arica related to climate change and variability compared to other parts of the world have not been fully researched (Getahun and Shefine 2015; Sonwa et al. 2017; Matata and Adan 2018). Africa is extremely endowed with biodiversity and is composed of an estimated one-fifth of mammals, plants, and birds, including one-sixth of herbs (Siegfried 1989).

Africa, on the other hand, hosts several of the world's most distinctive and biologically significant ecosystems threatened by negative effects of climate change, including savanna, tropical forests, coral reefs, marine, and freshwater habitats, and wetlands. These ecosystems in Africa provide very great benefits to communities, which are jointly called community services (MA 2005).

The services provided by the ecosystem are to meet the needs of various human lives, such as food, feed, wood, firewood provision, regulation (disease and climate regulation, etc.), and support (soil structure, nutrition supply, etc.). It is diverse, because it is necessary for cultural (e.g., recreation, ecotourism) services [Millennium Ecosystem Assessment (MA) 2005; Wangai et al. 2016].

Nevertheless, the very rich biodiversity in tropical Africa with its multiple roles of provision, regulating, and supporting communities is threatened by a change in the climate. Climate change is both a cause and a consequence of changes in Africa's biodiversity and ecosystems. (Thomas et al. 2004). In addition, there are anthropogenic stressors that further put pressure on this biodiversity. The various factors of climate change are expected to be the principal force responsible for biodiversity loss at all levels (Parmesan 2006). The decline in biodiversity resulting from climate change has, directly and indirectly, changed the patterns as well as the changes in the flow of energy and circulation of materials (Zhong and Wang 2017) that have largely affected the ecosystem in Africa as well the services provided. The diversity of species supported by the ecosystem will determine their ability to provide climate regulation services (Bellard et al. 2012).

In addition, climate change can occur through the process by which humans use biological resources to provide goods and services, especially the process of converting grasslands and forests into agricultural land (Lambin and Meyfroidt 2011). The numerous needs of man from the production of biological resources for foods, fuels, and fibers, including other activities that man engages directly affect

discharging of numerous greenhouse gases (Hector and Bagchi 2007; Burnham and MA 2015) responsible majorly for this change. Africa's climate is constantly changing, and the consequences of biodiversity loss continue to increase over time, posing a significant threat to the continent. The resulting impact is the change in the range and distribution arrangement of the ecosystem and a reduction in the services that they provide for the survival of the human race.

Climate change and biodiversity are closely linked and connected to human existence, as a change in one has an impact on the other either positive or negative and, in turn, affects the lives of humans. Rapid biodiversity loss and climate change are also closely linked, sharing and interacting with basic, direct, and indirect impetus, and having a rapid and challenging impact on human well-being. May threaten social goals (Diaz et al. 2019; IPBES 2019). Another important area where climate affects Africa's biodiversity is to reduce the number and availability of favorable habitats and elimination of other important species (Lovett and Barnard 2005; Hély et al. 2006; Doak and Morris 2010; Dawson et al. 2011). Species decline in ecosystems affects not only lost species but also dealing with other species and the shared ecological benefits expected from such interactions.

In the bid to safe guide whatever is left of the natural habitat, these habitats perform an essential function in the survival of species as well as a player in the storage of carbon for climate stabilization, more attention, therefore, needs to be given to the interaction that occurs within these natural ecosystems. When forested land is converted into an urban area or for agricultural purposes, the carbon that is contained in the vegetation most times enters the atmosphere during burning processes which are associated with land clearing in Africa. The burning process converts the carbon in vegetation to CO_2 in the atmosphere, a scenario that releases about one-third of greenhouse gases into the atmosphere yearly. Therefore, the protection of natural habitats from destruction can serve both as a prevention of species extinction and a significant contributor to controlling climate change.

16.4 Soil, Topography, and Biodiversity Nexus: African Perspective

Biodiversity has been an intriguing aspect to scientists for a very long time because of its complexity and distribution. Biodiversity is the biological variety and variability of plants and animal species on earth. Biodiversity deals with a variety of species on earth, including plants, bacteria, and fungi. The biodiversity of the earth is very rich, so not all species are recognized, even as others are threatened with extinction.

Biodiversity consists of genetic, species, and ecosystem diversity. It will be interesting to know that biodiversity in space is not uniform and is largely influenced by soil and topography among other factors, as shown in Fig. 16.2. Among the inanimate factors affecting biodiversity, topographical structure and soil variety

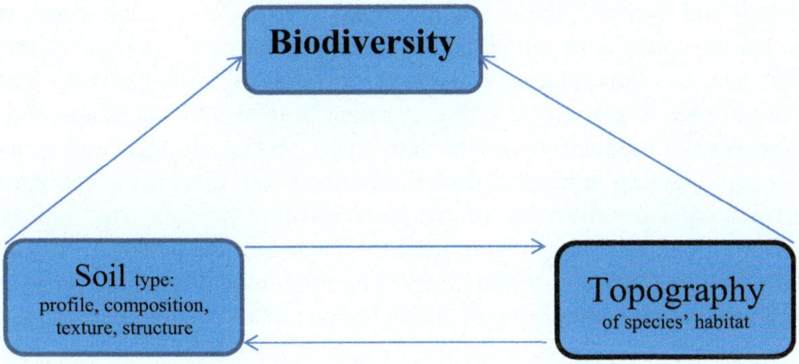

Fig. 16.2 Nexus between soil, topography, and biodiversity

constitute the most important in the characteristics of various habitats and, as a result, contribute to the somatic differentiation of vegetation (Oliveira-Filho and Ratter 2002; Osawaru et al. 2011, 2014, 2015a, 2015b, 2016; Omoigui et al. 2016). This chapter focuses on the influence of these abiotic factors on the distribution of biodiversity.

Soil is commonly defined as the upper or surface level of the earth's surface which supports the growth of the plant. However, the soil is a complex relationship between flora and fauna. A handful of soil can contain as many microbes as possible. The influence of soil type on vegetation distribution has been well-documented. Soil plays a necessary role in habitat formation and change, so it can bring adaptation in the form of vegetation and plant diversity. Soil variables were revealed to have affected both vegetation structure and micro- and macro-scale biodiversity (Nwaogu et al. 2019; Kerfahi et al. 2019; Song et al. 2019; Ikhajiagbe et al. 2020, 2022a, b). Habitats were structurally different and diversity was different between the savannah and the forest community. However, comparing the savannah habitats with each other revealed a greater difference. Results from studies on plant species and tropical vegetation productivity have a positive relationship with soil potency (Neri et al. 2012; Dybzinski et al. 2008; Poulsen et al. 2006). Some research, however, showed a non-positive relationship or influence between species productivity and soil fertility (Nadeau and Sullivan 2015; Enright et al. 1994); this, however, makes it necessary to carry out in-depth studies, especially for land in the tropical region.

Waide et al. (1999) did an intense literature review on the connection between vegetal productiveness and biodiversity in the desert, northern forests, tropical forests as well as wetland ecosystems. This study found that distribution patterns are inconsistent and that the relationship between the biodiversity of plants and productivity varies from habitat to habitat and can be impacted by diverse living and non-living components. In explicit research, Janssens et al. (1998) noted that there is a relationship between plant biodiversity and various soil chemical elements in different parts of temperate grassland ecosystems, plant biodiversity (abundance and diversity), and extractable P and K in soil and found a high degree of correlation with the concentration. No association has been established between plant biodiversity and various factors, such as total N and Ca concentrations, pH, and organic matter content. In temperate regions, Tilman et al. (1999) examined the effects of species abundance and diversity on ecosystem productivity. The availability of nitrogen in plant roots reduces the leaching loss of soil N and at the same time increases the productivity of the ecosystem, so that nitrogen will be available for faster root planting in higher species abundance and diversity plots. Partel et al. (2007) researched the diversity of plants concerning productivity, and 163 case studies from around the world were used to investigate the pattern of relationships between plant diversity and productivity. The monomodal relationship of low and very high productivity and excess in the midrange seemed to dominate the temperate region.

Topography is the shape and features of the earth's surface. It is otherwise seen as the high and low of the earth's surface. Ecologists are paying careful attention to the influence of topography on biodiversity knowing well that topography also plays a great role in the climate of a place. Topography greatly influences abiotic stipulations that, consequently, affect the framework, feature as well as dynamics in environmental communities (Jucker et al. 2018; Fortunel et al. 2018; Féret and Asner 2014; Baldeck et al. 2012). The topography of a geographical location, over time and space, influences local weather elements and also impacts diversification strategies (Antonelli et al. 2018; Badgley et al. 2017). However, until recently, the lack of high-resolution terrain records has constrained the ability to measure the organic relevance of fine terrain diversity. Technological improvement in remote sensing provides a new perspective for revealing the effects of subtle topographical homogeneities on ecosystem composition and dynamics (Féret and Asner 2014; Mascaro et al. 2014). Jucker et al. (2018) revealed that the relationship between topography, abiotic condition, and characteristics of species can lead to mutations in other forests, such as distribution patterns and biomass. This research, however, provides a good groundwork to grasp the intricate manner in which topography affects abiotic conditions. Notwithstanding, the effects of nearby topographically transmitted abiotic slopes can also depend on climatic conditions. For instance, the neighborhood fluctuations in soil moisture from valleys to ridges are more biologically widespread, as it may represent a change in availability of water compared to the average conditions of seasonally dry forests, as opposed to seasonally moist forests.

The relief shown on the scale in the United Kingdom may have had only a limited impact on plant life and fauna, and this impact may generally be revealed by climate. As a general rule of thumb, a 300 m rise corresponds to a 40F temperature drop. Therefore, in most cases in the United Kingdom, the problem is not that there is a big difference between vegetation and ecology as a result of changes in climate change, but the impact created by such changes. However, in areas such as the Pennines, Lake District, and Dartmoor, there are significant differences in elevation (up to 600 m) and geological structure/outcrops between adjacent lowland and highland areas, resulting in significant differences in mean temperature.

Early studies on ecology show the complexity of the relationship that exists between the physical and biological components of ecosystems and habitats. In this complex context, it is clear that geology performs an important role in explaining the ecology of the region, both in its contribution to soil structure and its impact on topography. However, what we are seeing today is not the definitive form of plant and animal distribution, nor the diversity of the UK's nation-state ecosystem but an impact that is creating a great concern. Climatic variations from south to north and west to east perform a significant role in the development of flora and faunal kinds and man's action has also contributed largely to defining its surplus, limitation, and distribution. The various attempts made to ascertain the relationship that exists between geology and flora and fauna (especially plant existence as a result of the direct relationship with the type of soil); nevertheless, this has proven to be difficult and complex, even though such a relationship exists with a burden on ecological security (Ogwu 2019a, b).

In addition, different authors have used different methods to show that topography is a significant factor influencing vegetation structure (Moeslund et al. 2013). Topography has been revealed clearly to affect climatic conditions (Grzyl et al. 2014) and pedogenesis (Ridolfi et al. 2008). Other factors that have a significant influence on the distribution of vegetation are infiltration, runoff, erosion, and migration of seeds. (Jiao et al. 2009; Yu et al. 2006) which are greatly influenced by a change in the climate and elevational gradients (Ogwu et al. 2019). Vegetation patterns are indirectly influenced by topography (Kirkpatrick et al. 2014; Ridolfi et al. 2008). Vegetation distribution is usually limited by hillside slopes (Lin et al. 2014) or outputs (Xu et al. 2008), but through regional elevations (Moeslund et al. 2013). Vegetation distribution under certain climatic conditions is usually influenced using one or more topographical elements (Fu et al. 2004). However, topography differs greatly from region to region, and therefore, effects are also different. Therefore, identifying key topographical factors that facilitate distribution under species conditions is an important part of sustainable forest management.

Elevation, aspect, slope, and surface curvature are typically used to assess topographical results for vegetation distribution (Laamrani et al. 2014; Ohwawa et al. 2008). As shown by Pabst et al. (2013), the gradient determines the surface current which in turn controls the path of the current, and the rise while affecting the tempo of solar radiation that determines the elevation zone of the soil. On the other hand, evaporative strength and surface curvature facilitate the movement and accumulation of water into the landscape by gravity. The combination of these terrain elements determines the conditions of the boom and, therefore, vegetation distribution to a large extent.

There have been several studies that focused on the relationships between soil, topography, land use, and the diversity of plants and animals (Garca-Aguirre et al. 2007; Laamrani et al. 2014; Nwaogu et al. 2018). Researchers have also identified key topographical elements that affect distribution at particular locations. For instance, the slope is the only non-soil aspect that correlates with plant species in Southeastern Oregon, USA (Davies et al. 2007), and also the dominant limitation of China's hot and dry Jinsha River valley. Altitude is the most important issue in the river basin in Japan (Matsuura and Suzuki 2013). Further research has found two or more topographical component that mostly affects vegetation distribution, exceptionally, in dry heat valleys in south-western China (Xu et al. 2008). Slopes and elevations have a major impact on the vegetation patterns of the Atlantic Forest in south-eastern Brazil (Eisenlohr et al. 2013). Aspect, as well as elevation, are closely related to the abundance of shrubs in the deciduous forest area of Beijing, China (Fu et al. 2004). Therefore, the basic topographical element that affects the distribution of vegetation differs from region to region.

Wanga et al. (2015), applied various multivariate analyses and identified a complex association that exists in the pattern of understory vegetation in Massoniana forests and the topography of areas, where red soil was dissolved in southern China. The distribution of vegetation depends on the topography of the area to a large extent. Most of the flora was found in the canyon, but there were few plants on the ridges. Studies also revealed that the low ridge vegetation (25.2%) and low species richness (5 per plot) were previously due to severe soil erosion. Aspect and ground curvature were the most important topographical component affecting distribution, but surface curvature is more significant than aspect (Wanga et al. 2015). In conclusion, the impact that topography plays in the distribution of biodiversity has been well-studied especially as it relates to the richness of species. However, more studies need to be carried out, especially with the adverse impact created by climate change in recent times.

16.5 Roles of Pests, Diseases, and Invasive Species on Biodiversity in Africa

The impacts of pests, diseases, and non-native species on the biological diversity of different fauna and flora species cannot be overemphasized. There have been several studies on the roles of the trio (pests, diseases, and non-native species) on biodiversity (Singh et al. 2016; Guy-Haim et al. 2018; IPBES 2019; Lubek et al. 2020; MacIvor et al. 2017; Chime et al. 2015, 2018; Gibson et al. 2015; Ghelardini et al. 2017; Defra 2021; Mitchell et al. 2022; Graziosi et al. 2020; Shackleton et al. 2020). As a result of intensified global commerce and climate change, the forest ecosystem is presently witnessing a rapid increase in non-indigenous tree pests/pathogens (Freer-Smith and Webber 2017). Consequently, significant ecological wreck, economic loss, and biodiversity loss have resulted (Boyd et al. 2013). There are strong interactions among biodiversity, invasive species, pests, diseases, man, and climate as the presence of one might impact another either positively or negatively (Fig. 16.3). For example, species invasions, pests, and diseases have caused severe effects on biodiversity and local, national, regional, and global levels.

In addition, the impacts of climate and anthropogenic activities on the loss of biodiversity can never be overemphasized. An increase in population growth has also been known as a major driver of biodiversity loss by exacerbating the effects of climate change (Ikhajiagbe et al. 2022b). Many farmers have suffered great losses, while several agricultural species have become extinct due to attacks by birds,

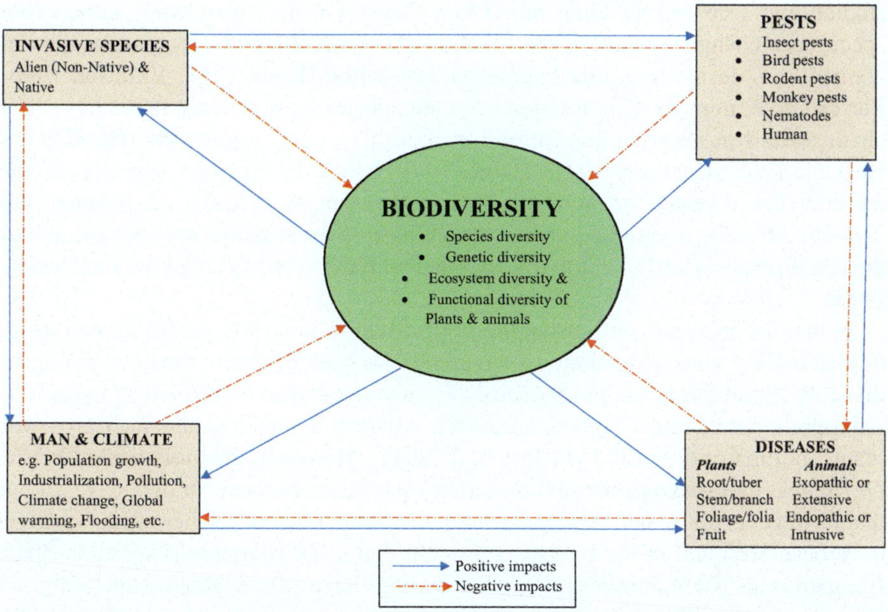

Fig. 16.3 Chart showing the relationships/interactions among biodiversity, invasive species, pests, diseases, man and climate

insects, rodents, and nematodes. A village farmland in Ikpo-Obibi community of Etche local government area of Rivers State, Nigeria has been abandoned for several decades due to the infestation of White-Curl grub (*Phyllophaga spp.*) larva (Nwaogu and Okweche 2019). On the other hand, functional and rich biodiversity can help in the mitigation of climatic impacts, help to ameliorate the impacts of invasive species, as well as support in the reduction of pests and diseases. Knowledge on the diversity within agroecosystems, and urban and home gardens, as well as a robust classification system, can help address some of these issues (Osawaru et al. 2013, 2015a, b; Ogwu et al. 2014a, b).

16.6 Impacts of Invasive Species on Biodiversity in Africa

Non-indigenous invasive species introduction and establishment are on the rise globally (Seebens et al. 2017), and this action is prompted by expanding global trade and exacerbated by climate change. Worldwide, invasive species are recognized as a key danger to biodiversity and the health of the ecosystem. Numerous non-indigenous flora and animals have been implicated in the destruction of indigenous local taxa and their biodiversity (Vitousek 1990).

Invasion of non-indigenous species is a major cause of species extinction on continents, specifically on islands. Although most biological invasions are not well-studied and recorded as some fail (Elton 1958), On the other hand, some exotic species can change the pattern and function of communities and ecosystems through competition, destruction, and habitat reconstruction (Elton 1958; Vitousek 1990). For example, impacts of non-indigenous ant species on native ant fauna have been characterized in many biological and ecological systems worldwide (Haskins and Haskins 1965; Erickson 1971; Haines and Haines 1978). The majority of this research has revealed an amazing loss in the species richness, abundance, and diversity of indigenous ants in regions dominated by non-native ant species, though the certain class of ant's seldom live together with the invaders (Haskins and Haskins 1965).

In another instance, an outstandingly devastating invasive species known as the dog-strangling vine (*Vincetoxicum rossicum*) has been of severe threat to biological diversity. In addition, the existence of *D. rossicum* is also confirmed as harmful to arthropods (Ernst and Cappuccino 2005), whereas it impedes and interferes with plants' pollinator diversity (MacIvor et al. 2017). However, the biological effects of *D. rossicum* are beyond the aboveground ecosystem, because studies have shown that its impacts transcend to belowground microbial dynamics, where many changes have been recorded in the biological system due to *D. rossicum* (Day et al. 2015; Bongard et al. 2013). Furthermore, other findings have attested that by operating as a fungal generalist, *D. rossicum* significantly stimulates a surplus of fungal plant–root colonizers (Day et al. 2015), especially those that of benefit to D. *rossicum,* and simply harm the development and reproduction of indigenous plant species (Day et al. 2016). Similarly, other empirical research stressed that D. *rossicum* could cause a decline or change in the form of soil bacterial and fungal biodiversity, by discharging allelochemicals including the toxin–antofine. An alteration is anti-biodiversity, since the microbial diversity becomes altered, thus hindering the development of native indigenous species (Gibson et al. 2015; Ogwu and Osawaru 2015).

The escalation of pests for the standing species of trees in the natural and/or non-natural forests is also increasing worldwide (Gibson et al. 2015). In North America, especially in the United States, about 400 insect pests in the forest were discovered between 1860 and 2006 (Aukema et al. 2010). In addition, in the European continent, more than 100 alien tree pathogens were added between 1800 and year 2009 (Aukema et al. 2010). Some other striking and worrisome cases are the unimaginable destruction in urban USA and Canada, as exotic emerald ash borer (*Agrilus planipennis)* causes severe damage to biodiversity.

The most worrisome truth is that the impacts of either pests, diseases, or invasive species on the plant or animal biodiversity tend to be infinity with geometric occurrences. For example, there is increasing evidence that the reduction in basic tree and animal species can affect hundreds of related species, not just those that depend on the basic species (Lubek et al. 2020; Mitchell et al. 2022). For example, Fraxinus excelsior is presently becoming extinct in Europe as a result of the non-native *ascomycete Hymenoscyphus fraxineus* (Kjær et al. 2012).

In addition, *Fraxinus excelsior* is threatened by the non-traditional emerald ash beetle *Agrilusplanipennis.* US trees (Herms & McCullough, 2014) extend beyond the Eurasian continent to Ukraine (Orlova Bienkowskaja et al., 2020). *Petraea*/ Rober is presently a non-indigenous pest, the moth of the oak procession (Tomlinson et al. 2015), the rapid decrease in oak tree engendered by native insects and bacteria (Doonan et al. 2020), due to climate change, endangered species (Brown et al. 2018), and various non-indigenous dusty mildew species (Lonsdale 2015). In addition, *Q. petraea*/robur is at risk of diminishing due to *Xylella fastidiosa,* as this bacterium becomes established in the United Kingdom (Defra, 2021).

16.7 Implications of Pests, Diseases, and Invasive Species on Flora Diversity in Africa

In the latest study by (Mitchell et al. 2022), it became clear that the total number of endangered species due to plant diseases affecting two sympatric hosts could exceed the total number of related species threatened by the decline of only one host. It was concluded that considering functional or lack of functional redundancy within the affected ecosystem, a complete assessment of the impact of biodiversity on plant or animal pests and diseases of related species is not possible (Mitchell et al. 2022). While the epidemic of invasive pests and diseases has increased rapidly (FreerSmith and Webber 2017), the increasing impact on related biodiversity has also increased substantially, resulting in serious biodiversity loss (Jonsson and Thor 2012).

In Africa, especially, several pests, including insect pests, have been reported to attack both indigenous and non-native plant species which consequently have adverse effects on the biodiversity of the affected species (Graziosi et al. 2020). For instance, across the countries of Africa, the black borer (*Apate monachus*) has a devastating effect on the richness and diversity of many tree species, including *Azadirachta indica, Terminalia* species (Graziosi et al. 2020). Besides having negative impacts on the host plant species, the animals which depend on the affected tree species were in turn affected.

The biodiversity and abundance of the West African tree T*rilochitonscleroxylon* (Stink bug), African teak (*Milicia excelsa),* and *Milicia regia* (Moraceae) are invaded by the aphids and *Phytolyma fusca*, respectively. It is also affected by sexual pests, such as Iroko goal bug (Graziosi et al. 2020). In Ghana and Malawi, a pest called *Lamprosema lateritialis* (Lepidoptera: Pyraloid moth) impedes the diversity and growth of the sensitive *Pericopsis elata* (Fabaceae) (IUCN 2018). Many indigenous insect species known as pests of native trees have been reported to expand their host range and destroy exotic trees.

Longhorn beetles such as the stalk beetle Analeptes and Paranareptes (Beetle: Cerambycidae) are capable of causing significant harm to a lot of non-native trees, such as African baobabs *Adansonia digitata* and *Bombax costatum* (Mallow), and also the non-native teak tree *T. grandis*, the eucalyptus tree and cashew nuts, and the *Anacardium occidentale* (Cashew nuts). (IUCN 2018). Eucalyptus longhorn beetles *Phoraanthase mipunctata* and Phoraanthare curva (Coleoptera: Longhorn beetle) were introduced in South Africa in the early 1900s and are now two stalk perforators affecting eucalyptus production in countries in the south and north of Africa (IUCN 2018; Watson et al. 1999). The cypress aphid Cinaracupressi has introduced about 30 years ago and is now found in all major cypress growing areas, reducing the biodiversity and fruitfulness of exotic cypress plantations. The invasion of the hinoki cypress Cinaracupressi also affects the endangered native African cedar Juniperus procera (Cupressaceae), a native species throughout Africa (IUCN 2018; Watson et al. 1999).

Apart from pests, diseases, and invasive species, the effects of microbial pathogens on species diversity cannot be ignored. Drastically, pathogens are seriously causing an impact on the health of natural as well as semi-natural forest ecosystems in Africa. For example, in the 1600s, the *Armillaria* root rot pathogens were brought to the continent of Africa from Europe (Watson et al. 1999). Initially, it was discovered that the disease was destroying planted pine as well as eucalypt in the early 1900s in South Africa. The impact of the pathogen later became critically native to the biodiversity of many forest plant species in Tanzania, Uganda, Kenya, Ethiopia, and other Eastern African countries (Graziosi et al. 2020). Nevertheless, eucalypts have been reported as important hosts of many global plant pathogens (Graziosi et al. 2020). Studies have shown that *Anthracnose* caused by *Colletotrichum spp*., comprising wide-ranging *C. gloeosporioides*, which is posing serious threats to the richness and diversity of many tree species in Africa, including eucalypt and others (Graziosi et al. 2020).

In 1992, across the East, North, South, and Western African countries, Leucaena psyllid (*Heteropsylla cubana*) was detected as an insect pest. It has been negatively impacting the biodiversity of some species, including *Leucaena, Mimosa* species, and others (Graziosi et al. 2020). Studies have also revealed that the Mahogany shoot borer (*Hypsipyla robusta*), Giant looper (*Ascotis selenaria*), and Tussock moth (*Orgyia basali*) have caused adverse biodiversity loss in Nigeria, Ghana, Benin Republic, Cameroun, and other west, East and southern African countries. In these countries and regions, these pests attack many species, such as Khaya, *Entandrophragma*, *Eucalyptus* spp., *Acacia mearnsii*, *Tectona grandis*, and *Pinus* spp.

16.8 Responses of Fauna Diversity in Africa to Pests, Diseases, and Invading Species

The effects of pests, diseases, and invading species on the biodiversity of many animal species have also been documented by many scientists inside and outside Africa (Marr et al. 2017; Zengeya et al. 2020; Measey et al. 2020a, b, Davies et al. 2020, Shackleton et al. 2020). For example, Tilapia fish in the Nile (Oreochromis species) such as the *Oreochromis mossambicuss* are facing danger from Oreochromis niloticus (Marr et al. 2017). In the same manner, smallmouth bass

(*Micropterus dolomieu*) transforms invertebrate community structures, locally exterminating and fragmenting native fish populations through competition and predation, impacting biodiversity in South Africa and other neighboring African countries (Marr et al. 2017). In the same regions of Africa, declines in biodiversity and cases of extinction of native invertebrates, frogs, and fish were reported due to predation by Brown Trout (Salmo trutta) (Zengeya et al. 2020). *Oncorhynchus mykiss* and *Salmo trutta* have also been reported as the main cause of the decrease of populations of *Hadromophryne natalensis* (Natal Cascade Frog) in UKhahlamba Drakensberg Park stream, (Karssing et al. 2012). They have also been known for the loss of diversity and extinction of sensitive fish *Amatolacypris trevelyani* (Border Barb) in headwater streams of the Keiskamma River in the Eastern Cape. Two alien abalone, *Tarebia granifera* and *Thebapisana*, have been reported to cause havoc in African biodiversity. The results show that these abalone are densely populated and are involved in swarming native snails and becoming the dominant element of invertebrate aggregates in the occupied territories (Perissinotto et al. 2014).

Some invasive amphibians such as *Duttaphrynus melanostictus* (Asian Common Toad) and *Sclerophrys gutturalis* (Guttural Toad) were reported to have potentially threatened the biodiversity of indigenous sensitive *Sclerophrys pantherina* (Western Leopard Toad) as well as causing the decline of local *Hyperolius horstocki* (Arum Lily Frog) (Measey et al. 2020a, b). In the bird kingdom, two species, *Anas platyrhynchos* (mallards) and *Pycnonotus jocosus* (red-whiskered bulbuls), are known to have a significant impact on their global invasion range. For example, *Anas platyrhynchos* damaged the genetic integrity of indigenous homologues by hybridization (Davies et al. 2020). A study has revealed that *Pycnonotus costs* have caused severe damage to different crops, spread weeds, preyed on indigenous species, and contends with them in South Africa and elsewhere. Other studies have studied the adverse biological effects of *Psittacula krameri* (Ring-necked Parakeet) having confirmed that it competes for breeding holes with bats and other birds which consequently leads to a population decline of affected species in most African countries (Zengeya et al. 2017; Shackleton et al. 2020).

Other studies have investigated the biological adverse effects of the Rose-ringed Parakeet and competition with bats and other birds for breeding holes, resulting in reduced populations of affected species in most African countries. (Zengeya et al. 2017; Shackleton et al. 2020). There are records of over 40 exotic mammalian species listed worldwide in South Africa and neighboring African countries (van Wilgen and Wilson 2018).

The presence of many invasive non-indigenous rodents, such as the black rat (*Rattus rattus*), the Norwegian rat (*Rattus norvegicus*), and the house mouse (*Mus musculus*), is associated with native invertebrates, birds, bats, etc. In addition, rodent species in some parts of the world inhabit continents and islands through a contest for food, survival of the fittest, disease as well as transmission. Similarly, some invading and wild species such as goats (*Capra hircus*), donkeys (*Equus asinus*), and pigs (*Sus scrofa*) compete for food and structure and composition of plant communities by browsing and rooting. Affected native species by changing (Campbell and Donlan 2005; Means and Travis 2007). The devastation, caused by this action, is widespread throughout the ecosystem, in particular, is the decline in the population of indigenous mammals, reptiles, and birds that has led to a reduction of biodiversity, loss of habitat, extinction of certain indigenous species, and increased soil erosion across Africa and beyond (Van Wilgen and Wilson 2018; Campbell and Donlan 2005; Means and Travis 2007).

16.9 Toward Biodiversity Sustainability in Africa

Sustainability rather than sustainable development could be described as the development that satisfies the desires of the present era without trading off the potential available for the coming generation to achieve theirs. Experts in all fields of human endeavor have seen sustainability as a multidisciplinary concept; hence, it has been globalized to incorporate different aspects including broadly the environmental factors and the socio-economic factors. Sustainability is an important issue in biodiversity at either gene, community, ecosystem, or local, regional, and/or global levels. Biodiversity is a principal component of the earth's natural resources. Therefore, mankind needs a full understanding of the processes, mechanisms, and practicability of sustainable development in the biodiversity dimension. Furthermore, the goals of biodiversity can only be fully met by a total adherence to the principles of sustainability. Humans have waited too long; now, it is time to leave no stone un-turn in conserving our earth—the gift from nature. We must not forget that preventing extinction is far better than restoration, because when there is extinction, there are no remedies.

16.10 Conclusions

The impacts of a changing climate and other environmental factors (such as topography, soil, pests, diseases, and nonindigenous species) on biodiversity and distribution of plants and animals in the world, especially in Africa, can never be overemphasized. Africa is highly vulnerable to biological attack owing to its location, economic status, sensitive agricultural ecosystem, defective border security, and low institutional competence. Thus, the growing acceptance and expansion of invasive species and extremely harmful pests, diseases, insects, and pathogens as well as invasive species are compounding financial challenges, food insecurity, and increasing biodiversity loss in Africa.

Although Africa is a highly diversified continent in land mass and species which have given the chances for numerous pests, diseases, and invasions, these should also ignite the need for us to curtail the related impacts to restore and maintain the spectacular rich biodiversity that our continent is known for. We are optimistic that this work will serve as a good basis for enhanced and improved management of biological pests, diseases, and invasions in Africa.

References

Adler PB, Leiker J, Levine JM (2009) Direct and indirect effects of climate change on a Prairie plant community. PLoS One 4(9):e6887

Aleman JC, Jarzyna MA, Staver AC (2018) Forest extent and deforestation in tropical Africa since 1900. Nat Ecol Evol 2:26–33

Anderson AR, Collinge JE, Hoffmann AA, Kellett M, McKechnie SW (2003) Thermal tolerance trade-offs associated with the right arm of chromosome 3 and marked by the HSR-omega gene in *Drosophila melanogaster*. Heredity 90:195–202

Antonelli A, Kissling WD, Flantua SGA, Bermúdez MA, Mulch A, Muellner-Riehl AN et al (2018) Geological and climatic influences on mountain biodiversity. Nat Geosci 11:718–725

Aukema JE, McCullough DG, Von Holle B, Liebhold AM, Britton K, Frankel SJ (2010) Historic accumulation of nonindigenous forests pests in the continental United States. BioScience 60(11):886–897

Badgley C, Smiley TM, Terry R, Davis EB, DeSantis LRG, Fox DL et al (2017) Biodiversity and topographic complexity: modern and geo historical perspectives. Trend Ecol Evolut 32:211–226

Baldeck CA, Harms KE, Yavitt JB, John R, Turner BL, Valencia R et al (2012) Soil resources and topography shape local tree community structure in tropical forests. Proc R Soc Lond B Biol Sci 280:2012–2532

Bellard C, Bertelsmeier C, Leadley P, Thuiller W, Courchamp F (2012) Impacts of climate change on the future of biodiversity. Ecol Lett 15:365–377

Biggs R, Simons H, Bakkenes M, Scholes RJ, Eickhout B, van Vuuren D et al (2008) Scenarios of biodiversity loss in southern Africa in the 21st century. Glob Environ Change 18:296–309

Blowes SA, Supp SR, Antão LH, Bates A, Bruelheide H, Chase JM et al (2019) The geography of biodiversity change in marine and terrestrial assemblages. Science 366(6463):339–345

Bongard CL, Navaranjan G, Yan W, Fulthorpe RR (2013) Fungal colonization of the invasive vine Vincetoxicum rossicum and native plants. Plant Ecol Evol 146:45–52

Bouwman AF, Van Vuuren DP, Derwent RG, Posch MA (2002) Global analysis of acidification and eutrophication of terrestrial ecosystems. Water Air Soil Pollut 141:349–382

Bowler DE, Bjorkman AD, Dornelas M, Myers-Smith IH, Navarro LM, Niamir A et al (2020) Mapping human pressures on biodiversity across the planet uncovers anthropogenic threat complexes. People Nature 2(2):380–394

Boyd IL, Freer-Smith PH, Godfray HCJ, Gilligan CA (2013) The consequence of tree pests and diseases for ecosystem services. Science 342:823–831

Brito-Morales I, García Molinos J, Schoeman DS, Burrows MT, Poloczanska ES, Brown CJ et al (2018) Climate velocity can inform conservation in a warming world. Trend Ecol Evolut 33(6): 441–457

Brook BW, Sodhi NS, Bradshaw CJA (2008) Synergies among extinction drivers under global change. Trend Ecol Evol 23:453–460

Brown N, Vanguelova E, Parnell S, Broadmeadow S, Denman S (2018) Predisposition of forests to biotic disturbance: predicting the distribution of acute oak decline using environmental factors. Forest Ecol Manag 407:145–154

Burnham M, Ma Z (2015) Linking smallholder farmer climate change adaptation decisions to development. Climat Dev 8:1–10

Butchart SHM (2010) Global biodiversity: indicators of recent declines. Science 328:1164–1168

Butler R (2006) Rainforest diversity – origins and implications. Retrieved 3/3/2022 from http://rainforests.mongabay.com/0301.htm

Campbell K, Donlan CJ (2005) Feral goat eradications on Islands. Conserv Biol 19:1362–1374

Ceballos G (2015) Accelerated modern human–induced species losses: entering the sixth mass extinction. Sci Adv 1(5):e1400253

Chaudhary C, Richardson AJ, Schoeman DS, Costello MJ (2021) Global warming is causing a more pronounced dip in marine species richness around the equator. Proc Natl Acad Sci U S A 118(15):e2015094118

Chime AO, Osawaru ME, Ogwu MC, Ogunlowo EO (2015) Preservation of fresh tomato fruits (*Solanum lycopersicon* L.) using leaf extracts of bitter melon (*Momordica charantia* L., Cucurbitaceae). J Phytopathol Plant Health 3:29–37

Chime AO, Aiwansoba RO, Ogwu MC (2018) Pathological status of plant germplasm and sustainable crop production and conservation. J Ener Nat Resour Manag 1:17–21

Clusella-Trullas S, Blackburn TM, Chown SL (2011) Climatic predictors of temperature performance curve parameters in ectotherms imply complex responses to climate change. Am Nat 177:738–751

Collier P, Conway G, Venables T (2008) Climate change and Africa. Oxford Rev Econ Pol 24(2): 337–353

Coumou D, Rahmstorf S (2012) A decade of weather extremes. Nat Clim Change 2:1–6

Davies KW, Bates JD, Miller RF (2007) Environmental and vegetation relationships of the Artemisia tridentata spp. wyomingensis alliance. J Arid Environ 70(3):478–494

Davies SJ, Jordaan MS, Karsten M (2020) Experience and lessons from alien and invasive animal control projects in South Africa. In: van Wilgen BW, Measey J, Richardson DM, Wilson JR, Zengeya TA (eds) Biological invasions in South Africa. Springer, Berlin, pp 625–660

Dawson TP, Jackson ST, House JI, Prentice IC, Mace GM (2011) Beyond predictions: biodiversity conservation in a changing climate. Science 332:53–58

Day NJ, Dunfield KE, Antunes PM (2015) Temporal dynamics of plant-soil feedback and root-associated fungal communities over 100 years of invasion by a non-native plant. J Ecol 103: 1557–1569

Day NJ, Dunfield KE, Antunes PM (2016) Fungi from a non-native invasive plant increase its growth but have different growth effects on native plants. Biol Invasion 18:231–243

Defra (2021) UK plant health risk register. DEFRA

Diagi B, Nwagbara M (2017) Potential benefits of climate change for crop production in Ebonyi state, South-Eastern Nigeria. J Environ Earth Sci 7(3):33–37

Diaz S, Settele J, Brondízio E, Ngo HT, Agard J, Arneth A et al (2019) Pervasive human-driven decline of life on earth points to the need for transformative change. Science (New York, NY) 366(6471):eaax3100. https://doi.org/10.1126/science.aax3100

Doak DF, Morris WF (2010) Demographic compensation and tipping points in climate-induced range shifts. Nature 467:959–962

Doonan JM, Broberg M, Denman S, McDonald JE (2020) Host–microbiota–insect interactions drive emergent virulence in a complex tree disease. Proc R Soc B Biol Sci 287(1933):20200956

Dybzinski R, Fargione JE, Zak DR (2008) Soil fertility increases with plant species diversity in a long-term biodiversity experiment. Oecologia 158:85–93

Eisenlohr PV, Alves LF, Bernacci LC, Padgurschi MCG, Torres RB, Prata E et al (2013) Disturbances, elevation, topography and spatial proximity drive vegetation patterns along an altitudinal gradient of a top biodiversity hotspot. Biodivers Conserv 22(12):2767–2783

Elton CS (1958) The ecology of invasions by animals and plants. Methuen and Co., London. https://doi.org/10.1007/978-1-4899-7214-9

Enright NJ, Miller BP, Crawford A (1994) Environmental correlates of vegetation patterns and species richness in the northern Grampians, Victoria. Austral Ecol 19:159–168

Erickson JM (1971) The displacement of native ant species by the introduced Argentine ant Iridomyrmex humilis Mayr. Psyche 78(4):257–266

Ernst CM, Cappuccino N (2005) The effect of an invasive alien vine, Vincetoxicum rossicum (Asclepiadaceae), on arthropod populations in Ontario old fields. Biol Invasion 7:417–425

European Commission (2008) The European Union's biodiversity action plan: halting the loss of biodiversity by 2010 – and beyond. European Commission, Belgium

Féret JB, Asner GP (2014) Microtopographic controls on lowland Amazonian canopy diversity from imaging spectroscopy. Ecol Appl 24:1297–1310

Forest Services Division (1996) The United States Department of Agriculture (USDA) forest services division report, Washington, DC

Fortunel C, Lasky JR, Uriarte M, Valencia R, Wright SJ, Garwood NC et al (2018) Topography and neighbourhood crowding can interact to shape species growth and distribution in a diverse Amazonian forest. Ecology 99(10):2272–2283

Freer-Smith PH, Webber JF (2017) Tree pests and diseases: the threat to biodiversity and the delivery of ecosystem services. Biodivers Conserv 26:3167–3181

Fu B, Liu S, Ma K, Zhu Y (2004) Relationships between soil characteristics, topography and plant diversity in a heterogeneous deciduous broad-leaved forest near Beijing, China. Plant Soil 261(1):47–54

Garca-Aguirre MC, Ortiz MA, Zamorano JJ, Reyes Y (2007) Vegetation and landform relationships at Ajusco volcano Mexico, using a geographic information system (GIS). For Ecol Manag 239(1–3):1–12

Garcia RA, Burgess ND, Cabeza M, Rahbek C, Araújo MB (2012) Exploring consensus in 21st century projections of climatically suitable areas for African vertebrates. Glob Chang Biol 18: 1253–1269

Garcia RA, Araújo MB, Burgess ND, Foden WB, Gutsche A, Rahbek C, Cabeza M (2014) Matching species traits to projected threats and opportunities from climate change. J Biogeogr 41(4):724–735

Getahun YS, Shefine BG (2015) Analysis of climate variability (Enso) and vegetation dynamics in Gojjam, Ethiopia. J Earth Sci Climat Chang 6(10):320. https://doi.org/10.4172/2157-7617.1000320

Ghelardini L, Luchi N, Pecori F, Pepori AL, Danti R, Della Rocca G et al (2017) Ecology of invasive forest pathogens. Biol Invasion 19(11):3183–3200

Gibson DM, Vaughan RH, Milbrath LR (2015) Invasive swallow-worts: an allelopathic role for -(–)antofine remains unclear. J Chem Ecol 41:202–211

Graziosi I, Tembo M, Kuate J, Muchugi A (2020) Pests and diseases of trees in Africa: a growing continental emergency. Plant People Plan 2:14–28

GRI (Global Reporting Initiative) (2007) Biodiversity: a GRI reporting resource. Global Reporting Initiative, Amsterdam

Grzyl A, Kiedrzyński M, Zielińska KM, Rewicz A (2014) The relationship between climatic conditions and generative reproduction of a lowland population of *Pulsatilla vernalis*: the last breath of a relict plant or a fluctuating cycle of regeneration? Plant Ecol 215(4):457–466

Guo D, Desmet PG, Powrie LW (2017) Impact of the future changing climate on the southern Africa biomes, and the importance of geology. J Geosci Environ Protect 5:1–9

Guy-Haim T, Lyons DA, Kotta J (2018) Diverse effects of invasive ecosystem engineers on marine biodiversity and ecosystem functions: a global review and meta-analysis. Glob Chang Biol 24: 906–924

Gyasi EA, Agyepong GT, Ardayfio-Schandorf E, Enu-kwesi L, Nabila JS, Owusu-Bennoah E (1995) Production pressure and environmental change in the forest-savanna zone of southern Ghana. Glob Environ Chang 5(4):355–366

Haines IH, Haines JB (1978) Pest status of the crazy ant, Anoplolepis longipes (Jerdon) (Hymenoptera: Formicidae), in the Seychelles. Bullet Entomol Res 68:627–638

Hall JB, Swaine MD (1981) Distribution and ecology of vascular plants in a tropical rain forests. Forest vegetation in Ghana. Geobotany l. Dr.W, Junk Hague

Hansen J, Sato M, Ruedy R (2012) Perception of climate change. Proc Natl Acad Sci U S A 109: E2415–E2423

Haskins C, Haskins E (1965) *Pheidole megacephala* and *Iridomyrmex humilis* in Bermuda, equilibrium or slow replacement? Ecology 46:736–740

Hawthorne WD, Abu-Juan M (1995) Forest protection in Ghana. IUCN Gland, Switzerland/ Cambridge

Hector A, Bagchi R (2007) Biodiversity and ecosystem multi functionality. Nature 448:188–190

Hély CL, Bremond-Alleaume S, Smith B, Sykes TM, Guiot J (2006) Sensitivity of African biomes to changes in the precipitation regime. Glob Ecol Biogeogr 15:258–270

Herms DA, McCullough DG (2014) Emerald ash borer invasion of North America: history, biology, ecology, impacts, and management. Annu Rev Entomol 59:13–30

Hoffmann AA, Chown SL, Clusella-Trullas S (2013) Upper thermal limits in terrestrial ectotherms: how constrained are they? Funct Ecol 27:934–949

Hulme M, Doherty RM, Ngara T, New MG, Lister D (2001) African climate change: 1900-2100. Clim Res 17:145–168

Ikhajiagbe B, Ogwu MC, Lawrence AE (2020) Single-tree influence of *Tectonia grandis* Linn. f. on plant distribution and soil characteristics in a planted forest. Bull Natl Res Centre 44:29. https://doi.org/10.1186/s42269-020-00285-0

Ikhajiagbe B, Chukwujama CA, Ikhajiagbe B (2022a) Intrinsic restoration of a ferruginous utisol using goosegrass (*Eleusine indica* L. [Gaertn.]) obtained from different land use areas. J Nat Resour Dev 12:1–18. https://doi.org/10.18716/ojs/jnrd2022.12.01

Ikhajiagbe B, Aituae W, Ogwu MC (2022b) Morpho-physiological assessment of oil palm (*Elaeis guineensis* Jacq.) seedlings exposed to simulated drought conditions. J Oil Palm Res 34(1): 26–34. https://doi.org/10.21894/jopr.202100018

IPBES (2019) In: Brondizio E, Settele J, Díaz S, Ngo H (eds) Global assessment report on biodiversity and ecosystem services of the intergovernmental science-policy platform on biodiversity and ecosystem services. IPBES Secretariat, p 1148

IPCC (2001) In: McCarthy JJ, Canziani OF, Leary NA, Dokken DJ, White KS (eds) Climate change 2001: impacts, adaptation and vulnerability. IPCC working group II, third assessment report. Cambridge University Press, Cambridge

IPCC (2013) Climate change 2013: the physical science basis. Contribution of working group I to the fifth assessment report of the intergovernmental panel on climate. Change Cambridge University Press, Cambridge/New York, NY

IUCN (2018) The IUCN red list of threatened species. Version 2018-2. http://www.iucnredlist.org

Janssens F, Peeters A, Tallowin JRB (1998) Relationship between soil chemical factors and grassland diversity. Plant Soil 202(1):69–78

Jiao J, Zou H, Jia Y, Wang N (2009) Research progress on the effects of soil erosion on vegetation. Acta Ecol Sinica 29(2):85–91

Jonsson MT, Thor G (2012) Estimating coextinction risks from epidemic tree death: affiliate lichen communities among diseased host tree populations of Fraxinus excelsior. PLoS One 7(9): e45701

Jucker T, Bongalov B, Burslem DFRP, Nilus R, Dalponte M, Lewis SL et al (2018) Topography shapes the structure, composition and function of tropical forest landscapes. Ecol Lett 21:989–1000

Karssing RJ, Rivers-Moore NA, Slater K (2012) Influence of waterfalls on patterns of association between trout and Natal cascade frog Hadromophryne natalensis tadpoles in two headwater streams in the Ukhahlamba Drakensberg Park World Heritage Site, South Africa. Afr J Aquat Sci 37:107–112

Kerfahi D, Tripathi BM, Slick JWF, Sukri RS, Jaafar S, Dong K, Ogwu MC, Kim H-K, Adams JM (2019) Soil metagenome of tropical white sand heath forests in Borneo: what functional traits are associated with an extreme environment within the tropical rainforest? Pedosphere 29(1): 12–23. https://doi.org/10.1016/S1002-0160(18)60054-2

Kirkpatrick JB, Green K, Bridle KL, Venn SE (2014) Patterns of variation in Australian alpine soils and their relationships to parent material, vegetation formation, climate and topography. Catena 121:186–194

Kjær ED, McKinney LV, Nielsen LR, Hansen LN, Hansen JK (2012) Adaptive potential of ash (Fraxinus excelsior) populations against the novel emerging pathogen Hymenoscyphus pseudoalbidus. Evolut Appl 5:219–228

Korcha D, Sorteberg A (2013) Validation of operational seasonal rainfall forecast in Ethiopia. Water Resource Res 49:7681–7697

Laamrani A, Valeria O, Bergeron Y, Fenton N, Cheng LZ, Anyomi K (2014) Effects of topography and thickness of organic layer on productivity of black spruce boreal forests of the Canadian Clay Belt region. For Ecol Manag 330:144–157
Lambin EF, Meyfroidt P (2011) Global land use change, economic globalization, and the looming land scarcity. Proc Nat Acad Sci 108:3465–3472
Leadley P, Pereira HM, Alkemade R, Fernandez-Manjarres JF, Proenca V, Scharlemann JPW et al (2010) Biodiversity scenarios: projections of 21st century change in biodiversity and associated ecosystem services. In: Secretariat of the convention on biological diversity (ed. Diversity SotCoB), Technical Series no. 50. Secretariat of the Convention on Biological Diversity, Montreal, pp 1–132
Lenoir J, Bertrand R, Comte L, Bourgeaud L, Hattab T, Murienne J et al (2020) Species better track climate warming in the oceans than on land. Natur Ecol Evol 4(8):1044–1059
Lin Y, Cui P, Ge Y, Chen C, Wang D (2014) The succession characteristic of soil erosion during different vegetation succession stages in dry-hot river valley of Jinsha River upper reaches of Yangtze River. Ecol Eng 62:13–26
Linder HP (2001) Plant diversity and endemism in sub-Saharan tropical Africa. J Biogeogr 28:169–182
Lonsdale D (2015) Review of oak mildew, with particular reference to mature and veteran trees in Britain. Arboricul J 37:61–84
Lovett JC, Midgely GF, Barnard PB (2005) Climate change and ecology in Africa. Afr J Ecol 43: 279–281
Lubek A, Kukwa M, Czortek P, Jaroszewicz B (2020) Impact of Fraxinus excelsior dieback on biota of ash-associated lichen epiphytes at the landscape and community level. Biodivers Conserv 29:431–450
MacIvor JS, Roberto AN, Sodhi DS, Onuferko TM, Cadotte MW (2017) Honey bees are the dominant diurnal pollinator of native milkweed in a large urban park. Ecol Evol 7:8456–8462
Marr SM, Ellender BR, Woodford DJ (2017) Evaluating invasion risk for freshwater fishes in South Africa. Bothalia 47:a2177
Mascaro J, Asner GP, Davies S, Dehgan A, Saatchi S (2014) These are the days of lasers in the jungle. Carbon Bal Manag 9:7. https://doi.org/10.1186/s13021-014-00070
Matata AC, Adan A (2018) Causes of climate change and its impact in the multi-sectoral areas in Africa-need for enhanced adaptation policies. Curr J App Sci Technol 27:1–10
Matsuura T, Suzuki W (2013) Analysis of topography and vegetation distribution using a digital elevation model: case study of a snowy mountain basin in north eastern Japan. Landscap Ecol Eng 9(1):143–155
McClean CJ, Lovett JC, Küper W, Hannah L, Sommer JH, Barthlott W et al (2005) African plant diversity and climate change. Ann Mo Bot Gard 92:139–152
McKechnie AE, Wolf BO (2010) Climate change increases the likelihood of catastrophic avian mortality events during extreme heat waves. Biol Lett 6:253–256
MEA (Millennium Ecosystem Assessment) (2005) Ecosystems and human Well-being: synthesis. Island Press, Washington, D.C.. https://www.cifor.org/knowledge/publication/1888/
Means DB, Travis J (2007) Declines in ravine-inhabiting dusky salamanders of the southeastern US coastal plain. Naturalist 6:83–96
Measey J, Hui C, Somers M (2020a) Terrestrial vertebrate invasions in South Africa. In: van Wilgen BW, Measey J, Richardson DM, Wilson JR, Zengeya TA (eds) Biological invasions in South Africa. Springer, Berlin, pp 113–150
Measey J, Robinson TB, Kruger N (2020b) South Africa as a donor of alien animals. In: van Wilgen BW, Measey J, Richardson DM, Wilson JR, Zengeya TA (eds) Biological invasions in South Africa. Springer, Berlin, pp 783–826
Meehl GA (2007) Global climate projections. In: Solomon S (ed) Climate change 2007: the physical science basis. Contribution of working group to the fourth assessment report of the intergovernmental panel on climate change. Cambridge University Press, Cambridge, UK, pp 747–845

Midgley G, Bond WJ (2015) Bond future of African terrestrial biodiversity and ecosystems under anthropogenic climate change. Nature Climate

Mitchell RJ, Bellamy PE, Broome A, Ellis CJ, Hewison RL, Iason GR et al (2022) Cumulative impact assessments of multiple host species loss from plant diseases show disproportionate reductions in associated biodiversity. J Ecol 110:221–231

Moeslund JE, Arge L, Bøcher PK, Dalgaard T, Odgaard MV, Nygaard B et al (2013) Topographically controlled soil moisture is the primary driver of local vegetation patterns across a lowland region. Ecospher 4(7):1–26

Mooney H, Larigauderie A, Cesario M, Elmquist T, Hoegh-guldberg O, Lavorel S et al (2009) Biodiversity, climate change, and ecosystem services. Curr Opinion Environ Sustain 1:46–54

Nadeau MB, Sullivan TP (2015) Relationships between plant biodiversity and soil fertility in a mature tropical forest. Costa Rica Int J For Res 2015:732946

Neri AV, Schaefer CEGR, Silva AF (2012) The influence of soils on the floristic composition and community structure of an area of Brazilian cerrado vegetation. Edinb J Bot 69:1–27

Niang I, Ruppel OC, Abdrabo MA, Dube P, Leary N, Schulte-Uebbing L, et al (2014). https://www.ipcc.ch/pdf/assessment-report/ar5/wg2/

Nwaogu C, Okweche SI (2019) Soil conservation, nutrient enrichment, and Pest control using allium cepa and allium sativa in a Long-term white-curl grub (*Phyllophaga Spp*) larva infested cassava farmland. Int J Rec Advance Multidisci Res 6:4728–4732

Nwaogu C, Okeke JO, Fashae O, Nwankwoala H (2018) Soil organic carbon and total nitrogen stocks as affected by different land use in an Ultisol in Imo watershed, southern Nigeria. Chem Ecol 34(9):854–870

Nwaogu C, Pechanec V, Vozenilek V (2019) Responses of soil and plants to spatio-temporal changes in landscape under different land use in Imo watershed, southern Nigeria. Arch Agron Soil Sci 65(10):1460–1476

Ogwu MC (2019a) Towards sustainable development in Africa: the challenge of urbanization and climate change adaptation. In: Cobbinah PB, Addaney M (eds) The geography of climate change adaptation in urban Africa. Springer Nature, Switzerland, pp 29–55. https://doi.org/10.1007/978-3-030-04873-0_2

Ogwu MC (2019b) Ecologic and economic significance of bryophyte. In: Rathoure AK, Chauhan PB (eds) Current state and future impacts of climate change on biodiversity. IGI Global, Pennsylvania, pp 54–78pp. https://doi.org/10.4018/978-1-7998-1226-5.ch004

Ogwu MC, Osawaru ME (2015) Soil characteristics, microbial composition of plot, leaf count and sprout studies of cocoyam (*Colocasia* [Schott] and *Xanthosoma* [Schott], Araceae) collected in Edo State, Southern Nigeria. Sci Technol Arts Res J 4(1):34–44. https://doi.org/10.4314/star.v4i1.5

Ogwu MC, Osawaru ME, Ahana CM (2014a) Challenges in conserving and utilizing plant genetic resources (PGR). Int J Genet Mol Biol 6(2):16–22. https://doi.org/10.5897/IJGMB2013.0083

Ogwu MC, Osawaru ME, Chime AO (2014b) Comparative assessment of plant diversity and utilization patterns of tropical home gardens in Edo state, Nigeria. Sci Afr 13(2):146–162

Ogwu MC, Takahashi K, Dong K, Song H-S, Moroenyane I, Waldman B, Adams JM (2019) Fungal elevational Rapoport pattern from a high mountain in Japan. Sci Rep 9:6570. https://doi.org/10.1038/s41598-019-43025-9

Ohwawa T, Saito Y, Sawada H, Ide Y (2008) Impact of altitude and topography on the genetic diversity of Quercus serrata populations in the Chichibu Mountains, central Japan. Flora 203:187–196

Oliveira-Filho AT, Ratter JA (2002) Vegetation physiognomies and woody flora of the cerrado biome. In: Oliveira PS, Marquis RJ (eds) The Cerrados of Brazil: ecology and natural history of a neotropical Aavanna. Columbia University Press, New York, NY, pp 91–120

Omoigui DO, Osawaru ME, Aiwansoba RO, Ogwu MC (2016) Morphological and phytodermological evaluation of Oka-Uselu (Maize- *Zea mays* L.). Appl Trop Agric 21(3):96–101

Orlova-Bienkowskaja MJ, Drogvalenko AN, Zabaluev IA, Sazhnev AS, Peregudova EY, Mazurov SG et al (2020) Current range of Agrilus planipennis Fairmaire, an alien pest of ash trees, in European Russia and Ukraine. Ann Sci 77:14

Osawaru ME, Ogwu MC (2014) Conservation and utilization of plant genetic resources. In: Omokhafe K, Odewale J (eds) Proceedings of 38th Annual Conference of The Genetics Society of Nigeria. Empress Prints Nigeria Limited, pp 105–119

Osawaru ME, Dania-Ogbe FM, Chime AO, Ogwu MC (2011) Epidermal morphology of West African Okra [*Abelmoschus caillei* (A. Chev.) Stevels] from South Western Nigeria. Sci World J 6(3):15–22

Osawaru ME, Ogwu MC, Imarhiagbe O (2013) Agro-morphological characterization of some Nigerian *Corchorus* (L.) species. Biol Environ Sci J Trop 10(4):148–158

Osawaru ME, Ogwu MC, Emokpare AA (2014) Preliminary assessment of the microanatomy of okra [*Abelmoschus* (L.)] wood. Egypt Acad J Biol Sci 5(1):39–54

Osawaru ME, Aiwansoba RO, Ogwu MC (2015a) Comparative micro-anatomical studies of the wood of two species of Okra [*Abelmoschus* species]. UNILAG J Med Sci Technol 3(2):1–13

Osawaru ME, Ogwu MC, Aiwansoba RO (2015b) Hierarchical approaches to the analysis of genetic diversity in plants: a systematic overview. Univ Maurit Res J 21:1–36

Osawaru ME, Ogwu MC, Aiwansoba RO (2016) Wood properties of *Abelmoschus caillei* [A. Chev.] Stevels. J Raw Mater Res 10(1):23–32

Pabst H, Kühnel A, Kuzyakov Y (2013) Effect of land-use and elevation on microbial biomass and water extractable carbon in soils of Mt.Kilimanjaro ecosystems. App Soil Ecol 67:10–19

Parmesan C (2006) Ecological and evolutionary responses to recent climate change. Ann Rev Ecol Evol Syst 37:637–699

Parmesan C, Yohe G (2003) A globally coherent fingerprint of climate change impacts across natural systems. Nature 421:37–42

Partel M, Laanisto L, Zobel M (2007) Contrasting plant productivity-diversity relationships across latitude: the role of evolutionary history. Ecology 88:1091–1097

Pecl GT, Araújo MB, Bell JD, Blanchard J, Bonebrake TC, Chen IC et al (2017) Biodiversity redistribution under climate change: impacts on ecosystems and human wellbeing. Science 355(6332). https://doi.org/10.1126/science.aai9214

Perissinotto R, Miranda NAF, Raw JL, Peer N (2014) Biodiversity census of Lake St Lucia, iSimangaliso Wetland Park (South Africa): gastropod molluscs. ZooKeys 440:1–43

Perrings C, Lovett J (1999) Policies for biodiversity conservation: the case of sub-Saharan Africa. Int Aff 75:281–305

Pimm SL, Jenkins CN, Abell R, Brooks TM, Gittleman JL, Joppa LN et al (2014) The biodiversity of species and their rates of extinction, distribution, and protection. Science 344:1246752. https://doi.org/10.1126/science.1246752

Pinsky M, Eikeset AM, McCauley DJ, Payne JL, Sunday JM (2019) Greater vulnerability to warming of marine versus terrestrial ectotherms. Nature 569(7754):108–111. https://doi.org/10.1038/s41586-019-1132-4

Poulsen AD, Tuomisto H, Balslev H (2006) Edaphic and floristic variation within a 1-ha plot of lowland Amazonian rain forest. Biotropica 38:468–478

Ridolfi L, Laio F, D'Odorico P (2008) Fertility island formation and evolution in dryland ecosystems. Ecol Soc 13(5):439–461

Rinawati F, Stein K, Lindner A (2013) Climate change impacts on biodiversity– the setting of a lingering global crisis. Diversity 5:114–123

Sala OE, Chapin FS, Armesto JJ, Berlow E, Bloomfield J, Dirzo R et al (2000) Biodiversity—global biodiversity scenarios for the year 2100. Science 287:1770–1774

Secretary of the Convention on Biological Diversity (SCBD) (2010) Global biodiversity outlook 3; secretary of the convention on biological diversity, Montréal, QC, Canada

Seebens H, Blackburn TM, Dyer EE, Genovesi P, Hulme PE, Jeschke JM et al (2017) No saturation in the accumulation of alien species worldwide. Nat Commun 8:14435. https://doi.org/10.1038/ncomms14435

Shackleton RT, Novoa A, Shackleton CM (2020) The social dimensions of biological invasions in South Africa. In: van Wilgen BW, Measey J, Richardson DM, Wilson JR, Zengeya TA (eds) Biological invasions in South Africa. Springer, Berlin, pp 697–726. https://doi.org/10.1007/978-3-030-32394-3_24

Siegfried WR (1989) Preservation of species in southern African nature reserves. In: Biotic diversity in southern Africa: concepts and conservation (Huntley, B.J. ed.). Oxford University Press Cape Town

Singh RP, Singh PK, Rutowski J, Hodson DP, He X, Jørgensen LN et al (2016) Disease impact on wheat yield potential and prospects of genetic control. Annu Rev Phytopathol 54:303–322

Song H-K, Singh D, Tomlinson K, Yang X, Ogwu MC, Slik JWF, Adams JM (2019) Tropical forest conversion to rubber plantation in Southwest China results in lower fungal beta diversity and reduced network complexity. FEMS Microbiol Ecol 95(7):fiz092. https://doi.org/10.1093/femsec/fiz092

Sonwa DJ, Dieye A, Mzouri EH, Majule A, Mugabe FT, Omolo N et al (2017) Drivers of climate risk in African agriculture. Clim Dev 9:383–398

Steinbauer MJ, Grytnes JA, Jurasinski G, Kulonen A, Lenoir J, Pauli H et al (2018) Accelerated increase in plant species richness on mountain summits is linked to warming. Nature 556:231–234

Stévart T, Dauby G, Lowry PP, Blach-Overgaard A, Droissart V, Harris DJ et al (2019) A third of the tropical African flora is potentially threatened with extinction. Sci Adv 5(11):eaax9444

Stocker TF, Qin D, Plattner GK, Tignor M, Allen SK, Boschung J et al (2013) Technical summary. In climate change. The physical science basis. Contribution of working group I to the fifth assessment report of the intergovernmental panel on climate change. Cambridge University Press, Cambridge/New York, NY

Thomas CD, Cameron A, Green RE, Bakkenes B, Beaumont LJ, Collingham YC et al (2004) Extinction risk from climate change. Nature 427:145–148

Tilman D, Wedin D, Knops J (1999) Productivity and sustainability influenced by biodiversity in grassland ecosystems. Nature 379(6567):718–720

Tomlinson I, Potter C, Bayliss H (2015) Managing tree pests and diseases in urban settings: the case of oak processionary moth in London, 2006–2012. Urban For Urban Green 14:286–292

Turner BL, Clark WC, Kates RW, Richard JF et al (1990) The earth as transformed by human action. Cambridge University Press, Cambridge

UNEP (United Nations Environment Programme) (2002) State of the environment and policy retrospective: 1972–2002. Earthscan, London

UNEP-WCMC (2016) The state of biodiversity in Africa: a mid-term review of progress towards the Aichi biodiversity targets. Cambridge, UNEP-WCMC; ISBN 978-92-807-3508-6

van Wilgen BW, Wilson JR (2018) The status of biological invasions and their management in South Africa 2017. South African National Biodiversity Institute, Kirstenbosch and DST-NRF Centre of Excellence for Invasion Biology, Stellenbosch

Vitousek PM (1990) Biological invasions and ecosystem processes: towards an integration of population biology and ecosystem studies. Oikos 57:7–13

Waide RB, Willig MR, Steiner CF (1999) The relationship between productivity and species richness. Ann Rev Ecol Syst 30:257–300

Wanga B, Zhang G, Duan J (2015) Relationship between topography and the distribution of understory vegetation in a Pinus massoniana forest in Southern China. Int Soil Water Conserv Res 3:291–304

Wangai P, Burkhard B, Muller F (2016) A review of studies on ecosystem services in Africa. Int J Sustain Built Environ 5:225–245

Watson GW, Voegtlin DJ, Murphy ST, Foottit RG (1999) Biogeography of the Cinara cupressi complex (Hemiptera: Aphididae) on Cupressaceae, with description of a pest species introduced into Africa. Bullet Entomol Res 89(3):271–283

Watson JE, Shanahan DF, Di Marco M, Allan J, Laurance WF, Sanderson EW et al (2016) Catastrophic declines in wilderness areas undermine global environment targets. Curr Biol 26(21):2929–2934

Wen J, Ickert-Bond SM, Appelhans MS, Dorr LJ, Funk VA (2015) Collections-based systematics: opportunities and outlook for 2050. J Syst Evol 53:477–488

World Bank (2011) Word development report 2011: conflict, security, and development. World Bank, Washington, DC

Wuver AM, Attuquayefio DK (2006) The impact of human activities on biodiversity conservation in a coastal wetland in Ghana. West Afri J Appl Ecol (WAJAE) 9

Xu X, Ma K, Fu B, Song C, Wen L (2008) Relationships between vegetation and soil and topography in a dry warm river valley, SW China. Catena 75(2):138–145

Yasuhara M, Wei CL, Kucera M, Costello MJ, Tittensor DP, Kiessling W (2020) Past and future decline of tropical pelagic biodiversity. Proc Nat Acad Sci U S A 117(23):12891. https://doi.org/10.1073/pnas.1916923117

Yu X, Shi S, Long R, Wang F, Chen B (2006) Research progress on effects of ecological factors on seed germination. Pratacult Sci 23(10):44–49. (in Chinese)

Zengeya T, Ivey P, Woodford DJ (2017) Managing conflict-generating invasive species in South Africa: challenges and trade-offs. Bothalia 47(2):a2160

Zengeya TA, Kumschick S, Weyl OLF (2020) An evaluation of the impacts of alien species on biodiversity in South Africa using different assessment methods. In: van Wilgen BW, Measey J, Richardson DM, Wilson JR, Zengeya TA (eds) Biological invasions in South Africa. Springer, Berlin, pp 487–512. https://doi.org/10.1007/978-3-030-32394-3_17

Zhong L, Wang J (2017) Evaluation on effect of land consolidation on habitat quality based on InVEST model. Trans Chin Soc Agric Eng 33:250–255

Chapter 17
Contamination of African Water Resources: Impacts on Biodiversity and Strategies for Conservation and Restoration

Baturh Yarkwan

Abstract The interplay of factors militating against water resources has created complicated situations with far-reaching ecological consequences. Increase in human population around the world, especially in Africa, has increased pressure on water resources on one hand. On the other hand, the impacts of climate change and attendant consequences of desertification in the Sahel savanna have increased demand for water resources. Lake Chad, sandwiched between Nigeria, Niger, Chad, and Cameroun, is fast diminishing, with a reported loss of at least 85% of its water content between 1963 and 2007. In South Africa, increasing demand for water; occasionally aggravated by drought has made this resource to be dubbed "liquid gold." This has led to increasing government expenditures to ensure adequate water supply to all populaces. In Algeria, very high temperatures combined with low precipitation creates a perennial water crisis, with several attendant consequences. Decrease in water supply coupled with growing population, industrial activities, increases demand on the available water bodies creating a greater tendency to further contaminate existing water sources. From the west to east, north to south, Africa is plagued with acute water crisis and continual contamination of existing water resources. This chapter reviews the water resources of Africa, in the context of its growing human population and attendant demands within households, as well as growing agricultural, and industrial activities. Water consumption is often accompanied by the discharge of wastewater, unfortunately, many urban communities in Africa lack municipal wastewater treatment plants, and hence, over 80% of wastewater is discharged into receiving water bodies (rivers, streams, and lakes) without treatment, leading to contamination of receiving water bodies. Sources of contaminants and impacts on biota have been reviewed in the light of published reports. The development of bacterial resistant strains and associated economic consequences has been pointed out. The physiological modulatory effects of organic contaminants on aquatic organisms have been highlighted as the initiation point toward biodiversity

B. Yarkwan (✉)
Department of Biochemistry, Joseph Sarwuan Tarkaa University, Makurdi, Nigeria

School of the Environment, University of Windsor, Windsor, ON, Canada

S. C. Izah, M. C. Ogwu (eds.), *Sustainable Utilization and Conservation of Africa's Biological Resources and Environment*, Sustainable Development and Biodiversity 32, https://doi.org/10.1007/978-981-19-6974-4_17

loss. This chapter concludes by proffering suggestions toward remedying already polluted waters as well as preventing further pollution of water resources to ensure both conservation of aquatic bodies and biodiversity for posterity.

Keywords Water resources · Water sustainability · Pollution · Emerging contaminants · Climate change

17.1 Introduction

Water pollution is a global problem confronting both developed and developing nations, threatening economic growth, challenging the environmental, and physical health of billions of people (FAO 2017). The central discussion on water around the world has often focused on the quantity of water for the teeming human population, such that impacts of climate change on the worlds' water resources have often been tabled in the context of quantity rather than quality. However, at least 80% of global municipal wastewater is released into receiving water bodies untreated; industries release tonnes of solvents, heavy metals, harmful sludge, and wastes into water bodies annually (WWAP 2017). On the other hand, improvements in agricultural practices are important for stemming the scourge of hunger confronting low-income countries. However, discharges containing agrochemicals, drug residues, organic matter, saline, and sediments are often emptied into water bodies, thereby contaminating surface water. Resultant pollution of water bodies constitutes great threats to aquatic organisms and soil/aquatic microflora, and facilitates biodiversity loss (Adeogun et al. 2019; Brauko et al. 2020; Chonova et al. 2017; Preena et al. 2020).

The population of Africa has increased about six times, from 227,794 million in 1950 to 1,340,598 billion in 2020 (UN 2021), while water levels in lakes and some rivers appear to be declining (Yunana et al. 2017; Beverly et al. 2020; Plisnier et al. 2018) owing to an interplay of climate change promoting factors, thereby creating water crisis and increased pollution of the available freshwater resources. Sub-Saharan Africa is greatly water stressed (Ngoran et al. 2015), exemplified by the near extinction state of Lake Chad (Eriegha et al. 2019) and desert encroachment. According to Vorster (2014), a water stress depicts a situation, where a population or country is using more than 20% of her renewable water resources.

Global water consumption per sector of economic activity has been on an increase since 1950, indicating a positive correlation between water consumption and population growth (Table 17.1). Agricultural sector remains the highest water consuming economic sector.

Several environmental factors are responsible for promoting global change (Sánchez-Bayo and Wyckhuys 2019). Among these factors, increased production and diverse applications of synthetic chemicals are the most effective, especially among low-income countries (Bernhardt et al. 2017), wherein all but few African nations belong. Synthetic chemicals (biocides) even when applied on cultivated fields, through water runoffs, they enter aquatic bodies, thereby polluting the water bodies. Domestic sewers contain these chemicals which are released into the

Table 17.1 Assessed and forecasted global water use (km^3/year) by sector of economic activity

Years parameter	1950	1995	2025[a]
Population (million)	2542	5735	7877
Agricultural (% sector)	722 (95.276)	1753 (92.978)	2252 (90.257)
Municipal (% sector)	16.7 (2.204)	49.8 (2.641)	74.1 (2.970)
Industrial (% sector)	19.1 (2.520)	82.6 (4.381)	169 (6.773)
Total	757.8	1885.4	2495.1
User quotient (Total population/Total sectoral use)	3.355	3.042	3.157

[a]Forecasted values based on economic activities
Table modified from Shiklomanov (2000)

environment either directly (WWAP 2017) or through wastewater treatment plants (WWTPs) (Ghernaout and Elboughdiri 2020; Galib et al. 2018).

The increasing environmental presence of diverse chemicals is a reason for major concern. From a systematic review of literature, the concern for chemical pollutants rest with their perceived bioactivity which could be as β-adrenergic receptors agonist blockers, endocrine disruptors, or can elicit oxidative stress, genetic alterations in DNA and epigenetic reprogramming via global DNA methylation, gene-targeted CpG methylation, and microRNA expression (Egbuna et al. 2021; Ojemaye and Petrik 2019). These physiological modulatory events may set the stage for reduced ability to cope in a highly competitive ecosystems leading to species extinction.

Of the regions of the world, it is predicted that Africa may face the direst consequences of climate change-induced hardships, ranging from hunger, immigration, ill health, and related social–economic consequences (Ndehedehe et al. 2016). Reduced crop yields have decreased farmers' income and aggravated food crisis and hunger for several farming communities on the continent (de Klerk 2014). Current geographical location-dependent extreme-opposite-ends scenario of water scarcity and flooding in some parts of the continent/nations and poor quality of domestic water (Ebele et al. 2020) owing to pollution of surface waters (Bessah et al. 2021) has dire social–economic and public health consequences. The solution lies in collaborative responses for effective interventions from all stakeholders (Beverly et al. 2020; Dudley et al. 2017; Yunana et al. 2017). This chapter is one such efforts to sound the trumpet call for urgent responses.

17.2 Water Resources in Africa: Past, Present, and Future

Water resources in Africa are largely stressed to a breaking point. The northern African states of Egypt, Morocco, Algeria, Tunisia, and Libya are susceptible to experiencing severe water scarcity under the prevailing climate change conditions with attendant severe social economic consequences (Schilling et al. 2020), including famine, emigration, and loss of livelihood. Over 36 years, increasing and decreasing trends in precipitation have been observed in two eco-regions of Ethiopia, suggesting a steady shift in yearly cycles of hydrologic regime (Gedefaw et al. 2019).

Lake Tanganyika, impacted by climate change, is under threats from anthropogenic activities culminating in shrinking water levels, contaminated water, evident in fish and biodiversity loss (Plisnier et al. 2018). The largest lake on African shores, Lake Victoria witnessed a drastic desiccation during the drought of 2010–2011 which affected at least ten million people in Kenya, Somalia, and Ethiopia (Beverly et al. 2020). Between 1984 and 2018, the lake lost at least 0.3% of its initial water levels showing obvious signs of drying up. Available data, future climate forecasts considering present precipitation levels, temperature surges, and rates of lake level fall, show that this source of White Nile could fall below outflow levels to the White Nile in the next 10 years and would recess beyond Kenyan shores within <400 years (Beverly et al. 2020). In Sub-Saharan Africa, dramatic water recession in Lake Chad shows water area has reduced from 25,000 km^2 in 1960s to barely 1350 km^2 in 2005 (Yunana et al. 2017) representing about 94% loss in water area. With the stressors still in place, the recession would continue.

Changes in precipitation and volume of water available will significantly impact on agricultural practices and crop yields, with dire consequences especially in areas, where local economic survival is dependent upon subsistent farming. This must have been one of the pillars de Klerk (2014) anchored his prediction of food scarcity in Africa. Chilunjika and Gumede (2021) submitted that Sub-Saharan Africa may be the global epicentre of food crisis, owing to threats arising from harsh conditions of induced extreme weather elements, injuries and deaths arising from flooding, water scarcity, and climate induced immigration. The decline in water levels in Lake Victoria between 2004 and 2007 exposed 34.5% Rwanjaba lakeshore community residents of Uganda to fish shortage, 12.7% of families broke down, 18.2% of responded residents engaged in sex for fish, while theft increased by 7.2% (Musoke and Boon 2012). Water scarcity would displace social–economic balances and heighten societal ills at unprecedented levels. The need to act by fighting the predisposing factors and building mitigatory strategies is critical to avert these challenges. The time to act is now!

17.3 Contamination of African Water Resources

Industrial processes have been at the centre of water pollution across Africa. From mining of metals, such as gold, silver, and copper in the gold-rich province of Guateng in South Africa (Minnaar 2020), to crude oil refining in the rich oil reserves region of Niger Delta, Nigeria (Babatunde 2020; Ejike et al. 2017; Elum et al. 2016) and the petrochemical industry in Algeria (Benlaribi and Djebbar 2020) have all contributed substantially to water pollution which has become a daily menace that Africans live with. The Nile, Egypt's life wire, is greatly threatened by the indiscriminate discharge of untreated industrial effluents into the river through Lake

Marriot, by at least 60% of more than 1250 industries situated along the Mediterranean coast (Abdel-Shafy and Aly 2002). Industrial activities have been fingered for driving water pollution in several African countries of Ethiopia (Gemeda et al. 2020; Lencha et al. 2021), Kenya (Kinuthia et al. 2020), Morocco (Cherif et al. 2019; Iavazzo et al. 2011), Namibia, Ghana (Duncan et al. 2020; Bessah et al. 2021) to copper smelting in DR of Congo (Shutcha et al. 2015). From across the four cardinals of the continent to the centre, no nation seems to be free of the scourge of water pollution and its attendant harmful effects on biodiversity. The illegal mining activities in African countries are worsening the situation (Duncan et al. 2020), since illegal miners usually focus on getting the product without an iota of concern on the health of the environment.

17.3.1 Sources of Water Contaminants

Pollutants identified in water bodies vary according to chemical properties, largely due to their source of origin. Notable classes of pollutants are pharmaceuticals and personal care products (PPCP), such as antibiotics, hormones, analgesics, lipid regulators, anti-epileptics, metabolites, betablockers, disinfectants, stimulants, fragrances, surfactants, antiretrovirals, personal care products, and preservatives which have been reported around the world (Segura et al. 2015; Hedgespeth et al. 2012; Miège et al. 2009; Tarpani and Azapagic 2018a, b; Biel-Maeso et al. 2019) and in Africa (Ojemaye and Petrik 2019; Egbuna et al. 2021; K'oreje et al. 2018; Sorensen et al. 2015). Other pollutants are heavy metals, (Ejike et al. 2017; Kinuthia et al. 2020), organic matter (FAO 2018; Sim et al. 2013), and more recently, microplastics (Egbuna et al. 2021) added to the list of pollutants because of emerging knowledge on its potential effects in the environment. Industrial effluents may contain an array of contaminants, which vary greatly depending on the raw materials and chemical substances used/produced within an industry (Kanu and Achi 2011) discharging the wastes. These contaminants can alter the physico-chemical parameters, such as carbon oxygen demand, biological oxygen demand (BOD), total suspended solids, turbidity, pH, conductivity, and total hardness of water. These parameters have the capacity of pushing to the extreme the survival tolerance of aquatic organisms by altering water chemistry and consequently distorting physiological processes in aquatic organisms. Castillo et al. (1999) surveyed effluents from tanneries, petrochemical plants, and textile industries and reported the presence of more than 50 compounds, including phenols, phthalates, esters, acids, and n-alkanes. Phenol is carcinogenic and a great source of threat to aquatic biota. Discharges from livestock farms contain organic matter, drug residues from administered pharmaceuticals, microbes arising from the litter, poultry droppings, and livestock excreta have considerable water polluting effects (Sim et al. 2013) and ultimately trigger public health concerns. Lake Victoria has its greatest BOD load on the Kenyan side, while domestic BOD exceeds industrial BOD, and nutrient input comes from atmospheric deposition and water run-off, together being responsible for 90% and

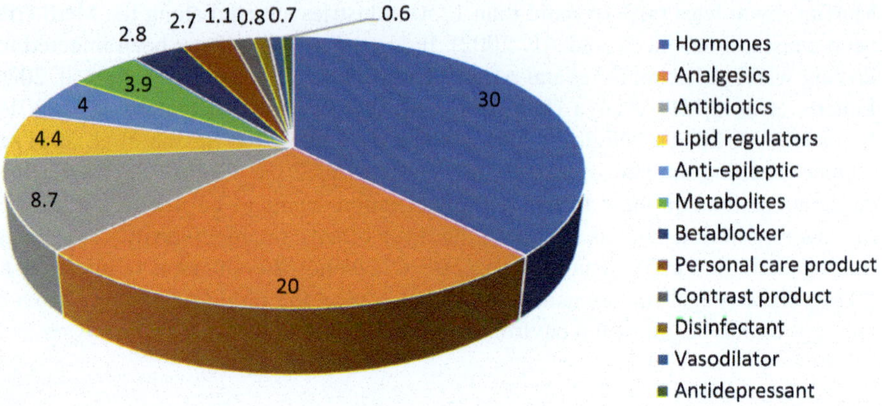

Fig. 17.1 Frequency (%) of most investigated groups of PPCPs in the environment. Modified from Miège et al. (2009)

94% of phosphorus and nitrogen, respectively input into the lake (Scheren et al. 2000). Chemical pollutants have also been detected in drinking water both from municipal treated water supply (Odendaal et al. 2015) as well as underground wells (Ejike et al. 2017). The direct contact of polluted water with diverse organisms beside human consumption raises ecological and public health concerns.

Human excreta are the major source of PPCPs release into the environment (Jones-Lepp and Stevens 2007), though discharges from hospitals (Kosma et al. 2010; Sim et al. 2013; Oliveira et al. 2015), pharmaceutical and related production companies (Tiwari et al. 2020; Kleywegt et al. 2019), and drug residues from livestock and aquaculture farms (Mo et al. 2015; Bustillo-Lecompte et al. 2016) also contribute substantially to the quantity of these pollutants in the environment. Contribution by chemical classes under PPCPs is summarized (Fig. 17.1). In each geographical location, the quantity and kinds of pollutants in a WWTP influent are determined by the population density and demography, lifestyle of the populace, disease burden of the inhabitants, seasonal changes, social events and local regulations, and the economic capability of the inhabitants (Adeleye et al. 2022; Segura et al. 2015; Golovko et al. 2014). Thiebault et al. (2019) reported a surge in analgesics and non-steroidal inflammatory agents during cold nights and a positive correlation between schools' activities and an increase in stimulants in WWTPs influents, characteristic of students' high intake of recreational drugs. This demonstrates the impact of seasonal weather variations and lifestyle on release of contaminants in wastewater discharges. The sensitivity of the analytical instruments/methodologies employed is also important in the reportage, since these chemicals must be analyzed before they can be published. Since majority of released effluents does not go through WWTPs, these factors could also determine the quantity of pollutants released directly into the environment.

Large numbers of chemicals have been reported in surface waters around the world. Egbuna et al. (2021) reported that a survey of literature for Nigeria showed the presence of over 250 emerging pollutants which belong to the above-mentioned classes, and in addition, polyaromatic hydrocarbons (PAHs), radionuclides, and particulate matter. In a national survey in South Africa, Wood et al. (2017) reported the presence of 99 pharmaceuticals in surface waters, the most frequently detected pollutants being Lamotrigine and Nevirapine. From extensive reviews, Miège et al. (2009) reported the presence of 184 PPCPs and showed their removal efficiencies through WWTPs. In Tianjin, Northern China, 227 chemicals were detected from an analytical survey targeting 1300 emerging pollutants; those with the highest frequency were bis(2-ethylhexyl) phthalate, siduron, lidocaine, and antipyrine with 100% detection (Kong et al. 2015).

From Spain, Biel-Maeso et al. (2019) reported a yearlong study monitoring 66 target emerging pollutants in influent and effluent of a WWTP. Oleivera et al. (2015) reported a total of 185 PPCPs were identified in 4 hospital WWTPs effluents, with 118 PPCPs detected in a single WWTP showing metformin at 720 μg/L and 33 of the identified PPCPs were associated with medium-to-high ecological risks.

Beside effluents from WWTPs and polluted freshwater bodies, underground water sources have also been reportedly contaminated with pollutants. Ejike et al. (2017) reported heavy metals concentrations significantly higher than the control groups, and far above the WHO recommendations, for Ar, Pb, and Cd. Deep borehole and shallow well water sources in Zambia were reported to contain several ECs, such as the antibiotic triclosan, chlorination by products (trihalomethanes), and surfactants (Sorensen et al. 2015). This indicates the ability of released pollutants to percolate deep into the ground and contaminate water sources away from their site of release. Since majority of Africans lack access to portable water, they obtain drinking water from wells. The contamination of underground water sources meant for domestic uses will expose the populace to higher pollutants uptake with the attendant consequences.

17.3.2 Factors Promoting Use of PPCPs Found in Municipal Discharges

The practice of lifestyle choices could minimize the exposure to such chemicals and the attendant consequences of environmental pollutions. Lesser contents of PPCPs in water could be because of the absence of the compounds due to lifestyle choices or non-availability/non-affordability of products containing such compounds. As observed across the world, with increases in purchasing power, humans often tend to be more complex with their consumption behaviors. Lenhart (2019) reported an expenditure increase between 10.5% and 20.3% of income on feeding following increase in financial fortunes of households, which may include diversifying from natural and locally sourced food items to processed foods containing preservatives, because of inherent desire to affirm one's new status, often as a competition to be at pal with counterparts, such as dwellers in low-income countries seeking to live like those in high-income nations (Chaudhary and Khatoon 2021). Those who depended on traditional medicines will access orthodox medicine, hence depending more on synthetic chemicals with consequent environmental discharges, thereby contributing to water pollution.

17.3.3 Impacts of WWTPs on Water Pollutants

Although Adeleye et al. (2022) reported that several PPCPs found in WWTPs influents are removed up to 90% during secondary treatment, this cannot be generalized for all compounds, since some due to their chemical nature may partition preferentially into the hydrophobic matrix of the sludge (Bielo-Maeso et al. 2019; Subedi and Kannan 2015) and end up into landfills or biosolids applied in agricultural fields (Biel-Maeso et al. 2019; Jones-Lepp and Stevens 2007). It was reported (Subedi and Kannan 2015) that over 50% total mass of ariprazole, norquetiapine, norsertraline, verapamil, propranolol, desmethyl citalopram, citalopram, and norverapamil in WWTPs influent were sorbed to particulate matter. The analyzed quantity of these chemicals in the effluents will be less by the sorbed and removed amounts. The portion partitioned in the sludge/biosolids will constitute a source of soil contamination wherever it is applied (Kodešová et al. 2019), indicating the link between water and soil pollution. The remaining quantity of PPCPs in effluents estimated at $\leq$10% of influent concentration (Adeleye et al. 2022) may be true for some compounds, such as acetaminophen, sulfamethoxazole, N-acetylsulfamethoxazol, ciprofloxacin, and caffeine which were reported to show between 95% and 99% removal following secondary treatment, while atelonol, carbamezipine, and clarithromycin were removed <90% (Al Qarni et al. 2016). The residual percentage is not sacrosanct and countered (Biel-Maeso et al. 2019; Sim et al. 2013) thus cannot be taken to apply for all PPCPs in influents of WWTPs with secondary treatment facilities. Since this is a percentage, it is only relative, and even at 95% removal, remaining concentration in effluent can still be high, as observed for acetaminophen, caffeine, ibuprofen, gabapentin, and naproxen at which residual concentrations, they can still elicit mild to high toxic effects on aquatic animals (Oliveira et al. 2015). For endocrine disruption chemicals (EDCs), environmental concentrations can elicit endocrine disruptive effects with great ecological consequences (Adeogun et al. 2019). Moreover, since 80% of the worlds' municipal wastewater is not even treated before discharge (WWAP 2017), this bulk constitutes a great menace in the environment.

Pollution with organic matter is detrimental, as it causes a decrease in water oxygen content, while phosphorus content has been increasing, positively correlating with decreased phytoplankton and macrobenthic community richness (Brauko et al. 2020). Pomati et al. (2017) also reported that a mixture of PPCPs showed capacity to reduce phenotypic diversity and response ability to natural phytoplankton communities.

17.4 Social–Economic Impacts of Water Contamination

The production of wastewater has tremendously increased around the world, owing to increased population, industrialization, besides changes in consumption patterns (Mishra et al. 2017). Released wastewater has increased in the last two decades; Egypt, Tunisia, China, and Nicaragua have increased wastewater generation by 97%, 50%, 158%, and 353% of their prior discharge capacities (FAO 2018). The practice of using wastewater, treated wastewater, ground water or a conjunction of either of these has been reported (Picó et al. 2019; Gomes et al. 2020) as a centuries old practice (Singh 2021) across several nations practicing irrigation farming, as a means of increasing food availability and financial income, especially in water-scarcity nations, such as Tunisia (Bedbabis et al. 2014). Beside water scarcity, irrigation with wastewater is desirable because of easy availability at low/no cost and due to the presence of nutrients in the wastewater (Akoto et al. 2015). Wastewater effluents arising from industrial facilities or municipal treatment plants are a great source of bioactive emerging contaminants, metabolic intermediates, transformation products (Picó et al. 2019), and heavy metals (Mohjoub et al. 2022; Cherfi et al. 2015) which are present owing to the failure of the treatment regimen to remove them, since WWTPs were not designed to cater for this class of contaminants (Ghernaout and Elboughdiri 2020). Crops irrigated with such polluted water tend to bioaccumulate the emerging contaminants (Singh 2021) and heavy metals (Berihun et al. 2021). The anti-epileptic Carbamazepine and its metabolites were found to have bioaccumulated in spinach leaves, telmisartan and sertraline accumulated preferentially in roots over other parts of the vegetable (Kodešová et al. 2019). Similarly, irrigation of maize under experimental conditions with glyphosate and ciprofloxacin resulted in the accumulation of the chemicals, and aminomethylphosphonic acid, an intermediate of glyphosate metabolism. Experimental evidence showed physiological changes capable of reducing per hectare yield of contaminated maize plants, beside the bioaccumulation of the chemicals in the plant leaves and kernels (Gomes et al. 2020). Reduced crop yields predispose farmers to famine and reduced monetary earnings. These chemicals are suspected to be toxic in experimental animals and humans. Where leaves are used for herbivores, livestock feeding and the maize eaten by humans even as staple food among farming communities, considering the suspicions of physiological modulations accompanying protracted exposures to the pollutants.

The presence of high concentrations of antibiotics in tributaries (K'oreje et al. 2018) could lead to the development of anti-resistant genes in pathogenic microbes leading to high cost of therapy for otherwise commonly treated bacteria-causing ailments. These would lead to loss of income, productive work hours and more discomfort and strains on family members who look after the sick person.

Water (environmental) contamination by pollutants may not only affect aquatic organisms and soil microflora. A recent study (Ding et al. 2022) has reported that smaller body sizes and higher morphological changes in wings, tarsus, and toe length in tree sparrows (*Passer montanus*) found in heavy metals (CU, Zn, Pb, and Cd) contaminated environments as observed in the liver, kidney, skeletons, and muscles as well as faces, corresponding to lower levels of calcium. On the flip side, the authors reported that tree sparrows in a non-polluted environment exhibited comparatively normally formed morphological features and higher levels of calcium. Morphological changes may not only be a good parameter in ecotoxicological studies but may well be one of several successive events culminating in biodiversity loss of such affected species.

Decline in water volume into the Kainji Dam in Nigeria and the Volta Lake in Ghana has been reported between 2002 and 2014 (Ndehedehe et al. 2016). The Kainji lake hosts Nigeria's national grid. Water recession in the lake could affect the already dwindling power generation from the lake. Moreover, reduced water levels could increase concentrations of contaminants per liter of water obtained from the dam, thereby increasing the tendency in uptake of pollutants by aquatic organisms inhabiting the lake, causing physiological modulations and ultimately species losses. This scenario may also aggravate the contamination of crops irrigated with the water. Lake Chad contributes at least 15% of annual fish consumed in Nigeria (Eriegha et al. 2019). Contamination of its water exposes consumers across Nigeria, especially Northern Nigeria, where the Lake is situated, beside the neighboring countries of Cameroun, Chad, and Niger to feeding on contaminants' bioaccumulated fish.

17.5 Impact of Water Pollution on Soil and Crops

Polluted water run-offs tend to transport pollutants from site of application to other locations (Dudley et al. 2017) especially down the slope, thereby making more soil area polluted. The pollution level of soils could depend on the soil characteristics, period of irrigation, and wastewater composition (Samia et al. 2013). Humic substances and clay soils bind strongly to Cr and Cu, thereby reducing their uptake into plants (Samia et al. 2013). The synergistic and additive effects of these metals in vivo in humans and other organisms suggest a total target hazard quotient exceeding 1, while the carcinogenic risk associated with these heavy metals at the levels found in African foods often points to the likelihood of consumers developing serious health risks, such as cancer (Atamaleki et al. 2021).

Heavy metals have been shown to be up taken by lettuce (*Lactuca sativa* L.) (Akoto et al. 2015; Agbenin et al. 2009) and turnip (*Brassica napus* L.) (Hashem et al. 2013) with preferential accumulation in shoots and roots. In lettuce and turnip, heavy metals caused lipid peroxidation, evident by the increase in number of polyamines and phenolic compounds which exhibit antioxidant activities in plants and caused significant reduction in leaf area, fresh and dry weights of shoots and roots, followed by a drastic reduction in chlorophyll content of leaves and damage to cell membranes (Hashem et al. 2013). For vegetables which are eaten raw, the high presence of heavy metals poses serious public health concerns (Berihun et al. 2021), especially in children, nursing, and pregnant mothers.

Microbial contamination of irrigated soils and consequently the crops been cultivated/harvested is another major concern for use of polluted wastewater in the environment. High contamination of irrigated soils with helminths of *Enterobius vermicularis* and *Toxocara* spp. at 11.41 and 29.26 eggs/g of soil, respectively, was observed as fecal coliforms, and *Staphylococci* and *Streptococcus* were observed at 1.35×10^4 CFU/mL, 8.92×10^6 CFU/mL and 0.27×10^7 CFU/mL, respectively, and reported in Taza city of Morrocco (Mherzi et al. 2021). Treated wastewater effluents used in irrigating vegetables contained varying loads of several microorganisms such as *Entamoeba coli, Enterobius vermicularis, Hymenolepsis nana, Ascaris lumbricoides,* and *Schistosoma mansoni* both in the water used to irrigate lettuce and were also subsequently isolated from a salad prepared with the lettuce (Nikaido et al. 2009). In South Africa, farming households feeding from wastewater irrigated vegetables showed high level of hookworm infections (42%) compared to 27% in other households (Gumbo et al. 2010). Therefore, consumption of vegetables irrigated with wastewater effluents poses serious health risks to the consumers, further impoverishing peasant farming families and stressing the fragile health care systems in low-income countries.

Increased infiltration of water, presence of micro-organisms from wastewater effluents used in irrigation of soil provides a rich source of micro-organisms which may facilitate the microbial mediated degradation of organic pollutants in soils with capacity to infiltrate oxygen (Samia et al. 2013). Thus, pollutants initial mass loadings unto agricultural sites may reduce over time owing to degradation or metabolism.

17.6 Ecological Toxicity of Water Pollutants

17.6.1 Toxicity of Antibiotics

Most toxicity studies often focus on a parameter in question. However, as shown above, myriad of parent pollutants, metabolic intermediates, and transformation by products exists in water. The additive or synergistic toxic effects of these mixtures on aquatic organisms are yet given a comprehensive study due to challenges in executing experiments of complex mixtures. Moreover, the environmental concentrations found often appear bearable, so studies may tend to focus on chronic exposures than acute assessments. Notwithstanding, there is ample information available to make informed decision over this topic. Ciprofloxacin and sulfamethoxazole have been shown at environmental concentrations to have risk quotient (RQ) high ($RQ > 1$) and moderate ($1 > RQ > 0.1$), respectively, for aquatic organisms (Daouk et al. 2016). These can hamper photosynthetic apparatus and alter carbon assimilation in aquatic algae (Daouk et al. 2016), a situation capable of de-establishing carbon flow through the carbon cycle for the local communities. Sengar and Vijayanandan (2022) studied 98 PPCPs and reported that 47% showed a possible risk (RQ > 1), two PPCPs showed the tendency for health hazards through groundwater, and yet others had an RQ > 1000 strongly signifying potential source of public health concerns, aquatic animals' toxicity, and probably death. Such chemicals could be prioritized for environmental monitoring due to their public health concerns.

17.6.2 Toxicity of Acetaminophen

Acetaminophen is a commonly patronized over-the-counter analgesic of first choice globally (Kristenssen et al. 2016). As its intake, so its presence as a water pollutant around the world (Miège et al. 2009) also in Africa (Ebele et al. 2020), where surface water mean concentration of 1.233 μg/L has been reported (Folarin et al. 2020). Acetaminophen triggers neurotoxic, cardiotoxic, and teratogenic effects in a dose-dependent manner in African catfish (*Clarias gariepinus*) embryo/larvae. In rainbow trout (*Oncorhynchus mykiss*), environmental relevant concentrations of acetaminophen initiated renal histological changes (movement and loss of nuclei, un-even nuclei size, and diminished tubular integrity); decreased oxygen intake during swimming and reduced swimming ability, while hepatic cells showed declined glycogen storage capacity (Choi et al. 2018). Diminished oxygen intake could affect ability to escape predators, leading to declined survival of the fish. These may affect cellular respiration with the accumulation of lactic acid in the muscles. The water flea, *Daphnia magma,* showed reduced body length, growth rate, and declined reproduction, which was dose dependent (Ding et al. (2020). In *Rhamdia quelen*, acetaminophen at environmental concentrations created reduced levels of hemoglobin and thrombocytes, triggered elevated leucocytes and thrombocytes, reduced testosterone levels and increased estradiol and superoxide dismutase activity and inhibition of ethoxyresorufin-O-deethylase and glutathione-S-transferase activities (Guiloski et al. 2017). Both EROD and GST are required for the detoxification of pollutants in fish and their inhibition is tantamount to asphyxiation of a terrestrial organism. Hence, the fish may experience oxidative stress, feminization of the males, and declined immunity over prolonged exposure. Fishes inhabiting such water bodies may become extinct owing to continual decline in population because of acetaminophen pollution. Although acetaminophen has a short half-life, its pseudo-persistence in the environment (Yarkwan 2021) is very inimical to aquatic organisms' survival.

17.6.3 Toxicity of EDCs

Studies on blackchin tilapia (*Sarotherodon melanotheron)* from waters contaminated with EDC (4-iso-nonylphenol, 4-t-octylphenol, monobutyltin cation, dieldrin) beside an array of heavy metals in Lagos, Nigeria produced molecular and histological evidence for intersex at 27.4% prevalence rate with evidence of testes-ova and ovo-testicular features, appearance of vitellogenin (vtg) in male fish and zona radiata proteins (zrp) mRNA levels being significantly higher in females than males when both observations were compared against the control (Adeogun et al. 2019). The African catfish is a highly cultivated, dominant freshwater fish and extremely patronized because of its delicious taste (Dauda et al. 2018). Males of this gonochoristic species posses an elongated urogenital papilla (UGP) having a pointed tip situated just below the anus and is an external secondary sexual feature, whose formation is determined by androgen, susceptible to modulation by EDCs. Water pollution with EDC causes developmental deformities in both gills and liver of the fish and altered gonadal structures (Pieterse et al. 2010). The alterations of the UGP coupled with microscopic evidence for gonadal intersex features reflect the effect of water pollution with EDCs on the male fish (Kruger et al. 2013).

17.6.4 Toxicity of Mixed Contaminants

Biofilms represent the normal microflora of an aquatic environment and can drive vital metabolic cycles. Their organization is affected by chemical, physical, and biological factors within the environment. Water contaminants have been shown to effect critical physiological changes in biofilms found near downstream of WWTPs effluents release point, compared to upstream (Chonova et al. 2017) despite the dilution in the receiving water bodies demonstrating the bioactivity of discharged chemical pollutants.

Tilapia (*Oreochromis* spp.) from Lake Tanganyika was reported to have higher bioaccumulated contents of synthetic emerging contaminants than those from Lake Victoria, in a manner that raises public health concern from human consumption (Polder et al. 2014). Lake Victoria from the Ugandan side receives large quantities of raw or poorly treated wastewater from municipal, industrial, and pharmaceutical sources, rich in emerging contaminants from Entebbe, Kampala and Jinja (Badamasi et al. 2019). The impact of these pollutants on selected fishes was investigated. Formation of liver lesions in marbled lungfish (*Protopterus aethiopicus*), Nile tilapia (*Oreochromis niloticus),* and Nile perch (*Lates niloticus)* was studied; 41.5% and 18.6% of sampled fishes showed moderate and severe lesions, respectively (Badamasi et al. 2019). Furadan and its derivatives (carbofuran, 3-hydroxycarbofuran, and 3-ketocarbofuran) were detected in Athi River, Kenya and were linked to the massive tragic loss of white-backed vultures (*Gyps africanus*) and hyenas in 2004 (Otieno et al. 2011). Member states of Eastern Africa (Cristiano et al. 2021) and in Egypt have also reported that water pollutants orchestrated toxic effects in aquatic organisms with deep concern over the propensity for biodiversity loss should the trend continue unchecked.

17.7 Implications of Physiological Modulations by Water Contaminants

The observed changes from biochemical, physiological, and morphological/histological characteristics of the above-mentioned aquatic organisms could be considered preludes to the onset of species decline culminating in biodiversity losses in several aquatic organisms. These could affect growth, reproduction, and other life processes (Parrish et al. 2019) and then create a decline in population which would be followed by a higher predator–prey quotient (more predators preying upon fewer preys). In an established ecosystem, the numbers of predators and preys remain constant, though variations in either of the organisms may be observed over the years in a manner that creates a regular cycle, without distorting the ecosystem balance. However, biotic factors (such as infectious pathogens) or abiotic factors may distort the ecological balance. Such sustained distortion, by way of continuous flow of toxic wastewaters may create feminization of male fish and tadpoles, development of intersex features (Adeogun et al. 2019; Kruger et al. 2013) and other physiological modulatory effects producing weaker species (Guiloski et al. 2017; Ding et al. 2020), thus creating a gradient scenario in which affected species transit over time to the edge of the slope and roll into irrecoverable extinction, producing biodiversity loss. These changes would affect carbon cycling, and established community food chain/webs (Nilsen et al. 2019).

17.7.1 Water Contaminants Induced Biodiversity Loss

Biofilm communities have been markedly affected by the release of effluents from WWTPs in a manner that varies between species, but generally, create genetic modifications which orchestrates shifts in bacterial community structures (Chonova et al. 2017). Water bodies polluted with atrazine, a herbicide, have been reported to cause feminization of frogs at concentrations less than 1 ppb leading to existential threat on frogs in farming communities using the herbicide (Beebee and Griffiths 2005). The extinction of four Californian anurans has been reported owing to water contamination with organophosphates and carbamates arising from agricultural fields (Davidson 2004). In Africa, ecological toxic effects including mortality, growth distortion and morphological changes on African anurans (Mediterranean painted frog, *Discoglossus pictus*, and African clowed frog, *Xenopus laevis*) have been documented, while the Mullers clawed frog (*Xenopus muellleri*) was shown to accumulate high concentrations of pesticides in its tissues Wolmarans et al. 2018). Acid rains are capable of lowering pH of host aquatic environment and amphibian embryos and larval are susceptible to both direct and indirect low pH (Beebee and Griffiths 2005). These occurrences aided by climate change and other abiotic factors could well account for biodiversity loss among amphibians, which has become more prominent in the last 40 years. A before–after–control–impact methodology was used to assess fish abundance, species richness, and water quality in river receiving effluents from a WWTP (Galib et al. 2018). After 8 years of WWTPs' effluents discharge, the authors reported a loss by >47% and >35% in fish abundance and species richness, respectively, compared to upstream sampling location, and 51% and 41%, respectively, when compared with pre-WWTP effluent discharge period. There is general biodiversity loss, which could be attributed to an interplay of biotic and abiotic factors, coupled with impacts of climate change (Beebee and Griffiths 2005). However, the loss is higher when stressors of chemical matrices in WWTPs effluents orchestrate a water contamination scenario. Phytoplankton exposed to a mixture of PPCPs at concentrations found in WWTPs effluents showed marked reduction in the rate of cell and population response by inhibition of cell–cell progression, which affected their life cycle (Pomati et al. 2017).

Studies focusing on PPCPs, PAHs, industrial chemicals and heavy metals identified in streams were accounted as major factor facilitating the loss of invertebrate populations especially on *Ephemeroptera*, *Plecoptera*, and *Trichoptera* species, while physico-chemical changes included reduction in oxygen concentration and elevated salinity levels (Stalter et al. 2013). Such changes in physico-chemical parameters would initiate imbalances demanding physiological adjustments to survive, which could be easy for larger organisms but tends to create a lag in growth phase (Ding et al. 2020) and ultimately a decline in populations unable to accommodate the change. Since organic pollutants are pseudo-persistent due to continual releases from pollution point-sources, they tend to alter the physiological processes for the community. Pollutants initiated that physiological changes may be slow but culminates in creating stressful environment through the alteration of abiotic factors which would obviously culminate in biodiversity loss.

Besides habitat loss and conversion to urbanization and intense agriculture, synthetic biocides applied in agricultural operations are the second most important factor responsible for biodiversity loss among entomofauna, which predictably would account for 40% loss of the worlds' insect species few decades away (Sánchez-Bayo and Wyckhuys 2019). The terrestrial ecosystem has suffered loss of species from taxa: *Lepidoptera*, *Hymenoptera*, and *Coleoptera* (dung beetles), while *Odonata*, *Plecoptera*, *Trichoptera*, and *Ephemeroptera* taxa in aquatic ecosystems have lost substantial numbers of species (Sánchez-Bayo and Wyckhuys 2019). This fate of both specific niche occupying species and generalists bestow dire consequences on ecological dynamics which may alter self perpetuation in affected ecosystems.

17.8 Water Quality Restoration

A majority of African countries do not have in place guidelines for wastewater effluents discharge into the environment, and where these guidelines exist, its implementation stands questionable (Fayiga et al. 2017). This no-guard–on-duty scenario has encouraged industries, municipalities, and other water users to act without regards for environmental safety. There is an urgent call for every African nation to develop legislations/policies guiding wastewater effluent discharges into receiving water bodies while also prohibiting surface release of wastewater into open lands in the neighborhood of manufacturing facilities, especially for small-scale industries. Where these policies are in place, there must be measures to ensure their implementation without fear or favor and apply sanctions where necessary to curtail continual environmental pollution.

Human excreta are a good contributing source PPCPs release into the environment. However, open defecation is a sad reality across African states. Water run-offs wash away such deposits into water bodies, thereby polluting aquatic bodies with PPCPs, micro-organisms and organic matter which would provide rich nutrients for algae growth. Many African nations are stepping up with zero-open defaecation policies. This must be encouraged to take roots and be established and enforced. Appropriate lavatory facilities must be put in place in towns, shopping malls, open market spaces, and at gas stations along highways and long-distance routes for sanitary needs of travelers. Urban boards must ensure that each house is served with and connected to a sewage system.

Multinationals operating on African shores must exercise the internal check measures to ensure that industrial operations are not synonymous with environmental pollutions as is the case of oil exploration in Nigerian Niger Delta region (Babatunde 2020), gold mining in South African Gauteng province (Minnaar 2020), and the petrochemical industries orchestrated pollution of Algerian environments (Benlaribi and Djebbar 2020). The liberty to be unethical while operating on African shores, contrary to what is obtainable in the developed nations housing the headquarters of these multinationals could sadly be envisaged as the modern-day master-slave divide still been perpetrated by foreign investors; sadly, often, with the active connivance of African elites/political class.

Since part of the water pollution witnessed in Africa is known to originate from industrial activities, some governments have imposed tax control measures. To curb industrial discharge menace in Olifants river basin, the most polluted, and third most stressed water basin in South Africa, authorities has introduced water pollution tax, which has been adjudged a huge success considering the goals set on curtailing discharges, protecting the ecosystem, and preserving biodiversity (Kyei and Hassan 2019). However, the other side of the coin is the tendency to scare off new companies from settling in such vicinities due to fear of increased operational cost; or even forcing existing companies to relocate to other regions, where they can maximize profit by paying lesser taxes (Kyei and Hassan 2019).

Low-income countries, to which class most African countries belong, do not have sufficient wastewater treatment operations in their towns leading to release of untreated wastewater, thereby contaminating the available freshwater resources (Segura et al. 2015). Therefore, to alleviate the danger posed to biodiversity and public health threats, these governments should make it a priority to construct WWTPs with at least secondary treatment processes and enact policies and create infrastructural facilities that will make it mandatory for all residences and institutions to discharge their wastes to the WWTPs. Depending on kind of discharges, industries should be encouraged to treat their wastewater. Cucumber cultivated by irrigation with WWTP effluents with secondary treatment facilities was analyzed for heavy metals and the content of Pb, Cr, Cu, and Zn (mg/kg) was 9.34, 4.67, 2.46 and 0.49, respectively, compared with 16.08, 9.12, 23.15 and 25.86, respectively, obtained from cucumber sampled from open market sources in Boumerdes, Algeria (Cherfi et al. 2015). This demonstrates the efficacy of optimized treatment regimen in mitigating heavy metal pollutants release into the environment. It also exposes the fact that Algerian soils are heavily contaminated with heavy metals among the farming communities. Though expensive, the gains of wastewater treatment far outweigh the cost of investments.

Municipalities should establish a database of commonly found water pollutants within their localities. This could serve as a basis of regular checks to ensure surface water quality is within acceptable standards. Discharges from institutions, industries, municipal WWTPs must be monitored to comply with established limits. Where possible, sanctions in form of environmental pollution tax should be applied to ensure that defaulting companies are not closed but made to pay the tax and effect remedial measures, where applicable.

Global water demand is forecasted to increase by 67–134% by 2050 (Hejazi et al. 2014). Since volume of consumption determines the volume of effluents discharged, it is necessary for African leaders to be proactive to prepare for adequate wastewater treatment and disposal or recycling techniques that would be implemented during this time. The foundation for a secured future must be laid today. Increased population density in the largest cities (Lagos, Nigeria; Cairo Egypt) and the fastest growing cities (Dares-salaam, Tanzania; Nairobi, Kenya; Kinshaha, Democratic Republic of Congo; Luanda in Angola, Addis Ababa, Ethiopia; and Abidjan, Cote D'ivoire) of Africa will put pressure on freshwater resources and surge the level of wastewater released to the environment. An effective method to treat and discharge the effluents is critical for a sustainable development.

Surface water and ground water contamination stem from the discharge of wastewater effluents from municipals and industrial activities into receiving water bodies, (Babatunde 2020; UNEP 2010), water run offs from agricultural lands polluted with biocides owing to field applications, often even without adherence to formulation requirements (Bessah et al. 2021), drainages from poorly situated sanitary and solid waste landfills, untreated effluents from domestic and industrial sources, and the flaring of gases in oil refineries—of which Algeria and Nigeria ranked fifth and sixth, respectively, among top seven gas flaring countries in the world (World Bank 2018) leading to the formation of acid rain, acidic soils, CO_2 emission, and attendant ecological consequences. In Africa, many countries do not have domestic wastewater collection and treatment procedures, leading to the discharge of such used waters on the earth's surface. In the Niger Delta region of Nigeria, years of oil spills, recording thousands of spilling events have contributed at least over three million barrels of crude oil into the surrounding water, constituting the single largest source of water pollution in the region (UNEP 2010) coupled with illegal oil refining (Duncan et al. 2020). Companies responsible for such spills should be held accountable for cleansing and remediation of polluted water bodies.

Such companies must also have environmental health monitoring unit which should be tasked to ensure regular surveillance and prevention/remediation of spills and discharges.

Assessment of toxicity must take a multi-techniques approach, employing molecular, histological, microscopic, ecological, and physical assessment tools to be able to arrest the onset of physiological modulatory events toward species decline and extinction, a position which was earlier recommended (Adeogun et al. 2019). Local/state/federal authorities should be careful to build databases on species richness and abundance within their jurisdiction which could serve as the reference point in future assessments. Such traditional studies conducted within 28 days may be good for invertebrates but are too limited to generate detailed information for larger organisms which may have better adaptive capabilities. Besides, to establish impacts of climate change—not climate variation—meticulous and long-term studies are mandatory (Plisnier et al. 2018). Galib et al. (2018) demonstrated biodiversity losses from aquatic species over 8 year period. Therefore, decades of meticulous monitoring activities are critical.

17.9 Conservation Strategy

It has been demonstrated that the practice of multiple-user-services (this approach highlights that houses use water both for domestic and productive purposes, with different productive uses of water being possible within this use context) using the form of managed aquifer recharge (MAR), a technique designed to enhance water infiltration into aquifers to encourage storage for environmental use or subsequent benefits, could harvest rainwater and retain run-off water into the dry season (Parker et al. 2022). Several types of MAR are available, such as in-channel modifications, excess irrigation, infiltration ponds, well injection, enhanced storage, and induced bank filtration. This practice will tend to make water available for multiple uses, such as irrigation, domestic washing, and animal husbandry typically obtainable in the rural settings of Africa.

Over harvesting of wild aquatic resources posses' threats to existence of the hunted species. Governments must encourage aquaculture to reduce dependence on wild species, and help the citizens to feed on less contaminated aquatic resources grown in the wild, which are exposed to uptake and bioaccumulation of heavy metals (Yarkwan and Apeh 2015), emerging contaminants (Adeogun et al. 2019) and also could hold antimicrobials-resistant genes from associated modified microbial strains (Duman et al. 2019; Katale et al. 2020; Preena et al. 2020) owing to previous exposures to antimicrobials-contaminated waters. Such occurrences have created a health crisis in East Africa (Katale et al. 2020) which needs serious monitoring to avoid a repeat elsewhere.

Careful management of uplands would imply safeguarding aquatic resources. In Lake Tanganyika, Britton et al. (2017) reported protected terrestrial shorelines correlated positively with a larger taxonomic and functional diversity of aquatic organisms and vice versa for unprotected upland areas. This suggests that upland activities are a key factor in determining aquatic biodiversity loss. This could be due to agricultural implements employed in cultivating uplands and other anthropogenic interruptions. In view of modern agricultural practices as a key driver of biodiversity loss, Dudley et al. (2017) postulated that six-step conservationists can employ to streamline biocides application and alleviate risks posed to biodiversity. They opined that, apt appropriation of land resources to conservation, farming, and other land use activities and drawing a balance between intense agricultural practices and sustainable agriculture employing biocides application, where necessary, must be within legal frameworks. They further stated its critical to minimize food waste and promote food efficiency through technical changes and informed choices besides developing frameworks that promote alternative to synthetic pesticides while outlawing harmful ones, and considerations should be given to landscapes in biocides application to minimize lowlands contamination due to water runoff from upper lands.

While these are good points to ponder, the debate between feeding a rapidly growing global population and climate change-induced changes in crop yields create a situation that must be carefully considered, namely: to fight hunger and feed the world or maintain environmental health and conservation of natural resources. The onus lies with governments of African countries at each level to work out sustainable and viable policies that will engage aquatic resources and till the earth to feed their populations in a sustainable manner.

17.10 Conclusions

Anthropogenic activities, prompted by a six-times growth in African population between 1950 and 2020, strongly supported by industrialization, agricultural practices, poor infrastructural designs, and climate change has created a severe water crisis globally and particularly in Africa. Droughts and flooding are happening simultaneously on different hydrological regimes. Water pollution has created social–economic problems ranging from health crisis, physiological modulation of aquatic organisms, and plants irrigated with polluted water. Soils contaminated with water runoffs hold residual contaminants which could be of organic, inorganic, microplastics or nanomaterials of anthropogenic origin and sometimes pollutants of natural origin. Impacts of water contamination affect all living things, from microorganisms to vertebrates. One of the gradual but sure impacts is biodiversity loss, witnessed on the continent and around the world. There is no single solution to this problem for all affected climes. While signing on to international treaties, African countries must look inward to develop feasible solutions which would promote a balance between uses of chemicals and environmental health toward a sustainable African continent. Investment in critical sanitary infrastructure is not optional. Careful monitoring of the environment and enactment of enabling legal tools and policies guiding the management of freshwater resources is very critical.

This should be followed by carefully monitoring the environment based on the legal frameworks to ensure that water resources are maintained in a sustainable manner. This could ease the burden on public health institutions, promote well-being, and improve quality of live for both humans and the myriad of biodiversity on our continental shores.

References

Abdel-Shafy HI, Aly RO (2002) Water issue in Egypt: resources, pollution and protection endeavors. Cent Eur J Med 8(2):3–21

Adeleye AS, Xue J, Zhao Y, Taylor AA, Zenobio JE, Sun Y et al (2022) Abundance, fate, and effects of pharmaceuticals and personal care products in aquatic environments. J Hazard Mat 424(Pt B):127284–127127. https://doi.org/10.1016/j.jhazmat.2021.127284

Adeogun AO, Ibor OR, Chukwuka AV, Regoli F, Arukwe A (2019) The intersex phenomenon in Sarotherodon melanotheron from Lagos lagoon (Nigeria): occurrence and severity in relation to contaminants burden in sediment. Environ Pollut 244:747–756. https://doi.org/10.1016/j.envpol.2018.10.091

Agbenin JO, Danko M, Welp G (2009) Soil and vegetable compositional relationships of eight potentially toxic metals in urban garden fields from northern Nigeria. J Sci Food Agric 89(1): 49–54. https://doi.org/10.1002/jsfa.3409

Akoto O, Addo D, Baidoo E, Agyapong EA, Apau J, Fei-Baffoe B (2015) Heavy metal accumulation in untreated wastewater irrigated soil and lettuce (*Lactuca sativa*). Environ Earth Sci 74:6193–6198. https://doi.org/10.1007/s12665-015-4640-z

Al Qarni H, Collier P, O'Keeffe J, Akunna J (2016) Investigating the removal of some pharmaceutical compounds in hospital wastewater treatment plants operating in Saudi Arabia. Environ Sci Pollut Res Inter 23(13):13003–13014. https://doi.org/10.1007/s11356-016-6389-7

Atamaleki A, Yazdanbakhsh A, Fallah S, Hesami M, Neshat A, Fakhri Y (2021) Accumulation of potentially harmful elements (PHEs) in lettuce (Lactuca sativa L.) and coriander (Coriandrum sativum L.) irrigated with wastewater: a systematic review and meta-analysis and probabilistic health risk assessment. Environ Sci Pollut Res Intern 28(11):13072–13082. https://doi.org/10.1007/s11356-020-12105-z

Babatunde AO (2020) Oil pollution and water conflicts in the riverine communities in Nigeria's Niger Delta region: challenges for and elements of problem-solving strategies. J Contemp Afri Stud 38(2):274–293

Badamasi I, Odong R, Masembe C (2019) Implications of increasing pollution levels on commercially important fishes in Lake Victoria. J Great Lakes Res 45(6):1274–1289. https://doi.org/10.1016/j.jglr.2019.09.024

Bedbabis S, Ben Rouina B, Boukhris M, Ferrara G (2014) Effect of irrigation with treated wastewater on soil chemical properties and infiltration rate. J Environ Man 133:45–50. https://doi.org/10.1016/j.jenvman.2013.11.007

Beebee TJ, Griffiths RA (2005) The amphibian decline crisis: a watershed for conservation biology? Biol Conserv 125(3):271–285. https://doi.org/10.1016/j.biocon.2005.04.009

Benlaribi R, Djebbar S (2020) Concentrations, distributions, sources, and risk assessment of polycyclic aromatic hydrocarbons in topsoils around a petrochemical industrial area in Algiers (Algeria). Environ Sci Pollut Res 27:29512–29529. https://doi.org/10.1007/s11356-020-09241

Berihun BT, Amare DE, Raju R, Ayele DT, Dagne H (2021) Determination of the level of metallic contamination in irrigation vegetables, the soil, and the water in Gondar City, Ethiopia. Nutr Diet Suppl 13:1–7. https://doi.org/10.2147/NDS.S283451

Bernhardt ES, Rosi EJ, Gessner MO (2017) Synthetic chemicals as agents of global change. Front Ecol Environ 15(2):84–90. https://doi.org/10.1002/fee.1450

Bessah E, Raji AO, Taiwo OJ, Agodzo SK, Ololade OO, Strapasson A, Donkor E (2021) Assessment of surface waters and pollution impacts in southern Ghana. Nord Hydrol 52(6): 1423–1435. https://doi.org/10.2166/nh.2021.051

Beverly EJ, White JD, Peppe DJ, Faith JT, Blegen N, Tryon CA (2020) Rapid Pleistocene desiccation and the future of Africa's Lake Victoria. Earth Planet Sci Let 530:115883. https://doi.org/10.1016/j.epsl.2019.115883

Biel-Maeso M, Corada-Fernández C, Lara-Martín PA (2019) Removal of personal care products (PCPs) in wastewater and sludge treatment and their occurrence in receiving soils. Water Res (Oxford) 150:129–139. https://doi.org/10.1016/j.watres.2018.11.045

Brauko KM, Cabral A, Costa NV, Hayden J, Dias CEP, Leite ES (2020) Marine heatwaves, sewage and eutrophication combine to trigger deoxygenation and biodiversity loss: a SW Atlantic case study. Front Mar Sci 7:590258. https://doi.org/10.3389/fmars.2020.590258

Britton A, Day J, Doble C, Ngatunga B, Kemp K, Carbone C et al (2017) Terrestrial-focused protected areas are effective for conservation of freshwater fish diversity in Lake Tanganyika. Biol Conserv 212:120–129. https://doi.org/10.1016/j.biocon.2017.06.001

Bustillo-Lecompte C, Mehrvar M, Quiñones-Bolaños E (2016) Slaughterhouse wastewater characterization and treatment: an economic and public health necessity of the meat processing industry in Ontario, Canada. J Geosci Environ Prot 4:175–186. https://doi.org/10.4236/gep.2016.44021

Castillo M, Barceló M, Pereira AS, Neto FRA (1999) Characterization of organic pollutants in industrial effluents by high-temperature gas chromatography–mass spectrometry. TrAC Trends Anal Chem 18(1):26–36. https://doi.org/10.1016/S0165-9936(98)00066-1

Chaudhary A, Khatoon S (2021) Impact of new middles class on consumer behavior: a case study of Delhi-NCR. J Asian Bus Econ Stud. ISSN:2515964X. https://www.emerald.com/insight/2515-964X.htm

Cherfi A, Achour M, Cherfi M, Otmani S, Morsli A (2015) Health risk assessment of heavy metals through consumption of vegetables irrigated with reclaimed urban wastewater in Algeria. Proc Safety Environ Prot 98:245–252. https://doi.org/10.1016/j.psep.2015.08.004

Cherif EK, Salmoun F, Nouayti N (2019) Sources of bathing water pollution in the west coast of Tangier, Morocco: effects of industrial zones. PRO 16(1):30. https://doi.org/10.3390/proceedings2019016030

Chilunjika A, Gumede N (2021) Climate change and human security in Sub-Saharan Africa. Afr Renaiss 2021(si1):13–37. https://doi.org/10.31920/2516-5305/v2021sin1a2

Choi E, Alsop D, Wilson JY (2018) The effects of chronic acetaminophen exposure on the kidney, gill and liver in rainbow trout (Oncorhychus mykiss). Aquatic Toxicol 198:20–29. https://doi.org/10.1016/j.aquatox.2018.02.007

Chonova T, Labanowski J, Cournoyer B, Chardon C, Keck F, Laurent É (2017) River biofilm community changes related to pharmaceutical loads emitted by a wastewater treatment plant. Environ Sci Pollut Res Intern 25(10):9254–9264. https://doi.org/10.1007/s11356-017-0024-0

Cristiano W, Giacoma C, Carere M, Mancini L (2021) Chemical pollution as a driver of biodiversity loss and potential deterioration of ecosystem services in eastern Africa: a critical review. South Afr J Sci 117(9-10):30–36. https://doi.org/10.17159/sajs.2021/9541

Daouk S, Chèvre N, Vernaz N, Widmer C, Daali Y, Fleury-Souverain S (2016) Dynamics of active pharmaceutical ingredients loads in a Swiss university hospital wastewaters and prediction of the related environmental risk for the aquatic ecosystems. Sci Tot Environ 547:244–253. https://doi.org/10.1016/j.scitotenv.2015.12.117

Dauda AB, Natrah I, Karim M, Kamarudin MS, Bichi AH (2018) African catfish aquaculture in Malaysia and Nigeria: status, trends and prospects. Fish Aquacul J 9(1):1. https://doi.org/10.4172/2150-3508.1000237

Davidson C (2004) Declining downwind: amphibian population declines in California and historical pesticide use. Ecol Appl 14:1892–1902

de Klerk B (2014) Baptism and the pollution of Africa's water. Hervormde Teologiese Stud 70(2): 1–8. https://doi.org/10.4102/hts.v70i2.2620

Ding R, Liu S, He C, Nie X (2020) Paracetamol affects the expression of *magna*. Detoxification- and reproduction-related genes and alters the life traits of Daphnia magna. Ecotoxicology 29: 398–406. https://doi.org/10.1007/s10646-020-02199-z

Ding J, Yang W, Wang S, Zhang H, Zhang Y (2022) Does environmental metal pollution affect bird morphometry? A case study on the tree sparrow Passer montanus. Chemosphere (Oxford) 295: 133947–133947. https://doi.org/10.1016/j.chemosphere.2022.133947

Dudley B, Zeldis J, Burge O (2017). New Zealand coastal water quality assessment. NIWA Riccarton, Christchurch. https://data.mfe.govt.nz/document/12907-dudley-et-al-2017. Assessed 27 Feb 2022

Duman M, Saticioglu IB, Altun S (2019) Molecular characterization and antimicrobial resistance profile of fecal contaminants and spoilage bacteria that emerge in rainbow trout (Oncorhynchus mykiss) farms. Biosci Microb Food Health 38(2):41–48. https://doi.org/10.12938/bmfh.18-007

Duncan AE, Rosa PR, Rosa Plá R (2020) The dangerous couple: illegal mining and water pollution—a case study in Fena River in the Ashanti Region of Ghana. J Chem 1–9. https://doi.org/10.1155/2020/2378560

Ebele AJ, Oluseyi T, Drage DS, Harrad S, Abdallah MA (2020) Occurrence, seasonal variation and human exposure to pharmaceuticals and personal care products in surface water, groundwater and drinking water in Lagos State, Nigeria. Emerg Contam 6:124–132

Egbuna C, Amadi CN, Patrick-Iwuanyanwu KC, Ezzat SM, Awuchi CG, Ugonwa PO (2021) Emerging pollutants in Nigeria: a systematic review. Environ Toxicol Pharmacol 85:103638–103638. https://doi.org/10.1016/j.etap.2021.103638

Ejike CECC, Eferibe CO, Okonkwo FO (2017) Concentrations of some heavy metals in underground water samples from a Nigerian crude oil producing community. Environ Sci Poll Res Intern 24(9):8436–8442. https://doi.org/10.1007/s11356-017-8524-5

Elum ZA, Mopipi K, Henri-Ukoha A (2016) Oil exploitation and its socioeconomic effects on the Niger Delta region of Nigeria. Environ Sci Poll Res Intern 23(13):12880–12889. https://doi.org/10.1007/s11356-016-6864-1

Eriegha OJ, Ovie SI, Ovie SO, Aminu AG (2019) Shrinking Lake Chad: initialization of culture-based fisheries for improved livelihood in Nigeria. Int J Fisher Aqua Stu 7(3):14–18

FAO (2017) Food and Agriculture Organization of the United Nations. ILRI International Livestock Research Institute, Washington, DC, p 72

FAO (2018) AQUASTAT—FAO's Global information system on water and agriculture. http://www.fao.org/nr/water/aquastat/data/query/index.html?lang=en. Accessed 21 Feb 2022

Fayiga AO, Ipinmoroti MO, Chirenje T (2017) Environmental pollution in Africa. Environ Dev Sust 20(1):41–73. https://doi.org/10.1007/s10668-016-9894-4

Folarin O, Otitoloju A, Amaeze N, Saliu J (2020) Occurrence of acetaminophen, amoxicillin, diclofenac and methylparaben in Lagos and Ologe Lagoons, Lagos, Nigeria. J Appl Sci Environ Manag 23(12):2143. https://doi.org/10.4314/jasem.v23i12.10

Galib SM, Mohsin AB, Parvez MT, Lucas MC, Chaki N, Arnob SS (2018) Municipal wastewater can result in a dramatic decline in freshwater fishes: a lesson from a developing country. Knowl Manag Aquat Ecosys 419:37. https://doi.org/10.1051/kmae/2018025

Gedefaw M, Yan D, Wang H, Qin T, Wang K (2019) Analysis of the recent trends of two climate parameters over two eco-regions of Ethiopia. Water (Basel) 11(1):161. https://doi.org/10.3390/w11010161

Gemeda FT, Guta DD, Wakjira FS, Gebresenbet G (2020) Occurrence of heavy metal in water, soil, and plants in fields irrigated with industrial wastewater in Sabata town, Ethiopia. Environ Sci Poll Res Intern 28(10):12382–12396. https://doi.org/10.1007/s11356-020-10621-6

Ghernaout D, Elboughdiri N (2020) On the treatment Trains for Municipal Wastewater Reuse for irrigation. Open Access Lib J 7:1–15. https://doi.org/10.4236/oalib.1106088

Golovko O, Kumar V, Fedorova G, Randak T, Grabic R (2014) Seasonal changes in antibiotics, antidepressants/psychiatric drugs, antihistamines, and lipid regulators in a wastewater treatment plant. Chemos (Oxford) 111:418–426. https://doi.org/10.1016/j.chemosphere.2014.03.132

Gomes MP, Rocha DC, Moreira de Brito JC, Tavares DS, Marques RZ, Soffiatti P (2020) Emerging contaminants in water used for maize irrigation: economic and food safety losses associated with ciprofloxacin and glyphosate. Ecotoxicol Environ Saf 196:110549. https://doi.org/10.1016/j.ecoenv.2020.110549

Guiloski IC, Ribasa JLC, Piancinib LDS, Dagostimb AC, Cirioc SM, Favorod LF (2017) Paracetamol causes enforcrone disruption and dhueptotoxicity inmale fish Rhamdia quelen after subchronic exposure. Environ Toxicol Pharm 53:111–120. https://doi.org/10.1016/j.etap.2017.05.005

Gumbo JR, Malaka EM, Odiyo JO, Nare L (2010) The health implications of wastewater reuse in vegetable irrigation: a case study from Malamulele, South Africa. Inter J Environ Health Res 20(3):201–211. https://doi.org/10.1080/09603120903511093

Hashem H, Hassanein R, El-Deep M, Shouman A (2013) Irrigation with industrial wastewater activates antioxidant system and osmoprotectant accumulation in lettuce, turnip, and tomato plants. Ecotoxicol Environ Saf 95:144–152. https://doi.org/10.1016/j.ecoenv.2013.05.030

Hedgespeth ML, Sapozhnikova Y, Pennington P, Clum A, Fairey A, Wirth E (2012) Pharmaceuticals and personal care products (PPCPs) in treated wastewater discharges into Charleston Harbor, South Carolina. Sci Total Environ 437:1–9. https://doi.org/10.1016/j.scitotenv.2012.07.076

Hejazi M, Edmonds J, Clarke L, Kyle P, Davies E, Chaturvedi V et al (2014) Long-term global water projections using six socioeconomic scenarios in an integrated assessment modeling framework. Technol Forecas Soc Change 81:205–226. https://doi.org/10.1016/j.techfore.2013.05.006

Iavazzo P, Ducci D, Adamo P, Trifuoggi M, Migliozzi A, Boni M (2011) Impact of past mining activity on the quality of water and soil in the high Moulouya Valley (Morocco). Water Air Soil Poll 223(2):573–589. https://doi.org/10.1007/s11270-011-0883-9

Jones-Lepp TL, Stevens R (2007) Pharmaceuticals and personal care products in biosolids/sewage sludge: the interface between analytical chemistry and regulation. Anal Bioanal Chem 387(4):1173–1183. https://doi.org/10.1007/s00216-006-0942-z

K'oreje KO, Kandie FJ, Vergeynst L, Abira MA, Van Langenhove H, Okoth M et al (2018) Occurrence, fate and removal of pharmaceuticals, personal care products and pesticides in wastewater stabilization ponds and receiving rivers in the Nzoia Basin, Kenya. Sci Total Environ 637-638:336–348. https://doi.org/10.1016/j.scitotenv.2018.04.331

Kanu I, Achi OK (2011) Industrial effluents and their impact on water quality if receiving Rivers in Nigeria. J Appl Tech Environ Sanit 1(1):75–86

Katale BZ, Misinzo G, Mshana SE, Chiyangi H, Campino S, Clark TG et al (2020) Diversity and risk factors for the transmission of antimicrobial resistance across human, animals and environmental compartments in East Africa: a review. Antimicrob Resist Infect Control 9:127. https://doi.org/10.1186/s13756-020-00786-7

Kinuthia GK, Ngure V, Beti D, Lugalia R, Wangila A, Kamau L (2020) Levels of heavy metals in wastewater and soil samples from open drainage channels in Nairobi, Kenya: community health implication. Sci Rep 10(1):8434–8434. https://doi.org/10.1038/s41598-020-65359-5

Kleywegt S, Payne M, Ng F, Fletcher T (2019) Environmental loadings of active pharmaceutical ingredients from manufacturing facilities in Canada. Sci Total Environ 646:257–264. https://doi.org/10.1016/j.scitotenv.2018.07.240

Kodešová R, Klement A, Golovko O, Fér M, Kočárek M, Nikodem A et al (2019) Soil influences on uptake and transfer of pharmaceuticals from sewage sludge amended soils to spinach. J Environ Manag 250:109407. https://doi.org/10.1016/j.jenvman.2019.109407

Kong L, Kadokami K, Wang S, Duong HT, Chau HTC (2015) Monitoring of 1300 organic micropollutants in surface waters from Tianjin, North China. Chemos (Oxford) 122:125–130. https://doi.org/10.1016/j.chemosphere.2014.11.025

Kosma CI, Lambropoulou DA, Albanis TA (2010) Occurrence and removal of PPCPs in municipal and hospital wastewaters in Greece. J Hazard Mat 179(1):804–817. https://doi.org/10.1016/j.jhazmat.2010.03.075

Kristenssen DM, Mazaud-Guitot S, Gaudriault P, Lesna L, Serrano T, Main KM, Jegou B (2016) Analgesic use – prevalence, biomonitoring and endocrine and reproductive effects. Nat Rev Endocrinol 12:381–393. https://doi.org/10.1038/nrendo.2016.55

Kruger T, Barnhoorn I, Jansen van Vuren J, Bornman R (2013) The use of the urogenital papillae of male feral African sharptooth catfish (*Clarias gariepinus*) as indicator of exposure to estrogenic chemicals in two polluted dams in an urban nature reserve, Gauteng, South Africa. Ecotox Environ Saf 87:98–107. https://doi.org/10.1016/j.ecoenv.2012.10.004

Kyei C, Hassan R (2019) Managing the trade-off between economic growth and protection of environmental quality: the case of taxing water pollution in the Olifants river basin of South Africa. Water Policy 21(2):277–290. https://doi.org/10.2166/wp.2019.190

Lencha SM, Ulsido MD, Muluneh A (2021) Evaluation of seasonal and spatial variations in water quality and identification of potential sources of pollution using multivariate statistical techniques for Lake Hawassa Watershed, Ethiopia. Appl Sci 11(19):8991. https://doi.org/10.3390/app11198991

Lenhart O (2019) The effects of income on health: new evidence from the Earned Income Tax Credit. Rev Econ Household 17:377–410. https://doi.org/10.1007/s11150-018-9429-x

Mherzi N, Lamchouri F, Toufik H (2021) Assessment of the effects of seasonal changes, urban discharges and leachates on the parasitological and bacteriological qualities of soil and water from Oued Larbaâ (north-eastern, Morocco). Environ Monit Asses 193(10):628–628. https://doi.org/10.1007/s10661-021-09326-1

Miège C, Choubert J, Ribeiro L, Eusèbe M, Coquery M (2009) Fate of pharmaceuticals and personal care products in wastewater treatment plants – conception of a database and first results. Environ Poll 157(5):1721–1726. https://doi.org/10.1016/j.envpol.2008.11.045

Minnaar A (2020) Water pollution and contamination from gold mines: acid mine drainage in Guateng Province, South Africa. In: Emam K, Segato L, Migliorini M (eds) Water, governance, and crime issues. Springer, Cham, pp 193–219. https://doi.org/10.1007/978-3-030-44798-4-12

Mishra BK, Regmi RK, Masago Y, Fukushi K, Kumar P, Saraswat C (2017) Assessment of Bagmati river pollution in Kathmandu Valley: scenario-based modeling and analysis for sustainable urban development. Sust Water Qual Ecol 9-10:67–77. https://doi.org/10.1016/j.swaqe.2017.06.001

Mo WY, Chen Z, Leung HM, Leung AOW (2015) Application of veterinary antibiotics in China's aquaculture industry and their potential human health risks. Environ Sci Poll Res Intern 24(10): 8978–8989. https://doi.org/10.1007/s11356-015-5607-z

Musoke SH, Boon EK (2012) Reduction in the water level of Lake Victoria and its implications on livelihoods in Rwanjaba lakeshore community, Uganda. Chin J Popul Resour Environ 10(1): 64–72. https://doi.org/10.1080/10042857.2012.10685061

Ndehedehe C, Awange J, Agutu N, Kuhn M, Heck B (2016) Understanding changes in terrestrial water storage over West Africa between 2002 and 2014. Adv Water Resour 88:211–230. https://doi.org/10.1016/j.advwatres.2015.12.009

Ngoran SD, Dogah KE, Xue XZ (2015) Assessing the impacts if climate change on water resources: the sub-Saharan African perspective. J Econ Sust Dev 6(1):185–193

Nikaido M, Tonani KAA, Julião FC, Trevilato TMB, Takayanagui AMM, Sanches SM (2009) Analysis of bacteria, parasites, and heavy metals in lettuce (Lactuca sativa) and rocket salad (Eruca sativa L.) irrigated with treated effluent from a biological wastewater treatment plant. Biol Trace Elem Res 134(3):342–351. https://doi.org/10.1007/s12011-009-8477-8

Nilsen E, Smalling KL, Ahrens L, Gros M, Miglioranza KS, Picó Y, Schoenfuss HL (2019) Critical review: grand challenges in assessing the adverse effects of contaminants of emerging concern on aquatic food webs. Environ Toxicol Chem 38(1):46–60. https://doi.org/10.1002/etc.4290

Odendaal C, Seaman M, Kemp G, Patterton H, Patterton HG (2015) An LC-MS/MS based survey of contaminants of emerging concern in drinking water in South Africa. S Afr J Sci 111(9/10): 160–106. https://doi.org/10.17159/sajs.2015/20140401

Ojemaye CY, Petrik L (2019) Occurrences, levels, and risk assessment studies of emerging pollutants (pharmaceuticals, perfluoroalkyl and endocrine disrupting compounds) in fish samples from Kalk Bay harbour, South Africa. Environ Poll 252(Pt A):562–572. https://doi.org/10.1016/j.envpol.2019.05.091

Oliveira J, Silveira M, Chacon D, Luchiari A (2015) The Zebrafish world colors and shapes: preference and discrimination. Zebrafish 12(2). https://doi.org/10.1089/zeb.2014.1019

Otieno PO, Lalah JO, Virani M, Jodinko IO, Schramm KW (2011) Carbofuran use and abuse in Kenya: residues in soils, plants, water courses and the African white-backed vultures (*Gyps africanus*) found dead. Environment 31(4):382–393. https://doi.org/10.1007/s10669-011-9350-9

Parker AH, Nyangoka J, Rodrigues I, Yadav B, Le Corre KS, Campo P et al (2022) The multiple uses of water derived from managed aquifer recharge systems in Kenya and India. J Water San Hyg Dev 12(2):208–216. https://doi.org/10.2166/washdev.2022.177

Parrish SC, Dormio SM, Richards SL, McCoy KA, McCoy MW (2019) Life in a contaminant milieu: PPCP mixtures generate unpredictable outcomes across trophic levels and life stages. Ecosphere 10(12):e02970. https://doi.org/10.1002/ecs2.2970

Picó Y, Alvarez-Ruiz R, Alfarhan AH, El-Sheikh MA, Alobaid SM, Barceló D (2019) Uptake and accumulation of emerging contaminants in soil and plant treated with wastewater under real-world environmental conditions in the Al Hayer area (Saudi Arabia). Sci Total Environ 652: 562–572. https://doi.org/10.1016/j.scitotenv.2018.10.224

Pieterse GM, Marchand MJ, Van Dyk JC, Barnhoorn IEJ (2010) Histological alterations in the testes and ovaries of the sharptooth catfish (*Clarias gariepinus*) from an urban nature reserve in South Africa. J Appl Ichthyol 26(5):789–793. https://doi.org/10.1111/j.1439-0426.2010.01557.x

Plisnier PD, Nshombo M, Mgana H, Ntakimazi G (2018) Monitoring climate change and anthropogenic pressure at Lake Tanganyika. J Great Lakes Res 44(6):1194–1208. https://doi.org/10.1016/j.jglr.2018.05.019

Polder A, Müller M, Lyche J, Mdegela R, Nonga H, Mabiki F et al (2014) Levels and patterns of persistent organic pollutants (POPs) in tilapia (Oreochromis sp.) from four different lakes in Tanzania: geographical differences and implications for human health. Sci Total Environ 488-489:252–260. https://doi.org/10.1016/j.scitotenv.2014.04.085

Pomati F, Jokela J, Castiglioni S, Thomas MK, Nizzetto L (2017) Water-borne pharmaceuticals reduce phenotypic diversity and response capacity of natural phytoplankton communities. PLoS One 12(3):e0174207–e0174207. https://doi.org/10.1371/journal.pone.0174207

Preena PG, Swaminathan TR, Kumar VJR, Singh ISB (2020) Antimicrobial resistance in aquaculture: a crisis for concern. Biologia 75:1497–1517. https://doi.org/10.2478/s11756-020-00456-4

Sánchez-Bayo F, Wyckhuys KA (2019) Worldwide decline of the entomofauna: a review of its drivers. Biol Conserv 232:8–27. https://doi.org/10.1016/j.biocon.2019.01.020

Samia K, Abdelkarim C, Yassin H, Teruo H (2013) The effect of long-term soil irrigation by wastewater on organic matter, polycyclic aromatic hydrocarbons, and heavy metals evolution: case study of Zaouit Sousse (Tunisia). Arab J Geosci 6:4337–4346. https://doi.org/10.1007/s12517-012-0666-7

Scheren PAGM, Zanting HA, Lemmens AMC (2000) Estimation of water pollution sources in Lake Victoria, East Africa: application and elaboration of the rapid assessment methodology. J Environ Manag 58(4):235–248. https://doi.org/10.1006/jema.2000.0322

Schilling J, Hertig E, Tramblay Y, Scheffran J (2020) Climate change vulnerability, water resources and social implications in North Africa. Reg Environ Chang 20:15. https://doi.org/10.1007/s10113-020-01597-7

Segura PA, Takada H, Correa JA, El Saadi K, Koike T, Onwona-Agyeman S et al (2015) Global occurrence of anti-infectives in contaminated surface waters: impact of income inequality between countries. Environ Intern 80:89–97. https://doi.org/10.1016/j.envint.2015.04.001

Sengar A, Vijayanandan A (2022) Human health and ecological risk assessment of 98 pharmaceuticals and personal care products (PPCPs) detected in Indian surface and wastewaters. Sci Tot Environ 807(Pt 1):150677–150677. https://doi.org/10.1016/j.scitotenv.2021.150677

Shiklomanov IA (2000) Appraisal and assessment of world water resources. Water Intern 25(1): 11–32. https://doi.org/10.1080/02508060008686794

Shutcha MN, Faucon M-P, Kissi CK, Colinet G, Mahy G, Luhembwe MN et al (2015) Three years of phytostabilisation experiment of bare acidic soil extremely contaminated by copper smelting using plant biodiversity of metal-rich soils in tropical Africa (Katanga, DR Congo). Ecol Eng 82:81–90. https://doi.org/10.1016/j.ecoleng.2015.04.062

Sim WJ, Kim HY, Choi SD, Kwon JH, Oh JE (2013) Evaluation of pharmaceuticals and personal care products with emphasis on anthelmintics in human sanitary waste, sewage, hospital wastewater, livestock wastewater and receiving water. J Hazard Mat 248-249:219–227. https://doi.org/10.1016/j.jhazmat.2013.01.007

Singh A (2021) A review of wastewater irrigation: environmental implications. Res Conserv Recycl 168:105454. https://doi.org/10.1016/j.resconrec.2021.105454

Sorensen JP, Lapworth D, Nkhuwa DC, Stuart M, Gooddy D, Bell R et al (2015) Emerging contaminants in urban groundwater sources in Africa. Water Res (Oxford) 72:51–63. https://doi.org/10.1016/j.watres.2014.08.002

Stalter D, Magdeburg A, Quednow K, Botzat A, Oehlmann J (2013) Do contaminants originating from state-of-the-art treated wastewater impact the ecological quality of surface waters? PLoS One 8(4):e60616–e60616. https://doi.org/10.1371/journal.pone.0060616

Subedi B, Kannan K (2015) Occurrence and fate of select psychoactive pharmaceuticals and antihypertensives in two wastewater treatment plants in New York State, USA. Sci Total Environ 514:273–280. https://doi.org/10.1016/j.scitotenv.2015.01.098

Tarpani RRZ, Azapagic A (2018a) Life cycle environmental impacts of advanced wastewater treatment techniques for removal of pharmaceuticals and personal care products (PPCPs). J Environ Manag 215:258–272. https://doi.org/10.1016/j.jenvman.2018.03.047

Tarpani RRZ, Azapagic A (2018b) A methodology for estimating concentrations of pharmaceuticals and personal care products (PPCPs) in wastewater treatment plants and in freshwaters. Sci Tot Environ 622-623:1417–1430. https://doi.org/10.1016/j.scitotenv.2017.12.059

Thiebault T, Fougère L, Destandau E, Réty M, Jacob J (2019) Impact of meteorological and social events on human-excreted contaminant loads in raw wastewater: from daily to weekly dynamics. Chemos (Oxford) 230:107–116. https://doi.org/10.1016/j.chemosphere.2019.04.221

Tiwari B, Drogui P, Tyagi RD (2020) Removal of emerging micro-pollutants from pharmaceutical industry wastewater. In: Varjani S, Pandey A, Tyagi RD, Ngo HH, Larroche C (eds) Current developments in biotechnology and bioengineering. Amsterdam, Elsevier, pp 457–480. https://doi.org/10.1016/B978-0-12-819594-9.00018-8

UNEP (2010) "Africa water atlas". Division of early warning and assessment (DEWA). United Nations Environment Programme (UNEP), Nairobi, Kenya

UN (2021) United Nations, Department of Economic and Social Affairs, Population Division (2021) World Population Prospects 2019, Online Edition. Rev. 1

Vorster M (2014) Climate change-related human security threats to border integrity and safeguarding for South Africa. Master's thesis, Faculty of Graduate Studies, University of Witwatersrand

Wolmarans NJ, Du Preez LH, Yohannes YB, Ikenaka Y, Ishizuka M, Smit NJ et al (2018) Linking organochlorine exposure to biomarker response patterns in anurans: a case study of Müller's clawed frog (*Xenopus muelleri*) from a tropical malaria vector control region. Ecotoxicology 27: 1203–1216. https://doi.org/10.1007/s10646-018-1972-y

Wood TP, Du Preez C, Steenkamp A, Duvenage C, Rohwer ER (2017) Database-driven screening of south African surface water and the targeted detection of pharmaceuticals using liquid chromatography - high resolution mass spectrometry. Environ Poll 230:453–462. https://doi.org/10.1016/j.envpol.2017.06.043

WWAP (2017) The United Nations World Water Development Report 2017: Wastewater, the untapped resource. United Nations World Water Assessment Programme (WWAP). United Nations Educational, Scientific and Cultural Organization (UNESCO), Paris

Yarkwan B (2021) Soybean Peroxidase Catalyzed Oligomerization of Acetaminophen and Selected Parabens for Wastewater Treatment (2021). Electronic Theses and Dissertations 8582. https://scholar.uwindsor.ca/etd/8582

Yarkwan B, Apeh D (2015) Assessment of heavy metals accumulation in tissues of *Tilapia zilli* and *Clarias gariepinus* found in lake Akpoko and River Benue, Nigeria. J Environ Earth Sci 5(10): 99–106

Yunana DA, Shittu AA, Ayuba S, Bassah EJ, Joshua WK (2017) Climate change and lake water resources in Sub-Saharan Africa: case study of Lake Chad and Lake Victoria. Nig J Techn 36(2):648. https://doi.org/10.4314/njt.v36i2.42

Chapter 18
Disease Outbreaks in Ex-Situ Plant Conservation and Potential Management Strategies

Matthew Chidozie Ogwu and Moses Edwin Osawaru

Abstract Plants are fundamental to directly or indirectly addressing most of humanity's contemporary challenges such as food scarcity and insecurity, availability of shelter, energy, and industrial raw materials as well as human and environmental health issues and mitigating climate change. However, around 40% of documented plant species are threatened with extinction due to anthropogenic pressure, especially from exponential population growth, overexploitation and mismanagement, habitat loss, modification and fragmentation, introduced alien species, pollution, and novel diseases. Plant diversity and their sustainable conservation and utilization underpin globalization and sustainable development as well as greater options for human innovation and adaptation. Plant conservation approaches are mainly either in situ or ex situ and they may be complementary to each other. Other little-known plant conservation approaches are linked to or derived from in situ and ex situ plant conservation and includes inter *situ*, *quasi* in situ, and circa *situm* plant conservation approaches. This chapter focuses on ex situ plant conservation. Ex situ plant conservation wholly seeks to sustain options for human innovation and adaptation by preserving plant species and taxa outside their native and historic range through the utilization of different techniques and conservation approaches, such as seed and gene banking, cryopreservation, gardens, and arboreta. However, certain factors tend to hinder or disrupt ex situ conservation techniques, including disease outbreaks, unstable weather and climatic conditions required for a variety of plant species, limited genetic diversity, plant stress from reintroduction and poor ex situ infrastructure, and inefficient management systems. For instance, *Zingiber* species held in field gene banks are known to be affected by soil-borne

M. C. Ogwu (✉)
Goodnight Family Department of Sustainable Development, Living Learning Center, Appalachian State University, Boone, North Carolina, USA
e-mail: ogwumc@appstate.edu

M. E. Osawaru
Department of Plant Biology and Biotechnology, Faculty of Life Sciences, University of Benin, Benin City, Nigeria

S. C. Izah, M. C. Ogwu (eds.), *Sustainable Utilization and Conservation of Africa's Biological Resources and Environment*, Sustainable Development and Biodiversity 32, https://doi.org/10.1007/978-981-19-6974-4_18

diseases caused by the bacteria *Ralstonia solacearum* which can spread further by sticking to human hands. Other examples of disease outbreaks under ex situ plant conservation include *Bogia coconut Syndrome'*, cassava viral diseases, and Alomoa-bobone viral disease of taro (*Colocasia esculenta*). This chapter seeks to highlight the potential and actual impacts of disease outbreaks under ex situ conservation using some examples. This chapter outlines how some of these diseases among and within plant species maintained in ex situ conservation can be prevented and/or managed optimally through various effective management strategies as well as the roles of different stakeholders involved in plant germplasm management.

Keywords Plant disease management · Disease outbreak · Plant conservation · Ex-situ conservation · Plant diversity · Germplasm management

18.1 Introduction

Plant conservation refers to a group of activities aimed at managing plant genetic resources and preventing threatened, endemic, and exceptional plant species and taxa from going extinct. These activities can assume a direct or indirect outlook and require the understanding of coupled natural systems and plant phenology (Osawaru et al. 2015a; Aiwansoba et al. 2019; Ogwu 2019a, b; McDonough et al. 2020; Ikhajiagbe et al. 2020, 2021a, b; Lortie et al. 2021). The works of Li et al. (2019) and Liu et al. (2021) showed that there is a correlative influence of coupled system outcomes on plant functional trait trade-offs that may influence plant diversity and coexistence within communities such as plant phenological shifting response to urbanization. Plant conservation may target wild plant populations, plant species within gardens or farms, education and research programs, management of invasive or introduced plant species, plant recovery and restoration work, research programs, using taxonomic and systematic knowledge and skills, survey and plant collections missions, field establishment or viability trials and germplasm banking. According to the IUCN Red List categorization, plant species or taxa may be considered vulnerable or endangered when exhaustive surveys within native or introduced distribution and historic range and habitat at appropriate times reveal minimal individuals (IUCN 2001). In this case, such plant species or taxa would require a focused conservation strategy that may be aimed at either a reintroduction to increase their population or maintenance away from their native or historic range. This is defined as ex situ conservation, and in both scenarios, the economic, ecological, diversity, and other value-based characteristics of the taxa may be restored over time.

Anthropogenic control of natural or managed plant dynamics varies and may include habitat modifications, alteration of reproductive cycles and habits, access restriction, and protection of plant germplasm (Rao et al. 2001; Osawaru et al. 2013; Ogwu et al. 2016). It, therefore, raises the need for the conservation of plant species to prevent them from extinction (Lindenmayer and Burgman 2005; Osawaru and Ogwu 2014; Ogwu et al. 2014a, b). Moreover, the exponential increase in global human population has also led to increased cultivation, processing, production, and consumption of plant-based food that exceeds production (Bussmann 2002; Ogwu 2019a, b, 2020; Osawaru and Ogwu 2020). Seed and gene banking is a major ex situ conservation technique that can avert such unforeseen emergencies. Plant materials held in seed and gene banks can be retrieved when needed or in emergencies and established on the farmlands or other human-controlled systems (Cohen et al. 1991; Maunder et al. 2004; Osawaru et al. 2015b). Hence, ex situ plant conservation refers to the off-site plant management, maintenance, preservation, and protection including outside of their natural, native, or historic diversity and distribution range either wholly or their germplasm—seed, flower, pollen, tissues, cell culture, etc. (Fazey et al. 2005). This system acts as a backup for plant resources in human-dominated systems and environments (Cohen et al. 1991). Besides seed and gene banks, the importance of other off-site plant conservation techniques such as botanical gardens, pollen banks, in vitro culture and cryopreservation (i.e., storage in liquid nitrogen at −196 °C), tissue culture, DNA bank, field bank, seed bank, etc. has been noted. In addition, biotechnological tools and techniques are becoming increasingly invaluable for the long-, medium- and short-term ex situ plant conservation and re-establishment of plant genetic resources, including micropropagation, protoplast culture, tissue culture, callus culture, somatic embryogenesis, organogenesis, micrografting, and cryopreservation (Khan et al. 2012; Chokheli et al. 2020; Coelho et al. 2020). In part, the current work focuses on seed bank management as an important technique for ex situ plant conservation. Plant species and taxa may require a specific ex situ approach in combination with seed banking to efficiently conserve their germplasm. For instance, some seeds may be intolerant, contain lots of water, and are short-lived and unsuitable for the dry and cold conditions of most seed banks especially recalcitrant seeds (Pence et al. 2017). On the other hand, plant in situ conservation approaches are considered complimentary of ex situ, but an understanding of their interrelationship is required for adoption and integration into sustainable environmental policies (Cohen et al. 1991; Volis 2017). The suitability of this complementary approach was verified in the conservation of *Trifolium thompsonii,* wherein it proved vital for monitoring mechanisms (such as genetic drift, gene flow, and natural selection) that may contribute to change within and between plant populations (Greene et al. 2014). In the view of Bellon et al. (2017), ex situ conservation is not perfect, because it only reflects diversity at the time of collection without accounting for biotic and abiotic evolutionary changes and may yet decline through genetic drift and improper storage and regeneration methods but remains invaluable. This value is connected to the roles of plants in global agriculture, food security, economy, and cheaper cost when compared to in situ conservation (Li and Pritchard 2009; Chime et al. 2015). On the other hand, the risks associated with ex situ plant conservation are connected to the species or taxa through longevity, their tolerance to ex situ conservation methods, length and condition of storage, and regularity of scientific assessment of held germplasm. Volis (2017) noted that these limitations make their adoption and utilization difficult and are necessitating the introduction of more flexible, integrated, or intermediated approaches, such as *quasi* in situ, *inter situs*, and circa *situm* approaches.

The main focus of this chapter is to highlight the potential and actual consequences of disease outbreaks in some ex situ plant conservation systems. The rate of disease emergence in conserved germplasm is alarming and the monitoring and control of such disease and disease outbreaks are ill-understood. Current methods include the surveillance of disease vectors and levels of exposure, but a framework targeting conservation education about disease spread and risk assessment within conserved individuals and populations may promote understanding of the disease risks and guide control and interventions (Ogwu 2009; Cooke et al. 2020). In addition, this chapter seeks to show that there is more to ex situ plant conservation than holding genetic materials by highlighting the need for efficient, effective, and quality management systems to monitor, prevent, and control disease outbreaks.

18.2 Plant Conservation Approaches

The core plant conservation approaches are either in situ or ex-situ.

In situ plant conservation: In in situ or on-site plant, conservation refers to the preservation, maintenance, protection, and recovery attempt targeted at ensuring viable populations of plant species within their natural habitats or historic range, neighboring areas, or secondary areas, where they have evolved distinctive properties and became naturalized. In situ plant management approaches can either be species or taxa specific, groups of species or taxa, or whole ecosystems with native or naturalized plants (Heywood and Dulloo 2005). Examples include crops that are of economic and ecological value or domesticated plant species that might have developed important biotic characteristics or transmittable traits and adapted to abiotic conditions including susceptibility and resistivity to various diseases in human-dominated environments (Myer et al. 2000). In in situ plant conservation, land areas such as nature reserves, national parks, sacred forests, sanctuaries, etc. may be set aside, where plant species are allowed to subsist within the ecosystems naturally or are properly managed within an ecological continuum by humans or systems set up by humans. Such reserves are useful for species that are endangered, threatened, or near extinction (Frankel and Hawkes 1975). However, for species that are widely distributed and utilized, the in situ conservation of their total genetic diversity may be difficult. Although species conserved within their natural habitats and historic distribution range have a high potential for the continued evolution of traits important for their survival undisturbed, as among other things, their populations are subjected to the influence of natural selection. However, some challenges associated with establishing this type of on-site system include social, economic, environmental, and cultural-like cost, size and management requirements, location, demand for the crop resources, infrastructure, accessibility, as well as political issues, and issues associated with the risk or loss of genetic resources such as wipe out from natural disasters such as fire, etc. Nonetheless, this conservation method is positively significant for the conservation of crop wild relatives of important economic and ecological plant species and several other categories of crop plants, including timber and non-timber forest plant species. The effectiveness of on-site plant conservation may also be connected to the availability of ex situ methods of plant conservation to complement it. Most in situ plant conservation programs are set up to cater to perennials especially those that are vegetatively propagated, and those with recalcitrant seeds that cannot survive the long cold and dry storage conditions of ex situ plant conservation programs. In addition, wild plant species and crop wild relatives are capable of best maintaining their original characteristics under in situ conservation, especially in native or historic ranges or habitats, where they are well-adapted. Therefore, there is a need to consider plant phenology, diversity, and distribution as well as the appropriateness of climatic, altitudinal, and latitudinal conditions zones when establishing in situ plant conservation sites (Pammenter and Berjak 1999).

Ex situ plant conservation: Ex situ or off-site plant conservation refers to the preservation of genes or whole plant germplasms outside their environment of natural occurrence (Hoyt 1988). It is defined as a set of techniques involving the transfer of endangered, endemic, threatened, or important plants or plant species away from their natural habitat or their center of origin to new sites under the care and control of humans to ensure their protection and preservation as well as the availability genetic materials for agriculture, education, research, breeding, and re-introduction. Ex situ plant conservation may be focused on the collection, maintenance, management, preservation, and breeding of endangered and threatened plants species under partially or wholly controlled conditions in specific areas. Examples of ex situ plant conservation systems include field banks, gene banks, botanical gardens, arboretum, pollen banks, tissue culture, seed banks, etc. The stress on plant diversity from competition for light, water, and other environmental resources and space can be avoided through ex situ conservation of plant genetic resources, wherein ideal conditions necessary for their growth, development, and reproductive viability are provided and monitored. Conservation of plants that are propagated by seed is relatively easy through ex situ conservation as well as plants that produce seeds with unorthodox storage behavior, for instance, seeds with viability that can only be maintained by drying or storage at (extremely) low temperature like under liquid nitrogen (−196 °C). In the case of orthodox seeds, i.e., desiccation-tolerant seeds, the lower seed moisture content is associated with an increase in their shelf life and viability albeit within certain limits (Costa et al. 2017; Pritchard et al. 2022). Some agro-horticultural crops have been reported to tolerate desiccation and this may have correlative effects on their productivity and other abiotic conditions outside ex situ conservation systems (Vílchez et al. 2016; Lamaoui et al. 2018; Liu et al. 2019). Certain economic crops such as cacao (*Theobroma cacao*) and rubber (*Hevea brasiliensis*), as well as many tropical fruits (e.g., pear (*Persea americana* and mango (*Mangifera* indica)) and many timber species [e.g., oak (*Quercus* species) and buckeye (*Aesculus* species)] produce recalcitrant seeds. These desiccation-sensitive seeds are killed when subjected to drying conditions below a critical moisture content value, which is usually between 12% and 35% moisture content. Physiological research on desiccation-sensitive

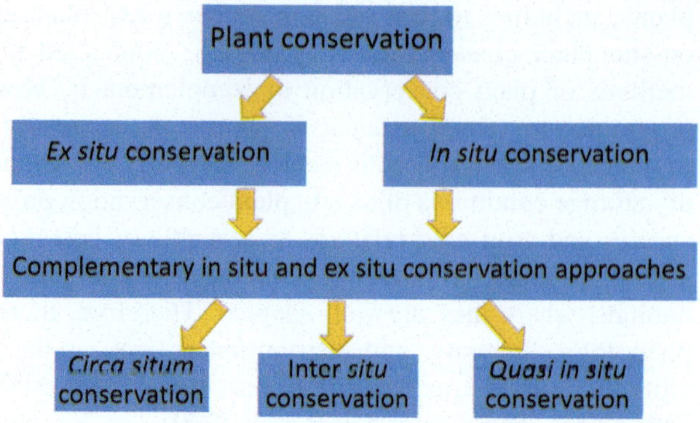

Fig. 18.1 Plant conservation approaches

seeds is currently very limited and key issues remain unresolved, especially pertaining to the effects of desiccation on their cells and tissues (Matilla 2021).

Other plant conservation approaches are largely considered complementary to and linked or derived from in situ and ex situ plant conservation and these include inter *situ*, *quasi* in situ, and circa *situm* plant conservation approaches (Fig. 18.1).

Inter *situ* plant conservation: This is a supplementary system of plant conservation that focuses on the restoration of endangered or threatened plant species and taxa with a significantly declining population within their distribution areas as well as in areas outside or near their historic or native areas of diversity and distribution range. This approach is considered complimentary because of the focus on plant maintenance, protection, and preservation of semi-wild habitat and area restoration—field gene banks in protected areas. Here, plant species survival may be linked to habitat restoration efforts and incorporates certain aspects of in situ as well as ex situ plant conservation techniques. The view of Volis (2016, 2017); Heywood (2018) suggests that inter *situ* conservation is a plant-based "conservation-restoration" approach, wherein there are no alternatives for the management of endangered or threatened plant taxa other than their reintroduction. An example is the restoration of Makauwahi Cave Reserve reported in Burney and Burney (2007).

***Quasi* in situ plant conservation**: *Quasi* in situ or near *situ* conservation is a complementary approach that is similar to inter *situ* conservation by focusing on the long-term prevention of genetic erosion and mitigation of conservation risks and environmental effects by keeping plants in human-managed ecosystems. Just like inter *situ,* the plants may be used for habitat restoration by incorporating aspects of in situ and ex situ plant conservation approaches. *Quasi* in situ conservation preserves plant genetic diversity close to their native range under natural or semi-natural conditions.

Circa *situm* plant conservation: This is a complementary conservation approach, wherein plants are held near or outside their native range under the control of humans including within farming systems to control ecosystem degradation and the decline of species' genetic resources.

18.3 Some Ex Situ Plant Conservation Methods

Conventional ex situ plant conservation methods include:

- **Gene banks,** e.g., Gene bank of the National Centre for Genetic Resources and Biotechnology, Nigeria, gene bank of the National Fruit Collection, UK, CGIAR Gene bank Platform in collaboration with the International Center for Tropical Agriculture, Sugarcane gene bank at the Centro Nacional de Pesquisa de Recurso Geneticos e Biotecnologia, Brazil, etc.
- **Botanical gardens,** e.g., Kirstenbosch National Botanical Garden Cape Town South Africa, Mauritius National Botanial Garden, Montreal Botanical Garden, Canada, Jardin Majorelle Marrakech, Morocco, National Botanical Garden Lucknow, India, Royal Botanic Garden UK, Jardim Botanico, Brazil, Sarius Palmetum and Botanical Gardens, Nigeria, etc.
- **Seed banks,** e.g., The Svalbard Seed Bank, Millenium Seed Bank, UK, Southern Cross Seed Bank USA, Seed Vault at PlantBank Australia, Ejere Farming Community Seed Bank Ethiopia, etc.
- **Field gene banks,** e.g., Fied gene banks of the International Institute for Tropical Agriculture Nigeria, International Centre for Tropical Agriculture Columbia, The National Plant Genetic Resources Centre of South Africa, The National Gene Bank of Kenya, etc.
- **Arboreta,** e.g., Kaya Cum Arboretum, Morton Arboretum, Trsteno Arboretum Dubrovnik Croatia, Arnold Arboretum, etc.

Biotechnological methods are commonplace in ex situ conservation and include:

- **Cryopreservation**: Storage plant tissues in liquid nitrogen at −196 °C for very long periods. This can be carried out via direct desiccation and freezing or desiccation of encapsulated tissues through vitrification. Vitrification is a type of cryopreservation that uses concentrated cryoprotectant cocktails to preserve plant tissues.
- **In-vitro conservation**: The use of sterile culture for vegetatively propagated plants either plants that do not produce seeds or produce recalcitrant seeds with long juvenile periods.
- **Slow growth method**: This technique seeks to minimize cell division and growth and increase plant longevity under storage without genetic changes.
- **Pollen storage**: Conservation of haploid plants and used to produce diploid plants.
- **Plant DNA storage or bank**: Regeneration of entire plants from stem cells or DNA (totipotency).

Others are micropropagation, protoplast culture, somatic embryogenesis, organogenesis, and micrografting under ex situ conservation; plant's genetic resources may be improved by marker techniques such as biochemical markers such as isozymes, phytochemical markers, and DNA-based marker techniques rapid amplified polymorphic DNA, inter simple sequence repeats, sequence characterized amplified region, simple sequence repeats, restriction fragment length polymorphism, amplified fragment length polymorphisms, selective amplification of microsatellite polymorphic loci, single-nucleotide polymorphism, expressed sequence tagged, single-strand conformation polymorphism, cleaved amplified polymorphic sequence, DNA barcoding, RNAi technologies, molecular pharming, etc. (Khan et al. 2012).

According to the Secretariat of the Convention on Biological Diversity (2009), some of the importance of ex situ conservation include:

- Prevent genetic erosion and support the survival and diversity of plant survival
- To safeguard plant populations from physical destruction by humans and as a cushion against natural disasters
- Promoting the plant species richness
- Easy access to plant materials for research, education, and breeding
- Ensuring viable reproductive units for plant species and recovery from threats
- Eradicates threats to biodiversity
- Ensure the provision of plant services to the human population.

The ex situ conservation of plant genetic resources contributes to prevent the extinction of plant while ensuring their perseverance (Hobbs and Gupta 2003). The diversity of plants is necessary for the functioning of natural and anthropogenic ecosystems, which will, in turn, provide fundamental support and benefits upon which life depends (Turnbull et al. 2016). These services and benefits include carbon sequestration, climate change mitigation, biogeochemical cycling of nutrients, food production, and pollination (Ogwu 2010; Ogwu et al. 2014a, b). Direct benefits from plant resources include food, clothes, shelter, medicine, and numerous raw materials for producing countless products and supporting the economic development of many countries. Plants are, therefore, an essential resource for environmental protection and human existence. Ex situ plant conservation is a key component of global, regional, and national biodiversity conservation efforts (Manning et al. 2006). Nonetheless, plant diversity is reportedly being lost at an unprecedented rate, and in the last centuries, the pace of human and societal development has caused around 25% of the 400,000 plant species documented in the world to be threatened with extinction mainly from habitat loss and degradation (BGCI 2020). Other sources of these threats are the introduction of invasive alien species, overexploitation, climate change, fertilizer overload, and pollution (Ogwu and Osawaru 2015; Osawaru and Ogwu 2015). Due to these multidimensional threats, only through coordinated action like the use of integrated and complementary conservation approaches will it be possible to halt or reduce the loss of biodiversity (Osawaru et al. 2014).

18.4 Disease Outbreak in Plants Maintained Under Ex Situ Conservation

Plant pathogens can affect the ex situ conservation of plant genetic resources. Most of these disease pathogens include fungi, bacteria, protozoa, and viral species. They may be infectious or non-infectious and enter the ex situ conservation facility from the plant accessions (where the pathogens may be in a resting phase), from other plant materials housed in the same unit within the ex situ facility, or introduced by humans or environmental changes. If undetected, they may be capable of annihilating plant genetic resources or cause debilitating effects in the ex situ conservation facility (Chime et al. 2018; Ogwu et al. 2018). It is important to note that some ex situ plant conservation storage conditions such as cryopreservation may discourage the proliferation or incidence of certain plant diseases. However, some ex situ conservation techniques and diseases that affect plants and genetic material held under these conservation techniques are presented below:

1. **Field Gene banks**: Field gene banks store crop plants wholly in fields under similar conditions as the species' natural habitat, preserving them and making these species available for reintroduction. Germplasm in field gene banks can be exposed to pathogenic microorganisms, insect vectors, bushfires, natural disasters, etc. (Evans and Youngquist 2004) that may lead to disease presentation. Taking the germplasm of the *Zingiber officinale* in a field gene bank as a case study. Research has shown that ginger plant is affected by soil-borne diseases which is a major biotic factor that can threaten this specie's existence. Plants in field gene banks are also propagated in the soil which makes them susceptible to soil-borne diseases (Nair 2013). A soil-borne disease that affects ginger germplasms in field gene banks includes bacterial wilt, which can also affect seeds and rhizome tissues. It is caused by the bacteria *Ralstonia solacearum yabuuchi,* which causes wilt in ginger within 5–10 days of infection (Kumar and Sarma 2005). Symptoms include linear streaks or water-soaked patches on the pseudostem. *Ralstonia solacearum* can spread easily by sticking to human hands, vehicle tires, and boots, as well as field equipment (Janse 1996). Outbreaks of bacterial wilt have also been recorded in sub-Saharan Africa, and Rwanda precisely especially at the local level amongst fields of *Ipomoea batatas* (Uwamahoro et al. 2018).
2. **Gene banks**: A plant gene bank is a biorepository for the preservation of plant genetic material. Plant species can be stored in gene banks in the form of sectioned tissue cultures as well as seed stockings under ambient temperatures. There are over 1240 gene banks worldwide. These gene banks conserve plant genetic material for long periods under conditions that do not alter their re-establishment or subsequent use. Gene banks aim at regenerating plant material using stored genetic information over some time. Gene bank activities, however, have been hindered by certain factors, one of which includes disease outbreaks. South Pacific Coconut gene bank which holds over 3200 coconut plant

materials with about 57 different varieties recorded an outbreak of the '*Bogia coconut Syndrome*' in 2012 caused by a bacteria, *Phytoplasma.* Cassava viral diseases have also been reported in African countries, such as Burundi, Rwanda, Tanzania, Uganda, DR Congo, and Kenya. They recorded the outbreak of cassava mosaic disease as well as the cassava brown streak disease, which is caused by potyviruses, and used survey standardized protocols in each of these countries (Sseruwagi et al. 2001; Legg et al. 2006).

3. **Botanical gardens**: Botanical gardens are specially protected locations established for the conservation of plant species. Some botanical gardens have seed banks and tissue culture facilities as well as other ex situ technologies. The main purpose of botanical gardens is plant resource development and utilization of living collections of plant material for the preservation of biodiversity as well as research. However, there are challenges faced by most botanical gardens arising from pests and diseases outbreak. For instance, the South China Botanical Garden has witnessed 2 bacterial and 13 fungal diseases in their ginger garden (Wu et al. 2002). Wu et al. (2002) suggested that these diseases can be controlled using fertile soil with proper drainage, regulating plant density, adopting rotation, irrigating moderately, shade, planting healthy seedlings and rhizomes, and pesticide use. In addition, O'Donnell and Sharrock (2018) reported the increase in fungal disease (Myrtle rust) incidence and impacts on numerous plant species from the Myrtaceae family in Australian and New Zealand Botanic gardens. Alomoa-bobone viral disease of taro (*Colocasia esculenta*) around botanic gardens in the Pacific Islands has been described as devastating (Dunn 2017). In Nigeria, a preliminary report implicates tomato diseases outbreak at a university botanical garden of tomatoes in Choba, Rivers State. Tomato cultivated in the garden was reported to have the following symptoms irregular leaf spots, dark brown lesions with chlorotic halos, and yellowing, which is likely due to a fungal infection.
4. **Arboretum**: An arboretum is a kind of garden that is used to cultivate and preserve trees and other woody plants. An arboretum can be likened to a botanical garden, but it is exclusively composed of forest trees, woody plants, and sometimes herbaceous or shrubby plant species. Arboretums around the world aim to improve forest species diversity by preserving living collections. An arboretum is, therefore, an ex situ conservation method that seeks to conserve threatened tree plant species. Various diseases as well as their causative organisms have been spotted in arboreta. For instance, the bur oak blight has affected most bur oak species (*Quercus macrocarpa*) in Morton Arboretum USA (The Morton Arboretum 2013). The organism responsible for this disease is a fungus known as *Tubakia iowensis* which causes purple coloration on the lower side of the bur oak leaves. The disease eventually spreads to the best part of the leaves and causes defoliation. In Holden Arboretum USA, the fungal disease agents *Cryptococcus fagisuga* and *Neonectria coccinea* have been reported to affect American beech (*Fagus grandifolia*).
5. **Cryopreservation**: Cryopreservation involves the preservation of living cells, tissues, and organs by cooling to very low temperatures, typically -196 °C in

Table 18.1 Seed-borne pathogen on cereal crops in Nigeria

Crop	Pathogen	Disease
Maize	*Ustilago* sp.	Cob smut
	Fusarium sp.	Cob rot
	Erwinia strewartii	Bacterial wilt
	Colletotrichum sp.	Anthracnose
Sorghum	*Aspergillus flavus*	Cob rot
	Aspergillus flavus	Maize wilt
Rice	*Sphacelotheca* spp.	Ergot
	Claviceps microcephala	Seed rot
	Colletotrichum sp.	Stalk rot/red leaf
	Fusarium sp.	Fusarium wilt
	Aspergillus spp.	Seed rot
	Drechslera oryzae	Seed rot

Source: Adapted from Amodu and Aku (2015)

liquid nitrogen (Tiersch 2008). Cryopreservation techniques apply to any type of plant tissue with regeneration potential with plant species in various forms, such as cell suspension, apices, and the somatic and zygotic embryo (Reed 2008). This modern biotechnology-based conservation method tends to preserve plant materials for over a very long period, making it the most advantageous method upholding the concept of protection of species until after factors threatening their existence have been eradicated. The extreme freezing temperatures of liquid nitrogen have been harnessed by scientists and used to preserve certain components of plant germplasm that can easily be lost in basic in vitro maintenance. However, the use of liquid nitrogen may be intolerable to certain plant genes and affect their resistance to certain diseases after storage. Therefore, there is a need to study plant characteristics before subjecting them to cryopreservation.

6. **Seed banks**: Seed banks as ex situ conservation facilities focus on keeping plant germplasms (seeds) in a controlled environment that is capable of prolonging and preserving seed viability and quality for subsequent use. Seed banking is aimed at conserving the genetic purity of seeds for regeneration (Agera and Dau 2012). Some seed-borne pathogens are capable of persevering and surviving seed processing processes before the seed is held in the seed bank. These pathogens can subsequently produce metabolites that affect seed metabolism at the cellular level within seed banks. Aflatoxins and other mycotoxins may reduce seed viability, germination, and vigor which ultimately affect seed yield (Marley 1996). In addition, the control of seedling diseases is considered a major priority in the seed bank system. This would require seed treatment to ensure systemic resistance against diseases during the reestablishment (Jensen et al. 1998). Some common seed-borne pathogens that may affect seed banking in Africa are presented in Table 18.1.

18.5 Management of Diseased Plants Held Under Ex Situ Conservation

The concept of ex situ plant conservation is to create a haven for plant species to thrive until factors affecting their existence in their center of origin or immediate habitat have been removed and immediate reintroduction is likely to occur successfully (Secretariat of The Convention on Biological Diversity 2009). Regular disease diagnosis and pathogen identification and subsequent disease management using cultural mechanical or physical, biological, chemical, and/or integrated approaches are necessary for every ex situ plant conservation facility to monitor and keep disease outbreaks under check (Chime et al. 2018). Some management strategies are discussed below:

Germplasm health: Infection and contamination of germplasms in gene banks with pathogens may cause several problems in field gene bank facilities and germplasm management. Seed longevity may be affected, and pathogens may spread within accessions and in the collections and affect susceptible species, which may spread further when the materials are distributed to new sites along with the infected germplasms. Therefore, adequate germplasm health management measures should be adopted by gene banks to eliminate infection and contamination or at least reduce it to the barest level (Sutherland et al. 2004). Samples need to be checked for the presence of diseases and disease vectors. In management principle, all samples intended for storage and subsequent distribution for use must be healthy to limit the spread of disease-causing organisms. This would require gene bank management to establish some set of detailed germplasm health monitoring strategies that would be species-appropriate and take into consideration global, regional, and national regulations. This will often entail the use of standard diagnostic tools for inspection of seed stocks to be stored for conservation and reintroduction (Frison and Jackson 1995). In addition, when involved in the international transfer of plant materials, other protocols might be applicable. However, ex situ protocols are preferentially designed for the country of origin of plant species or using temperate or tropical conditions.

Management of seed-borne pathogens in seed banks: Seedlings held under ex situ conservation such as in seed banks are often tested for viability involving field trials. If infected, this may lead to the reintroduction of a certain disease or disease vectors from the conserved plant material. Therefore, the control of seedling disease before, during, and before reestablishment should be considered a priority. According to Mortensen (1997); Cram and Fraedrich (2009); Amodu and Aku (2015) before seed treatment, identification, and detection of diseased materials may be achieved through direct plating, washing assay, destruction assay, enzyme link immunoabsorbent assay, radiographic method, polymerase chain reaction. After detection of diseased seedlings, seed bank institutions can now go-ahead to apply treatment before storage.

- **Chemical treatments of seed**: Some chemicals are widely used to control the spread of disease and to checkmate the activities of disease vectors. These include fungicides, bactericides, nematicides, avicides for fungi, bacteria, nematodes, birds, respectively, etc. However, in Africa, fungicides and insecticides are widely used for seed treatment.
- **Non-chemical seed treatment**: This includes biological or bio-based treatment of seeds. An example of biological treatment is the use of antagonistic biological agents, such as bacteria and fungi *Trichoderma* sp. (fungus), *Bacillus subtilis* (bacterium). These organisms multiply in the soil to protect root systems from soil-borne pathogens. In the case of bio-based treatment, renewable resources containing an active compound from a living agent or agents are used, e.g., plant extracts such as capsaicin from pepper plant (*Capsicum annum*). These treatments have little to no negative environmental impacts.

Management of diseases in arboreta: Like in other off-site conservation techniques, disease identification is followed by a counter activity to eradicate further spread, and this involves using various forms of methods such as dispersal of antimicrobial chemicals such as in the outbreak of '*bur oak blight',* the best options are to hamper further spread by introducing fungicides. Using the oak tree as a case study, recent outbreaks have shown that after defoliation and falling of the leaves, the petiole still carries the infection which could be a major challenge if further spread occurs; it may predispose the tree to other problems, such as *Armillaria* root rot as well as two-lined chest borer. Therefore, it would be in the best interest of arboreta to seek the help of a licensed professional for adequate injections.

Management of living collections in field gene banks: Living collections of various plant species are usually affected by natural soil-borne diseases, as they are cultivated on natural soil. As earlier stated, the *bacterial wilt* of the *Zingiber officinales* affects the collar region of the pseudostem of the plant. To halt the further spread of the bacterial infection, the pseudostem can easily be separated from the rest of the plant by gently pulling off or breaking off the diseased part at the base. On the other hand, this disease can also be prevented through chemical means by treating seeds and rhizomes with 'emisan' and plantomycin with an active dosage of 3 sprays for 30 min 15–30 days after planting. In addition, pouring 0.2% copper oxychloride on the soil can prevent or control the disease.

Management of botanical gardens: Most botanic gardens help to conserve a wide variety of plant materials both living germplasm, tissue culture as well as seedlings. Although botanical gardens are recognized as a source of visitors' attraction, they are one of the most important conventional methods of ex situ plant conservation. Like most scientific or research institutions, botanic gardens require effective management ranging from individual activity to implementation of government reforms. Reforms should be put in place to hamper the illegal or unauthorized collection of plant material from botanical sites, especially by tourists and visitors. Infrastructure to reduce the effects of biotic and abiotic stress on plant germplasm such as screen houses and pest repellant chemicals and equipment should also be provided. Soil-borne diseases tend to affect plant material in botanical gardens like in the case of 'wilt' in the *Lycopersicum esculentum.* To control the spread, action should be taken. Proper identification of the specific causative organism should be carried out by soil sample testing, then a proper mapping out of the area of soil affected should be carried out and segmented, growing or rather cultivation should not be carried out on that particular area to reduce the soil fungal activity over time, and fungicides are also an option. The use of naturally adapted or genetically modified disease-resistant cultivars can also be planted as a long-term disease management strategy.

Plant exclusion, pathogen eradication, and plant protection: To prevent the spread of disease or disease-causing agents, it is necessary to quarantine plant materials before housing them within ex situ conservation facilities. Plant materials at every point must be certified disease-free when infected should be treated and disinfected to prevent spreading to uncontaminated sections. This will likely reduce the inoculum density or eliminate it.

18.6 Recommendations and Conclusion

The knowledge of ex situ conservation is likened to a support system upholding the essence of biodiversity despite threatening factors. Species conservation by ex situ means has reduced the threat of extinction and addressed diverse sustainability issues about plant genetic resources. Nonetheless, there is a need for an adequate management system in most ex situ plant conservation facilities. In addition, ex situ conservation approach should be evaluated for suitability in achieving species-specific conservation objectives before adoption. Both conventional and biotechnological ex situ plant conservation strategies are still evolving and require innovation and organization of the species to be preserved about their natural environment, the center of origin, and distribution range. An efficient ex situ plant conservation strategy would seek to establish a relationship between species' natural or immediate habitat, native range, and intrinsic characteristics. An understanding of these indices would go a long way in eradicating the stress of adaptation and naturalization. In the case of recalcitrant plant materials such as the germplasm of *Mangifera indica, Hevea brasiliensis, Theobroma cacao, Persea americana,* etc. which tend to lose their viability if subjected to cryogenic procedures or cryopreservation, the adoption of a complementary approach such as conservation in gene banks (under tolerable temperature and other conditions) as well as their establishment in field gene bank or botanical gardens will be preferred.

Most cases of disease outbreaks in ex situ plant conservation facilities are a result of the introduction of new plant material(s), which have originally been predisposed to diseases in their natural habitat; therefore, standard procedures to manage new introductions will be to test for microbial presence and pretreatment before banking. For example, in most gene banks, in vitro conservation procedures will include a test to detect endophytic bacteria, and germplasms infected are treated with antibiotics (like Rifampicin 100 mg/l) before they are stored. This way plant materials can be preserved and also distributed to other facilities. The use of cryopreservation equipment and materials is also a reliable management strategy, as most microbes are intolerant to cryogenic temperatures. Even psychrophiles tend to be dormant and thawed at temperatures lower than −20 °C. Although cryopreservation treatment is expensive, it remains one of the most effective conservation methods as well as a solution to most plant diseases in ex situ facilities (Pegg 2007).

In conclusion, there is a need to establish a system to serve as a reservoir for the abundant natural resources such as plants held in ex situ plant conservation facilities, because human survival depends. It is also necessary to improve the efficiency of human activities that might affect plant diversity. Conservation entails preservation to counter inexhaustibility. Ex situ conservation is complementary to in situ conservation and creates an avenue or a situation for plant genetic materials to be properly safeguarded, thereby cancelling out extinction and extinction threats while allowing for an increase in plant diversity and ecosystems integrity. Ex situ conservation can be likened to refrigerating a perishable food item for future use. Giving us humans an opportunity to access resources unlimitedly from our immediate natural environment. The entirety of conservation is, however, based on the human ability for strategic management.

References

Agera SIN, Dau JH (2012) Seed processing, seed banks and seed conservation: a review. Nigeria J Edu Health Technol Res 3(2):8–18

Aiwansoba RO, Ogwu MC, Osawaru ME (2019) Assessing the relatedness of abelmoschus accessions using morphological characters. J Trop Biol Conserv 16:193–207

Amodu US, Aku BO (2015) Seed-borne disease and Nigerian agriculture. Sch J Agric Veterin Sci 2(3B):243–252

Bellon MR, Dulloo E, Sardos J, Thormann I, Burdon JJ (2017) In situ conservation-harnessing natural and human-derived evolutionary forces to ensure future crop adaptation. Evol Appl 10(10):965–977. https://doi.org/10.1111/eva.12521

BGCI (2020) Why conserve plants? https://www.bgci.org/about/about-plant-conservation

Burney DA, Burney LP (2007) Paleoecology and "inter-situ" restoration on Kaua'I, Hawai'i. Front Ecol Environ 5(9):483 490

Chime AO, Osawaru ME, Ogwu MC, Ogunlowo EO (2015) Preservation of fresh tomato fruits (*Solanum lycopersicon* L.) using leaf extracts of bitter melon (*Momordica charantia* L., Cucurbitaceae). J Phytopathol Plant Health 3:29–37

Chime AO, Aiwansoba RO, Ogwu MC (2018) Pathological status of plant germplasm and sustainable crop production and conservation. J Energy Nat Resour Manag 1:17 21

Chokheli VA, Dmitriev PA, Rajput VD, Bakulin SD, Azarov AS, Varduni TV, Stepanenko VV, Tarigholizadeh S, Singh RK, Verma KK, Minkina TM (2020) Recent development in micropropagation techniques for rare plant species. Plan Theory 9(12):1733. https://doi.org/10.3390/plants9121733

Coelho N, Gonçalves S, Romano A (2020) Endemic plant species conservation: biotechnological approaches. Plan Theory 9(3):345. https://doi.org/10.3390/plants9030345

Cohen JI, Williams JT, Plucknett DL, Shands H (1991) Ex situ conservation of plant genetic resources: global development and environmental concerns. Science 253(5022):866–872. https://doi.org/10.1126/science.253.5022.866

Cooke SJ, Madliger CL, Cramp RL, Beardall J, Burness G, Chown SL, Clark TD, Dantzer B, de la Barrera E, Fangue NA, Franklin CE, Fuller A, Hawkes LA, Hultine KR, Hunt KE, Love OP, MacMillan HA, Mandelman JW, Mark FC, Martin LB, Newman AEM, Nicotra AB, Robinson SA, Ropert-Coudert Y, Rummer JL, Seebacher F, Todgham AE (2020) Reframing conservation physiology to be more inclusive, integrative, relevant and forward-looking: reflections and a horizon scan. Conserv Physiol 8(1):coaa016. https://doi.org/10.1093/conphys/coaa016

Costa MD, Cooper K, Hilhorst HWM, Farrant JM (2017) Orthodox seeds and resurrection plants: two of a kind? Plant Physiol 175(2):589–599. https://doi.org/10.1104/pp.17.00760

Cram MM, Fraedrich SW (2009) Seed diseases and seed-borne pathogens of North America. Tree Planters Notes 53(2):35–44

Dunn CP (2017) Biological and cultural diversity in the context of botanic garden conservation strategies. Plant Divers 39(6):396–401. https://doi.org/10.1016/j.pld.2017.10.003

Evans J, Youngquist J (2004) Encyclopedia of Forest sciences. Elsevier, p 156

Fazey I, Fischer J, Lindenmayer D (2005) Who does all the research in conservation biology? Biodivers Conserv 14:917–934. https://doi.org/10.1007/s10531-004-7849-9

Frankel OH, Hawkes JG (1975) Crops genetic resources for today and tomorrow. Cambridge University Press, Cambridge

Frison EA, Jackson GVH (1995) Plant health and germplasm collectors. In: Guarino L, Ramanatha Rao V, Reid R (eds) Collecting plant genetic diversity. CAB International, Wallingford, pp 329–341

Greene SL, Kisha TJ, Yu LX, Parra-Quijano M (2014) Conserving plants in gene banks and nature: investigating complementarity with *Trifolium thompsonii* Morton. PLoS One 9(8):e105145. https://doi.org/10.1371/journal.pone.0105145

Heywood VH (2018) Conserving plants within and beyond protected areas - still problematic and future uncertain. Plant Divers 41(2):36–49. https://doi.org/10.1016/j.pld.2018.10.001

Heywood VH, Dulloo ME (2005) In Situ conservation of wild plant species, a critical global review of good practices. Food and Agriculture Org, Rome

Hobbs PR, Gupta RK (2003) Resource-conserving technologies for wheat in the rice–wheat system. In: Ladha JK, Hill JE, Duxbury JM, Gupta RK, Buresh RJ (eds) Improving the productivity and sustainability of rice-wheat systems: issues and impacts. American Society of Agronomy, Inc. https://doi.org/10.2134/asaspecpub65.c7

Hoyt (1988) Ex-situ plantconservation. https://www.sciencedirect.com/sciences/*ex-situ*conservation/ Accessed 14 Oct 2021

Ikhajiagbe B, Ogwu MC, Lawrence AE (2020) Single-tree influence of *Tectonia grandis* Linn. f. on plant distribution and soil characteristics in a planted forest. Bull Natl Res Centre 44:29. https://doi.org/10.1186/s42269-020-00285-0

Ikhajiagbe B, Aituae W, Ogwu MC (2021a) Morpho-physiological assessment of oil palm (Elaeis guineensis Jacq.) seedlings exposed to simulated drought conditions. J Oil Palm Res. https://doi.org/10.21894/jopr.202100018

Ikhajiagbe B, Ogwu MC, Ogochukwu OF, Odozi EB, Adekunle IJ, Omage ZE (2021b) The place of neglected and underutilized legumes in human nutrition and protein security. Crit Rev Food Sci Nutr 62(14):3930–3938. https://doi.org/10.1080/10408398.2020.1871319

IUCN (2001) IUCN red list categories and criteria, version 3.1. IUCN species survival commision. IUCN, Gland/Cambridge, p 30

Janse J (1996) Potato brown rot in Western Europe history, present occurrence and some remarks on possible origin, epidemiology and control strategies. Bull OEPP/EPPO 26:679–695

Jensen BD, Olumide Latunde Dada A, Hudson D, Lucas JA (1998) Protection of *brassica* seedlings against *downy mildew* and damping- off by seed treatment with CGA 245704, an activator of systemic acquired resistance. Pestic Sci 52(1):63–69

Khan S, Al-Qurainy F, Nadeem M (2012) Biotechnological approaches for conservation and improvement of rare and endangered plants of Saudi Arabia. Saudi J Biol Sci 19(1):1–11. https://doi.org/10.1016/j.sjbs.2011.11.001

Kumar A, Sarma YR (2005) Characterization of *Ralstonia solanacearum* causing bacterial wilt of ginger in India. Indian Phytopathol 57:12–17

Lamaoui M, Jemo M, Datla R, Bekkaoui F (2018) Heat and drought stresses in crops and approaches for their mitigation. Front Chem 6:26. https://doi.org/10.3389/fchem.2018.00026

Legg JP, Owor B, Sseruwagi P, Ndungwu J (2006) Cassava mosaic virus disease in east and Central Africa: epidemiology and regional management of a regional pandemic. Adv Virus Res 67:355–418

Li DZ, Pritchard HW (2009) The science and economics of ex situ plant conservation. Trends Plant Sci 14(11):614–621. https://doi.org/10.1016/j.tplants.2009.09.005

Li D, Stucky BJ, Deck J, Baiser B, Guralnick RP (2019) The effect of urbanization on plant phenology depends on regional temperature. Nat Ecol Evol 3(12):1661–1667. https://doi.org/10.1038/s41559-019-1004-1

Lindenmayer D, Burgman M (2005) Practical conservation biology. CSIRO Publishing, Collingwood, 624 p

Liu J, Moyankova D, Djilianov D, Deng X (2019) Common and specific mechanisms of desiccation tolerance in two gesneriaceae resurrection plants. Multiomics evidences. Front Plant Sci 10: 1067. https://doi.org/10.3389/fpls.2019.01067

Liu Y, Li G, Wu X, Niklas KJ, Yang Z, Sun S (2021) Linkage between species traits and plant phenology in an alpine meadow. Oecologia 195(2):409–419. https://doi.org/10.1007/s00442-020-04846-y

Lortie CJ, Filazzola A, Brown C, Lucero J, Zuliani M, Ghazian N, Haas S, Owen M, Butterfield HS, Nix E, Westphal M (2021) Facilitation promotes plant invasions and indirect negative interactions. Oikos 1:1–6. https://doi.org/10.1111/oik.08443

Manning AD, Fischer J, Lindenmayer DB (2006) Scattered trees are keystone structures – Implications for conservation. Biol Conserv 132(3):311–321

Marley PS (1996) The importance of fungi in cereal grain storage. NOMA 12:13–17

Matilla AJ (2021) The orthodox dry seeds are alive: a clear example of desiccation tolerance. Plants (Basel, Switzerland) 11(1):20. https://doi.org/10.3390/plants11010020

Maunder M, Havens K, Guerrant EO, Falk D (2004) Ex situ methods: a vital but underused set of conservation resources: supporting species survival in the wild. In: Guerrant EO, Havens K, Maunder M (eds) Ex situ plant conservation: supporting species survival in the wild. Island Press, pp 3–20

McDonough MC, Gallinat AS, Zipf L (2020) Low-cost observations and experiments return a high value in plant phenology research. Appl Plant Sci 8(4):e11338. https://doi.org/10.1002/aps3.11338

Mortensen CN (1997) Seed health testing for bacterial pathogens. Danish Government Institute of Seed Pathology for Developing Countries, Copenhagen. 66

Myer L, Wemik H, Gurino L, Smale M, Br AHD, Sadiki M, Sthapit B (2000). A training guide for in situ conservation on farm. International plant genetics resources institute

Nair KPP (2013) The agronomy and economy of tumeric and ginger. https://www.scribd.com/book/282548461/. Accessed 14 Oct 2021

O'Donnell K, Sharrock S (2018) Botanic gardens complement agricultural gene bank in collecting and conserving plant genetic diversity. Biopreserv Biobank 16(5):384–390. https://doi.org/10.1089/bio.2018.0028

Ogwu MC (2009) The significance and contribution of conservation education in Nigeria: an appraisal and a call for improvement. In: Babalola F (ed) Proceedings of the maiden seminar of Nigerian Tropical Biology Association. Nigerian Tropical Biology Association and Tropical Biology Association, pp 109–114

Ogwu MC (2010) Conserving biodiversity even in poverty: the African experience'. In: Babalola F (ed) Seminar Proceedings of Nigerian Tropical Biology Association. Nigerian Tropical Biology Association and Tropical Biology Association, pp 112–117

Ogwu MC (2019a) Towards sustainable development in Africa: the challenge of urbanization and climate change adaptation. In: Cobbinah PB, Addaney M (eds) The geography of climate change adaptation in urban Africa. Springer Nature, Switzerland, pp 29–55. https://doi.org/10.1007/978-3-030-04873-0_2

Ogwu MC (2019b) Lifelong consumption of plant-based GM foods: is it safe? In: Papadopoulou P, Misseyanni A, Marouli C (eds) Environmental exposures and human health challenges. IGI Global, Pennsylvania, pp 158–176. https://doi.org/10.4018/978-1-5225-7635-8.ch008

Ogwu MC (2020) Value of Amaranthus [L.] species in Nigeria. In: Waisundara V (ed) Nutritional value of Amaranth. IntechOpen, London, pp 1–21. https://doi.org/10.5772/intechopen.86990

Ogwu MC, Osawaru ME (2015) Soil characteristics, microbial composition of plot, leaf count and sprout studies of cocoyam (Colocasia [Schott] and Xanthosoma [Schott], Araceae) collected in Edo State, Southern Nigeria. Sci Technol Arts Res J 4(1):34–44. https://doi.org/10.4314/star.v4i1.5

Ogwu MC, Osawaru ME, Ahana CM (2014a) Challenges in conserving and utilizing plant genetic resources (PGR). Int J Genet Mol Biol 6(2):16–22. https://doi.org/10.5897/IJGMB2013.0083

Ogwu MC, Osawaru ME, Chime AO (2014b) Comparative assessment of plant diversity and utilization patterns of tropical home gardens in Edo State. Niger Sci Afr 13(2):146–162

Ogwu MC, Osawaru ME, Aiwansoba RO, Iroh RN (2016) Status and prospects of vegetables in Africa. In: Borokini IT, Babalola FD (eds) Conference proceedings of the joint biodiversity conservation conference of Nigeria Tropical Biology Association and Nigeria chapter of Society for Conservation Biology on MDGs to SDGs: toward sustainable biodiversity conservation in Nigeria. University of Ilorin, Nigeria, pp 47–57

Ogwu MC, Ahana CM, Osawaru ME (2018) Sustainable food production in Nigeria: a case study for Bambara groundnut (*Vigna subterranean* (L.) Verdc. Fabaceae). J Energy Nat Resour Manag 1:68–77

Osawaru ME, Ogwu MC (2014) Conservation and utilization of plant genetic resources. In: Omokhafe K, Odewale J (eds) Proceedings of 38th Annual Conference of The Genetics Society of Nigeria. Empress Prints Nigeria Limited, pp 105–119

Osawaru ME, Ogwu MC (2015) Molecular characterization of 36 accessions of two genera of cocoyam (Colocasia [Schott] and Xanthosoma [Schott], Araceae). Sci Technol Arts Res J 4(1): 27–33. https://doi.org/10.4314/star.v4i1.4

Osawaru ME, Ogwu MC (2020) Survey of plant and plant products in local markets within Benin City and environs. In: Filho LW, Ogugu N, Ayal D, Adelake L, da Silva I (eds) African handbook of climate change adaptation. Springer Nature, Switzerland, pp 1–24. https://doi.org/10.1007/978-3-030-42091-8_159-1

Osawaru ME, Ogwu MC, Ahana CM (2013) Current status of plant diversity and conservation in Nigeria. Niger J Life Sci 3(1):168–178

Osawaru ME, Ogwu MC, Ahana CM (2014) Study of the distinctiveness of two genera of cocoyam (Colocasia [Schott] and Xanthosoma [Schott]) using SDS – PAGE. Niger Soc Exper Biol J 14(3):168–179

Osawaru ME, Ogwu MC, Aiwansoba RO (2015a) Hierarchical approaches to the analysis of genetic diversity in plants: a systematic overview. Univ Maurit Res J 21:1–36

Osawaru ME, Ogwu MC, Imarhiagbe O (2015b) Principal component analysis (PCA) as an ideal tool for analysing on-farm research data. Biol Environ Sci J Trop 12(1):514–522

Pammenter NW, Berjak P (1999) A review of recalcitrant seed physiology in relation to desiccationtolerance mechanisms. Seed Sci Res 9:13–37

Pegg DE (2007) Principles of cryopreservation. Methods Mol Biol (Clifton, NJ) 368:39–57. https://doi.org/10.1007/978-1-59745-362-2_3

Pence VC, Finke LR, Chaiken MF (2017) Tools for the ex situ conservation of the threatened species, Cycladenia humilis var. jonesii. Conserv Physiol 5(1):cox053. https://doi.org/10.1093/conphys/cox053

Pritchard HW, Sershen, Tsan FY, Wen B, Jaganathan GK, Calvi G, Pence VC, Mattana E, Ferraz IDK, Seal CE (2022) Regeneration in recalcitrant-seeded species and risks from climate change. In: Baskin CC, Baskin JM (eds) Plant regeneration from seeds. Academic Press, pp 259–273. https://doi.org/10.1016/B978-0-12-823731-1.00014-7

Rao RV, Brown AHD, Jackson M (2001) Managing plant genetic diversity. In: Engels J, Ramanatha R, Brown A, Jackson M (eds) Plant breeding and genetic resources. CABI, Wallingford, p 512

Reed MS (2008) Stakeholder participation for environmental management: a literature review. Biol Conserv 141(10):2417–2431

Secretariat of The Convention on Biological Diversity. (2009). Convention on biological diversity plant conservation report. A Review of Progress in implementing the Global Strategy of Plant Conservation (GSPC) 48

Sseruwagi P, Sserubombwe WS, Legg JP, Ndugwu J, Thresh JM (2001) Methods of surveying the incidence and severity of casava mosaic disease and whitefly vector populations on cassava in Africa: a review. Virus Res 100:129–142

Sutherland WJ, Pullin AS, Dolman PM, Knight TM (2004) The need for evidence-based conservation. Trends Ecol Evol 19:305–308

The Morton Arboretum (2013) Collection of the Morton Arboretum. https://www.artsandculture.google.com/exhibit/what-is-an-arboretum. Accessed 14 Oct 2021

Tiersch TR (2008) Strategies for commercialization of cryopreserved fish semen. R Bras Zootechnol 37:15–19

Turnbull LA, Isbell F, Purves DW, Loreau M, Hector A (2016) Understanding the value of plant diversity for ecosystem functioning through niche theory. Proc Biol Sci 283(1844):20160536. https://doi.org/10.1098/rspb.2016.0536

Uwamahoro F, Berlin A, Bucagu C (2018) Potato bacterial wilt in Rwanda: Occurrence, risk factors, farmers knowledge and attitudes. Food Sec 10:1221–1235

Vílchez JI, García-Fontana C, Román-Naranjo D, González-López J, Manzanera M (2016) Plant drought tolerance enhancement by trehalose production of desiccation-tolerant microorganisms. Front Microbiol 7:1577. https://doi.org/10.3389/fmicb.2016.01577

Volis S (2016) Conservation meets restoration – rescuing threatened plant species by restoring their environments and restoring environments using threatened plant species. Israel J Plant Sci 63: 262–275

Volis S (2017) Complementarities of two existing intermediate conservation approaches. Plant Divers 39(6):379–382. https://doi.org/10.1016/j.pld.2017.10.005

Wu Y, Zulin D, Liu D, Gao Z (2002) The disease and its control on zingiberaceous plants in South China botanical garden. Zhong Yao Cai 25(11):773–775

Part IV
Integrated Conservation and Sustainable Utilization of Africa's Biological Resources and Environment

Chapter 19
Challenges of Sea Turtle Conservation in African Territorial Waters: The Way Out

Godfrey C. Akani, Luca Luiselli, Gijo A. Harry, Kia Tarela Jovita, and Grace N. Alawa

Abstract This paper presents a review of challenges and threats deterring effective marine turtle conservation on the continental shelf of Africa to identify gaps on a regional basis and offer recommendations for research, conservation, and management. The approach involved (i) a comprehensive literature search on marine turtle ecology, status, nesting, and foraging sites, (ii) ground-truthing studies conducted in the course of various coastal environmental impact assessments and PIAs along Nigerian coastlines between 2007 and 2021, and (iii) visits to oil spill sites, and a questionnaire administered to beach dwellers and fishers. Seven turtle species inhabit and nest around African waters. Around the continent, offshore archipelagos and island nations also provide suitable turtle nesting sites. Even the Kemp's ridley (*Lepidochelys kempi*), which is endemic to the North Atlantic and the Gulf of Mexico, has reportedly wandered into African territorial waters and beaches. The Australian flatback (*Natator depressus*) is the only sea turtle species that have never ventured into African coastlines. Significant gaps still exist in the nesting and foraging ecology of sea turtles in the Mediterranean and Red Seas, which happen to be important tourist destinations and oil and gas development areas. Although marine turtle research and conservation efforts have just begun in the East African coastline, parts of the Gulf of Guinea, and African Atlantic coast, Mediterranean and Red Sea coast, a lot more information is still needed for a full understanding of population dynamics, nesting, and foraging behavior of marine turtles on African coastlines. In general, incidental captures, poaching of meat, eggs, and shells for designing expensive tourist souvenirs or trophies; anthropogenic disturbance of nesting and foraging sites, oil industry activities, dredging operations, sand-mining,

G. C. Akani (✉) · G. N. Alawa
Department of Animal and Environmental Biology, Rivers State University, Port Harcourt, Rivers State, Nigeria

L. Luiselli
Centre of Environmental Studies "Demetra" s.r.l. and FIZV, Rome, Italy

G. A. Harry · K. T. Jovita
Department of Biological Sciences, Niger Delta University, Wilberforce Island, Bayelsa State, Nigeria

S. C. Izah, M. C. Ogwu (eds.), *Sustainable Utilization and Conservation of Africa's Biological Resources and Environment*, Sustainable Development and Biodiversity 32, https://doi.org/10.1007/978-981-19-6974-4_19

incessant oil spillage, and coastal developments on beaches are the major threats to sea turtle conservation in African territorial waters. Others are—ingestion of marine litter; delay and disorientation of hatchlings going to sea for the first time due to beach lighting, and series of vehicular ruts on beaches, and climate change impacts also threaten the conservation of sea turtles. Weak or lax governance also has a huge share of the sea turtle threat, and includes—weak turtle protection laws or their inadequate enforcement, lack of awareness, and poor land management and development systems, and pollution. Conservation and management strategies that will emphasize habitat protection, mandatory use of Turtle Excluder Device by trawlers, strengthening national laws on sea turtles; promoting community participation in turtle management; promoting huge national sensitization and enlightenment campaign; promoting national and regional cooperation; and sourcing fund for long-term turtle research and conservation are recommended to mitigate the multiple threats to the conservation of sea turtles. Emphasis on the problems precipitated by chelonitoxism in programmes may help salvage turtles from poaching.

Keywords African turtle diversity · Sea turtle · Biodiversity threats · Sea turtle conservation · African coastal regions · Sea turtle bycatch

19.1 Introduction

Sea turtles are a great asset to the world. They are the surviving members of an ancient group of reptiles. There are seven species of sea turtles that are morphologically well-adapted to sea life—the leatherback (*Dermochelys coriacea*), Green Turtle (*Chelonia mydas*), Hawksbill (*Eretmochelys imbricata*), Loggerhead (*Caretta caretta*), Olive Ridley (*Lepidochelys olivacea*), Kemp's Ridley (*Lepidochelys kempi*), and Flatback (*Natator depressus*). Their bodies are enclosed in a shield-like carapace, fused to its backbone, under it. They are graceful and fascinating swimmers, because they possess strong, flattened forelimbs and enlarged shoulders for swimming long distances, tear glands to remove excess salts, and streamlined carapace to improve hydrodynamic efficiency (Meylan and Meylan 1999; Muir and Sense 2005). Unlike the tortoises, they have heads and limbs that cannot be retracted into their shells or carapaces. They are found in tropical and subtropical seas throughout the world. Sea turtles are known to instinctively migrate long distances to the beaches, where they were hatched to lay their eggs. Using tagging and satellite tracking techniques, turtle behaviors may be monitored (Luschi et al. 1996; Papi et al. 1997). However, they would not lay their eggs until a suitable and safe site is located. Sea turtles are known to possess good sight and hearing ability and a good sense of smell. These adaptations and increased brain complexity have enabled them to thrive and survive most of the long voyages. Most species reach sexual maturity after 7 or 8 years and live 14 or more years. The long generation time of these marine chelonians, their ethnozoological values, the unnecessary consumption of their meat and eggs, and multiple environmental stress factors (Akani and Luiselli 2011) have placed all species on IUCN endangered species Red List (IUCN 1996). Appropriate nesting beaches, marine foraging grounds, and migratory corridors are preferential and important habitats that enhance the survival of sea turtles. Their foraging grounds consist primarily of seagrass beds, coral reefs, and hard-bottom habitats. Adult sea turtles have few natural predators due to their size and hard shells. However, sharks, crocodiles, large fish, and occasionally octopuses can attack them.

19.2 Ecosystem Services of Sea turtles

Sea turtles contribute to the health and stability of marine ecosystems. As keystone species, their elimination can disrupt the natural order of the marine ecosystem, as it will impact other creatures in different ways. For instance, the hawksbill turtles enhance the well-being of the reef system by grazing on sponges in the coral reef system. If left to grow unchecked, the sponges will eventually outgrow the corals, cover them up, and kill the reef. The dwindling numbers of turtles such as the hawksbills may be responsible for the inability of reefs to resist increasing pressure from pollution, algal blooms, overfishing, and climate change. In addition, the green turtles enhance the well-being and growth rate of seagrasses by constantly grazing on them. Leatherbacks forage in the open sea throughout their life, as top predators and check the population of the floating jellyfish. These jellyfish in turn feed on larval fish. It is, therefore, contended that if the number of leatherbacks continues to drop abysmally, jellyfish may increase locally and eat more larval fish, leaving a few fish to attain adulthood. Since a wide variety of marine creatures depend on these fish (especially commercial fisheries), they may end up suffering from this cascading effect.

As the female turtles return to the beaches to lay hundreds of eggs each season, they provide a source of nutrient which plays vital roles in the coastal ecosystem. Many predators feed on the eggs and hatchlings of the sea turtles. Even the eggshells and unhatched eggs from their nest provide nourishing nutrients to the vegetation cover (Hannan et al. 2007). Thus, the loss of such services will potentially affect the dynamics of nearshore ecosystems (Heithaus et al. 2005).

Marine turtles provide substrates for many epibionts (smaller marine creatures) that inhabit or permanently attach to their carapace. Since sea turtles are highly migratory and spend over 80% of their lives foraging, the epibionts serve as potential indicators of sea turtle migratory and foraging behaviors. Among the frequently recorded epibionts on sea turtles are algae, protozoans, nematodes, barnacles, microcrustaceans, anthozoans, etc. Of all the turtle species, loggerheads play host to the largest and most diverse communities of epibionts.

19.3 Economic and Cultural Significance of Sea Turtles

Apart from their role in marine ecosystem stability, sea turtles need to be conserved because of their cultural and economic values. All the values (which are inestimable) will be lost if they go extinct. For centuries, sea turtles have cultural, traditional, social, and economic significance not only to African coastal communities but also to Asia and Pacific regions. For example, as in Hindu mythology, the Indian deity Vishnu was said to have re-incarnated as "Kachhapa"—a turtle, bearing the burden of the world on its back. Huge amounts of sea turtle eggs are consumed in Southeast Asia because of the cultural belief linked to their aphrodisiac and medicinal properties. In many cultures today, the sea turtle has historically generated a lot of meanings, including deity, food, merchandise, medicine, scientific object, a protected animal, managed ecosystems, tourist attraction, and aesthetics (Vargas-Mena 2000; Akani and Luiselli 2011).

Turtle meat is known to enhance the human immune system, as it contains a lot of proteins and contains several micronutrients, including selenium, iron, potassium, calcium, phosphorus, thiamine, riboflavin, and zinc.

The shells of Hawksbill have been used to design jewelry, ornaments, musical instrument, utensils, and souvenirs, which find their way into international markets in the Asia Pacific region for centuries (Castro and Huber 2005). They are reportedly useful in some African and South American cultures, in the treatment of rheumatism, earache, and sore throat (Seixas and Begossi 2001; Alves 2006; Formia et al. 2003). In the Niger Delta region of Nigeria, particularly in Andoni culture, a state of emergency is announced when a town crier repeatedly plays the turtle carapace, and summons every man in the community to an emergency meeting in the chief's palace. Usually, failure to respond to such "turtle calls" attracts heavy penalties. From an economic point of view, sea turtles have important non-consumptive value especially and can be used to create revenues for coastal residents. Sea turtles are one of the most charismatic animals on the planet and virtually everybody loves them (Troëng and Drews 2009).

Although sea turtles are highly cherished in various coastal communities, they can be a route of food poisoning called chelonitoxism. It is a type of food poisoning that occasionally results from the consumption of contaminated sea turtle meat, in the region of the Atlantic, Pacific, and Indian Oceans. Four species of marine turtle have been associated with chelonitoxism—hawksbill, green turtle, loggerhead, and leatherback. Mostly implicated are the hawksbills which habitually eat poisonous sponges all over the world and other turtles that forage on poisonous blue–green algae during the hot season of November and March in the Indo-Pacific. In several places such as Tanzania's Pemba Island, coastal Madagascar, the Philippine Islands, and the Federated States of Micronesia, with this record, many persons were hospitalized, most of whom died because there is yet no antidote for chelonitoxism. Sea turtle meat may contain toxins, including heavy metals and polyaromatic hydrocarbons. In general, the symptoms of chelonitoxism manifest between 4 h and 5 days after consumption, and depending on the type of toxin and the quantity of

Table 19.1 Records of chelonitoxism in SWIO of Africa

Date	Locality/region	No. of victims hospitalized	No. of death
26/12/2021	Zanzibar	47	4
9/6/2021	Madagascar	–	19
–	Pemba, Tanzania	24	7
17/1/2018	Madagascar	–	8

contaminated meat consumed may be mild, moderate, or severe. Mild cases include those that have throat pains and dryness of mouth with or without diarrhea, vomiting, dizziness, malaise, and sweating. Moderate cases are those who develop mouth ulceration, and white-coated tongue or tongue fissures. Severe cases are those who developed neurological manifestations, including alternating phases of lethargy and agitations or a decrease in sensorium. Table 19.1 shows some recent records of chelonitoxism incidents in the South West Indian Ocean (SWIO) region of Africa.

19.4 Sea Turtle Threats on African Coastlines

The major threat to sea turtles on African coastlines is traceable to various anthropogenic activities and disturbances, on their nesting sites, foraging grounds, and migration corridors. Some of the human impingements are direct and some indirect, as listed below.

19.4.1 Fishing and Trawling Operations

Fishing is the major economic base of coastal communities in Africa. It has been estimated that the fishing industry accounts for the death of thousands of sea turtles every year. The greatest threat to sea turtle diversity and population is accidental capture known as bycatch. Virtually all gears including gillnet, beach seine net, longlines, and trawlers especially those with 45 mm mesh net entangle them, and as they fail to come back to the surface to breathe, they get drowned. The majority of the trawlers lack Turtle Exclusion Device (TED) which is supposed to spare them. In some localities, the fishermen deliberately set the nets and inspect them the following day. Others use special spears or harpoons to kill them.

19.4.2 Exploitation and Poaching

During the spawning season of sea turtles, most coastal communities engage in the exploitation and poaching of turtles. With their machetes, they hunt for the track of the turtles and trail them to their nests. Once located the nesting female is killed and the eggs are collected. Every year, many nesting turtles and their eggs are exploited for subsistence or trade. In addition, the high demand for sea turtle shell is increasing poaching in parts of Africa and globally. The shells are used in traditional medicine, aesthetic, jewelry, musical instrument, and souvenirs. Hawksbill turtles are mainly desired for these designs because of the striking details of the shell.

19.4.3 Habitat Loss Due to Coastal Developments

One of the major threats to the conservation of marine turtles in Africa is the rapid coastal developments going on in recent times. In the coastal states of Nigeria, for instance, various coastal settlements are urbanizing and many development projects (dredging operations, reclamation, and sand mining) and infrastructure—jetties, marinas, hotels, tourist beaches, oil rigs, estates, holiday resorts, shanties, fishing ports, etc., have been built—most of which encroach into sea turtle breeding sites. The implication now is that the female turtles are compelled to relinquish their quiet habitats for sub-optimal nesting sites, where hatchlings easily fall prey to predators. Habitat degradation that is exacerbated by vehicular traffic is exposing sea turtle to various hazards.

19.4.4 Artificial Lighting

Artificial lighting on beaches also contributes immensely to the threat of diminishing sea turtle populations. As soon as the hatchlings emerge from the nest, they instinctively move toward the brightest light. The bright light from numerous business centers (hotels, jetties, oil rigs, etc.) on the beach misdirects the hatchlings. Instead of moving toward the sea, they move in the direction of the artificial beach light and crawl endlessly toward them. In most cases, they fail to reach the sea, as they are exposed to several hazards, such as exhaustion, dehydration, and predators.

19.4.5 Marine Debris

Several million tons of plastic debris make their way into the ocean every year. On the coasts of Africa, like other coastal states of the world, amazing collections of plastic debris have become a great concern. The plastic bags floating in the ocean resemble jellyfish—a common food of sea turtles. All too often, the turtles mistake them for jellyfish and in the bid to swallow them, the plastics block their throat, making ingestion of the real food impossible. They eventually starve to death. Many dissected turtles have revealed the presence of other plastic debris-like toothbrushes inside their gastrointestinal tracts, and cases of sea turtles having straws in their noses are also known. Lost nets in the ocean (known as ghost nets) have severally been reported to trap sea turtles to the delight of fishermen (Adeyemi et al. 2019).

19.4.6 Dredging Operations

Because heavy ocean liners and trawlers ply the EEZ of African coastlines, maintenance dredging goes on, when necessary to ensure smooth navigation. This is particularly the case around Bonny, Brass, Eket, and Ibeano sites in Nigeria. The Bonny coastal waters are known to have been dredged from a depth of 8 m to 14 m to allow navigation of big vessels like the LNG boats. The turbidity plume and alteration of the bathymetry and hydrology arising from these operations are capable of affecting sea turtle migration to and from their spawning ground. Such maintenance dredging is also capable of stirring up toxic pollutants (e.g., heavy metals, radionuclides, etc.) trapped in the sediments for ages. Once in the water column, these pollutants can be taken by the sea turtles directly or indirectly through the food web. It has also been suggested that since sea turtles are highly visual predators, increased turbidity plumes would decrease prey capture efficiency, as observed in the Eastern Painted turtle (*Chrysemys pieta*). According to Goldberg et al. (2015), dredging presents a risk of injury and mortality on marine turtles through entrainment, vessel strike, or the effect of noise and vibrations. The severity of impact depends on the type of dredger in use, the hopper dredger being the most devastating (Dickerson et al. 2004).

One of the best ways to mitigate the impacts of dredging on marine turtles is to ensure that dredging operations do not take place during sea turtle spawning season. Another conservation strategy for avoiding turtle mortality is netting the sea turtles in the area of dredging and relocating them to safer areas, kilometers away.

Around Brass, Bonny, and Ibeano areas, oil companies and other companies have had to carry out dredging for various reasons. The more common reasons for dredging are:

- To improve navigation by widening channels or canals to allow passage of big vessels into the hinterland
- To obtain materials (sand, and gravel) for construction work
- For reclamation of beaches in areas, where space is lacking

Whatever be the reasons, beach dredging (as noted by Professor Wanless, Rosenthal School of Ocean and Atmospheric Science) causes a long list of environmental consequences, such as increased turbidity plume, re-suspension of toxic substances in the water column, beach profile anomalies, and instability of the ecosystem. Usually, the speed of resilience would depend on the magnitude of perturbation and the amount of clay in the sediment.

Shoreline dredging distorts the profile of the shoreline and discourages sea turtles that are returning to spawn. In such situations, spawning will be delayed until a suitable site is located. Furthermore, cutter-head dredges grind up the turtles, as they sweep through sediments, where turtles have embedded themselves.

19.4.7 Seismic Surveys

In oil-producing countries of Africa, such as Nigeria, Angola, Libya, etc., seismic activities have been going on in the onshore areas in search of oil reserves. Most times, the seismic crew does not recognize the fact that some of their shoreline operational areas include turtle spawning grounds, because the Environmental Impact Assessments (EIAs) that are supposed to highlight the presence of sea turtle nesting sites have hardly made mention of them (Akani and Luiselli 2011). The noise generated during the operations, the great numbers of the seismic crew, the series of cables that crisscross the area, the numerous shot holes, and the earthquake-like blasts disturb and put sea turtles off their spawning grounds.

Marine turtles show a strong avoidance response to the heavy sound produced by arrays of air-gun during the deep sea or offshore seismic surveys. In recent studies, Hawkins et al. (2014) observed that seismic air-gun exposure would inflict deadly injuries on sea turtles in a very close range of about 100 m or more to the sound source. Sea turtles are often entangled by anchor equipment during offshore seismic operations, e.g., eight Olive Kemp Ridley got entangled in ocean bottom gear, off the coast of Gabon (Nelms et al. 2015).

19.4.8 Oil Pollution

In oil-producing nations of Africa, which have onshore or offshore oilfields or transport oil by tankers or pipelines (such as Nigeria, Angola, and Libya) are prone to oil spill incidence. For instance, the incessant oil spill in Nigeria's Niger Delta region is known to affect air-breathing reptiles and amphibians that depend on clean water and clean beaches. Even though sea turtles are capable of holding their breathe on dives for long periods, they usually come to the surface to breathe several times per hour. In the process, they may come in contact with large oil slicks over and over again, and marine eco-toxicologists have posited that:

- The consumption of oiled food can lead to damage to the digestive tract and accessory organs and disrupt digestion.
- Physical tainting by oil can lead to irritation of nasal mucous membranes that may lead to organ inflammation and infection. The turtle may become narcotized, dazed, or moribund and become easy prey to sharks.
- Inhibition of embryonic development of the eggs. Whereas the eggs may be contaminated by oil from nesting sites.

Fig. 19.1 Segment of the beach at Okolo (Bonny–Andoni axis) that was devastated by an oil spill in 2014: resulting in the browning of *Nypa* fronds

Onshore and offshore oil spills get to the sea turtle nesting sites on the beach, through runoff or wave activities. A typical example is presented in Fig. 19.1, where a segment of the impacted shoreline vegetation, dominated by *Nypa fruticans* turned reddish-brown, following the Okolo Well Head spill on 22nd December 2014 in Nigeria. When an oil spill coincides with the spawning season of sea turtles, the population is affected, as both gravid females and hatchlings are oiled.

19.4.9 Climate Change Impacts

Years of oceanographic and marine biological research indicate that the abundance and distribution of food resources could occur as a result of the alterations in the thermal profile of the marine environment. As demonstrated in the herbivorous green turtle (*Chelonia mydas*), this, in turn, could lead to a shift in migratory patterns and ecological niches of some turtles.

Sex determination in sea turtles is dependent on the temperature of the beach sand at the time of spawning (Akani 2019). If the egg incubation temperature falls below 27.7 °C the hatchlings will be males, whereas the hatchlings will be females, if the eggs incubate above 31 °C. This means that female turtles are hatched in warmer incubation temperatures, while males are hatched in cooler incubation temperatures. However, temperatures that fluctuate between the two extremes will produce a combination of male and female hatchlings. The implication of temperature-dependent sex determination of sea turtles is that as global warming continues, an increase in temperature will result in the production of more female hatchlings, and if the temperature rise persists for decades, males may become extinct, leaving a monosexual population of female sea turtles. Then, sea turtles may become extinct because of the lack of males to fertilize the eggs. Warmer sea surface temperature can also lead to the loss of important foraging grounds for sea turtles (Atieno 2021).

19.4.10 Marine Turtle Disease

Sea turtles are prone to a disease known as fibropapillomatosis (FP). This is a debilitating disease that manifests through external tumors. The tumor arises mostly on the turtle's soft skin tissue, around the neck, at the base of flippers, and near the eye, and often grows too large hindering its ability to see, eat, swim, and escape from predators. These tumors have various shapes and colors and range in size from a pea to a grapefruit. FP is associated with infection by a herpes virus called, Chelonid Herpes virus 5. Blood analysis of most turtles infected with FP tumors indicates that they are typically anemic. Inexplicably, FP has been reported in all sea turtles, but the leatherbacks. The highest prevalence of FP is recorded in green turtles.

19.4.11 Vessel Strikes and Sea Turtle Stranding

Sometimes, a sea turtle collides with a travelling marine vessel resulting in instant death or severe injury. Several cases of sea turtle stranding have been recorded on African coastlines. A stranded turtle is an individual that is found washed ashore, dead, or alive. Three factors may cause a turtle to be stranded alive, namely, trauma arising from a boat strike, debility due to infectious diseases, such as FP, and crude oil pollution. A recent study by Flint et al. (2017) prompted by the increasing rate of marine turtles stranding in Queensland, Australia, has traced the phenomenon, primarily to the cumulative effect of freshwater discharge and several temperature-related factors. Turtles stranded alive are usually in a weakened or moribund state, because of sickness or injury, and best practices require that such live turtles be taken to the nearest rehabilitation clinic or marine biological station, where they can be resuscitated by experts (through warm-bathing, medication, and feeding) and subsequently released into a safe marine environment. Unfortunately, such rehabilitation homes are rare on African coastlines. Most times, the dead stranding is in fragments, and may consist of dismembered carcasses with broken necks, front flippers, chopped heads, or bleached carcasses with heads intact. It may also be a rotting carapace left by scavengers.

19.5 Sea Turtle Habitats in Africa—Regional Overview

Africa is endowed with numerous, conducive habitats for sea turtles on its Atlantic, East African, Mediterranean, and Red Sea coastlines. This has been attributed to the abundant wide sandy beaches in these sites that favor turtle nesting. Various authors have confirmed that five species of marine turtle annually visit the African coastlines (Amadi 1991; Fretey 2001; Fretey et al. 2002, 2007; Formia et al. 2003; Muir and Sense 2005; Akani and Luiselli 2011; Dallieau et al. 2020; Hochscheid and Toma 2021, etc.). They are: (i) Green Turtle (*Chelonia mydas*)—Linnaeus, 1758, (ii) Loggerhead (*Caretta caretta*)—Linnaeus, 1758, (iii) Olive Ridley, *Lepidochelys olivacea*)—Eschscholtz 1829, (iv) Hawksbill (*Eretmochelys imbricata*)—Linnaeus, 1766, and (v) Leatherback (*Dermochelys coriacea*)—Vandelli, 1761. They inhabit the coastal waters; forage and nest on specific beaches from Mauritania, through the Gulf of Guinea, down south to Angola on Africa's Atlantic coast, and from South Africa northward to Somalia on the southwest Indian Ocean. The offshore archipelagos and island nations associated with these regions also provide suitable turtle sites—these include the Canary Islands, Cabo Verde, Bijagos, Bioko, Sao Tome and Principe, Madagascar, Comoros, Mauritius, La Reunion, and Seychelles. Even the Kemp's ridley, *Lepidochelys kempi,* which is endemic to the North Atlantic and the Gulf of Mexico, has reportedly wandered into African territorial waters and beaches. The only species which has never been recorded on African coastlines is the Australian flatback (*Natator depressus*)—(seaturtlestatus.org; Agyekumhene et al. 2017).

19.5.1 Sea Turtles of East Africa and The Southwest Indian Ocean (SWIO) Region

Current reports of marine turtles in East Africa and the West Indian Ocean region indicate that this region harbors the world's most important turtle nesting site (Dallieau et al. 2020). The most abundant nesting sites are located on the islands of Seychelles, French Eparses Islands, and British Indian Ocean Territories. Decades of monitoring in this region have yielded positive trends in the number of turtles. Nesting also occurs, but to a lesser extent on virtually all the continental coasts of Kenya, Tanzania, Mozambique, and Madagascar. However, monitoring in the region had been inconsistent, due to the more difficult terrain, hence, the trends are inexplicable. Of the five species known in this region, the most abundant species is the green turtle, while olive Ridley is the least. In the SWIO region, hawksbill nesting sites are mainly located in the Seychelles archipelago at the level of the Granitic and Amirantes islands group as well as the Chagos. Strong conservation initiative implemented in Seychelles in the last two decades is benefiting the hawksbills. For loggerhead, the nesting centre is along the northeast coast of South Africa (in iSimangaliso Wetland Park, Kwazulu-Natal) and southern

Mozambique (south of Maputo) with occasional nesting records along the East African beaches. Loggerheads are also known to nest in the southeastern flank of Madagascar, but the activity is rated as low. As many as 1000 loggerhead females per annum can be counted at the main spawning ground (between Maputo and Cape Vidal).

The complicated pattern of the Indian ocean current on the coast of East Africa has a significant influence on the sea turtle movement. The north and south equatorial currents from the Indian Ocean flow east to west and break up on the continent north of Madagascar. On breaking, they rejoin to form the Somali and Agulhas currents which flow north and southward, respectively. The sea turtles, including juvenile loggerheads, use these currents as a migratory route from the islands onto the mainland or South or North, even up to the Gulf of Oman in the Middle East. Consequently, there is usually a mix of juvenile loggerhead populations from Madagascar, Reunion, and other islands in the neighborhood (Dallieau et al. 2020).

The leatherbacks are widely distributed on the beaches of the SWIO. The bulk of their nesting sites is located between Southern Mozambique and south-eastern Madagascar but at a reportedly low scale. Turtle threat in the SWIO region has been traced primarily to female poaching and mortality due to artisanal fishery. However, there are considerable conservation efforts for marine turtles in the region, because several of the regulatory mechanisms of international treaties strictly apply to sea turtles occurring in the area.

In Kenya, fishing and poaching are exacerbating sea turtle decline. Artisanal fishers use illegal fishing gear such as monofilament gillnets and spear guns to kill them in areas, where they are foraging in their numbers (Atieno 2021). Trawlers operate without TED, and fish near-shore waters, where there are lots of seagrasses—which is a delicacy for turtles. Sea turtle exploitation is mainly driven by their ethnozoological values. For instance, the cultural belief of the Mareremi people of Kenya is that turtle meat makes children grow stronger and healthy. Hence, pregnant women lookout for turtle meat to eat at all costs, irrespective of the fact that a small piece, sales from US$1 to US$1.5. During their breeding season, poachers use the turtle tracks on the beach to locate them, slaughter the meat, collect the eggs, and sell them around the villages. Some poachers simply set longlines overnight in water and inspect them the following day. Another driver of sea turtle decline in Kenya is the trade-in sea turtle oil, which is highly cherished because of its ethnomedicinal values. It is a common belief among locals that the oil treats asthma, impotence, infertility, water-borne diseases, ear arches, measles, and tuberculosis. The oil is more expensive, as midwives use it to induce quick placenta release.

19.5.1.1 United Republic of Tanzania

In Tanzania sea, turtle research and conservation are gaining momentum (Frazier 1976; Clark and Khatib 1993; Muir 2004a, b, Muir and Sense 2005, etc.). Five species are known in Tanzanian territorial waters. These include green, hawksbill, loggerhead, olive ridley, and leatherback. However, only two species—green and hawksbill—are known to nest in Tanzania. The green turtle which is most widespread, nests mainly between February and July, especially on the offshore islands of Zanzibar, Mafia, and the Songo–Songo archipelago with nesting grounds in Misali Island and Mafia Island between December and April. Information on the status and nesting behavior of other turtles is as little as they are rare. Among the major threats to turtles in Tanzania are: disturbance of nesting and foraging habitats, incidental bycatch, poaching of meat and eggs, inadequate enforcement of protection laws, limited awareness, land-based projects, and pollution (Muir and Sense 2005).

19.5.2 Sea Turtles of West Africa—Atlantic Coast of Africa

Reports accruing on the state of marine turtles in West Africa and the Atlantic coast of Africa (UNEP 2001) recognized the region as a hotspot for marine turtles, following abundant evidence of nesting and foraging sites identified.

Using secondary data collected from studies of several experts for over a century, Jacques Fretey, a seasoned Marine turtle expert on the French Committee of IUCN—The World Conservation Union, compiled an exquisite and comprehensive review on Biogeography and Conservation of the Atlantic Coast of Africa, highlighting the status and known nesting sites of the loggerhead, green, leatherback, hawksbill, Kemp's Ridley, and Olive Ridley turtle in every country from Morocco to South Africa.

19.5.2.1 Cape Verde

A significant nesting stock of loggerhead turtle was recently discovered on the Islands of Boa Vista and Sal in Cape Verde. Luis Felipe Lopez, a researcher at the University of the Canaries, Las Palmas, and his team estimated that the Cape Verde loggerhead population is the third-largest in the world. Turtle hunting in Cape Verde dates as far back as 1479. To date, turtle hunting has not ceased in the Islands; many are killed during the nesting season. Currently, developments along the coastline of Sal Island, mainly from the tourism industry, are reducing kilometers of conducive nesting sites for turtles. Every year thousands of kilograms of marine debris accumulate on the eastern coast of Sal, compelling female loggerheads to nest in non-optimal habitats.

19.5.2.2 Mauritania

Mauritania has two very important National Parks, namely, Banc d'Arguin and Levrier's Bay. The marine zone of Banc d'Arguin includes an exceptional reserve of seagrasses, which is the most attractive feeding ground for green turtles in West Africa, today. Other species also known in the region are loggerhead, leatherbacks, olive ridley, and hawksbill. Threats are also heavy and include poaching, by-catch, and stranding, all of which have both local and global concerns.

19.5.2.3 Guinea-Bissau

Guinea-Bissau represents one of the richest areas throughout the region for sea turtles. Of the seven extant turtle species, four are known to nest in Guinea Bissau, namely, green, olive ridley, hawksbill, and leatherback. The major threats include poaching, incidental bycatch, predation, degradation of nesting sites due to coastal erosion, flooding, and unbridled tourism practices. Some Biodiversity conservation organizations are making frantic efforts toward sensitizing the coastal communities about conserving the sea turtles of Guinea-Bissau.

19.5.2.4 Ghana

Potential nesting sites of sea turtles in Ghana are located on beaches from Princess Town to Busua, Senya Bereku to Accra, Prampram to Old Ningo, and Anloga to Denu. Four Species are commonly recorded on these beaches, namely, Olive ridley, leatherback, green turtle, and hawksbill. All turtle species are protected under Ghanaian law, but a deficiency in the enforcement of the laws has allowed the threats to linger, and these include incidental bycatch, poaching, pollution, coastal development, and climate change issues. Other threat factors include predators such as domestic pigs, and wild dogs, coupled with illegal sand-mining for construction, which undermine sea turtle habitats.

Following the multiple disturbance of sea turtles on Ghanaian beaches, an environment-concerned non-governmental organization in August 2019 during a weeklong celebration, launched a project at Gomoa Fetteh, (a fishing community) 50 km west of Accra to safeguard the sea turtles of Ghana. The enlightenment campaign was extended to 200 other fishing communities along the coast. They formed teams of coast guards who monitored turtle nesting grounds by night, debarring poachers of their nefarious business. The result was heartwarming as the team recorded 22 nesting turtles in the first season, and 26 nesting sites in the next, allowing the turtles to nest safely.

19.5.2.5 Benin

The presence of nesting sites for green, olive Ridley, and leatherback turtles has been confirmed in Benin. However, the continuous killing of females on the beaches and poaching of nests for commercial purposes raises concern. In Benin, the sales of turtle shells to tourists are a viable business. Leatherback fat is turned into oil, which is highly cherished for its medicinal properties.

19.5.2.6 Nigeria

In recent times, sea turtle migration to the Atlantic coast of Africa and Nigerian beaches at Akassa, Brass, Idama, Bonny, Andoni, and Ibeano, to nest has become more regular than ever; fishers in these areas captured them consecutively in 2005, 2006, 2007, and 2008 (Fretey 2001; Fretey et al. 2002, 2007; Formia et al. 2003; Akani and Luiselli 2011) and beyond. More nesting sites have been reported at Ilaje, in Ondo State (Ogunjobi and Surulere 2020) and parts of South West Nigeria (Adeyemi et al. 2019). Most of the nine Nigerian coastal states, namely—Akwa Ibom, Bayelsa, Cross River, Delta, Edo, Lagos, Ogun, Ondo, and Rivers—are veritable habitats for sea turtles. Of all the nesting grounds on Nigeria's 853 km coastline, that of Akassa attracts turtles more regularly, and the people of Akassa (a community of 19 villages) hunt them and collect their eggs for consumption. To stem this, the Tropical Research and Conservation Centre, about two decades ago, embarked on community-based conservation of marine turtles nesting on the Akassa coast, as one of its projects in southern Nigeria.

Several trawling boats (most of which are foreign fleets) spend quality time in the offshore waters of Akassa, Brass, Bonny, Idama, Andoni, Eket, and Ibeano, fishing, and shrimping. There are doubts as to whether these trawlers use TED, which is supposed to save or exclude trapped turtles. Fish trawlers are the main threat to marine turtles in the Niger Delta with as many as 20–30 trawlers sighted daily. Fishermen report the abundance of marine turtles in deep waters off the coast of Brass. According to their report, the seabed in that area is rocky and heavily covered with abundant seagrasses and colonies of corals, suggesting that it could be their foraging area. Three potential nesting sites are known here include the beach of Cape Formosa (west of Brass Island), the beach west and east of the Brass Airstrip between Ewoama and Okpoama, and the beach between Ilajekiri and Diema waterside. Other minor nesting sites for all the five species are the beaches between Brass River, St. Nicholas, Sangana fishtown, and Nun River estuaries. Since Nigeria is a signatory to the 1999 Memorandum of Understanding Concerning Conservation Measures for Marine Turtles of the Atlantic Coast of Africa, all the coastal states by dint of their proximity to sea turtle habitats, are obliged to protect these habitats.

19.5.2.7 Equatorial Guinea and Gabon

The Bay of Corisco is endowed with an exceptional population of seagrasses. Thus, it constitutes a major feeding ground for green turtles in the region. Other turtles such as hawksbill, olive ridley, and loggerhead (Tomas et al. 1999) also nest on the beaches south of Bioko Island in Equatorial Guinea, thus making the region an important turtle haven. Inhabitants of the Bay of Corisco especially the local Benga people have had turtle hunting as their way of life for generations. Recently, they have switched on to commercial hunting for the highly priced turtle product in the cities of Libreville and Bata. In addition to their bycatch, they enhance their catch of the turtles using harpoons or underwater guns. There are about 50 fishermen around the Bay of Corisco who specialized in harpooning sea turtles. Gabon is known to harbor huge reproductive stock of leatherback, one of the largest in the world. The systematic killing of female leatherbacks and egg collection has prompted payment of nest-guards, by the government to check this poaching.

19.5.2.8 Sao Tome and Principe'

Throughout the Atlantic coast of Africa, the Islands of Sao Tome and Principe' are renowned to harbor the greatest diversity of sea turtles, with males, females, and juveniles of green, olive ridley, hawksbill, and leatherback regularly visiting the coastal waters and beaches. The predominant species here is the hawksbill. The major drivers of over-exploitation include: the meat and egg trade on Bioko Island, coupled with the hawksbill shell-craft business in Sao Tome and Principe' (Castroviejo et al. 1994).

19.5.2.9 Sea Turtles of the African Mediterranean coastline (Morocco to Egypt)

The Mediterranean Sea is yet another region that hosts sea turtles. As a sea bordered by Europe on its northern shores, Asia to the east, and Africa to the south, sea turtles utilize this enclosed sea with over 150 million residents along its 20 coastal and two island nations. Endowed with beautiful scenic attractions, it is indeed a very busy sea which is by far the largest global tourism destination, which attracts almost a third of the world's international tourists every year. Consequently, sea turtles have a hard time, coping with multiple threats here which include incidental capture, high salinities over 3.8%, marine pollution, artificial beach lighting, cross-continental maritime traffic, and climate change impacts (Hochscheid and Toma 2021). To date, however, only three species are known throughout the Mediterranean Sea—the green, loggerhead, and leatherback.

19.5.2.10 Morocco

Three sea turtle species—loggerhead, leatherback, and green turtles—have been documented to forage on Morocco's 500 km coastline. Cases of stranding and bycatch are rampant and constitute a major threat in the region. Juvenile loggerhead is expensive in Morocco, as it is a source of the shell for designing ornamentals sold to tourists. However, the ban on turtle products in 2007 seems to have caused a crash in the business.

19.5.2.11 Algeria

In Algeria, incidents of sea turtle stranding and bycatch date as far back as the 1800s. Available records show that they are predominantly loggerhead (70%) and leatherback (30%). Up to date, there is no record of nesting by these turtles or any other turtle throughout the 1622 km coastline. This implies that they are all migrants.

19.5.2.12 Tunisia

Three species of sea turtle are known in Tunisia which forages and nest there. Loggerheads are most commonly encountered, followed by leatherbacks. Green turtles are very rare. The Gulf of Gabes in southern Tunisia, characterized by a remarkable wide continental shelf, is one of the most important foraging sites throughout the Mediterranean. The high frequency of incidental captures by various gears portrays a high population density of turtles in this area. Turtle stranding is also rampant in Tunisia, and this scenario has necessitated an active stranding network in Tunisia with a rescue center now located in the city of Monastir. The causes of these stranding are, however, not known. On Kurita Island is a regular loggerhead nesting site, where over 20 nests per year can be counted.

Libya—Of all the North African coastal countries bordering the Mediterranean, Libya has the longest sandy beach suitable for sea turtle nesting. It provides the oldest and most important nesting ground for loggerheads throughout the basin. No green turtle has been found nesting in the country. However, green turtles are known to use the Gulfs of Bomba and Sirte as foraging and overwintering habitats before they migrate to the Levantine coastlines (from Israel to Turkey including Cyprus), where they nest. Over the years, the Libyan Sea Turtle Program, supported by the Regional Activity Centre for Specially Protected Areas (RAC/SPA)—an arm of the United Nations Environment Programme's Mediterranean Action Plan (UNEP–MAP)—has assiduously been monitoring the migration patterns and population dynamics of the loggerheads, even in times of political instability. Up to date, the population of nesting females remains unknown (Hochscheid and Toma 2021). Threats to sea turtles in Libya include—incidental bycatch, human disturbance and coastal development, oil pollution, and 4x4 Jeep vehicle beach driving in summer, by holidaymakers. The ruts of this vehicle are deep, which slows down the movement of hatchlings going to the sea.

19.5.2.13 Egypt

The Egyptian coastline (approx. 1050 km) is an important foraging ground and migratory corridor for loggerheads and green turtles. The occurrence of a leatherback in the region has been established through stranding and bycatch data. Recently, the Bardawil lagoon has been identified as a possible site for wintering which deserves conservation attention. Compared with other Mediterranean sites, loggerhead and green turtle nesting in Egypt are generally low, although there are minor sites along the western flank of the Egyptian coastline. The best-known turtle nesting site is a stretch of about 20 km sandy beach in the Sinai Peninsula. A new nesting site located between Port Said and El Salum is currently being monitored by Egyptian authorities, aided by RAC/SPA to update information on nesting activities there. The threat to sea turtles in Egypt includes widespread habitat degradation, pollution, and by-catch. Illegal trade, for turtle products and consumption, has been a tradition in the area.

19.5.3 Marine Turtles of the Red Sea

Of the seven extant species of marine turtles throughout the globe, five species have been recorded in the Red Sea, namely—the green, hawksbill, loggerhead, Olive Ridley, and leatherback turtles. Presently, only the green and hawksbill turtles are known to forage and nest in the area. It has been estimated that about 450–550 female green turtles and 450–650 female hawksbills nest on the Red Sea coast annually (Mancini et al. 2015). No nesting activity has been reported for the loggerhead within the Red Sea. However, they are known to nest in great numbers outside the Gulf of Aden on the Socotra Islands of Yemen, in thc Middle East. For Olive Ridley, only one nest has been reported in the region, in 2006 from Eritrea. In the case of Leatherbacks, no nesting site has been found, but they are only seen as opportunistic feeders on jellyfish during the 'jellies' bloom season. Elaborate studies on the foraging niche and nesting site of Red sea turtles are lacking for most littoral countries on the Red Sea. It was only on the Egyptian Red Sea coastline that at least 453 green turtles were reportedly using 13 shallow bays as both feeding and nesting sites.

Sea turtles of the Red Sea are threatened by incessant bycatch and poaching, as well as habitat destruction, and oil pollution. Although all the coastal countries of the Red Sea have accorded legal protection to marine turtles, through national laws and international treaties, enforcement at sea and on nesting beaches are grossly inadequate and needs to be strengthened.

19.6 The Way Forward

To minimize sea turtle mortality along African and other coastlines, it is worthy to state that sea turtle conservation may be positively affected by the area of focus with the following main conservation and management considerations habitat protection, reduction of threat by incidental catch, strengthening national legislations, promoting community participation in conservation, sensitization, and education (Muir and Sense 2005).

19.6.1 Habitat Protection

- Local governments whose jurisdiction extends to the coast should set up agencies who should be raking and burning solid wastes from the beach, monthly or bimonthly. They may sort out the plastics and metals for recycling and incinerate the rest—such as papers and textile materials.
- Enforce, to the letter, coastal development regulations and ensure EIA proceeds any coastal development, such as harbor, dredging, seismic operations, pipeline constructions, refinery, etc. A detailed EIA proceeding such projects should be able to identify turtle nesting and foraging sites.
- Turtle sanctuaries should be created by government legislation in any identified nesting or foraging areas and ensure community participation in the protection of the area.
- The use of bright light (and reflectors) on the beach, where turtle nest should be prohibited by law. Only those lights of low intensity such as deem, blue, green, or sodium bulb lighting should be allowed.
- Governments of all African coastal states should establish Marine biology Research centres along the coast, to monitor the fate of sea turtles and other aquatic wildlife, when an oil spill occurs, and to rescue and rehabilitate affected victims.
- During offshore or deep-water seismic operations, the power of the air gun should be increased slowly to give sea turtles in the area the opportunity to relocate to safer places.
- All offshore oil concessionaires should always be ready to do clean-up, in the event of an oil spill from their rigs; booms should be quickly deployed around oil spill outfalls before oil spreads to the beach.

19.6.2 Reduction of Turtle Threat by Incidental Catch and Poaching

- To minimize the threat on marine turtles by incidental catch, Marine Police or Naval officers of every coastal state should be frequently drafted to inspect shrimp trawlers along African territorial waters and bring to book any trawlers operating without TED. TED consists of panels of large mesh webbings or metal grids inserted into the funnel-shaped trawl-net which has a vent through which a sea turtle can escape and avoid incidental drowning, while the prawns and other small animals are trapped. TED is the principal component of the United States sea turtle conservation programme,
- Poaching of turtles for meat could be minimized through programmes to discourage turtle consumption on health grounds and emphasize the problems of morbidity and mortality due to chelonitoxism.

19.6.3 Strengthening National Legislation

- Governments of coastal countries of Africa should strengthen all existing regulations on the conservation of sea turtle critical habitats and make it mandatory that all sea turtles caught in any fishing gear or stranded by the coast be returned to the sea immediately.
- Governments of all coastal countries of Africa should strengthen sea turtle legislation, and capacity building and provide logistic support, research equipment, and gadgets.
- Punitive or draconian measures (including arrests, prosecution, and heavy fines) should be meted out to culprits found handling or trading on sea turtle meat, eggs, carapace, or other products, irrespective of whose ox is gored. Such culprits should be paraded on TV to serve as a deterrent to others.

19.6.4 Promoting Community Participation in Sea Turtle Conservation Efforts

- Initiating beach monitoring teams for sea turtles and involving communities in their monitoring and management, such as the Beach Management Units enshrined in the 2003 Beach Tanzania Fisheries Act (Muir and Sense 2005) and the Pro-Natura International Turtle Conservation in Akassa, Nigeria.
- Conducting training programmes for community participants on sea turtle conservation issues and techniques, and supporting them with some stipends/incentives and PPEs.

19.6.5 Sensitization and Enlightenment Programmes

- Governments should undertake elaborate nationwide sensitization and education programmes using the mass media (through jingles, television programmes, etc.) and emphasizing the illegality of handling or trading on turtle meat, eggs and carapace, and other turtle products, and undermining sea turtle habitat.
- Colleges located in the coastal states should be encouraged to form "Sea Turtle Conservation Society" and organize seminars and TV talks on Sea turtle conservation on, World Wildlife Day (3rd March), International Day for Biological Diversity (22nd May), or World Environment Day (5th June).

19.7 Conclusions

The foregoing review has shown the behavioral adaptations sea turtles have evolved to enable them to spend quality time in the sea and come back to the beach of their birth to nest. It has also indicated that Nigerian territorial waters are one of the habitats in Africa for five species of marine turtles. They usually visit the coast, especially at Akassa, Brass, Idama, Bonny, Andoni, Eket, and Ibeano. Like all sea turtles of the world, they are endangered or vulnerable (IUCN 1996). The causes of the endangerment of these marine reptiles are virtually the same in the territorial waters of Africa and America. Their highly migratory disposition predisposes them to multiple pressure. Many are killed each year primarily as bycatch in fishing nets and trawlers; by poaching, and by coastal developments. Oil industry activities and oil spillage undermine and disturb the nesting sites. Artificial beach lighting causes disorientation of hatchlings going to sea, thereby exposing them to various hazards. Of these factors, the most devastating threats in African coastal waters are incidental net capture, poaching, and disturbance of nesting and foraging habitats. The lax attitude of some governments at enforcing conservation laws and funding research adds to the problem. Sea turtles are keystone species in the marine ecosystem and are of great economic and ecological importance. Despite the laws enacted to protect sea turtles in most countries, the illegal trade of their, eggs, meat, shell, and products such as oil, leather, and jewelry remains an ongoing threat to their survival in parts of Africa, Asia, and the Americas. A situation where some cultures eat them and others conserve them will not auger well for the sea turtles. All hands must be on deck to save the turtles. Emphasis on the problems of chelonitoxism may scare poachers and, therefore, help the conservation of sea turtles.

References

Adeyemi GA, Ayanda IO, Dedeke GA (2019) The interplay between sea turtle population and income generation in south-west Nigeria coastal environment. J Phys Conf Ser 1299(1):012127

Agyekumhene A, Aruna E, Airaud BF, Allman P, Ayissi I, Bourjea J, Dalleau M, Diagne T, Fallabrino A, Formia A, Fretey J (2017) The sea turtles of Africa. The State of the World's Sea Turtle Report, 14, pp. 14–23

Akani GC (2019) Wildlife and biodiversity conservation: an indispensable strategy for sustainable development: an inaugural lecture. Rivers State University, Nigeria, 200Pp

Akani GC, Luiselli L (2011) Diversity and distribution of Sea Turtles in the Niger Delta, Nigeria. Revue d' Ecol 64(4):369–374

Alves RRN (2006) Use of marine turtles in zootherapy in northern Brazil. Mar Turt Newsl 112:16–17

Amadi AA (1991) The Coastal and Marine Environment of Nigeria- Aspects of Ecology and Management. NIOMR Technical Paper 76

Atieno S (2021) How fishing and poaching are fueling sea turtle decline on Kenya's coast. Science Africa, Marereni

Castro P, Huber ME (2005) Marine biology, 5th edn. McGraw Hill Higher Education, Boston, MA, p 452

Castroviejo J, Juste BJ, Del Val Pérez J, Castelo R, Gil R (1994) Diversity and status of sea turtle species in the Gulf of Guinea islands. Biodivers Conserv 3(9):828–836

Clark F, Khatib AA (1993) Sea turtles in Zanzibar: status, distribution, management options, and local perspectives. Zanzibar Environ Study Ser 15:1–84

Dallieau M, Bourjea J, Nel R (2020) Sea Turtles in the East Africa and the West Indian Ocean Region. MTSG Annual Regional Report 2020. Report of the IUCN-SSC Marine Turtle Specialist Group. 78

Dickerson D, Wolters MS, Theriot CT, Slay C (2004) Dredging impacts on sea turtles in the southeastern USA: A historical review of protection. In Proceedings of World Dredging Congress XVII, Dredging in a Sensitive Environment (Vol. 27), Hamburg, Germany

Flint J, Flint M, Limpus CJ, Mills PC (2017) The impact of environmental factors on marine turtle stranding rates. PLoS One 12(8):e0182548

Formia A, Tiwari M, Fretey J, Billes A (2003) Sea turtle conservation along the Atlantic coast of Africa. Mar Turt Newsl 100(1):33–37

Frazier J (1976) Sea turtles in Tanzania. Tanzania Notes Records 77:11–20

Fretey J (2001) Biogeography and conservation of marine turtles of the Atlantic coast of Africa (CMS Technical Series Publication 6). UNEP/CMS Secretariat, Bonn

Fretey J, Meylan A, Tiwari M (2002) The occurrence of the hawksbill turtle (*Eretmochelys imbricata*) in West Africa. In Mosier, A., Foley, A., and Brost, B.(Comps.). Proceedings of the Twentieth Annual Symposium on Sea Turtle Biology and Conservation. NOAA Tech. Memor. NMFS-SEFSC-477 (pp. 95–96)

Fretey J, Billes A, Tiwari M (2007) Leatherback, Dermochelys coriacea, nesting along the Atlantic coast of Africa. Chelonian Conserv Biol 6(1):126–129

Goldberg DW, de Almeida DT, Tognin F, Lopez GG, Pizetta GT, Junior NDOL, Sforza R (2015) Hopper dredging impacts on sea turtles on the northern coast of Rio de Janeiro State, Brazil. Mar Turtle Newslett 147:16

Hannan LB, Roth JD, Ehrhart LM, Weishampel JF (2007) Dune vegetation fertilization by nesting sea turtles. Ecology 88(4):1053–1058

Hawkins AD, Popper AN, Fay RR, Mann DA, Bartol S, Carison TJ, Tavolga WN (2014) sound exposure guidelines for fishes and sea turtles. A technical report

Heithaus MR, Frid A, Wirsing AJ, Bejder L, Dill LM (2005) Biology of sea turtles under risk from tiger sharks at a foraging ground. Mar Ecol Prog Ser 288:285–294

Hochscheid S, Toma J (2021) Nesting range expansion of loggerhead turtles in the Mediterranean: phenology, spartial distribution, conservation implications. Global Ecology and conservation on line. Science Direct

IUCN (1996) 1996 IUCN Red list of threatened animals. IUCN, Gland

Luschi P, Papi F, Liew HC, Chan EH, Bonadonna F (1996) Long-distance migration and homing after displacement in the green turtle (Chelonia mydas): a satellite tracking study. J Comp Physiol A 178(4):447–452

Mancini A, Elsadek I, El-Alwany MA (2015) Marine turtles of the Red Sea. In: The Red Sea. Springer, Berlin, Heidelberg, pp 551–565

Meylan AB, Meylan PA (1999) Introduction to the evolution, life history, and biology of sea turtles. Res Manag Techn Conser Sea Turtles 4:3–5

Muir CE (2004a) An assessment of the status of turtles, dugongs and cetaceans in Mnazi Bay Ruvuma Estuary Marine Park and recommendations for a conservation strategy. Report to IUCN/MBREMP Project

Muir CE (2004b) Tanzania Turtle & Dugong Conservation & Research Programme – quarterly Progress Report, submitted to Commission for Science & Technology, Dar es Salam

Muir C, Sense S (2005) The status of marine turtles in the United Republic of Tanzania, East Africa. Sea sense (Tanzania Turtle & Dugong Conservation Programme). 35

Nelms SE, Piniak WE, Weir CR, Godley BJ (2015) Seismic surveys and marine turtles: An underestimated global threat? Biol Conserv 193:49–65

Ogunjobi JA, Surulere OC (2020) Sea turtles in Ilaje, Nigeria: status, sighting periods and conservation awareness. Nat Conserv Res 3 аповедная наука 5(2):28–35

Papi F, Luschi P, Crosio E, Hughes GR (1997) Satellite tracking experiments on the navigational ability and migratory behaviour of the loggerhead turtle Caretta caretta. Mar Biol 129(2): 215–220

Seixas CS, Begossi A (2001) Ethnozoology of fishing communities from Ilha Grande (Atlantic forest coast, Brazil). J Ethnobiol 21(1):107–135

Tomas J, Castroviejo J, Raga JA (1999) Sea turtles in the south of Bioko Island (Equatorial Guinea). Mar Turt Newsl 84:4–6

Troëng S, Drews C (2009) Money talks: economic aspects of marine turtle use and conservation. WWF. R-007.PDF (worldbank.org). OR cdoole@wwfint.org. Date of Retrieval 25th Jan 2022

UNEP (2001) New Insight into Africa's Threatened Sea Turtles. United Nation Press Release; UNEP/97 https://www.un.org. Date of retrieval 12th Feb 2022

Vargas-Mena E (2000) Signifacados culturales de la tortuga verde (*Chelonia mydas*) en el Caribe costarricense. In: Actitudes hacia la fauna Latinoamerica

Chapter 20
An Overview of Environmental Resources in Africa: Emerging Issues and Sustainable Exploitation

Adams Ovie Iyiola, Oluseun Adesola Akinsorotan, Abiola Elizabeth Ojeleye, and Abdulbasith Olabamiji Fajimolu

Abstract The natural environment is composed of both living and nonliving resources and their interaction dictates the condition of the environment. It provides food, shelter, economic returns, food source, and sustenance to diverse biota. The major components of the environment are the mountains, aquatic bodies, forests, and grasslands which provide a haven for diverse species of animals. Some of these components and areas serve as travel and tourist destinations, thereby providing foreign exchange which results in economic growth. The Northern African regions are well-established in these activities when compared with the sub-Saharan regions of Africa. These regions have potentials that can be exploited for revenue generation if policies are created to harness these resources. The synergy between atmospheric, water, and mineral resources in the environment must be preserved to enjoy the use and non-use benefits. Although, these resources are affected by human pressures in the form of overharvesting of natural resources, changes in a habitat-like fragmentation and modification, and the introduction of invasive alien species. Measures such as good governance and community participation by locals are important in resource management. To this end, this chapter reviews the environmental resources and approaches for sustainability.

A. O. Iyiola (✉)
Department of Fisheries and Aquatic Resources Management, Faculty of Renewable Natural Resources Management, Osun State University, Osogbo, Osun State, Nigeria

O. A. Akinsorotan · A. O. Fajimolu
Department of Wildlife and Ecotourism Management, Faculty of Renewable Natural Resources Management, College of Agriculture and Renewable Resources, Osun State University, Osogbo, Osun State, Nigeria

A. E. Ojeleye
Department of Agronomy, Faculty of Agricultural Production and Management, College of Agriculture and Renewable Resources, Osun State University, Osogbo, Osun State, Nigeria

S. C. Izah, M. C. Ogwu (eds.), *Sustainable Utilization and Conservation of Africa's Biological Resources and Environment*, Sustainable Development and Biodiversity 32, https://doi.org/10.1007/978-981-19-6974-4_20

Keywords Natural resources · Sustainability · Human population · Environmental policy · Community participation

20.1 Introduction

The resources of the environment are inestimable and can be in various forms of food from plants and animals. It provides services that are useful for sustainable living by providing various materials that have competing uses. These resources may be infinite (renewable) which is available for a long time, e.g., sun, or finite (non-renewable) which are gotten from the earth and cannot be re-used. The increasing human population has made the pressure on these environmental resources increase as compared with the pre-industrial era. Today, the transformation in environmental activities from the use of beast-of-burden and human strength to the use of machines, power plants, and cars which use coal, oil, or natural gas for their functions. The fuel to drive these types of machinery is gotten from the deposits in the environment and its use is not sustainable (United Nations Environment Programme [UNEP] 2016).

Africa relies heavily on natural resources to balance the ever-increasing human population. Land, water, plants, forests, renewable energy, and animal diversity are some of the resources which are fundamental in improving the livelihood of man and sustainable development (Sanginga et al. 2010). The dependence on natural resources is more on the rural population of Africa and poor economic growth policies are now placed as a key element by governments and planning institutions. There are very few areas in the world that are untouched or un-exploited. The resources in the forest can still meet the needs of humans, because it is a renewable resources that can grow rapidly and replace themselves in the environment (United Nations [UN] 2011; 2010). Of recent, forests are cleared for the harvest of hard-wood, farmland creation, and forage wood for cooking and heating. Despite all of these, it is important to find a sustainable way to conserve natural resources and this can be achieved in the following ways:

- Protection of tree species diversity
- Prevention of global climate change issues by preserving the vegetation that produces oxygen and stabilizing the chemistry of the atmosphere

The concept of environmental resource management is crucial at this time to strike a balance between the growth of the economy and prosperity with the need for sustainable use of resources (ILO 2011). For sustainable development, it is important to note that a reversal of natural resource exploitation that results in depletion must be addressed to reduce poverty (IMF 2010). The need to close the gap between consumption and supply rates is important. This chapter will, therefore, highlight the overview of classes, forms, and benefits of environmental resources, topical issues plaguing the diversity, and necessary sustainability measures.

20.2 Environmental Resources in Africa

Environmental resources are resources that are basic components in an environment and have great significance to humans (United Nations World Tourism Organization [UNWTO] 2009). The resources may be made up of organic or inorganic materials and are useful to plants, animals, and man. There are three major forms of environmental resources:

- Atmospheric resources
- Mineral resources
- Water resources (Fig. 20.1)

20.2.1 Atmospheric Resources

These are resources found in the atmosphere and have a direct effect on the weather conditions. Examples are gases, such as oxygen, nitrogen, carbon dioxide, water vapor, solar energy, and water. These resources have different functions:

- Oxygen is used for respiration by plants and animals, and it can support burning and can cause weathering by combining with water. It is used in industrial processes.
- Carbon dioxide is a principal element for photosynthesis and can combine with water to form a process known as carbonation. It also absorbs heat in the carbon cycle.
- Nitrogen is important to manufacture proteins in green plants. It can combine with water for plant growth and convert to nitrate for plant growth in the soil. The gas can be used for various industrial welding and processes.

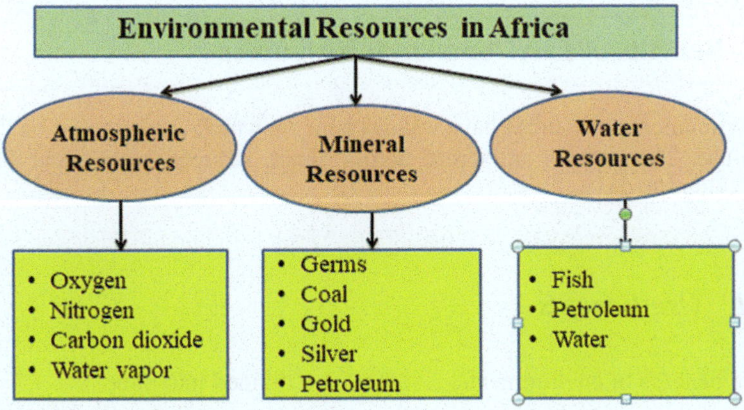

Fig. 20.1 Environmental resources in Africa

- Water vapor is very important in the formation of rain and the process of chemical weathering. It cools the atmosphere.
- Solar energy is derived from radiation from the sun. It is a principal component that influences photosynthetic activities by plants.

20.2.2 Mineral Resources

These are resources that are found embedded in rocks or the environment. Examples are germs, gold, silver, coal, and petroleum. The importance of mineral resources are:

- It is used in the beautification of structures, e.g., gold, silver, and germs.
- It is a source of fuel for combustion engines, e.g., coal and petroleum.
- It can serve as a source of foreign exchange from its sales and exports.
- It is used in industries.
- It provides raw materials for industries.
- It can provide employment.

20.2.3 Water Resources

This comprises all forms of resources that dwell in the aquatic bodies. Examples are water, fish, and petroleum. The importance of water resources includes:

- They can be exploited for their economic returns.
- For foreign exchange.
- Its nutritive value, e.g., fish.

20.3 Benefits of Environmental Resources

The environmental resources have various benefits which can be based on their use or non-use. The benefits are spread across direct, indirect, option, and existence benefits (Fig. 20.2).

20.3.1 Use Benefits

The use benefits of environmental resources are divided into three:

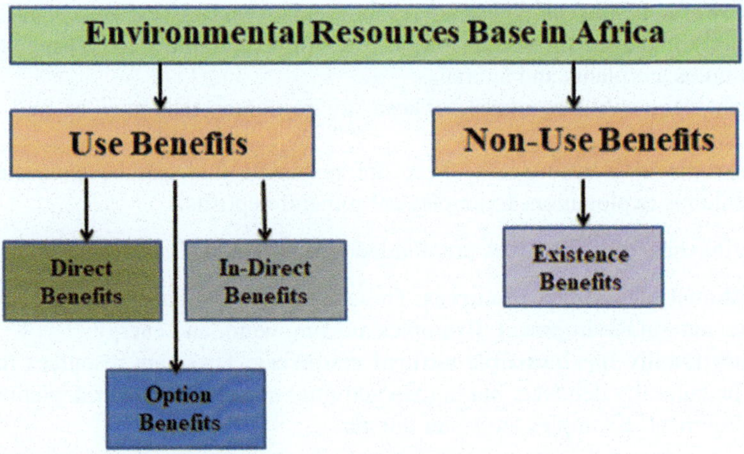

Fig. 20.2 Environmental resources based on Africa

- **Direct benefits**: It refers to when the environmental resources can be directly consumed or utilized by humans, such as food, minerals, and timber.
- **Indirect benefits**: It refers to the ecological benefits from the use of environmental resources, such as water protection, climate control, and carbon sink.
- **Option benefits**: It involves the focus placed on the sustainability of the resources and some may not be known presently.

20.3.2 Non-use Benefits

- **Existence benefits**: It is the intrinsic value of environmental resources regardless of their use. Examples are aesthetic and cultural significance (Serengeti National Park 2000).

20.4 Classification of Environmental Resources

Environmental resources are classified into two, namely:

- **The exhaustible natural resources**: These are resources that can be totally consumed and must be properly managed.
- **The inexhaustible natural resources**: These are resources that cannot be consumed, because they are components of the environment.

The exhaustible resources are divided into three, namely:

- **Renewable resources**: These are resources that cannot be exhausted and must be properly managed for sustainability. They have the ability to self-regeneration. Examples are plants and animals.
- **Partly renewable resources**: These are resources that may or may not be exhausted. An example is soil.
- **Non-renewable resources**: These are resources that can be exhausted after continuous exploitation. Examples are mineral deposits.

The inexhaustible resources are divided into two, namely:

- **Inexhaustible natural resources**: These are resources that can never be depleted in the natural environment. Examples are sun, wind, and energy.
- **Conditionally inexhaustible natural resources**: These are resources that may not be naturally depleted, but human activities can result in their depletion in the environment. Examples are water and air.

20.5 Contribution of Selected Sectors to Sustainable Environmental Development in Africa

The major sectors as components of the environment are (Fig. 20.3):

- Forest ecosystem.
- Biodiversity.
- Biotechnology practices.

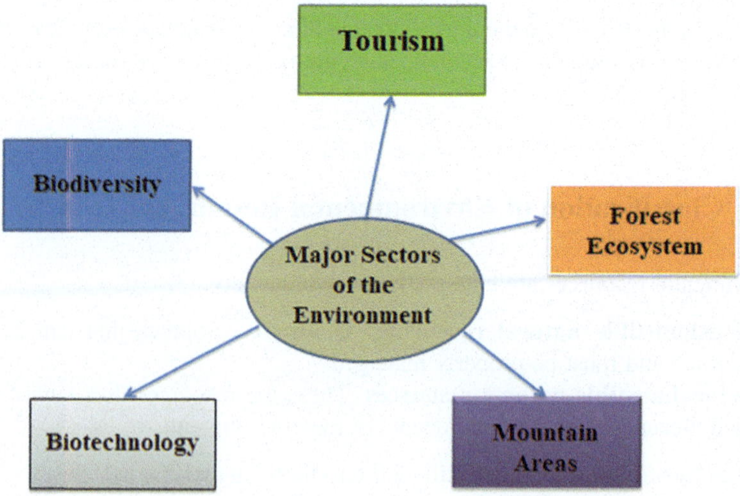

Fig. 20.3 Major sectors of the environment

- Mountain areas.
- Tourism (UNEP 2010; World Bank 2010).

20.5.1 Forests

Their presence is indispensable and management is very crucial in reducing poverty and development. It provides about 80% of energy in some countries and 6% of GDP in sub-Saharan Africa. There is a need to expand riparian and developmental areas, so that the increasing demand for forest products can be met and the issues of land degradation: climate change and desertification can be addressed. Diversification of forest products should be encouraged, so that the values derived from forest products and revenue generation can be increased (IMF 2010).

20.5.2 Biodiversity

It revolves around the people and the environment and their connections. There is an intimate connection between humans and the environment through their diets, energy, water, and recreational activities. They depend on the environment for food, medicine, and resources. For effective management, the equitability of sharing benefits from the environment must be strengthened to avoid land degradation, habitat loss, species, and genetic resource erosion. The need for national policy and plan harmonization is crucial. These policies can reduce the costs and undesirable results that may occur during the process of conservation (Kalaba et al. 2012).

20.5.3 Biotechnology

The significant improvement in food security was achieved through sustainable agriculture and more clean environmental processes. The interest of African countries in this research and development area is making huge progress and laws have been formulated to govern biotechnology. The approach must be ensured not to fall out of the regulatory framework. Public awareness of this new principle is important. Knowledge can be gained from accumulated information over the years to develop the biotechnology industry for improved environmental sustainability (UNEP 2010). The development of affordable, efficient, and regulatory regimes that are science-based should be developed by countries via their innovators without compromising the issues of biosafety. Political issues also have a major impact on emerging biotechnologies in the areas of scaling up investment levels and building physical, human, and institutional strategies (IMF 2010).

20.5.4 Mountains

Mountains are significant monuments in the African land mass and it is very crucial to sustainable development. They are rich envelopes of natural resources that when adequately harnessed; the development of Africa will be on the increase. Some contain minerals, and forests and can serve as a tourist attraction. They serve as the provision of water, clean energy, and habitat for biological diversity, and can serve as climate change indicators. Despite this, they are prone to threats from humans and degradation. The mountain areas suffer drastic damages from human activities because of their fragile nature, therefore, planning and adequate measures for mitigation are required for development in mountain areas. Population dynamics is a major issue that requires attention in the consideration of sustainable mountain areas. This is so because pressure from unsustainable activities, climate change, and human activities are on the increase. Therefore, strategic programmes, workshops, and actions must be developed in a manner that is participatory to involve the communities adjoining mountain areas (FAO/ITTO 2009).

20.5.5 Tourism

Tremendous growth has been recorded in the African tourism industry over the last two decades and has impacted significantly the economy. In 2010, Africa received about 63 million tourists which contributed 8.8% to economic growth. Issues of under-reporting were also identified because of the insufficient data capture at the state and national levels. Based on this, satellite monitoring and capturing systems can be employed for comprehensive reporting to capture the actual effects on the growth and development of the economy. Despite this wealth, the potential of the tourism industry has not been fully exploited (Miller 2012). There is a striking difference between tourism in North and sub-Saharan Africa. In the North, it is a major agenda in national policy and a resource for economic development and jobs, while in the sub-Saharan areas, it is yet to be fully capitalized as a resource (FAO 2010b). Africa should be able to explore the sub-Saharan areas to achieve the full benefits. There is a need for domestic and regional tourism which involves the identification of the significance of the growing middle class in travel and other touristic requirements. Policies should be created to diversify tourism and promote domestic and industrial tourism and travel (Burner et al. 2016).

20.6 Topical and Emerging Issues Plaguing Africa's Environmental Resources

Biological diversity is crucial to ecosystem structure and function. Human survival depends on the broad spectrum of goods and services provided by natural ecosystems (Leadley et al. 2010). Over 40% of the world economy is derived from biological resources (Travis 2003). Africa is immensely rich in biodiversity containing various species of plants, birds, and mammals (Raven et al. 2020). The continent of Africa remains the world's most diverse and biologically important ecosystem. These ecosystems include savannah, tropical forest, coral reef, marine and freshwater habitats, wetlands, and montane ecosystems (Dejene 2018). These regionally important ecosystems provide services (e.g., provisioning, supporting) to many African communities (MA 2005). In Africa, the temperature has increased and precipitation in the region has varied in most areas and many countries are already being affected (Funk et al. 2005). This increase in temperature has implications on the biological resources and their conservation, such as vegetative productivity and reproduction of tropical species (United Nations World Tourism Organization 2010, 2011, 2019; UNWTO 2010).

By and by, human activities have changed the ecosystem at an unprecedented rate (Owusu and Asumadu-Sarkodie 2016; Zelazowski et al. 2011). The change is being driven by the ever-increasing human population growth that led to increasing demand for basic nutritional and energy needs (Owusu and Asumadu-Sarkodie 2016). These needs are obtained from biological resources. The unsustainable extraction of these biological resources by humans has negative impacts that impede both the long-term well-being of humans and the resilience of the biophysical environment (Velenturf and Purnell 2017; Zhang et al. 2001). In the tropics, various environmental issues affect the persistence of the rainforest biomes. The major environmental issues in Africa are deforestation, land use changes or conversion, climate change, and natural resource exploitation with ecological and biodiversity consequences within tropical ecosystem (de Wasseige et al. 2015; Mallon et al. 2015; Song et al. 2019; Ikhajiagbe et al. 2020). Increasing human population growth with the demand for economic growth and diversification through diverse land use has taken a toll on the environment (Winfree and Kremen 2009). Furthermore, weak governance and enforcement of environmental laws could soon bring the region to a tipping point of rapid and unpredictable environmental change. The implications of all these environmental issues will not only be felt locally in the tropics but across the globe (Willis et al. 2013).

20.6.1 Global Warming and Climate Change

Presently, the whole world is facing serious challenges, such as global warming, poverty, international terrorism (UN 2007). Comparing this degree of threats, global warming and climate change pose by far the most serious challenge and hence, emerged as an important global concern (Arega 2018; Velenturf and Purnell 2017). Scientifically, it has been established those human activities such as the burning of fossil fuels in this era of industrialization increases the earth's average temperature which has consistently been on the rise (Dejene 2018; Ogwu 2019a; Osawaru and Ogwu 2014; Ogwu et al. 2014, 2016). In the tropics, the warming signal is dominant because of the relative importance of deforestation-induced evapotranspiration changes compared to albedo modifications in the high latitudes that can drive cooling in the annual mean (Perugini et al. 2017; UNFCCC 2009). This global warming is anthropogenic and the greenhouse gas (GHGs) (carbon dioxide) primarily originated from humans through deforestation (Mohammed 2013; Elke and Paul 2011; UNDP 2008). In addition, fluorinating gases are being released into the atmosphere from industrial processes, while other gases such as methane and nitrous oxides are originating from agriculture and waste (Arega 2018; Edwards et al. 2014). The implication of the rise in the global earth's temperature is the extreme weather events with its dramatic and devastating widespread extinction of many plant and animal species (Stępniewska and Kuźniar 2013; UN 2007; Townsend et al. 2000). Therefore, it is undeniably evident from findings, observations, and assertions from researchers that regardless of any adaptation and/or mitigation measures, the Earth's climate has changed, is changing, and will continue to change (Dantas-Torres 2015; Solomon et al. 2015; IPCC 2009; Smith 2009).

Climate change on the other hand is another major global threat (Dejene 2018). It refers to changes in climate over comparable periods, which may be due to natural variability or as a result of direct/indirect human activity. This change in climate is mainly caused by several natural and anthropogenic activities that increase the concentration of GHGs in the atmosphere (Arega 2018). Consequently, the conversion of grasslands and forests into croplands, and the production of biological resources for foods, fuels, and fibres both directly affect emissions of several GHGs and hence leads to climate change (Burnham and Ma 2015; Silverio et al. 2015; Lambin and Meyfroidt 2011; Vwioko et al. 2018; Ogwu 2019b, c, 2020). The global average temperature is rising, and a rise of about 0.7 °C has been documented with a prediction of continuous rise (Dejene 2018; Niang et al. 2014). It has been established that biodiversity and natural ecosystem are threatened and have been negatively impacted by climate change (Dejene 2018; Birdlife International 2013).

The dynamics of energy flow and material circulation patterns have directly and indirectly changed as the loss of biodiversity to climate change continues (Zhong and Wang 2017). This will have a great impact on the African ecosystem and ecosystem service. The supporting, regulating, and provisioning service capacity of the ecosystem depend on biodiversity and this may be lost to climate change (Pasquini and Shearing 2014; Saatchi et al. 2013; Bellard et al. 2012; Parmesan 2006; Thomas et al. 2004). Africa is endowed with biodiversity. However, the continent is highly vulnerable to climate change compared to other continents of the world. Overall, the African continent has warmed 0.7 °C over the twentieth century and warming is expected to continue at the rate of 0.2 °C and 0.5 °C per decade at low and high scenarios, respectively). As climate change surges over the next century, its synergetic effect on anthropogenic stressors is anticipated to be a major driver of the loss of African biodiversity (Midgley and Bond 2015; Bellard et al. 2012; Ikhajiagbe et al. 2022a). Thus, it is important to understand the link between biodiversity, ecosystem services, and climate change (Dejene 2018; Natural Environment Research Council 2012; Erinle et al. 2021).

20.6.2 *Resource Recovery Processes*

Ecosystems can be understood as the communities of various species' populations within their physical environment, and the flow of energy and matter between the living and non-living elements in the system (Bilitewski et al. 2012). A healthy environment is essential for human well-being. At the most fundamental level, people depend on the environment to meet basic human needs including access to clean water, food, and shelter (Velenturf and Purnell 2017; Muth and Bowe Jr 1998). Ecosystem stewardship recognizes the human species as an integral part of the ecosystem and argues that people carry a responsibility to manage the environment, such that society's use of resources is compatible with the ecosystem's capacity to sustain services (Murray-Brian et al. 2017; Chapin et al. 2009). The cause of the decline in the resources (flora and fauna species) in any ecosystem sometimes appeared obvious. However, actions to determine what is driving a species toward extinction require a successful diagnosis that depends on a logical series of steps to halt the decline toward species recovery (COMIFAC 2015; Pullin 2002).

The first step is to decide if a decline is taking place and to detect the rate of decline. Knowing the rate of decline establishes the time available in which to take remedial steps (Caughley and Gunn 1996). The greater the rate the more urgent the task determines what caused it (Hastenrath 2010). Sometimes, it is easier to detect and rectify declines, while a species is relatively abundant than when such species become rare. Therefore, regular monitoring of species with conservation interests must be conducted (Mohammed 2013). The second step is making lists of conceivable agents of decline toward diagnosing the cause of the problem/factors driving the decline. This can only be achieved by assembling evidence for and against possible cause(s) of the decline, conducting research, and concluding based on the findings to make recommendations (Herrmann and Khan 2008).

20.7 Factors Driving Environmental Resources Decline

Many researchers have identified various factors driving resource decline (Fig. 20.4). This is:

- Overharvesting.
- Habitat changes.

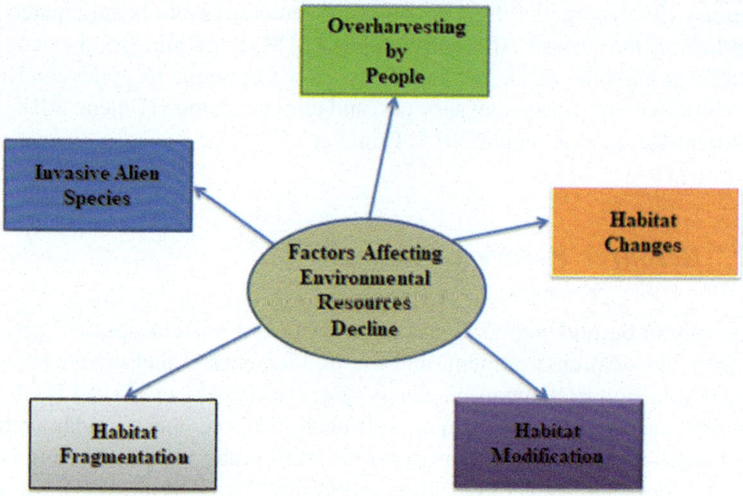

Fig. 20.4 Factors affecting environmental resources decline

- Habitat modification.
- Habitat fragmentation.
- Invasive alien species.

20.7.1 Overharvesting by People

Harvesting in this context means removing individuals from a population by hunting, fishing, and logging. Overharvesting is a common factor that drives population decline and, in extreme, leads to species extinction. Harvesting may be selective and unselective. It is selective if the targeted species is being harvested (Millennium Ecosystem Assessment 2005). Harvesting is unselective (by-catch) when other species are unintentionally forced into decline (ICIMOD 2011). Diagnosing overharvesting requires information on the harvesting rate, i.e., annual off-take of species of interest, and harvest types/levels (i.e., levels of subsistence and commercial harvesting). One common approach to gauging harvesting levels is to interview those involved in the business, e.g., Hunters, loggers, and gatherers (WEF 2011). Conducting such an interview could be tricky as the topic is sensitive and many people may feel unsafe if they give the true report/answer during the interview session. Hence, there will be underreporting of harvests especially when it is related to the illegal harvesting of resources. However, various studies have used the randomized response technique in obtaining vital information on such sensitive topics (Akinsorotan et al. 2019; St. John et al. 2015; COMIFAC 2014).

20.7.2 Habitat Changes

Habitat is a suite of resources (food and shelter) and environmental conditions (biotic and abiotic variables) that determine the presence, survival, and reproduction of a population. It can also refer to the composition of plants and animals in their natural environment Habitat change and loss are often advanced as the dominant cause of extinction (Lucas et al. 2020; Kohler and Maselli 2009).

20.7.3 Habitat Modification

Habitats can be modified by different factors which can dictate the survival of plants and animals in the environment. Most importantly, the change could be a result of the felling of trees and or change in fire regime which has negative effects on wildlife population density. Diagnosing the change in an element in a habitat is vital in preventing the species decline and a step toward species recovery (Lewis and Maslin 2015; Lovett 2015; Peacock 2014; Lewis et al. 2013).

20.7.4 Habitat Fragmentation

This is a man-made change imposed on natural habitat heterogeneity. Usually, the habitat is changed patch by patch into another land use and initially, those patches of modified habitat are embedded in the original habitat. Even at the early stages of fragmenting a habitat, relatively small areas of altered habitat can have a large effect on the ecology of the core area (Galetti and Dirzo 2013; IPCC 2013; Kuehl et al. 2009).

20.7.5 Impact of Invasive Alien Species

Alien species are referred to species that are not native to a particular environment. They are mostly introduced based on different reasons and can prey upon, compete with or disturb an indigenous species. Invasive alien species are disastrous and can lead to the loss of hundreds of different species (Fa et al. 2015; Feintrenie 2014; Crow et al. 2013).

20.8 Deforestation

Deforestation which is indiscriminate cutting down of trees is considered a potential factor causing climate change. It is the second-largest human-caused source of carbon dioxide in the atmosphere. Trees have aesthetic, recreational, economical, historical, cultural, and religious values to humans (Bommarco and Hallin 2018). The growing demand for forest products and agricultural land to feed the human population leads to deforestation (Damania and Wheeler 2015; Rudel 2013). Trees serve as a carbon sink in the environment. They are useful to humans especially the timber products (in making paper, houses, and buildings) and the non-timber products (fungi, nuts, bamboo, berries, medicinal compounds, dyes, and fabrics) (Mayaux et al. 2013). Hence, many people depend on the forest and its products. Economically, timber and other forest products are important locally and for exportation.

Deforestation or the death of trees reduces the amount of carbon stored and also releases carbon dioxide into the air. Consequently, the future generation will be affected by the extreme adverse effect of a large number of trees that are cut down or burnt from forested areas due to various reasons. These reasons vary and include:

- Demand for land for cultivation.
- The need for energy sources for cooking.
- Industrialization (Damania and Wheeler 2015; Hoare 2015; Mayaux et al. 2013; Rudel 2013).

The impact of deforestation on the environment is enormous. Besides the destruction of carbon sinks, deforestation makes soil prone to erosion and destroys animal habitats causing the disappearance of some endemic species of animals and plants (especially the ones with medicinal values).

20.9 Poaching Activities

Poaching remains one of the factors driving wildlife resources to extinction daily in Africa (Grey-Ross et al. 2010; Kideghesho 2009). Poaching includes all activities including hunting, trapping, and killing of animals in the conservation/protected areas established with extant laws and regulations to protect resources in the wild (William and Lemieux 2015). Despite the creation of protected areas to safeguard these wild animals, constant pressure from subsistence and commercial activities further reduces the population of these species. Pressure on forest resources is further exacerbated by the increased human population (Obour et al. 2016). The majority of these populations reside in communities neighboring conservation areas. For nutritional needs and economic empowerment, many inhabitants of these communities embark on poaching activities (Obour et al. 2016). Large proportions of their games are sold in the urban areas, where the demand for bush meat is higher (Obour et al.

2016). Hence, wild animals are trapped, hunted, and killed to be sold and consumed in urban areas making poaching an organized and lucrative business (Duveiller et al. 2018; Ijeomah et al. 2013).

Globally, the high commercial value of wildlife products poses threat to the wildlife resources in the protected areas. This includes poaching for ivory (examples from Kruger and Serengeti National Park of South Africa and Tanzania, respectively), where rhinoceros and elephants were targeted by groups of poachers, killed for their valuable horns, and tusks. The poaching activities were well-organized with the use of high-tech modern equipment, such as sedative drugs and night-vision goggles in their operations. Ivory poaching decimated the population of these iconic species with a dwindling population even in the protected areas. Nevertheless, meat poaching continued. Across African countries, hunting and trapping wild animals for meat have been reported by various researchers (Akinsorotan et al. 2019; Kim et al. 2016; Björn Schulte-Herbrüggen et al. 2013; Ganby 2012).

The drivers of bush meat hunting and consumption have been documented. These include:

- Individuals or households harvesting/consuming bush meat as alternative order to purchasing husbandry meat.
- It serves as a safety net function during a time of economic hardship.
- Affinity for the taste.
- Some are culturally inclined to eat bush meat.

These can further be grouped under poverty, cultural taste, household size, and taste preference purposes.

20.10 Other Issues Plaguing the Resources

It is undeniable that most of the changes observed in our environment are a result of massive human population growth (Georg et al. 2018; D'Amato et al. 2015; FAO 2014; Cubasch et al. 2001). In a quest to survive in this era, many habitats have continuously been destroyed. There are issues of lack of fresh water, overexploitation of natural resources, and species extinction (Darwall et al. 2018; Dudgeon et al. 2006). Our action is gradually destroying the lives of future generations. Aside from overpopulation, other issues affecting the environment are:

20.10.1 Pollution

Human activities have caused both water and air pollution. Globally, tons of sewage and agricultural and industrial wastes are released into the atmosphere and water bodies daily. This has been a growing problem as the effect of air and water pollution is causing several health problems to human, disrupts marine life, alter reproductive cycles, and increase mortality rates (Abernethy et al. 2016; Beyene et al. 2013; Aletti and Dirzo 2013; Crutzen 2002; Ikhajiagbe et al. 2022b).

20.10.2 *Waste Disposal*

There are tons of waste produced by the increasing human population. A large amount of methane is generated from bio-wastes. Methane is one of the worst GHGs and has the potential to create explosion hazards and global warming (Bright et al. 2017; Desclee et al. 2014; Dirzo et al. 2014; Collette et al. 2011; IPCC 2009).

20.11 Sustainable Exploitation of Environmental Resources in Africa

For sustainable exploitation of environmental resources, there is a need to harness the synergies necessary for sustainable growth and development. The resources in Africa can be managed through integrated approaches, because sustainable development is diverse, multi-scaled, and cuts across all governance and regimes (Nyenje et al. 2011). The line between natural resources management and institutional systems is very narrow and it is necessary to understand the effective management link for sustainable development across spatial and temporal spheres (van Drunen et al. 2009). Most opportunities are offered and more associated for effective responses are at the sectoral, national, and regional levels. It will require harmonization, monitoring, and understanding of the various linkages by integration of policy, participation of communities, and governance that is adoptive and innovational (UNEP 2010) (Fig. 20.5).

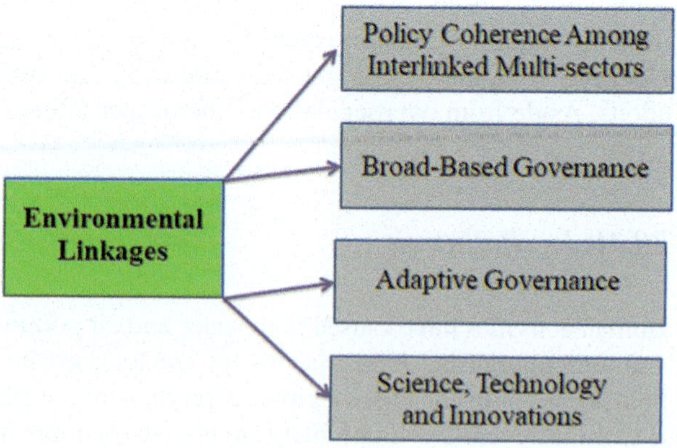

Fig. 20.5 Environmental linkages for sustainable development

20.11.1 *Promoting Policy Coherence amongst Integrated Multi-sectors*

In environments where goals, processes, and sectorial activities are linked, it is essential to strengthening the system of governance for sustainable environmental activities and development at all levels (Minnemeyer et al. 2011). Coherence and integration of policy must be built and must cover the economic, social, and environmental pillars of the environment while strengthening the processes of monitoring, accountability, and implementation. With this approach, overlapping conflicts and duplication of activities for managing forests, mountains, and biodiversity will be limited and technological advancement in biotechnology and tourism activities (Timko 2011). Stakeholders can benefit from this approach through national capacities for sustainable activities. This may be achieved by the government by forming policies that must be consistent and promote the involvement of stakeholders, relevant to environmental sustainability. With this, institutional mechanisms that support sustainable development will be in effect (World Bank 2010).

Successes have been recorded in seven West African countries in partnership with the Sahel and West Africa Club (SWAC) as reported by OECD (2011) who investigated the policy coherence in the fisheries sectors in Guinea, Cape Verde, the Gambia, Guinea-Bissau, Senegal, Mauritania, and Sierra Leone. It observed the need for multi-stakeholder dialogue on the coherence of policy as a prerequisite for sustainable development (Kiefer et al. 2010). The connection between the environmental resources such as forests, biodiversity, tourism, and mountains calls for policy coherence for trade, investment, agriculture, taxation, innovation, migration, security climate, and the environment (OECD 2011). A comprehensive approach is also required for policy coherence across the environment and sub-regions, thereby ensuring knowledge sharing and partnerships. It can be stimulated by analysis of the cost of incoherent policies, comprehensive assessment, and policy monitoring, identifying policy options, and synergy between trade-offs. Goals and targets of both international and national levels can be set, monitored, and harmonized (Karembu et al. 2009; Juma and Serageldin 2007).

20.11.2 *Broad-Based Governance*

Governance is a vital key to sustainable development and it is essential to building multi-linkages between all components of the environment in terms of biodiversity, forests, mountains, tourism, and biotechnology (Elson 2010). This approach can help to sustain all these components and promote social, economic, and infrastructural developments. Domestic mechanisms of accountability must be strengthened to address regulations, transparency, and management. The successes and achievements can be related to a well-structured consultative approach and activities by stakeholders in the social, environmental, and economic developmental process

(Cundill 2010). The consultation may target the actors who create the policies and are involved in the inter-linkage activities. The roles of governments, communities, civil societies, and partners are key to biodiversity management based on their roles and activities in the environment.

- The United Nations (UN) Commission on Private Sector and Development has recommended many countries to put plans for private sector engagement. The 2004 report of the UN Commission emphasized the best practices of identification, domestic policies, and strategic partnerships that can promote development in the national, international, public, and private sectors of the economy (Gobeze et al. 2009).
- The initiatives of UNDP stated partnerships between business, civil, government, and organizations to promote entrepreneurship between countries and neighbors to reduce poverty and increase sustainable development. This approach will also reduce poverty in these areas (Gondo 2011).

It has been shown from experience that environments that involve the active participation of private sectors and the government results in innovations and promote integration through the creation of structures and pro-poor policies. These can be achieved in the following ways:

- By creating and implementing environmental policies mainly targeted at agriculture, water, health, trade, and investments to stimulate economic growth.
- Development of partnership between stakeholders, government, private societies, civil societies, and actors with the aim of sustainable environmental management.
- By strengthening the social, ethical, and entrepreneurship activities that are sustainable and can implement global principles on environmental management.

20.11.3 Adaptive Governance

The underlying challenges of the environment such as inter-linkages among the forest ecosystem, tourism, mountain areas, biotechnology, and biodiversity which are associated with fragmentation, change, uncertainty, and various complexities must be adequately addressed. It was noted that these complexities can result in instability in governance (Biermann et al. 2010). The opportunities to transform these complexities are limited and it is complicated when there are conflicting national, institutional, or regional issues and interests by individuals. Adaptive management can be a possible solution to the issue of these complexities as recommended by Olsson et al. (2006). It involves the involvement of possible actors and individuals in the continuation and actions geared toward sustainable environmental development. This approach can enable all actors to be able to manage and be directly responsible for the environmental resources for development (Galaz et al. 2008).

20.11.3.1 Advantages of Adaptive Governance

The advantage of the adaptive governance approach to environmental sustainability is highlighted below:

- Sustainable development starts with existing organizations.
- It enables reaching out to other stakeholders (WTTC 2009).
- It can cascade and share responsibilities between stakeholders (WTTC 2011).
- It can broaden the knowledge base for sustainable development.
- It can enhance the social coordination application via networks (West Africa Discovery 2010).

By practicing this approach, policymakers can easily integrate efficiently fresh ideas, knowledge, and technologies to sustainable development. This will reduce drastically the cost of sustainable development and apply avenues that are appropriate for efforts in the future to address key environmental issues and inter-linkages.

20.11.3.2 Principles for Implementing Adaptive Management

The following principles are important in Africa for implementing a sustainable natural resource base for development.

- The dialogue between actors and stakeholders on the decision about the environment must be effective.
- Institutional layers of governance must be connected and well-nested and layered for sustainable development.
- An institutional mix that promotes integration and facilitates change, learning, and experimentation must be promoted.

20.11.4 Science, Technology, and Innovations (STI)

Development relies on information from various inter-linkages across all sectors of the environment. For poverty to be eradicated, individuals are required to key into science and technology for innovations. There must be investments in science and technology which focuses on management, knowledge, and information needs. In Africa, there are diverse initiatives that observe, monitor, network, and manage information for sustainable development challenges at various levels of governance. The African Union has a plan that is consolidated around science and technology in African systems of research and innovation. Presently, African countries are using STI principles to meet various needs, such as social, economic, and environmental challenges.

UNESCO (2011), political leadership is important in facilitation of innovation and entrepreneurship. Countries should learn from other countries on how needs were met and develop a technological and industrial base. Many African countries have provisions for STI harnessing which can contribute to human welfare and development. The use of Information and Communications Technology and bio-technology are areas, where skills can be developed and can be within universities, enterprises, and research and development institutions. In developed countries, innovation stimulates economic growth.

20.12 Possible Solutions to Decline in Environmental Resources

Nevertheless, one of the ways to halting resource/species decline and stimulate resource recovery is research-driven interventions. Considering the environment and human rights, research interventions targeted toward resource recovery must be approached with care. For instance, during recovery process to save indigenous species from the effects of invasive alien species, contaminants can escape into the ecosystem and negatively impacted the air, land, water, and the food chain through persistence/excessive use of chemicals (Bilitewski et al. 2012; Misana et al. 2012). The implication of these pollutants on the environment is less clear; however, it poses health risk to human (Slater et al. 2011). Critical assessment is necessary to ensure that resource recovery processes are truly sustainable. According to the United Nations (2010a), sustainable development is meeting the needs of the present without compromising the needs of future generations. To achieve this, the best strategy is to design a plan for resource recovery, so that after any incident, e.g., population decline, normal conditions can be restored (United Nations 2010b). Sustainable development creates a balance between capabilities of environment and economic growth and provides a suitable solution for the inhabitants of the earth without destroying the global ecosystem. Using sustainable development, mankind tries to achieve a better and healthier life (Iacovidou et al. 2017; Korcha and Sorteberg 2013; Buch et al. 2002; Ganesan 2000).

To achieve sustainable resource development, resource recovery is an effective strategy for environmental recovery (Mehrdad et al. 2016). Incidentally, all hands must be on deck as stakeholders (public, private, and international organizations and people) work together for this resource recovery. Commitment from the local and national organizations with their effective and integrated structure, skilled technical personnel and financial resources will go a long way in achieving sustainable management of the environmental resources (Ibisch et al. 2010). This will be achieved when local and national organizations are committed to it and have an effective, integrated structure, skilled technical personnel, and financial resources (Mehrdad et al. 2016). The importance of environmental and resource recovery process is to reduce future risks. Hence, policymakers, management of natural resources, and the public should be encouraged to base their policies/laws, interventions, food and energy production/consumption on research outcomes, innovative and green technologies (Mehrdad et al. 2016). Designing environmentally friendly structures should be considered in environmental management and recovery by authorities and the public (FAO 2012; United Nation Environmental Program 2010). Restoration of the environment after catastrophes can reduce underlying risks and their effects on the ecosystem. To this end and for environmental and resource recovery, in some cases, it is necessary to declare the affected area as a protected area (Mehrdad et al. 2016).

20.13 Conclusions

The ever-increasing human population has resulted to an increase in pressure on environmental resources in Africa. These resources are observed to be inestimable, although some are renewable, while others are non-renewable. The environment serves as sources for species biodiversity, tourism activities, timber and wood from forests and wildlife meat which are exploited at various levels. Based on this increased pressure, there is need for sustainability of these resources by protection and prevention of climatic change issues which have drastic effects on the environment and its resources. This can be achieved by striking a balance between resource use, exploitation, and recuperation. The atmospheric, mineral, and water resources are the three forms of environmental resources with each of them playing a significant role in the ecosystem. They offer various use and non-benefits spread between direct, indirect, option, and existence benefits. These benefits can be offered by major sector components of the environment which are forests, mountain areas, biotechnology, biodiversity, and tourism. The environment is also faced with various plaguing issues, such as climate change and resource recovery processes. Others are overharvesting, habitat changes, poaching, pollution, waste disposal, deforestation, habitat modification, habitat fragmentation, and invasive alien species. The solutions to these issues are resource recovery processes, policy, and legislation reforms and community participation which involves the actors, governments, or stakeholders and the immediate community. Environmental linkages must be encouraged which encompasses adaptive governance, broad-based governance, interlinked multi sectors, science, technology, and innovations.

References

Abernethy K, Maisels F, Lee JTW (2016) Environmental issues in Central Africa. Ann Rev Environ Resour 41:1–33

Akinsorotan OA, Oguntuase BG, Olaniyi OE, Nnamuka SS (2019) Evaluating rule breaking behavior in a Nigerian protected forest reserve area. J Appl Sci Environ Manag 23(6): 1075–1079

Aletti M, Dirzo R (2013) Ecological and evolutionary consequences of living in a defaunated world. Biol Conserv 163:1–6

Arega BB (2018) Global warming: a review of the debates on the causes, consequences and politics of global response Ghana. J Geogr 10(1):144–164

Bellard CC, Bertelsmeier P, Leadley WT, Courchamp F (2012) Impacts of climate change on the future of biodiversity. Ecol Lett 15:365–377. https://doi.org/10.1111/j.1461-0248.2012.01764.x

Beyene T, Ludwig F, Franssen W (2013) The potential consequences of climate change in the hydrology regime of the Congo River Basin. In: Haensler A, Jacob D, Kabat P, Ludwig F (eds) Climate change scenarios for the Congo Basin, pp 1–44

Biermann F, Betsill MM, Gupta J, Kanie N, Lebel L, Liverman D, Schroeder H, Siebenhuner B, Zondervan R (2010) Earth system governance: a research framework. Int Environ Agreem Polit Law Econ 10(4):277–298

Bilitewski B, Barbra RM, Barcelo B (2012) Global risk based management of chemical additives I: production, usage and environmental occurrence. Springer, Berlin

Birdlife International (2013) State of the World's birds: indicators for our changing world. Birdlife International, Cambridge

Björn Schulte-Herbrüggen et al. (2013) The importance of bushmeat in the livelihoods of west African cash-crop farmers living in a faunally-depleted landscape

Bommarco RGV, Hallin S (2018) Exploiting ecosystem Services in Agriculture for increased food security. Glob Food Sec 17:57–63. https://doi.org/10.1016/j.gfs.2018.04.001

Bright RM et al (2017) Local temperature response to land cover and management change driven by non-radiative processes. Nat Clim Chang 7:296–302

Bruner KM, Murray AJ, Pollack RA, Soliman MG, Laskey SB, Capoferri AA, Lai J, Strain MC, Lada SM, Hoh R, Ho YC, Richman DD, Deeks SG, Siliciano JD, Siliciano RF (2016) Defective proviruses rapidly accumulate during acute HIV-1 infection. Nat Med 22(9):1043–1049. https://doi.org/10.1038/nm.4156. Epub 2016 Aug 8. PMID: 27500724; PMCID: PMC5014606

Buch E, Masuku N, Mathee A (2002) Health and sustainable development meeting of senior officials and ministers of health. World Health Organization, Johannesburg

Burnham M, Ma Z (2015) Linking smallholder farmer climate change adaptation decisions to development. Clim Dev 8:1–10

Caughley G, Gunn A (1996) Conservation biology in theory and practice. Blackwell Science, Oxford

Chapin FS, Kofinas GP, Folke C (2009) Principles of ecosystem stewardship: resilience-based natural resource management in a changing world. Springer, New York, NY

Collette BB, Carpenter KE, Polidoro BA, Juan-Jorda MJ, Boustany A et al (2011) High value and long life—double jeopardy for tunas and billfishes. Science 333(6040):291–292

COMIFAC (2014) COMIFAC annual report 2014. COMIFAC, Yaounde, Cameroon. http://www.comifac.org/en/content/rapportannuel2014comifac

COMIFAC (2015) Plan de Convergence 2015–2025. COMIFAC, Yaounde, Cameroon

Crow MS, Tara O'Connor S, Stretesky PB (2013) Camouflage-collar crime: an examination of wildlife crime and characteristics of offenders in Florida. Deviant Behav 34:635–652

Crutzen PJ (2002) Geology of mankind: the Anthropocene. Nature 415:23

Cubasch U, Meehl GA, Boer GJ, Stoufer RJ, Dix M, Noda A, Senior CA, Raper S, Yap KS (2001) Projections of future climate change. Climate Change 2001: The Scientific Basis

Cundill G (2010) Monitoring social learning processes in adaptive co-management: three case studies from South Africa. Ecol Soc 15(3):28. http://www.ecologyandsociety.org/vol15/iss3/art28/

D'Amato G, Holgate ST, Pawankar R, Ledford DK, Cecchi L, Al-Ahmad M, Annesi-Maesano I (2015) Meteorological conditions, climate change, new emerging factors, and asthma and related allergic disorders. A statement of the world allergy organization. World Allergy Organ J 8(1):1–52

Damania R, Wheeler D (2015) Road improvement and deforestation in the Congo Basin countries. Policy Res. Work. Pap. 7274. World Bank, Washington, DC

Dantas-Torres F (2015) Climate change, biodiversity, ticks and tick-borne diseases: the butterfly effect. Int J Parasitol Parasites Wildl 4:452–461

Darwall W, Bremerich V, De Wever A, Dell AI, Freyhof J, Gessner MO, Weyl O (2018) The Alliance for freshwater life: a global call to unite efforts for freshwater biodiversity science and conservation. Aquat Conserv Mar Freshwat Ecosyst 28(4):1015–1022

de Wasseige C, Tadoum M, Eba'a Atyi R, Doumenge C (2015) The forests of the Congo Basin—forests and climate change. Weyrich, Belg.: Weyrich Ed.

Dejene WS (2018) Impact of climate change on biodiversity and associated key ecosystem services in Africa: a systematic review. Ecosyst Health Sustain 4(9):225–239. https://doi.org/10.1080/20964129.2018.1530054

Desclee B, Hansen M, Lola Amani P, Sannier C, Mertens B et al (2014) Evolution of forest cover at a national and regional scale and drivers of change. In: de Wassinge C, Louppe D, Hiol Hiol F, Mayaux P (eds) The forests of the Congo Basin—State of the forest, Weyrich USA, pp 21–46

Dirzo R, Young HS, Galetti M, Ceballos G, Isaac NJB, Collen B (2014) Defaunation in the Anthropocene. Science 345:401–405

Dudgeon D, Arthington AH, Gessner MO, Kawabata ZI, Knowler DJ, Lévêque C, Sullivan CA (2006) Freshwater biodiversity: importance, threats, status and conservation challenges. Biol Rev 81(2):163–182

Duveiller G et al (2018) The mark of vegetation change on Earth's surface energy balance. Nat Commun 9:679

Edwards D, Sloan S, Weng L, Dirks P, Sayer J, Laurance W (2014) Mining and the African environment. Conserv Lett 7(3):302–311

Elke U, Paul CS (2011) Public understanding of climate change in the United States. Am Psychol Assoc 66(4):315–328

Elson D (2010) Investing in locally controlled forestry: reviewing the issues from a financial investment perspective. Background paper for the Forests Dialogue's initiative on investing in locally-controlled forestry conference, London, 24–25 May 2010

Erinle KO, Ogwu MC, Evivie SE, Zaheer MS, Ogunyemi SO, Adeniran SO (2021) Impacts of COVID-19 on agriculture and food security in developing countries: potential mitigation strategies. CAB Rev 16(16):1–16. https://doi.org/10.1079/PAVSNNR202116016

Fa JE, Olivero J, Real R, Farfan MA, M'arquez AL et al (2015) Disentangling the relative effects of bush meat availability on human nutrition in Central Africa. Sci Rep 5:8168

FAO (2012) National forest programmes (NFPs) in practice. Ways to improve the implementation of national forest programmes. FAO, Rome

FAO (2014) Ethiopian country programme framework. FAO, Addis Ababa

FAO/ITTO (2009) Forest governance and climate. Policy Brief, Change Mitigation

Feintrenie L (2014) Agro-industrial plantations in Central Africa, risks and opportunities. Biodivers Conserv 23(6):1577–1589

Funk C, Senay G, Asfaw A, Verdin J, Rowland J, Michaelson J, Eilerts G, Korecha D, Choularton R (2005) Recent drought tendencies in Ethiopia and equatorial-subtropical Eastern Africa. FEWS-NET, Washington, DC

Galaz V, Olsson P, Hahn T, Svedin U (2008) The problem of fit among biophysical systems, environmental and resource regimes, and broader governance systems: insights and emerging challenges. In: Young OR, Schroeder H, King LA (eds) Institutions and environmental change: principal findings, applications and research frontiers. MIT Press, Cambridge, MA, pp 147–186

Galetti M, Dirzo R (2013) Ecological and evolutionary consequences of living in a defaunated world. Biol Conserv 163:1–6

Ganby Z (2012) To the rescue of endangered species. http://www.especesmenacees.ca/en/excessive-trade.php

Ganesan CT (2000) Environment and sustainable development; the role of the engineer. Paper presented at: strategies for a sustainable built environment; Aug. 23–25; Pretoria, South Africa

Georg LAG, Requier AF, Fijen T (2018) Complementarity and synergisms among ecosystem services supporting crop yield. Glob Food Sec 17:38–47. https://doi.org/10.1016/j.gfs.2018.03.006

Gobeze T, Bekele M, Lemenih M, Kassa H (2009) Participatory forestry management and its impacts on livelihoods and forest status: the case of Bonga forest in Ethiopia. Int For Rev 11(3): 346–358

Gondo PC (2011) Financing of sustainable forest management in Africa: an overview of the current situation and experiences. Southern Alliance for Indigenous Resources (SAFIRE), Harare

Grey-Ross R, Downs CT, Kirkman K (2010) An assessment of illegal hunting on farmland in KwaZulu-Natal, South Africa: implications for oribi (*Ourebia ourebi*) conservation. S Afr J Wildl Res 40(1):43–52

Hastenrath S (2010) Climatic forcing of glacier thinning on the mountains of Eastern Africa. Int J Climatol 30:146–152. https://doi.org/10.1002/joc.1866

Herrmann M, Khan HA (2008) Rapid Urbanization, Employment Crises and Poverty in African LDCs, paper prepared for UNU-WIDER Project Workshop "Beyond the Tipping Point: African Development in an Urban World". Cape Town, Geneva. http://mpra.ub.unimuenchen.de/9499/1/MPRA_paper_9499.pdf

Hoare A (2015) Tackling illegal logging and the related trade: what Progress and where next? Chatham House, The Royal Institute of International Affairs, London

Iacovidou E, Velis CA, Purnell P, Zwirner O, Brown A, Hahladakis J, Millward-Hopkins J, Williams PT (2017) Metrics for optimising the multi-dimensional value of resources recovered from waste in a circular economy: a critical review. J Clean Prod 166:910–938

Ibisch PL, Vega AE, Herrmann TM (eds.) (2010) Interdependence of biodiversity and development under global change. Technical Series No. 54

ICIMOD (2011) International Centre for Integrated Mountain Development (ICIMOD): Earth Observation and Climate Change; Sustainable Mountain Development; ICIMOD) No. 60; Autumn, 2011

Ijeomah HM, Ogogo AU, Ogbara D (2013) Analysis of poaching activities in Kainji lake National park of Nigeria. Environ Nat Resour Res 3(1):51

Ikhajiagbe B, Ogwu MC, Lawrence AE (2020) Single-tree influence of *Tectonia grandis* Linn. f. on plant distribution and soil characteristics in a planted forest. Bull Natl Res Cent 44:29. https://doi.org/10.1186/s42269-020-00285-0

Ikhajiagbe B, Ogwu MC, Ekhator PO, Omoayena II (2022a) Implications of soil nitrogen ehnacement on the yield performance of soybean (*Glycine max*) in cadmium-polluted soil. J Pollut Eff Contr 10(2):1–12. https://doi.org/10.1101/2021.11.22.469589

Ikhajiagbe B, Ogwu MC, Ekhator PO, Omoayena II (2022b) Implications of soil nitrogen enhancement on the yield performance of soybean (*Glycine max*) in cadmium-polluted soil. J Pollut Eff Contr 10(2):1–12. https://doi.org/10.1101/2021.11.22.469589

ILO (2011) Global employment trends for 2011: the challenge of a jobs recovery. International Labour Organization, Geneva

IMF (2010) Regional economic outlook: Sub-Saharan Africa: resilience and risks. International Monetary Fund World Economic and Financial Surveys, 0258-7440. Washington, DC

IPCC (2009) Climate change 2007: mitigation of climate change. In: Contribution of working group III to the fourth assessment report of the intergovernmental panel on climate change. Cambridge University Press, New York

IPCC (2013) Climate change 2013: the physical science basis. In: Stocker TF, Qin D, Plattner GK, Tignor M, Allen SK, Boschung J, Nauels A, Xia Y, Bex V, Midgley PM (eds) Contribution of working group I to the fifth assessment report of the intergovernmental panel on climate change. Cambridge University Press, Cambridge/New York, NY, USA

John FAS, Mai CH, Pei KJC (2015) Evaluating deterrents of illegal behaviour in conservation: carnivore killing in rural Taiwan. Biol Conserv 189:86–94

Juma C, Serageldin I (2007) Freedom to innovate: Biotechnology in Africa's development. A report of the High-level African Panel on Modern Biotechnology. African Union (AU) and New Partnership for Africa's Development (NEPAD). Addis Ababa and Pretoria

Kalaba FK, Quinn CH, Dougill AJ (2012) Contribution of forest provisioning ecosystem services to rural livelihoods in Copperbelt's Miombo woodlands, Zambia. SRI Papers No. 41. Sustainability Research Institute (SRI), School of Earth and Environment, The University of Leeds, Leeds
Karembu M, Nguthi F, Ismail H (2009) Biotech crops in Africa: the final frontier. ISAAA Afri-Center, Nairobi
Kideghesho JR (2009) The potentials of traditional African cultural practices in mitigating overexploitation of wildlife species and habitat loss: experience of Tanzania. Int J Biodiver Sci Manag 5(2):83–94
Kiefer I, Lopez P, Ramiarison C, Barthlott W, Ibish P (2010) Development, biodiversity conservation and global change in Madagascar
Kim et al (2016) The consumption of wild meat in Madagascar: drivers, popularity and food security
Kohler T, Maselli D (eds) (2009) Mountains and climate change. From understanding to action. Centre for Development and Environment (CDE), Institute of Geography, University of Berne Switzerland, Bern
Korcha D, Sorteberg A (2013) Validation of operational seasonal rainfall forecast in Ethiopia. Water Resour Res 49:7681–7697. https://doi.org/10.1002/2013WR013760
Kuehl HS, Nzeingui C, Yeno SL-D, Huijbregts B, Boesch C, Walsh PD (2009) Discriminating between village and commercial hunting of apes. Biol Conserv 142:1500–1506
Lambin EF, Meyfroidt P (2011) Global land use change, economic globalization, and the looming land scarcity. Proc Natl Acad Sci 108:3465–3472. https://doi.org/10.1073/pnas.1100480108
Leadley P, Pereira HM, Alkemade R, Fernandez Manjarrés JF, Proença V, Scharlemann JPW, Walpole MJ (2010) Biodiversity scenarios: projections of 21st century change in biodiversity and associated ecosystem services. Convention on Biological Diversity, Montreal, QC
Lewis SL, Maslin M (2015) Defining the anthropocene. Nature 519:171–179
Lewis SL, Brando P, Phillips OL, van der Heijden GMF, Nepstad D (2013) The 2010 Amazon drought. Science 331(6017):554
Lovett J (2015) Modeling the effects of climate change in Africa. Afr J Ecol 53:1–2
Lucas R, Zeppetello V et al (2020) Large scale tropical deforestation drives extreme warming. Environ Res Lett 15:084012
Mallon DP, Hoffman M, Grainger M, Hibert F, Van Vliet N, McGowan P (2015) An IUCN situation analysis of terrestrial and freshwater Fauna in west and Central Africa, vol 54. IUCN, Gland
Mayaux P, Pekel JF, Desclee B, Donnay F, Lupi A et al (2013) State and evolution of the African rainforests between 1990 and 2010. Philos Trans R Soc B 368:20120300
Mehrdad F, Dolatabadi ZA, Pakjouei S, Pouyesh V (2016) Review paper: approaches to post-disaster environmental recovery. Health Emerg Disast Quart 1(2):65–70
Midgley G, Bond WJ (2015) Bond future of African terrestrial biodiversity and ecosystems under anthropogenic climate change. Nat Clim Chang 5:823–829. https://doi.org/10.1038/nclimate2753
Millennium Ecosystem Assessment (MEA) (2005) Ecosystems and human well-being: synthesis. Island Press, Washington, DC. http://www.millenniumassessment.org/documents/document.356.aspx.pdf
Miller K (2012) Organizational communication: approaches and processes (6th ed.) In: Eckman M (ed) America Lyn Uhl Publishers, 338 pp
Minnemeyer S, Laestadius L, Sizer N, Saint-Laurent C, Potapov P (2011) A world of opportunity. World Resources Institute, Washington, DC. http://www.wri.org/restoringforests
Misana S, Sokoni C, Mbonile M (2012) Land-use/cover changes and their drivers on the slopes of Mount Kilimanjaro, Tanzania. J Geogr Reg Plann 5(6):151–164
Mohammed S (2013) Community perception and indigenous adaptive response to climate variability at Tehuledere Woreda, south Wollo zone. J Eng Int 1(2):18–27

Murray-Brian C et al (2017) How effective are us renewable energy subsidies in cutting greenhouse gases? American Economic Association. JSTOR 569-574. http://www.libproxy.uww.edu:2075/stable/pdf/42921000.pdf

Muth RM, Bowe JF Jr (1998) Illegal Harvest of Renewable Natural Resources in No National Tiger Conservation Authority (Undated). Project Tiger Authority, Ministry of Environment and Forests, Government of India. North America: Toward a Typology of the Motivations for Poaching. Soc Nat Res Int J 11:9–24

Natural Environment Research Council (NERC) (2012) Sustainable use of natural resources—theme action plan 3. Swindon, NERC. https://webarchive.nationalarchives.gov.uk/20120703112514/; https://www.nerc.ac.uk/research/themes/tap/documents/tap-natural-resources-p3.pdf

Niang I, Ruppel OC, Abdrabo M, Essel A, Lennard C et al (2014) Africa. In: Barros V, Field C (eds) Climate change 2014: impacts, adaptation, and vulnerability. Part B: regional aspects. Contribution of working group II to the fifth assessment report of the intergovernmental panel on climate change. Cambridge University Press, Cambridge, pp 1199–1265

Nyenje N, Kingamkono RR, Kullaya AK, Mneney EE (2011) Biotechnology for sustainable Agriculture, food security and poverty reduction in Africa. In: Pasternak C (ed) Access not Excess. pp 19–30

Obour R, Asare R, Ankomah P, Larson T (2016) Poaching and its potential to impact wildlife tourism: an assessment of poaching trends in the mole National Park in Ghana. Athens J Tour 3(3):169–192. https://doi.org/10.30958/ajt-3-3-1

OECD (2011) Better policies for development: recommendations for policy coherence. In: Organization for economic cooperation and development. Harvard Education PublishingGroup, USA, p 17

Ogwu MC (2019a) Towards sustainable development in Africa: the challenge of urbanization and climate change adaptation. In: Cobbinah PB, Addaney M (eds) The geography of climate change adaptation in urban Africa. Springer Nature, Switzerland, pp 29–55. https://doi.org/10.1007/978-3-030-04873-0_2

Ogwu MC (2019b) Understanding the composition of food waste: an "-omics" approach to food waste management. In: Gunjal AP, Waghmode MS, Patil NN, Bhatt P (eds) Global initiatives for waste reduction and cutting food loss. IGI Global, Pennsylvania, pp 212–236. https://doi.org/10.4018/978-1-5225-7706-5.ch011

Ogwu MC (2019c) Lifelong consumption of plant-based GM foods: is it safe? In: Papadopoulou P, Misseyanni A, Marouli C (eds) Environmental exposures and human health challenges. IGI Global, Pennsylvania, pp 158–176. https://doi.org/10.4018/978-1-5225-7635-8.ch008

Ogwu MC (2020) Value of Amaranthus [L.] species in Nigeria. In: Waisundara V (ed) Nutritional value of Amaranth. IntechOpen, London, pp 1–21. https://doi.org/10.5772/intechopen.86990

Ogwu MC, Osawaru ME, Ahana CM (2014) Challenges in conserving and utilizing plant genetic resources (PGR). Int J Genet Mol Biol 6(2):16–22. https://doi.org/10.5897/IJGMB2013.0083

Ogwu MC, Osawaru ME, Aiwansoba RO, Iroh RN (2016) Status and prospects of vegetables in Africa. In: Borokini IT, Babalola FD (eds) Conference Proceedings of the Joint Biodiversity Conservation Conference of Nigeria Tropical Biology Association and Nigeria Chapter of Society for Conservation Biology on MDGs to SDGs: Toward Sustainable Biodiversity Conservation in Nigeria. University of Ilorin, Nigeria, pp 47–57

Olsson P, Gunderson LH, Carpenter SR, Ryan P, Lebel L, Folke C, Holling CS (2006) Shooting the rapids: navigating transitions to adaptive governance of social-ecological systems. Ecol Soc 11(1):18

Osawaru ME, Ogwu MC (2014) Conservation and utilization of plant genetic resources. In: Omokhafe K, Odewale J (eds) Proceedings of 38th Annual Conference of The Genetics Society of Nigeria. Empress Prints Nigeria Limited, Nigeria, pp 105–119

Owusu PA, Asumadu-Sarkodie S (2016) A review of renewable energy sources, sustainability issues and climate change mitigation. Cogent Eng 3(1):1167990

Parmesan C (2006) Ecological and evolutionary responses to recent climate change. Annu Rev Ecol Evol Syst 37:637–699. https://doi.org/10.1146/annurev.ecolsys.37.091305.110100

Pasquini L, Shearing C (2014) Municipalities, politics, and climate change: an example of the process of institutionalizing an environmental agenda within local government. J Environ Dev 23(2):271–296

Peacock A (2014) The stern review: a dual critique. ANU Press, Australia, p 44

Perugini L, Caporaso L, Marconi S, Cescatti A, Quesada B, de Noblet-Ducoudré N, House JI, Arneth A (2017) Biophysical effects on temperature and precipitation due to land cover change. Env Res Lett 12:053002

Pullin AS (2002) Conservation biology. Cambridge University Press, Cambridge

Raven PH, Gereau RE, Phillipson PB, Chatelain C, Jenkins CN, Ulloa Ulloa C (2020) The distribution of biodiversity richness in the tropics. Sci Adv 6(37):eabc6228

Rudel TK (2013) The national determinants of deforestation in sub-Saharan Africa. Philos Trans R Soc B 368:20120405

Saatchi S, Asefi-Najafabady S, Malhi Y, Aragao L, Anderson L et al (2013) Persistent effects of a severe drought on Amazonian forest canopy. PNAS 110(2):565–570

Sanginga PC, Ochola WO, Bekalo I (2010) Natural resource management and development nexus in Africa. In: Ochola WO, Sanginga PC, Bekalo I (eds) Managing natural resources for development in Africa. Nairobi University Press, Nairobi

Serengeti National Park (2000) Park Management, Conservation and Research – Answers to conflict between man and Wildlife? Serengeti National Park. http://www.serengeti.org/main_serengeti.html

Silverio DV et al (2015) Agricultural expansion dominates climate changes in southeastern Amazonia: the overlooked non-GHG forcing. Environ Res Lett 10:104015

Slater FR, Singer AC, Turner S, Barr JJ, Bond PL (2011) Pandemic pharmaceutical dosing effects on wastewater treatment. FEMS Microbiol Lett 315:17–22

Smith SR (2009) Organic contaminants in sewage sludge (biosolids) and their significance for agricultural recycling. Philos Trans R Soc Lond A 367:4005–4041

Solomon A, Yihenew G, Getachew F, Birhanu G (2015) Time series trend analysis of temperature and rainfall in Lake tana sub-basin, Ethiopia. Environ Syst Res 4(25):1–12

Song H-K, Singh D, Tomlinson K, Yang X, Ogwu MC, Slik JWF, Adams JM (2019) Tropical forest conversion to rubber plantation in Southwest China results in lower fungal beta diversity and reduced network complexity. FEMS Microbiol Ecol 95:fiz092. https://doi.org/10.1093/femsec/fiz092

Stępniewska Z, Kuźniar A (2013) Endophytic microorganisms—promising applications in bioremediation of greenhouse gases. Appl Microbiol Biotechnol 97:9589–9596

Thomas CD, Cameron A, Green RE, Bakkenes M, Beaumont LJ, Collingham YC, Erasmus BFN et al (2004) Extinction risk from climate change. Nature 427:145–148. https://doi.org/10.1038/nature02121

Timko J (2011) HIV/AIDS, forests and futures in Sub-Saharan Africa, STEPS working paper 43. STEPS Centre, Brighton

Townsend CR, Begon M, Harper JL (2000) Essentials of ecology. Blackwell Publishing, Hoboken, NJ

Travis JMJ (2003) Climate change and habitat destruction: a deadly anthropogenic cocktail. Proc R Soc Lond Ser B 270:467–473. https://doi.org/10.1098/rspb.2002.2246

UN (2007) United Nations Expert Group Meeting on Population Distribution, Urbanization, Internal Migration and Development, United Nations population division, UN/POP/BGM-URB/2007/01

UN (2010) Population Division of the Department of Economic and Social Affairs of the United Nations Secretariat. In: World Population Policies 2009, New York

UNDP (2008) Fighting climate change solidarity in a divided world. Human Development Report

UNEP (2010) State of biodiversity in Africa. UNEP, Nairobi. http://www.cbd.int/iyb/doc/celebrations/iyb-egyptstate-of-biodiversity-inafrica.pdf

UNESCO (2011) 2011 Education for all global monitoring report. Document code: ED. 20011/WS/29. https://unesdoc.unesco.org/ark:/48223/pf0000212003
UNFCCC (2009) Climate change: impacts, vulnerabilities and adaptation in developing countries
United Nations (2010a) United Nations environment programme (UNEP). Opportunities in environmental management for disaster risk reduction: recent progress. United Nations Publishing, New York
United Nations (2010b) United Nations environment programme (UNEP). UNEP post-conflict capacity building programme in Liberia. United Nations Publishing, New York
United Nations (2011) Programme TUNE. Humanitarian action and the environment. United Nations Publishing, New York
United Nations Environment Programme (UNEP) (2016) Global Material Flows and Resource Productivity: Assessment Report for the UNEP International Resource Panel. http://unep.org/documents/irp/16-00169_LW_GlobalMaterialFlowsUNEReport_FINAL_160701.pdf
United Nations World Tourism Organization [UNWTO] (2010) Tourism highlights, 2010th edn. UNWTO, Madrid
United Nations World Tourism Organization (2011) Tourism Highlights 2011 Edition
UNWTO (2009) Tourism highlights, 2009th edn. UNWTO, Madrid
UNWTO (2010) Manual on tourism and poverty alleviation – practical steps for destinations, Madrid. http://pub.unwto.org/WebRoot/Store/Shops/Infoshop/4BEB/F726/BC4B/3784/9088/C0A8/0164/FDA7/100513_manual_pov_all_excerpt.pdf
Van Drunen M, Kuik O, Lutz C, Wiebe K (2009) Modeling future inter linkages, INDI-LINK – indicator based evaluation of inter linkages between different sustainable development objectives. Ther Deliv D2:5
Velenturf AP, Purnell P (2017) Resource recovery from waste: restoring the balance between resource scarcity and waste overload. Sustainability 9(9):1603
Vwioko DE, Okoekhian I, Ogwu MC (2018) Stress analysis of *Amaranthus hybridus* L. and *Lycopersicon esculentum* Mill. Exposed to Sulphur and nitrogen dioxide. Pertanika J Trop Agric Sci 41(3):1169–1191
WEF [World Economic Forum] (2011) The global competitiveness report 2011–2012. https://www.weforum.org/reports/global-competitiveness-report-2011-2012/
West Africa Discovery (2010) Responsible/Sustainable tourism development to create benefits for local communities. http://www.tripatini.com/group/westafricasustainableresponsibletourism/forum/topics/responsiblesustainable-tourism
William DM, Lemieux AM (2015) Poaching in Uganda: perspectives of law enforcement rangers. Deviant Behav 36(11):853–873. https://doi.org/10.1080/01639625.2014.977184
Willis K, Bennet K, Burrough S, Macias-Fauria M, Tovar C (2013) Determining the response of African biota to climate change: using the past to model the future. Philos Trans R Soc B 368:20120491
Winfree R, Kremen C (2009) Are ecosystem services stabilized by differences among species? A test using crop pollination. Proc R Soc B 276:229–237. https://doi.org/10.1098/rspb.2008.0709
World Bank (2010) World development report 2010: development and climate change. The International Bank for Reconstruction and Development/The World Bank, Washington, DC, p 192. http://go.worldbank.org/BKLQ9DSDU0
WTTC (2009) Overview of tourism to Africa, with reference to the Asian and Japanese out-bound market: a follow-up study to the fifth Africa –Asia business forum
WTTC (2011) The state of tourism in Africa-2011. http://www.wttc.org
Zelazowski P, Malhi Y, Huntingford C, Sitch S, Fisher JB (2011) Changes in the potential distribution of humid tropical forests on a warmer planet. Philos Trans R Soc Math Phys Eng Sci 3691934:137–160
Zhang H, Henderson-Sellers A, McGuffie K (2001) The compounding effects of tropical deforestation and greenhouse warming on climate. Clim Chang 49:309–338
Zhong L, Wang J (2017) Evaluation on effect of land consolidation on habitat quality based on InVEST model. Trans Chin Soc Agric Eng 33:250–255

Chapter 21
Touristic Value of African Environment: A Socio-economic Perspective

Opeyemi Pamela Babafemi, Adams Ovie Iyiola, and O. M. Ogundare

Abstract The African continent is rich in biodiversity which offers significant touristic value and holds great potential for continuous socio-economic advancement of the continent. Tourism, which is a major feature of the world economy, continues to play an important role in global development. It is a major driver of economic growth in emerging economies, as it is a key contributor of jobs and employment opportunities. Many countries (both developed and developing) have recognized the contribution of international tourism to their development. The tourism industry has experienced uninterrupted growth in international tourist arrivals over the last decade and Africa's natural wonders remain a source of attraction to many and, as such, continues to gain attention throughout the world. Some of the benefits derived from tourism include income generation, increased employment opportunities, foreign exchange, incentives to inward investment and regional development, as well as revenue creation. Tourism in Africa has vast potential and innovation, environmental consciousness, and sustainable management of touristic resources that can ensure that it remains one of the fastest-growing sectors in the global economy with huge socio-economic benefits that are not at the expense of the environment. This chapter reviews the touristic value of the African environment from the socio-economic perspective while giving attention to sustainability issues revolving around the tourism industry as well. This chapter starts with an overview of the various touristic resources presented in the African environment and their socio-economic value and highlights the present state of tourism in Africa. The socio-economic impacts of

O. P. Babafemi (✉)
Global Affairs and Sustainable Development Institute (GASDI), Osun State University, Osogbo, Nigeria

A. O. Iyiola
Department of Fisheries and Aquatic Resources Management, Faculty of Renewable Natural Resources Management, Osun State University, Osogbo, Osun State, Nigeria

O. M. Ogundare
Department of Tourism Studies, College of Humanities and Culture, Osun State University, Osogbo, Nigeria

S. C. Izah, M. C. Ogwu (eds.), *Sustainable Utilization and Conservation of Africa's Biological Resources and Environment*, Sustainable Development and Biodiversity 32, https://doi.org/10.1007/978-981-19-6974-4_21

tourism development in Africa and the environmental challenges that confront the tourism industry are also carefully examined. As the chapter progresses, the concept of biodiversity and sustainable environmental management is also explored. Case study scenarios of the application and relevance of tourism to sustainable biodiversity management are cited. This chapter concludes with a discussion on the sustainable development of Africa through sustainable ecotourism and the role of ecotourism in sustainable development. Some ecotourism projects across Africa are also identified and discussed. The importance of developing a resilient and sustainable tourism industry to boost GDP and investment, create employment opportunities, and reduce poverty rates is also duly highlighted.

Keywords Touristic resources · Tourism · Environment · Biodiversity · Socio-economic · Sustainable development · Ecotourism · Africa

21.1 Introduction

Tourism is a major feature of the world economy and it plays an important role in development (Dieke 2020). Many developed and developing countries have recognized the contribution of international tourism to their development (Jenkins 2015; Sharpley and Telfer 2015). Tourism, which can be briefly described as the act of moving away from home in pursuit of recreation, relaxation, and pleasure, while making use of the commercial provision of services during the time spent away, has especially become very vital to many African economies.

This chapter reviews the touristic value of the African environment from the socio-economic perspective while also giving attention to sustainability issues revolving around the tourism industry. Consequently, this chapter starts with an outline of the touristic resources presented in the African environment and their socio-economic value. It touches on fish as a touristic resource and identifies some fishing destinations in Africa. It then highlights the present state of tourism across the African continent and discusses the socio-economic and environmental impacts of tourism in Africa. Socio-economic impacts of the development of tourism in Africa and the environmental challenges that confront the tourism industry are carefully examined. This chapter explores the challenges and opportunities associated with the sustainable development of the tourism industry for economic growth, as well as key issues on sustainable management of touristic resources and biodiversity as a whole. As the chapter progresses, the concept of biodiversity and sustainable environmental management, as well as case study scenarios of the application of tourism to sustainable biodiversity management are discussed. This chapter concludes with a discourse on the sustainable development of Africa through sustainable ecotourism and the role of ecotourism to sustainable development. Some ecotourism projects across Africa are also cited and briefly discussed. The importance of developing a resilient and sustainable tourism industry to boost GDP and investment, create employment, and reduce poverty is also duly highlighted.

21.2 Touristic Resources and Socio-Economic Value of Africa's Environment

Touristic opportunities in Africa include cultural heritage tourism, beach tourism, safari tourism, nature/adventure tourism, etc. Several locations in Africa are benefitting immensely from their touristic potential which serves as a source of foreign exchange. The sites of attraction are usually a result of the beautiful masterpiece of nature and the work of art (Table 21.1). Countries such as Algeria, Egypt, Uganda, South Africa, Morocco, Tunisia, Ghana, and Tanzania depend heavily on tourism to increase both external and internal revenue (White Orange 2020). Egypt and South Africa are one of the biggest tourist locations in Africa. The African continent can be divided into three groups concerning tourism potential:

- Countries with well-developed tourism sectors (e.g., Morocco, South Africa, Egypt, and Tunisia).
- Countries with developing tourism industry and receive income steadily from tourism (e.g., Mauritius, Zimbabwe, and Swaziland).
- Countries that aspire to develop a tourism industry based on its prospects (e.g., Algeria and Burundi) (Otieno 2018).

21.2.1 Touristic Locations in Africa

21.2.1.1 Morocco

It is the most visited country in Africa with a total of 12.3million visitors and was the most visited country in the continent in 2019. This number exceeded the number in 2015 (9,409,000 tourists) and is not as high as the records from Egypt in 2015 (17,443,000 tourists). The countries have various amazing sites such as the drawcard, the colors, sight, and sound if the medina as well as the diverse food and pastries. The cities of Marrakesh, Fes, and Essaouira are sites to behold. The beautiful beach in Essaouira is a sight to behold.

Table 21.1 Countries and their tourist attractions

Countries	Tourist attractions
Morocco and Tunisia	Beautiful beaches and closeness to Europe
Egypt	Pyramids, artifacts, Red sand beaches, and Ancient Egypt town
South Africa and Kenya	Wild Safari cruise and expeditions

21.2.1.2 Egypt

When compared with 2015, the number of visitors in 2019 was reduced with 11.3 million tourists recorded and was the second most visited country in Africa. The sites are the ancient pyramids, tombs and tombstones and temples which can be dated many thousand years ago.

21.2.1.3 South Africa

Its sight is the safari and remains the most popular spot for safari. In 2019, a total of 10.5 million tourists were observed in the country. The numbers recorded in 2019 are higher than the records in 2015 (7,518,000 tourists).

21.2.1.4 Tunisia

The country is located in the Northern regions of Africa with 8 1.3 million tourists recorded. The country is an ancient city famous for the Sahara Desert and its hot temperature. The influx of visitors is also stimulated by the conducive policies for short distant flights or transit during extended flights.

21.2.1.5 Zimbabwe

This country is noted for the safari and Victoria Falls which is a known monument worldwide with 1708 m long. The country is land-locked in Southern Africa and is popular for safari.

21.2.1.6 Cote d'Ivoire

The country has numerous beaches and rainforests and is a popular travel destination for European tourists and travelers. The country is blessed with a beautiful temperature, sea shores, sunset and is an epitome of beauty. This location had a long history of French-colonial legacy and has been a traveling spot for sunset scenes.

21.2.1.7 Uganda

It is a spot for viewing gorillas that are domiciled in the mountain regions. The country has a lot of raw and natural beauty which has aesthetic values.

21.2.1.8 Kenya

It is the most visited country in East Africa with a lot of safaris. The Maasai Mara is a popular location for safari-viewing.

21.2.1.9 Mauritius

This country is also noticeable for its safari. It also has beautiful beaches, mountains, and diving activities which is world-class.

21.2.1.10 eSwatini (Formerly Swaziland)

This area is landlocked and located at the center of South Africa. It is an area for safari exploration, camping, and hikes.

21.3 Touristic Resources in African Environment

21.3.1 Safari

As elaborated by Beekwilder (2022), an analysis was carried out on safaris in Africa to determine their rankings in the continent. Safari Bookings carried out a survey and out of the 2373 reviews, Serengeti in Tanzania emerged as the finest safari park in Africa. It rated 4.92 out of 5 and this was also observed in the review of 2021. For the first top 50 safaris in the continent, 157 parks in 11 safari countries were considered. The countries are Kenya, Botswana, Malawi, Namibia, Rwanda, South Africa, Swaziland, Uganda, Tanzania, Zambia, and Zimbabwe. The 2373 reviews comprised 1363 reviews by safari tourists from around 83 countries and 1010 reviews from industrial experts. The top 10 are highlighted and the safaris in number 11 to 50 ranking are listed in Table 21.2.

- **Serengeti National Park**: It is located in Tanzania and it ranks first in Africa. It is known to house diverse safari, rare rhinoceros, numerous large cats, and the migration of wildebeests.
- **Mana Pools National Park**: It is located in Zimbabwe and it is a good spot for walking, canoe activities, and classic safaris. It houses four out of the five big and wild dogs, but no rhino is found in this park. It is ranked number two in Africa.
- **Mala Games Reserve**: It is located in South Africa and ranked third in Africa. The safari in this reserve is classic, and it is a private reserve and abundant in wildlife species, including the big five.
- **Okavango Delta**: It is located in Botswana and ranks fourth in Africa. It is known for canoe riding, boats, and classic safari activities. It houses the big five species.

Table 21.2 African Safaris in number 11–50 ranking in Africa

Position	Name of Safari	Country	Rating/5	Sights
11	Kidepo Valley National Park	Uganda	4.58	Classic safari, giraffe, cheetah, antelopes, the Big Five (no rhino)
12	Hwange National Park	Zimbabwe	4.57	Classic safari, the Big Five, and all major animals
13	Kgalagadi Transfrontier Park	South Africa	4.57	Classic safari, transfrontier park, all big cats are present
14	Chobe National Park	Botswana	4.53	Boat and classic safari, abundant elephants, big five are present (no rhino)
15	Ruaha National Park	Tanzania	4.52	Walking and classic safari, four out of the big five are present, cheetah, wild dogs present
16	Selous Game Reserve	Tanzania	4.51	Walking and boat classic safari, the big five are present, wild dogs present
17	Mahale Mountains National Park	Tanzania	4.48	Chimpanzee viewing, major safari animals present
18	Kruger National Park	South Africa	4.47	Classic safari, the big five are present
19	Madikwe Game Reserve	South Africa	4.45	Private reserve, classic safari, the big five are present
20	Phinda Game Reserve	South Africa	4.45	Classic safari, abundant big five
21	Amboseli National Park	Kenya	4.45	Classic safari, views of Mount Kilimanjaro, Big five are present
22	Etosha National Park	Namibia	4.43	Classic safari, wildlife abundant in the dry season, four out of the big five are present (no buffalo)
23	Central Kalahari Game Reserve	Botswana	4.40	Classic safari, big cats present, wildlife is scarce
24	Samburu National Reserve	Kenya	4.40	Classic safari, four out of big five presents (no rhino)
25	Ol Pejeta Conservancy	Kenya	4.39	Classic safari, chimpanzee present, big five present
26	Tsavo East National Park	Kenya	4.38	Classic safari, big five present, the density of wildlife is low
27	Hluhluwe-iMfolozi Game Reserve	South Africa	4.31	Classic safari, big five present, famous for rhino population
28	Shamwari Game Reserve	South Africa	4.25	Classic safari, big five are present, private reserve
29	Timbavati Nature Reserve	South Africa	4.25	Classic safari, big five present

(continued)

Table 21.2 (continued)

Position	Name of Safari	Country	Rating/5	Sights
30	Tarangire National Park	Tanzania	4.24	Classic safari, wildlife abundant in the dry season
31	Murchison Falls National Park	Uganda	4.23	Classic safari, boat safari, wildlife Is abundant, four out of big five are present (rhino absent)
32	Makgadikgadi Pans National Park	Botswana	4.23	Classic safari, the big five are present (rhino is scarce)
33	Tsavo West National Park	Kenya	4.11	Classic safari, big five presents, wildlife density is low
34	Lake Nakuru National Park	Kenya	4.02	Classic safari, white rhino present, it is a friendly park
35	Katavi National Park	Tanzania	4.00	Classic safari, the big five are present, private reserve
36	Queen Eliza-beth National Park	Uganda	4.00	Classic safari, boat safari, four out of the big five are present, no giraffe is present
37	Pilanesberg Game Reserve	South Africa	3.98	Classic safari, the big five are present
38	Addo Elephant National Park	South Africa	3.94	Classic safari, the big five are present, elephants are seen occasionally
39	Kafue National Park	Zambia	3.94	Classic safari, walking, boat safari, chee-tah present, four out of the big five are present, there is no giraffe
40	Aberdare National Park	Kenya	3.87	Classic safari, the big five are present
41	Nairobi National Park	Kenya	3.83	Classic safari, big five are present, it is close to the city, does not have wilderness
42	Lake Manyara National Park	Tanzania	3.83	Classic safari, big cats are present, four out of the big five are present (no rhino)
43	Nxai Pan National Park	Botswana	3.79	Classic safari, the big five are present, wildlife is scarce
44	Hell's Gate National Park	Kenya	3.77	Cycling, walking, Classic safari, lion (sighting is rare), and antelope are abundant
45	Augrabies Falls National Park	South Africa	3.75	Animals are scarce, some species of leop-ard, giraffe, antelopes, and good scenery.
46	Mountain Zebra National Park	South Africa	3.71	Classic safari, mountain zebra, black rhino, antelopes, lion, and cheetah are present
47	Arusha National Park	Tanzania	3.68	Classic safari, antelope found, hiking, black colobus.
48	Meru National Park	Kenya	3.63	Classic safari, four out of the big five are present

(continued)

Table 21.2 (continued)

Position	Name of Safari	Country	Rating/5	Sights
49	Mikumi National Park	Tanzania	3.54	Walking and Classic safari, four out of the big five are present
50	Karoo National Park	South Africa	3.50	Classic safari, black rhino, and lion are absent

- **Lower Zambezi National Park**: It is located in Zambia and ranks fifth in Africa. It is known for canoe and classic safari activities. It houses the big five wildlife species and no rhino or giraffe.
- **Moremi Game Reserve**. It is located in Botswana and ranks sixth in Africa. It is a spot for a classic safari, and the big five are present, but the rhino is rare. It is partly located in Okavango Delta.
- **South Luangwa National Park**: It is located in Zambia and ranks seventh in Africa. It is known for walking and classic safari, giraffe sighting, and four out of the big five are present (no rhino). Wildlife is abundant in this park.
- **Ngorongoro Crater**: It is located in Tanzania and ranks eighth in Africa. It is a spot for a classic safari, abundant in wildlife, and houses the big five wildlife species.
- **Sabi Sand Game Reserve**: It is located in South Africa and ranks ninth in Africa. The reserve has multiple private reserves and a good spot for viewing the big five wildlife species.
- **Masai Mara National Reserve**: It is located in Kenya and is a spot for a classic safari, the big five are present (rhino is rare), wildebeest, and many big cats are present.

21.3.2 Housing

- **Nzulezo**: This place is located in Ghana and its striking appearance is the entire village that is built on stilts on Lake Tandane in the western part of Ghana. This area shows the illustration of life entirely on water.

21.3.3 Monuments

- **African Renaissance Movement**: This statue is higher than the statue of liberty in New York. It is a bronze statue located in the city of Dakar. It is the tallest in Africa and it is very breathtaking when the monument is observed.

- **Pyramids of Giza**: This monument is located in Egypt and it is an ancient wonder of the world. It has been a tourist attraction to the Romans for over 2000 years ago.
- **Sahara Dunes**: This dune is located in Morocco and it is an entry point into the Sahara Desert.
- **Dara Valley**: It is a place, where hikers and trekkers are exposed to beautiful scenery. Date palms and Kasbahs are found in the valley.
- **Mount Kilimanjaro**: It is located in Tanzania and has breathtaking altitudes and views. It is a good spot for hikes and the most coveted spot in the world.
- **Nyiragongo Volcano**: It is located in the Virunga National Park in the Democratic Republic of Congo. The volcano is an active one and has a very enticing and fascinating sight to behold the awesome power of geologic forces.
- **Fish River Canyon**: This canyon is one of the largest in Africa and it is located in Namibia.
- **Mount Mulanje**: It is located in Malawi and is an area for bird watching and hiking.
- **Sphinx**: It is located in Egypt and this is a monument of the ancient colossus with a lot of mystery. The monument has a body of a lion and a human head.

21.3.4 Nature

- **Lekki Conservation Centre**: It is located in Lagos, Nigeria, and is characterized by the canopy walkway which is the longest canopy walk in Africa. This conservation center is a point of leisure and relaxation away from the hustling of the commercial city of Lagos, Nigeria.
- **Cape Verde Archipelago**: This spot is a good blend of African, Brazilian, and Liberian influences. They are areas for hikes, water sports, and scenic beaches. This spot is located in Cape Verde Islands.
- **Nzla Pan National Park**: It is located in northeast Botswana and houses a large number of giraffe herds in the African continent. The park is exquisite with the works of nature at its peak.

21.3.5 Wildlife Environments

- **Rhinos at Solio Reserve**: It is located in Kenya and it houses a collection of endangered creatures that are found on the African plains.
- **Bazaruto Archipelago**: It is located in Mozambique and it houses dolphins, whales, and dugongs of the Indian ocean.
- **Flamingos**: It is located in Lake Nakuru National Park in Kenya about 170 km from the northwest of Nairobi. This Lake is a haven for Flamingos. Rhinos, monkeys, and other animals are spotted in the park.

- **Riding safari**: It is located in Kenya and has diverse Zebra collections. Their site is very mesmerizing and there are several horseback activities.

21.3.6 Aquatic Environments

- **Lake Retba**: It is located in Senegal and has one of the highest levels of salinity in the world. The color of the lake is dictated by bacteria in the water which produces the green pigments that absorbs sunlight.
- **The Nile**: It is located in Egypt and it is the life of the country. It is the longest river in Africa and has years of history.
- **Lake Malawi**: The water of the lake is clear and houses diverse species of colorful fish. The coloration of fish species in this lake is second to none on the Earth and it has beautiful and relaxing scenery.
- **Lower Zambezi**: It is located in Zambia and the lower water is a haven for elephants, crocodiles, and diverse animals.

21.3.7 Fishing Destinations in Africa

Asides from the wildlife, and savannahs, Africa has a lot of diverse unique areas for fishing lot of unique areas for fishing that can serve as tourist destinations. The continent is surrounded by the Atlantic and Indian Oceans, and the Red and Mediterranean Seas with amazing water for fishing. Aleksandra (2020) elaborated on the major fishing spots in Africa:

- **Lake Nasser**: It is located in the southern part of Egypt and it is a spot with a great history of fishing in the country. The lake is man-made and it is one of the largest reservoirs in the world which houses diverse abundant freshwater fish species. *Nile perch* is the most popular fish species in the lake. Lizards, crocodiles, and other 30 species of fish are found in the Lake.
- **Aberdare Mountains**: It is located about 120 miles north of Nairobi, Kenya. The mountain has increased the reputation of the country as one of the best spots for sport fishing in Africa. Rainbow and Brown Trout are the most abundant for fishing adventures. They were introduced into the Kenya Mountain streams and metamorphosed over the years. Buffalos and leopards are also found around the mountain. It also houses a famous Treetops Lodge at the edge of the park which once hosted Queen Elizabeth II.
- **Cape Verde Islands**: It is a major spot, where deep fishing activities can occur in the Atlantic. The island is also called Cabo Islands. It is an island that lies 280 miles over the coast of Senegal and has deep spots for fishing and game fishing. The water houses the fastest fish in the world, the Blue Matlin which are extremely hard to catch and other species such as Sailfish, *Mahi-mahi,* Yellowfin

Tuna, big eye Tuna, Spearfish, and different types of sharks. The area is also a good spot for a beach holiday destination, hiking, and a nice tan. No matter the time of year, the island is visited, and there is always an activity for relaxation.

- **The Mnyera and Rhudji Rivers**: This spot is located in Tanzania and is known for saltwater fishing activities. The rivers are home to the freshwater monster called the Tigerfish. The fish has sharp teeth with great strength. Fish species such as Yellowfish Tuna and Vundu Catfish are found in the rivers. The fishing activities in the rivers are bound by strict guiding protocols by the Tourette Fishing and Kilombero North Safaris. Per week, only eight anglers are allowed to fish within 120 km a week with a catch-and-release policy in place. No fishing is allowed in spawning areas and no fishing is done in a particular river twice in a week. The Kilombero North Safari camp, Samaki camp, and Tengo camp are located around the rivers. Wildlife such as buffalos, lions, leopards, warthogs, pukus, zebras, and elephants are common in this area.
- **Luanda**: This spot is located on the Angola coastline, where the warm tropical water meets the cold Benguela current from the south Atlantic. This mixture results in diverse fish species and sardines which attracts huge Sailfish, *Mahi-mahi,* and big blue marlin. Luanda metropolis is an area that is developing and fast-growing with major developments all around the world. The area has a fortress of Sao Miguel which is the oldest building in Luanda and a UNESCO world heritage site.
- **Seychelles**: It is a spot for fine and big game fishing activities on the planet. The island houses the Blue and Black Marlin, Yellowfin Tuna, Wahoo, Giant Trevally, and Bonefish. The plateau in the area has numerous reefs and holes which can attract diverse fish species and become food for predator fish species, such as Tuna, wahoos, Sailfish, Bonito, and *Mahi-mahi.* The area has beautiful white shores and plenty of turquoise water. Hiking, scuba diving, snorkeling, and golfing are activities that are done along the shoreline and serve as activities for relaxation.
- **Durban**: It is located in South Africa. The country has several fishing sports and charters which offer diverse fish species for anglers. Marlin, Yellowfin, Skipjack Tuna, and Bludger Travellay are fish species found in this area all year round. Boat launching is a major activity for relaxation in this spot and these areas house the Golden Mile beachfront which is known for visitation by runners, and cyclers. Surfers and water spot enthusiasts.
- **Mauritius**: This area is a paradise for anglers with numerous fishing spots. Places such as Morne Brabant peninsula is an area with numerous local fishing legends and the Pamplemousses District which has the largest concentration of charter boats. The Flacq District is an area with numerous Sharks, Bonito, *Mahi-mahi,* or Marlin. This area is also a favorite spot for honeymoon with diverse fish species and beautiful scenery.
- **Lake Kariba**: It is located in Zimbabwe and it is the largest man-made lake in the world. The lake contains diverse fish species, such as Bream, Eel, Nkuoe, Squeaker, Cornish jack, Chessman, and the famous Tiger fish. The lake holds the Tiger fishing tournament, because the lake has one of the mightiest sizes in the

world. The Kariba Invitation Tigerfish Tournament is the largest in the world, because people travel to the Lake from all over the world. The Lake has beautiful and unforgettable scenery.

21.4 Tourist Arrivals and Income Generated in African Countries

The list of African countries and their tourist influx abundance and income generated according to the World Tourism Organization in 2015 cited by Rogerson (2017) is presented in Table 21.3. Egypt had the highest influx of tourists with 17,443,000 and Sao Tome and Principe was the least with 11,000 tourists.

The total number of tourists that arrived in Africa as of May 2021 was 1.04 million which was higher than the number of international tourists recorded in May 2020 by 533 thousand. The continent recorded the highest number of international tourists in January 2020 with over 5.3 million people and it was slightly higher than the number recorded in 2019 during this same period (Table 21.4).

21.5 African Countries with Tourism and Travel Ready Economies

The economy of Africa has a huge tourism potential, thereby creating employment, growth, and economic development. Tourism in Africa has a projected growth of 4.9% which amounts to about 9% of the GDP of the continent. According to the travel report of 2015, the 10 most tourism-ready countries in sub-Saharan Africa out of 141 countries are listed in Fig. 21.1.

From the figure, it can be seen that South Africa is the first in Africa and the 48th in the world. This is so because infrastructures put in place by the government during the World Cup tournament in 2010 is still in good condition and still attracts tourists to date. The stadia and hotels are still in good condition, thereby increasing business relationships and economic growth in the tourism industry. Despite these rich natural and rich resources, the continent is still at the developmental stages in aspects of tourism. The challenges are associated with infrastructure, hygiene, and security. Most countries in Africa are yet to discover or maximize their environmental and biodiversity potential to catalyze sustainable development, and in some cases, this is connected to packaging and branding issues (Ogwu 2019a, 2019b; Evivie et al. 2020). Two elements can relate to the policies of traveling and tourism. They are:

- Facilitation of travel: Most African countries have travel restrictions when they are within the same region. Countries such as South Africa are even tightening their policies on visa application and travels. Africa has the largest percentage

Table 21.3 Arrivals and Income generation in countries in Africa

Country	Arrivals in 2015	Income ($) in 2020
Algeria	4,244,000	
Angola	210,000	
Botswana	1.559,000	562,000,000
Burundi	148,000	2,000,000
Cameroon	210,000	
Cape Verde	198,000	123,000,000
Democratic Republic of Congo	61,000	
Djibouti	30,000	
Egypt	17,443,000	16,851,000,000
Eritrea	83,000	66,000,000
Gambia	111,000	
Guinea	45,000	
Kenya	–	879,000,000
Lesotho	304,000	30,000,000
Malawi	–	26,000,000
Mali	143,000	
Mauritius	934,827 (2020)	
Morocco	9,409,000	4,617,000,000
Mozambique	–	130,000,000
Namibia	–	348,000,000
Reunion	–	384,000,000
Sao Tome and Principe	11,000	
Senegal	769,000	
Seychelles	129,000	192,000,000
Sierra Leone	40,000	83,000,000
South Africa	7,518,000	7,327,000,000
Sudan	–	89,000,000
Swaziland	830,000	
Tanzania	–	4,468,000,000
Togo	81,000	
Tunisia	6,378,000	2,063,000,000
Uganda	1,468,000 (2017)	1,400,000,000
Zimbabwe	1,559,000	1,559,000

Countries not listed do not have available data

among countries that give visas on arrival to visitors (31%) and require a traditional visa for 58% of countries in the world.

- Policy changes: Members of the Economic Community of West African States (ECOWAS) have a visa-free policy for all its member states. This will enable easy movement and travels and can attract international tourists.

Table 21.4 Number of international tourist arrivals (thousands) in 2019–2021

Months	2019	2020	2021
January	5309	5309	990
February	4823	5823	850
March	5347	5347	964
April	5539	5539	1044
May	4775	4775	1043
June	5545	5545	
July	7384	7384	
August	7194	7194	
September	5452	5452	
October	5514	5514	
November	5270	5270	
December	6310	6310	
Total	68,462	21,387	4891

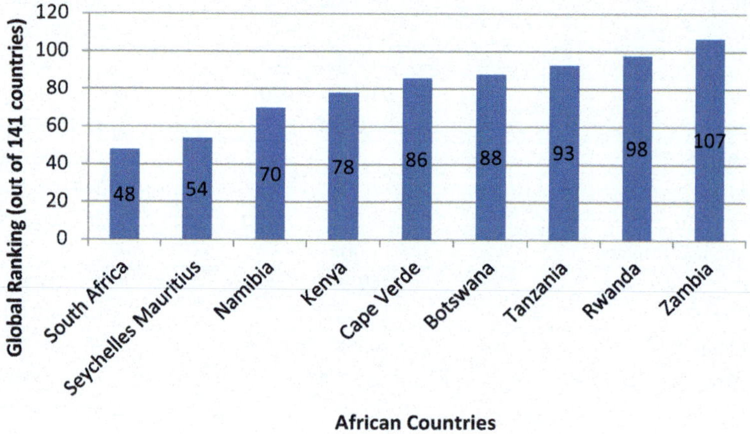

Fig. 21.1 Ten most ready economies for travel and tourism

21.6 Potentials of Tourism for Socio-economic Growth

Given the continent's natural splendor and abundance of resources, tourism industry has considerable potential in Africa. Between 1998 and 2005, industries such as tourism have experienced exponential growth across Africa. Consumer expenditure on tourism, hospitality, travel, and recreation in Africa is anticipated to reach about \$261.77 billion by 2030. This value is \$137.87 billion more than in 2015. However, tourism in most parts of Africa is still in need of further development, especially in the aspects of infrastructure and security. Being aware of the potential to develop their tourism sector to make its share of the economy larger, many countries have come up with strategic plans that can help catalyze development. Some of these countries putting meaningful efforts into advancing their tourism sector include the

Gambia, Tanzania, Kenya, and South Africa. Botswana, Rwanda, and Mauritius are especially working hard toward improving their business environment for tourism investment. Morocco has employed tourism as a tool to strengthen their economic growth by prioritizing their tourism industry (Signé 2018).

The tourism industry continues to offer numerous opportunities to invest in Africa's local communities, generate economic activity, and create employment opportunities for women and youths. To get the most of the potential tourism holds for socio-economic growth and development, policymakers and business leaders must prioritize investments that encourage innovations that could attract more tourists to their countries.

Policymakers should also endeavor to drive a much more coordinated effort across agencies to simplify regulations, visa requirements, and other administrative processes, thus removing the barriers to tourism development. This improvement in governance would help attract investors and promote touristic destinations. The ease of doing business in the tourism sector is another important aspect in maximizing the potential of tourism. The development of infrastructures and allied facilities for transport and communication that meets or exceeds the global standards are also ways to develop the travel and tourism industry. Locally sourcing for goods and services while promoting the inclusion of women and youths, and fostering environmental sustainability, should be highly encouraged among investors and entrepreneurs.

21.7 State of Tourism in Africa

In the 1990s, the African continent witnessed a remarkable increase in international tourist arrivals (Signé 2019) and has since been one of the world's fastest-growing tourism regions. About 300% increase culminated in 26.2 million arrivals in 2000 and by 2014, despite the Ebola crisis, international arrivals to the African continent rose to 65.3 million with income from international tourism running into an estimated US$ 43.6 billion and Egypt, Morocco, and Cote d'Ivoire being the countries that experienced the strongest growth (Africa Tourism Monitor 2015). A 0.64% increase saw international tourism arrivals in Africa grow to 62.9 million in 2016, with international tourism receipts in Africa totaling US$ 36.2 billion. Compared to 2015, Sierra Leone, Nigeria, and Burundi were among the countries that experienced the strongest increase in international arrivals with a whopping 126% increase in Sierra Leone and a 50.5% and 42.7% increase in Nigeria and Burundi, respectively. Eritrea, Togo, and Madagascar also witnessed a significant increase in the numbers of international arrivals (24.6%, 23.8%, and 20% increase, respectively (Africa Tourism Monitor 2018).

In 2016, Africa held a 5.1% and a 3.0% share of worldwide tourism arrivals and global tourism receipts, respectively. South Africa also surpassed ten million arrivals for the first time in the year 2016, while Morocco, a thriving tourism country, sustained more than ten million arrivals for the fourth consecutive year. Africa

Table 21.5 Countries and their income from Tourism in US dollars equivalent (2020)

Country	Receipts in US dollars (2020)
Egypt	16,851,000,000
Kenya	879,000,000
Morocco	4,617,000,000
South Africa	7,327,000,000
Uganda	1,400,000,000
Tunisia	2,063,000,000
Tanzania	4,468,000,000

Source: United Nations World Tourism Organization

continues to record a significant increase in the number of tourist arrivals, and at present, the continent welcomes over growing millions of annual visitors' countries such as South Africa, Kenya, and Tanzania have improved their travel and tourism infrastructure to accommodate the increasing number of tourists.

Generally speaking, South Africa, Egypt, Morocco, and Tunisia are among some of the countries in Africa that have a developed tourism industry that yields consistent economic benefits. Both natural and man-made attractions put these countries on the tourism map. These attractions and points of interest include climate, wildlife, historic sites, landscapes, etc. South Africa continues to attract tourists and wildlife lovers from around the world who undertake safari expeditions. The success of tourism in Egypt derives from its rich ancient history as seen in the presence of pyramids and artifacts. Morocco and Tunisia, for example, are placed at an advantage due to their picturesque beaches as well as their proximity to Europe. Other countries such as Kenya, Zimbabwe, Mauritius, and Swaziland can be considered as developing and in need of expansion (Otieno 2018).

The economic conditions of tourists generating regions such as the United States and Europe play an important role in tourism. In a buoyant economy, there is usually a correlation between disposable per capita income and the inclination to travel. Economic downturns and uncertainty prevent people from committing themselves to travel expenditures, even though there is evidence that consumers prioritize travel plans and potential holiday expenditures in their annual budgets (Dieke 2020).

The table below (Table 21.5) presents data as gotten from the World Tourism Organization regarding the income from tourism in US dollars equivalent in the year 2020.

Unfortunately, the recent COVID-19 pandemic gave way to an unparalleled crisis for Africa's tourism and related sectors like the agricultural and food industries with various supporting sectors that depend on tourism for employment and income suffering a huge impact (Erinle et al. 2021). In July 2020, it was reported that the tourism industry lost an estimated value of $55bn in revenues, and about two million jobs were lost within the first few months of the pandemic. African airlines are not left out in the loss of revenues. According to the African Airlines Association, an estimated $10.21bn is said to have been lost in passenger revenue in 2020. In addition, according to the International Monetary Fund, the real GDP among African countries reliant on tourism dropped by 12% in 2020. Employing about two million people, Egypt's travel and tourism industry, which is one of Africa's largest tourism markets, accounted for about 12% of the country's GDP before the COVID-19 pandemic. However, the industry's contribution to the country's economy decreased to $14.4bn in 2020 from a value of $32bn in 2019. In 2019, the country welcomed about 13 million tourists, which is an improvement from the 11.3 million tourist arrivals in 2018. However, in 2020, only 3.6 million visitors were received (Mitchell 2021).

While the COVID-19 pandemic rocked tourism in Africa a bit leading to the overall economic slowdown in other developing and emerging regions of the world, beaches, safari parks and heritage sites in the African region are once again bustling with tourists. This tourism market rebound is a relief to many countries. In October, lockdown restrictions were eased and international travels started to normalize in places such as South Africa, Africa's second-largest tourism industry. By September, over 93,000 international arrivals were recorded at Johannesburg's international airport. This figure still pales in comparison with the 400,000 international arrivals received monthly in 2019; nevertheless, it denoted a significant recovery in the tourism industry (Mitchell 2021).

21.8 Social, Economic, and Environmental Impacts of Tourism in Africa

Tourism impacts Africa in different ways. There exist the social and economic aspects of it which bring about notable growth and economic development in the region, as well as the environmental concerns that continually beset various stakeholders in the industry as regards the preservation of biological resources of touristic value that are present in the continent. The multifaceted impacts of tourism in Africa will be discussed under the following subheadings:

21.8.1 Socio-economic Impact of Tourism Development in Africa

21.8.1.1 Contribution to Employment

Tourism has long been known to create jobs and stimulates enterprise growth. The African travel and tourism industry generates both direct and induced employment running into millions of jobs that support the industry. The World Travel and Tourism Council puts the value of the number of jobs created by tourism at 20.5 million. In Seychelles, Cape Verde, and Mauritius, tourism makes up 57.3%, 35.8%, and 24.2% of their total employment, respectively (Africa Tourism Monitor 2015).

The tourism industry has experienced uninterrupted growth in international tourist arrivals over the last decade and Africa's natural wonders remain a source of attraction and continue to gain attention throughout the world. Some of the benefits derived from tourism include income generation, more employment opportunities, foreign exchange, incentives to inward investment and regional development as well as revenue creation (Dieke 2020). According to the World Tourism organization, tourism accounts for about 10% of the world's GDP and is considered an avenue for the creation of millions of jobs and, as such, increased levels of employment. Not only does tourism drive employment but it also encourages economic inclusion for women and young people. Tourism-related employment in Africa makes up about 71% of the continent's total employment. In 2019, Africa's tourism industry was reported to have generated over 21 million jobs. Tourism, therefore, remains a major driver of economic growth in emerging economies (Signé 2019), as it is a major contributor to jobs and employment opportunities.

In 2017, 2.5 million tourism jobs in North Africa and 6.8 million in sub-Saharan Africa saw direct travel and tourism) employment in Africa rises to 9.3 million which is an 11.2% increase from the previous year (Africa Tourism Monitor 2018). This trend expands economic opportunities for many especially the youths.

21.8.1.2 Contribution to Business Growth and Development Opportunities

Since the adoption of the Tourism Action Plan by the African Union Development Agency, tourism has become integral to economic development policies. Several African countries are considered to hold huge potential for attracting tourists, entrepreneurs, or investors thereby positioning Africa as a global destination for entrepreneurship and investment. The tourism industry remains one of the most vibrant sectors with an incredible potential to boost economic growth across Africa, thereby improving livelihoods and reducing poverty (Africa Tourism Monitor 2014).

To judge how well African countries are exploiting the opportunities presented by tourism, African countries were categorized into four performance categories by a recent World Bank study. These four categories consist of "pre-emergent," "potential," "emerging," and "consolidating" tourism destinations. Some performance indicators include factors such as the international arrival per head of population, the tourism receipts per arrival, and the calculated growth in tourism arrivals (Signé 2019).

Though security poses a challenge to the tourism industry, the level of government commitment to the tourism industry is one of the reasons that some developing countries such as Sudan and Somalia are yet to experience growth. Other countries such as Ethiopia and Gabon show some promise but still battle government challenges preventing the prioritization of the tourism industry and stalling competitiveness.

21.8.1.3 Environmental Implications of Tourism in Africa

Emerging countries in Africa are gradually shifting away from dependence on agriculture and trade in commodities toward advancing industries, such as tourism. This has placed tourism in the spotlight of national development. While tourism can contribute positively to socio-economic development, when unrestrained, its growth can lead to problematic environmental issues. In this section, the adverse effects of tourism on the environment are explored in sufficient detail under the following subheadings.

21.8.1.4 Air Pollution

Tourism is not possible without travel, be it primary transportation to the actual place or city of attraction (mostly involving international travels by air) or domestic transportation within the location. Thousands of air travel are undertaken annually for touristic purposes (see Table 21.6, showing the number of passengers carried by air transport between 2015 and 2019). In 2017, the Ethiopian Airlines Group experience such tremendous growth that it transported about nine million

Table 21.6 Number of passengers carried by air transport between 2015 and 2019 (air passengers carried include both domestic and international aircraft passengers of air carriers registered in the country)

Countries	Number of air passengers	Countries	Number of air passengers
Algeria	30,915,564	Malawi	40,962
Angola	7,056,147	Mauritania	1,713,022
Burkina Faso	805,431	Mauritius	8,206,747
Cabo Verde	2,688,862	Morocco	40,963,868
Cameroon	1,310,817	Mozambique	3,005,692
Congo, Democratic Republic of the	2,990,235	Namibia	2,550,582
Congo, Republic of the	2,620,372	Niger	45,000
Cote d'Ivoire	3,692,223	Nigeria	27,838,314
Egypt	60,048,716	Rwanda	5,285,097
Equatorial Guinea	2,180,436	Senegal	709,239
Eritrea	413,224	Somalia	76,767
Eswatini	104,377	South Africa	109,759,438
Ethiopia	49,015,731	Sudan	2,438,333
Gabon	214,651	Togo	2,603,724
The Gambia	315,535	Tunisia	19,675,929
Ghana	2,496,631	Uganda	149,992
Kenya	26,849,964	Zambia	536,389
Libya	6,710,099	Zimbabwe	1,314,046
Madagascar	2,472,606		

passengers, a rise from three million in 2010 (Africa Tourism Monitor 2018). As revealed by a Norwegian research study, tourist travel is a major cause of serious environmental problems (Høyer 2000), and solving the problem of sustainable transportation represents a major challenge for the future development of tourism. Cohen et al. (2014) have projected that by the year 2050, given that other economic sectors have greatly lowered their CO_2 emissions, tourism is likely to be generating 40% of global carbon emissions. The United Nations World Tourism Organization (UNWTO) report of 2014 suggests that the preferred mode of travel for most tourists is air transport as this includes both movements from outside and within the continent. The alarmingly high rate of air transportation, however, has grievous impacts on the environment and biological resources by promoting stress in the latter and degradation in the former (Vwioko et al. 2018; Ikhajiagbe et al. 2021, 2022). Aviation, for instance, accounts for 55% of CO_2 emissions from transportation which is about 40% of tourism's total, and when considering the effect of all greenhouse gas emissions, aviation alone could account for up to 75% of tourism's impact on climate. Tourism's impact on climate is due to its heavy reliance on fossil fuels. About 72% of tourism's CO_2 emissions come from transportation, 24% from accommodations, and 4% from local activities (Peeters and Dubois 2010).

Transportation, both local and international, is a crucial part of modern living in every part of the world, but we seem to be ever leaning toward gratification and comfort at the expense of longer term implications, such that the tourism transport emission challenge remains largely unattended to. In general, due to climate change concerns as a result of CO_2 emissions, fossil fuel phase-out actions are being taken by some nations around the world, but one of the challenges accompanying the phase-out of fossil fuels is the heavy reliance that the world still has on them. Some modes of transportation, e.g., electric powered, and solar-powered, are already in use in some developed nations. However, tourist transport which usually involves high velocity and long-distance transport and utilizes petroleum-based fuel remains a big concern. Achieving radical emission reductions in all sectors of the economy is an urgent and critical challenge that needs quick attention for climate change mitigation.

A paper by Peeters and Dubois (2010) describes an inventory giving the CO_2 emission generated by global tourism for the year 2005, presenting a 30-year projection and a 45-year simulation. The study found that tourists cause about 4.4% of global CO_2 emissions, and these emissions are anticipated to grow at an average rate of 3.2% per year up to 2035. This is quite problematic as a global reduction in emissions of up to 3–6% is required to avoid 'dangerous' climate change. Employing the use of contemporary scenario techniques, it appeared difficult to find a "future tourist travel system" consistent with CO_2 emission reductions of up to 70% by 2050 wiconcerning005. Gossling et al. (2010), in their paper on tourism growth and climate policy, concluded that given the projected emission growth rates, technology and management will not be sufficient to achieve even modest absolute emission reductions in the tourism sector. It is, therefore, rightly noted that "radical shifts" will be needed in tourism transport if the tourism industry is to adapt to a lower carbon future (Dickinson 2010). Unfortunately, it has become a norm to satisfy short-term pleasure at the expense of longer term negative effects on the environment, and to aim at managing the effect tourism has on the world climate, developing potential solutions to tourist transport's reliance on fossil fuels will play a key role of in realizing climatically sustainable tourism on both domestic and international levels.

21.8.1.5 Destruction of Ecology

In many areas, negative impacts of tourism on the environment often arise alongside the notable positive socio-economic impacts of the development of tourism. The Okavango Delta in Botswana is one example of such a places. Being home to a principal tourist destination with scenic beauty and rich wildlife resources, the Okavango region has witnessed the development of a range of infrastructure and facilities, such as hotels, lodges, tarred roads, airports, and communication facilities. Tourism in this region has also enhanced the establishment of both wholesale and retail businesses that supply various goods to the tourist industry. A huge increase in employment opportunities for the local community has also come about as a result of the tourism industry that not only thrives in the region but is also a major source of foreign exchange for Botswana. However, despite the noteworthy socio-economic impacts of tourism development in this region, the industry is beginning to have negative environmental impacts in the area such as the destruction of the ecology in the area. Driving outside of the prescribed trails, poor waste management, and noise pollution are some of the environmental concerns in the Okavango Delta (Mbaiwa 2003).

21.9 Sustainable Tourism Development in Africa

The African Union and sub-regional communities, having recognized the potential of tourism for economic growth and development, have prioritized the development of the industry by endorsing the Tourism Action Plan aimed at making Africa the tourism destination of the twenty-first century A visa policy introduced by the members of the ECOWAS enables free movement of people across member states, thereby offering a larger market to international travelers, increasing travel rates, and further advancing the African tourism industry. However, potential social, cultural, and environmental impacts of such development endeavor have to be carefully considered, as these potential impacts may compromise the sustainability of the sector.

The United Nations World Tourism Organization describes 'sustainable tourism' as tourism that meets the needs of present tourists and their host regions while protecting and enhancing opportunities for the future. It is a principle that underpins all tourism activities and is integral to all aspects of tourism development and management. The clear objective of sustainable tourism is to retain the economic and social advantages of tourism and tourism development while reducing or mitigating any undesirable impacts on the natural, historic, cultural, or social environment. This is achieved by balancing the needs of tourists with those of the destination.

Sustainable Tourism, thus, is the level of tourism activity that can be maintained over the long term without jeopardizing or undermining its benefits for the social, economic, cultural, and natural conditions of the area in which it takes place. When the impacts of tourism are neither permanent nor irreversible, it can be said to be environmental, socio-culturally, and economically sustainable.

21.9.1 Pillars of Sustainable Tourism

Sustainability, which implies meeting our present needs without compromising the ability of future generations to meet their own needs, is majorly characterized by environmental conservation, social responsibility, and economic development. Sustainable tourism is, therefore, deeply concerned with people and the well-being of communities. It is devoted to make a negligible impact on the environment while creating an enriching experience for the local people, tourism operators or companies, and tourists as well. Improving and supporting the development of the tourism sector with a minimal environmental impact would involve both tourism operators and tourists. Local populations and community stakeholders need to be engaged as well in the mission to achieve sustainable tourism in the dynamic travel and tourism sector in Africa.

There are three important aspects of sustainability that needs to be considered for tourism to truly be sustainable. They are the environmental, economic, and socio-cultural aspects that deal with environmental integrity, economic development, and social justice.

21.9.1.1 Environmental Sustainability

Environmental sustainability, otherwise known as ecological sustainability, means preserving the environment for use by the future generation (Ogwu 2019b; Ikhajiagbe and Ogwu 2021). The environment must be responsibly protected and supported by tourists and locals alike and being environmentally conscious plays a good role in the preservation of the environment.

21.9.1.2 Economic Sustainability

Economic development is an aspect of sustainability that is quite important. It refers to practices that support long-term economic growth without impacting the environmental, social, and cultural aspects of the community negatively. Economic sustainability revolves around money and is a key to make a tourism venture sustainable. When a community is involved in tourism and share in the financial benefits, this helps in keeping the money local and contributes much to the local economy.

21.9.1.3 Socio-cultural Sustainability

Tourists are bound to influence their host community socially and culturally. There may be increased congestion in towns, cities, and the introduction of new values and languages. Socio-cultural sustainability implies promoting cultural exchange and preserving local traditions by encouraging positive impacts and minimizing negative ones. Getting the locals involved in the tourism industry is one way of achieving this, because tourists will have a more genuine experience and locals will be able to see tourism in a positive light, as they will feel involved and part of it.

21.9.2 Tourism for Sustainable Environmental and Biodiversity Management

21.9.2.1 The Concept of Biodiversity

Biodiversity is the biological variety and variability of life on earth, and it occurs at three levels the genetic diversity within species and the diversity of the ecosystem (Cho 2011). It is concerned with the variety of individual species in an ecosystem. However, the most used definition of 'biodiversity' was given by the Convention of Biological Diversity which defines biodiversity as the variability among living organisms from all sources, which include terrestrial, marine and aquatic ecosystems (CBD 1992). The International Union for the Conservation Nature also uses the term biological diversity to describe the variety of life on earth, which includes the number of species, variety, and variability of living organisms, such as animals, plants, fungi, microbes etc. Biodiversity is very beneficial to the existence of man. It provides functioning ecosystems that supply oxygen, clean air and water, pollination of plants, pest control, wastewater treatment, and many numerous ecosystem services (Bradley et al. 2012). Valuable nature of biodiversity can be categorized into two according to Tackacs (1996).

21.9.2.2 The Intrinsic Value of Biodiversity

This refers to the innate value of bio-resources to exist concurrently with humans (Ogwu 2020). It is part of the natural world and is important for the maintenance of natural ecological processes. It is the duty of humans to leave it for posterity without any alteration, as it was handed down to the current generation by their forebears. It is not just the economic incentives that can be obtained from the biodiversity in nature that should be the driving agent for their conservation but also, their admirability and spiritual or religious reason (Huber et al. 2002).

21.9.2.3 Anthropocentric Value of Biodiversity

This refers to conservation of biodiversity resources for their economic value. The economic benefits human that can benefit from biodiversity can be categorized into direct benefits and indirect benefits. The direct benefits that humans derive from biodiversity include the food, clothing, shelter, fuel, medicine, raw materials for industries, etc. Attuquayefio and Fobil (2005) opined that some biodiversity species have the potential of providing new as well as proactive medicines for the management or possible cure of intractable diseases, such as HIV, AIDS or cancer. The indirect benefits which living organisms gain from bio-resources in nature cannot be easily recognized like the direct benefits. These indirect benefits are numerous but difficult to quantify, and they refer to ecological services which can be sub-divided into provisional services, regulatory services, cultural services, supporting services, etc. (Tackacs 1996). These benefits are threatened by human modification of the environment as well as their over exploitation of biological resources (Ogwu et al. 2014, 2016; Osawaru and Ogwu 2014).

21.9.2.4 Contributions of Tourism for Biodiversity Management

Mankind is living in ecosystems that can be recognized at many different levels, ranging from a small forest to the entire globe; natural ecosystems are dynamically stabilized based on balanced inputs and outputs (Abdulla et al. 2009). Tourism like every other industry is accompanied by impacts. These impacts either support or degrade the environment. The environment in this context refers to the physical component which can be categorized as the natural surroundings and physical structures. It is the ability of the ecosystem to absorb the impacts which does not exclude the tourist and commercial exploitation of the ecosystem. Ecotourism, which is a major tool for sustainable tourism, does not set limits on tourist numbers but only controls the extent of their activity (Tyler and Dangerfield 1999).

21.9.2.5 Positive Contributions of Tourism to Biodiversity Management

Tourism can positively influence conservation of protected areas and decreases deforestation and hunting rate. It can also ensure the conservation of the physical and natural structures that it depends on. For instance, most of the ecotourism destination provides economic benefits to the host community. Its goals are to develop an ecologically sound and culturally sensitive tourism program that maximizes benefits to local villages and discourage the conversion of forest to agricultural and pastoral land. It is, thus, one of the preferred tools for conservation and community development in many rural areas (Stem 2003).

Raising of environmental awareness is also one of the core benefits of tourism especially with respect to natural environment. It promotes an increase in the public understanding and appreciation of nature. Most ecotourism destinations encourage this by exposing tourists to nature through education via eco-tour guiding, eco-lodge, signage and even design and operate environmental conscious activities that tourist can participate in (Cusack and Dixon 2006). In addition, tourism, as have been discussed in earlier sections of this chapter, is an important source of employment for the people in the territory. Local residents benefit through employment into the tourism industry which includes hotels, guesthouses, restaurants, catering services, and bar management Boxill and Severin 2004). It serves as an important way to increase employment and reduce pressure on the biodiversity. The revenue from tourism could also be used to help the local people with access to civilization and education.

21.9.2.6 Negative Impact of Tourism on Biodiversity Management

However, negative environmental impacts of tourism on biodiversity still abounds. Environmental degradation is a major concern when it comes to the tourism industry. Tourism can have negative effects on species, communities, and populations by influencing their feeding, reproductive, and social behaviors, and is an industrial activity that has series of impacts that are similar to most other industrial activities (Cole and Landres 1995). The construction of tourism attractions and infrastructures such as road and other activities in tourism destination brings about degradation of the environment. During construction, lands are cleared and transformed which then displaces the natural occupants of such places (Kafle 2015). Tourism, when not managed efficiently, can cause long-term damage to wildlife habitat.

Disruption or of wildlife breeding cycles and behaviors is also another impact of tourism on biodiversity that is worth highlighting. Tourism activities, especially mass tourism, disrupt the breeding cycles of wildlife and their natural behavior and intensive development of facilities in a protected area can change migratory routes of wildlife, for example, the wildebeests (Haysmith and Hunt 1995). Species are often particularly vulnerable to the effects of disturbance during their period and during their juvenile stages. Any disruption of courtship and mating behavior or care for offspring reduces overall success and, therefore, is a threat to population maintenance and survival (Haysmith and Hunt 1995).

Finally, in areas with high concentrations of tourist activities and appealing natural attractions, disposal is usually a serious problem and improper disposal can be a major despoiler of the natural environment, river, scenic areas, and roadsides (Joseph 2007) and pollution of the natural environment does not contribute positively to environmental management.

21.10 Sustainable Development of Africa Through Sustainable Ecotourism

There exists an agreement that tourism development should be sustainable, but how this can be accomplished continues to be a thing of continuous deliberation amongst many scholars, as the idea of sustainable tourism is closely connected to a number of key themes in sustainability. Tourism, being one of the largest industries in the world, is one whose growth demands that tourists seek to promote sustainable tourism by being open to avenues to protect the environment and make tourism, as an ever-expanding industry, sustainable. With their remarkable landscapes and ample natural resources, many regions in Africa are positioned for massive growth in tourism, enabling them to empower people by providing abundant opportunities for employment, innovation, entrepreneurship, and sustainable livelihoods (Africa Tourism Monitor 2015). Components of tourism include accommodation, entertainment, shopping, food, and recreation, and are by no means less important in matters surrounding sustainability in tourism. The general approach to sustainable tourism is to conserve resources utilizing the least possible amount of resources, especially non-renewables, and by being environmentally conscious.

21.10.1 Sustainable Ecotourism: An Overview

Global Ecotourism Network (2020) defines ecotourism as a form of tourism involving responsible travel (sustainable transport) to natural areas, conserving the environment, and improving the well-being of the local people. Ecotourism actively embraces the principles of sustainable tourism and contributes to the conservation of nature and preservation of the cultural heritage of indigenous communities. The purpose ecotourism that aims to serve is the education of the traveler on the natural environment and local surroundings, promotion of ecological conservation, fostering respect for different cultures, as well as driving economic development in local communities, while ensuring that future generations can experience preserved locations relatively untampered by human intervention (Honey 2008). Ecotourism, therefore, is an industry in which tour operators seek to conserve elements of biodiversity in localities, as one of the major attractions for ecotourists is biodiversity and many are less likely to travel to degraded places or areas that has been deforested. Returns on investment also relies on the maintenance of the rich attributes in such locations; hence, this serves as an incentive for conservation as well. In addition, the definition of ecotourism is the effort to produce economic opportunities for the locals while making conservation of natural resources easy and possible.

Ecotourism can help generate huge revenues for tourism operators and local and indigenous communities. However, it can very difficult to ensure sustainability, as the tourism industry generally tends to exert pressures on biodiversity. There is the increase in population in the local communities consisting of tourists and the employees providing various services to them, which inadvertently contributes to pollution, habitat fragmentation, reduced water quality, and in some cases, the influx of exotic plants and animals. Ecotourism relies greatly on the presence of biodiversity and these problems negatively impact and causes the decline of the animals and plants that the tourism industry is built around. These losses in biodiversity threaten the sustainability of the industry (Beattie et al. 2005).

21.10.2 Role of Sustainable Ecotourism in Sustainable Development

Sustainable tourism fundamentally involves visiting a place as a tourist and trying to make positive impact on both the environment and society, as well as the economy. It establishes a suitable balance between the environmental, economic, and socio-cultural aspects of tourism development and plays a vital role in promoting biodiversity conservation. It also seeks to minimize the impact of tourism on the environment and the local culture while sustainably promoting the generation of income and employment, and the conservation of local ecosystems (UNWTO). Consequently, sustainable ecotourism eagerly promotes the positive contribution of tourism to biodiversity conservation, raises awareness on responsible tourism, and fosters positive behavioral changes for biodiversity conservation amongst tourists. It also provides incentives for habitat protection, improves livelihoods and fights poverty, and promotes the achievement of sustainable development goals. Revenues accrued from tourists' expenditures are often directed back into conservation initiatives or program directed at capacity building for the purpose of managing protected areas. These all serve to further the sustainable development agenda.

21.10.3 Some Sustainable Ecotourism Projects in Africa

When it comes to sustainable ecotourism projects, there are certain characteristics to look out for. Such projects would usually entail deliberate steps to minimize impact, building of environmental awareness, provision of financial incentives for conservation, provision of financial benefits and empowerment for the local people, and respect for local culture. The typical functions of ecotourism projects include: protection of ecological zones or areas, education and awareness, revenue generation, and local participation. Some notable projects across Africa are stated in the table below (Table 21.7).

Table 21.7 Case scenarios of sustainable environmental/biodiversity management in tourism

Project Name and Country	Prominent features of sustainable environmental or biodiversity management and tourism development
Campi ya Kanzi Luxury ecolodge, Kenya	– Transformative tourism that strives to protect wildlife and the environment, including the pristine Kuku wilderness, Chyulu Hills National Park, and Mzima Springs – Provision of economic benefits to the local people (the Maasai tribe) who are involved in management and running of the lodge. Most of the staff are Maasai people who received vocational training and economic empowerment – Encourages eco-friendly practices and works toward minimization of carbon footprint by deriving their energy and heat from solar panels, using eco-friendly stoves for all cooking and cropping water from rainfall – Educates guests about sustainable practices to encourage the conservation of resources both on the camp and when they return to their respective destinations
Kicheche Laikipia Camp, Kenya	– Concerned with environmental conservation, preservation of resources, and solid waste management – Compliant with relevant legal requirements for sustainability in areas of their operations – Has achieved noteworthy reduction on organic, plastic, glass and metallic waste within the premises – Promote responsible tourism to protect natural habitats – Employs innovation in resource conservation
Isoitok Camp Manyara, Tanzania	– Eco-friendly camp that promotes sustainable experiences that benefits both guests and hosts – Provides sustainable and responsible accommodation that blends with the natural environment
Mombasa Beach Operators' Livelihood Project, Kenya	– Concerned with promoting sustainable long-term tourism support for beach operators along the Kenyan coast – Strengthens the livelihoods of beach operators, e.g., the local people working as safari guides, sellers, boat operators, photographers, Samburu warriors and curio sellers, by helping to improve their interaction with tourists, and lessen tourist hassle
Mashovela Bush Lodge, South Africa	– Known for its eco-friendly accommodation, and educational initiatives that supports the arts and crafts of the local community – Employs local staff thereby driving economic benefits in the local community – Possesses strong commitment to fair and responsible tourism which includes the promotion of fair wages and working conditions – Promotes an equitable distribution of benefits – Prioritize respect for culture and the environment
The Berjaya Beau Vallon Bay Resort and Casino, Seychelles	– Involved in rainwater harvesting to provide water for use and to maintain and manage the landscape during drought – Promotes staff awareness on sustainable practices – Keen on reducing wastage and managing water, as well

(continued)

Table 21.7 (continued)

Project Name and Country	Prominent features of sustainable environmental or biodiversity management and tourism development
	as educating guests on the benefits of water conservation – Utilization of 100% LED lights in public areas and minimization of energy consumption in guest rooms through a key card assist system

21.11 Recommendations

The economic advantages of tourism are so much stressed at the expense of its inescapable impacts on the environment. These impacts when not duly considered inevitably undermine the sustainability of the tourism sector. To unlock Africa's tourism potential, certain areas need to be improved in order to accelerate the development pf tourism, attract investors, and promote socio-economic growth. This would help facilitate the achievement of Sustainable Development Goals and African Union's Agenda of target dates 2030 and 2063, respectively (Signé 2018). As such, governments across the continent should prioritize tourism in their development agendas and programs. They should also advance investment in infrastructure and improve the business atmosphere for tourism to thrive. Importantly, the continent should foster cooperation for sustainable tourism through partnerships.

The tourism sector should also play a key role in raising awareness on the devastating impacts of wildlife crime among both policy-makers and tourists. The industry should also help in financing anti-poaching initiatives (Africa Tourism Monitor 2015). In addition, to achieve sustainable tourism, tourism operators and tourists should take into consideration the role of sustainable transportation in sustainable tourism and thereby seek to reduce the impact of tourism by creating and adopting sustainable strategies.

21.12 Conclusions

The tourism industry in Africa holds huge potential for massive growth and development for its populations across the continent. More needs to be done to maximize the potentials of African touristic resources for sustainable socio-economic growth and development. Some of the challenges faced by the tourism industry and that is responsible for its inability to bloom in some of the countries in Africa have been sufficiently discussed in this chapter.

Tourism is clearly one of the chief industries driving continuous change and development in several regions on the continent and it provides manifold opportunities for socio-economic growth which promotes improved livelihoods and poverty reduction. Job creation, regional economic development, infrastructural development, increased local consumption, promotion of cultural heritage, environmental preservation, and empowerment of women and youth are among the positive contributions and value of Tourism in Africa. The World Tourism Organization predicts that African tourism industry will continue to expand. The future of the tourism sector in Africa is, thus, very promising. By 2030, the number of International tourist arrivals in Africa is estimated to have reached 134million—a figure that is more than double its present value. This kind of growth forecast presents an enormous opportunity to promote socio-economic development. However, it should also inspire responsibility in protecting Africa's natural resources, her rich biodiversity, and the host communities (UNWTO).

References

Abdulla FA, Abu-Dieyeh MH, Qnais E (2009) Human activities and ecosystem health. In: Laboy-Nieves EN, Schaffner F, Abdelhadi A, Goosen MFA (eds) Environmental management, sustainable development and human health. Chapter 28. Taylor & Frances, London

Africa Tourism Monitor (2014) Leveraging Africa's Human Capacity for Tourism Development. African Development Bank Group (AfDB), Africa House at NYU, and Africa Travel Association Volume 2 Issue 1. November 2014

Africa Tourism Monitor (2015) Africa tourism monitor-unlocking Africa's tourism potential. Volume 3 Issue 1. October 2015

Africa Tourism Monitor (2018) The High 5s – Tourism as a Pathway to Industrialization, Integration, Quality of Life, Agriculture, and Powering Up Africa. African Development Bank Group (AfDB), Africa House at NYU, and Africa Travel Association. Volume 5 Issue 1 July 2018

Aleksandra (2020) 9 Best Fishing Destinations in Africa Our Top Picks. https://www.fishingbooker.com/blog/category/captains-corner/

Attuquayefio DK, Fobil JN (2005) An overview of biodiversity conservation in Ghana: challenges and prospects. West Afr J Appl Ecol 7:1–18

Beattie A, Barthlott W, Kate K, Elisabetsky E, Farrel R, Kheng CT et al (2005) New products and industries from biodiversity. In: Ecosystems and human well-being: current state and trends : findings of the Condition and Trends Working Group of the Millennium Ecosystem Assessment. Island Press, Washington, DC, pp 271–296

Beekwilder J (2022) Top 10 Best African Safari Parks and Destinations of 2022. https://www.safaribookings.com/blog/top-10-best-african-safari-parks-and-destinations

Boxill I, Severin FO (2004) An exploratory study of tourism development and its impact on the Caribs of Dominica. Int J Hosp Tour Adm 5(1):1–27

Bradley C, Emmet DJ, Gonzalez A (2012) Biodiversity loss and its impact on humanity. Nature 486:59–67

Cho R (2011) What you can do to protect biodiversity? http://www.earth.columbia.edu

Cohen S, Higham J, Peeters P, Gossling S (2014) Why tourism mobility behaviours must change. Ch. 1 in: Understanding and Governing Sustainable Tourism Mobility: Psychological and Behavioural Approaches

Cole DN, Landres PB (1995) Wildlife and recreationists. Island Press, Washington, DC

Convention on Biological Diversity (CBD) (1992) Convention on biological diversity, viewed 16 June 2016. http://www.biodiv.org

Cusack D, Dixon L (2006) Community-based ecotourism and sustainability. J Sustain For 22(1–2): 157–182

Dickinson J (2010) Tourism and climate change - an introduction. J Transp Geogr 18(3):445–446

Dieke PUC (2020) Tourism in Africa: issues and prospects. In: Baum T, Ndiuini A (eds) Sustainable human resource management in tourism. geographies of tourism and global change. Springer, Cham. https://doi.org/10.1007/978-3-030-41735-2_2

Erinle KO, Ogwu MC, Evivie SE, Zaheer MS, Ogunyemi SO, Adeniran SO (2021) Impacts of COVID-19 on agriculture and food security in developing countries: potential mitigation strategies. CAB Rev 16(16):1–16. https://doi.org/10.1079/PAVSNNR202116016

Evivie SE, Ogwu MC, Ebabhamiegbebho PA, Abel ES, Imaren JO, Igene JO (2020) Packaging and the Nigerian food industry: challenges and opportunities. In: Ogunlade CA, Adeleke KM, Oladejo MT (eds) Food technology and culture in Africa. Reamsworth Publishing, Ibadan, pp 28–99

Global Ecotourism Network (GEN) (2020) What is (not) Ecotourism? | Global Ecotourism Network (GEN). https://www.globalecotourismnetwork.org/what-it-is-not-ecotourism/; www.globalecotourismnetwork.org. Retrieved 8 July 2020

Gossling S, Hall M, Peeters P, Scott D (2010) The future of tourism: can tourism growth and climate policy be reconciled? A climate change mitigation perspective. Tour Recreat Res 35(2): 119–130

Haysmith L, Hunt JD (1995) In: Knight RL, Dutzwiller KJ (eds) Nature tourism: impacts and management. Chapter 12. Wildlife and recreationists. Island Press, Washington, DC

Honey M (2008) Ecotourism and sustainable development: who owns paradise? 2nd edn. Island Press, Washington, DC. ISBN 978-1-59726-125-8

Høyer K (2000) Sustainable tourism or sustainable mobility? The Norwegian case. J Sustain Tour 8(2):147–160. https://doi.org/10.1080/09669580008667354

Huber H, Hohn M, Rachel R, Fuchs T, Wimmer V, Stetter K (2002) A new phylum of Archaea represented by a nanosized hyperthermophilic symbiont. Nature 417:63–67

Ikhajiagbe B, Ogwu MC (2021) Growth performance of ferulic acid composition of *Amaranthus* [L.] species and *Mitracarpus villosus* (Sw.) DC. in competition with weed species. Environ Sustain. https://doi.org/10.1007/s42398-021-00163-z

Ikhajiagbe B, Ogwu MC, Lato NF (2021) Growth and yield response of soybean (*Glycine max* [L.] Merr.) accessions after exposure to cadmium. Vegetos 34(1):107–118. https://doi.org/10.1007/s.42535-021-00189-y

Ikhajiagbe B, Ogwu MC, Ekhator PO, Omoayena II (2022) Implications of soil nitrogen ehnacement on the yield performance of soybean (*Glycine max*) in cadmium-polluted soil. J Poll Effects Control 10(2):1–12. https://doi.org/10.1101/2021.11.22.469589

Jenkins CL (2015) Tourism policy and planning for developing countries: some critical issues. Tour Recreat Res 40(2):144–156

Joseph PD (2007) Tourism principles, policies and practices. MBTM 3001, Directorate of Distance Education, Pondicherry University. pp. 206

Kafle S (2015) Prospects of ecotourism: a case study of Gaurishankar Conservation Area, Dolakha, Central Nepal, in Central Department of Environmental Science. Tribhuvan University, Kathmandu

Mbaiwa JE (2003) The socio-economic and environmental impacts of tourism development on the Okavango Delta, north-western Botswana. J Arid Environ 54(2):447–467, ISSN 0140-1963. https://doi.org/10.1006/jare.2002.1101

Mitchell J (2021) Western tourists have returned to Africa. Will business travellers follow?—Investment Monitor. https://www.investmentmonitor.ai/sectors/tourism-business-activities/western-tourists-africa-business-travellers-covid

Ogwu MC (2019a) Towards sustainable development in Africa: the challenge of urbanization and climate change adaptation. In: Cobbinah PB, Addaney M (eds) The geography of climate change adaptation in Urban Africa. Springer Nature, Switzerland, pp 29–55. https://doi.org/10.1007/978-3-030-04873-0_2

Ogwu MC (2019b) Ecologic and economic significance of bryophyte. In: Rathoure AK, Chauhan PB (eds) Current state and future impacts of climate change on biodiversity. IGI Global, Pennsylvania, pp 54–78. https://doi.org/10.4018/978-1-7998-1226-5.ch004

Ogwu MC (2020) Value of *Amaranthus* [L.] species in Nigeria. In: Waisundara V (ed) Nutritional Value of Amaranth. IntechOpen, London, pp 1–21. https://doi.org/10.5772/intechopen.86990

Ogwu MC, Osawaru ME, Ahana CM (2014) Challenges in conserving and utilizing plant genetic resources (PGR). Int J Genet Mol Biol 6(2):16–22. https://doi.org/10.5897/IJGMB2013.0083

Ogwu MC, Osawaru ME, Aiwansoba RO, Iroh RN (2016) Status and prospects of vegetables in Africa. In: Borokini IT, Babalola FD (eds) Conference Proceedings of the Joint Biodiversity Conservation Conference of Nigeria Tropical Biology Association and Nigeria Chapter of Society for Conservation Biology on MDGs to SDGs: Toward Sustainable Biodiversity Conservation in Nigeria. University of Ilorin, Nigeria, pp 47–57

Osawaru ME, Ogwu MC (2014) Conservation and utilization of plant genetic resources. In: Omokhafe K, Odewale J (eds) Proceedings of 38th Annual Conference of The Genetics Society of Nigeria. Empress Prints Nigeria Limited, Nigeria, pp 105–119

Otieno V (2018) Tourism in Africa information: Facts, Statistics, Tourism in Africa by Country 2018. TINA Magazine https://tinamagazine.com/2018/07/tourism-in-a frica-facts.html

Peeters P, Dubois G (2010) Tourism travel under climate change mitigation constraints. J Transp Geogr 18(3):447–457. https://www.sciencedirect.com/science/article/abs/pii/S0966692309001422?via%3Dihub

Rogerson C (2017) Reviewing Africa in the global tourism economy United Nations World Tourism Organization:24(3)

Sharpley R, Telfer D (2015) Tourism and development: concepts and issues, 2nd edn. Channel View Publications, Bristol. http://scholar.google.com/scholar_lookup?title=Tourism%20and%20development%3A%20Concepts%20and%20issues&publication_year=2015

Signé L (2018) Africa's tourism potential Trends, drivers, opportunities, and strategies. Report Brookings December 3, 2018. https://www.brookings.edu/wp-content/uploads/2018/12/Africas-tourism-potential_LandrySigne1.pdf

Signé L (2019) Africa's tourism: A global destination for investment and entrepreneurship. Africa in focus Brookings March 5, 2019

Stem CJ (2003) How eco is ecotourism? A comparative case study of ecotourism in Costa Rica. J Sustain Tour 11(4):322–347

Tackacs D (1996) The idea of biodiversity: philosophies of paradise. John Hopkins University Press, Baltimore, MD

Tyler D, Dangerfield JM (1999) Ecosystem tourism: a resource-based philosophy for ecotourism. J Sustain Tour 7(2):146–158

Vwioko DE, Okoekhian I, Ogwu MC (2018) Stress analysis of *Amaranthus hybridus* L. and *Lycopersicon esculentum* Mill. Exposed to sulphur and nitrogen dioxide. Pertanika J Trop Agric Sci 41(3):1169–1191

WhiteOrange (2020) Homepage. Ghana Tourism Authority

Chapter 22
Environmental Sustainability: Relevance of Forensic Insects and Other Ecosystem Services in Africa

Maduamaka Cyriacus Abajue

Abstract Greater percentage of African people depend on exploring the natural resources for agriculture, herding, fishing, and hunting as a means of livelihood. The manner and rate that these practices are carried out are not sustainable in line with the United Nations sustainable development goals that center on getting natural sources of livelihood cautiously and maintain a quality and sustainable environment. The environment is man's surroundings in which he lives and carries out his day-to-day activities and, perhaps, does not live in an isolation; hence, he is biologically surrounded by plants and animals. Environmental sustainability requires awareness and attitudinal behavior to ones need for natural resources and bearing in mind that any abuse on them today will take revenge on the future generation. It appreciates population growth, sustainable cultivation, and yield, sustainable disposal of wastes as well as economic development and industrialization that are anchored on renewable resources. Insects among the animal groups play a useful role in plant reproduction as part of ecological services that are germane for the survival of human beings man and animals and help to maintain and sustain the environment for the benefit of man. The relevance of insects as regards crops' propagation, biological monitoring of polluted ecosystems, organic matter decomposers, food for man and animals is presented in this chapter. Farming activities, burning of fossil fuels, flaring of gases, methods of disposing urban wastes, and climate change are highlighted as human distortions and extortions that are degrading African environment against environmental sustainability for insects of special interest. The use of insects' of forensic relevance in criminal investigation as a vanguard for sustainable environment in Africa is advocated for as a mitigation approach in sustaining insects' biodiversity of African environment.

M. C. Abajue (✉)
Department of Animal and Environmental Biology, Faculty of Science, University of Port Harcourt, Port Harcourt, Nigeria
e-mail: maduamaka.abajue@uniport.edu.ng

S. C. Izah, M. C. Ogwu (eds.), *Sustainable Utilization and Conservation of Africa's Biological Resources and Environment*, Sustainable Development and Biodiversity 32, https://doi.org/10.1007/978-981-19-6974-4_22

Keywords Environmental sustainability · Insects · Forensic entomology · Ecosystem services · Environmental degradation · Africa

22.1 Introduction

Insects play inestimable roles in the maintenance and sustenance of the environment for the benefit of man and animals including the insects themselves. Insects provide ecosystem services such as pollination of flowers for plant propagation and food production. They serve as food for man and other animals that is a key component in animals' food chain. Insects' presence or absence in aquatic and terrestrial environments at a time of check is a probable indication that pollution is paramount. Plants and animals are made up of organic compounds and at death; the compounds are returned to the earth with the help of insects playing a major role to their decomposition. Among the insects that regularly participate in decomposition of animals are very useful in forensic evaluation of time since death. The environment is a man's surroundings in which he lives and carries out his daily activities. Perhaps, man does not live in an isolation; hence, he is biologically surrounded by plants and animals. The plants support live by trapping carbon dioxide that are used in making food and oxygen for animals' survival. Conversely, the plants need other nutrients to survive with which the animals provide. Thus, the relationship between plants and animals are inseparable, and where the interactions exist is known as the environment. Regrettably, the insatiable wants of man among the animals have altered and continue to alter any environment, where the interactions of plants and animals exist. Man is continuously increasing in number globally and his wants and activities continue to grow; hence, the environment is gradually not sustaining his activities for the continued mutual interactions to exist. Sustainability of an environment in this context, therefore, encompasses plant and animal interactions with minimal or without man's interference from man for perpetual continuity. It requires that man should take the responsibility of conserving the nature resources for his immediate and future benefits. In this chapter, the relevance of insects in the environment as it relates to pollination of crops and propagation of wild plants, serving as food for man and animals, being a tool for environmental bio-monitoring and agent for decomposition of organic matters shall be discussed. The activities that degrade African environment such as farm practices, use of fossil fuel, flaring of gases, and methods of waste disposal are also highlighted as distortions and extortions to environmental sustainability for insects of special interest. The sustainability of insects of forensic relevance is promoted as a tactic to sustain insect biodiversity in African environment.

22.2 Relevance of Insects in African Environment

22.2.1 Insects as Crops and Plants Pollinators

Insects play a useful role in plant reproduction as part of ecological services that is germane for survival of man and animals. Among the animal pollinators, insects constitute about 98% of them and about 250,000 of plant species depend on insects for pollination. This ecological service of pollination of agricultural crops is provided mostly by wild and domesticated bees. Domesticated bees alone pollinate almost 15% of species of plants (Ingram et al. 1996; Tscharntke 2007). An insect is tagged a pollinator when it consistently moves pollen lodged in the male gamete of a flower to a female gamete of the same species of a flower. Insect pollinators are liable for providing orchard, agricultural crops, and horticultural crops with fertilization. The sustainability of any ecosystem and its biodiversity is sustained by insect pollinators through propagation of wild plant species. Insect pollinators are, therefore, relevant in food production and expansion of crop yields and values.

Apis mellifera (bee), for instance, has continued to be the most economically treasured pollinators of crops globally, and without them fruits, seeds and nut crops yield will decrease by more than 90% (Klein et al. 2007). In the absence of wild bees, cultured honeybees are dependent on to pollinate agricultural farms especially mono-cropping farms. Bees in the wild and other insects are capable of pollinating several crops. In Africa, this singular ecosystem service lacks documentation, but has been progressively being documented in the developed countries (Kevan and Phillips 2001; Klein et al. 2003; Slaa et al. 2006). It is pertinent that studies in Africa should envisage managing local agro-ecosystems and devise means of conserving the habitats for survival of the bees and other insect pollinators. There are values associated with insect pollinators that are beyond agriculture in terms of food production. Insects and their habitats offer environmental, social, financial, health, human, and cultural values. Insect pollinators enrich reproduction and genetic uniformity of most plant species, especially dioecious ones that are completely dependent on insects for pollination.

Insect pollinators exploit floral parts of plants for food that has a form of nectar and pollen. Some insects collect oils, resins, and use the plants for shelter and breeding sites (Simpson and Neff 1983; Sakai 2002; Hembry and Althoff 2016). The interaction between plants and insects that visit their flowers sustain plant diversity and diversity of other species of animals estimated at 350,000 with insects dominating the group (Ollerton 2017). The accessibility of active pollinators, especially insects in plant communities, is a clear prerequisite for fruitful seed production in most plant species. Experiment on selective exclusion of a group of active pollinators in which plants are already adjusted to has ended in failure of the plants from producing fruits or seeds (Ratto et al. 2018) and a connection between pollinator diversity and seed production are, therefore, being constituted for plants with an overall list of pollination system (Albrecht et al. 2012). Note, pollinators include different animals but dominated by insects, particularly bees. Others comprise some species, such as flies, wasps, beetles, moths, butterflies, thrips, weevils, ants, and gnats, while non-insect pollinators include birds, bats, marsupials, primates, rodents, and reptiles. About 90% of crops globally are visited by bees. Currently, little species of bees that are reared in a developed world, such as western honey bee (*Apis mellifera*), eastern honey bee (*Apis cerana*), some bumble bees, stingless bees, and little solitary bees. Some pollinators restrict their visit to a minor range of flowering plants; meanwhile, others visit a wide range of plants (Klein et al. 2007). Thus, this vital ecological service of pollinating crops and plants by insects warrant that Africa should engage in practices that would make African environment sustainable for insects' habituation and survival.

22.2.2 *Insects as Food for Man and Animals*

Consumption of insects by man in Africa is known in history, and to use them as animal feed is being presently advocated for (van Huis and Oonincx 2017). Insects add to food security, because they are regularly used as food at home and serve as a means of generating money to households when they are sold in the market (Kalaba et al. 2013). The cost of purchasing edible grasshoppers and palm weevils is at times costlier than other meat products in Africa (Agea et al. 2008; Ayemele et al. 2016). Interestingly, about 2100 edible insect species are collected from their natural habitats in the tropics including Africa (Jongema 2017). Insects are among the alternative protein resources proposed for food for man and feed for animals (van Huis and Oonincx 2017). Harvesting insect species regarded as pests of field crops would be very beneficial in Africa as food and feed and would certainly lessen their population in the field and saves money meant for buying pesticides for their control without jeopardizing the environmental health (van Huis and Oonincx 2017). A forest annexed for captive rearing of wild insect species with their host plants has been advocated (Vantomme et al. 2004) for edible insects as a quest for a sustainable environment in Africa.

Nutritionally, edible insect species are very crucial in the food security of developing countries especially in Africa (Kelemu et al. 2015; Sogari et al. 2017), as they are key natural resource to rely on during food shortage or famine (De Foliart 1997; Mutungi et al. 2019). They provide relevant socio-economic and ecological securities for people of African countries (Alamu et al. 2013; Nonaka 2009). Insects that are consumed in Africa are estimated at 500 species (Kelemu et al. 2015), though 212 insect species in nine orders that are edible were documented between years 2000 and 2020 (Hlongwane et al. 2020). The order Lepidoptera at 41% is the most consumed followed by Orthoptera 23%, Coleoptera 15%, Blattodea/Isoptera 12%, Hemiptera 4%, Hymenoptera, Diptera, Mantodea less than 1%. The nutritional composition of insect species that are edible vary; with a highest protein content found in Lepidoptera between 12% and 79% and Orthoptera has between 12% and 73%, while the lowest found in Blattodea with 0% and 39%. The Coleoptera has highest crude fiber between 2% and 28%, Lepidoptera 2% and 16%, while the crude fiber content has the lowest in Hemiptera 0% and 5%. For moisture content, Lepidoptera has between 3% and 86%, Blattodea has as low as 2.8% and 3% moisture content. Coleoptera has between 15% and 52% carbohydrate content more than Orthoptera with 15–47% while Blattodea has 0% and 32%. For fat content, Lepidoptera has 2–55% and lowest was Orthoptera with 2–16% (Hlongwane et al. 2020). Insect species that are edible based on insect orders have been reported to contain minerals at various compositions in milligram at one hundred gram (mg/100 g). For instance, Orthoptera contains about 0.3 to 910 mg/100 g of iron, Blattodea with 27–332 mg/100 g and Hemiptera 0–20 mg/100 g. For calcium content, Blattodea has 18–132 mg/100 g, Lepidoptera 8–15 mg/100 g for phosphorus content, Lepidoptera has 100–730 mg/100 g, Orthoptera 106–125 mg/100 g. Magnesium content was found in Lepidoptera at 1–160 mg/100 g and Blattodea has 0.1–0.3 mg/100 g. Comparatively, 79% protein content of edible insects, was higher than the quantity found in chicken at 43% and beef at 54%, respectively (Teffo et al. 2007; Malinga et al. 2018). Similarly, Vitamins B1, B2 and B3 that are contained in a house fly were higher than that found in chicken, beef, or salmon (Bukkens 1997), and palatable crickets comprise twice more than Vitamin B12 in beef (Akhtar and Isman 2018). *Microtermes nigeriensis* has a higher content of Niacin, Thiamine, Vitamin A, and C (Igwe et al. 2012). Grasshoppers and termites are also good, but are seasonal and not feasible to domesticate (van Huis et al. 2013). The larvae of black soldier fly (*Hermatia illucens*) are mainly found in their numbers on organic litters, such as coffee bean pulp, vegetables, fish offal, and wastes from distillers. These insects are commercially used to reduce environmental problems associated with manure and several organic wastes by decreasing manure mass, moisture content, and unpleasant odors. They are also being very relevant as feedstuff for cattle, pig, poultry, and fish (Newton et al. 2005).

22.2.3 Insects in as Bio-Monitoring Agents

Some insect species are bio-monitors that assist to monitor the health status of a particular environment. Some selected insect species are more suitable to use when carrying out environmental impact assessment, because they are highly diversified, easy to survey in their habitats, taxonomically stable, easy to identify, respond rapidly upon any alteration in their environment, and highly tolerant to contaminants (da Rocha et al. 2011). Major insects that are relevant in bio-monitoring are dragonflies, damselflies, mayflies, caddis-flies, stoneflies, midges, honey bees, beetles, collembollans, lepidopterans, ants, and syrphid fly.

Environmental bio-indicators are used for quantitative and qualitative checks and are capable of displaying changes in the environment, where physical, biological, chemical, or human phenomena are assessed together in the composite dynamics of the environment (Martos et al. 1997). Anthropogenic activities in the environment require that bio-indicators that are capable of reflecting changes in the environment are sought after (Zaghloul et al. 2020). Among the bio-indicators, insects are very relevant in practically assessing the level of environmental degradation (Lopes 2008). Insects for bio-monitoring are predominant, because they constitute more than half of animal species and their uniformity makes them easier to assess the difference between different habitats on a tolerable spatio-temporary scale (Tylianakis et al. 2004).

Some aquatic insects such as dragonflies were very sensitive during changes that affect their habitat, mainly lakes and flooded drainage areas (da Rocha et al. 2011). Many other species in the families of Gyrinidae, Dytiscidae, Hydrophilidae (Coleoptera), Notonectidae, Veliidae (Heteroptera) and Plecoptera and Ephemeroptera are highly adaptive in most environments and are found throughout the year hence, showcasing ecological and geographical changes. Some aquatic organisms are tolerant to heavy metals, as metallothioneins formation in many aquatic organisms has been elucidated. Hence, insect species in the genus *Halobates* are appropriate for bio-monitoring cadmium and mercury (Nummelin et al. 2007). Insects that mirror environmental changes are taken as sentinels in aquatic environments; hence, they are used in bio-monitoring programs of anthropogenic pollution. Chironomids and caddisflies tolerate some level of metals in the water, but species of Ephemeroptera in the family Heptageniidae are very sensitive to metals (Guimarães et al. 2019).

On land some coleopterans such as carabid beetles are bio-monitors of different environmental changes in forest and agricultural areas, where human activities have led to pollution from oil, sulfur, herbicides, CO_2, insecticides, radioactive substances, and phosphorus (Louzada and Lopes 1997; Davis 2000). The lepidopterans (moths and butterflies) are sensitive to heavy metals and carbon dioxide in an environment closer to industrial and urban areas. Copper, iron, nickel, cadmium, sulfuric acid, and substances that were used in fertilizers have been assessed on some lepidopteran families (Geometridae, Noctuidae, Eriocraniidae, Nymphalidae) (Fajer et al. 1989; Klein et al. 1992; Koricheva and Haukioja 1992; Heliovaara and Vaisanen 1990). The collembolans are sensitive to anthropogenic changes in a soil and diversity reduction is an evidence that heavy metals, agricultural pesticides, and soil water acidification by organic pollutants and wastes may have impacted the soil (Rusek 1998).

The ants (Hymenoptera) are sensitive to human impact and are used as soil bio-monitors that play a functional role of replenishing a degraded and reforested areas (Mejer 1984). They are environmental indicators of different terrestrial ecosystems (Folgarait 1998; Peck et al. 1998). Ants have strong resistance to radioactive and industrial pollutants (Peck et al. 1998). The influence of ants in soils is shown by leaf cutters found in the tropics as agents that improve physical and chemical quality of soils (Cherret 1989). Honeybees (*Apis mellifera*) are useful pollinators but are as well considered as bio-indicators capable of monitoring an environment that is chemically impaired when higher mortality rate is noted in them after visiting flowers for nectar collections (Ghini et al. 2004). Honeybees are one of the adaptable and able bio-indicators, because they are useful in monitoring metals in urban environments, radioactivity substances, pesticides, and industrial wastes (Urbini et al. 2006). The larval frass of wasps in the genus *Polistes* and some other social wasps as predators accumulate lead substance, thus presenting them as agents of monitoring lead pollution (Urbini et al. 2006).

Among the dipterans, some species of sarcophagid family are bio-monitors of heavy metals, asbestos fibers and waste chemicals, and are sensitive to insecticides and herbicides. The syrphid family is a bio-indicator, because its larvae have different environmental conditions from the adults (Sommaggio 1999). Deformities noted in larvae of some chironomids genera (e.g., *Procladius*, *Chironomus,* and *Cryptochironomus*) are indications of polluted sediments (Servia et al. 1998), while the gerrid family is used to detect iron, and manganese concentrations (Nummelin et al. 2007).

22.2.4 Insects as Organic Decomposers

Insects play an immeasurable role in degrading organic wastes and dung beetles have been widely reported in relation to plant growth, improvement of soil quality, and soil recuperation (Kaleri et al. 2020). Both larval and adult beetles, flies, ants and termites shred dead plant and animal matters until they are suitable for fungi and bacteria to consume. This process makes minerals and nutrients of dead organisms become readily obtainable in the soil for uptake by plants. Animal carcasses are eaten by fly maggots and beetles, while dung beetles of different species play a crucial role in degrading animal manure. They have the potential to colonize dung wastes within a short time, thereby inhibiting flies from establishing on them. Dungs contribute up to 80% nitrogen to the atmosphere when they are not acted upon on soil's surface by dung beetles. Dung beetles help to recycle carbon and minerals back into the soil for plants absorption. Dung beetles that are acting on ungulate dungs in Africa, Europe, and Hawii have been useful in Australia to help degrade cattle dungs which marsupial dung beetles could not degrade. Thus, about 23 species of cattle dung beetles helping to degrade dungs in Australia are alien beetles (Hinton 1945; Bornemissza 1976).

Organic manure is a cherished resource; however, its management such as discharging in lagoons, composting, dispersal, and others was ineffective, environmentally perplexed, and is not sustainable (Adhikari et al. 2005; Chang and Janzen 1996; Edmonds et al. 2003; Grossman 2014; Kyakuwaire et al. 2019; Welch and O'Hagan 2010). Among the recent procedures advocated for managing manure to ameliorate its influence on the environment and to sustain animals' production is to involve insects in the management of organic manure. Insects such as dung beetles (Coleoptera: Scarabaeidae) that feed on animals' manure are referred to as manure managers (Nichols et al. 2017) as they recycle organic wastes and improve soil health (Bertone et al. 2006; Nichols et al. 2008). Fly species (Diptera) are solely found in manure and many are erroneously referred to as pests hence man always result to toxic chemicals to control them without considering the effects of the chemical on biodiversity and environmental health (Axtell 1986; Sheppard 1983). Insects such as house flies (*Musca domestica*) decompose large volumes of organic manure within a small time frame. The larvae of house fly are capable of digesting 100 kg of swine manure in 7 days (Zhang et al. 2012) and about 400 kg in 14 days (Čičková et al. 2012). Feeding activity of fly larvae on organic manure reduces odors (Wang et al. 2013) and moisture content (Wang et al. 2013), aerate the manure to allow an environment to remain aerobic in order to accelerate loss of gases and water (Beard and Sands 1973). Fly larvae provide bio-fertilizer that is less toxic to the environment, as their feeding on organic manure reduces antibiotic resistant genes found in animal's manure (Wang et al. 2015, 2017a, b). Fly larvae can moderate swine manure mass to 70% when wet (Wang et al. 2013), 31–35% when dry (Miranda et al. 2020; Wang et al. 2013) and nitrogen to 78% while and phosphorous to 30% (Wang et al. 2013).

The larva of black soldier fly (Stratiomyidae) is capable of consuming more manure than house fly (Čičková et al. 2015), because it possesses a diverse digestive enzymes (Kim et al. 2011). The ability of black soldier fly to digest organic manure has resulted to greater reduction of odors from organic manure (Beskin et al. 2018). The larvae of black soldier fly do not only aerate manure and promote the growth of beneficial saprophytic microbes (Sarpong et al. 2019) but reduce pathogens found in organic wastes when they feed on them. As copraphagous insects, they reduce bacteria that would be detrimental to other animals including man (Erickson et al. 2004). The ability of insects to turn organic wastes into wealth, especially the fly larvae, has mediated the industrial cultivation of black soldier fly.

22.3 Activities that Impact African Environment Against Insect Sustainability

22.3.1 Farming Activity

Warming of the climate is significant in Africa and about 78.7% causes of biodiversity decline is attributed to agricultural intensification, urbanization, deforestation, and pesticide pollution account for 21.3% (Adebimpe and Kayode 2018; Sanchez-Bayo and Wyckhuys 2019; Skendžić et al. 2021). Destroying of insect habitat due to farming activity is one major causes of biodiversity loss (Sanchez-Bayo and Wyckhuys 2019). Intensification of agriculture leads to the use of a broad-based pesticide which is considered an important source of insect decline and biodiversity loss (Schulz et al. 2021). The use of herbicides kills weeds and wild plants, which serve as food and shelter for insects and other arthropods (pests and it natural enemies). Hence, fluctuations in insect biomass are relevant for an ecological functioning of any ecosystem (Hallmann et al. 2017). In aquatic environment, mayflies, caddisflies, and dragonflies are abundantly affected by pyrethroid insecticides, while Neonicotinoids affect pollinators, such as honeybees and bumblebees, especially when they are used as post-bloom spray on perennial trees and field crops.

Industrial pollution also contributes immensely to insect decline and fertilizer account for 10% (Sanchez-Bayo and Wyckhuys 2019).

Insect biomass is used as a proxy for measuring a biodiversity, but the checklist of insects in Africa, where many of the insect diversities are, has not been copiously recorded. Consequently, many of the insect species in Africa may disappear without being noted and in most cases before being identified and documented (Wagner et al. 2021). Agricultural crops occupy about 11% of earth's land area and more than 30% of land are used for grazing and most of which are still relatively natural, and nearly, the land stocking densities are eroding and/or unsustainable causing continuous species lost (Newbold et al. 2015). In the tropics, especially Africa, agriculture is continually expanding very fast leading to destruction of natural vegetation. Increasing numbers of persons contribute to clearing of vegetation for agriculture and for other purposes (Curtis et al. 2018), ensuing in the swift loss of biodiversity.

Insects are declining very fast, and it has been noted in western and northern Europe, and the major stressor is linked to intensification of agriculture (Bell et al. 2020; Habel et al. 2019; Habel and Linsenmair 2019; van Klink et al. 2020; Powney et al. 2019). Butterflies are the most common and best evaluated terrestrial insects, but their declining in number are connected to agriculture as their principal driver (van Strien et al. 2019). Agriculture in Africa has changed tremendously as pesticides, fertilizers, and tractors are now available, making way for increased industrialization of farming methods. Farmlands in Africa are bigger in scale and are more of a monoculture farm practices that mainly dependent on fertilizer, insecticide, and herbicide application (Raven and Wagner 2021). Intensification of agriculture appears to be a major driver in insect population decline (Seibold et al. 2021; Wagner 2020). The production of about 13 crops that are wholly reliant on pollinators for fruits production would be critically affected by pollinator loss through agricultural intensification (Kremen et al. 2007). The grassland insects such as butterflies and noctuid moths (Lepidoptera); ants, bees, and wasps (Hymenoptera); scarab and ground beetles (Coleoptera); crickets, grasshoppers, and katydids (Orthoptera); leaf and plant hoppers, seed bugs, and their relatives (Heteroptera) are facing elevated rates of loss. Of these groups, only butterflies have been well-documented especially in Europe (Habel et al. 2019; Habel and Linsenmair 2019; Warren et al. 2021) in addition to numbers reports of declines of wild bees (Biesmeijer et al. 2006; Grab et al. 2019). In Africa, the clean clearing of woodlands for crops, pasture, was happening at a disturbing rate. For instance, from 2001 to 2015, a mean value of five million acres of tropical woodland were lost per annum to industrial-scale agriculture (Curtis et al. 2018; Stokstad 2018).

22.3.2 *Use of Wood as Fossil Fuel*

The entire continent of Africa is still on a journey to economic development and, thus, relies so much on her irreplaceable natural resources for survival. Utilization of fossil fuels mainly woods in Africa for energy among other natural resources contribute immensely to stock of carbon dioxide (CO_2), a greenhouse gas that is continually released into the atmosphere (Delucia et al. 2008) contributing to a remarkable upsurge in global temperature (IPCC 2007). Upsurge in global temperature and CO_2 are reshuffling the communities of plants and animals in their habitats, thus affecting food supplies globally (Parmesan 2006). The use of fossil fuels in Africa is a catalyst that affects the quantity and diversity of insects that are damaging plants because of the continued discharge of CO_2 and elevated temperature (Currano et al. 2008) leading to increased insect herbivory (Delucia et al. 2008). Elevation in CO_2 and artificial rise in average annual temperature from 10.5° to 20.1 °C increased the percentage and the diversity of leaves damaged by insects from 38% to 57% (Currano et al. 2008). An upsurge in CO_2 largely rises the ratio of carbon to nitrogen in plant tissues (Lincoln et al. 1984), thus causing a reduction of the nutritional quality for protein for insects. This compels insects to increase their consumption on leaves to augment for low leaf nitrogen content (Coviella and Trumble 1999; Knepp et al. 2005). Similarly, beetles (*Epilachna varivestis*) have shown preferential feeding on leaves with higher CO_2 level (Hamilton et al. 2005).

The continued usage and demand for fossil fuels in Africa not only result in greater insects damaging valuable crops and forests (Currano et al. 2008) but would accentuate obnoxious use of pesticides by farmers to control them hence, resulting to environmental degradation and loss of insect fauna diversity. Fossil fuels produce harmful substances that are not natural constituents of the environment. Harmful substances emitted by fossil fuels include carbon monoxide (CO), oxides of sulfur (SO_2, SO_3), oxides of nitrogen (NO, NO_2), and particulate matters which consist of soot and ash particles. These pollutants are capable of interacting with the environment to produce further harmful effects to plants and animals. For instance, these elements react with atmospheric moisture to produce acid rain and smog that are injurious to plants and animals in any environment.

22.3.3 *Flaring of Gas*

Flaring of gases causes atmospheric emissions that adversely affect the environment and people (Mafimisebi and Nkwunonwo 2015; Morrison 2006) in major oil and gas producing communities. Its impact on the environment is not limited to deposit of black dust and or soot that settles on surfaces, foods, and clothes, but also compromises quality of life by causing itching and skin rashes on people within the flare areas. Gas flaring impacts health and safety of people and causes acid rain which makes lakes and streams acidic and damages crops and vegetation. Flaring of gas is burning of crude gases in connection to oil production in a controlled chamber (Nwaugo et al. 2006). This occurs when an initial separation of unrefined oil into gas, oil, and water has taken place. The separated oil undergoes fractional distillation, while gas is burnt or flared and the water discarded into the environment (Nwaugo et al. 2006).

Gas flares are synonymous within the landscape of oil producing communities in Africa, especially the Niger Delta areas of Nigeria. Many flare stations in Nigeria have burn daily without quenching for over 50 years. The natural darkness at night in the communities closer to where the flare stations are no longer enjoyed, while the impacts of the gas flareing on biodiversity of ecosystems contiguous to the flares are affected. Flaring of gas is associated with enormously high levels of carbon dioxide (CO_2) and methane (CH_4) gases that are emitted into the atmosphere, thus impacting the climate patterns (Steiner 2010). Methane (CH_4) which probably due to low efficiency has a much higher global warming potential than CO_2 (Eweje 2006). Gas flaring has been reported to deplete an ozone layer, cause climate change, global warming, acid rain, and level of sea rise (IPCC 2014). Acid rain due to gas flaring is affecting communities in Nigeria, Algeria, Angola and Libya, where oil production is taken place in Africa. Carbon dioxide emitted during gas flaring with other gases trap long-wave infrared radiation in the atmosphere to form greenhouse gases that warm the earth (Nkwunonwo and Mafimisebi 2013). Production of CO_2 is mainly from combustion of fuels (oil, coal, and gas and decay of organic matters) and is continually being deposited in the atmosphere at 280 parts per million (ppm) in the year 1880 to 355 ppm in 1992 (IPCC 1992).

22.3.4 Urban Waste Disposal

Globally urban cities are having problems of waste disposal and its management. The obsolete and unlawful ways of disposing urban wastes affect resident communities in all countries (Fazzo et al. 2017; Marsili et al. 2009). In African cities, only approximately 20% of the urban wastes are likely to be landfilled, while the remaining ones are deposited at unapproved dumpsites. It is also noted that Africa is one of the main hub for unlawful trans-boundary trade of urban and harmful wastes known as electronic wastes (e-wastes) from developed countries (Marsili et al. 2009). E-wastes contain known harmful substances that could be directly released after deposit or during the reprocessing cycle. Unsafe reprocessing methods of e-wastes in Africa, where 75% of EU and 80% of USA e-wastes, respectively, are unlawfully traded affect workers seriously especially children and women (Perkins et al. 2014) and affects soil organisms (Ewuim et al. 2014).

Paper waste, leftover food, and several other toxic substances such as sewage sludge, healthcare waste, batteries, paint, asbestos, and radioactive substances are generally constituted as waste (Pohl et al. 2008; Samson et al. 2011; Shankar 2017; Qasima et al. 2020). These wastes are capable of polluting both the environmental surface and groundwater (Taylor and Allen 2006). Animals and vegetable wastes are mostly shredded by insects prior to further decomposition by bacteria and fungi (Kunz 2009). Waste is produced through a day-to-day activities of man and is grouped into three categories: solid, liquid and gas (Taylor and Allen 2006). Both gas and liquid wastes contaminate a nearby environment and harmfully affect human health. Some solid wastes decompose faster and cause environmental contamination by emitting offensive smell and secondarily attracting arthropod vectors, such as insects (Zhao et al. 2011; Alam and Ahmade 2013; Vidyavathy 2018; Nor Faiza et al. 2019). Many synanthropic dipteran insects reproduce in organic wastes and are referred to as pests.

Cockroach invasion is highest near waste dumps (Koehler et al. 1987) and mosquitoes (*Aedes aegypti*) are comfortably breeding in containers found in waste dumps (Pérez-Guerra et al. 2009).

Among insect vectors (Diptera, Dictyoptera and Coleoptera) found in waste dumps, dipteran families; Blattidae, Serabidae, Muscidae, Fannidae, Culicidae, Calliphoridae, and Psychodidae are very abundant in the urban cities (Montoya et al. 2009) dominating the population with over 80% (Ahmed 2011). About 50 species of dipteran flies are responsible for human diseases, 21 species of them are associated with organic wastes causing digestive diseases in man because of their synanthropy, unrestrained behavior and desire to always visit organic wastes (Nurita and Hassan 2013) just as cockroaches are normally found in hospital waste and spreading diseases to humans (Pai et al. 2003). Very interestingly, but not generally known is that blow flies in the family Calliphoridae has about 1500 species in about 150 genera globally (Verves 2005; Bharti 2011; Kosmann et al. 2013). Among which, some are vectors of various human diseases (Sawabe et al. 2011; Adenusi and Adewoga 2013; Hemmati et al. 2018), and some fly larvae cause serious infestations on man and animals by feeding on their tissues, a condition known as myiasis (Chan et al. 2005; Wang et al. 2012; Ruiz-Zapata et al. 2019). Some blow fly species are noted to parasitize humans (Hemmati et al. 2018), while some are forensically relevant in medical and criminal (forensic) fields (Yadong et al. 2011; Wang et al. 2017a, b).

In Africa, management of urban waste is poor and very far from cueing into the sustainable practices of managing wastes to curb the propagation of harmful insects. Urban wastes in many cities of Africa are hubs for diversity of insects of medical importance (Abajue et al. 2020). The diversity of insects associated with waste dumps is detrimental to man and his environments, thus impeding sustainable development at all sectors of social and economic fronts. The implication of proliferations of waste dumps in the cities of Africa is that peoples' health is threatened by insects and other arthropods-borne diseases. African environments are usually inundated with pesticides used against pests which also indirectly affect flora and fauna diversity in favor of pests' resurgence, resistance, secondary outbreak, and establishment of invasive alien species (Shine et al. 2000).

22.3.5 Drought, Flooding, and Climate Change in Africa

The climate is a critical element that is described as a phenomenon, which includes environmental factor variations (temperature, humidity, and precipitation) over several years. Due to rise in temperatures, climate excesses, increased carbon dioxide and further greenhouse gases in addition to different precipitation designs, global food production would be on a severe threat (Shrestha 2019) which is already taking tolls in Africa. The basic reason for global warming is elevated concentrations of greenhouse gases (carbon dioxide, methane, and nitrous oxide) in atmosphere which are prompted by human activities, such as burning of fossil fuels and land-use change (Yoro and Daramola 2020). Increased carbon dioxide in the atmosphere, higher temperature, and lowered soil moisture as the leading causes of climate change could meaningfully impact the population dynamics of insects and this could favor insect pests at the detriment of agricultural crops (Fand et al. 2012). Because of climate change; new ecological niches for insect pests are created, providing chances for pests to build and spread to new geographical regions (FAO 2008).

Rainfall fluctuations, volume, and incidence are vital indicators of changing climate. In Africa, the occurrence of rainfall has reduced, but intensity has increased. This kind of rainfall has led to droughts and floods respectively. Insect species that hide in soils are directly impacted by rainfall. In Africa, rainfall regularly causes flooding and extended stagnation of water that affects insects' diapause. Heavy rains and flooding wash insect eggs and larvae away from their habitats (Shrestha 2019). Rapid growth of wireworm populations in upper part of soil is due to increase of seasonal rainfall which has been noted (Staley et al. 2007) and herbivorous insects are also affected by drought via noted mechanisms. For instance, dry climate offer environmental conditions suitable for development of herbivorous insects, while drought-stressed plants entice certain insect species.

Drought-stressed plants are more vulnerable to insect attack because of a decrease in production of secondary metabolites (Yihdego et al. 2019). Altered temperature and rainfall could form a predictable change in climate which will define the distribution, survival, and reproduction of insect species within its future (Fand et al. 2012). This can lead to spread of insect pests toward new areas and a shift to areas, where their host plants are found. In Africa, farmers have witnessed unadorned pest problems in addition to issues such as poor soil properties and environmental structure that would be impacted by climate change (Lastuvka 2010). Climate change causes higher risk of aggressive alien insect species that are habitually field, stored-product, forestry, household, or organizational pests as well as vectors of various diseases of plants and animals (Ward and Masters 2007).

Flooding, on the other hand, negatively affects termites (Forschler and Henderson 1995; Schuurman 2005) and other wood-dwelling arthropods (Ballinger et al. 2010), and this, in turn, has the potential to further influence decomposition rates. In temporary pools, mosquito larvae (e.g., *Culiseta longiareolata*) are typically statistically dominant during a stable season at little elevations, but species richness drops abruptly after a flood washes out invertebrate residents dominated by insects in pools, thus leading to species richness close to zero, but rises as the season progresses (Ward and Blaustein 1994; Duchet et al. 2017). Insects as a tool for sustainable environment in Africa in (Fig. 22.1) shows they are relevant as pollinators of crops and other plants, serve as food for man and animals, act as bio-monitoring agents and organic decomposers while farming activity, use of fossil fuel, flaring of gas, urban waste disposal, drought flooding and climate change are factors that impact environmental sustainability.

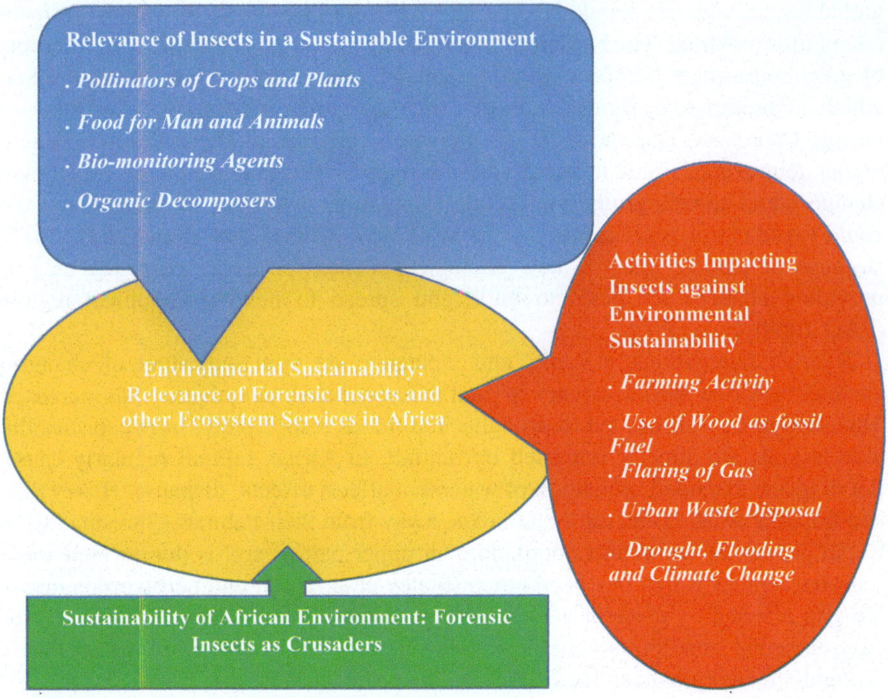

Fig. 22.1 Insects as a tool for sustainable environment in Africa

22.4 Environmental Sustainability in Africa: Forensic Insects as Crusaders

The aquatic and terrestrial environment in Africa is a source of sustenance for a greater percentage of her citizen, as they depend on exploring the natural resources for agriculture, herding, fishing, and hunting as a means of livelihood. These practices are not sustainable in line with the United Nations sustainable development goals that center on getting natural sources of livelihood cautiously and maintain a quality and sustainable environment (Todaro and Smith 2015; Oyebanji et al. 2017). Environmental sustainability requires awareness and attitudinal behavior to one's needs bearing in mind that any abuse on the environment today will take revenge on the future generation. Environmental sustainability appreciates population growth, sustainable cultivation, and yield, sustainable disposal of wastes as well as economic development and industrialization that are anchored on renewable resources. Concerned organizations have continuously placed a premium on any developmental agenda to ensure environmental sustainability by adhering to clean water and sanitation, sustaining healthy ecosystems, reduce activity culminating to climate change, and refining other aspects of livelihood (UNDP 2014; World Bank 2014, 2015; FAO 2016).Technically, environmental sustainability tries to envisage how different nations exert pressure on their environment via industrial progressions that are linked to developmental and consumption activities in relation to human populations. Beyond social and political contemplations; the realities of the impacts of climate change and environmental degradation have started showing its ugly face in Africa. These include increased problems of shortage of water, floods, droughts, and desertification that are threatening food security. In the past few years, many urban and rural communities have experienced floods, a case study of Nigeria in the year 2012 (Odermorho 1993; Adejumo and Adejumo 2014), and currently unaccounted lives and properties that are lost this year in Nigeria (Anambra, Bayelsa, Delta, Kogi, and other states). Notwithstanding the enormous natural resources in Africa, the prospects for environmental sustainability and sustainable growth are unsatisfied, because its future is compromised by environmental degradation and socio-political instability. Environmental degradation generally signifies damage caused on the biosphere due to hazardous and despicable activities of human beings. This culminates to a situation, whereby natural resources are consumed faster than it can replenish itself and when human beings are degrading the ecosystem in the process of development.

Having noted the causes and consequences of anthropogenic activities on biodiversity, it is imminent that different mitigation approaches geared toward showcasing, the importance of biodiversity to man and the reason to conserve them should be sought after. In Africa, insects and other related arthropods are seen as pests, nuisance, and vectors of different diseases with a negligible number recognizing their usefulness in food security (van Huis 2015). In this section, the relevance of insects in criminal investigation related to questionable death is promoted as a means of informing African people the importance of insects in our environment, hence their conservation in a sustainable environment. The choice of insects of forensic relevance as a crusader for environmental sustainability among other ecosystem services played by insects in our environment is that every individual needs security of lives and properties. Hence, every individual will be concerned to safe and secure environment devoid of criminality culminating to death. Environmental sustainability which denotes preservation and lucid exploitation of nature and its resources in a way to preserve the veracity of every ecosystem and ensure to protect biodiversity and cautiously avoid degrading an environmental (Gbenda 2012) should be anchored on all life forms. The novelty of insects in criminal investigation is solidly growing and has become relevant in the developed world, while it is still at the elementary stage in Africa. The use of insects in forensic study is referred to as forensic entomology. Forensic entomology evaluates the relationship between dead animals including human beings and insects found on them to substantiate cases related to murder, suicide, neglect, accident, and pouching (Villet 2011). Remarkably, only few scientific researchers related to insects study know the potentials of forensic entomology and just few cases have been used in prosecutions in South Africa, Zimbabwe, and Sudan only (Blair 1979; Louw and van der Linde 1993; Gifford 2005a, b; Williams and Villet 2006a, b; Gurafi and Mohamed (2013)). Insects collected from dead animals is used in forensic investigations to determine an estimate of time when the animal or person died. The time is usually referred to as postmortem interval in medicine and time of colonization (TOC) or period of insect activity in entomology (Smith 1986; Byrd and Castner 2010).

Forensic entomology is grouped into three (forensic urban entomology, forensic stored products entomology, and criminal entomology). Criminal entomology generally referred to as forensic entomology is interested in dead animals or human beings to ascertain when, where and what killed them using insects and other arthropod relatives found on their bodies as pieces of evidence that would be useful in law court. Thus, forensic entomology plays a vital role in the delivery of justice and help curb criminality in the society which is one of the components of sustainable development goals (SDGs) of the United Nations in 2015. The current status of environmental degradation in Africa contributing to global climate change requires that the benefits of insects and other arthropods should be publicized. In view of this, insects that are relevant to forensic investigation are sieved out as the vanguard for a sustainable environment, in Africa. Interestingly, some studies geared toward identifying the diversity of forensic insects in Africa is gradually spreading in all the five regions of Africa: Central Africa, Eastern Africa, Northern Africa, Southern Africa, and Western Africa (Table 22.1). Thus, there is hope that using the diversity of insects of forensic relevance in the law courts of these regions, Africa will soon be a major stakeholder in the crusade for sustainable environment.

22.5 Conclusion

Insects are very relevant in every ecosystem, as they pollinate crops and vegetables, serve as food for man and animals, and shred organic matters that are locked in dead plants and animals for soil nutrients enrichment. Insects remain the dominant fauna that are saddled with decomposition of animal carcasses, including human cadavers. Their arrival and succession on cadavers are very useful in estimating time of cadaver's death. This singular role has been outstanding in forensic investigation relating to questionable decease cases. In African, diversity of insects are abound, but its environment is threatened by farming activities, burning of fossil fuels, flaring of gas, and poor disposal of urban wastes. These activities are orchestrated by human beings, and are seriously contributing to global climate change which culminate to biodiversity loss. To restore, or at least reduce further degradation of African environment, especially against diversity of beneficial insects, the importance of sustainable use of natural resources should be paramount in all structural developmental projects. Thus, this paper campaigns for sustainable environment in Africa by publicising the relevance of insects in forensic investigation of homicides to the people, especially at the grassroots. Because every individual is interested in security of lives and properties; as enshrined in the United Nations Sustainable Development Goals, they will be curious about the knowledge and acceptance of insects as a tool for curbing criminality. The inspiration will make them to lessen the activities that degrade the ecosystems against biodiversity. Definitely, Africa as enthusiastic users

Table 22.1 Diversity of insects of forensic relevance in Africa

Insect species	CA	EA	NA	SA	WA	Insect species	CA	EA	NA	SA	WA
Angionychus lividus					+	*Monopsis argillacea*					+
Anthomyia sp.			+		+	*Morellia nilotica*					+
Aphaenogaster senilis	+		+			*Morellia prolectata*					+
Aphids sp.	+					*Musca domestica*	+		+	+	+
Aplomya sp.			+		+	*Musca sorbens*			+		+
Apogonia nitidula					+	*Musca* sp.			+	+	+
Atherigona occidentalis					+	*Muscina stabulans*	+		+		+
Atherigona soccata				+		*Muscina* sp.					
Atherigona sp.					+	*Mymicaria striata*					+
Attagenus fasciatus			+			*Myremacaris senaarensis*					+
Bledius incertus					+	*Myrmecia* sp.					+
Brachymeria sp.			+			*Myrmeleon* sp.			+		
Buphonella sp.					+	*Nasonia* sp.			+		
Calliphora sp.					+	*Necrobia ruficolis*			+		+
Calliphora vicina	+			+	+	*Necrobia rufipes*			+	+	+
Calliphora vomitoria	+		+		+	*Necrobia* sp.					+
Camonutus sp.					+	*Necrodes littoralis*			+		
Camponotus maculatus					+	*Necrophorus* sp.					+
Camponotus periisi					+	*Nicrophorous tomentus*					+
Camponotus sericeus					+	*Nicrophorus* sp.					+
Carpophilus hemipterus			+			*Ocypus raffrayi*					+
Ceratocoris bucephallus					+	*Ocypus* sp.					+
Chalcis sp.			+			*Oecophylla longinoda*	+				+
Chlorichaeta albipennis			+			*Oecophylla smaragdina*					+
Chrysis sp.			+			*Omosita colon*			+		

(continued)

Table 22.1 (continued)

Insect species	CA	EA	NA	SA	WA	Insect species	CA	EA	NA	SA	WA
Chrysomya albiceps			+	+	+	*Ontholestes* sp.			+		
Chrysomya bezziana			+			*Onthophagus crassicollis*				+	
Chrysomya chloropyga				+	+	*Onthophagus taurus*					+
Chrysomya laxifrons	+					*Onthophagus vacca*				+	
Chrysomya marginalis		+	+	+	+	*Ophyra aenescens*					+
Chrysomya megacephala			+	+	+	*Ophyra* sp.	+				
Chrysomya putoria			+	+	+	*Ophyro ignava*			+		+
Chrysomya regalis					+	*Pachylister inaequalis*					+
Chrysomya rufifacies			+		+	*Parasarcophaga aegyptiaca*			+		
Chrysomya sp.	+				+	*Paratrechina longicornis*	+				
Chrysomyza africana					+	*Paratrechina* sp.					+
Chrysosoma tenuipenna					+	*Phaenicia (=L) sericata*					+
Coelalysia sp.			+			*Pheidole megacephale*				+	+
Conficata sp.					+	*Pheidole* sp.	+				+
Cosmina griscoviridis					+	*Philonthus* sp.			+		+
Cremastogaster megaponera					+	*Philonthus stragulatus*			+		
Cremastogaster sp.					+	*Phormia regina*			+		+
Creophilus maxillosus			+			*Phortica* sp.					+
Culex pipiens					+	*Physiphora alceae*			+		
Cynomyopsis cadaverina					+	*Phytomyza* sp.			+		
Cynomyopsis cadaverina					+	*Piophila casei*	+		+		+
Dermestes ater			+	+		*Platysoma* sp.					+
Dermestes atomarieus			+			*Pododula ancisa*		+			+
Dermestes frichii			+		+	*Protophormia terraenovae*					+
Dermestes lardarius					+	*Psychoda* sp.			+		

Dermestes maculatus	+		+	+	+	*Pullimosina heteroneura*			+		
Dermestes peruvianus	+		+			*Rhinia apicalis*					+
Dermestes sp.			+	+	+	*Rhizopertha dominica*					+
Dermestes undulates			+			*Rhycomyia aenea*					+
Diasemopsis sp.					+	*Rhyncomya* sp.					+
Dichaetochilus vagans					+	*Riptortus dentipes*					+
Diopsis apicalis					+	*Sacrophaga bercaea*			+		
Discomyza incurve					+	*Sagarizera analis*					+
Discomyza sp.					+	*Saprinus aeneus*			+		
Dorylus affinis					+	*Saprinus bicolor*				+	
Drosophila replete			+			*Saprinus bonnairei*			+		
Drosophila sp.				+		*Saprinus caerulescens*			+		
Empoasca sp.			+			*Saprinus chalcites*			+		
Endustomus senegalensis					+	*Saprinus cruciatus*				+	
Eristalis sp.			+			*Saprinus furvus*			+		
Fannia canicuaris					+	*Saprinus semipunctatus*			+		
Fannia sp.			+		+	*Saprinus semistriatus*			+		
Graphomyia maculta			+			*Saprinus splendens*				+	
Gymnogrylus lucens.					+	*Saprinus* sp.			+	+	
Gymnopleurus sp.					+	*Sarcophaga africa*			+		+
Haematobia minuta,					+	*S. argyrostoma*	+		+		
Hemipyrellia fernandica	+				+	*Sarcophaga calcifera*				+	
Hemipyrellia sp.	+					*Sarcophaga canaria*	+				+
Hermatia illucens					+	*Sarcophaga cruentata*					+
Hetronychus mosambicus					+	*Sarcophaga destructor*			+		
Hipocacculus sp.					+	*Sarcophaga exuberans*					+
Hister cadaverinus					+	*S. haemorrhoidalis*					+

(continued)

Table 22.1 (continued)

Insect species	CA	EA	NA	SA	WA	Insect species	CA	EA	NA	SA	WA
Hister monitor					+	*Sarcophaga hertipes*			+		
Hister quadrinotatus					+	*Sarcophaga inzi*					+
Hister sp.			+		+	*Sarcophaga muscaria*			+		
Hister unicolor			+		+	*Sarcophaga nodosa*		+			+
Hycleus lunatus				+		*Sarcophaga villa*					+
Hydrotae adentipes			+			*Scatophaga* sp.			+		+
Hydrotaea sp.				+		*Sepedomyia* sp.					+
Hyparpalus sp.					+	*Sepsis* sp.					+
Hypocacculus buqueti					+	*Silpha odscura*	+				
Hypocacculus sp.					+	*Silpha rugosa*			+		
Hypocalus sp.					+	*Staphylinus violaceous*					+
Iridomyrmex purpureus					+	*Stomorhina cribrata*					+
Ischnotrachelus sp.					+	*Stomoxys calcitrans*			+		+
Isomyia dubiosa					+	*Stomoxys evanida*					+
Isomyia sp.					+	*Synthesiomyia nudiela*			+		
Larinus haroldi					+	*Synthesiomyia nudiseta*			+		
Larra sp.					+	*Synydas* sp.					+
Lasioderma sp.			+			*Tachyphanes* sp.					+
Leistus sp.			+			*Tenebrio molitor*					+
Lepisma saccharina			+			*Teritriosoma* sp.					+
Leptothorax acervorum	+					*Tetramorium* sp.	+				
Liosarcophaga emmrichiana				+		*Thanatophilus micans*				+	
Lucilia caesar	+				+	*Thanatophilus mutilates*				+	
Lucilia cuprina			+	+		*Thanatophilus rugosus*			+		
Lucilia illustris	+				+	*Thanatophilus sinuatus*			+		

Lucilia infernalis					+	*Thaumastocera* sp.					+
Lucilia sericata	+		+		+	*Thelairosoma* sp.					+
Lucilia Silvarum			+			*Tinearia alternate*			+		
Lucilia sp.	+				+	*Trachyderma hispidae*			+		
Margarinotus brunneus			+		+	*Trigona carbonaria*					+
Margarinotus ventralis			+			*Trirhithum* sp.					+
Medetera sp.			+			*Trox cadaverinus*					+
Megaselia scalaris			+		+	*Trox hispidus*			+		
Megaselia sp.					+	*Trox sabulosus*			+		
Melipona beecheii					+	*Trox scaber*			+		
Mesostina puncticollis			+			*Trox sceber*			+		+
Messor sp.					+	*Wasmannia auropuunctata*					+
Messorgalla sp.					+	*Wolfahartia nuba*			+		+
Monomonium minimum					+	*Wolfhartia* sp.		+			+
Monomorium pharaonis	+					*Zophosis abbreviate*		+			+
Monomorium sp.	+		+			*Zophosis quadrilineata*					+
						Zophosis sp.					+

Authors: Abajue et al. (2013), Abajue et al. (2014), Abajue et al. (2015), Abd El-bar et al. (2016), Abd El-Bar and Sawaby (2011), Ahmed and Joseph (2016), Ahmed et al. (2016), Akpa et al. (2021), Benkenana Naima (2020), Bensaada et al. (2014), Daniel et al. (2017), Dao et al. (2019), Ekanem and Dike (2010), Ekrakene and Iloba (2011), El-bassiony (2020), El-Mehy et al. (2015), Ewuim (2007), Feugang et al. (2011), Feugang et al. (2012), Guerroudj and Berchi (2019), Iloba and Odigie (2006), Iloba and Odo (2020), Joshua and Noutcha (2019), Koffi et al. (2018), Kpama-Yapo et al. (2020), Mabika et al. (2014), Mashaly et al. (2019), Ndueze et al. (2013), Njau et al. (2016), Odo (2020), Odo and Iloba (2020), Onyishi et al. (2020), Padonou et al. (2017), Saifi et al. (2017), Salem et al. (2015), Sawaby et al. (2018), Sebastião and Castro (2019), Severin et al. (2021), Tantawi et al. (1996), Tantawi et al. (2010), Tembe et al. (2021), Tembe and Mukaratirwa (2021), Williams and Villet (2006a, b)

Note: *CA* Central Africa, *EA* Eastern Africa, *NA* Northern Africa, *SA* Southern Africa, *WA* Western Africa, + insect species present

of natural resources will start to be cautious in utilizing them in view to conserve nature, and be part of the vanguards, for environmental sustainability.

References

Abajue MC, Ewuim SC, Akunne CE (2013) Insects associated with decomposing pig carrions in Okija, Anambra State, Nigeria. Bioscience 1(1):54–59

Abajue MC, Ewuim SC, Akunne CE (2014) Preliminary checklist of flies associated with pig carrions decomposition in Okija, Anambra State, Nigeria. Anim Res Int 11(1):1899–1904

Abajue MC, Ewuim SC, Akunne CE (2015) Preliminary checklist of beetles associated with pig carrions decomposition in Okija, Anambra State, Nigeria. Anim Res Int 12(2):2166–2170

Abajue MC, Babatunde BB, Okwelogu SI (2020) Composition of resident macro-invertebrates in refuse dumps at University of Port Harcourt, Nigeria. Int J Sci Technol 8(1):21–31

Abd El-bar MM, Sawaby RF (2011) A preliminary investigation of insect colonization and succession on remains of rabbits treated with an organophosphate insecticide in El-Qalyubiya governorate of Egypt. Forensic Sci Int 208(2011):e26–e30. https://doi.org/10.1016/j.forsciint.2010.10.007

Abd El-bar MM, Sawaby RF, El-Hamouly H, Hamdy R (2016) A preliminary identification of insect successive wave in Egypt on control and zinc phosphide-intoxicated animals in different seasons. Egypt J Forensic Sci 6:223–234. https://doi.org/10.1016/j.ejfs.2016.05.004

Adebimpe AO, Kayode EP (2018) Urban growth issues and environmental sustainability in Nigeria. Coven J Res Built Environ 6(2):2000–2008

Adejumo V, Adejumo O (2014) Prospects for achieving sustainable development through the millennium development goals in Nigeria. Europ J Sustain Dev 3(1):33–46

Adenusi AA, Adewoga TOS (2013) Human intestinal parasites in non-biting synanthropic flies in Ogun State, Nigeria. Travel Med Infect Dis 11:181–189

Adhikari M, Paudel KP, Martin NR Jr, Gauthier WM (2005) Economics of dairy waste use as fertilizer in Central Texas. Waste Manag 25:1067–1074. https://doi.org/10.1016/j.wasman.2005.06.012

Agea JG, Biryomumaisho D, Buyinza M, Nabanoga GN (2008) Commercialization of Ruspolia nitidula (Nsenene grasshoppers) in Central Uganda. Afr J Food Agric Dev 8:319–332. https://doi.org/10.4314/ajfand.v8i3.19195

Ahmed A (2011) Insect vectors of pathogens in selected undisposed refuse dumps in Kaduna town, northern Nigeria. Sci World J 6:21–26

Ahmed A, Joseph SG, Naman K (2016) Dry season study of necrophagous insects associated with *Cavia porcellus* (Guinea pig) carcass in Kaduna, Nigeria. Int J Entomol Res 04(03):99–106

Ahmed A, Joseph SG (2016) Estimation of postmortem interval using the blowfly *Phaenicia* (*Lucilia*) *sericata* (Diptera: Calliphoridae) in Kaduna, northern Nigeria. Int J Res Med Sci 4(8):3417–3420. https://doi.org/10.18203/2320-6012.ijrms20162304

Akhtar Y, Isman MB (2018) Insects as an alternative protein source. Food Sci Nutr 2:263–288

Akpa HO, Tongjura JDC, Amuga GA, Ombugadu RJ (2021) Postmortem evaluation off rabbit carcasses using insect populations in Keffi Nasarawa state, Nigeria. Europ J Biol Biotechnol 2(6):6–9

Alam P, Ahmade K (2013) Impact of solid waste on health and the environment. Int J Sustain Dev Green Econ 2:165–168

Albrecht M, Schmid B, Hautier Y, Mueller CB (2012) Diverse pollinator communities enhance plant reproductive success. Proc R Soc B-Biol Sci 279:4845–4852

Alamu OT, Amao OA, Nwokedi IC, Oke AO, Lawa O (2013) Diversity and nutritional status of edible insects in Nigeria: a review. Int J Biodivers Conserv 5:215–222

Axtell RC (1986) Fly management in poultry production: cultural, biological, and chemical. Poult Sci 65:657–667

Ayemele AG, Muafor FJ, Levang P (2016) Indigenous management of palm weevil grubs (Rhynchophorus phoenicis) for rural livelihoods in Cameroon. J Insects Food Feed 3:43–50. https://doi.org/10.3920/JIFF2016.0002

Ballinger A, Mac Nally R, Lake PS (2010) Decay state and inundation history control assemblage structure of log-dwelling invertebrates in floodplain forests. River Res Appl 26:207–219

Beard RL, Sands DC (1973) Factors affecting degradation of poultry manure by flies. Environ Entomol 2:801–806

Bensaada F, Nadir BA, Saifi M, Doumandji S (2014) First data on insects to forensic interest in the region of Gouraya, Algeria. Int J Agric Innovat Res 3(2):461–466

Beskin KV, Holcomb CD, Cammack JA, Crippen TL, Knap AH, Sweet ST, Tomberlin JK (2018) Larval digestion of different manure types by the black soldier fly (Diptera: Stratiomyidae) impacts associated volatile emissions. Waste Manag 74:213–220. https://doi.org/10.1016/j.wasman.2018.01.019

Bertone MA, Green JT, Washburn SP, Poore MH, Watson DW (2006) The contribution of tunneling dung beetles to pasture soil nutrition. Forage Grazing Lands 4:1–12

Bharti M (2011) An updated checklist of blowflies (Diptera: Calliphoridae) from India. Halteres 3: 34–37

Biesmeijer JC, Roberts SP, Reemer M, Ohlemuller R, Edwards M, Peeters T, Schaffers AP, Potts SG, Kleukers RJ, Thomas CD, Settele J (2006) Parallel declines in pollinators and insect-pollinated plants in Britain and the Netherlands. Science 313(5785):351–354

Blair B (1979) Insects of use and of nuisance to a policeman. Outpost 57(5):24–26

Bornemissza GF (1976) The Australian dung beetle project 1965-1975. Austral Meat Res Committ Rev 30:1–30

Bell JR, Blumgart D, Shortall CR (2020) Are insects declining and at what rate? An analysis of standardised, systematic catches of aphid and moth abundances across Great Britain. Insect Conserv Divers 13:115–126

Bukkens SGF (1997) The nutritional value of edible insects. Ecol Food Nutr 36:287–319

Byrd JH, Castner JL (2010) Forensic entomology: the utility of arthropods in legal investigations, 2nd edn. CRC Press, Boca Raton, FL, p 681

Chan JCM, Lee JSW, Dai DLK, Woo J (2005) Unusual cases of human myiasis due to Old World screwworm fly acquired indoors in Hong Kong. Transc R Soc Trop Med Hyg 99:914–918

Chang CCY, Janzen HH (1996) Long-term fate of nitrogen from annual feedlot manure applications. J Environ Qual 25:785–790

Cherret JM (1989) Leaf-cutting ants. Biogeographical and ecological studies. Ecosystem of the World, Tropical rain forest ecossystem. Elsevier, New York, pp 473–488

Čičková H, Newton GL, Lacy RC, Kozanek M (2015) The use of fly larvae for organic waste treatment. Waste Manag 35:68–80. https://doi.org/10.1016/j.wasman.2014.09.026

Čičková H, Pastor B, Kozanek M, Martinez-Sanchez A, Rojo S, Takac P (2012) Biodegradation of pig manure by the housefly, *Musca domestica*: a viable ecological strategy for pig manure management. PLoS One 7:e32798. https://doi.org/10.1371/journal.pone.0032798

Coviella CE, Trumble JT (1999) Effects of elevated atmospheric carbon dioxide on insect-plant interactions. Conserv Biol 13(4):700–712

Curtis PG, Slay CM, Harris NL, Tyukavina A, Hansen MC (2018) Classifying drivers of global forest loss. Science 361:1108–1111

Currano ED, Wilf P, Wing SL, Labandeira CC, Lovelock EC, Royer DL (2008) Sharply increased insect herbivory during the Paleocene–Eocene thermal maximum. Proc Natl Acad Sci U S A 105:1960–1964

Daniel CA, Midgley JM, Villet MH (2017) Determination of species and instars of the larvae of the Afrotropical species of *Thanatophilus* Leach, 1817 (Coleoptera, Silphidae). Afr Invertebr 58(2): 1–10. https://doi.org/10.3897/AfrInvertebr.58.12966

Dao H, Louis Aboua LRN, Yalamoussa TUO, Koffi AF, Yapo CEY (2019) Diversity of the entomological fauna associated with PIG carcasses (*Sus Scrofa domesticus* L.) exposed in the open air in the Sudano-Guinean and Sub-Sudanese zones of Côte d'Ivoire. J Entomol Zool Stud 7(2):1338–1344

da Rocha JRM, Almeida JR, Lins GA, Durval A (2011) Insects as indicators of environmental changing and pollution: a review of appropriate species and their monitoring. HOLOS Enviro 10(2):250–262

Davis AJ (2000) Does reduced-impact logging help preserve biodiversity in tropical rainforests? A case study from Borneo using dung beetles (Coleoptera: Scarabaeoidea) as indicators. Environ Entomol 29(3):467–475

De Foliart GR (1997) An overview of the role of edible insects in preserving biodiversity. Ecol Food Nutr 36:109–132

Duchet C, Moraru GM, Segev O, Spencer M, Hayoon AG, Blaustein L (2017) Effects of flash flooding on mosquito and community dynamics in experimental pools. J Vector Ecol 42(2): 254–263

DeLucia EH, Castee CL, Nabity PD, O'Neill BF (2008) Insects take a bigger bite out of plants in a warmer, higher carbon dioxide world. PNAS 105(6):1781–1782. https://doi.org/10.1073/pnas.0712056105

Edmonds L, Gollehon N, Kellogg RL, Knight L, Lander C, Lemunyon J, Meyer D, Moffitt DC, Schaefer J (2003) Costs associated with development and implementation of comprehensive nutrient management plans, part I – nutrient management, land treatment, manure and wastewater handling and storage, and recordkeeping. United States Department of Resources Natural Resources Conservation Service, Washington, DC

Ekanem MS, Dike MC (2010) Arthropods succession on pig carcasses in southeastern Nigeria. Pap Avuls de Zool 50(35):561–570

Ekrakene T, Iloba BN (2011) One death, many insect species, yet one insect generation. J Entomol 8:27–39

El-Mehy SA, Soliman E, Hassan NA, Alrouf TA (2015) Application of entomology in some medicolegal issues. Austin J Forens Sci Criminol 2(4):1036

El-Bassiony GM (2020) Blowflies (Diptera: Calliphoridae) as forensic indicators in Egypt with special reference to the development data of *Lucilia Cuprina* (Wiedemann). Sumerianz J Biotechnol 3(7):43–51

Erickson MC, Islam M, Sheppard C, Liao J, Doyle MP (2004) Reduction of *Escherichia coli* O157: H7 and *Salmonella enterica serovar* Enteritidis in chicken manure by larvae of the black soldier fly. J Food Prot 67:685–690. https://doi.org/10.4315/0362-028x-67.4.685

Eweje G (2006) Environmental costs and responsibilities resulting from oil exploration in developing countries: the case of the Niger Delta of Nigeria. J Bus Ethics 69:27–56. http://www.mendeley.com/catalog/environmental-costs-responsibilities-resulting-oil-exploitationdeveloping-countries-case-niger-delta-nigeria/

Ewuim SC, Akunne CE, Abajue MC, Nwankwo EN, Faniran OJ (2014) Challenges of E-waste pollution to soil environment in Nigeria. A review. Anim Res Int 11(2):1976–1981

Ewuim SC (2007) Dipteran fauna of an abattoir and its contiguous fallow plot in a Guinea savannaecosystem. Anim Res Int 4(1):626–629

Fajer ED, Bowers MD, Bazzaz FA (1989) The effects of enriched carbon dioxide atmospheres on plant-insect herbivore interactions. Science 243:1198–1199

Fand BB, Kamble AL, Kumar M (2012) Will climate change pose serious threat to crop pest management: a critical review. Int J Sci Res 2:1–14

FAO (2008) Climate related transboundary pests and diseases. http://www.fao.org/3/a-ai785e.pdf. Accessed on 13 May 2022

FAO (2016) The state of food insecurity in the world 2015. Food and Agriculture Organization of the United Nations, Rome. http://www.fao.org/hunger/en/

Fazzo L, Minichilli F, Santoro M, Ceccarini M, Seta MD, Bianchi F, Comba P, Martuzzi M (2017) Hazardous waste and health impact: a systematic review of the scientific literature. Environ Health 16(107):1–11. https://doi.org/10.1186/s12940-017-0311-8

Feugang YFD, Djieto-Lordon C, Ateba AJC, Bilong BCF (2011) A preliminary study of arthropod associated with carrion in Yaounde, Cameroon: a first step in forensic entomology in Central Africa. J Ecol Nat Environ 3(6):215–220

Feugang YFD, De Coninck E, Bilong BCF, Hubrecht F, Djiéto-Lordon C, Braet Y, Bourguignon L (2012) First records on five species of Calliphoridae (Diptera) reared from maggot collected on rat carrions corpse during a forensic entomology experiment in the campus of the University of Yaounde I-Cameroon. Int J Biosci 2(3):75–80

Folgarait PJ (1998) Ant biodiversity and its relationship to ecosystem functioning: a review. Biodivers Conserv 7:1221–1244

Forschler BT, Henderson G (1995) Subterranean termite behavioral reaction to water and survival of inundation: implications for field populations. Environ Entomol 24:1592–1597

Gbenda J (2012) A keligo-cultural respective on ecological sustainability in Nigeria. African dynamics of. Soc Sci Res 2 2(2):1–8

Ghini S, Fernandez M, Pico Y, Marin R, Fini F, Manes J, Girotti S (2004) Occurrence and distribution of pesticides in the province of Bologna, Italy, using honeybees as bioindicators. Arch Environ Contam Toxicol 47:479–488

Gifford G (2005a) It's science vs Moodley. Bullets, flies, spiders and chicken livers—all used to see if Leigh's killer is lying. The Star July 28:1

Gifford G (2005b) Small black spider tells Leigh's story. Pretoria News 1

Grab H, Branstetter MG, Amon N, Urban-Mead KR, Park MG, Gibbs J, Blitzer EJ, Poveda K, Loeb G, Danforth BN (2019) Agriculturally dominated landscapes reduce bee phylogenetic diversity and pollination services. Science 363(6424):282–284

Grossman E (2014) As dairy farms grow bigger, new concerns about pollution. Yale Environment 360. https://e360.yale.edu/features/as_dairy_farms_grow_bigger_new_concerns_about_pollution

Guerroudj FZ, Berchi S (2019) Composition of the necrophagous fauna on a rabbit corpse in Constantine, Northeastern Algeria. Energy Proc 157:1083–1088

Guimarães RMS, Juliano J, Corbi JJ, Jacobucci GB (2019) Aquatic insects as bioindicators of heavy metals in sediments in Cerrado streams. Limnetica 38(2):575–586. https://doi.org/10.23818/limn.38.33

Gurafi LMA, Mohamed EAE (2013) Insects as forensic indicators: analysis of some cases. Sudan J Sci 5(2):15–20

Habel JC, Trusch R, Schmitt T, Ochse M, Ulrich W (2019) Long-term large-scale decline in relative abundances of butterfly and burnet moth species across southwestern Germany. Sci Rep 9: 14921

Habel JC, Linsenmair KE (2019) Arthropod decline in grasslands and forests is associated with landscape-level drivers. Nature 574(7780):671–674

Hallmann CA, Sorg M, Jongejans E, Siepel H, Hofland N, Schwan H et al (2017) More than 75 percent decline over 27 years in total flying insect biomass in protected areas. PLoS One 12: e0185809

Hamilton JG, Dermody O, Aldea M, Zangerl AR, Rogers A, Berenbaum MR, DeLucia EH (2005) Anthropogenic changes in tropospheric composition increase susceptibility of soybean to insect herbivory. Environ Entomol 34:479–485

Heliovaara K, Vaisanen R (1990) Heavy-metal contents in pupae of *Bupalus piniarus* (Lepidoptera: Geometridae) and *Panolis flammea* (Lepidoptera: Noctuidae) near an industrial source. Environ Entomol 19:481–485

Hembry DH, Althoff DM (2016) Diversification and coevolution in brood pollination mutualisms: windows into the role of biotic interactions in generating biological diversity. Am J Bot 103: 1783–1792

Hemmati S, Afshar AA, Mohammadi MA, Afgar A, Nasibi S, Harandi MF (2018) Experimental and field investigation of non-biting flies as potential mechanical vectors of Echinococcus granulosus eggs. Exp Parasitol 189:43–48

Hlongwane ZT, Slotow R, Munyai TC (2020) Nutritional composition of edible insects consumed in Africa: a systematic review. Nutrients 12(2786):1–28

Hinton HE (1945) A monograph of the beetles associated with stored products Volume I. pp. 443

Igwe C, Ujowundu C, Nwaogu L (2012) Chemical analysis of an edible African termite, *Macrotermes nigeriensis*; a potential antidote to food security problem. Biochem Anal Biochem 1:1–4

Iloba BN, Odigie OO (2006) Arthropod succession on buried carrion of the African giant rat (*Crycetomys gambianus*). Afr Sci 7(2):53–59

Iloba BN, Odo PE (2020) Forensic entomology: arthropods collected on decomposing pig carrions in Warri, Delta state, Nigeria. Nigerian Ann Pure Appl Sci 3(3):8–19

Ingram M, Nabhan GC, Buchmann SL (1996) Impending pollination crisis threatens biodiversity and agriculture. Tropinet 7:1

Intergovernmental Panel on Climate Change (IPCC) (2007) Fourth Assessment Report of the Intergovernmental Panel on Climate Change. www.ipcc.ch/ipccreports/assessments-reports.htm

IPCC (1992) Intergovernmental Panel on Climate Change 1992 IPCC supplement of the fifth session of the IPCC (IPCC-V). www.ipcc_90_92_asse. . .cc_suppl(2).pdf

IPCC (2014) Summary for policymakers. In climate change 2014: impacts, adaptation, and vulnerability, part a: global and sectoral aspects; contribution of working group ii to the fifth assessment report of the intergovernmental panel on climate change, vol 2014. Cambridge University Press, Cambridge/New York, NY, pp 1–32

Jongema Y (2017) List of edible insect species of the world. Laboratory of Entomology, Wageningen University, Wageningen. http://www.entwurnl/UK/edible+insects/worldwide+species+list

Joshua I, Noutcha MAE (2019) Succession patterns and diversity of arthropods associated with decomposing domestic rabbit (*Oryctolagus cuniculus* L, 1758) in different habitats. Environ Ecol Res 7(6):303–312. http://www.hrpub.org. https://doi.org/10.13189/eer.2019.070601

Kalaba FK, Quinn CH, Dougill AJ (2013) Contribution of forest provisioning ecosystem services to rural livelihoods in the Miombo woodlands of Zambia. Popul Environ 35:159–182. https://doi.org/10.1007/s11111-013-0189-5

Kaleri AR, Ma J, Jakhar AM, Hakeem A, Ahmed A, Napar WF, Ahmed S, Han Y, Abro SA, Nabi F, Tan C, Kaleri AH (2020) Effects of dung beetle-amended soil on growth, physiology, and metabolite contents of Bok Choy and improvement in soil conditions. J Soil Sci Plant Nutr 20:2671–2683. https://doi.org/10.1007/s42729-020-00333-8

Kelemu S, Niassy S, Torto B, Fiaboe K, Affognon H, Tonnang H, Maniania NK, Ekesi S (2015) African edible insects for food and feed: inventory, diversity, commonalities and contribution to food security. J Insects Food Feed 1:103–119

Kevan PG, Phillips T (2001) The economics of pollinator declines: assessing the consequences. Conserv Ecol 5:8. http://www.consecol.org/vol5/iss1/art8

Kim W, Bae S, Park K, Lee S, Choi Y, Han S, Koh Y (2011) Biochemical characterization of digestive enzymes in the black soldier fly, *Hermetia illucens* (Diptera: Stratiomyidae). J Asia Pac Entomol 14:11–14. https://doi.org/10.1016/j.aspen.2010.11.003

Klein A, Vaissière BE, Cane JH, Steffan-Dewenter I, Saul A, Cunningham SA, Kremen C, Koricheva J, Haukioja E (1992) Effects of air pollution on host plant quality, individual performance, and population density of *Eriocrania miner* (Lepidoptera: Eriocraniidae). Environ Entomol 26(6):1386–1392

Klein AM, Steffan-Dewenter I, Tscharntke T (2003) Fruit set of highland coffee increases with the diversity of pollinating bees. Proc R Soc B 270:955–961

Klein A, Vaissière BE, Cane JH, Steffan-Dewenter I, Cunningham SA, Kremen C, Tscharntke T (2007) Importance of pollinators in changing landscapes for world crops. Proc R Soc B 274: 303–313. https://doi.org/10.1098/rspb.2006.3721

Koehler PG, Patterson RS, Brenner RJ (1987) German cockroach (Orthoptera: blattellidae) infestations in lowincome apartments. J Econ Entomol 80:446–450

Koffi AF, Aboua LRN, Kone BA, Fofana D, Dao H, Djodjo M (2018) Dynamics of the main necrophagous Diptera populations of forensic interest, in the Guinean zone of Côte d'Ivoire. Int J Entomol Res 3(2):11–22

Koricheva J, Haukioja E (1992) Effects of air pollution on host plant quality, individual performance, and population density of Eriocrania miner (Lepidoptera:Eriocraniidae). Environ Entomol 26(6):1386–1392

Kosmann C, de Mello RP, Souza ÉSH, Luz JRP (2013) A list of current valid blow fly names (Diptera: calliphoridae) in the Americas South of Mexico with key to the Brazilian species. Entomo Brasilis 6:74–85

Knepp RG, Hamilton JG, Mohan JE, Zangerl AR, Berenbaum MR, DeLucia EH (2005) New Phytol 167:207–218

Kpama-Yapo YCE, Dao H, Koffi AF, Aboua LRN (2020) Influence of the conditions of exposure of pigs carcasses (*Sus scrofa domesticus* L.) on the egg-laying delay of necrophagous Diptera in the Guinean zone of Ivory Coast. Am J Entomol 4(4):66–73. https://doi.org/10.11648/j.aje.20200404.12

Kremen C, Williams NM, Aizen MA et al (2007) Pollination and other ecosystem services produced by mobile organisms: a conceptual framework for the effects of land-use change. Ecol Lett 10(4):299–314

Kunz RG (2009) Solid waste. Environmental calculations. John Wiley and Sons, Inc., Chichester, pp 375–414

Kyakuwaire M, Olupot G, Amoding A, Nkedi-kizza P, Ateenyi BT (2019) How safe is chicken litter for land application as an organic fertilizer? A review. Int J Environ Res Public Health 16: 3521

Lincoln DE, Sionit N, Strain BR (1984) Growth and feeding response of Pseudoplusia includens (lepidoptera: Noctuidae) to host plants grown in controlled carbon dioxide atmospheres. Environ Entomol 13(6):1527–1530

Louw SVDM, van der Linde TC (1993) Insects frequenting decomposing corpses in Central South Africa. Afr Entomol 1:265–269

Louzada JNC, Lopes FSA (1997) Comunidade de Scarabaeidae copro-necrófagos (Coleoptera) de um fragmento de Mata Atlântica. Rev Brasil Entomol 41(1):117–121

Lopes BGC (2008) Levantamento da entomofauna bioindicadora da qualidade ambiental em diferentes áreas do Alto Jequitinhonha – Minas Gerais. 2008. 47f. Trabalho de Conclusão de Curso - Escola Agrotécnica Federal de Inconfidentes

Lastuvka Z (2010) Climate change and its possible influence on the occurrence and importance of insect pests. Plant Prot Sci 45:S53–S62

Mabika N, Masendu R, Mawera G (2014) An initial study of insect succession on decomposing rabbit carrions in Harare, Zimbabwe. Asian Pac J Trop Biomed 4(7):561–565

Mafimisebi OP, Nkwunonwo UC (2015) Environmental risk: exploring organisational resilience and robustness. Int J Sci Eng Res 6(1):1103–1115

Malinga GM, Valtonen A, Lehtovaara VJ, Rutaro K, Nyeko ROP, Roininen H (2018) Diet acceptance and preference of the edible grasshopper Ruspolia di_erens (Orthoptera; Tettigonidae). Appl Entomol Zool 53:229–239

Martos HL, Maia NB, Brown-Jr S (1997) Indicadores Ambientais. Sorocaba USP 266

Marsili D, Fazzo L, Comba P (2009) Health risks from hazardous waste disposal: the need for international scientific cooperation. Eur J Oncol 14:151–159

Mashaly A, Al-Ajmi R, Rady A, Al-Musawi Z, Farrukh A (2019) Species richness of scavenger insects on different carcass types. Trop Biomed 36:630–639

Mejer JD (1984) The influence of ants on seeding operations in northern Australian mined areas. Reclam Reveg Res 2:299–313

Montoya ALG, Sanchez JD, Wolff EM (2009) Synanthropy of Calliphoridae (Diptera) from La PintadaSynanthropy of Calliphoridae (Diptera) from La Pintada, Antioquia-Colombia. Rev Colomb Entomol 35: 73–82

Miranda CD, Cammack JA, Tomberlin JK (2020) Mass production of the black soldier fly, *Hermetia illucens* (L.), (Diptera: Stratiomyidae) reared on three manure types. Animals 20:1243

Morrison J (2006) The international business environment, 2nd edn. Palgrave Macmillan, New York

Mutungi C, Irungu FG, Nduko J, Mutua F, Affognon H, Nakimbugwe D, Ekesi S, KKM F (2019) Postharvest processes of edible insects in Africa: a review of processing methods, and the implications for nutrition, safety and new products development. Crit Rev Food Sci Nutr 59: 276–298

Naima B (2020) Insects involved in decomposing corpses in the Constantine region-Algeria. Open J Trauma 4(1):022–024. https://doi.org/10.17352/ojt.000026

Ndueze O, Noutcha MAE, Umeozor OC, Okiwelu SN (2013) Arthropods associated with wildlife carcasses in lowland rainforest, Rivers state, Nigeria. Europ J Exp Biol 3(5):111–114

Newton L, Sheppard C, Watson DW, Burtle G (2005) Using the black soldier fly, *Hermetia illucens,* as a value-added tool for the management of swine manure. North Carolina, North Carolina State University. www.cals.ncsu.edu/wastemgt/smithfield_projects/phase2report05/cd,web%20files/A2.pdf

Nichols E, Alarcon V, Forgie S, Gomez-Puerta LA, Jones MS (2017) Coprophagous insects and the ecology of infectious diseases of wildlife. ILAR J 58:336–342. https://doi.org/10.1093/ilar/ilx022

Nichols E, Spector S, Louzada J, Larsen T, Amezquita S, Favila ME (2008) The Scarabaeinae research network: ecological functions and ecosystem services provided by Scarabaeinae dung beetles. Biol Conserv 161:1461–1474

Newbold T, Hudson LN, Hill SL, Contu S, Lysenko I, Senior RA, Börger L, Bennett DJ, Choimes A, Collen B, Day J (2015) Global effects of land use on local terrestrial biodiversity. Nature 520(7545):45–50

Nonaka K (2009) Feasting on insects. Entomol Res 39:304–312

Nor Faiza MT, Hassan NA, Mohammad FR, Edre MA, Rus RM (2019) Solid waste: its implication for health and risk of vector borne diseases. J Wastes Biomass Manag 1:14–17

Njau DG, Muge KK, Kinyanjui PW, Omwandho COA, Mukwana S (2016) STR analysis of human DNA from maggots fed on decomposing bodies: assessment of the time period for successful analysis. Egypt J Forens Sci 6:261–269

Nummelin M, Lodenius M, Tulisalo E, Hirvonen H, Alanko T (2007) Predatory insects as bioindicators of heavy metal pollution. Environ Pollut 145:339–347

Nkwunonwo UC, Mafimisebi OP (2013) Responding to the challenges of global climate change in Nigeria through GIS mapping of carbon iv oxide gas emission. Int J Sci Eng Res 4(12):765–774

Nwaugo VO, Onyeagba RA, Nwachukwu NC (2006) Effect of gas flaring on soil microbial spectrum in parts of Niger Delta area of Southern Nigeria. Afr J Biotechnol 5(19):1824–1826

Nurita A, Hassan AA (2013) Filth flies associated with municipal solid waste and impact of delay in cover soil application on adult filth fly emergence in a sanitary landfill in Pulau Pinang, Malaysia. Bull Entomol Res:1–7

Odermorho F (1993) Flood control in a third world city: Benin City, Nigeria: some environmental and policy issues. Geo J 29(4):391–396

Ollerton J (2017) Pollinator diversity: distribution, ecological function, and conservation. In: Futuyma DJ (ed) Annual review of ecology, evolution, and systematics, vol 48, pp 353–376

Odo PE (2020) Forensic entomology: a dry season study of insects collected on pig carrions in Neke, Enugu State, Nigeria. Int J Sci Res Multidiscipl Stud 6(9):29–36

Odo PE, Iloba BN (2020) Insects' succession on the decomposing pig carrion during the wet season at the college of education, Delta State, Nigeria. SAU Sci-Tech J 5(1):83–96

Onyishi GC, Osuala F, Aguzie IO, Okwuonu ES, Orakwelu CH (2020) Arthropod succession on exposed and shaded mammalian carcasses in Nsukka Nigeria. Anim Res Int 17(3):3869–3877

Oyebanji IJ, Adeniyi B, Khobai H, Le Roux P (2017) Green growth and environmental sustainability in Nigeria. Int J Energy Econ Policy 7(4):216–223

Padonou GG, Gnanguenon V, Ossè R, Oussou O, Odjo E, Akinro B, Akogbeto M (2017) First evidence of forensic entomology revealed the presence of arthropods on rabbit carrion in Cotonou, Benin (West Africa). Int J Entomol Res 2:94–98

Pai HH, Ko Y, Chen E (2003) Cockroaches (*Periplaneta americana* and *Blattella germanica*) as potential mechanical disseminators of Entamoeba histolytica. Acta Trop 87:355–359

Parmesan C (2006) Ecological and evolutionary responses to recent climate change. Annu Rev Ecol Evol Syst 37:637–669

Peck SL, Mcquaid B, Cambell CL (1998) Using ant species (Hymenoptera: Formicidae) as a biological indicator of agroecosystem condition. Environ Entomol 27(5):1102–1110

Pérez-Guerra CL, Zielinski-Gutierrez E, Vargas-Torres D, Clark GG (2009) Community beliefs and practices about dengue in Puerto Rico. Rev Panam Salud Pública 25:218–226

Perkins DN, Drisse BMN, Nxele T, Sly PD (2014) E-waste: a global hazard. Ann Glob Health 80: 286–295. https://doi.org/10.1016/j.aogh.2014.10.001

Pohl HR, Tarkowski S, Buczynska A, Fay M, De Rosa CT (2008) Chemical exposures at hazardous waste sites: experiences from the United States and Poland. Environ Toxicol Pharmacol 25:283–291. https://doi.org/10.1016/j.etap.2007.12.005

Powney GD, Carvell C, Edwards M, Morris RK, Roy HE, Woodcock BA, Isaac NJ (2019) Widespread losses of pollinating insects in Britain. Nat Commun 10:1018

Qasima M, Xiaoa H, Hea K, Nomanc A, Liua F, Chena M, Hussaind D, Jamale ZA, Lia F (2020) Impact of landfill garbage on insect ecology and human health. Acta Trop 211(105630):1–6. https://doi.org/10.1016/j.actatropica.2020.105630

Ratto F, Simmons BI, Spake R, Zamora-Gutierrez V, MacDonald MA, Merriman JC, Tremlett CJ, Poppy GM, Peh KS, Dicks LV (2018) Global importance of vertebrate pollinators for plant reproductive success: a meta-analysis. Front Ecol Environ 16(2):82–90

Raven PH, Wagner DL (2021) Agricultural intensification and climate change are rapidly decreasing insect biodiversity. PNAS 118(2):e2002548117. https://doi.org/10.1073/pnas.2002548117

Ruiz-Zapata JD, Figueroa-Gutiérrez LM, Mesa-Franco JA, Moreno-Gutierrez PA (2019) Umbilical myiasis by Cochliomyia hominivorax in an infant in Colombia. Front Med 6:292

Rusek J (1998) Biodiversity of collembola and their functional role in the ecosystem. Biodivers Conserv 7:1207–1219

Saifi M, Berrouane FZ, Bensaade F, Doumandji S (2017) Preliminary study of necrophagous diptera associated with boar corpse (*Sus scrofa*) in a semi-arid environment in eastern Algeria: implications in forensic entomology. Adv Environ Biol 11(3):1–8

Sakai S (2002) A review of brood-site pollination mutualism: plants providing breeding sites for their pollinators. J Plant Res 115:161–168

Salem AM, Adham FK, Picard CJ (2015) Survey of the genetic diversity of forensically important *Chrysomya* (Diptera: Calliphoridae) from Egypt. J Med Entomol 52(3):320–328. https://doi.org/10.1093/jme/tjv013

Samson O, Oluwole A, Abimbola S (2011) On the physical composition of solid wastes in selected dumpsites of Ogbomoso land, South-Western Nigeria. J Water Resour Prot 3:7521

Sanchez-Bayo F, Wyckhuys KAG (2019) Worldwide decline of entomofauna: a review of its drivers. Biol Conserv 232:8–27

Sarpong D, Oduru-Kwarteng S, Gyasi SF, Buamah R, Donkor E, Awuah E, Baah MK (2019) Biodegradation by composting of municipal organic solid waste into organic fertilizer using the black soldier fly (*Hermetia illucens*) (Diptera: Stratiomyidae) larvae. Int J Recycl Organ Waste Agric 8:45–54

Sawabe K, Hoshino K, Isawa H, Sasaki T, Kim KS, Hayashi T, Tsuda Y, Kurahashi H, Kobayashi M (2011) Blow flies were one of the possible candidates for transmission of highly pathogenic H5N1 avian influenza virus during the 2004 outbreaks in Japan. Influenza Res Treat 2011: 652652

Sawaby RF, Hamouly HE, Abo-El Ela RH (2018) Diagnosis and keys of the main dipterous families and species collected from rabbit and guinea pig carcasses in Cairo, Egypt. J Basic Appl Zool 79(10):1–14. https://doi.org/10.1186/s41936-018-0018-6

Schulz R, Bub S, Petschick LL, Stehle S, Wolfram J (2021) Applied pesticide toxicity shifts toward plants and invertebrates, even in GM crops. Science 372:81–84

Schuurman G (2005) Decomposition rates and termite assemblage composition in semiarid Africa. Ecology 86:1236–1249

Sebastião M, Castro CP (2019) A preliminary study of carrion insects and their succession in Luanda, Angola. J Med Entomol 56(2):378–383. https://doi.org/10.1093/jme/tjy203

Seibold S, Gossner MM, Simons NK, Blüthgen N, Müller J, Ambarlı D, Ammer C, Bauhus J, Fischer M, Severin D, Barankanira E, Gasogo A (2021) Necrophagous insect's on corpse in Africa Great Lakes Region. East Afr Schol J Agric Life Sci 4(4):96–100

Servia MJ, Cobo F, Gonzalez MA (1998) Deformities in larval *Prodiamesa olivacea* (Meigen, 1818) (Diptera, Chironomidae) and their use as bioindicators of toxic sediment stress. Hydrobiology 385:153–162

Severin D, Barankanira E, Anastasie G (2021) Necrophagous insects' succession on corpse in African Great Lakes Region. East Afr Scholars J Agric Life Sci 4(4):96–100

Shankar S (2017) Management and remediation of problem soils, solid waste and soil pollution. In: Principles and applications of environmental biotechnology for a sustainable future. Springer, pp 143–171

Sheppard C (1983) House fly and lesser fly control utilizing the black soldier fly in manure management systems for caged laying hens. Environ Entomol 12:1439–1442

Shine C, Williams N, Gündling LA (2000) Guide to designing legal and institutional frameworks on alien invasive species, environmental policy and law paper no. 40. In: IUCN, Gland/Cambridge/Bonn

Shrestha S (2019) Effects of climate change in agricultural insect pest. Acta Sci Agric 3:74–80

Skendžić S, Zovko M, Živković IP, Lešić V, Lemić D (2021) The impact of climate change on agricultural insect pests. Insects 12:440. https://doi.org/10.3390/insects12050440

Slaa EJ, Sánchez CLA, Malagodi-Braga KS, Hofstede FE (2006) Stingless bees in applied pollination. Pract Perspect Apidol 37:293–315. https://doi.org/10.1051/apido:2006022

Simpson BB, Neff JL (1983) Evolution and diversity of floral rewards. In: Jones CE, Little RJ (eds) Handbook of experimental pollination biology. Van Nostrand Reinhold Co., New York, pp 142–159

Smith KGV (1986) A manual of forensic entomology. British Museum (Natural History), London, p 205

Sogari G, Menozzi D, Mora C (2017) Exploring young foodies' knowledge and attitude regarding entomophagy: a qualitative study in Italy. Int J Gastron Food Sci 7:16–19

Sommaggio D (1999) Syrphidae: can they be used as environmental bioindicators? Agric Ecosyst Environ 74:343–356

Staley JT, Hodgson CJ, Mortimer SR, Morecroft MD, Masters GJ, Brown VK, Taylor ME (2007) Effects of summer rainfall manipulations on the abundance and vertical distribution of herbivorous soil macro-invertebrates. Eur J Soil Biol 43:189–198

Steiner R (2010) Double standard: Shell practices in Nigeria compared with international standards to prevent and control pipeline oil spills and the Deepwater horizon oil spill. Milieu Defense, Amsterdam. http://www.milieudefensie.nl/publicaties/rapporten/double-tandard

Stokstad E (2018) New global study reveals the 'staggering' loss of forests caused by industrial agriculture. Science. https://www.sciencemag.org/news/2018/09/scientists-reveal-how-much-world-s-forests-being-destroyed-industrial-agriculture

Tantawi TI, El-Kady EM, Greenberg B, El-GHhaffari HA (1996) Arthropod succession on exposed rabbit carrion in Alexandria, Egypt. J Med Entomol 33(4):566–580

Tantawi TI, Williams KA, Villet MH (2010) An accidental but safe and effective use of *Lucilia cuprina* (Diptera: Calliphoridae) in maggot debridement therapy in Alexandria, Egypt. J Med Entomol 47(3):492–494

Taylor R, Allen A (2006) Waste disposal and landfill: information needs. In: Schmoll O, Howard G, Chilton J, Chorus I (eds) Protecting groundwater for health: managing the quality of drinking-water sources. IWA, London

Teffo LS, Toms RB, Eloff JN (2007) Preliminary data on the nutritional composition of the edible stink-bug, *Encosternum delegorguei* Spinola, consumed in Limpopo province, South Africa. S Afr J Sci 103:434–436

Tembe D, Malatji MP, Mukaratirwa S (2021) Molecular identification and diversity of adult arthropod carrion community collected from pig and sheep carcasses within the same locality during different stages of decomposition in the KwaZulu-Natal province of South Africa. PeerJ 9:e12500. https://doi.org/10.7717/peerj.12500

Tembe D, Mukaratirwa S (2021) Insect succession and decomposition pattern on pig carrion during warm and cold seasons in Kwazulu-Natal Province of South Africa. J Med Entomol 58(6): 2047–2057. https://doi.org/10.1093/jme/tjab099

Todaro M, Smith S (2015) Development economics. Pearson Education, Upper Saddle River, NJ

Tscharntke T (2007) Importance of pollinators in changing landscapes for world crops. Proc R Soc B 274:303–313. https://doi.org/10.1098/rspb.2006.3721

Tylianakis J, Veddeler D, Lozada T, López RM, Benítez P, Klein AM, Koning GHJ, Olschewski R, Eldkamp E, Navarrete H, Onore G, Tscharntke T (2004) Biodiversity of land-use systems in coastal Ecuador and bioindication using trap-nesting bees, wasps, and their natural enemies. Lyonia 6(2):7–15

UNDP (2014) Sustaining human progress: reducing vulnerabilities and building resilience (2014 human development report). United Nations Development Programme, New York. http://hdr.undp.org/sites/default/files/hdr14-report-en-1.pdf

Urbini A, Sparvoli E, Turillazzi S (2006) Social paper wasps as bioindicators: a preliminary research with *Polistes dominulus* (Hymenoptera: Vespidae) as a trace metal accumulator. Chemosphere 64:697–703

van Huis A (2015) Edible insects contributing to food security? Agric Food Sec 4:20. https://doi.org/10.1186/s40066-015-0041-5

van Huis A, Oonincx DGAB (2017) The environmental sustainability of insects as food and feed. A review. Agron Sustain Dev 37:43

van Huis A, Itterbeeck JV, Klunder H, Mertens E, Halloran A, Muir G, Vantomme P (2013) Edible insects: future prospects for food and feed security. Food and Agriculture Organization of the United Nations, Rome

van Klink R, Bowler DE, Gongalsky KB, Swengel AB, Gentile A, Chase JM (2020) Meta-analysis reveals declines in terrestrial but increases in freshwater insect abundances. Science 368(6489): 417–420. https://doi.org/10.1126/science.aax9931

van Strien AJ, van Swaay CA, van Strien-van Liempt WT, Poot MJ, Wallis De Vries MF (2019) Over a century of data reveal more than 80% decline in butterflies in the Netherlands. Biol Conserv 234:116–122

Vantomme P, Göhler D, N'Deckere-Ziangba F (2004) Contribution of forest insects to food security and forest conservation: the example of caterpillars in Central Africa. Odi Wildlife Pol Brief 3(4):345–352. https://www.odi.org/sites/odi.org.uk/files/odiassets/publications-opinion-files/3306.pdf

Verves YG (2005) A catalogue of oriental Calliphoridae (Diptera). Int J Dipterolo Res 16:233–310

Vidyavathy K (2018) A study on solid waste management and vector borne diseases (dengue) control in polluted areas (Warangal City). Pharma Innovat J 7:259–263

Villet H (2011) African carrion ecosystems and their insect communities in relation to forensic entomology. Pest Technol 5(1):1–15

Wagner DL (2020) Insect declines in the Anthropocene. Annu Rev Entomol 65:457–480

Wagner DL, Grames EM, Forister ML, Berenbaum MR, Stopak D (2021) Insect decline in the Anthropocene: Death by a thousand cuts. Proc Natl Sci U S A 118(2):e2023989118. https://doi.org/10.1073/pnas.2023989118

Wang X, Zhong M, Wen J, Cai J, Jiang H, Liu Y, Aly SM, Xiong F (2012) Molecular characterization and expression pattern of an odorant receptor from the myiasis-causing blowfly, *Lucilia sericata* (Diptera: Calliphoridae). Parasitol Res 110:843–851

Wang Y, Ma MY, Jiang XY, Wang JF, Li LL, Yin XJ, Wang M, Lai Y, Tao LY (2017a) Insect succession on remains of human and animals in Shenzhen, China. Forensic Sci Int 271:75–86

Wang H, Zhang Z, Czapar GF, Winkler MK, Zheng J (2013) A full-scale house fly (Diptera: Muscidae) larvae bioconversion system for value-added swine manure reduction. Waste Manag Res 31:223–231

Wang H, Li H, Gilbert JA, Li H, Wu L, Liu M, Wang L, Zhou Q, Yuan J, Zhang Z (2015) Housefly larva vermicomposting efficiently attenuates antibiotic resistance genes in swine manure, with concomitant bacterial population changes. Appl Environ Microbiol 81:7668–7679

Wang H, Sangwan N, Li HY, Su JQ, Oyang WY, Zhang ZJ, Gilbert JA, Zhu YG, Ping F, Zhang HL (2017b) The antibiotic resistome of swine manure is significantly altered by association with the *Musca domestica* larvae gut microbiome. ISME J 11:100–111

Ward D, Blaustein L (1994) The overriding influence of flash floods on species-area curves in ephemeral Negev Desert pools: a consideration of the value of island biogeography theory. J Biogeogr 21:595–603

Ward NL, Masters GJ (2007) Linking climate change and species invasion: an illustration using insect herbivores. Glob Chang Biol 13:1605–1615

Warren MS et al (2021) The decline of butterflies in Europe: problems, significance, and possible solutions. Proc Natl Acad Sci U S A 118(2):e2002551117. https://doi.org/10.1073/pnas.2002551117

Welch C, O'Hagan M (2010) Officials seek cause of Snohomish dairy-manure spill. The Seattle Times April: 14. https://www.seattletimes.com/seattle-news/officials-seek-cause-of-snohomish-dairy-manure-spill/

Williams KA, Villet MH (2006a) A history of southern African research relevant to forensic entomology. S Afr J Sci 102:59–65

Williams KA, Villet MH (2006b) A new and earlier record of *Chrysomya megacephala* in South Africa, with notes on another exotic species, *Calliphora vicina* (Diptera:Calliphoridae). Afr Invertebr 47:347–350

World Bank (2014) Risk and opportunity: managing risk for development. Washington, DC. https://siteresources.worldbank.org/EXTNWDR2013/Resources/8258024-1352909193861/8936935-1356011448215/89869011380046989056/WDR-2014_Complete_Report.pdf

World Bank (2015) World development report 2015: mind, society, and behaviour. Washington, DC. http://www.worldbank.org/en/publication/wdr2015

Yadong G, Jifeng C, Zhenchu T, Xiong F, Zhang L, Fu Y, Jianbo L, Yaoqing C, Fanming M, Jifang W (2011) Application of *Aldrichina grahami* (Diptera, Calliphoridae) for forensic investigation in central-South China. Romanian J Legal Med 19:55–58

Yihdego Y, Salem HS, Muhammed HH (2019) Agricultural pest management policies during drought: case studies in Australia and the state of Palestine. Nat Hazards Rev 20:05018010

Yoro KO, Daramola MO (2020) Carbon dioxide emission sources, greenhouse gases, and the global warming effect. In: Rahimpour MR, Farsi M, Makarem MA (eds) Advances in carbon capture, 1st edn. Wood Head Publishing, Sawston, pp 3–28

Zaghloul A, Saber M, Gadow S, Awad F (2020) Biological indicators for pollution detection in terrestrial and aquatic ecosystems. Bull Natl Res Cent 44(127):1–11

Zhang Z, Wang H, Zhu J, Suneethi S, Zheng J (2012) Swine manure vermicomposting via housefly larvae (*Musca domestica*): the dynamics of biochemical and microbial features. Bioresour Technol 118:563–571. https://doi.org/10.1016/j.biortech.2012.05.048

Zhao Y, Christensen TH, Lu W, Wu H, Wang H (2011) Environmental impact assessment of solid waste management in Beijing City, China. Waste Manag 31:793–799

Chapter 23
Intrinsic Values of the African Environment: A Sustainable Perspective

Adams Ovie Iyiola and John Valbo-Jørgensen

Abstract Africa is a continent with immerse renewable and non-renewable resources. The African continent has an excess of 30 million km^2 of natural resources base which is spread across the mountains, deserts, lakes, coasts, rivers, etc. These resources most especially non-renewable resources such as coal, oil, gas, metals, and non-metals contribute significantly to the economic returns of African countries. The continent is known as a major producer of various natural resources, including minerals and non-metals, which are found in various countries within the continent. Minerals and non-metals are concentrated majorly in the southern, northern, and western regions with a minor deposit in the eastern regions. South Africa is the continent's leading producer of minerals which are divided into energy, precious, industrial, ferrous, and non-ferrous minerals. It houses diverse biodiversity of importance to the economy. The rivers are enormous with huge volumes of water with the Nile, Congo, Niger, Zambezi, Orange, Limpopo, Volta, and Okavango as the major rivers. The resources are majorly threatened by anthropogenic activities. Other issues such as pollution, climate change, sedimentation, and the introduction of alien species are also threats. Management and governance issues are paramount to the sustainability of natural resources based on a framework covering natural resource policy, fiscal regimes, sustainable development, and regulatory framework as implementation strategies for sustainable development. Natural resources have various importance, such as natural capital, source of energy, food, materials for industries, medicinal value, scientific studies, and shelter. This chapter, therefore, elaborates on the African natural resources base, sustainability, and management framework for resource conservation.

Keywords Sustainability · Africa · Natural Resources · Management framework

---

A. O. Iyiola (✉)
Department of Fisheries and Aquatic Resources Management, Faculty of Renewable Natural Resources Management, Osun State University, Osogbo, Osun State, Nigeria

J. Valbo-Jørgensen
Fisheries Division, Food and Agricultural Organization (FAO) of the United Nations, Vialedelle Terme di Caracalla, Rome, Italy

S. C. Izah, M. C. Ogwu (eds.), *Sustainable Utilization and Conservation of Africa's Biological Resources and Environment*, Sustainable Development and Biodiversity 32, https://doi.org/10.1007/978-981-19-6974-4_23

23.1 Introduction

The continent of Africa has diverse natural resources which are either renewable or non-renewable. These resources range from the environment, plants, and animals (Fig. 23.1). About 30% of the mineral reserves in the world, 8% of natural gas, and 12% of the oil reserves in the world are located in Africa. In terms of minerals, Africa has 40% of gold, 90% of platinum and chromium, and the largest deposits of diamonds, cobalt, and uranium in the world. It holds 65% of the world's arable land and 10% of the planet's internal renewable fresh water source. The natural capital of countries in Africa is about 30–50% of the total wealth of the continent. In sub-Saharan Africa, about 70% of the populace living in this region depends directly on the woodlands and forests for sustenance. The land is a socio-economic resource that enhances economic development but a large part of it is degraded by illegal activities, thereby reducing the benefits of the resource. It was estimated that the African continent loses in its natural capital about $195 billion every year from activities, such as illegal mining, illegal wildlife trade and logging, unregulated and illicit financial flow, illegal, unregulated, and unreported fishing, and the degradation of the environment.

Presently, the continent is creating funds available for environmental restoration and this will go a long way in achieving prosperity, economic returns, and future growth. This will also enhance climate resilience and better exploitation of resources for sustainability. Like renewable resources, countries are also benefitting from non-renewable natural resources. Revenues are generated from these activities and it has stimulated the middle class in the urban areas to develop but not an increase in physical wealth. It was reported by AfDB (2013) that poverty has reduced and the

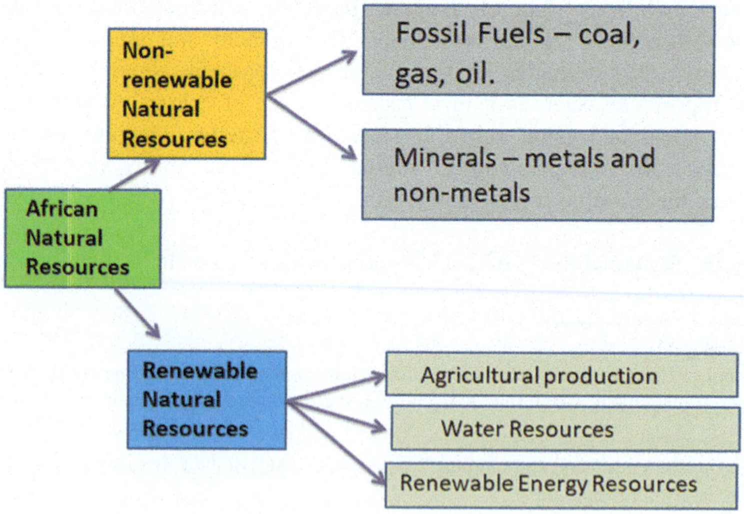

Fig. 23.1 African natural resources

indicators such as infant mortality, enrollment in school, life expectancy, and issues of malnourishment have improved. This also was evident in statistics with a 34% increase in the middle class of the African population in 2010 as against 27% in 2000 (AfDB 2011.

23.2 Renewable Resources in Africa

A renewable resource is a natural resource that can be regenerated or replenished by natural means or a period after consumption or utilization. These forms of resources are part of the earth and are of various types:

- Agricultural production (e.g., food crops—green plants and non-food crops—wood).
- Water resources (e.g., freshwater).
- Renewable energy resources (e.g., solar, wind, and thermal energy) (Park and Allaby 2017).

23.3 Non-renewable Resources in Africa

Non-renewable resources are resources that when exploited theta cannot be replaced or regenerated by natural means. There are two classes of non-renewable resources in Africa, namely:

- Fossil fuels (coal, gas, and oil).
- Minerals (metals and non-metals).

These resources can contribute to the economy of African countries and the top five exports in Africa are mineral-related (Table 23.1).

23.3.1 Coal

It is majorly found around the southern and western parts of Africa around Botswana, the Democratic Republic of Congo, Nigeria, South Africa, Mozambique, and Zimbabwe. It is found in smaller quantities around Niger, Senegal, Benin, Ethiopia, Somalia, Zimbabwe, Tanzania, Malawi, Egypt, Madagascar, and Morocco (TradeMap 2007). South Africa is the largest reserve of coal in Africa with an estimate of 34 billion tons and is the country with the sixth-largest coal reserve in the world. In the country, it meets the need for energy production (77%) and electricity (90%) (DME 2007).

Table 23.1 First five African exports and their values

Country	Products	Value in \$ USD (2005)	Percentage in export
Algeria, Chad, Egypt, Libya, Morocco, Tunisia, Nigeria, Angola	Petroleum oil from crude oil	133,534,293	17
Libya, Gabon, Egypt, Chad, Tunisia, and Morocco	Petroleum oil, not crude oil	19,283,817	5
Algeria, Mozambique, Libya, Niger, Morocco, Nigeria, Rwanda, Ghana, Egypt, Tunisia, Tanzania, and Namibia	Liquefied Natural Gas	7,968,296	18
Botswana, South Africa, Angola, Namibia, Democratic Republic of Congo	Diamond	7,494,806	22
Botswana, Democratic Republic of Congo, Nigeria, Mozambique, South Africa, Zimbabwe, Senegal, Benin, Niger, Ethiopia, Somalia, Tanzania, Egypt, Morocco, and Madagascar	Bituminous coal	3,180,845	8

23.3.2 *Crude Oil*

The major areas with huge deposits are around the northern part of Africa, such as Tunisia, Morocco, Libya, Egypt, Chad, and Algeria. Minor deposits are located in West Africa in places, such as Nigeria and Angola. Africa contributed to 11.6% of the world's crude oil supply with increased production in countries, such as Libya, Chad, Sudan, Equatorial Guinea, Algeria, Nigeria, and Angola. These countries formed the Organization of Petroleum Exporting Countries in Africa. The oil production in Africa is expected to increase but is dependent on discoveries of more oil fields and a reduction in conflicts in oil-producing areas. However, it should be noted that crude oil if not well-managed can affect environmental and biological resources in the continent (Osawaru et al. 2013a, b; Ikhajiagbe and Ogwu 2020).

23.3.3 *Natural Gas*

Algeria and Mozambique have the largest deposit of gas in Africa with a substantial amount in Libya, Morocco, Niger, Nigeria, Rwanda, Ghana, Tunisia, Egypt, and off South African, Tanzania, and Namibia coasts. Algeria is the biggest producer of natural gas in Africa with 54%. Africa exported 37 million tons in 2005 with 79% exported as liquefied gas. The production of dried natural gas is expected to increase in Africa (DTI 2005).

23.3.4 *Minerals*

The continent has diverse minerals, such as the non-ferrous, precious, ferrous, and industrial types. Africa is the largest producer of gold and platinum. South Africa is the leading producer of chromite and ferrochromium, gold, platinum, and vanadium and the second-largest producer of manganese and ferromanganese in the world. Minerals are formed from geological processes which is a natural phenomenon. It has its chemical and physical composition and ranges from pure elements and simple salts to complex silicates which have diverse forms. Minerals can be divided into various forms:

- Energy minerals (e.g., uranium).
- Precious metals and minerals (e.g., gold, platinum group metals, diamonds).
- Ferrous metals (e.g., chrome, manganese).
- Industrial minerals (e.g., fluorspar).
- Non-ferrous metals and minerals (e.g., aluminum, copper, nickel, zinc).

23.4 Management of Non-renewable Natural Resources

The non-renewable resources in Africa are enormous and are currently tapped with diverse mineral resources undiscovered and untapped. There is increased interest in these resources and it is, therefore, essential to manage these activities for sustainability. The management involves:

- The process of ensuring that the resource is available for exploitation and extraction.
- Allocating the resources between the local and the international miners in competition for the natural resource.
- Ensuring that the resources from mining activities are of high integrity.
- Developing policies that can adequately manage the resources.
- Limiting the impacts of the resource exploitation on the environment.
- Ensuring that the health and safety measures are in place and guaranteed when exploiting the resources.
- The process of converting the resources to the sustainable development of the economy through appropriate linkages.

23.5 African Natural Resources Base

With an area above 30 million km^2, Africa is second to only Asia in size among the world's continents. The Equator bisects Africa in two parts of approximately equal size, and it is the only continent reaching both the northern temperate and the

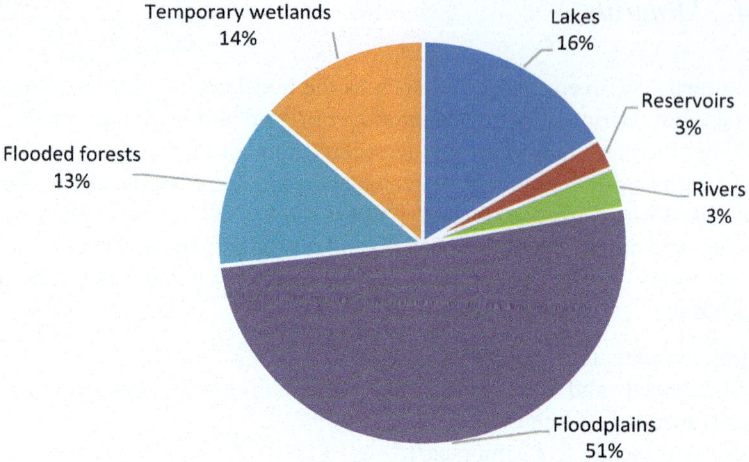

Fig. 23.2 Percentage contribution of different freshwater ecosystems in Africa

southern temperate zones—a distance of 8000 km (Shahin 2003). The vast landmass is covered by grasslands, mountains, deserts, rainforests, coasts, lakes, rivers, and wetlands and is home to thousands of species of plants and animals—including 1.2 billion people of more than 3000 ethnic groups and cultures. The earliest human relatives evolved in Africa millions of years ago, where the natural bounty provided by these ecosystems allowed them to survive. About 75% of Africa is located in the tropical zone (Shahin 2003), and next to Australia, Africa is the driest continent (FAO 2003), but aquatic ecosystems, nevertheless, cover 1.3 million km^2 (Lymer et al. 2016) and are fundamental parts of Africa's natural and cultural heritage. The relative contribution of the different types of aquatic ecosystems to the overall area is shown in Fig. 23.2. Modern civilization had its origin in the fertile plains of the Nile, and the importance of the services delivered by aquatic ecosystems to the African populations to this day cannot be overestimated.

23.5.1 River Basins

Africa can be divided into the northern and western sedimentary and upland plains located between 150 and 600 m above sea level known as "Low Africa" a region bordered by the Nile and the upper Congo Basin, in contrast to the land to the East and South "High Africa" is a table land 1000 m or more above sea level (Lévêque and Paugy 2010). The relief of the continent means that most of the major rivers enter the sea to the west, the exception being the Nile running north and Zambesi, Limpopo, and the smaller Juba running to the Indian Ocean. Africa hosts 18 rivers longer than 1000 km including the longest river in the world, the Nile, which runs for 6700 km in a south–north direction crossing the equator. It also harbors the world's

Table 23.2 Major rivers and their basins in Africa

Rivers	Length (km)	Size of Basin (million km^2)	Population around the Basin (million people)
Nile	6700 (1[a])	3.1 (3[a])	259.0
Congo	4700 (10[a])	3.8 (2[a])	100.6
Niger	4200 (15[a])	2.1 (9[a])	126.0
Zambesi	2600	1.4	45.3
Orange	2400	1.0	19.0
Limpopo	1800	0.4	18.7
Volta	1600	0.4	27.6
Okavango	1100	0.7	1.2

[a]Indicates a ranking in the world

Table 23.3 Some major African wetlands

Wetlands in Africa	Basin	Area (million ha)
Ngiri–Tumba–Maindombe	Congo	6.5
Congo River and tributaries	Congo	5.9
Sudd	Nile	5.7
Okavango delta	Okavango	5.5
Bahr Anouk and Salamat floodplains	Chad	4.9
Niger inner delta	Niger	4.1

second-largest river basin, the Congo, which drains an area larger than India, with a discharge of more than two times the amount of water transported by the Mississippi and is responsible for 30% of the continent's runoff (FAO 2003). There are 12 major river basins and approximately 80 other smaller basins in Africa, most of which are shared by several countries (Ainsworth et al. 2021; Aquastat 2007; Frenken 2005). Details of some major rivers are found in Table 23.2.

23.5.2 *Wetlands*

Many of the major river systems are associated with enormous wetlands (Table 23.3) home to extraordinary biodiversity. Floodplains are ephemeral wetlands, intermediate between terrestrial and aquatic ecosystems. Their production cycles are driven by annual flood pulses that lead to favorable conditions for different sets of organisms in the respective phases of the flood. Floodplains are among the most productive natural ecosystems. Aquatic organisms, including fish, move into the floodplains with the advance of the water during the flood and quickly build up enormous biomass, which serves as the basis for other species feeding on these including humans. During the drawdown, and the subsequent dry season, fish become isolated in a shrinking environment; at that time, the fishes are easy targets for their predators and mortality is high, but the species have life strategies that allow them to cope with this and are, therefore, also particularly resilient to fisheries exploitation.

23.5.3 *Lakes*

Lakes are important features of the African landscape and range from small ponds to water bodies stretching hundreds of kilometers (Table 23.4). Lakes are formed, where water accumulates in depressions in the terrain, for example, carved by ice or created through volcanic or tectonic activity such, for instance, the large rift lakes (Victoria, Tanganyika, and Malawi) in eastern Africa. Lake Victoria is the second-largest lake in the world, and with an area of 68,000 km^2 is roughly the size of Latvia, Lithuania, or Sri Lanka. Lake Tanganyika is almost 1500 m deep and is the second-deepest lake in the world. All the rift lakes are millions of years old and are natural laboratories, where evolution has allowed extraordinary assemblages of fish species to appear and an enormous variety of fish species particularly cichlids, e.g., an estimated 1000 species in Lake Malawi, to evolve. Many large lakes also have large populations of fast-growing small pelagic fish species living in the open water. In years when it rains so many, nutrients will be washed out of the soil in the upland and carried into the lake by the tributaries and will cause a bloom in plankton production. Small pelagic fish species have a short life span and may spawn several times in a year. Their populations will, therefore, respond quickly to such changes in productivity which can be termed rainfall years, hence, equal good fish years. Larger growing species with later maturity will take longer to respond to such changes and the relationship is less strong as other factors will also interfere (Ainsworth et al. 2021).

Another type of lake is the so-called endorheic lake without an outlet, where all the water entering the lake eventually evaporates. These lakes vary enormously in

Table 23.4 Major African lakes and their surface areas

Lakes	Location	Area (km^2)
Lake Victoria	Tanzania and Uganda	68,000
Lake Tanganyika	Borders of Tanzania, Zambia, Burundi, and the Democratic Republic of Congo	32,900
Lake Malawi/Nyasa/ Niassa	Between Malawi, Tanzania, and Mozambique	30,900
Lake Turkana	Northwestern Kenya and Southwestern Ethiopia	6750
Lake Albert	The border between Congo (Kinshasa) and Uganda	5300
Lake Rukwa	Southwestern Tanzania	5760
Lake Mweru	Between Zambia and Congo (Kinshasa)	5120
Lake Tana	Northwestern Ethiopia	3150
Lake Kivu	Between Western Congo (Kinshasa) and Eastern Rwanda	2370
Lake Edward	Between the border of Congo (Kinshasa) and Uganda	2325

size over the year and between years due to seasonal weather patterns and climate variability. These changes greatly affect habitat conditions for aquatic organisms and, therefore, the composition of fish assemblages and ultimately the fisheries. There are many such basins in Africa, including Lake Chad and Lake Chilwa. Depending on the geochemistry in the watershed associated with these lakes, their water may become salty affecting the fauna and flora. The examples above demonstrate the extraordinary variety of the African natural resources base.

23.5.4 *Reservoirs*

Reservoirs (Table 23.5) are man-made lakes constructed by interrupting the flow of a river or stream by a dam to store water for a variety of purposes and people have constructed them for thousands of years. Most African dams have been constructed since 1980 for irrigation (52%), supply water to municipalities (20%), electricity generation (6%), and flood control (1%)—20% were found to have multiple purposes (Aquastat 2007). According to Kolding and van Zwieten (2006), there were 176 large reservoirs in Africa around 2000 of which 40% were located in South Africa. The total reservoir area associated with large dams in Africa was around 54,600 km^2 corresponding to 80% of the area of Lake Victoria (Kolding and van Zwieten 2012; Welcomme and Lymer 2012).

Lake Volta, Cahora Bassa, Lake Kariba, and Lake Nasser all have important fisheries; however, small reservoirs are more productive with yields of up to 329 kg/ha/year (Kolding et al. 2014). On the negative side, dams have fragmented rivers preventing migratory fish from reaching their spawning sites and have interrupted the natural flood pulse and thus the seasonal dynamics that drive ecosystem productivity. Reservoirs are similar to lakes in that they are standing water bodies, but reservoir depth and area are often more variable over the years than in a natural lake

Table 23.5 Major reservoirs in Africa

Reservoirs	Location	Area (km^2)	Purpose
Volta	Ghana	8500	Hydropower
Nasser	Southern Egypt and Northern Sudan	6500	Irrigation, flood control, hydropower
Kariba	Between Zambia and Zimbabwe	5100	Hydropower
Cahora Bassa	Western Mozambique	2000	Irrigation, flood control, hydropower
Roseires Reservoir	Sudan	2000	Irrigation, hydropower
Koussou	Cote d'Ivoire	1800	Hydropower
Grand Ethiopian Renaissance Dam (not completed)	Ethiopia	1600	Hydropower
Kainji	Nigeria	1300	Hydropower

of a similar size, and drawdown may happen out of season and will thus affect the ecology of the system. When a reservoir has been created, the species present will be the ones that used to live in the river, well-adapted to life in running water, and few of them may be fit to live in a reservoir (for example, due to the lack of suitable spawning grounds or nursery areas for the fry). It is, therefore, a common practice to introduce species well-suited to life in standing water when new reservoirs have been created. Small pelagic clupeids have been introduced in Lake Kariba, Cahora Bassa, Lake Kainji, and Lake Volta, where they have formed self-sustaining populations, which contribute to most of the catches. Catches have, for example, increased five-fold in Lake Kariba after the introduction of *Limnothrissa modern* from Lake Tanganyika creating a million-dollar industry apparently without causing any damage to populations of native species (Kolding and van Zwieten 2006, 2012). However, such practices are, nevertheless, risky and may have serious consequences for the ecosystem, some of which may be felt well beyond the place, where the species were introduced, since fishes are mobile organisms.

23.5.5 Species Diversity

Estimated 16,789 species of fauna have been identified from freshwaters in sub-Saharan Africa (Balian et al. 2008) including around 2945 species of African freshwater fish species from 48 families (Lévêque et al. 2008) and more will surely be discovered in the future. Many of these species are endemic, e.g., 95% of the cichlids from the great lakes are only known from the same lake (Snoeks 2000). There are only a few fish species that move systematically between the sea and freshwater in Africa, but some fishes migrate for hundreds of kilometers from lower river reaches upstream, and others from lakes into tributaries, to complete their lifecycles, and most species displace themselves laterally between rivers and floodplains as a response to the flood cycle.

The *Tilapias* constitute a conspicuous group of African cichlids that play an important role as food fish, and they are caught in fisheries throughout the continent, but in addition, several of these species are now bred and grown in captivity and have been introduced for that purpose to tropical and subtropical regions around the globe, and they are among the three most-produced species (groups) in aquaculture. Another African species that are encountered in tropical aquaculture around the world is the catfish *Clarias gariepinus* which is famous for its hardiness and its fast growth. Ironically Africa is the last continent to take advantage of these resources as aquaculture remains little developed (see below). Other famed African fishes include the long-snouted strange looking and highly intelligent elephant fishes (family Mormyridae), the voracious tiger fish (*Hydrocynus* spp.) that are favored targets for sport fishers, the electric catfish (*Malapterus* spp.) that stuns its prey with an electric shock, the air-breathing bichir (*Polypterus* spp.) that survives the dry season deeply buried in the mud, and many others.

23.5.6 Ecosystem Services

African lakes, rivers, and wetlands ecosystems are important for the services that they provide to millions of rural communities who depend directly on them for their livelihoods. These services include navigation, water purification, groundwater recharge, sediment trapping, fodder, fuel wood, building materials, flood- and drought-buffering capacity, carbon sequestration, chemical regulation, nutrient recycling, habitat for aquatic and semi-aquatic species, food and nutrition for people, cultural and recreational opportunities, and eco-tourism. Freshwater ecosystems also indirectly influence valuable coastal and marine fisheries through regulating freshwater flows nutrient and sediment loads in outflowing water and the exchange of species. Brugere et al. (2015) made a crude estimate that the value of the services provided by African aquatic ecosystems could be in the order of magnitude of USD 10.8 trillion.

23.5.7 Fisheries

Inland capture fisheries are a fundamental ecosystem service that provides a broad range of benefits for development and contributes directly to the United Nations' Sustainable Development Goals (SDGs). Inland fisheries sustain the livelihoods of millions of Africans, because they are diverse and widely distributed including in remote rural locations, where communities lack access to nutrient and protein-rich foods. In these areas, inland fisheries are typically integrated with farming and other economic activities, and there may be few alternative income-generating opportunities seasonally when the crops require less attention. During times of crop failure, economic downturns or civil unrest fishing may provide a safety net, as inland fisheries require only minimal investment and training to participate. More than 3.6 million people are employed in fisheries in Africa, representing 0.8% of the economically active population on the continent. How many of them work in inland fisheries is quite difficult to determine, because there are many part-time fishers, and inland fisheries are often integrated with farming and other rural activities to form a multi-faceted livelihood strategy for rural households. In many parts of Africa, fishers seasonally move between fishing grounds according to the availability of fish. This is a strategy that is well-adapted to a resource with a highly variable abundance.

African inland fisheries are largely small-scale, including the sub-sectors of catching, transportation, trade, gear manufacture, and fish processing, where large numbers of women are working (Neiland et al. 2005). The majority of fishing vessels are canoes (un-decked) with some motorization provided by outboard engines. It is virtually impossible to determine the exact number of canoes given the dispersed and isolated nature of many fisheries (Neiland et al. 2005). Africa is responsible for roughly 25% of global catches from inland fisheries; however, when converted to annual catch per capita (2.56 kg/capita/yr), the contribution of fish from inland fisheries is far higher in Africa than in any other continent mainly due to the poorly developed aquaculture sector. The 3.3 million tons of fish caught annually in African inland fisheries would meet the full dietary animal protein of around 40 million people for 1 year. Fish consumption in Africa with 8.3 kg fish/capita/year is less than half the global average and has remained static for years. Fish nevertheless contributes on average 21% of the daily protein intake for African countries, and nutrients and minerals from fishes are essential for ensuring proper child development and preventing disease and poor health conditions (Craviari et al. 2008; O'Meara et al. 2021). The role of inland fisheries as a supplier of fish in countries with abundant water resources such as Tanzania and Congo is indisputable, in 19 low-income food-deficit African countries (as defined by FAO), more than 50% of the fish produced comes from inland fisheries, and in 9 land-locked countries, more than 90% of the fish produced comes from inland fisheries (Fig. 23.3) (FAO FishStat 2021). With few exceptions, most notably in Egypt, the contribution of aquaculture to fish production in Africa is marginal. Marine small-scale fisheries are important, and there is higher production in countries with easy access to the sea. However, even in some of those countries, the production from inland fisheries can be considered as in the case of Nigeria, Mozambique, and Côte d'Ivoire where inland fisheries still contribute more than 25% of national fish production. This should be noted that inland fisheries are more difficult and expensive to monitor than aquaculture and marine fisheries, and therefore, landings are more likely to be underestimated. Even in countries with relatively small inland fisheries production, it may seasonally contribute significantly to people's livelihoods. It should also be emphasized that inland fisheries concentrate around major water bodies, rivers, or wetlands and the local importance is not reflected in national aggregate statistics.

Most of the production of inland fish is found in the large river and lake basins, e.g., Lake Victoria with one million tons, Niger River basin with 300 thousand tons, Lake Volta with 270 thousand tons, and Lake Tanganyika with 200 thousand tons (Neiland et al. 2005). It is, however, important to recall the important role of small wetlands and water bodies that are not covered by official fisheries statistics, they are more productive per unit area than large lakes and reservoirs, and because they are so numerous, they jointly constitute a much larger surface area (Kolding et al. 2016). For countries such as Chad, Mali, and Uganda, inland fisheries contribute as much as 5–10% to Gross Domestic Product (GDP) (Neiland et al. 2005). Poorly developed facilities at the landing sites include a lack of ice, and deficient transportation infrastructure limit the distance in which fish can be transported fresh, and they are, therefore, mostly traded and consumed locally. Small pelagic are dried and form the basis for a considerable regional trade penetrating areas well beyond the water bodies, where they were caught. The only major export-oriented capture fisheries are the Nile perch fishery in Lake Victoria, the export value of fresh and frozen Nile perch fillets was USD240 million in 2008 (Kolding et al. 2014). Other regional trade in dried and smoked fish products, for example, in the Niger and Lake Chad basins is less formal. The total value of African inland fisheries was estimated to be USD 6.3 billion in 2011 (de Graaf and Garibaldi 2014).

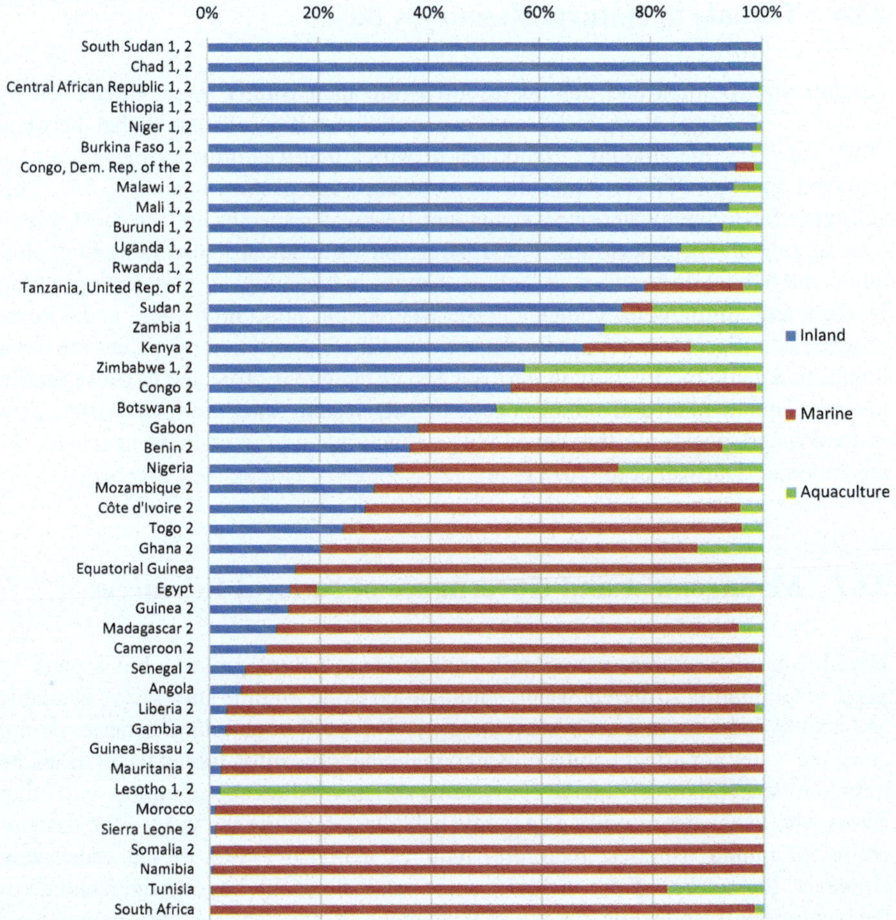

Fig. 23.3 Relative contribution of inland fisheries, marine fisheries, and aquaculture to fish production in African countries reporting inland fisheries catches. (1) Landlocked countries; (2) low-income food-deficit countries

In addition, recreational fisheries are of growing importance and a source of income in several African countries, e.g., South Africa, Namibia, and Botswana, although the lack of statistics does not allow to quantify their contribution to their national economies; finally, it was estimated the export of African aquarium fish to Europe represent 2.5 million USD (Lévêque and Paugy 2010).

23.6 Threats to Natural Resources Base

Africa's environment and natural resources are increasingly being threatened by escalating pressures from fast-growing populations. It is estimated that between 1999 and 2008, the amount of land area in Africa that was utilized for agricultural purposes rose by 30.7%, while the amount used for pasture rose by 8.5%. This strongly affects freshwater ecosystems and freshwater species are the most threatened among all species groups. Floodplains and wetlands are vulnerable to exploitation and habitat destruction, as well as changes in hydrological functioning caused by damming, draining of wetlands, water abstraction, flood protection, and a range of other activities. Pollution, sedimentation, and the introduction of species are other threats to aquatic biodiversity in Africa. Climate change is projected to cause further losses of aquatic plant and animal species and ecosystem functioning. In Africa, 21% of freshwater species are threatened with extinction, and out of those species, 91% are endemic (Darwall et al. 2011).

23.7 Management and Governance of Natural Resources

Inland fisheries resources are not only vulnerable to fishing pressure but depend on good water quality, sufficient water, natural flood cycle (in time and space), available and healthy habitats, and ecosystem integrity. Thus, although inland fisheries do not consume water nor affect habitat or water quality significantly, inland fisheries are an important stakeholder in aquatic ecosystems and are in direct competition with other actors who use water, modify or degrade habitats, or pollute the water. The sustainability of inland fisheries, therefore, requires a holistic view of the ecosystem. However, the control of non-fisheries activities is normally beyond the mandate of the authorities responsible for the management of the fisheries, and achieving a sustainable outcome will, therefore, require substantial efforts to coordinate with the entities responsible for water and land management, and broader developments related to land use (FAO 2019a). Sustainability thus requires a shift away from the traditional management approach narrowly focusing on fish toward critical balancing between the various human needs, i.e., food, income, and livelihoods, as well as the need to sustain the aquatic ecosystem's functioning capacity to deliver services. The objective should always be improving human well-being and equitable sharing of costs and benefits among the various stakeholders. This will require reaching compromises and some aspects of a production system may have to be sacrificed to accommodate another (FAO 2019a). The fisheries authority should act in the sector's best interest by drawing attention to its needs and raising awareness of the benefits it generates, and should actively seek involvement in developing strategies for other sectors that are using water resources or have an impact on them. This will require a comprehensive sectoral analysis of the vulnerability, pressures/drivers, and how inland fisheries relate and interact with other sectors (FAO 2019a).

Planning and management will need to operate at different spatial and political scales if development is to be pursued sustainably while addressing the effects of climate change in the continent (Ogwu 2019; Ikhajiagbe et al. 2022). One way of doing this could be a basin approach, i.e., divide a major basin into manageable units of different sizes (e.g., fisher group, community, stream, sub-basin, and basin) (Hoggarth et al. 1999a, b), although this may not necessarily coincide with the way planning is done in the sub-national context (e.g., village, district, and provincial government). A further complication stems from the fact that most of the major basins in Africa are shared between several countries, i.e., 68 basins, for example, Congo, Niger, Nile, Lake Chad, and the Zambezi all have eight or more countries in their watersheds, each country will have its priorities and legal framework in which it operates (McCracken and Wolf 2019). In fewer than 23 African agreements, a basin committee or commission has been established to oversee that the agreement is respected by the parties, and/or to constitute a forum for discussion and consultation on basin development (Schulze 2015). A permanent secretariat has been established by the most effective basin organizations, which can institute a common vision for the development of the relevant basin and assist the countries in formulating projects and in obtaining donor funding for programs or projects. At the continental level, inland fisheries production in Africa has increased by more than 50% in the last 20 years, but there are still places where the fisheries can be further developed and areas where better management of water and natural resources may lead to an increase in fish production.

In most countries in Africa, fisheries management is centralized with a fisheries department mandated to monitor the fish and set and enforce rules and regulations. The objectives of management have been to maximize fish production using the maximum sustainable yield as the reference point and controlling, and the way to achieve this has been through licensing and other ways to control effort (Neiland et al. 2005). Historically, the performance of these management approaches has been poor, fishing pressure is increasing, many fisheries are in reality open-access, and some fish stocks are threatened by overexploitation (Neiland et al. 2005). These failures are frequently due to the fisheries department's lack of staff and resources to enforce the regulations. Meaning that the fishers themselves become the de-facto managers, therefore, the fishers have been actively involved in management decisions and implementation; these efforts are likely to fail. Imposed regulations coming from the central level are often locally inappropriate and can cause the fishers to be unwilling to comply with them. Management objectives, plans, and measures should rather be developed in partnership with local people, and build on and integrate their knowledge and traditions. One way to achieve this would be through the use of an Ecosystem Approach to Fisheries (EAF) as advocated by the FAO.

The EAF is a holistic and flexible planning framework that promotes decision-making processes that balance ecological, human, and societal well-being, within improved governance frameworks. It addresses the multiple needs and desires of societies, without jeopardizing sustainability. During the implementation of the EAF, a management plan is developed through consultations with the primary stakeholders. The plan will address the concerns and problems that have been identified and will consider the risks assessed. Objectives, indicators, and management measures will be agreed upon. The plan should be implemented in small steps using precautionary principles and should be adaptive and incorporate periodic revisions and adjustments (FAO 2019a, b).

23.8 Framework for Natural Resources Management

The framework may include a description and the characteristics of natural resources which will form the implementation of policies and strategies to address the issues. The framework involves four basic criteria (Fig. 23.4):

- Natural resource policy.
- Legal and regulatory framework.
- Fiscal regimes.
- Sustainable development.

23.8.1 Natural Resource Policy

It is a document that clearly states and addresses the objectives of all stakeholders and actors in the exploitation of natural resources. The purpose of this policy is to:

- Coordinate the use of these resources by clearly stating the role of individuals, stakeholders, and the government in exploratory activities of the resources.

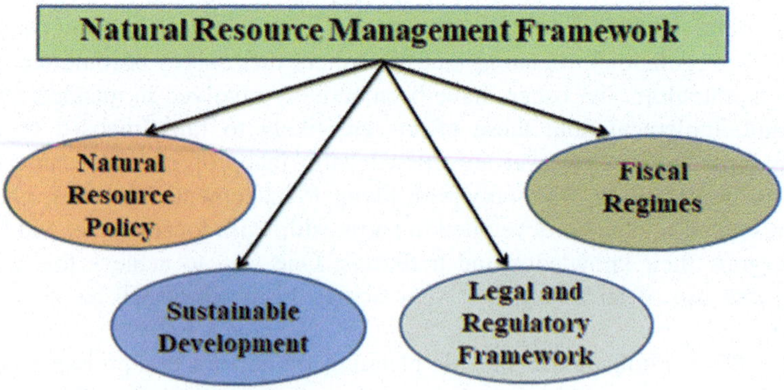

Fig. 23.4 Framework for natural resources management

- Ensure sustainable developmental practices are followed when the resources are being exploited.
- Ensure revenues from the exploitation of natural resources are used for appropriate investments and programs as stated.

23.8.2 Legal and Regulatory Framework

It outlines all the necessary and relevant elements of the natural resource policy and puts in place the legal and regulatory frameworks, procedures, and actions to be taken during the exploration of the natural resources. The purpose of this framework is to:

- Provide a framework that is legislative and administers the exploitation of the natural resources.
- Provide licenses, permits, and rights to miners for the exploration of the natural resource.

23.8.3 Fiscal Regimes

It outlines and illustrates in detail all the fiscal terms and policies that are related to the exploitation of natural resources. The purpose of this is to:

- Provide fiscal terms that can be applied to all commercial activities during the exploitation of the resource.
- The non-renewable resources should have a higher level of taxation than other resources principally because they cannot be replaced.

23.8.4 Sustainable Development

It illustrates and promotes sustainable explorative activities and the exploitation of natural resources. It covers the social, environmental, and economic issues of exploitation. The unstable nature of non-renewable resources had made a difference in the management approach of non-renewable and renewable natural resources. The purpose of this is to:

- Ensure that the principles and policies of sustainable development are respected and obeyed. This will enable socio-economic and environmental development.
- Preservation of the natural biota and environment.

23.9 Sustainable Development in Africa

The natural wealth of Africa which lies in the environment has social and economic benefits which require urgent attention. Issues such as the management, and the economic and environmental impacts must be addressed (Fig. 23.5).

The conservation measures of the African government are supported by the United Nations Environmental Program Africa Office and they are in charge of the following:

- Translate statements taken into practical actions.
- Translate decisions into practical actions.
- Wealth and job creation for locals.
- Food security.
- Generation of revenue.
- Social equity.
- A healthy environment.

There are three major advantages of sustainable development:

- The life of present and future generations is improved.
- The environmental impact of pollutants in the air, soil, and water can be reduced.
- Long-term growth of the economy can be achieved.

The contribution of Africa to the mineral resources in the world is 30% and 57% of the earnings is from the exportation of hydrocarbons. The oil reserves in Africa have expanded by 150% from 1980 to 2012 (Al-Jazeera 2018) and the third-largest budget for projects that are exploitation oriented was received by Africa in 2020. This reduced in 2020 to 10% and it was recorded as the lowest when compared with the previous 4 years (Heiberg and Reid 2021). The discovery of Gold was a major account in West Africa from 2009 to 2019 and had a great contribution to the economy and foreign exchange (Jamasmie 2021).

Fig. 23.5 Issues that affect sustainable development

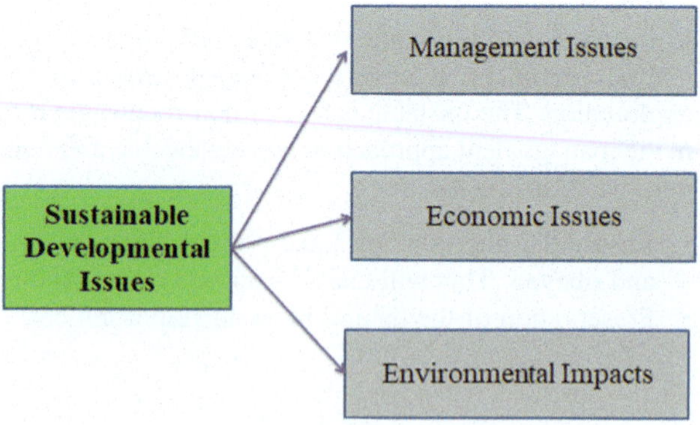

23.10 State of Natural Resources in Africa

A critical resource in Africa is Land which is very abundant but globally it is poorly managed and highly degraded. The land is a major key in food production and nutrition for humans and measures to protect that it is pertinent. It is used for agriculture which employs 60–70% population and 30–40% of the wealth of Africa. Issues such as insecurity, conflicts, lack of agricultural services, lack of infrastructure, and extreme weather conditions have affected the nutrition of about 22.9% of Africans (FAO 2013). In Africa, 66% are deserts (Liniger et al. 2011), while 29% are arable land (NEPAD 2002). Despite these resources, there is under-exploitation of resources with Africa having the lowest yield, because it uses less than 30% of land for agriculture. Soil degradation is a major issue that is affected by various factors presented in Table 23.6. The loss from these factors is estimated to be 25% and 6.6% for cropland and pastureland, respectively, and it also impacts forests, freshwater, and biodiversity.

Water scarcity is a major issue in Africa and it is caused by increased temperatures resulting in high evaporation and the fluctuating patterns of rains. It was reported that nearly 400 million people living around the 36 largest river basins in Africa experience scarcity of water at least once a month each year (AfDB and WWF 2012; Unmusig and Cramer 2008). Countries in the East, Southwest, and Northern part of Africa experience water scarcity, because they exhaust their reserves. In sub-Saharan Africa, only 60% have access to water that is clean and safe in 2010, and has increased by 11% in the last 20 years, because progress is very slow (WHO and UNICEF 2010). In Africa, only 4% of the arable land uses an irrigation system, and the climate change factor is projected to increase water scarcity by 2025 (Juel 2013; Somorin 2010). Anthropogenic factors and resultant pollutants from these factors degrade the natural catchment, thereby reducing its usefulness (UNEP 2013). Mismanagement of effluents from agricultural, domestic, or industrial sources is the major cause of water stress (UNEP 2013; FAO 2010).

Africa accounts for one-third of the biodiversity globally and these resources are disappearing very quickly, affecting the future development of the environment. The continent has 8 biodiversity hotspots out of the 34 found in the world and these provide goods, services, and medicine for over 80% of the population (UNEP 2013). The Eastern and Southern parts of Africa are rich ecotourism destinations and travels. Crop improvements by developing disease and pests' resistant varieties are offered by Africa's genetic diversity. Pharmaceutical products and climate-resistant crops can also be genetically produced. However, biodiversity is affected by the

Table 23.6 Issues affecting soil degradation in Africa

S/N	Issues	Contribution (%)
1	Overgrazing	50
2	Poor management of agriculture	24
3	Removal of vegetation	14
4	Overexploitation	13

destruction and fragmentation of habitats, overharvesting, illegal trade, and invasive species.

23.11 Importance of Natural Resources

Natural resources have diverse contributions to the natural economy and people who depend on them directly for one function or the other. Some of the importance are itemized (Fig. 23.6):

- Provision of natural capital.
- Source of energy.
- The basis for further scientific research.
- Source of food.
- Source of raw materials for industries.
- Medicinal values.
- Shelter.
- Employment opportunities.
- National development.
- Ecosystem services.

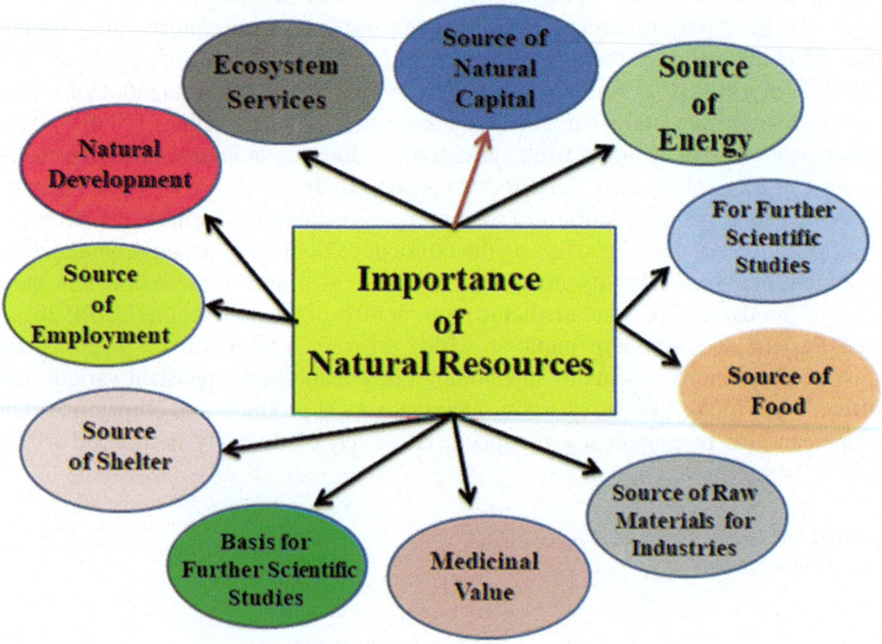

Fig. 23.6 Importance of natural resources

23.11.1 Energy Source

The provision of energy for life processes is one of the most important functions of natural resources. Energy for use can be generated from resources, such as water, wind, tides, petroleum, solar radiation, etc. The sun is the principal source of energy for life processes in the natural environment. It governs photosynthetic activities and food production. Almost 100% of the energy used by man is from the natural resources in the environment (Grieg-Gran 2008).

23.11.2 Natural Capital

These describe all the products and assets provided by the environment which is free and essential in life processes. Examples are water, soil, air, and living biota. The natural capital is seen as the world's stock of all-natural resources available in the environment. Despite this gift of nature, the resources must be exploited sustainably.

23.11.3 Basis for Further Scientific Studies

The dynamics of the natural environment and its resource composition are a source of inspiration for scientists who are hungry for biodiversity exploration and sustainability. They study these parameters and report significant information necessary for the management, utilization, and conservation of the environment. Such information is of great benefit to individuals and the government at large. For instance, petroleum and cotton have been converted via scientific research into energy sources and cotton products, respectively. This further elaborates that the value placed on natural resources is very pivotal.

23.11.4 Food Source

Nature is very rich in food materials necessary for healthy living. The environment contains diverse resources which are important sources for all classes of food required for a balanced diet by humans and animals. Plants are not left out, as the environment provides all nutrients required for their growth and production. The sources of food in the natural environment are plants, organisms, and other animals. No living thing can survive without food and this emphasizes the great importance of the natural environment.

23.11.5 Sources of Raw Materials for Industries

Raw materials are required for production processes by industries and ancillary services. These resources can be provided by the natural environment. For instance, the raw material required by the petroleum industry is crude oil produced by fossil fuels under the ground. The textile industries may use materials such as hides and skin gotten from the environment for the production of cotton materials. It must be noted that these raw materials must be exploited sustainably for a balance in nature.

23.11.6 Medicinal Value

The forests can provide items that are used to cure diseases and treat various ailments in the world. The majority of the population in the world today depends on the environment for treatment. Drugs can be made from products gotten from the tree bark, leaves, roots, grasses, and other plant parts and used as vaccines against diseases. Some minerals can also be used for medicinal purposes; for instance, titanium which is an element concentrated in the earth's crust can be used in prosthetics (Juel 2013).

23.11.7 Further Scientific Studies

The existence of natural resources in the environment has inspired and still inspires so many enthusiastic scientists. These have enabled them to undergo further studies and make products that can benefit mankind and the world at large. This includes petroleum; which turned into a popular source of energy through scientific research, and cotton; which now serves as the most popular raw material for cloth production, with all these values; the importance of natural resources cannot be overemphasized.

23.11.8 Provision of Shelter

Materials used for the construction of buildings such as timber, limestone, mud, gravel, etc. are gotten from natural materials in the environment. The provision of shelter is mostly brought into existence by natural resources. Conscious efforts must be geared toward maintaining a balance between deforestation and a sustainable environment. The trees help in carbon sequestration, and when depleted, the concentration of greenhouse gases may increase leading to climate change.

23.11.9 Provision of Employment Opportunities

Over 80% of the world's workforce is involved in the developmental stages of natural resources. They are participatory in the processing of the raw materials from natural resources into finished products. Most of the processes in the development of raw materials have a series of stages that can serve as potential employment areas and millions of jobs are created. A typical example is the exploration of oil; it involves processes, such as the mapping of the site, test drilling, construction of tanks, laying of pipes, buildings, and refining. Each of these stages are areas, where employees can be granted (Walter 2011).

23.11.10 National Development

Countries with well-established laws and policies have their GDP values on the increase. This is because the natural resources are properly managed and it translates to economic development. Natural resources have contributed over 90% of revenue in African countries and this indicates the great importance of the resource. The increased levels of corruption and bad governance have resulted in countries being poor despite the numerous and diverse natural resources that they have. Those in the arm of governance see it as a means to accrue wealth rather than economic development. To reduce this act, there is a need to improve cooperation between the financial institutions, the country, and the private sector companies (Buncombe 2006).

23.11.11 Ecosystem Services

This refers to all benefits that are derived from the natural environment and these services can contribute to the natural resources in the environment. The ocean is an important resource for the hydrological cycle (Ross 2001). The services include:

- Supporting services: these include photosynthesis, water, and nutrient cycle in the environment.
- Regulatory services: these include water purification, pollination, and climate regulation activities.
- Cultural services: these include aesthetics.
- Provision services: these include the provision of food, shelter, and water for life processes in the ecosystem.

23.12 Sustainable Management of African Natural Resources: Approaches and Challenges

23.12.1 Challenges

The sustainable management of African natural resources centers around Sustainable Land Management (SLM) which is yet to be applicable across the African continent. SLM as defined by Liniger et al. (2011) is the adoption of land-use systems that can enable land users to maximize benefits from the land through appropriate managerial practices and ecological support of resources. SLM approach is not new in most African countries, because it recognizes the developmental plans (TerrAfrica 2009). The main challenges faced in sustainability are (Fig. 23.7):

- Demographic pressures.
- The consequent needs for land.
- Insufficient technical, institutional, and financial capacities.
- Inadequate incentives, laws, and policy instruments including economic, regulatory, and social marketing instruments that are implementable to scale.
- Lack of secured tenure rights that are usually required for farmers to undertake long-term investments on their land (TerrAfrica 2009).

In the African continent, policies on water and its use have been concentrated on infrastructure expansion instead of managing the water resources. The framework of governance and enforcement of the law is very weak and the financial means to ensure sustainable management and use of resources (Sachs and Warner 2001). The Integrated Water Resources Management (IWRM) adopted a policy in the year 2000 which circled the development, coordination, and management of water to ensure sustainability. The implementation of the IWRM was poor and international basin authorities for the continent were created, because major lakes and rivers were situated at the border of most countries (Juel 2013). Their mission which was to facilitate a sustainable and efficient management system for water resources amongst

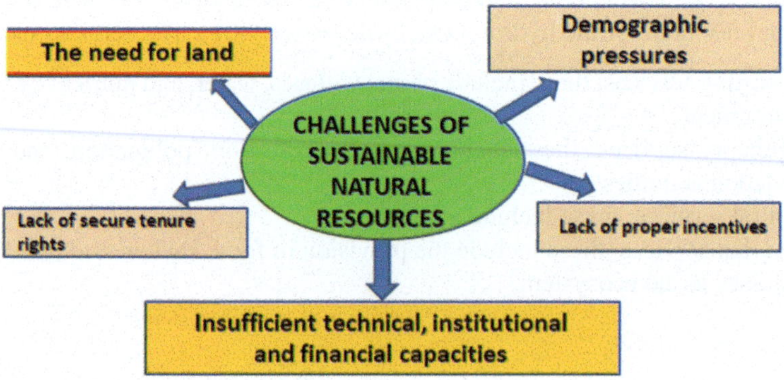

Fig. 23.7 Challenges of sustainable natural resources

the neighboring countries could not be carried out, because the authorities lacked the appropriate capacity to achieve this (WWC/CONAGUA 2006; FAO 2005).

23.12.2 Approaches

The strategies for sustainable management in Africa have recently relied on three main approaches (Fig. 23.8):

- The protected area.
- Sustainable timber exploitation.
- Community-based forest management.

Forests were principally created for conserving biodiversity and forests situated in protected areas were more conserved and at 13.4% of the total forest area (FAO 2010; Song et al. 2019; Kerfahi et al. 2019; Ikhajiagbe et al. 2020). Forest area conservation have increased by five million hectares since 1990 and this has increased due to appropriate forest management plans and actions. Many African countries are proposing decentralizing the management of forest areas to local communities. These measures will benefit the local populace, and help synergy between the conservation of resources and local development. With the slow development of community-based NRM in sub-Saharan Africa, protected areas have increased to 11.7% of land and 5.8% of territorial waters. This area has also received the lowest levels of human resources and investments and the results were low in terms of management and effective reduction in the extinction of biodiversity (FAO 2010). The lack of financial investments for sustainable management as well as protected areas is threatened. The government of Africa allocates a minor budget to this sector and allows investors and donors to make the largest investment contribution (Grieg-Gran 2008).

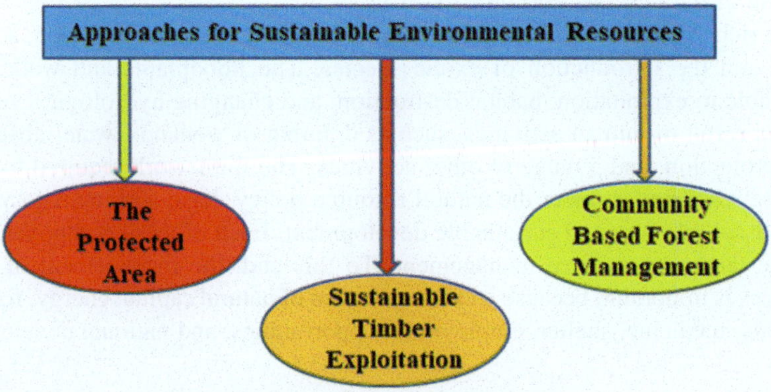

Fig. 23.8 Approaches for sustainable environmental management

The allocation of funds and spending of most African countries on conservation is less than 1% of their GDP. The countries that benefit from tourism spend more of their funds on it. Budget allocation and implementation must ensure to cover management and compensation for locals (James et al. 2001). About US$ 1.9 billion/year is required for the conservation of forests and it covers the implementation of policies, programs that may bring changes and compliance of locals (Juel 2013).

23.13 Conclusions

The African environment is rich in diverse natural resources which can be harnessed for various purposes ranging from aesthetics to income generation for the economy. The continent is a major exporter of various metals and non-metals, precious stones, and a major contributor to crude oil production. The focus of countries is on the exploitation of non-renewable resources, because it commands more price, although they cannot be recuperated by natural processes. Unlike renewable resources, there is less pressure on them and they can regenerate through natural processes. Fossil fuels such as crude oil, gas, and oil are the major components of this category. Minerals which comprise energy minerals, precious metals and minerals, ferrous and non-ferrous metals, and industrial minerals are all exploited from the natural environment. These resources must be managed for sustainability; most especially the non-renewable resources by ensuring optimal exploratory activities, developing policies, and ensuring the health guarantee of exploiters. These will also strengthen the African natural resources base which contains more freshwater with floodplains (51%) having the largest component. Other components are the forests (13%), rivers (3%), reservoirs (3%), lakes (16%), and temporary wetlands (14%). All of these have various effects on the diversity of fish species and the services delivered by these aquatic ecosystems to the African population cannot be overestimated. Fish species such as tilapia, *Clarias gariepinus* (catfish), elephant fish, Tiger fish (*Hydrocynus* spp.), electric fish (*Malapterus* spp.), and the *Polypterus* spp. have contributed to the boosting of the fisheries economy of Africa.

The rich African resources are threatened by issues, such as pollution, sedimentation, and the introduction of exotic species. The floodplains and wetlands are vulnerable to exploitation, habitat destruction, and changing hydrological functioning as a result of human activities, such as draining of wetlands, water abstraction, flood protection, and a range of other activities. The framework required to sustain all these resources is under the natural resource policy, legal and regulatory framework, fiscal regimes, and sustainable development. Each of these components has a specific target in resource management. To this end, the management of natural resources is important, because it can be a source of natural capital, energy, food, raw materials, medicine, shelter, employment opportunities, and national development.

References

African Development Bank (AfDB) (2013) African Development Report 2012. Towards Green Growth in Africa. http://www.afdb.org/fileadmin/uploads/afdb/Documents/Publications/African_Development_Report_2012.pdf

African Development Bank (AfDB) (2011) Organization for Economic Cooperation and Development (OECD), United Nations Development Program (UNDP), and United Nations Economic Commission for Africa (UNECA) (2011) African Economic Outlook 2011: Africa and its Emerging Partners. Paris/Tunis: AfDB and OECD. http://www.afdb.org/fileadmin/uploads/afdb/Documents/Generic-Documents/Media_Embargoed_Content/EN-AEO_2011_embargo%206%20Juin.pdf

African Development Bank (AfDB) and World Wide Fund for Nature (WWF) (2012) Africa Ecological Footprint Report. http://d2ouvy59p0dg6k.cloudfront.net/downloads/africa_efr_english_low_res_1.pdf

Ainsworth R, Cowx IG, Funge-Smith SJ (2021) A review of major river basins and large lakes relevant to inland fisheries. https://www.fao.org/documents/card/en/c/cb2827en/

Al-Jazeera (2018) Mapping Africa's natural resources. 20 February 2018

Aquastat (2007) Dams and agriculture in Africa. FAO, Rome. https://www.fao.org/3/a-bc815e.pdf

Balian EV, Segers H, Lévèque C, Martens K (2008) The freshwater animal diversity assessment: an overview of the results. Hydrobiologia 595:627–637. https://doi.org/10.1007/s10750-007-9246-3

Brugere C, Lymer D, Bartley DM (2015) Ecosystem services in freshwater fish production systems and aquatic ecosystems: recognizing, demonstrating and capturing their value in food production and water management decisions. TEEB Agriculture & Food. UNEP, Geneva

Buncombe A (2006) African bio-resources 'exploited by West'. The Independent

Craviari T, Pettifor JM, Thacher TD, Meisner C, Arnaud J, Fischer PR, the Rickets Convergence Group (2008) Rickets: an overview and future directions, with special reference to Bangladesh. J Health Popul Nutr 26(1):112–121

Darwall WRT, Smith KG, Allen DJ, Holland RA, Harrison IJ, Brooks EGE (eds) (2011) The Diversity of Life in African Freshwaters: Under Water, Under Threat. An analysis of the status and distribution of freshwater species throughout mainland Africa. IUCN, Cambridge/Gland, p 347. Annexes

de Graaf G, Garibaldi L (2014) The value of African fisheries, FAO Fisheries and Aquaculture Circular. No. 1093. FAO, Rome, p 76

DME (2007) The South African Department of Minerals and Energy (DME). http://www.dme.gov.za

DTI (2005) Department of Trade and Industry (DTI) (SA) Economic Database http://www.dti.gov/za/econdb/

FAO (2019a) Regional Conference on river habitat restoration for inland fisheries in the Danube River basin and adjacent Black Sea areas. Conference Proceedings, 13–15 November 2018, Bucharest, Romania. FAO Fisheries and Aquaculture Proceedings. No. 63. FAO, Rome

FAO (2019b) Ecosystem approach to fisheries management training course (Inland fisheries)—Volume 1: Handbook for trainees. Rome, FAO

FAO (2013) Equitable payments for watershed services: financing conservation and development. http://www.fao.org/fileadmin/user_upload/pes-project/docs/FAO_RPEPES_WWF_Tanzania.pdf

FAO (2010) Global Forest Resources Assessment 2010. http://www.fao.org/docrep/013/i1757e/i1757e.pdf

FAO (2005) AQUASTAT Information System on Water and Agriculture, Country Profiles

FAO (2003) Review of world water resources by country. FAO, Rome, p 71

FAO FishStat (2021) FAO global fisheries and aquaculture statistics. https://www.fao.org/fishery/statistics

Frenken K (2005) Irrigation in Africa in figures. AQUASTAT Survey – 2005. FAO, Rome, p 649

Grieg-Gran M (2008) The cost of avoiding deforestation: Update of the report prepared for the Stern Review of the Economics of Climate Change. Paper commissioned by the Office of Climate Change as background work to its report Climate Change: Financing Global Forests (The Eliasch Review). IIED, London. http://pubs.iied.org/pdfs/G02489.pdf

Heiberg T, Reid H (2021) Africa's pandemic-hit mining sector faces exploration challenge. Reuters

Hoggarth DD, Cowan VJ, Halls AS, Aeron-Thomas M, McGregor JA, Garaway CA, Payne AI, Welcomme RL (1999a) Management guidelines for Asian floodplain river fisheries. Part 1. A spatial, hierarchical and integrated strategy for adaptive co-management, FAO Fisheries Technical Paper. No. 384/1. FAO, Rome, p 63

Hoggarth DD, Cowan VJ, Halls AS, Aeron-Thomas M, McGregor JA, Garaway CA, Payne AI, Welcomme RL (1999b) Management guidelines for Asian floodplain river fisheries. Part 2. Summary of DFID research, FAO Fisheries Technical Paper. No. 384/2. FAO, Rome, p 117

Ikhajiagbe B, Ogwu MC (2020) Hazard quotient, microbial diversity and plant composition of spent crude oil polluted-soil. Beni-Suef Univ J Basic Appl Sci 9(26). https://doi.org/10.1186/s43088-020-00052-0

Ikhajiagbe B, Chukwujama CA, Ikhajiagbe B (2022) Intrinsic restoration of a ferruginous utisol using goosegrass (*Eleusine indica* L. [Gaertn.]) obtained from different land use areas. Journal of Natural Resources and Development 12:1–18. https://doi.org/10.18716/ojs/jnrd2022.12.01

Ikhajiagbe B, Ogwu MC, Lawrence AE (2020) Single-tree influence of *Tectonia grandis* Linn. f. on plant distribution and soil characteristics in a planted forest. Bull Natl Res Centre 44:29. https://doi.org/10.1186/s42269-020-00285-0

Jamasmie C (2021) Exploration boom in West Africa challenged by lack of services — report. MINING.COM

James A, Gaston K, Balmford A (2001) Can we afford to conserve biodiversity? Bioscience 51:1. http://mail.cbd.int/doc/articles/2002–/A-00486.pdf

Juel M (2013) Scoping review, Renewable Natural Resources in Africa. Report for the African Development Bank

Kerfahi D, Tripathi BM, Slick JWF, Sukri RS, Jaafar S, Dong K, Ogwu MC, Kim H-K, Adams JM (2019) Soil metagenome of tropical white sand heath forests in Borneo: What functional traits are associated with an extreme environment within the tropical rainforest? Pedosphere 29(1): 12–23. https://doi.org/10.1016/S1002-0160(18)60054-2

Kolding J, van Zwieten PAM (2006) Improving productivity in tropical lakes and reservoirs, Challenge Program on Water and Food - Aquatic Ecosystems and Fisheries Review Series 1. Theme 3 of CPWF. Cairo, Egypt, WorldFish Center, p 139

Kolding J, van Zwieten P (2012) Relative Lake level fluctuations and their influence on productivity and resilience in tropical lakes and reservoirs. Fish Res 115–116:99–109

Kolding J, Medard M, Mkumbo O, van Zwieten P (2014) Status, trends and management of the Lake Victoria fisheries. In: Welcomme RL, Valbo-JØrgensen J, Halls AS (eds) Inland fisheries evolution and management – case studies from four continents, FAO Fisheries and Aquaculture Technical Paper No. 579. FAO, Rome, p 77

Kolding J, van Zwieten P, Marttin F, Poulain F (2016) Fisheries in the drylands of sub-Saharan Africa – "Fish come with the rains". Building resilience for fisheries-dependent livelihoods to enhance food security and nutrition in the drylands, FAO Fisheries and Aquaculture Circular No. 1118. FAO, Rome. http://www.fao.org/3/a-i5616e.pdf

Lévêque C, Paugy D (2010) Animal resources and diversity in Africa, Freshwater fishes in Africa. Encyclopedia of Life Support Systems (EOLSS). Developed under the auspices of the UNESCO. Eolss Publishers, Paris

Lévêque C, Oberdorff T, Paugy D, Stiassny MLJ, Tedesco PA (2008) Global diversity of fish (Pisces) in freshwater. Hydrobiologia 595:545–567. https://doi.org/10.1007/s10750-007-9034-0

Liniger HP, Mekdaschi SR, Hauert C, Gurtner M (2011) Sustainable Land management in practice – guidelines and best practices for Sub-Saharan Africa. Terr Africa, World overview of

conservation approaches and technologies (WOCAT) and food and agriculture organization of the United Nations (FAO)
Lymer D, Marttin F, Marmulla G, Bartley DM (2016) A global estimate of theoretical potential annual Inland capture fisheries harvest. Freshwater, fish, and the future: Proceeding of the global cross-sectoral conference. FAO/American Fisheries Society
McCracken M, Wolf A (2019) Updating the Register of International River Basins of the world. Int J Water Resour Dev 35(5):732–782. https://doi.org/10.1080/07900627.2019.1572497
Neiland AE, Chimatiro S, Khalifa U, Ladu BMB, Nyeko D (2005) Inland fisheries in Africa key issues and future investment opportunities for sustainable development. technical review paper – Inland Fisheries NEPAD-Fish for all summit. NEPAD, Abuja
NEPAD (2002) Comprehensive Africa Agriculture Development Programme. http://www.nepad.org/system/files/caadp.pdf
Ogwu MC (2019) Towards sustainable development in Africa: the challenge of urbanization and climate change adaptation. In: Cobbinah PB, Addaney M (eds) The geography of climate change adaptation in Urban Africa. Springer Nature, Switzerland, pp 29–55. https://doi.org/10.1007/978-3-030-04873-0_2
O'Meara L, Cohen PJ, Simmance F, Marinda P, Nagoli J, Teoh SJ, Funge-Smith S, Mills DJ, Thilsted SH, Byrd KA (2021) Inland fisheries critical for the diet quality of young children in sub-Saharan Africa. Glob Food Sec 28:100483. https://doi.org/10.1016/j.gfs.2020.100483
Osawaru ME, Ogwu MC, Braimah L (2013a) Growth responses of two cultivated Okra species (*Abelmoschus caillei* (A. Chev.) Stevels and *Abelmoschus esculentus* (Linn.) Moench) in crude oil contaminated soil. Niger J Basic Appl Sci 21(3):215–226
Osawaru ME, Ogwu MC, Chime AO (2013b) Assessment of growth performance of two Okra species (*Abelmoschus esculentus* [L.] Moench and *Abelmoschus caillei* [A. Chev.] Stevels) exposed to crude oil contaminated soil. Niger J Biotechnol 26:11–20
Park C, Allaby M (2017) A dictionary of environment and conservation. Oxford University Press. https://doi.org/10.1093/acref/9780191826320.001.0001
Ross M (2001) Does Oil Hinder Democracy. World Polit 53(3)
Sachs JD, Warner AM (2001) Natural resource abundance and economic growth, development discussion paper no. 517a. Harvard institute for international development, Cambridge
Schulze SBC (2015) Environmental Change and Adaptation Capacities of River Basin Organizations in Southern Africa. Dissertation. Leipzig University
Shahin M (2003) Hydrology and water resources of Africa, vol 41. Water Science and Technology Library/Springer, Dordrecht. https://doi.org/10.1007/0-306-48065-4_1
Snoeks J (2000) How well known is the ichthyodiversity of the large East African lakes? In: Rossiter A, Kawanabe H (eds) Ancient lakes: biodiversity, ecology and evolution, Advances in Ecological Research, vol 31, pp 17–38
Somorin O (2010) Climate impacts, forest-dependent rural livelihoods and adaptation strategies in Africa: a review. Afr J Environ Sci Technol 4(13):903–912. http://www.ajol.info/index.php/ajest/article/view/71407/60354
Song H-K, Singh D, Tomlinson K, Yang X, Ogwu MC, Slik JWF, Adams JM (2019) Tropical forest conversion to rubber plantation in southwest China results in lower fungal beta diversity and reduced network complexity. FEMS Microbiol Ecol 95:fiz092. https://doi.org/10.1093/femsec/fiz092
TerrAfrica (2009) Policy and Financing for Sustainable Land Management in Sub Saharan Africa. http://www.caadp.net/pdf/Policy%20and%20Financing%20for%20SLM%20in%20Sub-Saharan%20Africa%201.0.pdf
TradeMap (2007) International Trade Database Online http://www.trademap.org
UNEP (2013) Africa Environment Outlook 3: Summary for Policy Makers. http://www.unep.org/pdf/aeo3.pdf
Unmusig B, Cramer S (2008) Climate Change in Africa. GIGA Focus 2, German Institute of Global and Area Studies. Institute of African Affairs, Hamburg. http://www.boell.de/sites/default/files/assets/boell.de/images/download_de/worldwide/giga2_hamburg_afrika.pdf

Walter R (2011) How Europe underdeveloped Africa. Baltimore. ISBN 9781574780529. OCLC 815651132

Welcomme R, Lymer D (2012) An audit of inland capture fishery statistics – Africa. FAO Fisheries and Aquaculture Circular No. 1051. Rome, FAO. 61.

WHO/UNICEF (2010) Progress on sanitation and drinking-water 2010 update. http://whqlibdoc.who.int/publications/2010/9789241563956_eng_full_text.pdf

WWC/CONAGUA (2006) Local Actions for a Global Challenge, Regional Document: Middle East and North Africa. 4 World Water Forum, Mexico 2006. http://www.worldwaterforum4.org.mx/uploads/TBL_DOCS_109_21.pdf

Chapter 24
Toward Sustainable Biological and Environmental Policies in Africa

Smith Etareri Evivie and Ejiroghene Ruona Evivie

Abstract The African continent is endowed with significant biodiversity and environmental resources, including terrestrial and aquatic species, water bodies, natural mineral reserves, and an enormous and diverse landmass. This makes it a prime investment hub for some of the world's biggest corporations. Therefore, effective and sustainable policies on managing these resources are crucial to balance the need for and pursuit of development with preserving biodiversity and environmental resources. This chapter gives an overview of the policies regulating the biological and environmental resources on the continent. It begins by presenting Africa's biological and ecological resources based on available literature and then highlights the policy structure currently adopted by some countries within the continent. The policy and legal frameworks of some of the continent's leading and emerging economies—Nigeria, South Africa, Egypt, and Ethiopia—are also discussed. To enhance the current policy structure, African countries need to understudy the constantly evolving legal frameworks of model countries, such as China and Australia, which have made significant progress in framing and constantly updating their biodiversity and ecological policies. Some key challenges of sustainable biological and environmental policymaking in Africa include inadequate government funding, the government's inability or unwillingness to prioritize sound

---

S. E. Evivie (✉)
Department of Animal Science, Faculty of Agriculture, University of Benin, Benin City, Nigeria

Department of Food Science and Human Nutrition, Faculty of Agriculture, University of Benin, Benin City, Nigeria

Key Laboratory of Dairy Science, Ministry of Education, Northeast Agricultural University, Harbin, China
e-mail: smith.evivie@uniben.edu

E. R. Evivie
Division of Plant Science, Research School of Biology, College of Science, Australian National University, Acton, Australia

Department of Plant Biotechnology, College of Life Sciences, Northeast Agricultural University, Harbin, China

S. C. Izah, M. C. Ogwu (eds.), *Sustainable Utilization and Conservation of Africa's Biological Resources and Environment*, Sustainable Development and Biodiversity 32, https://doi.org/10.1007/978-981-19-6974-4_24

policymaking, and scarce literature on progress made by several African countries in supporting evidence-based policymaking and enacting. We recommend that political support be galvanized for sound policymaking and revisions for sustainability purposes. It is pertinent that evidence-based research informs scientifically clear policy drafts and that this should replace the conventional, politically driven, and ineffective ones. The potential of digital platforms to raise awareness of policymakers about emerging biological and environmental concerns is also encouraged.

Keywords Africa · Biodiversity · Environment · Depletion · Policy · Sustainability

24.1 Introduction

The vast continent of Africa is endowed with an array of biological and environmental resources (Asare et al. 2013; Negash 2021). From marine biodiversity in the north (Mona et al. 2019), medicinal plants in the south (Van Wyk and Prinsloo 2018), and forest reserves in the east/central region (Karame 2015; Richardson et al. 2020), to a mix of biological and environmental resources in the sub-Saharan west (Adom 2018; Izah et al. 2018), Africa is indeed a diversity hub for several ecological and biological matrices. These have played integral roles in sustaining the continent for centuries. Still, as several recent reports suggest, these resources are at risk of being depleted below threshold levels (Beirne et al. 2019; Lal and Stewart 2019; Orimoloye et al. 2020).

Climate variations, particularly in the past two decades, have raised environmental activists' and researchers' concerns about its devastating effects on biotic and abiotic resources across the continent (Ogwu 2019a; Raimi et al. 2021). Notable examples include human activities that deplete biodiversity resources in Rwanda (Venuste et al. 2017), Egypt (Elsaid Saeed and Bedair 2021), Nigeria (Osawaru et al. 2013a, 2013b, 2013c; Imarhiagbe et al. 2020), South Africa (Sink et al. 2012), and Morrocco (Benamar et al. 2021). In addition to political and economic instabilities, inconsistencies in policy enactment have been identified as a deficit factor across several governments and establishments (Ordway 2015). Government policy lays out the goals that need to be achieved by the government, backed by empirical data, and the pathways by which stipulated objectives will be met. It clarifies a government's political position and priority in a fast-changing world (Lukey and Hall 2020).

It is pertinent that sustainable policies bordering on biological and environmental resources in Africa be enacted in the shortest possible time, drawing from recent commentaries showing the detrimental effects of pollutants in various parts of the continent (Ikhajiagbe and Ogwu 2020; Olisah et al. 2020). In addition, recent findings indicate that biological resource depletion in Africa can have far-reaching economic implications (Diagne et al. 2021). This chapter thus sets out to give an overview of Africa's current position on steps taken to safeguard and sustain her natural and biodiversity resources. Specifically, it will explore the policy structure of several African countries, the diversities in these policies, commendable models to consider adopting and where governments should be headed in the future. It is anticipated that this will be a renewed call to wake up the sleeping giants and point government departments and policymakers in the right direction regarding concrete and sustainable policy development.

24.2 Africa's Biological and Environmental Resources

As a continent, Africa represents a growing global investment location, attracting some of the most lucrative actors in the oil, gas, tourism, biological, environmental and financial sectors (Habimana 2022; Signé and Heitzig 2022). This section will take a cursory look at some of the continent's major environmental and biological hubs and how their diversity plays strategic roles in positioning the countries for future development potential. Egypt, known historically as an integral point in human civilization, harbors some of the richest algae and photosynthetic organisms on the entire continent, with marine macroalgae being one of the most biologically active natural resources (Levine and Fleurence 2018). These algae and other lower plant species contain bioactive ingredients that have been reported to be bactericidal, anti-inflammatory, and anticoagulant (Ogwu 2019b; Barzkar et al. 2019; Khalid et al. 2018). In addition, macroalgae have beneficial renoprotection and dietary properties. China and Japan are among the world's highest algae consumers (Leandro et al. 2019). The Nile River is a major water source for a significant proportion of the population for agricultural, industrial, and personal use (Abdelhafez et al. 2020). However, it has been reported that using wastewater and seawater desalination make up non-conventional water sources in Egypt (Djuma et al. 2016).

Nigeria is Africa's largest economy and is generally regarded as the continent's giant. Although the country is known globally as a major oil producer, it also has various environmental and biodiversity resources in its northern, southern, eastern, and western regions (Jekayinfa et al. 2020). From different aquatic species in the riverine areas, the famous rivers Niger and Benue confluences (Kogi–Niger region), to mineral resources (gold, tin, iron ore, columbite, among others), Nigeria has a significant reserve of these resources enough to attract long-term investors (Olade 2021). The vast forest and grassland resources in the country's rainforest and savanna ecological zones are home to some of its rare biological treasures (Merem et al. 2012). Section 20 of the Federal Republic of Nigeria Constitution contains the country's environmental objectives "to protect and improve the environment and safeguard the water, air, land, forest and wildlife" (Kankara et al. 2013). This has not prevented these hubs' systematic degradation and depletion, raising concerns about their future lifespan (Galadima et al. 2011).

South Africa is considered one of the most biodiverse countries in the world. The country, with a land area of 1219 million km^2 (World Data 2022), is endowed with migratory bird species (Dean 1997), medicinal plants (van Wyk and Prinsloo 2018), water resources (Rankoana 2020), diverse livestock and aquatic species (Beinart 2008; Ngxumeshe et al. 2020), and land for agriculture and industry (Greenberg 2010; Pradhan and Mbohwa 2014). Recent studies indicate that South Africa has a constantly evolving policy stricture that seeks to safeguard abundant and endangered biodiversity elements (Lukey and Hall 2020; Shih and Mabon 2018) and protect the environment by establishing protected areas (Iorember et al. 2021). Similarly, West Central countries like Cameroun also have policies regulating agricultural, industrial, freshwater, and biodiversity resources (Beckline et al. 2022; Ntoko and Schmidt 2021). In addition, a recent study advocated that the country's science–policy interface be strengthened to encourage higher stakeholder participation in climate change adaptation initiatives (Nkiaka and Lovett 2019).

Lesotho has a luxuriant blend of wetlands, biodiversity and ecosystem services (Chatanga and Seleteng-Kose 2021; Sieben and Chatanga 2019). The country recently proposed its first biosphere reserve as a green economy initiative (Seleteng-Kose et al. 2021). Liberia is at the 'horn' of West Africa, endowed with freshwater and other aquatic resources at its coast areas (Tweh et al. 2018). It also has non-timber forest products scattered across the country (Hwang et al. 2020). With its parks and reserves, Madagascar is a growing tourist centre (Waeber et al. 2020). A recent study revealed that its agricultural sector contributes significantly to its GDP (Máiz-Tomé et al. 2018). It is thus a constellation of natural, freshwater, and agricultural resources. Fish and fisheries are integral to Malawi's biodiversity component with recent case studies of Lake Malawi (Weyl 2019) and Lake Malombe (Makwinja et al. 2022). Mali's marine biodiversity (N'Souvi et al. 2021) and vast expanse of rangeland (Flintan et al. 2022) reveal its potential to become a leading hub in agriculture aquaculture technology and renewable energy production. In Table 24.1, an overview of several African countries' biological and environmental resources available from literature as far back as three decades ago sources from Google Scholar is presented to show how blessed the continent is and why sustainable policies must be implemented to prevent further loss.

24.3 Diversity in Biological and Environmental Policies in Africa

As previously established, biological and environmental resource policies vary from one African country to another, depending on the prevailing circumstances. This section builds on this premise to highlight the biological and ecological approaches currently adopted in some of the continent's leading economies and overview their key similarities and differences. It must be emphasized that their differences do not translate to a disadvantage but a synergy that individual countries are encouraged to develop further.

South Africa has been a critical player in developing policies regarding biological invasions, traced as far back as the 1860s, presenting it as one of the continent's

Table 24.1 Overview of biological and environmental resources in Africa

Country	Biological and/or environmental resources	References
Algeria	Large fish sp. base (endemic commercial, thermophilic and tropic sp), diverse plants and trees, and a potential investment hub for renewable energy technology	Maamar et al. (2018), Parisi et al. (2022), Bélaïd and Youssef (2017)
Benin	A riparian forest epicentre, inland valleys for potential intensive farming, and backyard farming are strongly encouraged. Presence of diverse poultry breeds.	Djagba et al. (2019), Natta et al. (2002), Orounladji et al. (2021), Salako et al. (2014)
Botswana	Diverse aquatic and wildlife species and a growing ecotourism destination. Favorable land area for leafy vegetable cultivation	Alonso and Nordin (2004), Chweya and Eyzaguirre (1999), Mbaiwa (2015), Wheelwright et al. (1996)
Burundi	Riverine resources, wetland habitat, and large land expanse for silviculture and agroforestry programs. Has a large reserve of unused wild medicinal plant seeds	Anatole et al. (2022), Harding and Devisscher (2009), Niyukuri et al. (2021)
Cameroun	High Guinea and Sudan–Sahelian savannas, enormous forest areas. A significant population of livestock, e.g., Djallonke sheep (*Ovis aries*)	Adoum (2022), Baenyi et al. (2018), Nkiaka and Lovett (2018)
Central African Republic	Landlocked country with savanna regions and diverse wildlife species. It also has a vibrant timber sector from a luxurious rainforest region	Cerutti et al. (2018), Orijemie (2022), Scholte et al. (2022)
Cote D'Ivoire	Large forest resource base, cocoa production is intense in certain parts of the country, and coastal areas are rich in aquatic resources, making fishing a common vocation	Amoutchi et al. (2021), Dosso et al. (2012), Kouassi et al. (2021)
Democratic Republic of Congo	It has large timber and biodiversity zones in the Congo basin rainforests and has the potential for sustainable hydropower projects	Abbott et al. (2018), Aquilas et al. (2022), Bele et al. (2015), Nasi et al. (2012), Winemiller et al. (2016)
Djibouti	Rich benthic fauna and other biodiversity habitats in the north Alboran Sea. Currently being explored as a potential renewable energy production hub	Dadashi et al. (2022), Rueda et al. (2021), Souhaid (2021)
Gabon	Rich biodiversity and natural resources base contribute to the country's GDP. A national biodiversity portal was recently launched and has collected over 4500 vertebrate information.	Abbott et al. (2018), Cutler (2019), Tobi et al. (2021)
Guinea Bissau	It has a mangrove rice biodiversity hotspot with potential valorization initiatives. Suitable for growing cashew and a diurnal raptor habitat	Pereira et al. (2022), Rodrigues et al. (2020), Tesio et al. (2021)

(continued)

Table 24.1 (continued)

Country	Biological and/or environmental resources	References
Kenya	Large swamp lakes with diverse fish and snail species, a geo-environmental resource hub, wetland habitat	Aloo (2003), Gurarie et al. (2017), Paron et al. (2013), Harding and Devisscher (2009)
Mauritius	Land and aquatic biodiversity were recently reported to have the potential for nutraceutical product development	Iranah et al. (2018), Ramjane et al. (2021), Tee et al. (2019)
Mozambique	This country is endowed with enormous forest cover and landmass.	Mucova et al. (2018), Niza et al. (2021)
Rwanda	Riverine resources, wetland habitat	Bolson and Patzek (2022), Harding and Devisscher (2009), Richard and Singh (2020)
Senegal	The country has rainforest vegetations with luxuriant marine and coastal environment	Amara et al. (2019), Mathieu et al. (2018), Wade et al. (2018)
Somalia	Has extensive agro-ecological resources, the coastal region is a commercial fish hub	Farah (2021), Manzelli et al. (2006)
Sudan	It has a vast plant and fish biodiversity, has land for potential sustainable energy production	Bakhiet and Barshem (2020), Hegazy et al. (2021), Saeed (2020)
Togo	It has significant biodiversity levels despite its small size	Kokou et al. (2018), Raoufou and Kouami (2018)
Zambia	A vast array of forestry, fisheries and wildlife resources	Chabwela (1990), Mabeta et al. (2018), Siachoono (2018)
Zimbabwe	Rich biodiversity base and freshwater food resources, large land area for gaming reserves	Mutasa and Ndebele-Murisa (2015), Pritchard et al. (2020)

leading forces in taking concrete steps in safeguarding natural and biodiversity resources within its borders (Lukey and Hall 2020). These have culminated in the establishment of biological and environmental protection hubs like the Centre for Invasive Biology, a leading institute in invasion biology, with broad research focus leading to a diverse research program that has produced many integrated products and the production of policy- and management-relevant research products arising from the dynamic nature of research it conducted (van Wilgen et al. 2014; Richardson et al. 2020). As recently opined, the social and ecological issues that threaten biological niches can draw a leaf from South Africa's extensive history in combating them. It can also guide other nations, especially within Africa, as they strive to develop meaningful and sustainable policies and related action plans to safeguard biological and environmental resources. The country also has strong collaborative ties with organizations like the Intergovernmental Science–Policy Platform on Biodiversity and Ecosystem Services and the International Union for Conservation of Nature (van Wilgen et al. 2020). The Biodiversity Policy and the

Biodiversity Act are tangible examples of efforts made by the government to mitigate biological and environmental degradations (King et al. 2018; Kumschick et al. 2020). While much progress has been made regarding policymaking and implementation against natural and biological threats, researchers have stated that it will be a long-term campaign involving systematic management programmes and intervention pathways to create 'buffer zones' around national parks (Cole et al. 2018; Foxcroft et al. 2011, 2019).

Nigeria is one of the continent's leading economies in Africa. In addition, it has various policies for biological, agricultural, environmental, and work health safety for professionals in these fields. As a participant of the United Nations Conference on Environment and Development, Nigeria is obligated under customary international law to ensure that environmental and biodiversity resources within its borders are safeguarded (Nwafor et al. 2018). To achieve this, the Federal Government has set up various legal frameworks in place, such as the Forestry Ordinance, the National Parks Decree, the Federal Environmental Impact Assessment Decree, and the Environmental Impact Assessment, among others (Adewoye et al. 2019; Okafor and Anikelech 2019). Literature regarding previous and current policies has been scarce, suggesting that they are either not easily assessable or impeded by poor implementation of these policies (Kankara et al. 2013).

Despite these formative years in the country's biological and environmental regulation measures, concerns have been raised in the last three decades about several aspects of degradation, suggesting that deficits in clear scientific criteria on how biological and ecological threats should be managed (Chokor 1993; Adelegan 2002). In addition, the country's gas flaring rates have been considerably high, representing around 45% of the continent's gas use (Olujobi 2020). Some devastating effects on communities, biological life and the environment have been reported across the country and surrounding regions (Akpan and Bassey 2020; Raimi et al. 2021). In addition, oil pollution incidences have been rampant in the Niger Delta region of the country. Although some efforts have been made to conduct gradual bioremedial procedures using biological resources, no considerable progress has been made (Osawaru et al. 2013b, 2013c; Ukhurebor et al. 2021). Adelegan (2002) previously argued that these were largely qualitative legalized policies that were not clearly outlined and customized to meet changing environmental and biodiversity goals.

Like many developing countries, Egypt has its fair share of biological and environmental challenges. These include air and water pollution (Goher et al. 2019; Al Naggar et al. 2018) and sediment quality, which measures moderate- and long-term metal accumulation (Hasaballah et al. 2019; Singovszka et al. 2017). In 1991, the government initiated an environmental action plan managed by the government, research and non-government units (Anwar 2003). Concerns were recently raised about diminishing sediment quality in the Suez Bay due to accumulating wastes from industrial products, plant fertilizers and sewage (El-Sikaily et al. 2021). Similar patterns have been observed at the River Nile, believed to be the product of industrialization (Getachew et al. 2021). In Ethiopia, organic and inorganic chemical pollutants are increasing, partly stemming from the country's elevated fertilizer use. This has raised serious concerns about the ecological integrity and biodiversity safety in regions like Lake Ziway (Merga et al. 2021), and similar trends have been reported in other sub-Saharan reservoirs (Teklu et al. 2018; Wenaty et al. 2019). The Ethiopian government has adopted the climate-resilient green economy initiative, which implements the protection and re-establishment of forest areas, improving livestock and crop practices, expanding green energy generation by 2025 (Zegeye 2018). A summary of biodiversity and environmental policies have been given in Table 24.2.

We first observed that there appears to be a challenge in policy information available for several African countries, partly because they have not been digitalized. This made a more comprehensive policy diversity appraisal more challenging to conduct. The second is that it is unclear as to who does what. Who are the policymakers, and what informs some of the policy drafts they advocate? Unlike South Africa, which appears to encourage an evidence-based policy formulation approach, other countries are somewhat ambiguous in approaching the constantly changing factors and threats to the environment and biodiversity. Third, there appears to be a 'research deficit', where carefully planned studies are either not carried out, or are insufficient, in helping the relevant government and private partners understand the constantly changing situation regarding natural and biological factors. These will be discussed subsequently.

24.4 Development of Sustainable Biological and Environmental Policies for Africa

Policies at any clime are never static, as biodiversity and nature constantly change. African countries must thus develop and continuously review biological and environmental policies robustly to ensure that individual governments can stay in touch with and effectively combat new and existing threats. Previous sections have attempted to present a robust picture of the policy frameworks across several African countries set up to manage biodiversity and environmental spheres. Here, models currently implemented successfully by some countries across the globe will be highlighted in anticipation that governments will take a cue as to how best to enhance policy implementation and sustenance efforts significantly. Some of the guiding questions for African governments and policymakers to ruminate on are: have policy development pathways changed? Are some of the current policies obsolete?

This section begins with China, the globe's second-largest economy, with a 2020 GDP of USD14.72 trillion and a population of around 1.402 billion people. In addition, it has one of the most vibrant and threatened biodiversity in the world, owing to its unprecedented development rates and climate change (Zhang et al. 2022). Although China's governance system is shrouded in uncertainties and controversies, some recent studies indicate that it has constantly reviewed wildlife and

Table 24.2 Diversity of biodiversity and environmental policies in Africa[a]

Country	Policy structure and regulations on biodiversity and environmental resources	References
Botswana	The tribal grazing land policy of 1972 plays a pioneering role in biodiversity and grassland conservation. A recent call to review existing policies has been made	Maude and Reading (2010), Tsinda et al. (2018)
Burkina Faso	Along with Kenya and Zambia have, livestock policies are framed in conjunction with the program of accompanying research for agricultural innovation (PARI), financed by Germany's Federal Ministry of Economic Cooperation and Development (BMZ). Preliminary deliberations into national climate change adaptation policymaking have been reported	Kariuki et al. (2022), Theokritoff and Lise D'haen (2021)
Cameroun	Government has a 1996 national policy for forest management and biodiversity conservation. Policy for creating protected areas have also been enacted. 'Operation green Sahel' was relaunched in 2008	Nkiaka and Lovett (2018)
Egypt	Policy construct managing biodiversity and urbanization trends, recently enacted healthy ecosystems for rangeland development (HERD) policy system	Rizk and Saifelnasr (2020), Zetter and Hassan (2002)
Gambia	Recently developed a policy-backed program led by the UNDP to safeguard key ecological zones and biodiversity niches enforced by the Nagoya protocol. This project is also supported by the global environment facility (GEF)	Gambia (2010), Zhongming et al. (2021)
Guinea–Bissau	Several industrial policies regulate the production and export of cashew, its main cash crop. Potential policy initiatives toward the valorization of mangrove rice biodiversity have been reported	Lundy (2021), Tesio et al. (2021)
Ghana	There are environmental conservation policies (over 120 documents) to ensure sustainability compliance levels. The country has a draft national biodiversity policy in line with the UN CBD. Deliberations toward strengthening eco-innovation policies have been reported	Botchway et al. (2021), MEST (2021), MoES (2002), Ozor and Nyambane (2021), Yesutanbul et al. (2021)
Kenya	Along with Burkina Faso and Zambia, have livestock policies framed in conjunction with the program of accompanying research for agricultural	Kariuki et al. (2022), Ozor and Nyambane (2021)

(continued)

Table 24.2 (continued)

Country	Policy structure and regulations on biodiversity and environmental resources	References
	innovation (PARI) financed by Germany's Federal Ministry of Economic Cooperation and Development (BMZ). Deliberations toward strengthening eco-innovation policies have been reported	
Malawi	Deliberations toward strengthening eco-innovation policies have been reported	Ozor and Nyambane (2021)
Zambia	Have livestock policies framed in conjunction with the program of accompanying research for agricultural innovation (PARI), financed by Germany's Federal Ministry of Economic Cooperation and Development (BMZ). Deliberations toward strengthening eco-innovation policies have been reported	Kariuki et al. (2022), Ozor and Nyambane (2021)
Zimbabwe	The country has a national environmental policy draft providing the legal framework which outlines specific strategies on how the environment should be protected. An ethical structure was later included	Kurebwa (2022), Mangena (2014)

[a]While several previous and recent studies abound regarding environmental and biodiversity concerns, available literature regarding policy responses to them is relatively scarce. In addition, due to space constraints, several other relevant studies were not included in this section

environmental laws, culminating in how its sustainability policies have informed efficient ecosystem services (Ali et al. 2018; Li et al. 2022). Given the enormous opportunities and challenges in biodiversity and environmental spheres, China is an avid supporter of the Global Biodiversity Plan, being one of the first countries to sign and ratify the Convention on Biological Diversity. The China Biodiversity Conservation was released in October 2021, announcing that the cause has become a national strategy. The white paper details specific ecosystem targets aligned with the 2030 Sustainable Development Goals (SDG). As the COP15 host country, China discussed how it plans to manage its biodiversity plans with the SDG goals seamlessly through a coordinated blueprint detailing cooperation between different government tiers (Zhang et al. 2022). One notable factor that has driven most of China's success in ensuring a (near) balance in development and biodiversity conservation is unrivalled political support. This driver is lacking in many African countries. This chapter's challenges and recommendations sections will highlight some weaknesses and opportunities in China's biodiversity policies.

Australia is a megadiverse country in Oceania, boasting a 2020 GDP of 1.331 trillion USD. Even with a relatively small population of around 25 million, it is known for its strong governance arrangements and high-functioning institutional capacities (Evans 2016). However, Australia faced significant deforestation levels at the start of the twenty-first century, noted as among the top ten on the planet (ACF 2001; FAO 2001). Deforestation removes vegetation, erodes the topsoil contents, and raises salinity levels, which are key drivers in the migration of woodland birds, reptiles, and mammals (Archer et al. 2017). Although deforestation policies are regulated at the state level, a nationally coordinated plan was also in place to increase native vegetation through funding and frameworks from the Australian Federal Government (Guglyuvatyy 2022; Evans 2016). Since 2001, the Federal Government's Environmental Protection and Biodiversity Conservation Act 1999 (EPBC Act) has listed land clearing as a critical, threatening process and thus prohibited by law (Lindenmayer 2005; Macintosh 2004). From 2010 to 2016, there have been strategic policy responses to deforestation at the Australian state levels. For instance, in 2011, the New South Wales government announced that it was going to 'strike the right balance between sustainable agriculture and protecting the environment, by conducting a statutory review of the 2005 *Native Vegetation Regulation* under the *National Vegetation Act (NVA) 2003* (Condon and Bryant 2013). Although this was welcomed by landholders, a review of this initiative was conducted, where the emerging report recommended the repeal of the *NVA 2003,* arguing that it should be included with other biodiversity policies into one single *Biodiversity Conservation Act* (Byron et al. 2014). Similar policy reviews have also been conducted by the governments of Victoria (Department of Environment and Primary Industries 2013) and Western Australia (Standing Committee on the Environment and Public Affairs 2015). In Queensland, more liberal policies were enacted, allowing landholders to clear 'high value' land unrestricted, a move that has been strongly criticized (Taylor 2015; WWF International 2015). Thankfully, a newly elected Queensland government has since reversed this in 2015 by clearing restrictions to apply retrospectively from March 17, 2016 (Chambers 2016; State of Queensland 2016). Compared to many African countries, it is interesting to note that environmental and biodiversity policy space has been very active in Australia, involving policymakers, researchers, landholders and activists. It again reminds us that policies are formulated based on emerging economic and socio-political changes, not a static, informed procedure.

24.5 Challenges in Policy Implementation in Africa

Although some progress has been made in mechanized agriculture in some countries, Olisah et al. (2020) recently reported that the continent still relies heavily on pesticide use from 1990 to 2016, accounting for about 2.1% of global pesticide usage. This trend poses a serious problem to the sustainable use of environmental factors such as the atmosphere (Tudi et al. 2021), water (Teklu et al. 2021), sediments and soils (Alengebawy et al. 2021), and biological factors such as animals (Kumar et al. 2021; Riyaz et al. 2021), humans (Benbrook et al. 2021), and food products (Umapathi et al. 2022). Many African governments see pesticides as a less expensive route to sustaining agricultural productivity and have either not assessed their long-term deleterious effects or blatantly ignored it (Holt-Giménez 2019).

Several studies have also indicated that funding is a significant constraint in policy implementation in many African countries (Leal Filho et al. 2018). While some countries like Comoros, Guinea–Bissau, and Sao Tome and Principe may not have large economies and GDP to finance the level of biological and environmental management that specific policies require (Ahmed et al. 2021; Patel et al. 2021), other countries (e.g., Nigeria, South Africa, and Egypt) have not fared better either (Jalam et al. 2021; Olawuyi and Olusegun 2018; Ramaano 2021). In some instances, the budgeted resources are embezzled. Nothing is often done to recover the funds or instill severe penalties backed by law (Ogwang and Vanclay 2021; Onimisi 2021). Olawuyi and Olusegun (2018) discussed other vital challenges, including the outdated and incoherent legal regime, non-availability of a post-2025 biodiversity agenda, no political will, and inadequate stakeholder participation in biodiversity management.

It is sad to note that even with the biological and environmental resource depletion being reported across the globe, particularly in the last two decades, several African countries have still not made the conservation of biological and ecological resources a priority in annual plans and budgetary allocations (Canavan et al. 2021). This generally stems from a perceived unwillingness to change the status quo, and calls for action are interpreted as politically motivated (Pocock et al. 2019). The critical point to note here is that unless national governments in Africa genuinely assess the implications of not protecting its natural hubs and enact relevant policies to ensure their sustainability, no significant changes will be seen. It would only be a matter of time before some of the continent's major biological and environmental resources dry up (Mazur-Panasiuk et al. 2019; Vera et al. 2022). It is also noteworthy that the coronavirus 19 (COVID-19) pandemic may significantly impact many African countries' ability to strengthen policy frameworks further, as economies are still recovering from the shock wave created almost 3 years ago (Cumming et al. 2021; Kideghesho et al. 2021).

Several African countries either have deficient biodiversity and environmental policies or lack good enforcement mechanisms, as earlier reported in Nigeria (Ojo 2020), reiterating the lack of coordination within its government tiers (Kankara et al. 2013). Several African countries such as Tanzania (Brüntrup et al. 2018), Kenya (Joseph Kanyua 2020), Gabon (Abbott et al. 2018), Ghana (Botchway and Hlovor 2019), and Ethiopia (Yami and Mekuria 2022) follow this trend. The Belt Road Initiative (BRI) is arguably one of the most ambitious projects in history, estimated to cost around eight trillion USD, with projects already earmarked in several African countries (Smith 2018; Zhang et al. 2022). It has been argued that although the initiative will foster cooperation, trade, and create employment in the participating countries, the scale of biodiversity loss and environmental degradation may be much larger than estimated (Lee et al. 2022). Emerging studies indicate that African partner countries will need to take a step back and conduct a broadscale risk assessment on the long-term effects on the environment and biodiversity (Mengdi and Wang 2021).

24.6 Future Perspectives

Several previous and recent studies across Africa have given strategic recommendations to governments and NGOs alike to improve and sustain its biological and environmental resources (Schweizer et al. 2021; Wassie 2020). However, as consistently opined by these studies, a major constraint is not the lack of funding but non-prioritization (Kimbu 2011; Omo and Etuvoata 2021). Many African countries already have drafted policies toward biological and environmental resources management. Still, they have been hindered by several socio-economic and political factors (Mkonda and He 2017; Rihoy and Maguranyanga 2007). In Nigeria, for example, Omolola (2013) has listed potential funding sources for environmental conservation, including oil revenue, ecotourism returns, donor agencies, and federal government budgetary allocations. For international projects like the BRI, African countries must, as a matter of urgency, work with China to draft policies safeguarding the environment and biodiversity that intensive and long-term construction activities would significantly impact. These policies should be informed by carefully conducted studies showing the project's long-term detrimental effects in Africa. While it is crucial to encourage bilateral cooperation, participating African countries must also be aware of the toll these will have on the environment and biological activities.

It is no news that many African countries are battling environmental degradation incidences (Hardoy et al. 2013; Herbig 2019). These countries should initiate policy responses to curb pollution problems by ensuring that government and non-government stakeholders abide by a highly transparent operations procedure. Policymakers across the continent must keep abreast with recent environmental and biodiversity threat reports from their various countries and enact policies regulating the activities of local and foreign exploring corporations (Materu et al. 2018; Rochell et al. 2022). While such procedures may take considerable time, it is recommended that they be initiated in the shortest possible time. NGOs also have a role in policy implementation for biological and environmental resources conservation. While it should be acknowledged that some NGOs are making progress in this regard, others are still experiencing bottlenecks that impede the realization of their core objectives (Bendjebbar and Fouilleux 2022; Ola and Benjamin 2019). It is recommended that NGOs prompt relevant lawmakers to propose biological and environmental protection bills that will ensure their smooth running. It is no news that the government alone cannot get everything done. Another advantage is that this conservation cause can become a viable investment that other actors would want to have a shared interest in, becoming an additional income stream for the national government and the people living in that region. Eventually, though, it should be emphasized that the conservation of biological and environmental resources is the responsibility of everyone. As custodians, humans owe biodiversity and the environment a lifetime task to preserve and enact policies to enhance the environment. Social media and other ICT tools have digitalized the globe. Hey can be a powerful means with which Africans (in Africa or the diaspora) can draw attention to either what they are doing to save the environment and biodiversity or an opportunity to reach relevant government agencies on pressing policies to implement to protect the environment. While scrutinizing each paper cited in this chapter, the authors acknowledge that not all the relevant works were included due to space constraints, but the most relevant ones were captured. A more comprehensive follow-up study should focus on policy effectiveness, especially in the last two decades.

24.7 Conclusion

Through this chapter, an overview of the biological and environmental resources of several African countries and the existing policies regulating them sourced from available literature are presented. Some African countries have made more progress than others in sustainable policy awareness. However, the general performance as a continent is still less than desired. As a guide to develop sustainable policies that evolve with global climate change indices, two countries—China and Australia—were presented as models that African countries are encouraged to understudy and, afterward, develop robust policies that mitigate biodiversity loss and environmental degradation. The challenges and recommendations toward sustainable biodiversity and ecological policymaking in Africa have also been highlighted.

References

Abbott P, Tsinda A, Mugisha R (2018) Review of policies for biodiversity informatics in Central Africa: case studies of the Democratic Republic of Congo (DRC) and Gabon. Working Paper 3, Centre for Global Development, University of Aberdeen, January 2018

Abdelhafez AA, Metwalley SM, Abbas HH (2020) Irrigation: water resources, types and common problems in Egypt. In: Technological and modern irrigation environment in Egypt. Springer, Cham, pp 15–34

ACF (2001) Australian land clearing, a global perspective: latest facts & figures. Australian Conservation Foundation (ACF), Melbourne

Adelegan J (2002) Environmental policy and slaughterhouse waste in Nigeria. The 28th WEDC Conference, Calcutta, India

Adewoye OA, Oyewole OD, Lawal OO (2019) Environmental management policies in Nigeria. In 1st International Conference On Engineering and Environmental Sciences (p. 415)

Adom D (2018) Traditional biodiversity conservation strategy as a complement to the existing scientific biodiversity conservation models in Ghana. Environ Nat Resour Res 8(3):1–24

Adoum D (2022) The influence of periodic fires on the high Guinean savannas and the carbon stock in Cameroun. Am J Environ Res Econ 7(1):15–24

Ahmed F, Houessenou P, Nikiema A, Zougmore R (2021) Transforming agriculture in Africa's Small Island developing states: lessons learnt and options for climate-smart agriculture investments in Cabo Verde, Guinea-Bissau and Seychelles. Food & Agriculture Org

Akpan CO, Bassey SA (2020) The quandary on water pollution in Nigeria's Niger delta: an environmental ethical analysis. Bull Pure Appl Sci Geol 2

Al Naggar Y, Khalil MS, Ghorab MA (2018) Environmental pollution by heavy metals in the aquatic ecosystems of Egypt. Open Acc J Toxicol 3:555603

Alengebawy A, Abdelkhalek ST, Qureshi SR, Wang MQ (2021) Heavy metals and pesticides toxicity in agricultural soil and plants: ecological risks and human health implications. Toxics 9(3):42

Ali M, Kennedy CM, Kiesecker J, Geng Y (2018) Integrating biodiversity offsets within circular economy policy in China. J Clean Prod 185:32–43

Alonso LE, Nordin LA (Eds.). (2004) A rapid biological assessment of the aquatic ecosystems of the Okavango Delta, Botswana: High Water Survey. Conservation International

Aloo PA (2003) Biological diversity of the Yala swamp lakes, with special emphasis on fish species composition, in relation to changes in the Lake Victoria Basin (Kenya): threats and conservation measures. Biodivers Conserv 12(5):905–920

Amara R, Diop M, Diop C, Ouddane B (2019) The Senegalese coastal and marine environment. In World seas: an environmental evaluation. Academic Press, Cambridge, pp 855–873

Amoutchi AI, Mehner T, Ugbor ON, Paul KE (2021) Fishermen's perceptions and experiences toward the impact of climate change and anthropogenic activities on freshwater fish biodiversity in Côte d'Ivoire. Discov Sust 2(1):1–18

Anatole B, Masharabu T, Rose A, Tite V, Jacques N (2022) Indigenous versus exotic tree species used in silviculture and agroforestry: an overview from Burundi seed Centre data. East Afr J Sci Technol Innov 3

Anwar WA (2003) Environmental health in Egypt. Int J Hyg Environ Health 206(4–5):339–350

Aquilas NA, Mukong AK, Kimengsi JN, Ngangnchi FH (2022) Economic activities and deforestation in the Congo Basin: an environmental Kuznets curve framework analysis. Environ Challeng:100553

Archer SR, Andersen EM, Predick KI, Schwinning S, Steidl RJ, Woods SR (2017) Woody plant encroachment: causes and consequences. In: Rangeland systems. Springer, Cham, pp 25–84

Asare RA, Kyei A, Mason JJ (2013) The community resource management area mechanism: a strategy to manage African forest resources for REDD+. Philos Trans R Soc B Biol Sci 368(1625):20120311

Baenyi P, Meutchieye F, Basengere RA, Bwihangane BA, Karume K, Mushagalusa NG, Ngoula F (2018) Biodiversity of indigenous Djallonké sheep (Ovis aries) in Sudano Guinean region in Cameroon. GABJ 2(2):1–10

Bakhiet HHA, Barshem AA (2020) Fish biodiversity and anatomical weight composition of fish from Al-Rahad Turda, North Kordofan State in Sudan

Barzkar N, Jahromi ST, Poorsaheli HB, Vianello F (2019) Metabolites from marine microorganisms, micro, and macroalgae: immense scope for pharmacology. Mar Drugs 17(8):464

Beckline M, Manan A, Dominic N, Mukete N, Hu Y (2022) Patterns and challenges of forest resources conservation in Cameroon. Open Acc Lib J 9(5):1–13

Beinart W (2008) The rise of conservation in South Africa: settlers, livestock, and the environment 1770–1950. Oxford University Press, Oxford

Beirne C, Meier AC, Mbele AE, Menie GM, Froese G, Okouy J, Poulsen JR (2019) Participatory monitoring reveals village-centered gradients of mammalian defaunation in Central Africa. Biol Conserv 233:228–238

Bélaïd F, Youssef M (2017) Environmental degradation, renewable and non-renewable electricity consumption, and economic growth: assessing the evidence from Algeria. Ener Pol 102:277–287

Bele MY, Sonwa DJ, Tiani AM (2015) Adapting the Congo Basin forests management to climate change: linkages among biodiversity, forest loss, and human well-being. For Pol Econ 50:1–10

Benamar L, Bennas N, Hassoun M, Millán A (2021) Threatened endemic water beetles from Morocco. J Insect Conserv 25(3):465–477

Benbrook C, Kegley S, Baker B (2021) Organic farming lessens reliance on pesticides and promotes public health by lowering dietary risks. Agronomy 11(7):1266

Bendjebbar P, Fouilleux E (2022) Exploring national trajectories of organic agriculture in Africa. Comparing Benin and Uganda. J Rural Stud 89:110–121

Bolson N, Patzek T (2022) Evaluation of Rwanda's energy resources. Sustainability 14(11):6440

Botchway TP, Hlovor IK (2019) Mitigating the challenges related to the implementation of the convention on biological diversity in Ghana. J Sust Dev 12(3):91–102

Botchway M, Davis RE, Merchant AT, Appiah LT, Moore S (2021) Diabetes-related stigma and its influence on social networks, social support, and HbA1c in Ghana. Ethn Dis 31(1):57

Brüntrup M, Schwarz F, Absmayr T, Dylla J, Eckhard F, Remke K, Sternisko K (2018) Nucleus-outgrower schemes as an alternative to traditional smallholder agriculture in Tanzania–strengths, weaknesses and policy requirements. Food Sec 10(4):807–826

Byron N, Craik W, Keniry J, Possingham H (2014) A review of biodiversity legislation in NSW: final report. State of NSW and the Office of Environment and Heritage

Canavan K, Paterson ID, Ivey P, Sutton GF, Hill MP (2021) Prioritization of targets for weed biological control III: a tool to identify the next targets for biological control in South Africa and set priorities for resource allocation. Biocont Sci Technol 31(6):584–601

Cerutti PO, Mbangolo JD, Nakoe GP, Vandenhaute M (2018) State of the timber sector in Central African Republic (2016)

Chabwela HN (1990) The exploitation of wildlife resources in Zambia: a critique. In Utilizing local resources for development. Proceedings of the 9th PWPA conference, Eastern, Central and Southern Region, held at Musungwa Lodge, Zambia, July 1988. (pp. 54–65). Professors World Peace Academy of Zambia

Chambers G (2016) Land-clearing laws retrospective. The Australian Newspaper, 29 March 2016

Chatanga P, Seleteng-Kose L (2021) Montane palustrine wetlands of Lesotho: vegetation, ecosystem services, current status, threats and conservation. Wetlands 41(6):1–19

Chokor BA (1993) Government policy and environmental protection in developing world. Environ Manag 17(1):15–30

Chweya JA, Eyzaguirre PB (1999) The biodiversity of traditional leafy vegetables. University of Nairobi

Cole N, van Wilgen NJ, Foxcroft LC, et al. (2018) Invasive alien species management. South African National Parks policy document Reference number: 17/P-CSD/Pol/AIS (09–17) v1

Condon M, Bryant S (2013) NSW farmers welcome change in native veg policy. ABC Rural, 14 June 2013. https://www.abc.net.au/news/rural/2013-06-14/native-vegetation-nsw-friday-stoner-nationalsconfernece/4755492. Accessed 18 Dec 2015

Cumming T, Seidl A, Emerton L, Spenceley A, Kroner RG, Uwineza Y, van Zyl H (2021) Building sustainable finance for resilient protected and conserved areas: lessons from COVID-19. Parks 27:149–160

Cutler JS (2019) Biodiversity, biogeography, and conservation of freshwater fishes in Gabon. University of California, Santa Cruz

Dadashi Z, Mahmoudi A, Rashidi S (2022) Capacity and strategies of energy production from renewable sources in Arab countries until 2030: a review from renewable energy potentials to environmental issues. Environ Sci Pollut Res Int:1–30

Dean W (1997) The distribution and biology of nomadic birds in the Karoo, South Africa. J Biogeo 24(6):769–779

Department of Environment and Primary Industries (2013) Reforms to Victoria's native vegetation permitted clearing regulations: overview. The State of Victoria Department of Environment and Primary Industries, Melbourne

Diagne C, Turbelin A, Moodley D, Novoa A, Leroy B, Angulo E et al (2021) The economic costs of biological invasions in Africa: a growing but neglected threat? NeoBio 67:11–51

Djagba JF, Zwart SJ, Houssou CS, Tenté BH, Kiepe P (2019) Ecological sustainability and environmental risks of agricultural intensification in inland valleys in Benin. Environ Dev Sustain 21(4):1869–1890

Djuma H, Bruggeman A, Eliades M, Lange MA (2016) Non-conventional water resources research in semi-arid countries of the Middle East. Desalin Water Treat 57(5):2290–2303
Dosso K, Yéo K, Konaté S, Linsenmair KE (2012) Importance of protected areas for biodiversity conservation in Central Côte d'Ivoire: comparison of termite assemblages between two neighboring areas under differing levels of disturbance. J Insect Sci 12(1):131
Elsaid Saeed MA, Bedair R (2021) Monitoring land degradation and soil productivity in Bilqas District, Dakahlia Governorate, Egypt. Egyptian J Soil Sci 61(3):311–321
El-Sikaily A, Ghoniem DG, Emam MA, El-Nahrery EM (2021) Biological indicators for environmental quality monitoring of marine sediment in Suez Gulf, Egypt. Egypt J Aqua Res 47(2): 125–132
Evans MC (2016) Deforestation in Australia: drivers, trends and policy responses. Pac Conserv Biol 22(2):130–150
FAO (2001) Global forest resources assessment 2000. United Nations Food and Agriculture Organization (UNFAO)
Farah LI (2021) Analysis of the Most common commercial fish species found in Mogadishu Coast, Somalia. Fen ve Ziraat Bilimleri 127
Flintan FE, Diop A, Coulibaly M (2022) Opportunities for participatory rangeland management (PRM) in the great Green Wall initiative in Mali and Senegal. ILRI Project Report
Foxcroft LC, Jarošík V, Pyšek P et al (2011) Protected area boundaries as a natural filter of plant invasions from surrounding landscapes. Conserv Biol 25:400–405
Foxcroft LC, Spear D, van Wilgen NJ et al (2019) Assessing the association between pathways of alien plant invaders and their impacts in protected areas. NeoBio 43:1–25
Galadima A, Garba ZN, Leke L, Almustapha MN, Adam IK (2011) Domestic water pollution among local communities in Nigeria-causes and consequences. Eur J Sci Res 52(4):592–603
Gambia FP (2010) Forestry sub-sector policy Republic of the Gambia. Forest policy of the Gambia
Getachew B, Manjunatha BR, Bhat HG (2021) Modeling projected impacts of climate and land use/land cover changes on hydrological responses in the Lake Tana Basin, upper Blue Nile River Basin, Ethiopia. J Hydrol 595:125974
Goher ME, Ali MH, El-Sayed SM (2019) Heavy metals contents in Nasser Lake and the Nile River, Egypt: an overview. Egypt J Aqua Res 45(4):301–312
Greenberg S (2010) Status report on land and agricultural policy in South Africa
Guglyuvatyy E (2022) Australian forests and climate change. In: Climate change, forests and federalism. Springer, Singapore, pp 63–88
Gurarie D, King CH, Yoon N, Wang X, Alsallaq R (2017) Seasonal dynamics of snail populations in coastal Kenya: model calibration and snail control. Adv Wat Res 108:397–405
Habimana V (2022) Rationalization of international trade and investment agreements: case of East African community (Doctoral dissertation, Anglia Ruskin University)
Harding B, Devisscher T (2009) Review of the economic impacts of climate change in Kenya, Rwanda and Burundi: ecosystems chapter, Kenya
Hardoy JE, Mitlin D, Satterthwaite D (2013) Environmental problems in an urbanizing world: finding solutions in cities in Africa, Asia and Latin America. Routledge
Hasaballah AA, Hegazy T, S Ibrahim M, A El-Emam D (2019) Assessment of water and sediment quality of the River Nile, Damietta Branch, Egypt. Egypt J Aqua Biol Fisher 23(5):55–65
Hegazy AK, Hosni HA, El-Sheikh MA, Abo-El-Kassem A, Badawi ESM, Lovett-Doust L (2021) Plant biodiversity and soils in the Jebel Marra region of Darfur, Sudan. Arid Land Res Manag 35(2):127–161
Herbig FJ (2019) Talking dirty-effluent and sewage irreverence in South Africa: a conservation crime perspective. Cog Soc Sci 5(1):1701359
Holt-Giménez E (2019) Can we feed the world without destroying it? John Wiley & Sons, New York
Hwang LC, Juliani HR, Govindasamy R, Simon JE (2020) Traditional botanical uses of non-timber Forest products (NTFP) in seven counties in Liberia. In African natural plant products,

volume III: discoveries and innovations in chemistry, bioactivity, and applications. American Chemical Society, pp 3–43
Ikhajiagbe B, Ogwu MC (2020) Hazard quotient, microbial diversity and plant composition of spent crude oil polluted-soil. Beni-Suef Univ J Basic Appl Sci 9(26). https://doi.org/10.1186/s43088-020-00052-0
Imarhiagbe O, Egboduku WO, Nwankwo BJ (2020) A review of the biodiversity conservation status of Nigeria. J Wild Biodiv 4(1):73–83
Iorember PT, Jelilov G, Usman O, Işık A, Celik B (2021) The influence of renewable energy use, human capital, and trade on environmental quality in South Africa: multiple structural breaks cointegration approach. Environ Sci Pollut Res 28(11):13162–13174
Iranah P, Lal P, Wolde BT, Burli P (2018) Valuing Visitor Access to Forested Areas and Exploring Willingness to Pay for Forest Conservation and Restoration Financethe Case of Small Island Developing State of Mauritius
Izah SC, Aigberua AO, Nduka JO (2018) Factors affecting the population trend of biodiversity in the Niger Delta region of Nigeria. Int J Avian Wild Biol 3(3):199–207
Jalam AM, Sharaai AH, Ariffin M, Zainudin N, Musa HD (2021) Closing the policy-practice gaps in Nigeria's desertification interventions: a qualitative document analysis of sustainable practice. J Environ Pol Plan 23(3):381–398
Jekayinfa SO, Orisaleye JI, Pecenka R (2020) An assessment of potential resources for biomass energy in Nigeria. Resources 9(8):92
Joseph Kanyua M (2020) Effect of imposed self-governance on irrigation rules design among horticultural producers in peri-Urban Kenya. Sustainability 12(17):6883
Kankara AI, Adamu GK, Tukur R, Ibrahim A (2013) Examining environmental policies and laws in Nigeria. Int J Environ Eng Manag 4:165–170
Karame P (2015, August) Sustaining Practical Biodiversity Informatics Initiatives for Decision Support in Natural Resources Management sector, Rwanda. In TDWG 2015 Annual Conference
Kariuki J, Yameogo V, Daum T, Birner R, Chagunda M (2022) Do African livestock policies address sustainability trade-offs? Evidence from Kenya, Zambia, and Burkina Faso. Hohenheim Working Papers on Social and Institutional Change in Agricultural Development. 0022022. University of Hohenheim
Khalid S, Abbas M, Saeed F, Bader-Ul-Ain H, Suleria HAR (2018) Therapeutic potential of seaweed bioactive compounds. IntechOpen, London
Kideghesho JR, Kimaro HS, Mayengo G, Kisingo AW (2021) Will Tanzania's wildlife sector survive the COVID-19 pandemic? Trop Conserv Sci 14:19400829211012682
Kimbu AN (2011) The challenges of marketing tourism destinations in the central African subregion: the Cameroon example. Int J Tour Res 13(4):324–336
King ND, Strydom HA, Retief FP (2018) Environmental management in South Africa, 3rd edn. Juta, Cape Town
Kokou K, Afiademanyo KM, Elikplimabotsi K, Segniagbeto H, Gomina M, Amévoin K (2018) Biodiversity in Togo. Global Biodiversity: Selected Countries in Africa 3: 341
Kouassi CJA, Khan D, Achille LS, Omifolaji JK, Kebin Z (2021) Forest resources depletion: an ecological model for biodiversity preservation and conservation in cote D'Ivoire. Open J Ecol 11(12):870–890
Kumar R, Sankhla MS, Kumar R, Sonone SS (2021) Impact of pesticide toxicity in aquatic environment. Bioint Res Appl Chem 11(3):10131–10140
Kumschick S, Foxcroft LC, Wilson JR (2020) Analyzing the risks posed by biological invasions to South Africa. In: van Wilgen BW, Measey J, Richardson DM, Wilson JR, Zengeya TA (eds) Biological invasions in South Africa. Springer, Berlin, pp 569–592
Kurebwa J (2022) Gender roles and utilization of indigenous knowledge systems in the management of forest biodiversity: the case of Madondo Communal Lands in Zimbabwe. Int J Soc Ecol Sust Dev 13(1):1–11
Lal R, Stewart BA (eds) (2019) Soil degradation and restoration in Africa. CRC Press, Boca Raton, FL

Leal Filho W, Balogun AL, Ayal DY, Bethurem EM, Murambadoro M, Mambo J et al (2018) Strengthening climate change adaptation capacity in Africa-case studies from six major African cities and policy implications. Environ Sci Pol 86:29–37

Leandro A, Pereira L, Gonçalves AM (2019) Diverse applications of marine macroalgae. Mar Drugs 18(1):17

Lee HS, Arestis P, Chong SC, Yap S, Sia BK (2022) The heterogeneous effects of urbanization and institutional quality on greenhouse gas emissions in belt and road initiative countries. Environ Sci Pollut Res 29(1):1087–1105

Levine I, Fleurence J (eds) (2018) Microalgae in health and disease prevention. Academic Press, London

Li Y, Kong J, Ji J (2022) Environmental regulation, technological innovation and development of marine fisheries—evidence from ten coastal regions in China. Fishes 7(1):20

Lindenmayer D (2005) Vegetation loss and degradation. In: Lindenmayer D, Burgman M (eds) Practical conservation biology. Chapter 9. Melbourne, CSIRO Publishing

Lukey P, Hall J (2020) Biological invasion policy and legislation development and implementation in South Africa. In: van Wilgen BW, Measey J, Richardson DM, Wilson JR, Zengeya TA (eds) Biological invasions in South Africa. Springer, Berlin, pp 513–548

Lundy BD (2021) Cashews in conflict: the political ecology of cashew pomiculture in Guinea-Bissau. In Political ecology of industrial crops. Routledge, pp 124–149

Maamar B, Nouar B, Soudani L, Maatoug M, Azzaoui M, Kharytonov M et al (2018) Biodiversity and dynamics of plant groups of Chebket El Melhassa region (Algeria). Biosyst Divers 26(1): 62–70

Mabeta J, Mweemba B, Mwitwa J (2018) Key drivers of biodiversity loss in Zambia. Policy Brief 3 (February 2018)

Macintosh A (2004) Why the environment protection and biodiversity conservation act's referral, assessment and approval process is failing to achieve its environmental objectives. EPLJ 288

Máiz-Tomé L, Sayer C, Darwall WRT (2018) The status and distribution of freshwater biodiversity in Madagascar and the Indian Ocean islands hotspot. IUCN, Gland

Makwinja R, Mengistou S, Kaunda E, Alamirew T (2022) Complex interactions between benefits, ecosystem services and landscape dynamics: a synthesis of Lake Malombe, Malawi. Lakes Reser Res Manag 27(1):e12392

Mangena F (2014) Environmental policy, management and ethics in Zimbabwe, 2000-2008. J Pan Afr Stud 6(10):224–240

Manzelli M, Benedettelli S, Vecchio V (2006) Agro-biodiversity in subsistence farming systems of South Somalia–Collection and Agronomic Assessment of Somali Sorghum (Sorghum bicolor (L.) Moench). Germplasm Tropicult 24(4):213–220

Materu SF, Urban B, Heise S (2018) A critical review of policies and legislation protecting Tanzanian wetlands. Ecosys Health Sust 4(12):310–320

Mathieu G, Taïbou BA, Aminata DS (2018) Biodiversity in Senegal: terrestrial ecosystems Flora and Great Fauna. Glob Biodiver Sel Count Afr 3:173

Maude G, Reading RP (2010) The role of ecotourism in biodiversity and grassland conservation in Botswana. Great Plains Res 20(1):109–119

Mazur-Panasiuk N, Żmudzki J, Woźniakowski G (2019) African swine fever virus–persistence in different environmental conditions and the possibility of its indirect transmission. J Vet Res 63(3):303

Mbaiwa JE (2015) Ecotourism in Botswana: 30 years later. J Ecotour 14(2–3):204–222

Mengdi Y, Wang CN (2021) Debt-for-nature swaps: a triple-win solution for debt sustainability and biodiversity finance in the belt and road initiative (BRI)? Green BRI Center, International Institute of Green Finance (IIGF), Beijing

Merem EC, Wesley J, ATwumasi Y, Richardson C, Romorno C (2012) Tropical forest landscape change and the role of agroforestry systems in Southern Nigeria. Br J Environ Clim Change 2(3):288

Merga LB, Mengistie AA, Alemu MT, Van den Brink PJ (2021) Biological and chemical monitoring of the ecological risks of pesticides in Lake Ziway, Ethiopia. Chemosphere 266:129214

Ministry Environment Science and Technology (MESTI) (2021) Draft National Biodiversity Policy. MESTI, Accra, Ghana

Ministry of Environment and Science (MoES) (2002) National biodiversity strategy for Ghana

Mkonda MY, He X (2017) Yields of the major food crops: implications to food security and policy in Tanzania's semi-arid agroecological zone. Sustainability 9(8):1490

Mona MH, El-Naggar HA, El-Gayar EE, Masood MF, Mohamed ESN (2019) Effect of human activities on biodiversity in nabq protected area, South Sinai, Egypt. Egypt J Aqua Res 45(1): 33–43

Mucova SAR, Leal Filho W, Azeiteiro UM, Pereira MJ (2018) Assessment of land use and land cover changes from 1979 to 2017 and biodiversity & land management approach in Quirimbas National Park, Northern Mozambique, Africa. Glob Ecol Conserv 16:e00447

Mutasa M, Ndebele-Murisa M (2015) Biodiversity and human development in Zimbabwe. Beyond the Crises: Zimbabwe's Prospects for Transformation, 199–241

Nasi R, Billand A, van Vliet N (2012) Managing for timber and biodiversity in the Congo Basin. For Ecol Manag 268:103–111

Natta AK, Sinsin B, van der Maesen LJG (2002) Riparian forests, a unique but endangered ecosystem in Benin. Botan J Syst Pflanz Pflanzeng 124(1):55–69

Negash L (2021) A selection of African native trees: biology, Uses, Propagation and Restoration Techniques. 621

Ngxumeshe AM, Ratsaka M, Mtileni B, Nephawe K (2020) Sustainable application of livestock water footprints in different beef production systems of South Africa. Sustainability 12(23): 9921

Niyukuri J, Raiti J, Ntakarutimana V, Hafidi A (2021) Lipid composition and antioxidant activities of some underused wild plants seeds from Burundi. Food Sci Nutr 9(1):111–122

Niza H, Bento M, Lopes LF, Cartaxana A, Correia AM (2021) A picture is worth a thousand words: using digital tools to visualize marine invertebrate diversity data along the coasts of Mozambique and São Tomé & Príncipe. Biodiv Data J 9:e68817

Nkiaka E, Lovett JC (2018) Mainstreaming climate adaptation into sectoral policies in Central Africa: insights from Cameroun. Environ Sci Pol 89:49–58

Nkiaka E, Lovett JC (2019) Strengthening the science-policy interface for climate adaptation: stakeholder perceptions in Cameroon. Reg Environ Chang 19(4):1047–1057

N'Souvi K, Sun C, Zhang H, Broohm DA, Okey MKN (2021) Fisheries and aquaculture in Togo: overview, performance, fisheries policy, challenges and comparative study with Ghana, Mali, Niger and Senegal fisheries and aquaculture. Mar Pol 132:104681

Ntoko VN, Schmidt M (2021) Indigenous knowledge systems and biodiversity conservation on Mount Cameroon. For Trees Live 30(4):227–241

Nwafor N, Alozie OJ, Onyeabor E (2018) Lessons from legal transubstantiation: analyzing the Nigerian environmental protection regime. Afr Naz Univ Law J 6(1):66–91

Ogwang T, Vanclay F (2021) Resource-financed infrastructure: thoughts on four Chinese-financed projects in Uganda. Sustainability 13(6):3259

Ogwu MC (2019a) Towards sustainable development in Africa: the challenge of urbanization and climate change adaptation. In: Cobbinah PB, Addaney M (eds) The geography of climate change adaptation in Urban Africa. Springer Nature, Switzerland, pp 29–55. https://doi.org/10.1007/978-3-030-04873-0_2

Ogwu MC (2019b) Ecologic and economic significance of bryophyte. In: Rathoure AK, Chauhan PB (eds) Current state and future impacts of climate change on biodiversity. IGI Global, Pennsylvania, pp 54–78. https://doi.org/10.4018/978-1-7998-1226-5.ch004

Ojo OO (2020) Enforceability of environmental rights: a possible panacea to man's environmental imbroglio–Nigeria's Niger-delta as a case study. Afr Focus 33(1):125–134

Okafor PC, Anikelech IG (2019) Legal framework for environmental management in Nigeria: focus on plant and animals. J Nat Build Pol Stud 3(2):149

Ola O, Benjamin E (2019) Preserving biodiversity and ecosystem services in West African forest, watersheds, and wetlands: a review of incentives. Forests 10(6):479

Olade MA (2021) Mineral deposits and exploration potential of Nigeria. Prescott books. 368

Olawuyi DS, Olusegun OO (2018) Achieving the United Nations sustainable development goals on biological diversity in Nigeria: current issues and future directions. Glob J Comp Law 7(1): 37–60

Olisah C, Okoh OO, Okoh AI (2020) Occurrence of organochlorine pesticide residues in biological and environmental matrices in Africa: a two-decade review. Heliyon 6(3):e03518

Olujobi OJ (2020) Analysis of the legal framework governing gas flaring in Nigeria's upstream petroleum sector and the need for overhauling. Soc Sci 9(8):132

Omo EE, Etuvoata A (2021) Efficacy of the legal framework on the environmental impact of agricultural technology in Nigeria. Law Soc Just Rev 2(1)

Omolola O (2013) Why Nigeria is not yet sustainably developed. APCBEE Proc 5:383–387

Onimisi T (2021) Corruption and the implementation of sustainable development goals in Nigeria: an overview. Univ Led Knowl Innov Sustain Dev 9:92–103

Ordway EM (2015) Political shifts and changing forests: effects of armed conflict on forest conservation in Rwanda. Glob Ecol Conserv 3:448–460

Orijemie EA (2022) Human behaviour and climate-linked fluctuations in the rainforests of West-Central Africa. Philosoph Trans R Soc B 377(1849):20200488

Orimoloye, I. R., Kalumba, A. M., Mazinyo, S. P., Nel, W. (2020). Geospatial analysis of wetland dynamics: wetland depletion and biodiversity conservation of Isimangaliso Wetland, South Africa. Journal of King Saud University-Science, 32(1), 90-96.

Orounladji BM, Tozo SK, Chrysostome CAAM (2021) Morphobiometric characteristics and biodiversity of indigenous Guinea fowl (Numida meleagris) in Benin. J World's Poult Res 11(1):136–150

Osawaru ME, Ogwu MC, Ahana CM (2013a) Current status of plant diversity and conservation in Nigeria. Niger J Life Sci 3(1):168–178

Osawaru ME, Ogwu MC, Braimah L (2013b) Growth responses of two cultivated Okra species (*Abelmoschus caillei* (A. Chev.) Stevels and *Abelmoschus esculentus* (Linn.) Moench) in crude oil contaminated soil. Niger J Basic Appl Sci 21(3):215–226

Osawaru ME, Ogwu MC, Chime AO (2013c) Assessment of growth performance of two Okra species (*Abelmoschus esculentus* [L.] Moench and *Abelmoschus caillei* [A. Chev.] Stevels) exposed to crude oil contaminated soil. Niger J Biotechnol 26:11–20

Ozor N, Nyambane A (2021) Eco-innovation policies for sustainable development in Africa: case studies from Botswana, Ghana, Kenya, Malawi, Nigeria, and Zambia. African Technology Policy Studies Network (ATPS) Technopolicy Brief No 61

Parisi C, De Marco G, Labar S, Hasnaoui M, Grieco G, Caserta L et al (2022) Biodiversity studies for sustainable lagoon: thermophilic and tropical fish species vs. endemic commercial species at Mellah lagoon (Mediterranean, Algeria). Water 14(4):635

Paron P, Olago DO, Omuto CT (2013) Kenya: a natural outlook: geo-environmental resources and hazards. Elsevier B.V, Netherlands, p 363

Patel S, Steele P, Kelly L, Adam JP (2021) Innovative financing for Africa: Harnessing debt for climate and nature. Economics: Climate Change. October 2021 Issue paper. United Nations Economic Commission for Africa

Pereira SC, Lopes C, Pedroso JP (2022) Mapping Cashew Orchards in Cantanhez National Park (Guinea-Bissau). Rem Sens Appl Soc Environ 26:100746

Pocock MJ, Roy HE, August T, Kuria A, Barasa F, Bett J et al (2019) Developing the global potential of citizen science: assessing opportunities that benefit people, society and the environment in East Africa. J Appl Ecol 56(2):274–281

Pradhan A, Mbohwa C (2014) Development of biofuels in South Africa: challenges and opportunities. Renew Sust Ener Rev 39:1089–1100

Pritchard R, Grundy IM, van der Horst D, Dzobo N, Ryan CM (2020) Environmental resources as 'last resort'coping strategies following harvest failures in Zimbabwe. World Dev 127:104741

Raimi MO, Vivien OT, Oluwatoyin OA (2021) Creating the healthiest nation: climate change and environmental health impacts in Nigeria: a narrative review. Scholink Sust Environ 6(1):61–122

Ramaano AI (2021) Potential of ecotourism as a mechanism to buoy community livelihoods: the case of Musina municipality, Limpopo, South Africa. J Bus Socio-Eco Dev 1(1):47–70

Ramjane H, Bahorun T, Ramasawmy B, Ramful-Baboolall D, Boodia N, Aruoma OI, Neergheen VS (2021) Exploration of the potential of terrestrial and marine biodiversity for the development of local nutraceutical products: a case for mauritius. Am J Biopharm Pharmac Sci 1:3

Rankoana SA (2020) Climate change impacts on water resources in a rural community in Limpopo province, South Africa: a community-based adaptation to water insecurity. Int J Clim Change Strat Manag 12(5):587–598

Raoufou R, Kouami K (2018) Usefulness of plant biodiversity in the cities of Togo. Selected studies in biodiversity. IntechOpen, London

Richard N, Singh P (2020) Biodiversity business innovations: a case study of Akagera National Park, Rwanda. In: The 2nd International Conference on Tourism And Entrepreneurship (ICTE) 2020 (p. 1653)

Richardson DM, Abrahams B, Boshoff N et al (2020) South Africa's Centre for Invasion Biology: an experiment in invasion science for society. In: van Wilgen BW, Measey J, Richardson DM, Wilson JR, Zengeya TA (eds) Biological invasions in South Africa. Springer, Berlin, pp 875–912

Rihoy E, Maguranyanga B (2007) Devolution and democratization of natural resource management in southern Africa: a comparative analysis of CBNRM policy processes in Botswana and Zimbabwe

Riyaz M, Shah RA, Sivasankaran K (2021) Pesticide residues: impacts on Fauna and the environment. In Biodegradation, IntechOpen, Rijeka

Rizk R, Saifelnasr E (2020) Policies and Interventions Related to Sustainable Rangeland Management & Land Degradation Neutrality in Egypt Arab Organisation for Agriculture Development (AOAD) 158

Rochell K, Xie L, Fisher R, Griffin K (2022) Contextual factors for transnational municipal network's local environmental action: a study of ICLEI Africa's LAB Wetlands SA Programme. Loc Environ:1–18

Rodrigues P, Mirinha M, Palma L (2020) Diurnal raptors of West Africa woodland-farmland mosaics: data from walking-transects in eastern Guinea-Bissau. Avian Biol Res 13(1–2):18–23

Rueda JL, Gofas S, Aguilar R, Torriente ADL, García Raso JE, Lo Iacono C et al (2021) Benthic fauna of littoral and deep-sea habitats of the Alboran Sea: a hotspot of biodiversity. In: Alboran Sea-ecosystems and marine resources. Springer, Cham, pp 285–358

Saeed TM (2020) Sustainable energy potential in Sudan. J Eng Comp Sci (JECS) 20(3):1–10

Salako VK, Fandohan B, Kassa B, Assogbadjo AE, Idohou AFR, Gbedomon RC et al (2014) Home gardens: an assessment of their biodiversity and potential contribution to conservation of threatened species and crop wild relatives in Benin. Gen Res Crop Evol 61(2):313–330

Scholte P, Pays O, Adam S, Chardonnet B, Fritz H, Mamang JB et al (2022) Conservation overstretch and long-term decline of wildlife and tourism in the Central African savannas. Conserv Biol 36(2):e13860

Schweizer D, van Kuijk M, Ghazoul J (2021) Perceptions from non-governmental actors on forest and landscape restoration, challenges and strategies for successful implementation across Asia, Africa and Latin America. J Environ Manag 286:112251

Seleteng-Kose L, Kobisi K, Pool-Stanvliet R, Mohapi K (2021) A rapid biodiversity assessment of Lesotho's first proposed biosphere reserve: a case study of Bokong nature reserve and Tsehlanyane national park. Both Afr Biodiver Conserv 51(2):1–34

Shih WY, Mabon L (2018) Land-use planning as a tool for balancing the scientific and the social in biodiversity and ecosystem services mainstreaming? The case of Durban, South Africa. J Environ Plan Manag 61(13):2338–2357

Siachoono SM (2018) Biodiversity in Zambia. Global Biodiversity: Selected Countries in Africa. 3: 387

Sieben EJ, Chatanga P (2019) Ecology of palustrine wetlands in Lesotho: vegetation classification, description and environmental factors. Koedoe Afr Prot Area Conserv Sci 61(1):1–16

Signé L, Heitzig C (2022) Effective engagement with Africa: capitalizing on shifts in business, Technology, and Global Partnerships

Singovszka E, Balintova M, Demcak S, Pavlikova P (2017) Metal pollution indices of bottom sediment and surface water affected by acid mine drainage. Metals 7(8):284

Sink KJ, Holness S, Harris L, Majiedt PA, Atkinson L, Robinson T et al (2012) National biodiversity assessment 2011: technical report, Volume 4: marine and coastal component. South African National Biodiversity Institute, Pretoria

Smith J (2018) China's belt and road initiative: strategic implications and international opposition. Background 3331:2018–2008

Souhaid ACS (2021) Appraisal environmental and social review summary (ESRS)-horn of Africa initiative: Djibouti regional economic corridor project-P174300

Standing Committee on the Environment and Public Affairs (2015) Petition no. 42—request to repeal the environmental protection (environmentally sensitive areas) notice 2005. Government of Western Australia, Perth

State of Queensland (2016) Vegetation Management (Reinstatement) and Other Legislation Amendment Bill 2016

Taylor MFJ (2015) Bushland destruction rapidly increasing in Queensland. WWF–Australia, Sydney

Tee LKAS, Puchooa D, Bhoyroo V, Appadoo C (2019) A review of nudibranch (Mollusca: Euthyneura) diversity from the Republic of Mauritius: status and future work. West Ind Ocean J Mar Sci 18(1):83–93

Teklu BM, Hailu A, Wiegant DA, Scholten BS, Van den Brink PJ (2018) Impacts of nutrients and pesticides from small-and large-scale agriculture on the water quality of Lake Ziway. Ethiopia. Environ Sci Pollut Res 25(14):13207–13216

Teklu BM, Haileslassie A, Mekuria W (2021) Pesticides as water pollutants and level of risks to environment and people: an example from central Rift Valley of Ethiopia. Environ Dev Sustain 24:5275–5294

Tesio F, Camerini F, Maucieri G, Bertini C, Cerise S (2021) Mangrove rice biodiversity valorization in Guinea Bissau. A bottom-up approach. Exp Agric 57(4):244–254

Theokritoff E, Lise D'haen SA (2021) How is science making its way into national climate change adaptation policy? Insights from Burkina Faso. Clim Dev:1–9

Tobi E, Djiembi GAN, Feistner A, Iponga DM, Liwouwou JF, Mabala C et al (2021) Creating a National Biodiversity Database in Gabon and the challenges of mobilizing natural history data for francophone countries. Biodiv Inform Sci Stand 5:e75643

Tsinda A, Abbott P, Mugisha R (2018) Review of policies for biodiversity informatics in southern Africa: case studies of Botswana and South Africa. SSRN 3110889

Tudi M, Daniel Ruan H, Wang L, Lyu J, Sadler R, Connell D et al (2021) Agriculture development, pesticide application and its impact on the environment. Int J Environ Res Pub Health 18(3): 1112

Tweh CG, Kouakou CY, Chira R, Freeman B, Githaiga JM, Kerwillain S et al (2018) Nest counts reveal a stable chimpanzee population in Sapo National Park, Liberia. Prim Conserv 32:141–152

Ukhurebor KE, Athar H, Adetunji CO, Aigbe UO, Onyancha RB, Abifarin O (2021) Environmental implications of petroleum spillages in the Niger Delta region of Nigeria: a review. J Environ Manag 293:112872

Umapathi R, Ghoreishian SM, Sonwal S, Rani GM, Huh YS (2022) Portable electrochemical sensing methodologies for on-site detection of pesticide residues in fruits and vegetables. Coord Chem Rev 453:214305

van Wilgen BW, Davies SJ, Richardson DM (2014) Invasion science for society: a decade of contributions from the Centre for Invasion Biology. S Afr J Sci 110(7–8):1–2. https://doi.org/10.1590/sajs.2014/a0074

van Wilgen BW, Wilson JR, Wannenburgh A et al (2020) The extent and effectiveness of alien plant control projects in South Africa. In: van Wilgen BW, Measey J, Richardson DM, Wilson JR, Zengeya TA (eds) Biological invasions in South Africa. Springer, Berlin, pp 593–624

Van Wyk AS, Prinsloo G (2018) Medicinal plant harvesting, sustainability and cultivation in South Africa. Biol Conserv 227:335–342

Venuste N, Olivier H, Valens N (2017) Knowledge, attitudes and awareness of pre-service teachers on biodiversity conservation in Rwanda. Int J Environ Sci Edu 12(4):643–652

Vera I, Goosen N, Batidzirai B, Hoefnagels R, van der Hilst F (2022) Bioenergy potential from invasive alien plants: environmental and socio-economic impacts in eastern cape, South Africa. Biom Bioene 158:106340

Wade TI, Ndiaye O, Mauclaire M, Mbaye B, Sagna M, Guissé A et al (2018) Biodiversity field trials to inform reforestation and natural resource management strategies along the African Great Green Wall in Senegal. New For 49(3):341–362

Waeber PO, Rafanoharana S, Rasamuel HA, Wilmé L (2020) Parks and reserves in Madagascar: managing biodiversity for a sustainable future. Protected Areas Nat Parks Sust Fut. Bakar AN, Suratman MN (eds.) 89–108

Wassie SB (2020) Natural resource degradation tendencies in Ethiopia: a review. Environ Sys Res 9(1):1–29

Wenaty A, Mabiki F, Chove B, Mdegela R (2019) Assessment of persistent organochlorine compounds contamination on the Lake Victoria water and sediments: a case study in Tanzania. African J Aquat Sci 44(3):281–290

Weyl O (2019) Lake Malawi–home to unique fish species. Wat Wheel 18(2):28–29

Wheelwright NT, Dangerfield JM, Flyman M, Tjibae M (1996) Strengthening biological theory in wildlife research and management: a case study for Botswana. South Afr J Sci 92(4):173–177

Winemiller KO, McIntyre PB, Castello L, Fluet-Chouinard E, Giarrizzo T, Nam S et al (2016) Balancing hydropower and biodiversity in the Amazon, Congo, and Mekong. Science 351(6269):128–129

World Data (2022) South Africa: country data and statistics. https://www.worlddata.info/africa/south-africa/index.php#:~:text=Geography,of%20the%20area%20of%20Texas. Accessed 26 May 2022

WWF International (2015) Chapter 5: saving forests at risk. In 'WWF living forests Report'. WWF International, Geneva

Yami M, Mekuria W (2022) Challenges in the governance of community-managed forests in Ethiopia. Sustain 14(3):1478

Yesutanbul NA, Ogutu AO, Ossei-Wusu P, Offosu P (2021) Forest destruction, agriculture practices and biodiversity conservation at Lake Bosomtwe Basin. In: Policy brief, community-based independent reporting project, Lake Bosomtwe, Ghana FIDEP foundation Ghana. Spiritan University, Ejisu, Kumasi, Ghana

Zegeye H (2018) Climate change in Ethiopia: impacts, mitigation and adaptation. Int J Res Environ Stud 5(1):18–35

Zetter R, Hassan AM (2002) Urban economy or environmental policy? The case of Egypt. J Environ Pol Plan 4(2):169–184

Zhang S, Zhou Y, Yu R, Xu X, Xu M, Li G et al (2022) China's biodiversity conservation in the process of implementing the sustainable development goals (SDGs). J Clean Prod 338:130595

Zhongming Z, Linong L, Xiaona Y, Wangqiang Z, Wei L (2021) The Gambia invests in valuing and safeguarding biodiversity. Glob S&T Dev Trend Analy Plat Res Environ

Correction to: Sustainable Utilization and Conservation of Africa's Biological Resources and Environment

Sylvester Chibueze Izah and Matthew Chidozie Ogwu

Correction to:
S. C. Izah, M. C. Ogwu (eds.), *Sustainable Utilization and Conservation of Africa's Biological Resources and Environment*, Sustainable Development and Biodiversity 32, https://doi.org/10.1007/978-981-19-6974-4

This book published under the series 'Sustainable Development and Biodiversity' was inadvertently published without the volume number. Volume number 32 has been updated with this correction.

The updated version of the book can be found at
https://doi.org/10.1007/978-981-19-6974-4

S. C. Izah, M. C. Ogwu (eds.), *Sustainable Utilization and Conservation of Africa's Biological Resources and Environment*, Sustainable Development and Biodiversity 32, https://doi.org/10.1007/978-981-19-6974-4_25

Index

A

Africa, 2, 37, 62, 105, 120, 152, 178, 204, 236, 272, 296, 324, 354, 400, 414, 443, 470, 503, 523, 544, 572, 605, 636, 666
African coastal regions, 522
African religious, 187–189
African turtle diversity, 523
Air pollution, 379–383, 405, 412–418, 557, 589–591
Anthropogenic effects, 412
Aquatic environment, 99, 111, 273, 274, 419, 420, 422–429, 432–434, 481, 482, 580, 608, 610

B

Biodiversity, 2, 36, 87, 120, 152, 180, 204, 237, 270, 344, 355, 378, 442, 470, 504, 532, 548, 572, 604, 641, 666
  conservation, 4, 16, 24, 35–57, 61–80, 100, 110, 140, 155, 171, 172, 187, 225–228, 237, 241, 242, 244–246, 272, 284–289, 405, 504, 532, 597, 673–675
  management, 8–11, 14–15, 50–52, 80, 560, 572, 593–595, 598, 599
  resources, 227, 594, 666–668, 671

C

Climate change, 2, 49, 68, 100, 139, 154, 204, 248, 271, 356, 382, 414, 443, 470, 504, 521, 544, 590, 613, 648, 668
Community forest, 131, 135, 139, 141, 144
Community participation, 17, 56, 286, 536–538, 563
Conservation, 2, 36, 62, 100, 121, 154, 177, 205, 235, 271, 312, 324, 359, 401, 421, 443, 487, 498, 524, 549, 579, 617, 652, 670
Conservation strategies, v, 13, 23, 50, 61–80, 85–111, 151–172, 180, 185, 197–199, 285, 288, 486–487, 498, 510, 525
Crop storage, 364–368
Cryopreservation, 259–260, 499, 503, 505–507, 510, 511
Cultivation, 5, 14, 213, 258, 281, 298, 313, 324, 340–343, 356, 359, 362, 364, 397, 498, 510, 556, 610, 616, 669
Cultural beliefs, 187–189, 196–197, 325, 343, 522, 530

D

Disease outbreak, 75, 497–511

E

Ecosystem services, 2, 237, 238, 243, 251, 272, 273, 278, 279, 283, 284, 521, 552, 553, 603–624, 645, 654, 657, 668, 670, 674
Ecotourism, 41–48, 56, 57, 64, 80, 143, 195, 230, 244, 255, 258, 279, 443, 447, 572, 594–599, 653, 669, 677
Emerging contaminants, 477, 481, 486
Endemic, 6, 7, 44, 67, 100, 120, 121, 124, 140, 143, 144, 152, 154, 155, 164, 172, 240, 254, 257, 275, 281, 328, 335, 498, 501, 529, 556, 644, 648, 669

S. C. Izah, M. C. Ogwu (eds.), *Sustainable Utilization and Conservation of Africa's Biological Resources and Environment*, Sustainable Development and Biodiversity 32, https://doi.org/10.1007/978-981-19-6974-4

Environment, 2, 36, 62, 87, 120, 152, 180, 204, 236, 271, 295, 337, 355, 378, 412, 442, 471, 501, 527, 544, 572, 604, 636, 667
Environmental degradation, v, 2, 21, 62, 168–169, 271, 340, 357, 362, 398, 400, 411–437, 442, 595, 608, 612, 617, 618, 671, 676, 677
Environmental laws, 51, 171, 199, 551, 674
Environmental laws and policies, 15, 50
Environmental policy, 72, 198, 401, 499, 560, 665–678
Environmental protection, 13, 56, 69, 71, 416, 504, 670, 675
Environmental sustainability, 2, 4, 163, 167, 172, 344, 364, 401–402, 404, 549, 559, 561, 592, 603–624
Ex-situ conservation, 5, 74, 76–77, 248, 255–261, 498, 499, 501, 503–507, 510, 511

F

Forensic entomology, 617, 618

G

Germplasm management, 508

H

Heritage sites, 41, 45, 48, 251, 253, 581, 587
Human activities, v, 65, 68, 69, 100, 101, 104–108, 111, 120, 158, 184, 238, 378, 379, 385, 387, 393, 397, 398, 405, 415, 416, 446, 511, 548, 550–552, 557, 608, 615, 660, 666
Human population, 2, 16, 21, 73, 154, 161, 211, 218, 224, 240, 247, 273, 340, 359, 389, 392, 397, 470, 498, 504, 544, 551, 556–558, 563

I

Insects, 44, 65, 77, 88, 123, 155, 161, 165, 168–170, 204, 206, 246, 280, 332, 336, 342, 365, 366, 368, 394, 415, 426, 443, 453–456, 458, 483, 505, 603–624
In situ and ex situ, 38, 72, 248, 502
Integrated systems approach, 15
Invasive species, 98, 166, 247, 254, 273, 274, 283, 452–456, 458

L

Legislation, 2, 10, 13–15, 49–51, 53, 55, 171, 226, 286, 340, 484, 536–538, 563
Local food crops, 353–369

M

Macrophytes, 85–111, 167, 414, 422, 434
Management, 2, 39, 63, 100, 154, 179, 205, 236, 272, 313, 325, 355, 385, 421, 451, 486, 498, 536, 544, 572, 609, 639, 670
Marine, 3, 39, 67, 86, 152, 226, 240, 270, 390, 411, 444, 520, 551, 593, 645, 666
Medicinal plants, 2, 208, 209, 221, 250, 281, 282, 296, 323–344, 358, 666, 667, 669
Medicinal uses, 261

N

Natural resources, 2, 36, 69, 178, 207, 236, 271, 325, 378, 423, 443, 487, 511, 544, 596, 606, 636, 667

O

Oil spill, 71, 105, 172, 212, 220, 412, 429, 435, 485, 526, 527, 537
Open grazing, 171

P

Pests, 222, 237, 247, 259, 280, 282, 364, 365, 368, 426, 427, 443, 452–458, 506, 509, 593, 606, 609, 610, 614, 615, 617, 653
Phytoplankton, 88–91, 95, 96, 100, 101, 105–109, 156, 389, 392, 414, 423, 434, 476, 483
Plant conservation, 162, 497–511
Plant disease management, 508–510
Plant diversity, 167, 275, 340, 449, 450, 498, 501, 504, 511
Plant protection, 11, 510
Plant utilization, 357
Pollution, 48, 64, 104, 154, 182, 205, 237, 272, 378, 411, 443, 470, 504, 521, 557, 591, 604, 648, 671
Post-harvest treatments, 356, 364, 365

R

Rationale, 13, 235–261
Resource utilization, 8, 195

S
Seasonal bush burning, 167
Sea turtle, 7, 241, 248, 520–539
Sea turtle conservation, 519–539
Socio-economic, 364, 369, 400, 403, 458, 571–600, 606, 636, 651, 677
Soil, 69, 97, 154, 205, 237, 274, 298, 342, 356, 378, 412, 442, 470, 505, 545, 608, 642, 675
Soil pollution, 385–389, 405, 430, 476
Sustainability, 2, 36, 63, 95, 163, 180, 205, 237, 270, 324, 355, 401, 442, 510, 544, 572, 604, 636, 673
Sustainable development, 11, 13, 18–23, 36, 39, 51, 54, 57, 63–64, 71, 72, 207, 226, 228, 253, 269–289, 299, 324, 354, 355, 401, 402, 458, 485, 544, 550, 558–562, 572, 582, 596–599, 616, 618, 639, 650–652, 660
Sustainable tourism, 35–57, 572, 591–597, 599

T
Terrestrial, 3, 6, 8, 43, 87, 93, 96, 99, 104, 109, 169, 204, 251, 253, 270, 274, 282, 385, 388, 394, 414, 419, 426–428, 444, 480, 483, 486, 593, 608, 611, 641
Terrestrial environment, 93, 428, 437, 604, 616
Therapeutic potential, 295–314
Threats, 19, 38, 65, 109, 120, 156, 188, 205, 246, 271, 313, 339, 355, 405, 413, 443, 470, 504, 523, 550, 595, 615, 648, 671
Topography, 38, 448–452
Tourism, 3, 17, 19, 21, 23, 35–57, 64, 101, 143, 207, 215, 241, 244, 251, 253, 287, 288, 531, 532, 534, 545, 549–551, 559, 560, 563, 572, 573, 582, 584–600, 660, 667
Touristic resources, 572–582, 599
Toxicity, 105, 108, 243, 311–313, 421, 479–482, 486
Traditional medicine, 162–163, 221, 247, 323–325, 328, 330–332, 334, 335, 340, 342, 476, 524

U
Utilization patterns, v, 4, 354, 355, 357, 360, 363, 364

W
Water pollution, 108, 158, 378, 389–392, 395, 397, 405, 412, 418–423, 434, 470, 472, 473, 476, 478–479, 481, 484, 485, 487, 557, 671
Water resources, 109, 110, 157, 159, 165, 281, 469–488, 545, 546, 563, 637, 646, 648, 658, 668
Wildlife, 5, 10, 12–14, 16, 17, 23, 24, 36–38, 40–45, 47–51, 55–56, 67, 69, 71–75, 79, 120, 128, 131, 141, 142, 144, 155, 161, 162, 167–169, 178, 180, 188, 189, 192, 197, 198, 204, 206, 209, 214, 215, 219, 222, 223, 227, 230, 237, 244, 245, 247, 248, 250, 252, 253, 256, 257, 261, 273, 402, 537, 538, 555–557, 563, 575–581, 586, 591, 595, 598, 599, 636, 667, 669, 670, 672

Z
Zooplankton, 88–90, 92, 94–96, 100, 106–108, 156, 166, 423, 434

GPSR Compliance

The European Union's (EU) General Product Safety Regulation (GPSR) is a set of rules that requires consumer products to be safe and our obligations to ensure this.

If you have any concerns about our products, you can contact us on ProductSafety@springernature.com

In case Publisher is established outside the EU, the EU authorized representative is:

Springer Nature Customer Service Center GmbH
Europaplatz 3
69115 Heidelberg, Germany

The manufacturer's authorised representative in the EU is Springer Nature Customer Service Centre GmbH, Europaplatz 3, 69115 Heidelberg, Germany. If you have any concerns regarding our products, please contact ProductSafety@springernature.com

Printed and bound by CPI Group (UK) Ltd, Croydon, CR0 4YY
01/10/2026
02225901-0013